美国西北理工大学
财经学院张雅芳：

执手软件世界牛耳，
舍我龚益之孙其谁?!

作者：Jaypi G
龚继忠
2012年3月12日

New Software Engineering Paradigm Based on Complexity Science

Jay Xiong

New Software Engineering Paradigm Based on Complexity Science

An Introduction to NSE

Springer

Jay Xiong
1545 Jackson St. #103
Oakland, CA 94612, USA
jay@nsesoftware.com

Additional material to this book can be downloaded from http://extras.springer.com

ISBN 978-1-4419-7325-2   e-ISBN 978-1-4419-7326-9
DOI 10.1007/978-1-4419-7326-9
Springer New York Dordrecht Heidelberg London

Library of Congress Control Number: 2011921248

© Springer Science+Business Media, LLC 2011
All rights reserved. This work may not be translated or copied in whole or in part without the written permission of the publisher (Springer Science+Business Media, LLC, 233 Spring Street, New York, NY 10013, USA), except for brief excerpts in connection with reviews or scholarly analysis. Use in connection with any form of information storage and retrieval, electronic adaptation, computer software, or by similar or dissimilar methodology now known or hereafter developed is forbidden.
The use in this publication of trade names, trademarks, service marks, and similar terms, even if they are not identified as such, is not to be taken as an expression of opinion as to whether or not they are subject to proprietary rights.

Printed on acid-free paper

Springer is part of Springer Science+Business Media (www.springer.com)

Preface

Why This Book?

Today software has become the driving force for the development of all kinds of businesses, engineering, sciences, and the global economy. As pointed by David Rice, "like cement, software is everywhere in modern civilization. Software is in your mobile phone, on your home computer, in cars, airplanes, hospitals, businesses, public utilities, financial systems, and national defense systems. Software is an increasingly critical component in the operation of infrastructures, cutting across almost every aspect of the global, national, social, and economic function. One cannot live in modern civilization without touching, being touched by, or depending on software in one way or another" [Ric08].

But unfortunately, software itself is not well engineered. The total economic cost of insecure software is very high: $180 billion a year in the USA [Ros08].

As Dr. Lyle N. Long pointed out, "the list of software disasters grows each year. Some of the best-known include the following: the Ariane 5 rocket (Flight 501), the Federal Bureau of Investigation Virtual Case File system, the Federal Aviation Administration Advanced Automation System, the California Department of Motor Vehicle system, the American Airlines reservation system, and many, many more. The F-22 aircraft also had problems initially due to its complex software systems. Software disasters cost the United States billions of dollars every year, and this may only get worse since future systems will be more complex. Boeing spent roughly $800 million on software for the 777, and they might need to spend five times that on the 787. Aerospace systems will also include some levels of autonomy, accompanied by an entirely new level of software complexity" [Lon08].

Since the term *software engineering* first appeared in the 1968 NATO Software Engineering Conference it has been more than 40 years past. Although many software process models, software development methodologies, software engineering techniques and tools have been innovated and broadly applied in practices, such as the Object-Oriented software development techniques, the Agile software development methods, RUP (Rational Unified Process), CMMI (*Capability Maturity Model Integration*), and the Component-Based Software Development technology, software are still not well engineered – many fundamental issues still exist.

The Fundamental Issues Exist with Today's Software Engineering Paradigm

There are many critical issues existing with today's software engineering paradigm:

(a) It is still unclear what should be the right foundation for software engineering.
(b) Software disasters happen more often now.
(c) It is unreliable – "Major software projects have been troubling business activities for more than 50 years. Of any known business activity, software projects have the highest probability of being canceled or delayed. Once delivered, these projects display excessive error quantities and low levels of reliability." [Jon06].
(d) It is unmaintainable – "Over three decades ago, software maintenance was characterized as an 'iceberg.' We hope that what is immediately visible is all there is to it, but we know that an enormous mass of potential problems and cost lies under the surface. In the early 1970s, the maintenance iceberg was big enough to sink an aircraft carrier. Today, it could easily sink the entire navy!" [Pre05-P841], "The fundamental problem with program maintenance is that fixing a defect has a substantial (20–50%) chance of introducing another" [Bro95-P122].
(e) The software project success rate is still very low: about 30% – it is not acceptable in any other industry.
(f) **"No Silver Bullet"** – pointed by Professor Frederick P. Brooks Jr., "**There is no single development, in either technology or management technique, which by itself promises even one order-of-magnitude improvement within a decade in productivity, in reliability, in simplicity.**" [Bro95-P179], "Of all the monsters who fill nightmares of our folklore, none terrify more than werewolves, because they transform unexpectedly from the familiar into horrors. For these, we seek bullets of silver that can magically lay them to rest. **The familiar software project has something of this character (at least as seen by the nontechnical manager), usually innocent and straightforward, but capable of becoming a monster of missed schedules, blown budgets, and flawed products.**" [Bro95-P180]. "**Not only are there no silver bullets now in view, the very nature of software makes it unlikely that there will be any – no inventions that will do for software productivity, reliability, and simplicity what electronics, transistors, and large-scale integration did for computer hardware. We cannot expect ever to see twofold gains every two years.**" [Bro95-P181].

It seems that having those critical problems is normal to software products and software engineering.

A Sudden Realization

I have been working in the field of software engineering for more than 20 years since I established my first company, Advanced Software Automation, Inc. (ASA) in Silicon Valley in 1987. At that time, I realized that automation should be the direction for the development of software engineering. ASA's first product, Hindsight designed by me and implemented by me and my colleagues with many automated functions in software testing and visualization was chosen by Sun Microsystems as the test suite for its many software products except the operating systems. In 1992, I established my second software company, International Software Automations, Inc. (ISA) in Silicon Valley. As the designer of ISA's first product, Panorama, I extended the automated capability from the back-end to include the support for the front-end of software engineering. About Panorama, Professor Roger S. Pressman stated that "Panorama: developed by International Software Automation, Inc. encompasses a complete set of tools for object-oriented software development, including tools that assists test case design and test planning." [Pre05-P409].

Later on, I realized that although automation is important to software engineering, it cannot be used to solve the major critical issues existing with software engineering – low quality and productivity, and high cost and risk.

Where is the outlet of software engineering?

One day in the summer of 2005, in a book store I accidentally found a book introducing complexity science. After reading it curiously, I suddenly realized that it is what I am looking for! Yes, complexity science will be the powerful means to solve the all critical issues existing with today's software engineering, because complexity science is the science studying complex systems with many interactive components. Complexity science offers holistic and global approaches rather than partial and local approaches to handle complex systems. That day I bought five different books on complexity science.

"The next century will be the century of complexity" (Stephen Hawking, January 2000). Complexity science is the driving force for the development of sciences, engineering, and business in the twenty-first century. Complexity science explains how holism emerges in the world, and more. Definitions of complexity are often tied to the concept of a complex system – something with many parts that interact to produce results that cannot be explained by simply specifying the role of each part. This concept contrasts with traditional machine or Newtonian constructs, which assume that all parts of a system can be known, that detailed planning produces predictable results, and that information flows along a predetermined path.

What is Wrong with Today's Software Engineering Paradigm?

After I changed my standing point from traditional Newtonian constructs to complexity science, I realized that almost all of the components of the existing software engineering paradigm (except the technologies for database, operating systems, and programming languages) are wrong or outdated:

(a) **The foundation of today's software engineering paradigm is wrong** – Software is a nonlinear complex system. "The complexity of software is an essential property, not an accidental one….Many of the classical problems of developing software products derive from this essential complexity and its nonlinear increases with size" [Bro95-P183], but unfortunately, the existing software engineering paradigm is based on linear thinking, reductionism, and superposition principle that the whole of a system is the sum of its parts, so that almost all tasks/activities are performed linearly, partially, and locally.

(b) **The process models are wrong** – They are all linear ones (no matter if it is a waterfall-like model, an incremental development model **which is "a series of Waterfalls" [*GSAM03*]**, or an iterative development model in which each time of the iteration is a waterfall) with which there is only one track in a forward direction – no upstream movement at all, and the work flow is always going forward from the upper phases to the lower phases. Those models require that the developers always do all things right without making any mistake or wrong decision – it violates the nature of human beings. The result is that defects introduced in the upper phases easily propagate to the lower phases to make the defect removal cost increase tenfold many times.

(c) **The software development methodologies are outdated** – They are based on linear thinking, reductionism, and Constructive Holism principle to complete the components of a software product first, then, as CMMI states, "*Assemble the product from the product components, ensure the product, as integrated, functions properly and deliver the product.*" **[CMMI1.1]** – they handle a logic software product created by people as a machine which can be **assembled**. Regarding the quality assurance, those methodologies are test driven – mainly depending on software testing after production – it is too late.

(d) **The existing software modeling approaches are outdated**, because they are outcomes of reductionism and superposition principle, use different sources for human understanding and computer understanding of a software system separately with a big gap between them. The obtained models are not traceable for static defect removal, not executable for debugging, and not testable for dynamic defect removal, not consistent with the source code after code modification, and not qualified as the road map for software development.

(e) **The software testing paradigm is outdated** – Most software defects are introduced to a software product in requirement development phase and the product design phase, but the existing software testing paradigm can only be dynamically used after production, so that NIST (National Institute of Standards and Technology) concluded that "Briefly, experience in testing software and systems has shown that testing to high degrees of security and reliability is from a practical perspective not possible. Thus, one needs to build security, reliability, and other aspects into the system design itself and perform a security fault analysis on the implementation of the design." (**"Requiring Software Independence in VVSG 2007: STS Recommendations for the TGDC," November 2006**, http://vote.nist.gov/DraftWhitePaperOnSIinVVSG2007-20061120.pdf).

(f) **The quality assurance paradigm is outdated** – Current software quality is ensured mainly through inspection and dynamic testing after production, it violates W. Edwards Deming's product quality principle that *"Cease dependence on inspection to achieve quality. Eliminate the need for inspection on a mass basis by building quality into the product in the first place."* [Dem86].
(g) **The software maintenance paradigm is wrong** – with it, software maintenance is performed blindly, partially, and locally without the capability to prevent the side effects in the implementation of requirement changes or code modifications, making the maintained software product unstable day by day.
(h) **The software visualization paradigm is outdated** (see Chap. 2).
(i) **The documentation paradigm is outdated** (see Chap. 2).
(j) **The project management paradigm is outdated** (see Chap. 2).
(k) **The "Software" definition is outdated** (see Chap. 1).
(l) **The entire software engineering paradigm is outdated** (see Chap. 2).
(m) **The "No Silver Bullet" conclusion is outdated** – it is an outcome of linear thinking, reductionism, and superposition principle, only suitable to the old-established software engineering paradigm (see Chap. 2 for more detailed description).

What Is the Root Cause for Those Critical Issues Existing with Today's Software Engineering?

The root cause for those critical issues comes from the wrong foundation of the software engineering paradigm that software and the software engineering paradigm are complex nonlinear systems, and should be handled with complexity science to comply with the essential principles of complexity science, particularly the Nonlinearity principle and the Holism principle to make all tasks and activities being performed holistically and globally rather than partially and locally.

The Difficulty in Solving Those Critical Issues

As described above, there are many components with software engineering paradigm. According to complexity science, the behaviors and characteristics of the whole of a complex system emerge from the interaction of its components, and cannot be inferred simply from the behavior of any individual part, so that only improving its one or two components such as focusing the improvement of software engineering process and the software management process only will not be able to make significant improvement to the whole of the software engineering paradigm – it could be the main reason why the failure rate of the implementation of CMM/CMMI is about 70% [Nia09].

The difficulty in solving those critical issues comes from two major steps – step 1: bring revolutionary changes to the all major components of the software engineering paradigm; step 2: after the revolutionary changes of the all major components, make revolutionary changes of the whole of the software engineering paradigm emerge from the interaction of all of its components changed revolutionarily – **it is how NSE (Nonlinear Software Engineering paradigm) is established and implemented, and why this book comes**.

The Major Features of This Book

The major features of this book are listed as follows:

(a) **New** – This book introduces many new concepts, ideas, algorithms, models, methods, techniques, and tools.
(b) **Original** – Almost all of the new concepts, ideas, algorithms, models, methods, techniques, and tools introduced in this book are innovated by me and implemented by me and my colleagues, not collected from others' contributions or other books. Those innovations include the following:

1. The new definition of "software" – see Chap. 1.
2. The FDS (Five-Dimension Synthesis Method) general paradigm-shift framework for various industry revolutions from the old-established paradigm based on linear thinking and superposition principle to a revolutionary paradigm based on nonlinear thinking and complexity science (not only for software engineering) – see Chap. 4.
3. Many new software engineering techniques innovated for the implementation of NSE – see Chap. 6.
4. The NSE visualization paradigm and the interactive and traceable J-Chart, J-Diagram, and J-Flow diagram used to make an entire software development process and the work products visible – see Chap. 7.
5. The NSE process model which is a nonlinear, incremental, and parallel model with multiple-tracks for bidirectional iteration – see Chap. 8.
6. The facility for automated and self-maintainable traceability among documents and test cases and the source code through the use of Time Tags for data mapping between test cases and the source code, and some special keywords to indicate the document formats, the file paths, and the bookmarks for opening the traced documents from the specified locations – see Chap. 9.
7. The NSE software development methodology complying with the Generative Holism principle (rather than Constructive Holism principle), which is driven by defect prevention and five types of bidirectional traceabilities – see Chap. 10.
8. The Holistic, Actor–Action and Event–Response driven, Traceable, Visual, and Executable technique (HAETVE) used for Source Code Driven Dynamic Software Modeling and Engineering (see Chap. 11). Here "Dynamic Software Modeling" means:

Using only one kind of source (source code) for both human understanding of a complex software in diagrams automatically generated from the code, and computer understanding of the software in textual format, through forward engineering using dummy programs (a dummy module has an empty body or only a list of function call statements) or reverse engineering using regular programs (Top-down + Bottom-up). Since the diagrams/models are generated from the source code, they are always consistent with the code.

The generated diagrams/models are executable directly or indirectly through the corresponding code.

The generated diagrams/models not only can represent the static properties of a software product but can also represent the dynamic properties of a software product, such as the code test coverage and the percentage of the execution time spent in each module.

The generated diagrams/models are interactive and traceable.

The most important feature of Dynamic Modeling is that the generated diagrams/models no longer statically exist – they dynamically exist ("alive") – the generated diagrams/models, the generators of the diagrams/models, and the interfaces for accepting users' commands (using the diagrams/models themselves), are three in one: when a diagram/model is shown, its generator is always working and waiting for a user's command through the diagram/model (acting as the interface) – after receiving a user's command, the generator will dynamically respond to it such as generating a subtree (see Fig. 7.11), printing out a chart (see Fig. 7.23), or performing untested path analysis and automatically highlighting a "best" one with the most untested branches and automatically extracting the execution conditions to help users design the most efficient test case.

The generated diagrams/models and the corresponding source code are no longer separated; instead, they are combined together to form a powerful union to help users develop a software product better, understand a software product better, test a software product better, and maintain a software product better. For instance, clicking on a module-box from the generated call graph to directly edit the source code of that module as shown in Fig. 11.31, or clicking on a module from the generated control flow diagram to trace the corresponding test cases and directly play the captured GUI test operations back dynamically as shown in Fig. 11.32.

9. The NSE software testing paradigm and the Transparent-box testing method, which combines functional testing and structural testing together seamlessly with the capability to establish bidirectional traceability among documents and test cases and source code, and can be used dynamically in the entire software development lifecycle including the requirement development phase and software design phase (because having an output is no longer a condition to use this kind of testing method

and tools dynamically – to each test case, it checks whether the output (if any, can be none) is the same as what is expected, and checks whether the execution path covers the expected path specified, and then establishes bidirectional traceability to help users remove the inconsistency defects, plus many other ways for defect prevention and inspection using traceable documents and traceable source code.) – see Chap. 16.

10. The NSE quality assurance paradigm based on defect prevention and defect propagation prevention through dynamic testing, software visualization, and semiautomated inspection and review using traceable documents and source code diagrammed in the entire software development lifecycle – see Chap. 17.

11. The NSE maintenance paradigm which is systematic, disciplined, and quantifiable with the capability to prevent side effects for the implementation of requirement changes and code modifications supported by various traceabilities – see Chap. 18.

12. The NSE documentation paradigm with which the documents and the source code are managed together with bidirectional traceability to keep them consistent – see Chap. 19.

13. The NSE project management paradigm combining the software development process and software project management process together to make software project management documents also traceable with the implementation of requirements and the source code – see Chap. 20.

14. The new algorithms innovated to support NSE – see Chap. 21.

15. Many automated tools and the support platform, Panorama++, designed for supporting NSE – see Chap. 22.

(c) **Based on complexity science** – Almost all of the new concepts, ideas, algorithms, models, methods, techniques, and tools innovated are based on complexity science, complying with the essential principles of complexity science, particularly the Nonlinearity principle and the Holism principle.

(d) **The described new concepts, ideas, algorithms, models, methods, techniques are commercially implemented** – All of them are supported by the Panorama++ platform for software development, testing, and maintenance.

(e) **Complete** [Xio09-1], [Xio09-2] – It covers almost all aspects in software engineering to offer a holistic and global solution for software engineering, rather than a partial and local solution, and also offers all required tools to support the applications of NSE to form a complete solution.

(f) **Detailed** – It not only introduces the concepts or ideas but also introduces the implementation algorithms step by step.

(g) **Easy to read and understand** – It describes the contents with several hundred graphics, most of which are screenshots from real application examples; **easy to try** – trial versions of the NSE support platform Panorama++ are provided with application examples (see the "**Toolkits Provided for This Book**" section); **and easy to use** – NSE (with its support platform Panorama++) can be applied for new software product development, or a product being developed using any other method – in this case, the users only need to rewrite the test cases according to NSE's simple rules, and set the corresponding bookmarks to the related documents – other work can be performed automatically by the NSE support platform Panorama++ in which many easy-to-use automated tools are integrated.

(h) **Beneficial** – Preliminary applications of NSE and the support platform Panorama++ introduced in this book show that compared with the old-established software engineering paradigm, **it is possible for NSE with its support platform Panorama++ to help software organizations double their software productivity, halve their cost, greatly reduce the risks, remove 99.99% of the defects in their products, and double their project success rate because**

- With NSE, almost all tasks/activities are performed nonlinearly, holistically, and globally, rather than linearly, partially, and locally.
- The quality is ensured through defect prevention and defect propagation prevention performed in the entire lifecycle from the first step down to maintenance through dynamic Transparent-box testing and semiautomatic inspection using traceable documents and traceable source code.
- Software requirement changes or code modifications are responded to in real time with side-effects prevention through various traceabilities.
- The Software maintenance process is combined with the software development process and performed holistically and globally with side-effect prevention. The regression testing after code modification is performed with test case efficiency measurement and test case minimization and intelligent test case selection through backward traceability. Because the NSE nonlinear process model is followed and the quality of a software product is ensured from the first step down to maintenance, the defects propagated to the maintenance phase is greatly reduced. Even if the product maintenance team is different from the product development team, according to the new software definition with NSE and the support platform, the database built through static and dynamic measurement of the product and a set of Assisted Online Agents will also be delivered to the customer to form almost the same conditions as the product development site for maintaining the product. So, the effort and cost spent in software maintenance will be almost the same as the effort and cost spent in the software development process – it means about half of the total effort and cost can be reduced (usually with the old-established software engineering paradigm, software maintenance takes 75% or more of the total effort and total cost in a software product development. With NSE, software maintenance will take the total effort and cost almost the same as the development process – only 25% of the total effort and total cost, it means about 50% of the total effort and total cost can be saved).
- The entire process of a software development, testing, and maintenance is visible through the applications of the NSE software visualization paradigm,

which generates interactive and traceable J-Chart, J-Diagram, and J-Flow diagrams automatically.
- The software documents are traceable with the source code to keep consistency among them, and stored virtually without huge disk and memory space.
- With NSE, the project management process is combined with the product development process closely, making the project management documents traceable with the implementation of requirements and the source code.

The Scope of This Book

Considering that complexity science is the driving force for the development of sciences, engineering, and business in the twenty-first century, and software is becoming the foundation of modern civilization, it means that both are closely related to the future of mankind and the economic development of the world.

Today, more and more industries are becoming increasingly aware that traditional approaches to design and engineering are failing to keep up with the increasing scale of systems [Mck99]. The foundation of those traditional approaches is based on linear thinking and established science complying with the reductionism and superposition principle that **the whole of a system is the sum of its parts**. But, in fact, all people problems and issues are nonlinear which do not comply with the superposition principle because they exist in a dynamic and changeable environment, rather than a static one [Lim05].

Although there are many ways proposed for the applications of complexity science, none of them aims for a new round of industrial revolution. I believe I am the first person to not only realize that complexity science can be efficiently applied in a new round of industrial revolution but also innovated a corresponding paradigm-shift framework, the Five-Dimensional Structure Synthesis Method (FDS, see Fig. 1), and successfully use it to complete the paradigm-shift of the software industry – the most difficult one to handle. It proves that FDS is useful and operational. Since complexity science and the FDS paradigm-shift framework can be successfully used to revolutionarily complete the paradigm shift of the software industry from that based on linear process, reductionism, and superposition principle to that based on nonlinear process and complexity science, why can't other industries do the same?

I also realize that directly applying complexity science to handle the problems of an individual complex system in an industry without shifting the entire paradigm from the old-established one (consisting of many components including the process models, the development methodologies, the algorithms, the technologies, the quality standards, and the tools) based on linear process and reductionism principle to a new one based on nonlinear process and complexity science in that industry will be very difficult – if not impossible, because the "Sunlight" of complexity science cannot directly "Reach" the target without removing the big "Umbrella" in the middle – the old-established paradigm. I suggest that the application of complexity science should follow two major steps: (1) the first step is to complete the paradigm shift from the old one based on linear process and reductionism principle to a new one based on nonlinear process and complexity science; (2) then, after the paradigm has been shifted, the second step is to apply complexity science to efficiently handle the problems of an individual complex system. The two-step approach is also shown in Fig. 1.

The relationships among the five elements represented in the five axes of FDS are shown in Fig. 2.

For the detailed description about FDS, see Chap. 4.

When FDS is used for the paradigm shift of an industry, it is required to comply with the essential principles of complexity science (including the Nonlinearity principle, the Holism principle, the Dynamics principle, the Self-Organization principle, the Self-Adaptation principle, the Openness principle, and more) to redefine the process model, reinnovate the methodology, redesign the tools and platform, reestablish the quality assurance methodology and the standard, and so on in order to establish a complete new paradigm in that industry. It is clear that, for instance, a waterfall-like process model will not be redefined because it does not comply with the Nonlinearity principle and the Holism principle of complexity science. After paradigm-shift is done, FDS can also be used for handling the problems of an individual complex system.

It is why this book is written not only for people in the field of software engineering and computer science but also for people in all other fields who want to

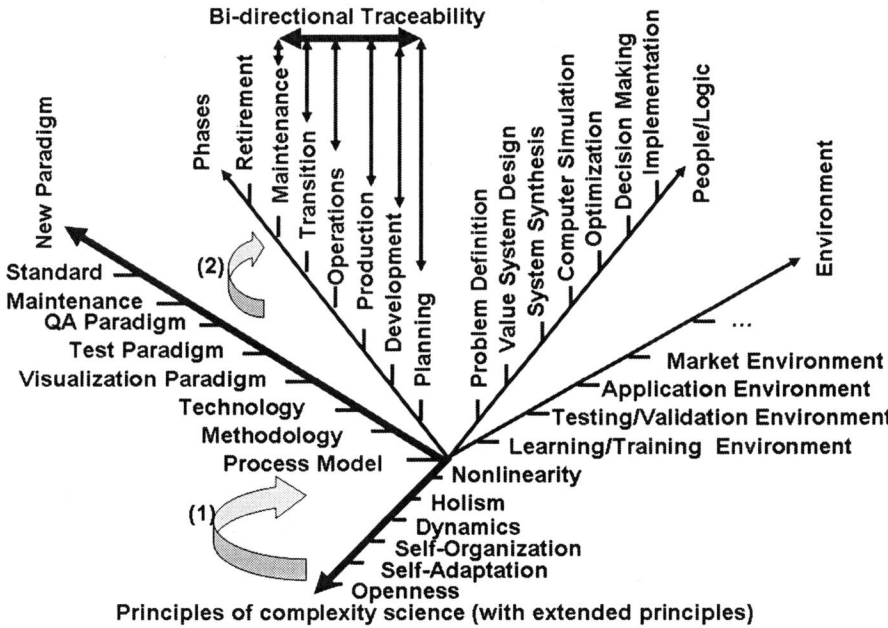

Fig. 1 The innovated FDS (five-dimensional structure synthesis) framework

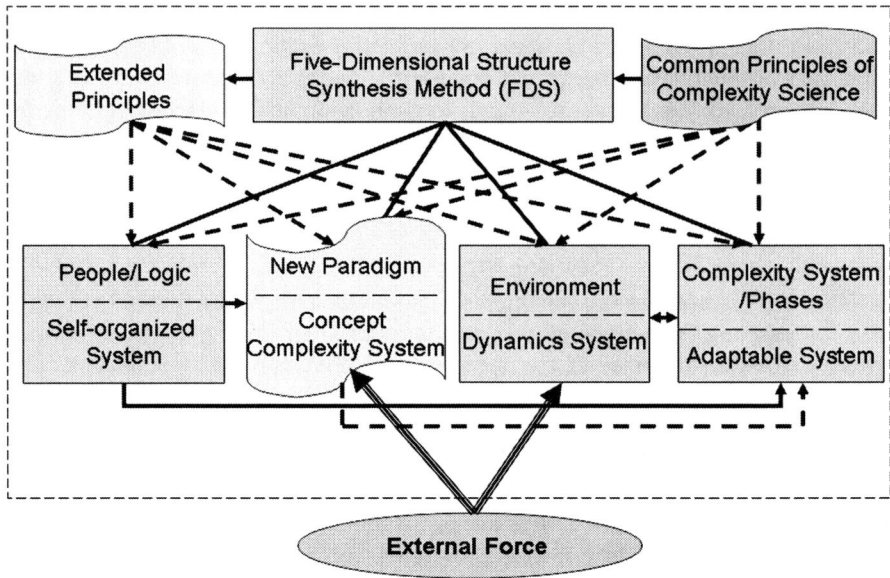

Fig. 2 The five elements of FDS and their relationships

apply complexity science as a powerful means to perform a revolutionary paradigm-shift from the old one based on linear process and superposition principle to a new one based on nonlinear process and complexity science through the general paradigm-shift framework, FDS. For more related information, see Chap. 3 titled "**Foundation for Establishing NSE: Complexity Science**" and Chap. 4 titled "**Prediction and Practices: A New Round of Industrial Revolution Driven by Complexity Science, and a General Paradigm-Shift Framework (FDS).**"

Who Should Read This Book

People Working in the Field of Software Engineering or Computer Science

This book is for perplexed software and management professionals who want to know and use a revolutionary software engineering paradigm based on complexity science to help software organizations to dramatically solve the most critical problems with today's software engineering at the same time – to double their productivity and their project success rate, halve their cost, greatly reduce the risk, and improve the quality of their product tenfold several times, compared with the existing software engineering paradigm with the same level of the resource. If you want to know what critical issues exist with today's software engineering paradigm, why those critical issues exist for more than 40 years without being solved, what are the root causes of those critical issues, what is complexity science, how complexity science can be applied to solve those critical issues, how a revolutionary software engineering paradigm (NSE – Nonlinear Software Engineering paradigm) is established, how can NSE help software organizations, and how can you try the NSE support platform Panorama++, this book is for you:

- Executives and Project managers should read this book to know what is complexity science, how it can help your organization in the software development, what are the major differences between NSE and the old-established software engineering paradigm, how a project can be developed and managed holistically and globally, and whether the productivity can be doubled, the cost can be reduced to half, and the quality can be improved greatly at the same time, and how the project management documents can be traced automatically to get the first-hand information.
- Software developers should read this book to know what is complexity science, what is NSE, how it can help for you to perform your jobs better, how software testing can be dynamically performed in the entire software development lifecycle, how documents and test cases and the source code can be made traceable, how software maintenance can be done with side-effect prevention, and how NSE can help you for your career.
- Computer science and software engineering researchers should read this book to consider whether it is a new direction to apply complexity science on software engineering research, review NSE, compare NSE with the old-established software engineering paradigm, then find possible research topics and make contributions to the future of software engineering.
- Computer science and software engineering students should read this book to learn what is complexity science, what is NSE and the major differences between NSE and the old-established software engineering paradigm, and try the demo program of NSE.
- Customers should read this book to particularly know how requirement changes can be implemented through bidirectional traceability to prevent side effects, how a software product can be maintained in your site with almost the same conditions as that in the product development site – with NSE, a **software** (software product) is redefined as and delivered to the customer with (1) a computer program (a regular program, or a cloud computing program, or a program developed through the internet) with the source code, (2) the data used, (3) all of the related documents (including the test case scripts too) traceable to and from the source code, plus (4) the database built though static and dynamic measurement of the program, and (5) a set of Assisted Online Agents (automated and intelligent tools working with the program and the database) for handling the issue of complexity and supporting the testability, visibility, changeability, conformity, reliability, and

traceability – making the software product adaptive and truly maintainable in the new working environment at the customer site, and that the requirement validation and the acceptance testing can be done dynamically in a fully automated way with mouse clicks only.

Recommended Courses Using This Book as a Textbook in the Computer Science Department of a University and a Software Engineering College

The twenty-first century is the century of complexity science. Compared with the old-established software engineering paradigm, the major advantages of NSE can be summarized in one sentence: with NSE, almost all software engineering tasks and activities are performed nonlinearly, holistically, and globally rather than linearly, partially, and locally. Therefore, although this book describes a complete revolution in software engineering based on complexity science, it is also suitable as a textbook in the computer science department of a university or a software engineering college:

1. It is organized hierarchically according to software engineering workflows, such as in the following chapters; Chap. 11 introduces requirement engineering under NSE, Chap. 12 introduces software design under NSE, Chap. 13 introduces software coding under NSE, and so on.
2. Several hundred detailed illustrations are provided.
3. Detailed application examples are provided (see Chap. 1) – people work well through examples.
4. In each chapter, there is a "Summary" section designed.
5. In each chapter, there is a section of "Points and Questions to Ponder" designed.
6. The hints for answering the "Points and Questions to Ponder" for each chapter are provided in Appendix D.
7. Trial Versions of the NSE support platforms are provided (see the "**Toolkits Provided for This Book**" section) for students to get hands-on experience in using the powerful tools to design, evaluate, test, validate, and maintain their own learning projects.
8. A detailed Tutorial is also provided to help students to apply NSE and the support platform Panorama++ in practice, step by step.

Recommended course titles:

    (a)   Nonlinear Software Engineering Paradigm Based on Complexity Science
    (b)   Advanced Software Engineering
    (c)   The Future of Software Engineering

Suggested Level + Length:

1. Undergraduate (seniors), 2 semesters (28–30 weeks)
2. Master program, 1 semester (14–15 weeks)
3. Postgraduate course, 8 weeks

I believe that to meet the urgent needs of the software industry and raising the competition power in the near future, the earlier the computer science departments of a university or a software engineering college to offer NSE courses, the better for them and their students to win over their competition.

Note: Besides universities and software engineering colleges that teach their students internally, it is also welcome for an individual or an organization to work with us to offer co-held training courses for software engineers, programmers, and employees working in a software-related company. For ensuring the quality of the courses on NSE with the use of the trial versions of the NSE support platform Panorama++, the instructors of the courses should take a corresponding exam to get the authority certificate first. If you are interested in offering a co-held training course on NSE (the corresponding certificates for trainees will also be provided), please send an email with your proposal to me (jayxiong@yeah.net and jay@nsesoftware.com).

People Working in Other Fields Who Want to Know How Complexity Science and the FDS Framework Can Be Used to Complete the Paradigm-Shift Revolutionarily in Their Industries

This book is written for you too! Please ignore Chaps. 1 and 2 (options), pay more attention to Chaps. 3 and 4, and consider other chapters as an application example of complexity science and the FDS paradigm-shift framework in the establishment of NSE, a revolutionary new paradigm for software engineering.

How to Read This Book

For easy comparison of the old-established software engineering paradigm and the new software engineering paradigm, NSE, to be introduced in detail in this book, it is strongly recommended for readers to install and try the NSE-CLICK toolkit through an application example (a calculator software product, see Chap. 1), while reading this book. After that try the S_Panorama (for C/C++) or S_Panojava (for Java language) product designed for students to learn NSE with small projects (less than 1,501 lines of the source code). About how to get those toolkits, see **"Toolkits Provided for This Book" section below.**

Organization of This Book

This book is organized as follows:

- Chapter 1 is an introduction to this book.
- Chapter 2 concludes that the old-established software engineering paradigm is outdated.
- Chapter 3 introduce the Foundation for establishing NSE: Complexity Science.
- Chapter 4 describes prediction and practices : a new round of industrial revolution driven by complexity science, and a general paradigm-shift framework.
- Chapter 5 is the outline of NSE Paradigm.
- Chapters 6–19 introduce the body of NSE, including the nonlinear NSE process model, the NSE software development methodology complying with the Generative Holism principle of complexity science, NSE software visualization paradigm generating interactive and traceable charts and diagrams which are holistic and virtual, NSE software testing paradigm based on the innovated Transparent-box testing method combining functional and structural testing together seamlessly, the NSE software quality assurance paradigm driven by defect prevention and defect propagation prevention, the NSE documentation paradigm to make software documents traceable to and from the source code, the NSE software maintenance paradigm with side-effect prevention in the implementation of requirement changes or code modifications.
- Chapter 20 introduces the NSE project management paradigm working closely with the software development process to make the management materials traceable with the requirement implementation and the source code.
- Chapter 21 introduces the algorithms innovated for establishing NSE.
- Chapter 22 describes the NSE support tools and support platforms.
- Chapter 23 introduces NSE applications – NSE not only can be used for new software development but also can be used for a software product being developed using other methodologies in any stage by rewriting the test cases and set bookmarks to the related documents (other documents can automatically be generated) for improving the development process, testing and ensuring the product quality, or efficiently maintaining the product with side-effect prevention.
- Chapter 24 summarizes the entire NSE software engineering paradigm, compares it with the old-established software engineering paradigm, and proposes three Candidates of "Silver Bullet" – the NSE automated and self-maintainable traceability, the NSE software testing paradigm, and the entire NSE software engineering paradigm.
- Appendix A provides a template for requirement specification.
- Appendix B shows an example about how to realize 100% MC/DC (Modified Condition/Decision Coverage) test coverage for a program unit.
- Appendix C describes how to control/simulate the return values to a program unit being tested.
- Appendix D provides hints for answering the "Points and Questions to Ponder" in each chapter.
- Glossary provides a list of specialized terms with definitions.

Toolkits Provided for This Book

It is strongly recommended for readers to install and try the NSE-CLICK and other toolkits provided (on Springer Extras at http://extras.springer.com/ and then use this book's ISBN).

After downloading the file (NSE_Panorama.rar) and unzipping it, you will find the following files and directories as shown in Table-P1.

Table P1 The files and directories included in the NSE_Panorama Tool Package

Type	Name	Description
File	readme.doc	The first document to read
File	license_agreement.txt	License agreement
File	installation.doc	Installation guide (NSE support platform and tools are green software without complicated installation operations)
File	NSE_CLICK_J_Tutorial.pdf	A tutorial for using NSE_CLICK_J.
File	NSE_CLICK_Tutorial.pdf	A tutorial for using NSE_CLICK
File	NSE_J_Tutorial.pdf	A tutorial for using Pano_java product
File	NSE_Tutorial.pdf	A tutorial for using Panorama++ product
Directory	floating_license	The directory with files regarding the use of floating license of the regular Panorama++ products
Directory	isa_common_tools	The directory including all Assisted Online Agents to be delivered with a software product developed using NSE
Directory	isa_examples	The directory including some application examples, particularly a calculator software product used to show all the major features of NSE and the support platform Panorama++
Directory	isa_NSE	A trial version of Panorama++ for C/C++ products (for learning NSE)
Directory	NSE_CLICK	The directory including the NSE-CLICK toolkit and the Interface – a demo product for fully automated product acceptance testing of a C/C++ product
Directory	NSE_CLICK_J	The directory including the NSE-CLICK_J toolkit and the Interface – a demo product for fully automated product acceptance testing of a Java product
Directory	Pano_java	A trial version of Panojava for Java products (for learning NSE)

Acknowledgments I would like to thank Hamid R. Arabnia, Ph.D., a Professor of Computer Science, Graduate Coordinator, who invited me to offer a tutorial titled "Complete Revolution in Software Engineering Based on Complexity Science" to WORLDCOMP'09 where I got a lot of useful feedback to improve the NSE paradigm. I would also like to thank Professor Ni Guangnan, academician of the Chinese Academy of Engineering, for his insightful suggestions. Thanks to professor Zheng Renjie from Tsinghua University of China for sharing his thought on the old-established software engineering paradigm and his valuable suggestions. Thanks to Michael Zhao, Jonathan Xiong, and more than 50 of my colleagues of International Software Automation, Inc. (ISA US) and ISA Shanghai, Ltd for their support in the implementation of NSE and the development of the NSE support platform Panorama++ and SilverBullet (both consist of about 10,000 function points with about one million lines of source code). Special thanks to Brett Kurzman from Springer for his great help in the planning, organization, and publishing of this book.

Oakland, California, US   Jay Xiong

References

[Bro95-P122] Brooks FP Jr (1995) The mythical man-month. Addison-Wesley, Reading, p 122
[Bro95-P179] Brooks FP Jr (1995) The mythical man-month. Addison-Wesley, Reading, p 179
[Bro95-P180] Brooks FP Jr (1995) The mythical man-month. Addison-Wesley, Reading, p 180
[Bro95-P181] Brooks FP Jr (1995) The mythical man-month. Addison-Wesley, Reading, p 181
[Bro95-P183] Brooks FP Jr (1995) The mythical man-month. Addison-Wesley, Reading, p 183
[CMMI1.1] Phillips M (2002) CMMI V1.1 and appraisal tutorial. http://www.sei.cmu.edu/cmmi/
[Dem86] Deming WE (1986) Out of the crisis. MIT Press, Cambridge
[GSAM03] Department of the Air Force Software Technology Support Center (2003) Condensed GSAM Handbook, Chapter 2. CrossTalk
[Jon06] Jones C (2006) Social and technical reasons for software project failures. CrossTalk, June Issue
[Lim05] Lindberg C (2005) Complexity, the science of relationships. Nursing, the profession of relationships. Plexus Institute, Allentown, NJ, 14 November 2005
[Lon08] Long LN (2008) The critical need for software engineering education. CrossTalk, Jan Issue
[Mck99] McKenzie CA (1999) MIS327 – systems analysis and design. Course Schedule, 1999
[Nia09] Niazi M (2009) Software process improvement implementation: avoiding critical barriers. CrossTalk, Jan Issue
[Ric08] Rice D (2008) Geekonomics: the real cost of insecure software. Addison-Wesley, Upper Saddle River
[Ros08] Rosenberg D (2008) Total economic cost of insecure software: $180 billion a year in the U.S. http://news.cnet.com/8301-13846_3-9978812-62.html
[Pre05-P409] Pressman RS (2005) Software engineering: a practitioner's approach. McGraw-Hill, New York, p 409
[Pre05-P841] Pressman RS (2005) Software engineering: a practitioner's approach. McGraw-Hill, New York, p 841
[Xio09-1] Xiong J (2009) Tutorial, A complete revolution in software engineering based on complexity science. In: WORLDCOMP'09, Las Vegas, 13–17 July 2009
[Xio09-2] Xiong J, Xiong J (2009) A complete revolution in software engineering based on complexity science. In: WORLDCOMP'09 – SERP (Software Engineering Research and Practice 2009), pp 109–115

Contents

1 **Introduction** 1
   1.1   What Is Software? 1
   1.2   What Is Software Engineering? 29
   1.3   The Major Activities/Tasks to Be Performed
         in Software Engineering 31
   1.4   The Popular Lifecycle/Process Models with the Existing
         Software Engineering Paradigm 32
         1.4.1   The Waterfall Model 32
         1.4.2   The Incremental Development Models 34
         1.4.3   The Iterative Models 36
         1.4.4   More Popular Process Models 39
         1.4.5   General Comments to All Process Models Existing
                with the Old-Established Software Engineering Paradigm 43
   1.5   Why the Current Software Is Not Sufficiently Engineered
         at This Time to Fulfill the Role of "Foundation" 45
   1.6   What Does a Revolution Mean? 47
         1.6.1   Three Phases of Scientific Revolutions 47
         1.6.2   Progress Through Revolutions 48
   1.7   What Is NSE? 48
   1.8   Summary 57
   1.9   Points and Questions to Ponder 58
   1.10 Further Reading and Information Source 58
   References 58

2 **Is the Old-Established Software Engineering
Paradigm Entirely Out of Date?** 61
   2.1   The *20 Famous Software Disasters* Reported 65
         2.1.1   Very High Project Failure Rate Reported 67
   2.2   What Is the Root Cause for Software Disasters
         and Very High Software Project Failure Rate? 67
   2.3   The "Software" Definition Is Outdated 69

	2.4	The Current Software Development Process Models Are Out of Date 70
	2.5	Current Software Development Methodologies Are Out of Date 71
	2.6	The Existing Software Modeling Approaches Are Outdated 72
	2.7	Current Software Testing Paradigm Is Out of Date 72
	2.8	Current Software Quality Assurance Paradigm Is Out of Date 72
	2.9	Current Software Visualization Paradigm Is Out of Date................... 73
	2.10	Current Software Documentation Paradigm Is out of Date 73
	2.11	Current Software Maintenance Paradigm Is Out of Date 73
	2.12	Current Software Project Management Paradigm Is Out of Date 74
	2.13	"The Mythical Man-Month" Is an Outcome of Linear Thinking; The "No Silver Bullet" Conclusion Is Out of Date 74
	2.14	Summary... 76
	2.15	Points and Questions to Ponder 77
	2.16	Further Reading and Information Source 77
	References... 77

3 Foundation for Establishing NSE: Complexity Science 79

3.1		The Basis of Complexity Science... 79
	3.1.1	Linear and Nonlinear.. 80
	3.1.2	Reductionism.. 80
	3.1.3	Chaos Theory 80
	3.1.4	System 81
	3.1.5	System Categories 81
	3.1.6	Linear System... 81
	3.1.7	Nonlinear System and Complex System................................. 81
	3.1.8	Feedback 82
	3.1.9	Fractal... 82
	3.1.10	Fractal Dimension 82
	3.1.11	Dynamical System 82
	3.1.12	Dissipation Structure.. 82
	3.1.13	Li–Yorke Theorem: Period Three Theorem.......................... 83
	3.1.14	Self-Organization 83
	3.1.15	Synergetics 83
	3.1.16	Catastrophe Theory.. 83
	3.1.17	Complex Adaptive System... 84
	3.1.18	Meta-Synthesis... 84
	3.1.19	Cellular Automata 84
	3.1.20	Genetic Algorithm.. 85
	3.1.21	Soliton 86
3.2		Linear Thinking and Nonlinear Thinking.. 86
3.3		The Essential Principles of Complexity Science 87
3.4		Applications of Complexity Science 88

	3.5	Complexity Science and NSE 89
	3.6	Summary 89
	3.7	Points and Questions to Ponder 89
	3.8	Further Reading and Information Source 89
	References 90

4 Prediction and Practices: A New Round of Industrial Revolution Driven by Complexity Science and a General Paradigm-Shift Framework 91

    4.1 Prediction: A New Round of Industrial Revolution Driven by Complexity Science Is Coming 91
    4.2 The Contribution and Limitation of Hall's Systems Engineering Framework 92
    4.3 The Background for the Innovation of FDS 93
    4.4 The Objectives of Innovating FDS 93
    4.5 The Description of FDS 94
        4.5.1 The "Principles of Complexity Science" Axis 94
        4.5.2 The "Environment" Axis 96
        4.5.3 The "People/Logic" Axis 96
        4.5.4 The "New Paradigm" Axis Modified from the "Knowledge/Skills" Axis in Hall's Framework 97
        4.5.5 The "Phases" (Workflows) Axis 97
    4.6 The Major Features of FDS 98
    4.7 Applications of FDS 99
    4.8 Bringing Feedback to the Research and Development of Complexity Science 100
    4.9 Summary 101
    4.10 Points and Questions to Ponder 101
    4.11 Further Reading and Information Source 101
    References 101

5 Outline of the NSE Paradigm 103

    5.1 A Tree Will Not Fall at One Blow: The Difficulty in Software Engineering Revolution 103
    5.2 The Objectives for Establishing NSE 105
    5.3 The Strategy to Achieve the Objectives of NSE 106
    5.4 The Establishment of NSE 106
    5.5 The Structure of NSE 107
    5.6 The Components of NSE 107
    5.7 The Major Feature and Characteristics of NSE 109
    5.8 Summary 112
    5.9 Points and Questions to Ponder 112
    5.10 Further Reading and Information Source 112
    References 113

6 The Techniques Innovated to Support NSE 115
- 6.1 Definitions 115
- 6.2 Holistic, Virtual, and Traceable Diagram Generation Technique 117
- 6.3 Virtual and Traceable Documentation Technique 119
- 6.4 Holistic and Intelligent Version Comparison Technique 121
- 6.5 Holistic and Dynamic Traceability Technique 122
- 6.6 Comprehensive Software Testing Technique Mainly Based on the Transparent-Box Method 122
- 6.7 Defect Prevention Driven Quality Assurance Technique 123
- 6.8 Test Case Efficiency Analysis and Test Case Minimization Technique 125
- 6.9 Refactoring Technique with Defect Prevention 126
- 6.10 Holistic MC/DC Test Coverage Analysis and Graphical Representation Technique 127
- 6.11 Assisted Test Case Design Technique 128
- 6.12 Intelligent Regression Test Case Selection Technique 128
- 6.13 Holistic, Actor–Action and Event–Response Driven, Traceable, Visual, and Executable Technique for Requirement Development 130
- 6.14 Synthesis Design and Incremental Growing Up (Implementation and Integration) Technique 131
- 6.15 Holistic, Global, and Side-Effect-Prevention Based Software Maintenance Technique 133
- 6.16 Summary 133
- 6.17 Points and Questions to Ponder 134
- 6.18 Further Reading and Information Source 134
- References 134

7 NSE Software Engineering Visualization Paradigm 135
- 7.1 The Old-Established Software Engineering Visualization Paradigm Is Outdated 135
- 7.2 The Revolutionary Solution Offered by NSE 137
- 7.3 The 3J graphics (J-Chart, J-Diagram, and J-Flow) 138
- 7.4 J-Chart 138
- 7.5 J-Diagram 140
- 7.6 J-Flow 148
- 7.7 Entire Software Life Cycle Visualization with NSE 153
- 7.8 Rich Options for Generating 3J Graphics 155
    - 7.8.1 For J-Chart Generation 155
    - 7.8.2 For J-Diagram and J-Flow Generation 160
- 7.9 The Major Features of NSE Software Visualization Paradigm 160
- 7.10 Applications 180
- 7.11 Self-Documenting 191

Contents   xxvii

	7.12	Summary	195
	7.13	Points and Questions to Ponder	196
	7.14	Further Reading and Information Source	196
	References		197

8 NSE Process Model 199

	8.1	Some Experts' Expectations	199
	8.2	All of the Existing Software Engineering Process Models Are Outdated	201
	8.3	Outline of the Revolutionary Solution Offered with NSE	202
	8.4	The Driving Forces and The Support Techniques	203
	8.5	The Graphical Representation of the NSE Process Model	204
		8.5.1 The Objectives of the Preprocess	206
		8.5.2 The Objectives of the Main Process	207
		8.5.3 The Objective of the Support Facility for Automated and Bidirectional Traceability	208
	8.6	The Major Steps of the Preprocess	208
	8.7	The Major Steps of the Main Process	213
	8.8	The Support Facility for Automated and Bidirectional Traceability	224
	8.9	The Manifestation of the Essential Principles of Complexity Science in the NSE Process Model	225
	8.10	The Major Features and Characteristics of the NSE Process Model	226
	8.11	Summary	234
	8.12	Points and Questions to Ponder	235
	8.13	Further Reading and Information Source	236
	References		236

9 The Facility for Automated and Self-Maintainable Traceability 237

	9.1	The Importance of Requirement Traceability	238
	9.2	The Problems Addressed	238
	9.3	The Solution Offered with NSE	239
		9.3.1 Part 1	240
		9.3.2 Part 2	240
	9.4	How It Works	242
		9.4.1 Bidirectional Traceability Between the Test Cases and the Source Code Modules or Branches	245
		9.4.2 Extending the Bidirectional Traceability to Include All Related Documents	246
	9.5	The Major Features	249
		9.5.1 Automated	249
		9.5.2 Self-Maintainable	250

		9.5.3 Methodology-Independent 250
		9.5.4 Nonlinear, Bidirectional, and Parallel 250
		9.5.5 Accurate 250
		9.5.6 Precise 251
		9.5.7 Extended to Include Software Project Management Documents 251
		9.5.8 Extended to Include Web Pages 251
		9.5.9 Extended for Multiproject Support 251
		9.5.10 Dynamic 252
		9.5.11 Easy to Add on at Any Time, In Any Status 253
	9.6	Application 254
	9.7	Summary 254
	9.8	Points and Questions to Ponder 255
	9.9	Further Reading and Information Source 256
	References 256

10 NSE Software Development Methodology Driven by Defect Prevention and Traceability 257
    10.1 Almost All Existing Software Development Methodologies Are Outdated 257
    10.2 Outline of the Revolutionary Solution Offered by NSE 259
    10.3 The Driving Forces for the Innovation of the NSE Software Development Methodology 263
    10.4 The Related NSE Software Engineering Process Model 265
    10.5 Graphical Presentation of the NSE Software Development Methodology 267
    10.6 Application 270
        10.6.1 Some Suggestions About the Applications of the NSE Software Development Methodology 270
    10.7 The Major Features of the NSE Software Development Methodology 271
    10.8 Summary 271
    10.9 Points and Questions to Ponder 272
    10.10 Further Reading and Information Source 272
    References 272

11 Requirement Engineering Under NSE: Source Code Driven Dynamic Software Modeling 273
    11.1 Are All the Existing Software Modeling Approaches Outdated? 273
    11.2 Outline of the Revolutionary Solution Offered by NSE 276
    11.3 Description of the HAETVE Technique 279
    11.4 Applications of HAETVE 286
    11.5 How to Make a Hard Copy of a Graphical Requirement Document 303

	11.6	Suggestions for the Requirement Documentation Design............304
	11.7	The Major Features of HAETVE..306
	11.8	More About Dynamic Modeling...309
	11.9	Summary..311
	11.10	Points and Questions to Ponder...311
	11.11	Further Reading and Information Source....................................311
	References..312

12 Design Engineering Under NSE 313
- 12.1 The Major Problem Addressed..313
- 12.2 Outline of the Solution for Software Design with NSE..............314
- 12.3 Description of the Innovated "Synthesis Design and Incremental Growing Up" Technique.........................315
  - 12.3.1 Basic Ideas...315
  - 12.3.2 What is Synthesis? What is Analysis?..........................316
  - 12.3.3 Recommendation for Graphic Document Creation/ Generation..318
  - 12.3.4 Self-Documenting..320
  - 12.3.5 Detailed System Hierarchy Design................................321
  - 12.3.6 Static Defect Prevention and Defect Propagation Prevention Through Traceability..................321
  - 12.3.7 Dynamic Defect Prevention and Defect Propagation Prevention..321
  - 12.3.8 Data Structure Design...323
  - 12.3.9 Detailed Logic Design of the Modules..........................323
- 12.4 Application...325
- 12.5 The Major Features of the Software Synthesis Design Technique..336
- 12.6 Summary..337
- 12.7 Points and Questions to Ponder...337
- 12.8 Further Reading and Information Source....................................338
- References..338

13 Coding Engineering with NSE..339
- 13.1 The Problems Addressed...339
- 13.2 The Solution: Software Coding Engineering with NSE Using the Synthesis Design and Incremental Integration Technique........341
- 13.3 Unit Testing and Integration Testing Support.............................349
- 13.4 MC/DC Test Coverage Measurement Support...........................353
  - 13.4.1 Conclusion...361
- 13.5 Semiautomated Inspection Support..362
- 13.6 Defect Prevention Driven Quality Assurance in Programming......364
- 13.7 Quality Measurement for an Entire Software Product and Each of Its Components..366
- 13.8 Application...367

13.9 The Major Features 368
13.10 Summary 368
13.11 Points and Questions to Ponder 368
13.12 Further Reading and Information Source 369
References 369

14 The Basis of Software Testing 371
14.1 The Purpose of Software Testing 371
14.2 Functional Testing and the Black-Box Method 373
14.3 Structural Testing and the White-Box Method 373
   14.3.1 Test Coverage Metrics 374
   14.3.2 Instrumentation Methods 374
14.4 Gray-Box Testing 375
14.5 Performance Testing and the Testing Method 376
14.6 Other Nonfunctional Testing 377
14.7 Unit Testing, Integration Testing, and System Testing 378
14.8 Regression Test After Code Modification 378
14.9 Object-Oriented Software Testing 378
14.10 Web Application Testing 380
14.11 Embedded Software Testing 381
14.12 GUI Operation Capture and Playback 382
14.13 Acceptance Testing 383
14.14 Why Should Software Testing Tools Be Used 383
14.15 The Major Drawback of the Major Existing
       Software Testing Paradigm and the Solution 383
14.16 Summary 384
14.17 Points and Questions to Ponder 384
14.18 Further Reading and Information Source 384
References 384

15 Software Test Case Design 387
15.1 What Is a Test Case? 387
15.2 The Basis of Test Case Design 388
   15.2.1 Equivalence Class Partition
           and Boundary Value Analysis 388
   15.2.2 State Transition Analysis 389
   15.2.3 Conditions Combination Method 389
15.3 Semiautomated Test Case Design 390
15.4 Test Case Efficiency Measurement 391
15.5 Test Case Minimization 391
15.6 NSE Test Case Design with HAETVE Technique
       for Both Functional Testing and Structural Testing 396
15.7 Automated Test Case Selection with Automated
       Test Case Execution 405
15.8 Summary 406

	15.9	Points and Questions to Ponder 407
	15.10	Further Reading and Information Source 407
	References 407	

**16 The NSE Software Testing Paradigm Based
on the Transparent-Box Method** 409
- 16.1 The Major Existing Software Testing Methods,
  Techniques, and Tools Are Outdated 409
- 16.2 The Transparent-Box Testing Method 411
- 16.3 The New Software Testing Paradigm Based
  on the Transparent-Box Testing Method 413
- 16.4 The Major Features of the New Software Testing Paradigm 417
- 16.5 A General Comparison Between the New Software
  Testing Paradigm and the Old One 429
- 16.6 Summary 432
- 16.7 Points and Questions to Ponder 432
- 16.8 Further Reading and Information Source 432
- References 432

**17 NSE Software Quality Assurance Paradigm Driven
by Defect Prevention** 433
- 17.1 The Old-Established Software Quality
  Assurance Paradigm Is Outdated 433
- 17.2 Outline of NSE Software Quality
  Assurance Paradigm (NSE-SQA) 435
- 17.3 Description of NSE Software Quality Assurance Paradigm 436
  - 17.3.1 The Foundation of NSE-SQA 436
  - 17.3.2 The Framework for Establishing NSE-SQA 436
  - 17.3.3 The Purpose of NSE-SQA 437
  - 17.3.4 Definitions 437
  - 17.3.5 The Quality Assurance Strategy of NSE-SQA 439
  - 17.3.6 The Implementation of the Quality Assurance
    Strategy of NSE-SQA 439
- 17.4 Application of NSE-SQA 460
- 17.5 The Major Features of NSE-SQA 460
- 17.6 Summary 463
- 17.7 Points and Questions to Ponder 464
- 17.8 Further Reading and Information Source 464
- References 464

**18 NSE Software Maintenance Paradigm:
Systematic, Disciplined, and Quantifiable** 467
- 18.1 The Existing Software Maintenance Engineering
  Paradigm Is Outdated 467
- 18.2 Outline of the NSE Software Maintenance Paradigm 470

	18.3	Description of NSE Software Maintenance Engineering Paradigm 476
	18.4	Application 477
	18.5	The Major Features 485
	18.6	Summary 487
	18.7	Points and Questions to Ponder 487
	18.8	Further Reading and Information Source 487
	References 488

19 NSE Documentation Paradigm: Virtual, Traceable, and Consistent with the Source Code 489

	19.1	The Old-Established Software Documentation Paradigm Is Outdated 489
	19.2	Outline of NSE Documentation Paradigm 491
	19.3	Description of the NSE Documentation Paradigm 494
		19.3.1 The Critical Issues with the Old-Established Software Documentation Paradigm 494
		19.3.2 The Solution Offered with NSE 495
		19.3.3 The Objectives of the NSE Documentation Paradigm 496
		19.3.4 Working with Dummy Programming 497
		19.3.5 Working with NSE Software Visualization Paradigm 497
		19.3.6 Working with HAETVE Requirement Development Technique 497
		19.3.7 How It Works 500
		19.3.8 Making a Software Product Visible in Multiple-Views 500
	19.4	The Major Features of NSE Documentation Paradigm 505
	19.5	Application 510
	19.6	Summary 510
	19.7	Points and Questions to Ponder 512
	19.8	Further Reading and Information Source 514
	References 515

20 NSE Project Management Paradigm: Seamlessly Combined with the Project Development Process 517

	20.1	The Old-Established Software Project Management Paradigm Is Outdated 517
	20.2	Outline of the NSE Project Management Paradigm 518
	20.3	The Foundation of NSE Project Management Paradigm 519
	20.4	The Strategy of NSE Project Management Paradigm 520
	20.5	People Oriented 521
	20.6	Focusing on Maintenance 522
	20.7	More Method and Tool Support 523
	20.8	Combination of Product Development and Project Management 524
	20.9	Finding Problems Early and Solving the Problems in Time 528

20.10	Quality Management	528
20.11	Multiple-Project Management	528
20.12	Summary	528
20.13	Points and Questions to Ponder	529
20.14	Further Reading and Information Source	530
References		530

21 Algorithms Innovated for Establishing NSE 531

21.1 The Algorithm for Realizing Modified Condition/Decision Coverage Test Coverage Measurement 532
- 21.1.1 The Requirements 532
- 21.1.2 The Basic Idea 532
- 21.1.3 The Major Steps 533
- 21.1.4 Application 533

21.2 The Algorithm for Test Case Efficiency Analysis and Test Case Minimization 533
- 21.2.1 The Requirements 533
- 21.2.2 The Basic Idea 534
- 21.2.3 The Major Steps 535
- 21.2.4 Application 536

21.3 The Algorithm for Performance Analysis 536
- 21.3.1 The Requirements 536
- 21.3.2 The Basic Idea 537
- 21.3.3 The Major Steps 538
- 21.3.4 Application 539

21.4 The Algorithm for Cyclomatic Complexity Analysis 540
- 21.4.1 The Requirement 540
- 21.4.2 The Basic Idea 540
- 21.4.3 The Major Steps 541
- 21.4.4 Application 541

21.5 The Algorithm for Tracing the Execution Path of a Runtime Error 542

21.6 The Algorithm for the Layout of the Call Graph of a Program Using J-Chart Notations 543

21.7 The Algorithm for Holistic Version Comparison of a Software Product 543

21.8 The Algorithm for Memory Leak and Usage Violation Analysis 543

21.9 The Algorithm for Realizing the Traceability of the Diagrammed Source Code 546

21.10 The Algorithm for Dynamic Traceability 549

21.11 Summary 553

21.12 Points and Questions to Ponder 553

21.13 Further Reading and Information Source 555

References 555

22 NSE Support Tools and NSE Support Platforms 557
- 22.1 Full Software Development Lifecycle Support 557
- 22.2 The Product Development History 557
  - 22.2.1 The First Generation: Hindsight 557
  - 22.2.2 Second Generation: Panorama 558
  - 22.2.3 Panorama++ 560
- 22.3 Automated Tools Integrated with Panorama++ 560
- 22.4 Panorama++ Product Installation 560
- 22.5 A Guided Tour of Panorama++ for C/C++ 565
- 22.6 Network Floating License Support 574
- 22.7 The Major Features of Panorama++ 575
- 22.8 Applications 575
- 22.9 Summary 575
- 22.10 Points and Questions to Ponder 575
- 22.11 Further Reading and Information Source 575
- References 576

23 NSE Applications 577
- 23.1 The Whole and Its Components: A General Comparison Between NSE and Other Approaches 577
- 23.2 What Makes NSE Special? 579
- 23.3 Applications in New Software Development 579
  - 23.3.1 Benefits 579
  - 23.3.2 Recommended Process 580
- 23.4 Applications in a Software Product Being Developed Using Other Approaches 583
- 23.5 Possible Combination with UML 583
  - 23.5.1 About the Future of UML 583
  - 23.5.2 Question to the Future of UML 583
  - 23.5.3 Possible Combination with UML (NSE-UML?) 584
  - 23.5.4 Possible Combination with CMMI (NSE-CMMI?) 585
- 23.6 Possible Combination with Agile Software Development Approaches 587
  - 23.6.1 Possible Combination with XP (NSE-XP?) 592
- 23.7 Possible Combination with RUP (NSE-RUP?) 593
- 23.8 Support for CBSE 593
- 23.9 Summary 594
- 23.10 Points and Questions to Ponder 595
- 23.11 Further Reading and Information Source 595
- References 595

24 Candidates of "Silver Bullet" 597
- 24.1 Is "The Mythical Man-Month" an Outcome of Linear Thinking, Reductionism, and Superposition Principle? 597
  - 24.1.1 A Great book 598
  - 24.1.2 Limitation 599

24.2	Is the "No Silver Bullet" Conclusion Outdated?		599
24.3	The First Candidate of "Silver Bullet"		602
24.4	The Second Candidate of "Silver Bullet"		604
24.5	Can the "Silver Bullet" Defined by Brooks Slay the "Werewolves" Defined by Him?		605
24.6	What Kind of "Silver Bullet" Can be Used to Slay the "Werewolves" Defined by Brooks?		607
24.7	The Third Candidate of "Silver Bullet": The Entire NSE Paradigm		609
	24.7.1	What Is NSE: The Whole and Its Components	610
	24.7.2	The Components of NSE	614
	24.7.3	The Major Features and Characteristics of NSE	616
	24.7.4	The Major Differences Between NSE and the Old-Established Software Engineering Paradigm	618
	24.7.5	Qualification as a Candidate of "Silver Bullet" for Slaying Software "Werewolves"	628
24.8	Summary		647
24.9	Points and Questions to Ponder		648
24.10	Further Reading and Information Source		649
References			649

Appendix A: Software Requirements Specification Template To Be Used with NSE 651

Appendix B: An Example About How to Realize 100% MC/DC (Modified Condition/Decision Coverage) for a Program Unit 675

Appendix C: How to Control/Simulate the Return Values of a Program Unit Being Tested 699

Appendix D: Hints for Answering the "Points and Questions to Ponder" in Each Chapter 703

Glossary 727

Index 731

Chapter 1
Introduction

> *Software is becoming the foundation of modern civilization; software constitutes or will control the products, services, and infrastructure people will rely on for a wide variety of daily activities from the vital to the trivial. ... software is not sufficiently engineered at this time to fulfill the role of "foundation."*
>
> David Rice

Today software is becoming the foundation of modern civilization. It is playing an important role in the development of all kinds of businesses in the world. It affects almost all aspects of our lives and our everyday activities.

1.1 What Is Software?

With the existing software engineering paradigm, software is defined as follows:

1. Instructions (computer programs) that when executed provide desired features, function, and performance
2. Data structures that enable the programs to adequately manipulate information
3. Documents that describe the operation and use of the programs [Pre95-p4]

But this software definition is outdated because

- The program(s) and the documents are provided without describing how they are managed together with bidirectional traceability among them.
- The documents are often inconsistent with the source code after code modification is done again and again in the software development process.
- The history and the results of the static and dynamic program measurement are missing or ignored.
- The program(s) is not represented graphically, making it hard to read and understand.

- The working conditions at the customer's site are quite different from the product development site, making the product acceptance testing and product maintenance hard to perform.
- The software product as defined is not adaptive to its new working environment in the customer site.

With NSE (Nonlinear Software Engineering paradigm, will be described in detail from Chaps. 3 to 24 in this book), a **software** (software product) is redefined as and delivered to the customer with

(a) A computer program (a regular program, or a cloud computing program, or a program developed through the internet) with the source code
(b) The data used
(c) All of the related documents (including the test case scripts too) traceable to and from the source code, plus
(d) **The database built though static and dynamic measurement of the program**
(e) **A set of Assisted Online Agents (automated and intelligent tools working with the program and the database) for handling the issue of complexity and supporting the testability, visibility, changeability, conformity, reliability, and traceability – making the software product adaptive and truly maintainable in the new working environment at the customer site, and that the requirement validation and the acceptance testing can be done dynamically in a fully automated way with mouse clicks only**

Why should a software product be delivered to the customer with the database plus a set of intelligent agents too? The main objective is to make the software product easy to understand, test, and truly maintainable at the customer site.

For comparing the old and new software definition and getting hands-on experience with the new software definition to know how software acceptance testing can be done automatically and how software maintenance can be performed globally and holistically with side-effect prevention, it is strongly recommended for readers to install and try the NSE-CLICK toolkit provided (see Preface) through an application example (a calculator software product). This example shows what a customer will get for his/her software product developed by a third party (or through outsourcing software development) applying the NSE paradigm and the support platform Panorama++: the customer will get the program with the source code, the data used, the documents with bidirectional traceability to the source code to keep consistency with each other, plus the database built through static and dynamic measurement, and (after signing a maintenance contract) an end-user license to use (but not own) a set of Assisted *Online Agents*, including.

1.1 What Is Software?

- The NSE-CLICK interface
- The OO-Browser for generating interactive and traceable call graphs or class inheritance charts shown in J-Chart notations (see Chap. 7) innovated by me
- The OO-Diagrammer for generating interactive and traceable logic diagrams shown in J-Diagram notations or control flow diagram in J-Flow notations (see Chap. 7) innovated by me
- The OO-Test for performing software testing using Transparent-box method combining functional testing and structural testing (for Modified Condition/Decision (MC/DC) test coverage analysis) together seamlessly with the capability to establish bidirectional traceability among the related documents and test cases and source code (see Chap. 16) innovated by me
- The OO-V&V for Requirement Validation and Verification through bidirectional traceability
- The OO-SQA for software quality measurement
- The OO-Analyzer for dynamic and static program measurement
- The OO-MemoryCheck for checking memory leaks and usage violations
- The OO-Performance for performance measurement
- The OO-DefectTracer for tracing each runtime error to the execution path
- The OO-MiniCase for test case efficiency analysis and test case minimization in order to perform regression testing efficiently after code modification
- The OO-Playback for GUI operation capture and playback after code modification
- The OO-CodeDiff for holistic and intelligent software version comparison, etc. for supporting testability, visibility, changeability, conformity, traceability, and maintainability.

With these *Assistant Online Agents*, the acceptance testing can be done in a fully automated way as follows:

- Start the NSE-CLICK (double click the e_NSE_CLICK.exe file from the home directory of NSE-CLICK after the installation), the NSE-CLICK interface will show up (see Fig. 1.1).
- Load the database of the calculator (from C:\isa_examples\English_examples\analyzed_for_review\cal) as shown in Fig. 1.2.
- View the program structure shown in J-Chart notations with some overall measurement results using OO-Browser (see Figs. 1.3–1.8) – the operations and the results.

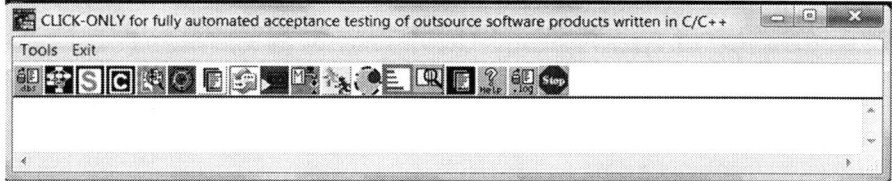

Fig. 1.1 The NSE-CLICK interface (the original icons are shown in different colors)

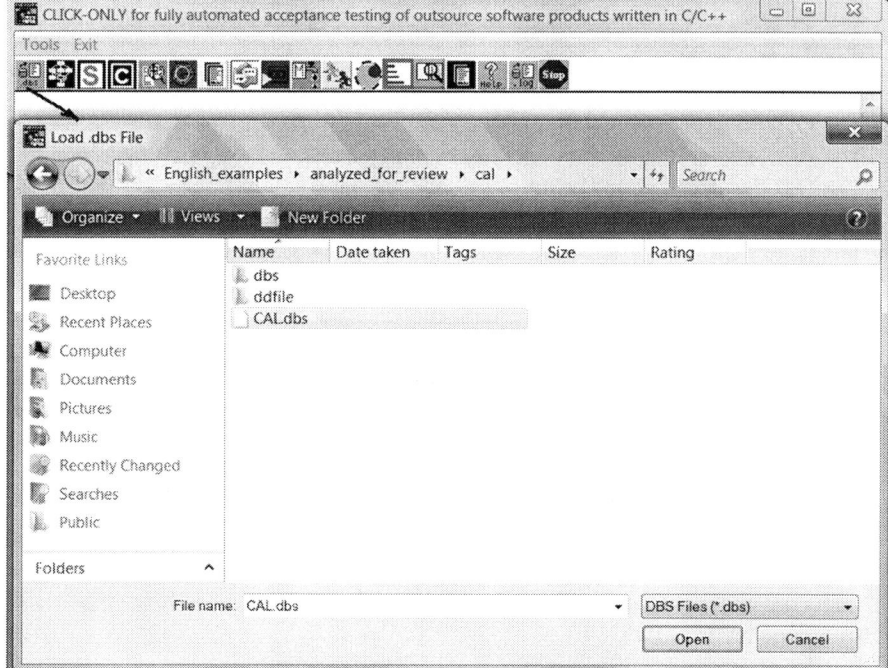

Fig. 1.2 The operations for loading the database of the calculator program

- View the program logic diagram using J-Diagram notations and control flow diagram using J-Flow diagram with automated traceability as shown in Figs. 1.1–1.14 – the operations and the results:
- Select the corresponding options to view the MC/DC (Modified Condition/ Decision Coverage) test coverage measurement result with untested conditions and branches highlighted, as shown in Fig. 1.12.
- **Try to use the traceability automatically established, as shown in Fig. 1.13.**
- **View the performance measurement result as shown in Figs. 1.15 and 1.16 – the operations and the result.**

1.1 What Is Software? 5

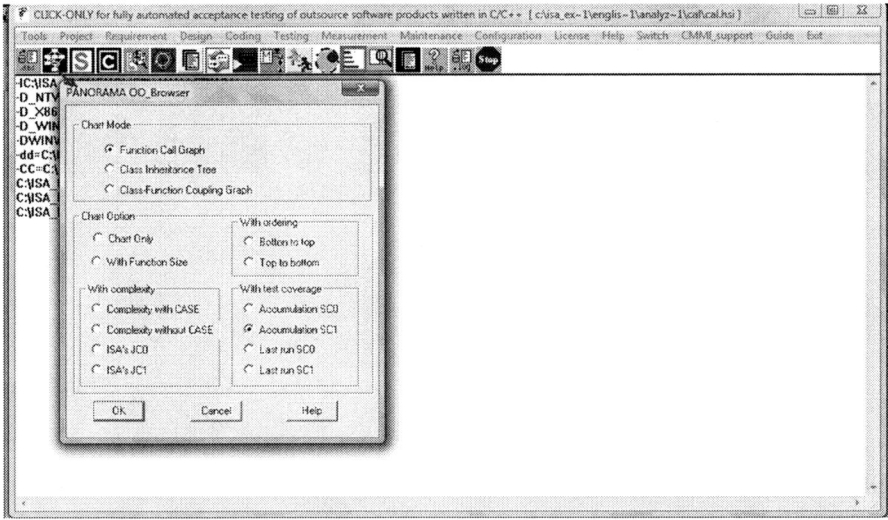

Fig. 1.3 Start the OO-Browser

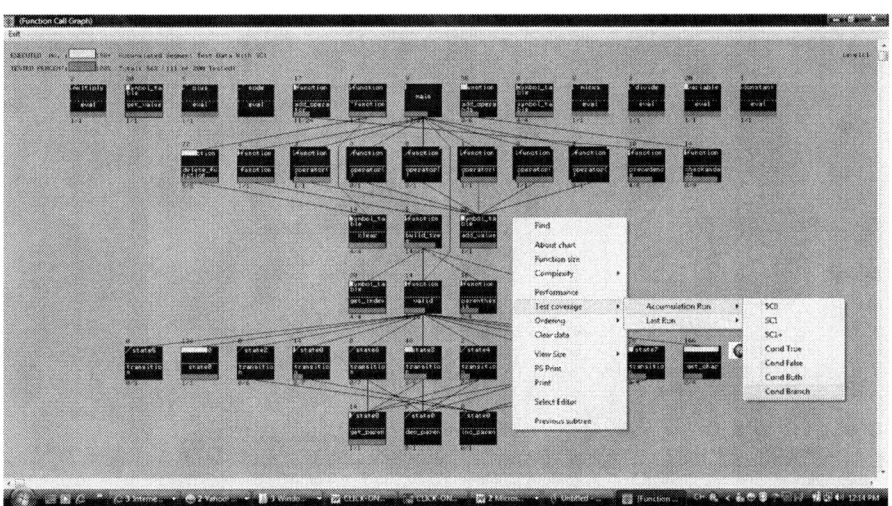

Fig. 1.4 View the overall test coverage measurement result (a *bar graph* at the *bottom* of each module-box shows the percentage of the elements tested)

6   1 Introduction

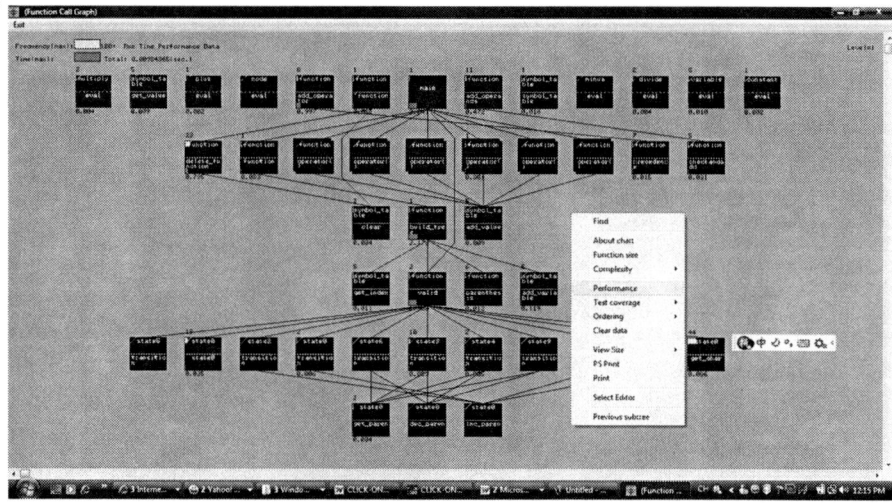

Fig. 1.5 View the performance measurement result for locating bottlenecks (a *bar* at the *bottom* of each module-box shows the percentage of the time used)

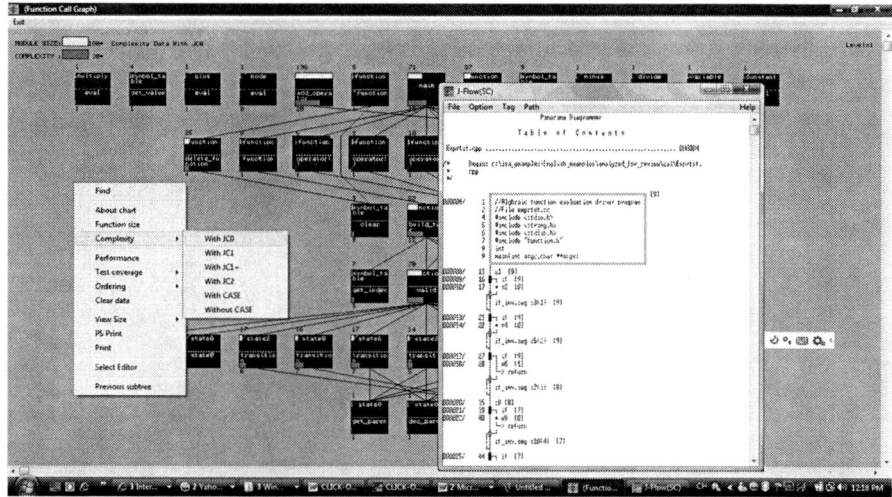

Fig. 1.6 View the measurement result of the cyclomatic complexity (the number of decision statements such as "if," "for," "while," "do," "switch")

1.1 What Is Software?

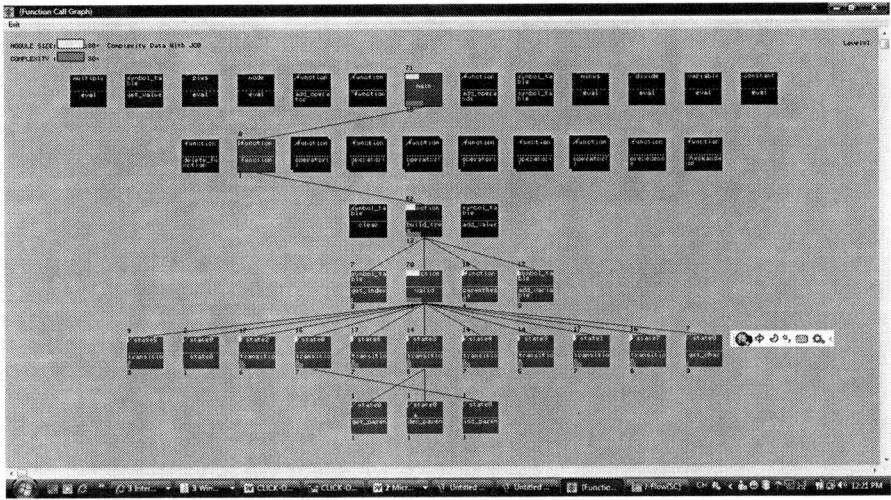

Fig. 1.7 Highlight/trace a module with all of the related modules calling and called by it

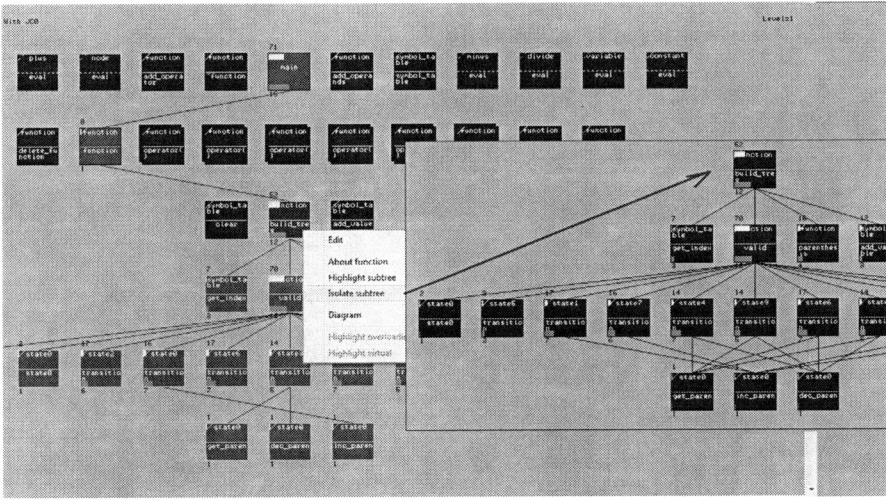

Fig. 1.8 Use a module as the root to generate a sub-call-graph

8   1 Introduction

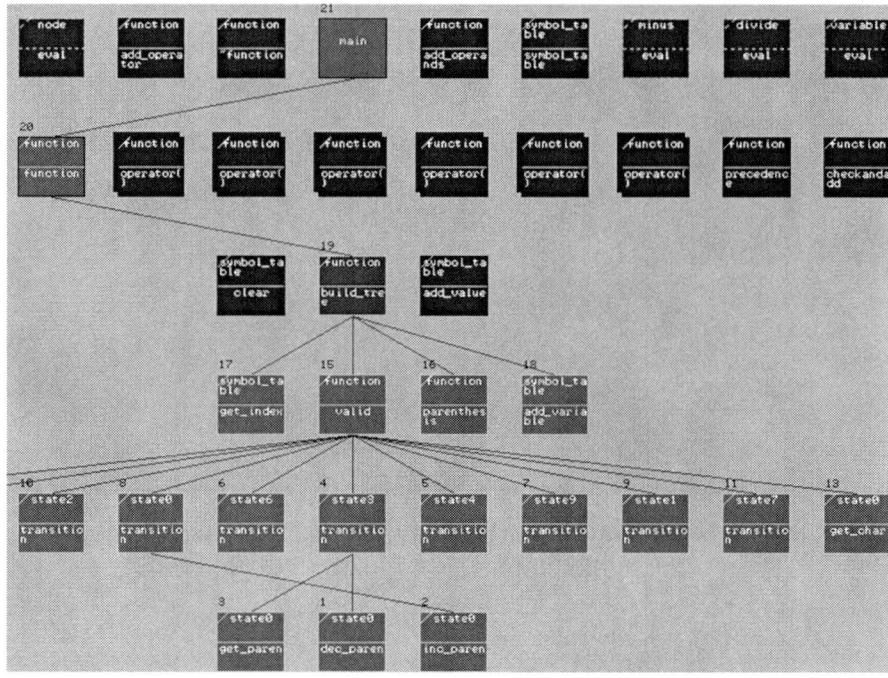

Fig. 1.9 Assign bottom-up order for incremental unit coding and unit testing for a critical path or the entire product to support reverse engineering

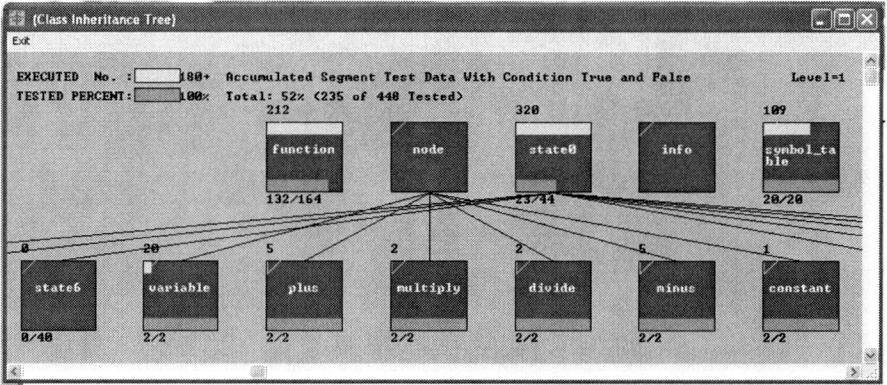

Fig. 1.10 View the class inheritance chart with test coverage data (note: a class cannot be tested directly, so the class test coverage data are collected from their instances)

1.1 What Is Software?

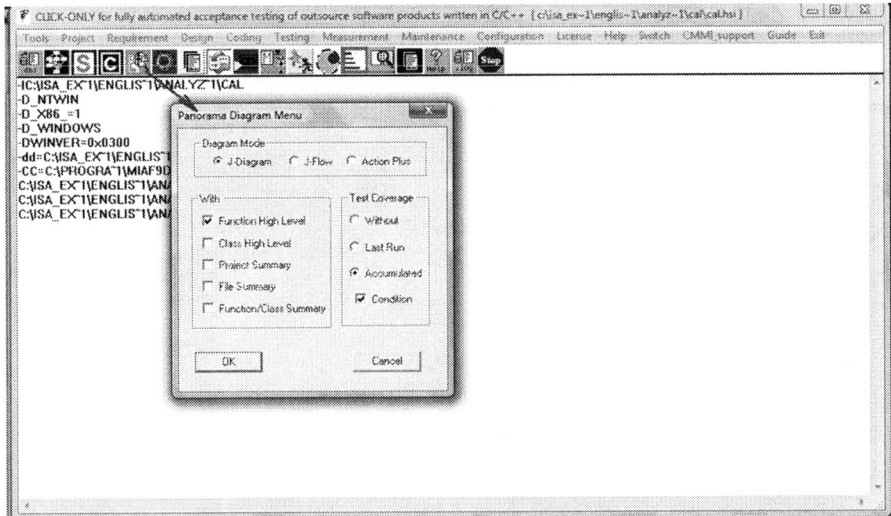

Fig. 1.11 Start OO-Diagrammer

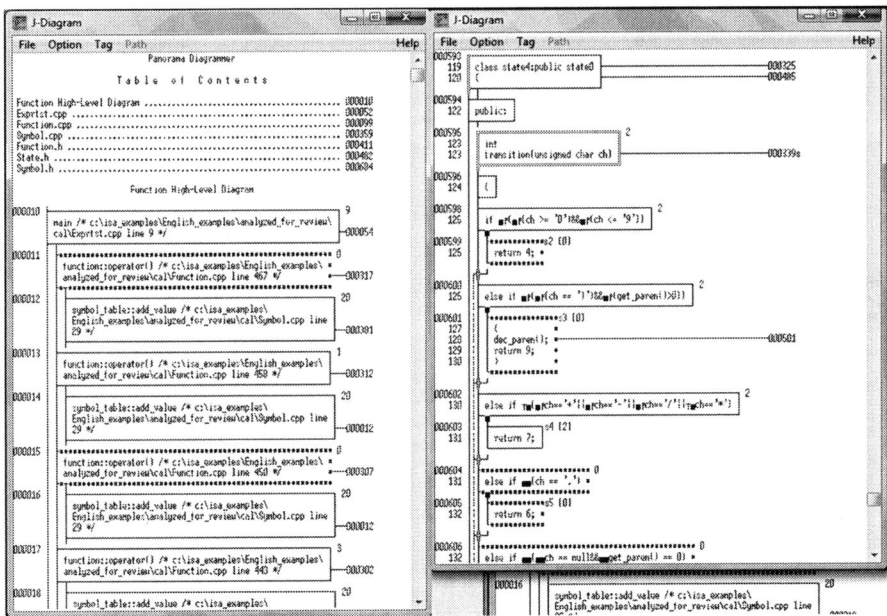

Fig. 1.12 The MC/DC (Modified Condition/Decision Coverage) test coverage measurement result

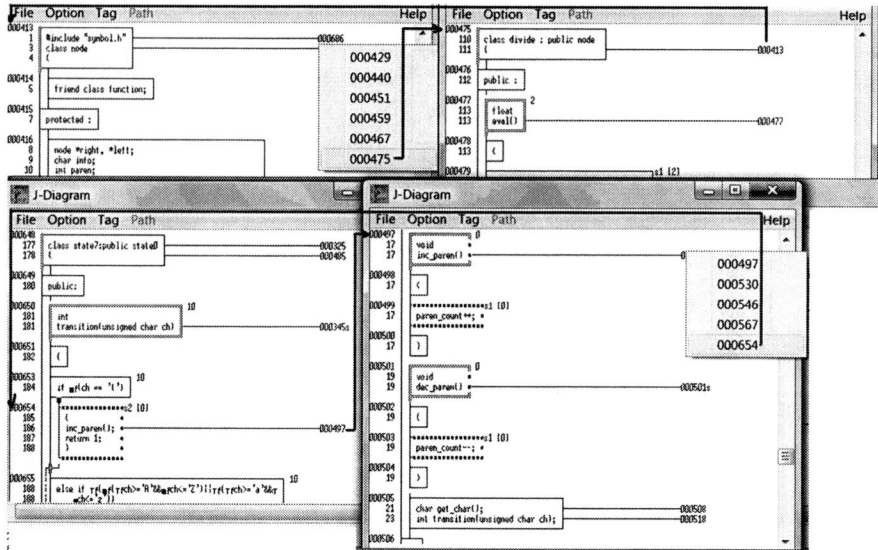

Fig. 1.13 The traceabilities established automatically

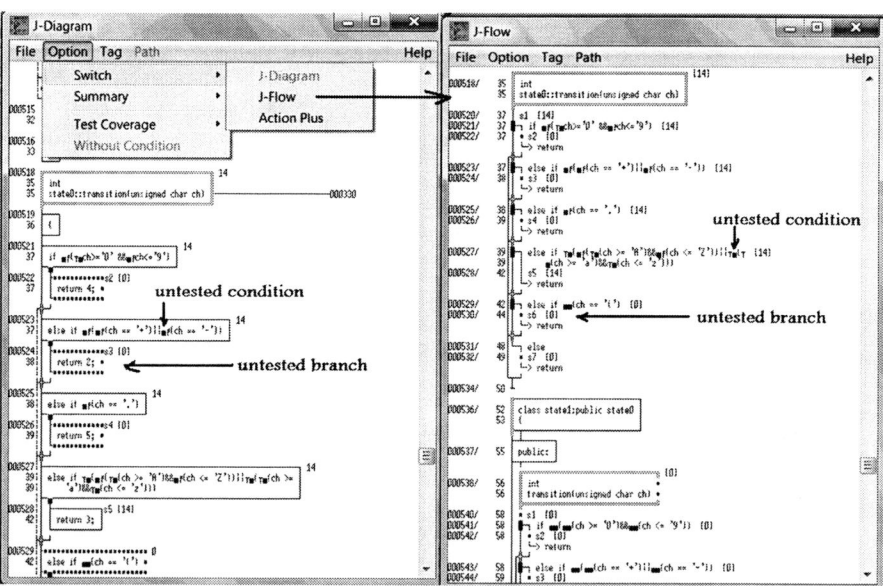

Fig. 1.14 Convert J-Diagram (showing MC/DC test coverage measurement result with untested conditions and branches highlighted) to J-Flow diagram

1.1 What Is Software?

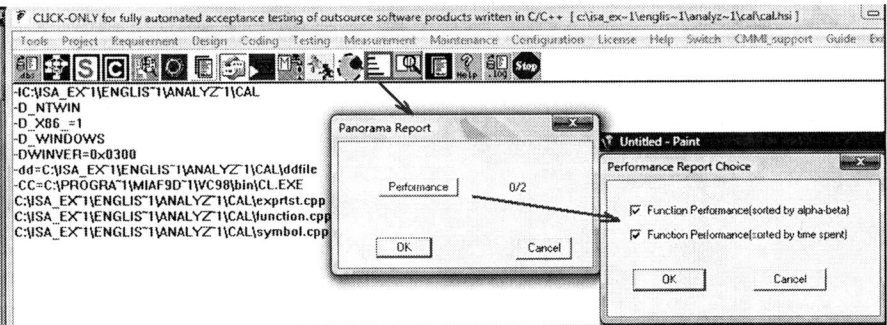

Fig. 1.15 Start the OO-Performance tool

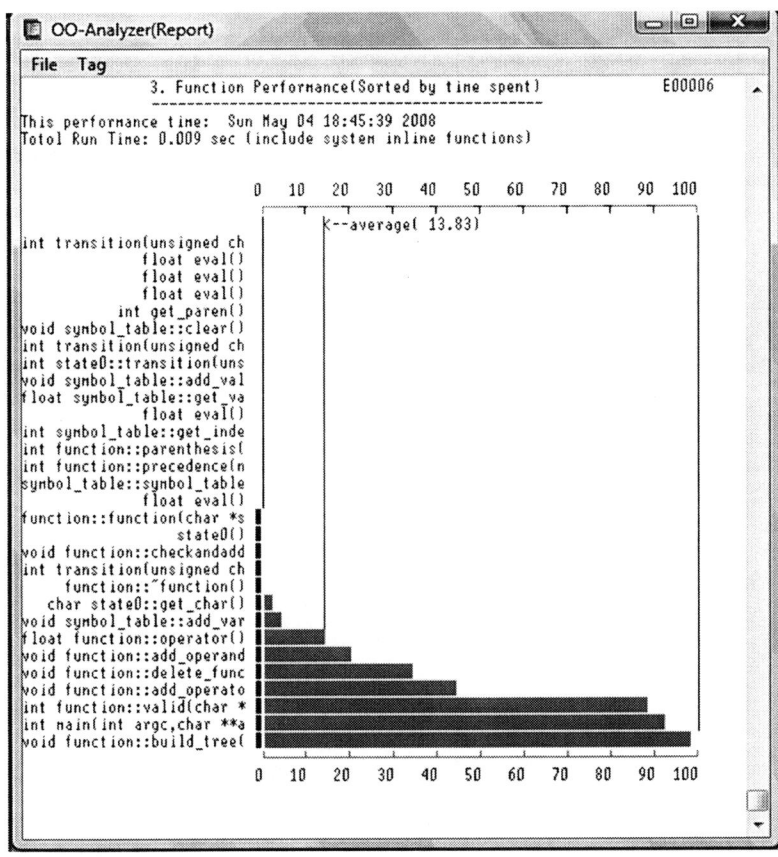

Fig. 1.16 The performance measurement result

- View the general static and dynamic measurement reports as shown in Figs. 1.17 and 1.18 – the operations and the results.
- View the quality measurement result as shown in Figs. 1.19 and 1.20 – the operations and the results.
- View the execution path for a runtime error as shown in Fig. 1.21 – the operation and the result.
- View the memory leak measurement result as shown in Fig. 1.22.
- View the measurement results of test case efficiency analysis and test case minimization for efficient regression testing after software modification shown in Figs. 1.23–1.25 – the operations and the results.
- View the support for bidirectional traceability (established through Time Tags automatically inserted into both the test cases and the code test coverage measurement database for mapping them, see Chap. 9), requirement validation and verification in Figs. 1.26–1.32 – the operations and the results:
  - View the traceability between test cases and the source code as shown in Figs. 1.27–1.29 – the operations and the results.
  - View the traceability extended to include all related documents and test cases and source code as shown in Figs. 1.30–1.34 – the operations and the results.

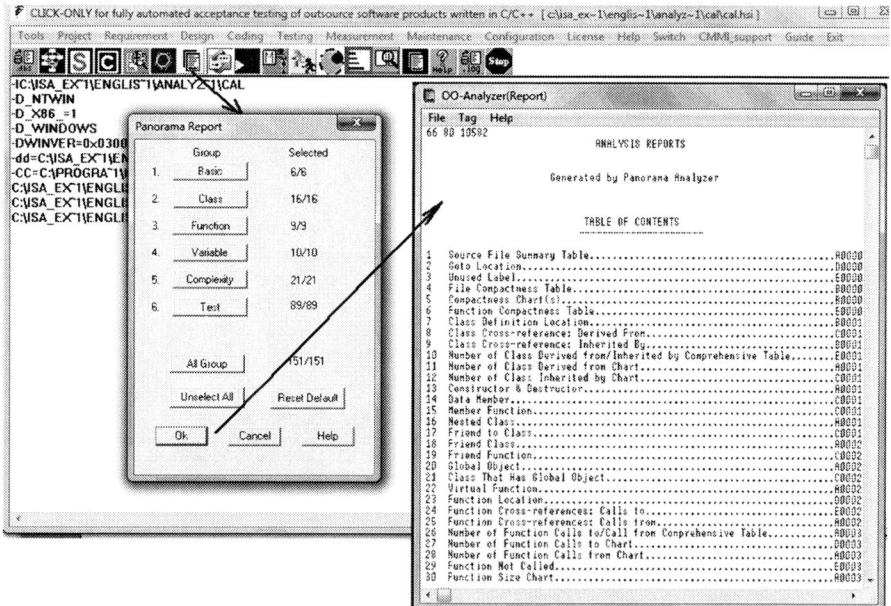

Fig. 1.17 Start the OO-Analyzer tool which generates about 150 dynamic and static measurement reports

1.1 What Is Software?

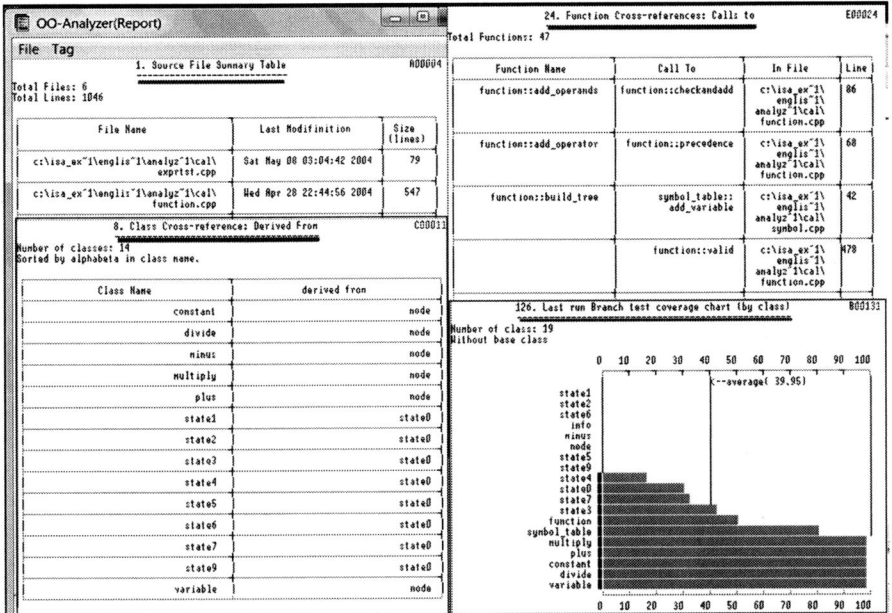

Fig. 1.18 Partial reports

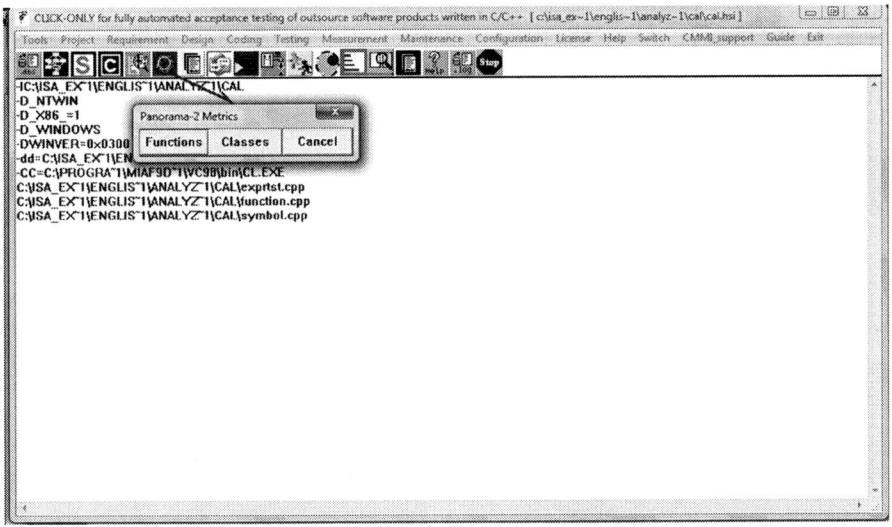

Fig. 1.19 Start the OO-SQA tool

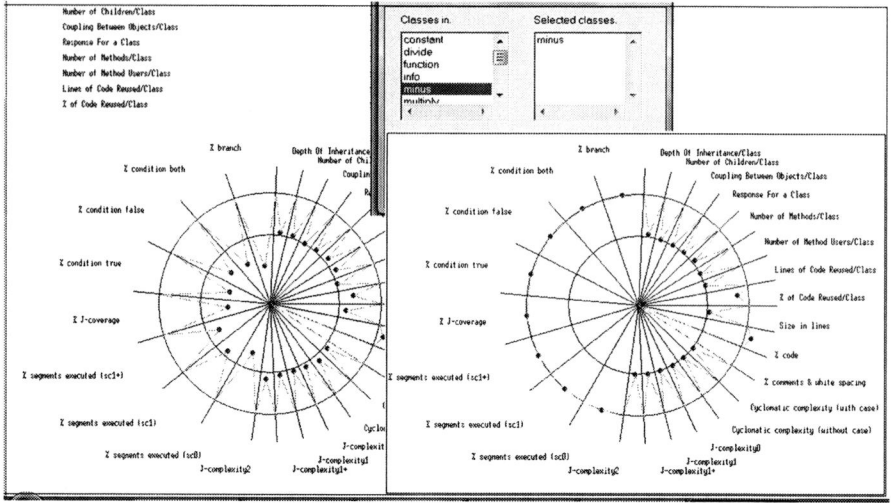

Fig. 1.20 View the total quality measurement result for the entire product or each individual class/function shown with a Kiviat diagram

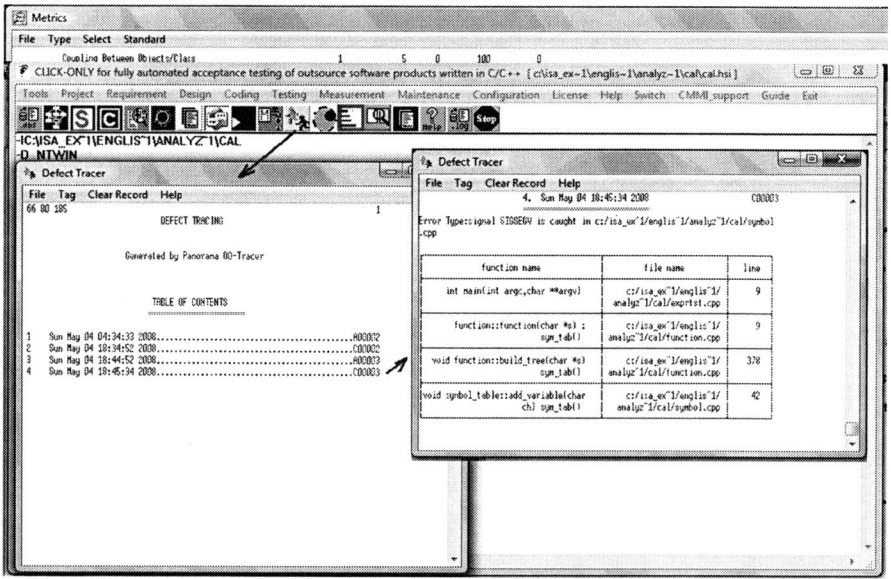

Fig. 1.21 View the execution path traced from a runtime error to easily locate the defect

1.1 What Is Software?

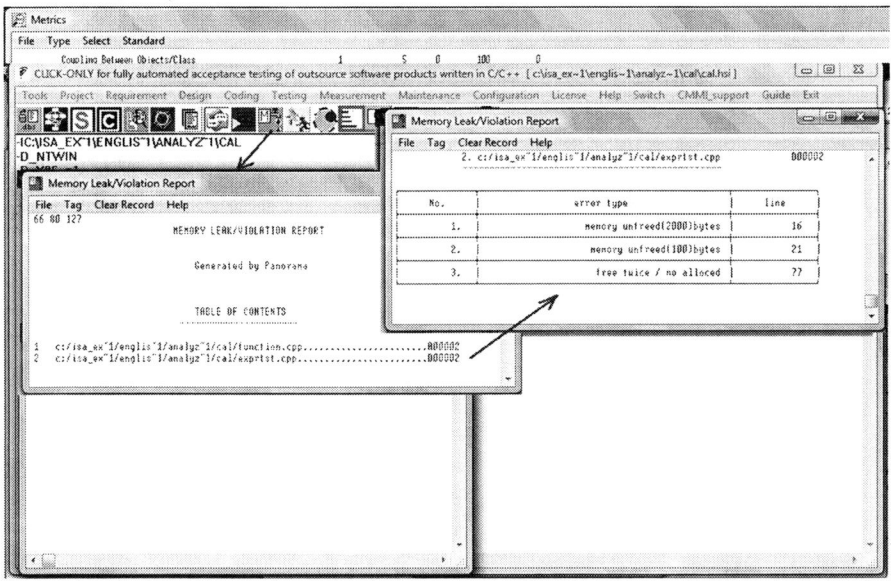

Fig. 1.22 Start the OO-MemoryCheck tool and view the measurement result

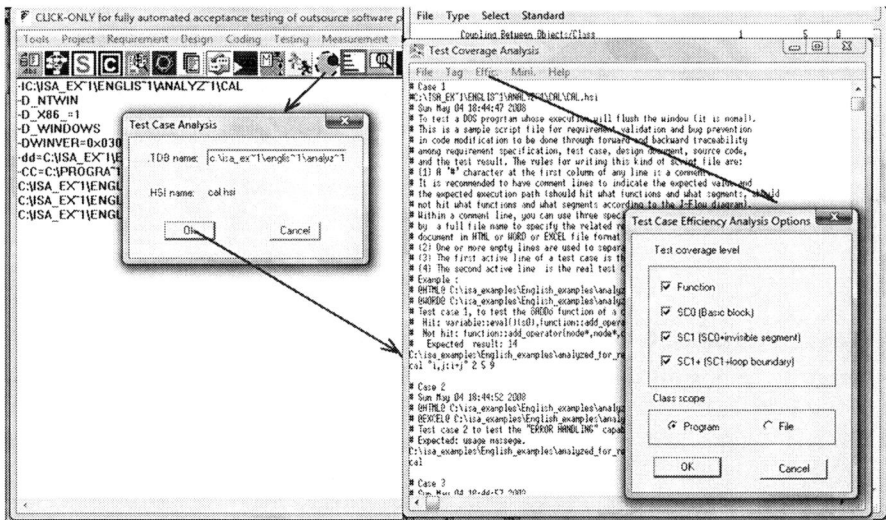

Fig. 1.23 Start the OO-MiniCase tool

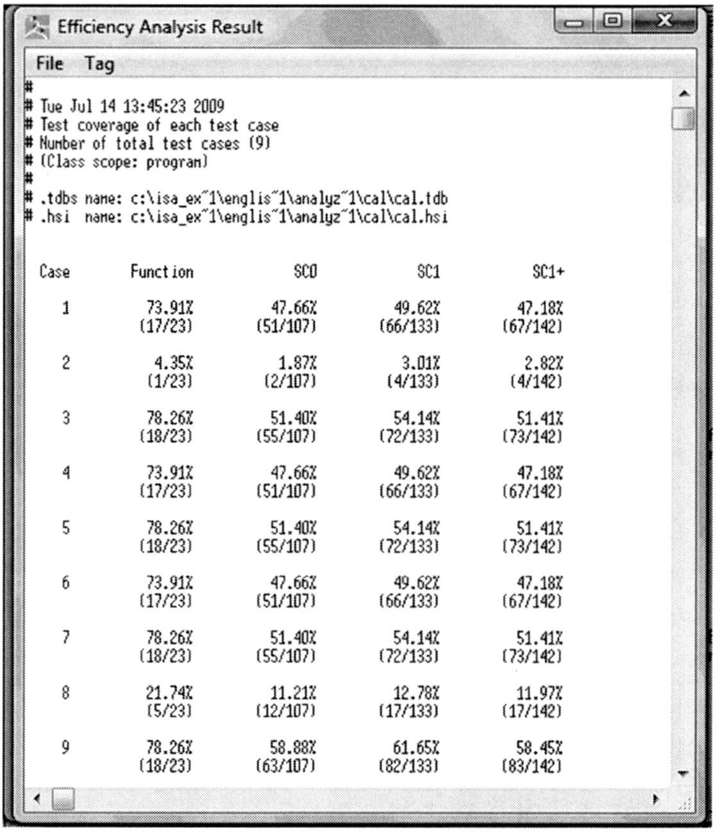

Efficiency Analysis Result

File Tag

```
#
# Tue Jul 14 13:45:23 2009
# Test coverage of each test case
# Number of total test cases (9)
# (Class scope: program)
#
# .tdbs name: c:\isa_ex~1\englis~1\analyz~1\cal\cal.tdb
# .hsi  name: c:\isa_ex~1\englis~1\analyz~1\cal\cal.hsi
```

Case	Function	SC0	SC1	SC1+
1	73.91% (17/23)	47.66% (51/107)	49.62% (66/133)	47.18% (67/142)
2	4.35% (1/23)	1.87% (2/107)	3.01% (4/133)	2.82% (4/142)
3	78.26% (18/23)	51.40% (55/107)	54.14% (72/133)	51.41% (73/142)
4	73.91% (17/23)	47.66% (51/107)	49.62% (66/133)	47.18% (67/142)
5	78.26% (18/23)	51.40% (55/107)	54.14% (72/133)	51.41% (73/142)
6	73.91% (17/23)	47.66% (51/107)	49.62% (66/133)	47.18% (67/142)
7	78.26% (18/23)	51.40% (55/107)	54.14% (72/133)	51.41% (73/142)
8	21.74% (5/23)	11.21% (12/107)	12.78% (17/133)	11.97% (17/142)
9	78.26% (18/23)	58.88% (63/107)	61.65% (82/133)	58.45% (83/142)

Fig. 1.24 The measurement result of the test case efficiency analysis – here, SCO means visible branch test coverage, SC1 means visible and invisible branch test coverage, SC2 means enhanced branch test coverage (a loop statement is considered as three branches)

- **View the version comparison result of a more complex program (a GNU project: bison) in system level, file level, and statement level as shown in Figs. 1.35–1.44 – the operations and the results.**
- **To a program with a GUI (sortdemo), selectively play the test case back dynamically for efficient regression testing – the operations and the result (see Figs. 1.45–1.47).**

1.1 What Is Software?

Fig. 1.25 View the test case minimization result (the related algorithm will be introduced in Chap. 23)

There are more functions available (After trying NSE-CLICK, it is recommended to try the S_Panorama/S_Panajava product (see Preface) designed for students to learn NSE with your small project (less than 1,501 lines of the source code)).

The above operations and results show that with the new definition, a software is much easier to understand, test, maintain, and that the acceptance testing can be done automatically.

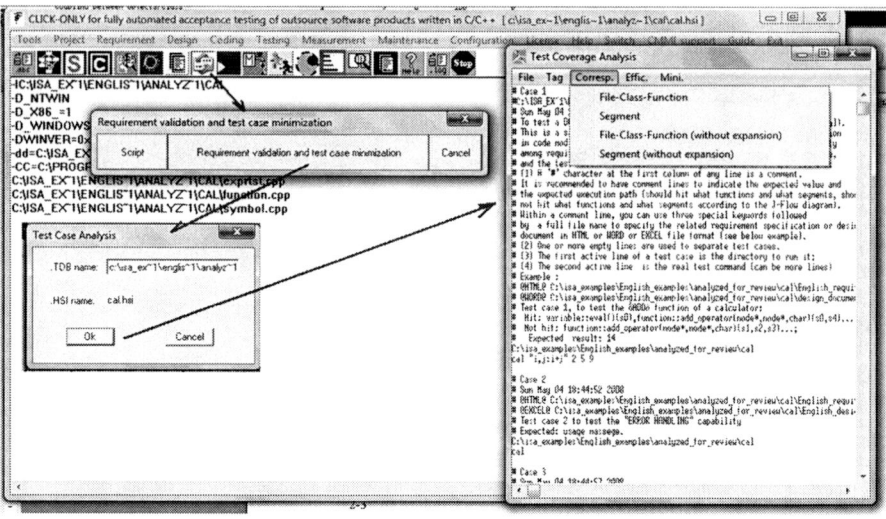

Fig. 1.26 Start the OO-V&V (Requirement Validation and Verification) Tool

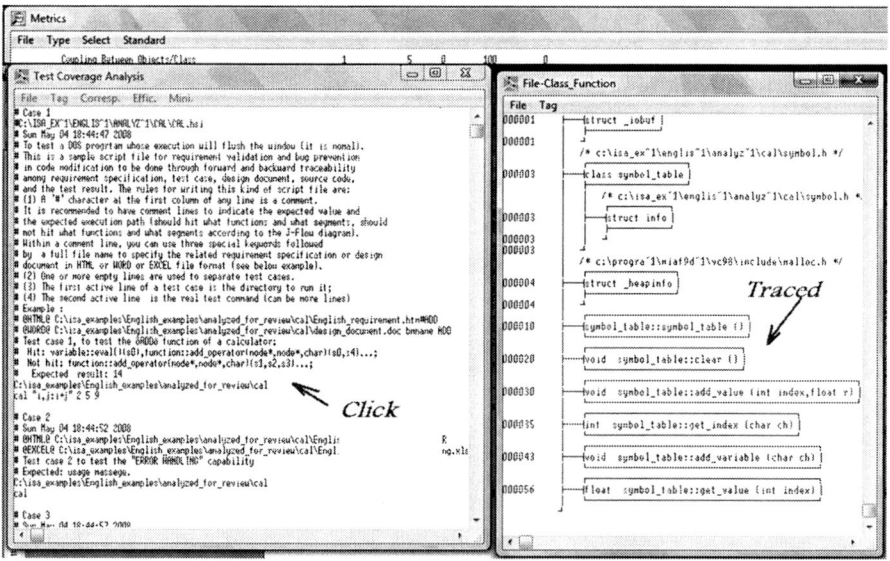

Fig. 1.27 Perform forward tracing at the module level – click a test case to trace the modules tested

1.1 What Is Software?

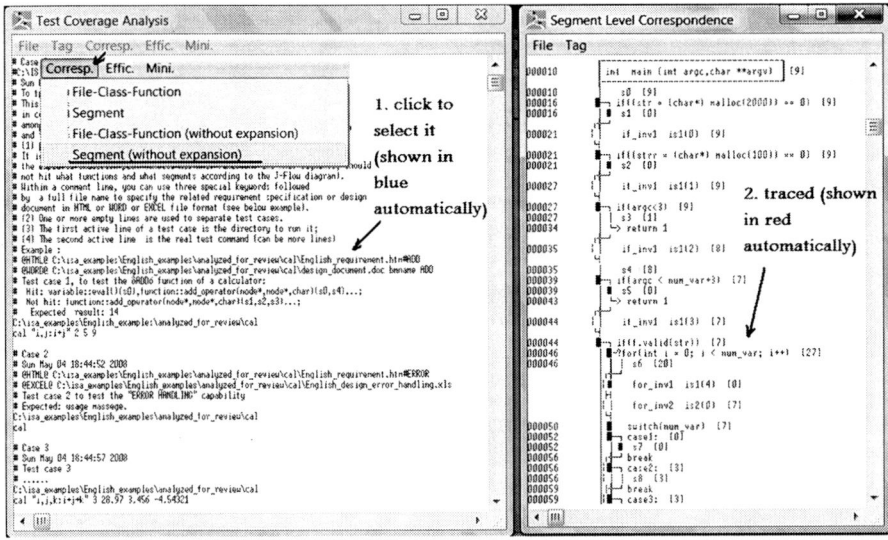

Fig. 1.28 Perform forward tracing at the segment (a set of statements with the same execution conditions) level – click a test case to trace the code segments tested

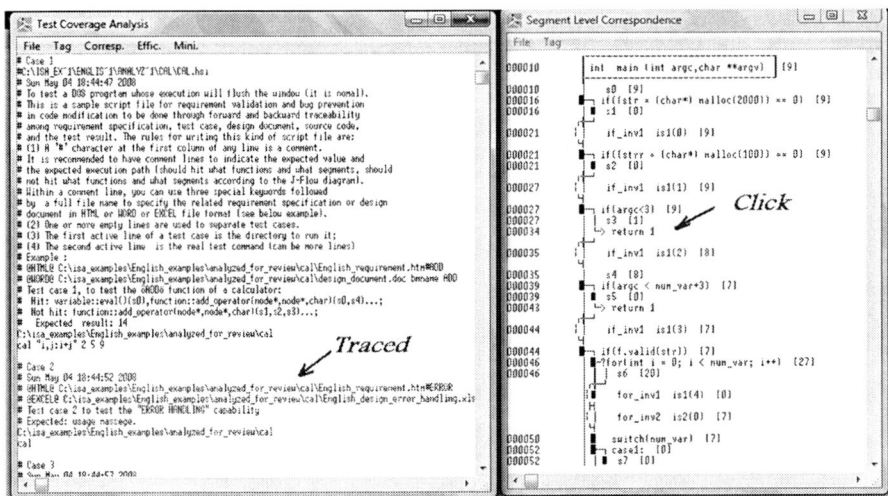

Fig. 1.29 Perform backward tracing from a code segment to find what test cases can be used to test the segment

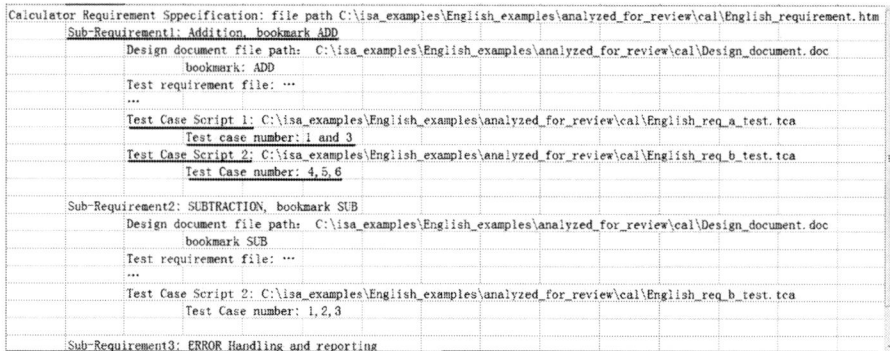

Fig. 1.30 View the simple framework showing the relationship between requirement specifications and the related documents including the test scripts and test cases (file name: C:\isa_examples\English_examples\analyzed_for_review\cal\document_relationship.jpg)

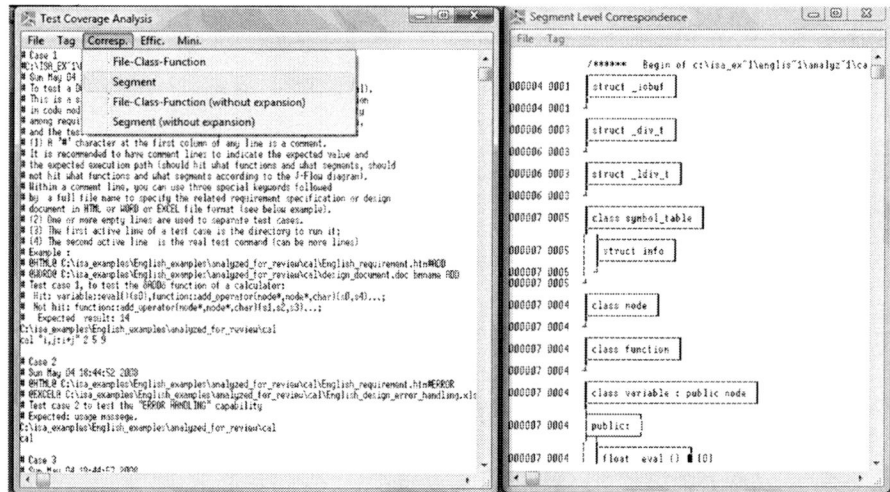

Fig. 1.31 View the options for selecting the traceability type

1.1 What Is Software?

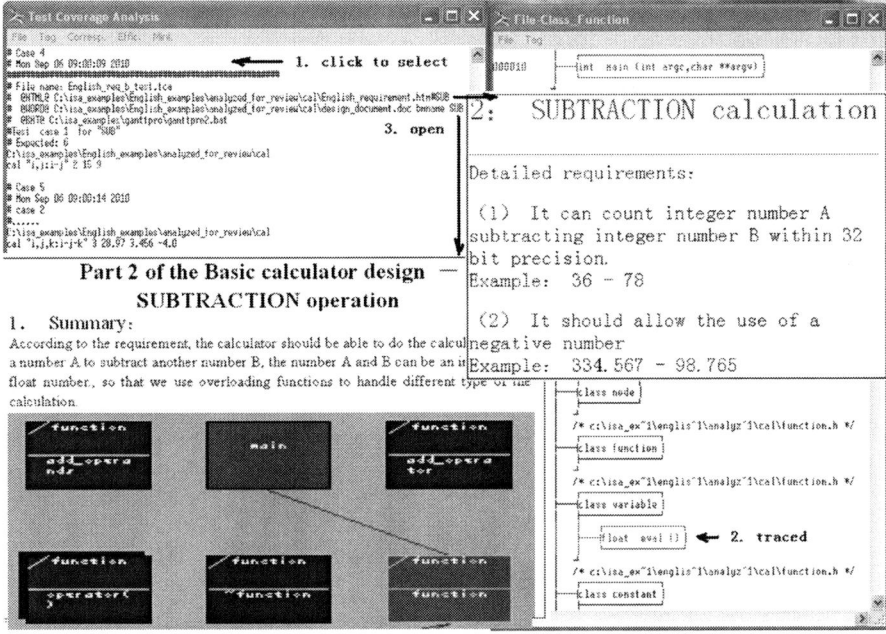

Fig. 1.32 From the subrequirement ADDITION, a related test case is selected to perform forward tracing (the requirement specification file is opened and shown from the location pointed by the ADD bookmark)

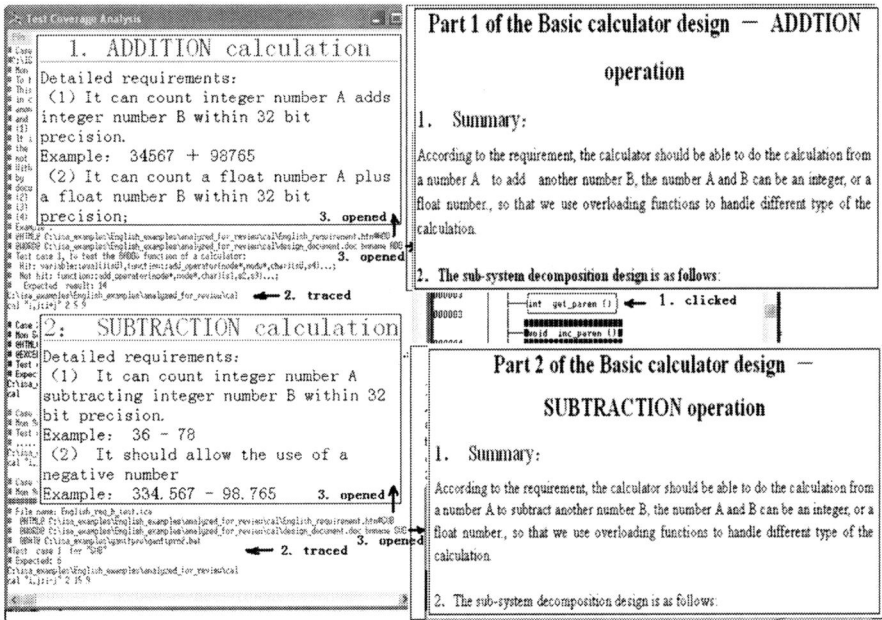

Fig. 1.33 Backward tracing from a module. In this example, several test cases and two subrequirements (ADDITION and SUBTRACTION) are traced – it means this module is used for the implementation of both subrequirements – if it needs to be modified, it should satisfy both requirements

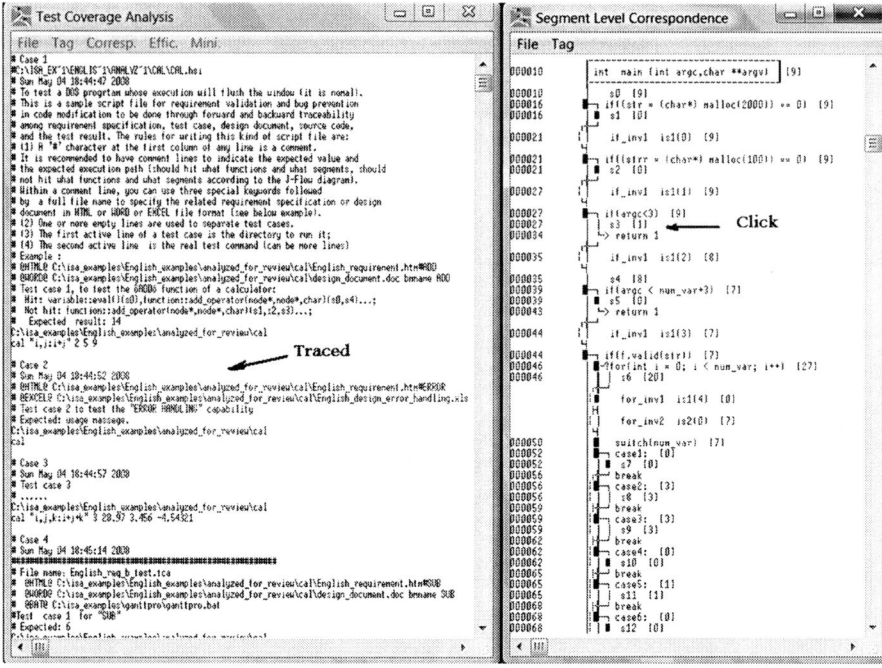

Fig. 1.34 Intelligent test case selection for efficient regression testing after code modification: when a module or code branch is modified, click it to find what test cases can be used to retest it through backward tracing – in this example, for retesting segment s3, only one test case is useful

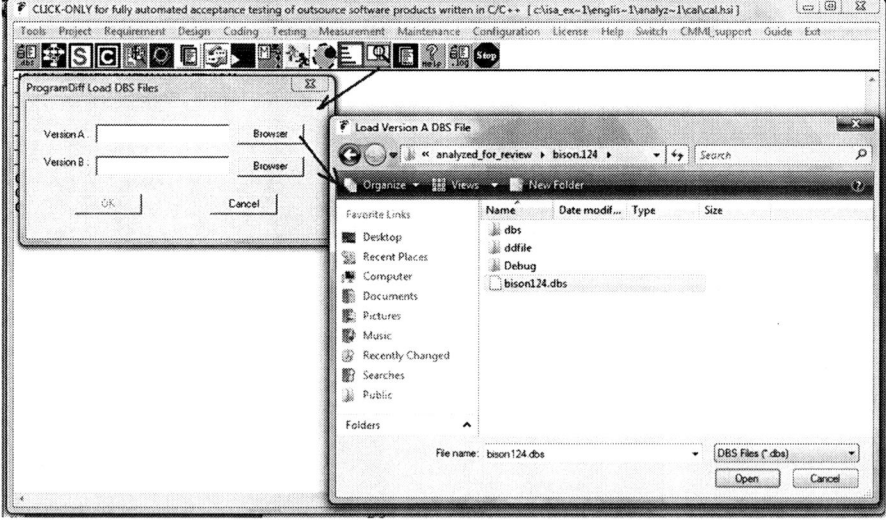

Fig. 1.35 Start the OO-CodeDiff tool

1.1 What Is Software?

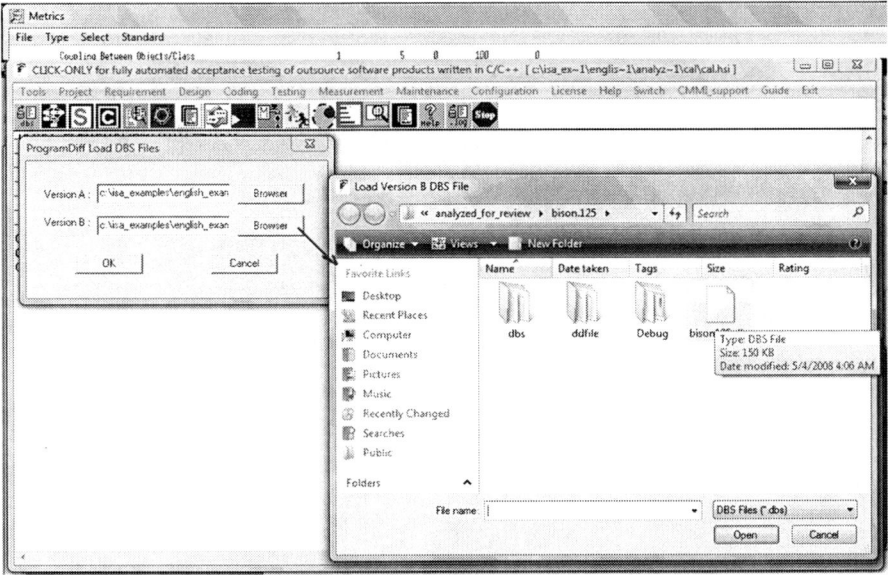

Fig. 1.36 Load the databases of the two versions of the program (bison1.24 and bison1.25) separately

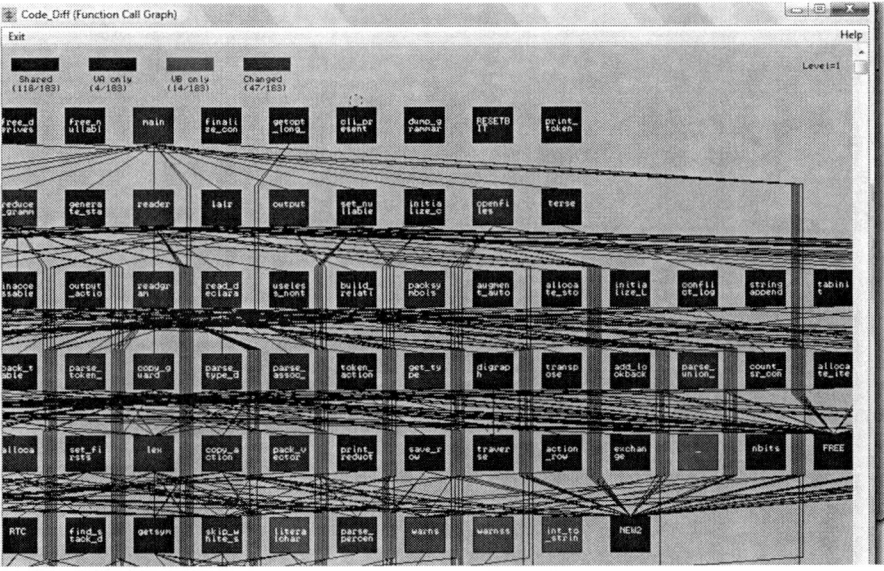

Fig. 1.37 View the system-level differences (in the original figure, shown in color rather than black and white, the unchanged modules are shown in *blue*, changed modules are shown in *red*, deleted modules are shown in *brown*, and new added modules are shown in *green*)

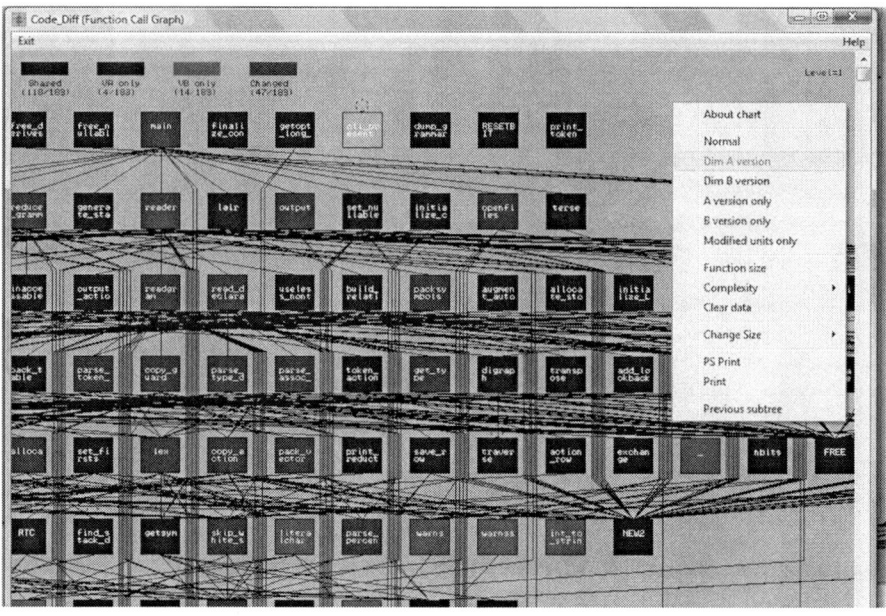

Fig. 1.38 View the new version only

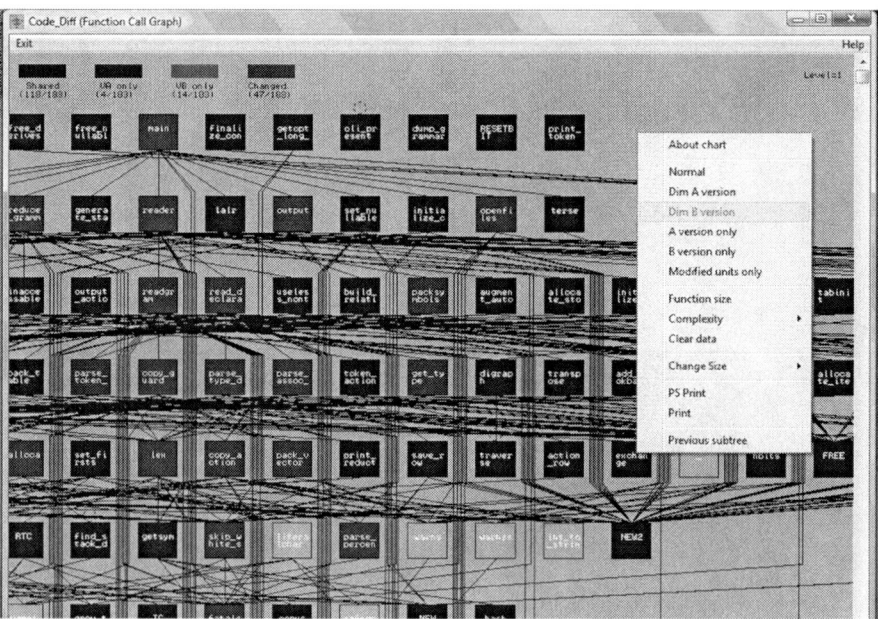

Fig. 1.39 View the old version only

1.1 What Is Software?

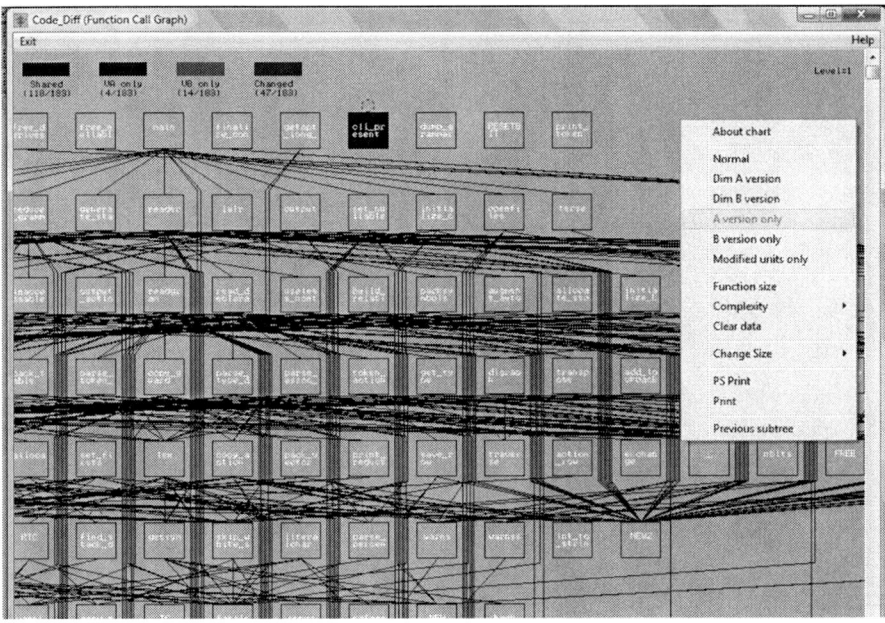

Fig. 1.40 View the modules deleted from the old version

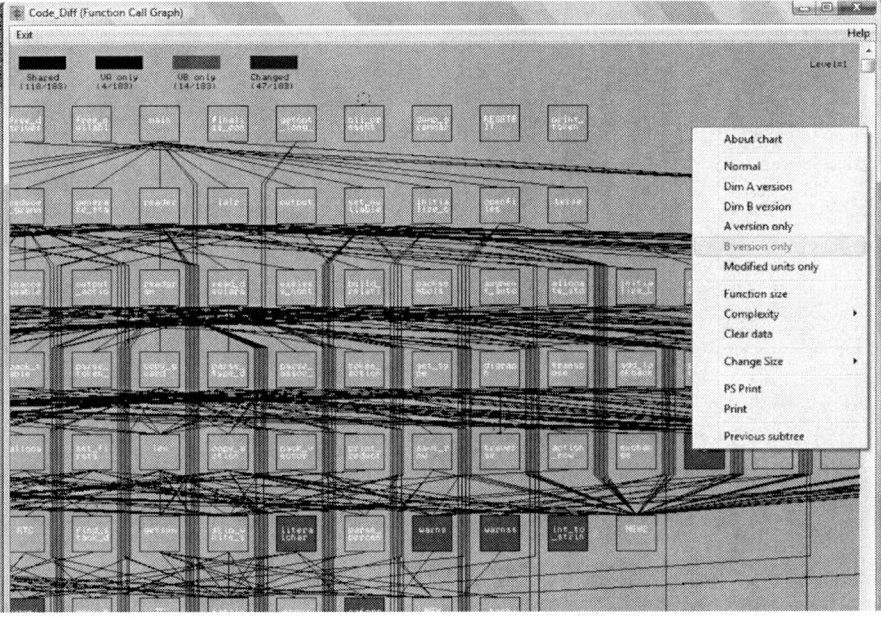

Fig. 1.41 View the modules added into the new version

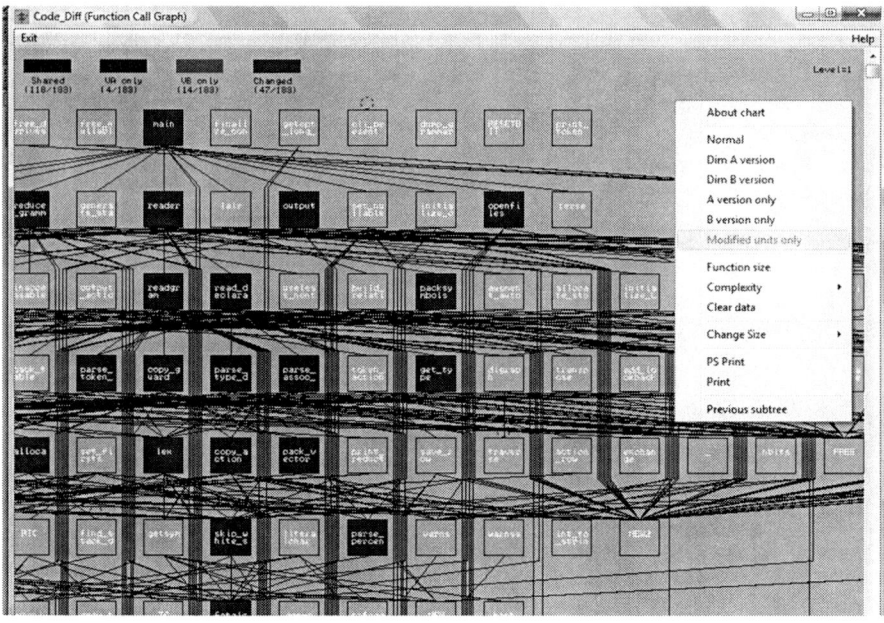

Fig. 1.42 View the changed modules

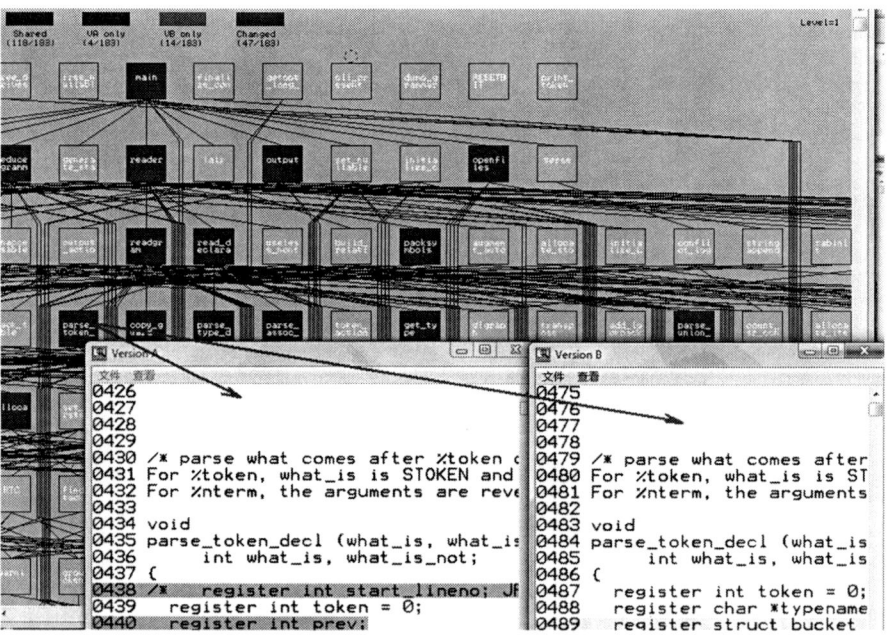

Fig. 1.43 View the statement-level differences of a changed module

1.1 What Is Software? 27

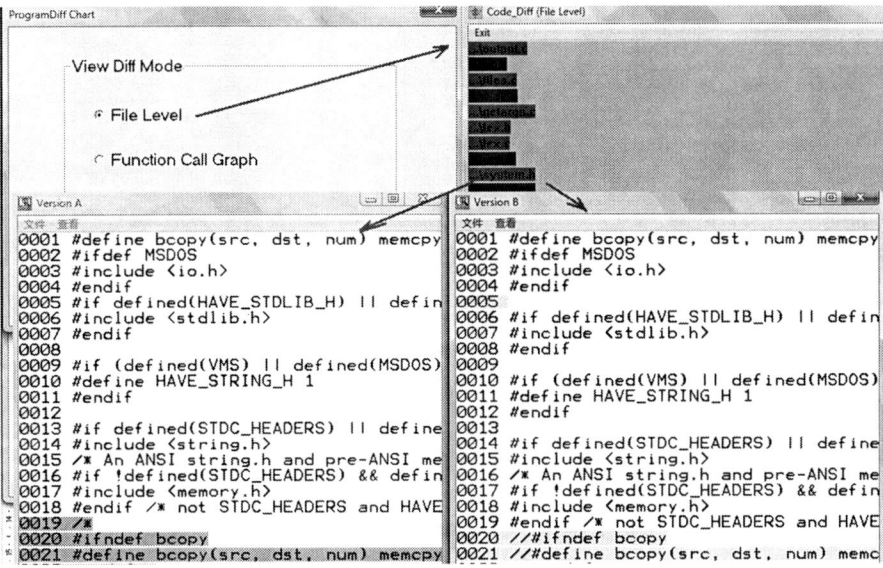

Fig. 1.44 View the file-level differences and the statement differences of a modified source file

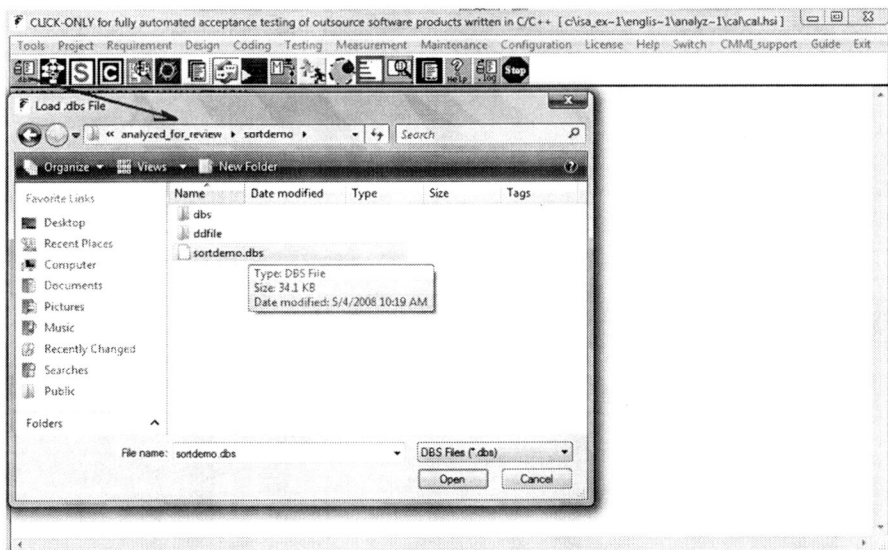

Fig. 1.45 Load the database of the sortdemo program

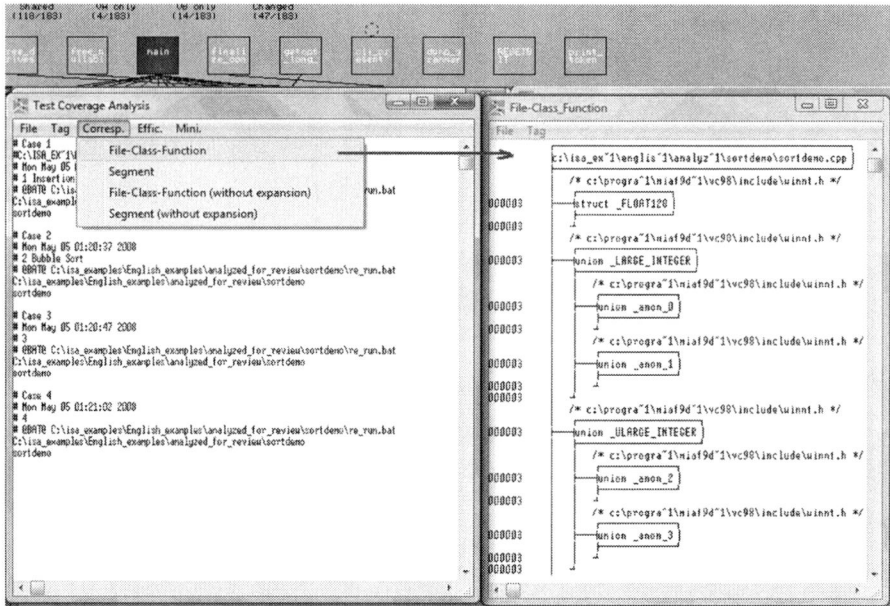

Fig. 1.46 Choose the traceability type

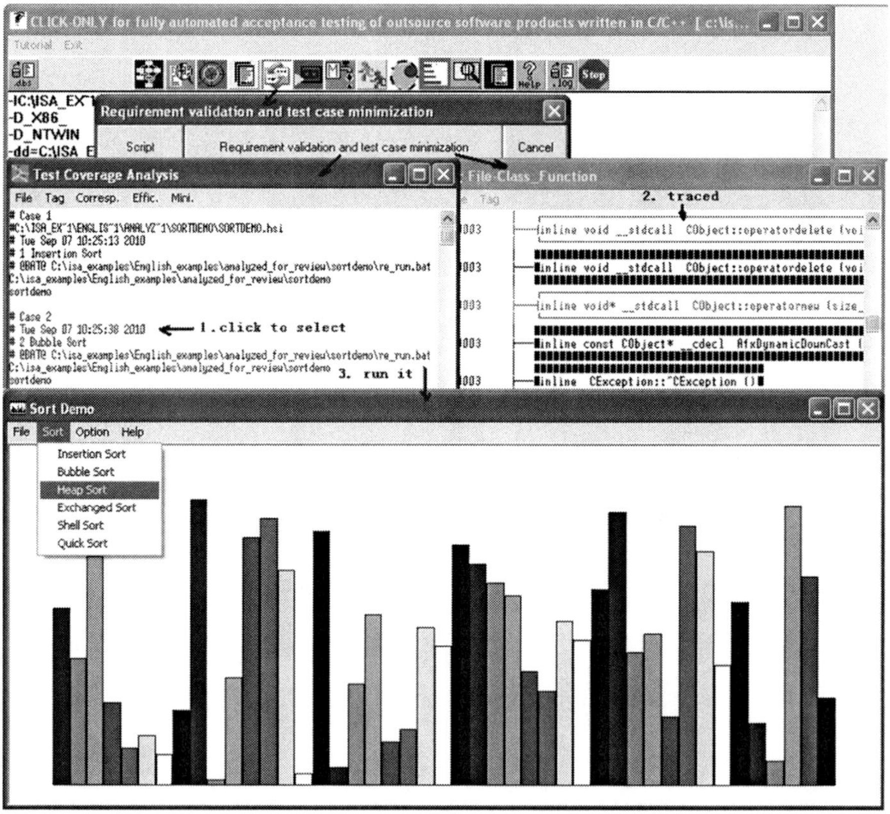

Fig. 1.47 Click a test case to selectively play the test operations back (although there is only one of the GUI record file and the batch file is the same for all test cases in the script file, we can still selectively play the different operations back through the different test cases with different Time Tags)

1.2 What Is Software Engineering?

The term *software engineering* first appeared in the 1968 NATO Software Engineering Conference and was meant to provoke thought regarding the "software crisis" at the time.

According to the *IEEE Standard Computer Dictionary*, 610, ISBN 1-55937-079-3, 1990, "Software Engineering" is defined as "The application of a systematic, disciplined, quantifiable approach to development, operation, and maintenance of software; that is, the application of engineering to software."

But, for instance, do we really have a systematic, disciplined, quantifiable approach to the maintenance of software with the existing software engineering paradigm? The answer is no.

As Scott Ambler pointed out, "The Unified Process suffers from several weaknesses. First, it is only a development process… it misses the concept of maintenance and support…. It's important to note that development is a small portion of the overall software life cycle. The relative software investment that most organizations make is allocating roughly 20% of the software budget for new development, and 80% to maintenance and support efforts" [Amb05].

With the existing software engineering paradigm, the maintenance of software is performed partially, locally, and blindly without the means to prevent the side effects for the implementation of a requirement change or code modification. In fact, not only RUP (Rational Unified Process) but almost all existing process models and methods do not really support software maintenance because of the lack of various kinds of bidirectional traceabilities. As Professor Brooks pointed out in his book of "The Mythical Man-Month":

> "Two Steps Forward and One Step Back
> …The fundamental problem with program maintenance is that fixing a defect has a substantial (20-50 percent) chance of introducing another. So the whole process is two steps forward and one step back.
>
> Why aren't defects fixed more cleanly? First, even a suitable defect shows itself as a local failure of some kind. In fact it often has system-wide ramifications, usually nonobvious. Any attempt to fix it with minimum effort will repair the local and obvious, but unless the structure is pure or the documentation very fine, the far reaching effects of the repair will be overlooked. Second, the repairer is usually not the man who wrote the code,…
>
> One Step Forward and One Step Back
>   …All repairs tend to destroy the structure, to increase the entropy and disorder of the system. Less and less effort is spent on fixing original design flaws; more and more is spent on fixing flaws introduced by early fixes. As time passes, the system becomes less and less well-ordered. Sooner or later the fixing ceases to gain any ground. Each forward step is matched by a backward one" [Bro95-p120]….
>
> "Clearly, methods of designing programs so as to eliminate or at least illuminate side effects can have an immense payoff in maintenance cost" [Bro95-p122].

In my opinion, for truly supporting software maintenance, a model or a software development method must satisfy the following conditions:

(a) Being able to help users perform software maintenance holistically and globally
(b) Being able to greatly reduce the amount of defects introduced into the software product and propagated to software maintenance phase through defect prevention and defect propagation prevention performed from the first step to the entire software development process
(c) Being able to help users prevent the side effects for the implementation of requirement changes or code modifications
(d) Being able to provide the necessary means to help users greatly reduce the time, resources requested, and cost in regression testing after the implementation of requirement changes or code modifications, such as the capability for test case efficiency analysis and test case minimization, or automated, efficient, and intelligent test case selection
(e) Being able to help the customer side to maintain a software product with almost the same conditions as if the software product is maintained by the product development side

Different from all existing models and methods, the NSE (with its support platform Panorama++) model and methodology based on nonlinear thinking and complexity science satisfies these five conditions – it brings revolutionary changes to not only software development but also to software maintenance (see Chap. 18).

It is possible for NSE to help software developers double their productivity and halve their cost by reducing about two-third of the effort and cost spent in software maintenance. For the detailed information about the differences in software maintenance between NSE and the existing models and methods, please read Chap. 18.

It is important to point out that with NSE there is no major difference between the software development process and the software maintenance process, because

- Both processes support requirement changes and code modifications with side-effect prevention for the implementation of requirement changes or code modifications through various bidirectional traceabilities.
- The NSE nonlinear process model is followed and the quality of a software product is ensured from the first step down to maintenance through defect prevention and defect propagation prevention, so that the defects propagated to the maintenance phase is greatly reduced.
- Even if the team for the development of a software product is different from the team for the maintenance of the software product, as described before in this chapter with the new software definition, the working conditions (with the program and the source code, the data used, the documents traceable to the source code, the database built through static and dynamic measurement, and a set of Assisted Online Agents) for the product maintenance are almost the same as that for the product development.

1.3 The Major Activities/Tasks to Be Performed in Software Engineering

The major activities/tasks to be performed in software engineering are listed as follows:

- **Software requirements engineering** – defines needed information, function, behavior, performance and interfaces, mainly including

    - Requirement elicitation/gathering – the practice of obtaining the requirements of a system from customers (or users/stakeholders).
    - Functional decomposition of functional requirements (often applying Use Case approach) and the description of nonfunctional requirements. With NSE, it is performed using a Holistic, Actor–Action and Event–Response driven, Traceable, Visual, and Executable technique (HAETVE, see Chap. 11) to replace the Use Case approach, which is not a holistic one, and the obtained results are not traceable and not directly executable for defect removal.
    - Requirement analysis – a modeling activity where the objective is to understand what the customer really wants. With NSE, it is done using the HAETVE technique, dummy programming technique, and visual diagramming techniques (see Chap. 11).
    - Specification – a complete description of the behavior of the system to be developed. With NSE, a template is provided for preventing errors of something missing.
    - Validation – tests to ensure that the software conforms to customers' requirements. With NSE, it is done through forward traceability that is automatically established.

- **Software design engineering** – an activity that translates the requirements model into a more detailed model that is the guide to implementation of the software, including the data structures, software architecture, interface representations, and algorithmic details. It is usually done based on the **Constructive Holism** principle with Computer-Aided Software Engineering (CASE) tools and use standards for the format, such as the Unified Modeling Language (UML). With NSE, it is done mainly through dummy programming and visual diagramming techniques (see Chap. 12) based on the **Generative Holism** principle.
- **Software coding** – the construction of software through the use of programming languages. With NSE, it is done incrementally with defect prevention.
- **Software testing** – a set of activities conducted with the intent of finding errors in software. With NSE, it is done by applying the Transparent-box method innovated by me, which combines functional and structural testing together seamlessly and can be dynamically used in the entire software development lifecycle (see Chap. 16).
- **Software quality assurance (SQA)** – means of monitoring the software engineering processes and methods used to ensure quality. The methods by which this

is accomplished are many and varied, and may include ensuring conformance to one or more standards, such as ISO 9000 or CMMI (Capability Maturity Model Integration). With NSE, it is done mainly through defect prevention and defect propagation prevention (see Chap. 17).
- **Software deployment and support** – Software deployment is an evolving collection of interrelated processes such as release, install, adapt, reconfigure, update, activate, deactivate, remove, and retire. The connectivity of large networks, such as the Internet, is affecting how software deployment is performed (Richard S. Hall, Dennis Heimbigner, Alexander L. Wolf, *A Cooperative Approach to Support Software Deployment Using the Software Dock*, Software Engineering Research Laboratory, University of Colorado, Boulder, CO 80309-0430, USA). The software developed should be delivered to the customers and supported.
- **Software maintenance** – Software systems often have problems and need enhancements for a long time after they are first completed. With NSE, it is done using a systematic, disciplined, quantifiable approach with side-effect prevention for the implementation of a requirement change or code modification (see Chap. 18).
- **Software configuration management** – Software systems are very complex, their configuration (such as versioning and source control) have to be managed in a standardized and structured method. With NSE, it is done with CVS (a GNU product) plus intelligent version comparison technique in system level, file level, module level, and statement level.
- **Software engineering management** – The management of software systems borrows heavily from project management, but there are nuances encountered in software not seen in other management disciplines. With NSE, it is done by combining the project development process and the project management process together seamlessly, making the management documents traceable with the implementation of requirements and the source code (see Chap. 20).
- **Software development process** – The process of building software is hotly debated among practitioners with the main paradigms being agile or waterfall or other paradigms such as NSE. With NSE, a nonlinear process model based on complexity science is applied (see Chap. 8).

1.4 The Popular Lifecycle/Process Models with the Existing Software Engineering Paradigm

There are several popular lifecycle/process models with the existing software engineering paradigm.

1.4.1 The Waterfall Model

A waterfall model is shown in Fig. 1.48, a modified version of the waterfall model with feedback is shown in Fig. 1.49. A waterfall model is a typical linear model.

1.4 The Popular Lifecycle/Process Models with the Existing Software Engineering   33

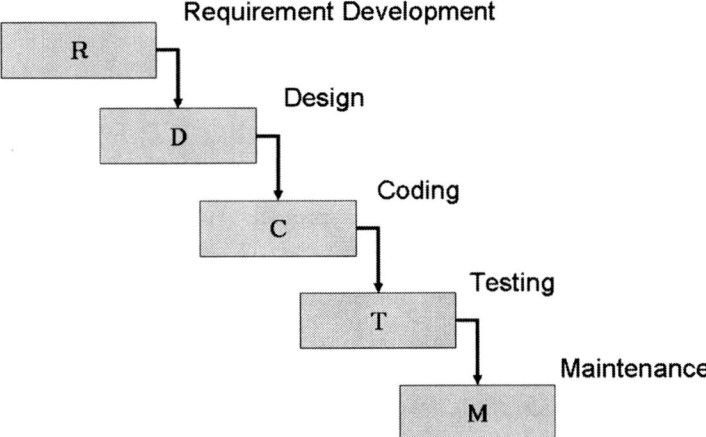

Fig. 1.48 The waterfall model

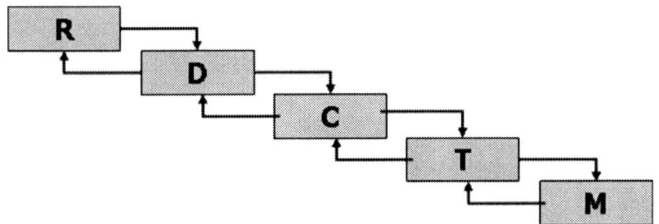

Fig. 1.49 A modified waterfall model with feedback

A waterfall model with feedback is still a linear model. As pointed out by Roger S. Pressman, "Although the original waterfall model proposed by Winston Royce made provision for 'feedback loops,' the vast majority of organizations that apply the process model treat it as if it were strictly linear." It is clear that in a modified waterfall model with feedback there is only local feedback at the transition between phases with no real upstream movement. A nonlinear process model must be supported with forward and backward traceabilities.

As shown in Figs. 1.48 and 1.49, the waterfall model divides the software lifecycle into five main processes: requirement analysis, design, coding, testing, and maintenance. Each lower phase begins after the upper phases are completely finished.

The first formal description of the waterfall model is often cited to be an article published in 1970 by Winston W. Royce (1929–1995) [Roy70], although Royce did not use the term "waterfall" in this article. Royce was presenting this model as an example of a flawed, nonworking model.

Compared with other linear process models, the **advantages** of the waterfall models are listed as follows:

(a) System is well documented.
(b) It is easy to understand.
(c) It can be applied with or without tools.
(d) "Big Design Up Front" can be done to avoid some kind of rework.
(e) It is suitable for fixed requirement projects such as science computing projects.

The **disadvantages** of the waterfall models are as follows:

(a) All risks must be dealt with in a single software development effort.
(b) Before a previous phase (such as the requirement analysis) is completed, the next process (such as the design) phase cannot be performed – it must wait. It will waste time and human resources.
(c) Requirement changes are hard to perform efficiently.
(d) A working product is not available until late in the project – the customer must wait for the product evaluation until the product is produced.

1.4.2 The Incremental Development Models

Incremental development is a staging and scheduling strategy in which various parts of the system are developed at different times or rates and integrated as they are completed.

[Coc08]. The incremental development model is shown in Fig. 1.50.

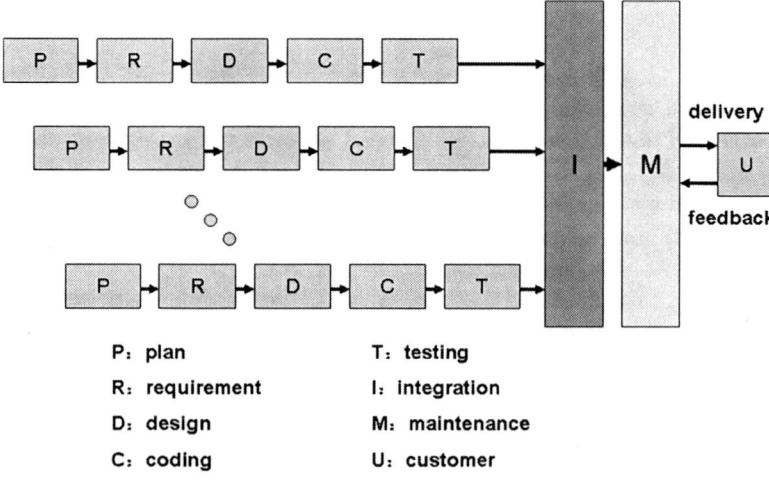

P: plan   T: testing
R: requirement   I: integration
D: design   M: maintenance
C: coding   U: customer

Fig. 1.50 Incremental development model

As the name suggests, an incremental software development process model guides the requirement implementation incrementally rather than totally at one time. It is also called a Micro-Waterfall Model which divides the total requirements into some subrequirements and implements and integrates them incrementally. In this way it reduces the waiting time and also reduces the risk, where the customer can partially evaluate the product early and find the possible problems early. But it does not remove the other major drawbacks of the linear process models, such as the defects introduced in upper phases will still easily propagate to the lower phases to make the defect removal cost increase tenfold several times.

Compared with other linear process models, the advantages and disadvantages of the incremental development methods are as follows [GSAM03]:

1.4.2.1 Advantages

- Provides some feedback, allowing later development cycles to learn from previous cycles.
- Requirements are relatively stable and may be better understood with each increment.
- Allows some requirements modification and may allow the addition of new requirements.
- It is more responsive to user needs than the waterfall model.
- A usable product is available with the first release, and each cycle results in greater functionality.
- The project can be stopped any time after the first cycle and leave a working product.
- Risk is spread out over multiple cycles.
- This method can usually be performed with fewer people than the waterfall model.
- Return on investment is visible earlier in the project [Mck95].
- Project management may be easier for smaller, incremental projects [Mck95].
- Testing may be easier on smaller portions of the system.

1.4.2.2 Disadvantages

- The majority of requirements must be known in the beginning.
- Formal reviews may be more difficult to implement on incremental releases than on a complete system [Ree95].
- Because development is spread out over multiple iterations, interfaces between modules must be well-defined in the beginning [Ree95].
- Cost and schedule overruns may result in an unfinished system.
- Operations are impacted as each new release is deployed.
- Users are required to learn how to use a new system with each deployment.

1.4.3 The Iterative Models

Iterative development is a rework scheduling strategy in which time is set aside to revise and improve parts of the system [Coc08]. There are several different iterative software development models.

1.4.3.1 The Prototype Models

Two prototype models are shown in Figs. 1.51 and 1.52.

1.4.3.2 Spiral Model

Defined by Barry Boehm, the **spiral model** (see Fig. 1.53), also called the **spiral lifecycle model,** which combines the features of the prototyping model and the waterfall model together, is often used in large project development.

Compared with other linear process models, the advantages and disadvantages of the iterative development models/methods are as follows [GSAM03]:

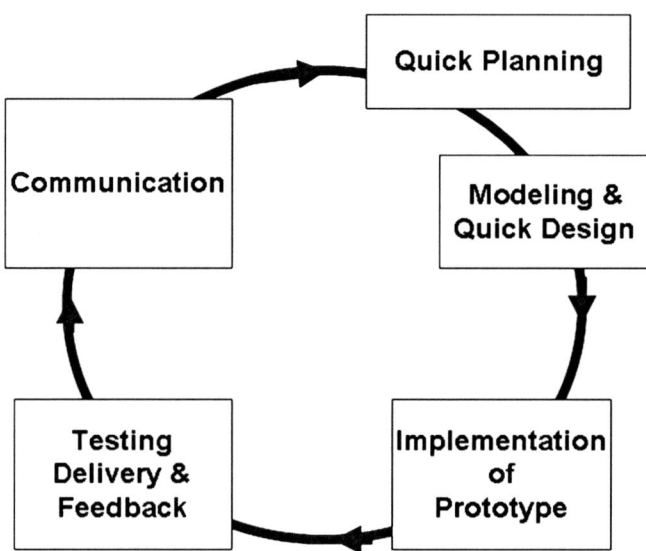

Fig. 1.51 A typical prototype model

1.4 The Popular Lifecycle/Process Models with the Existing Software Engineering

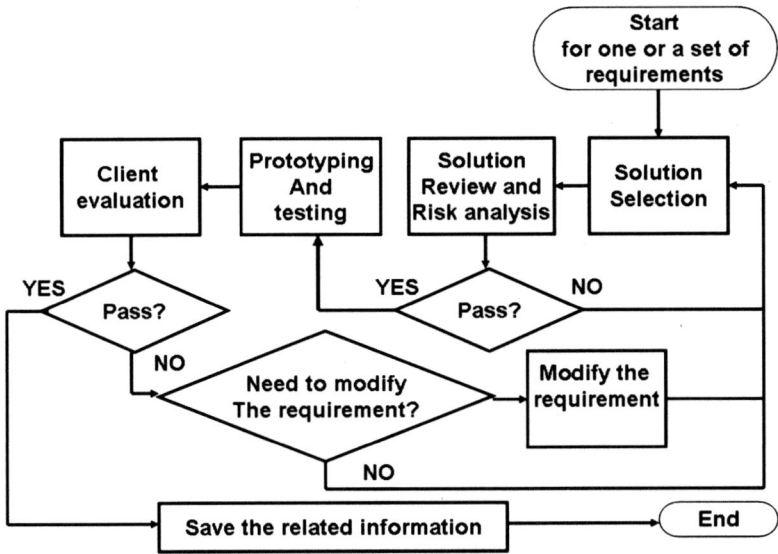

Fig. 1.52 The prototype model used with the preprocess of the NSE model

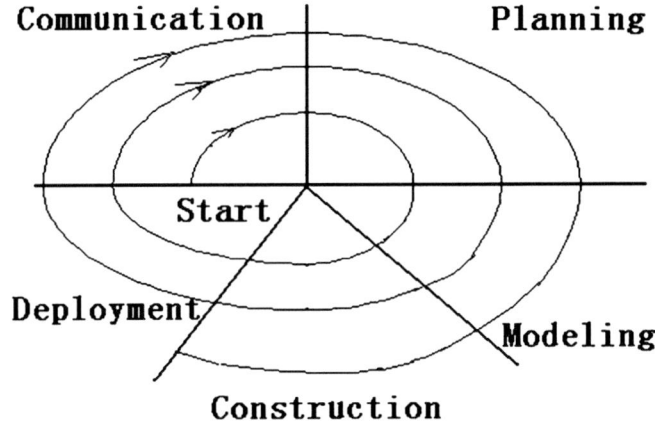

Fig. 1.53 Spiral model

1.4.3.3 Advantages [GSAM00]

- Project can begin without fully defining or understanding requirements.
- Final requirements are improved and more in line with real user needs.
- Risks are spread over multiple software builds and controlled better.
- Operational capability is achieved earlier in the program.
- Newer technology can be incorporated into the system as it becomes available during later prototypes.
- Documentation emphasizes the final product instead of the evolution of the product [Ree95].
- This method combines a formal specification with an operational prototype [Ree95].

1.4.3.4 Disadvantages [GSAM00]

- Because there are more activities and changes, there is usually an increase in both cost and schedule over the waterfall method.
- Management activities are increased.
- Instead of a single switch over to a new system, there is an ongoing impact to current operations.
- Configuration management activities are increased.
- Greater coordination of resources is required.
- Users sometimes mistake a prototype for the final system.
- Prototypes change between cycles, adding a learning curve for developers and users.
- Risks may be increased in the following areas:
  - Requirements – temptation to defer requirements definition.
    - (a) Management – Programs are more difficult to control. Better government/ contractor cooperation needed.
    - (b) Approval – vulnerable to delays in funding approval, which can increase schedule and costs.
  - Architectural – Initial architecture must accommodate later changes:
    - (a) Short term benefits – Risk of becoming driven by operational needs rather than program goals.
  - Risk avoidance – Tendency to defer riskier features until later:
    - (a) Exploitation by suppliers – Government bargaining power may be reduced because initial contract may not complete the entire task, and subsequent contracts are not likely to be competed.
    - (b) Patchwork quilt effects – If changes are poorly controlled, the product quality can be compromised.

1.4.4 More Popular Process Models

There are three more popular process models, CMMI, Agile, and RUP.

1.4.4.1 CMMI

According to *GSAM Handbook* [GSAM03], originally, there were several different versions of capability maturity models: one for software, one for system engineering, and one for software acquisition. Recently, these separate models have been integrated into a single model, the CMMI. As shown in Fig. 1.54, two different representations are available for the CMMI, a continuous representation and a staged representation previously used by both the Software and Software Acquisition CMMs. The staged representation shows progress as a series of five levels. Each of these levels is described by certain attributes characterizing its level of competency. Each level is associated with process areas, and each process area is described in terms of common practices that support that level's goals. These levels, descriptions, and process areas are shown in Fig. 1.55.

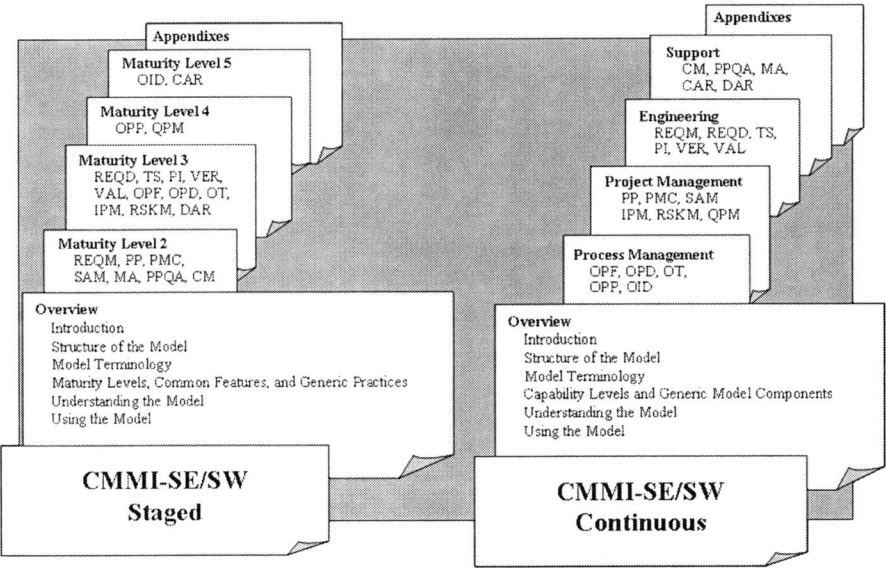

Fig. 1.54 The two different representations of CMMI

Process Areas by Maturity Level

Level	Focus	Process Areas
5 Optimizing	*Continuous process improvement*	Organizational Innovation and Deployment Causal Analysis and Resolution
4 Quantitatively Managed	*Quantitative management*	Organizational Process Performance Quantitative Project Management
3 Defined	*Process standardization* (SS) (IPPD) (IPPD)	Requirements Development Technical Solution Product Integration Verification Validation Organizational Process Focus Organizational Process Definition Organizational Training Integrated Project Management Integrated Supplier Management Risk Management Decision Analysis and Resolution Organizational Environment for Integration Integrated Teaming
2 Managed	*Basic project management*	Requirements Management Project Planning Project Monitoring and Control Supplier Agreement Management Measurement and Analysis Process and Product Quality Assurance Configuration Management
1 Performed		

Fig. 1.55 The process areas by maturity level

1.4.4.2 Agile Software Development Model

A typical agile software development model, XP (Extreme Programming), is shown in Fig. 1.56 [Pre05-p78].

The Agile Manifesto

- *Individuals and interactions* over processes and tools
- *Working software* over comprehensive documentation
- *Customer collaboration* over contract negotiation
- *Responding to change* over following a plan

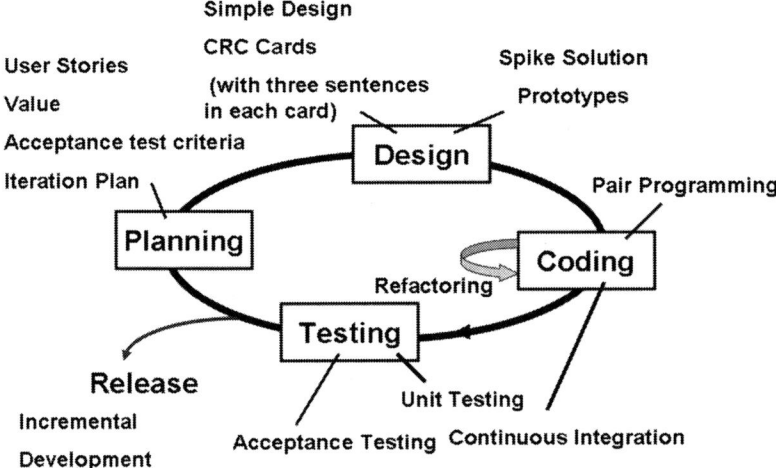

Fig. 1.56 XP model

Twelve Agile Principles

1. Our highest priority is to satisfy the customer through early and continuous delivery of valuable software.
2. Welcome changing requirements, even late in development. Agile processes harness change for the customer's competitive advantage.
3. Deliver working software frequently, from a couple of weeks to a couple of months, with a preference to the shorter time scale.
4. Business people and developers must work together daily throughout the project.
5. Build projects around motivated individuals. Give them the environment and support they need, and trust them to get the job done.
6. The most efficient and effective method of conveying information to and within a development team is face-to-face conversation.
7. Working software is the primary measure of progress.
8. Agile processes promote sustainable development. The sponsors, developers, and users should be able to maintain a constant pace indefinitely.
9. Continuous attention to technical excellence and good design enhances agility.
10. Simplicity – the art of maximizing the amount of work not done – is essential.
11. The best architectures, requirements, and designs emerge from self-organizing teams.

12. At regular intervals, the team reflects on how to become more effective, then tunes and adjusts its behavior accordingly.

1.4.4.3 Agile Methods (Table 1.1)

1.4.4.4 Rational Unified Process

Rational Unified Process is graphically shown in Fig. 1.57.

With RUP, in each iteration, a micro-waterfall model is applied as shown in Fig. 1.58.

As shown in Fig. 1.57, the unified process groups iterations into four phases: Inception, Elaboration, Construction, and Transition.

- Inception identifies project scope, risks, and requirements at a high level but in enough detail that work can be estimated.
- Elaboration delivers a working architecture that mitigates the top risks and fulfills the nonfunctional requirements.

Table 1.1 Agile methods (Rich Mironov CMO, Enthiosys, Mitigating Risk with Agile Development, http://www.enthiosys.com/wp-content/uploads/2009/09/agile_mironov_fairfax.pdf)

Scrum	Extreme Programming (XP)	Agile Project Management Framework (APM)	Crystal methods	Dynamic Systems Development Model (DSDM)
Rational Unified Process (RUP)	Feature Driven Development (FDD)	Lean development	Rapid Application Development (RAD)	–

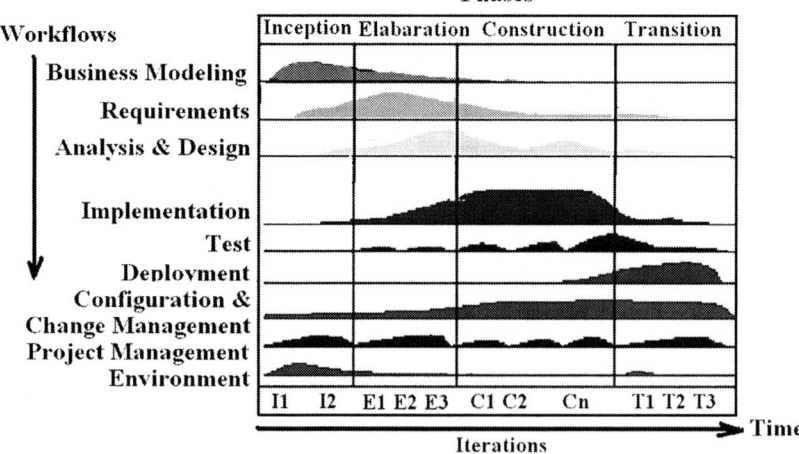

Fig. 1.57 Rational Unified Process

1.4 The Popular Lifecycle/Process Models with the Existing Software Engineering

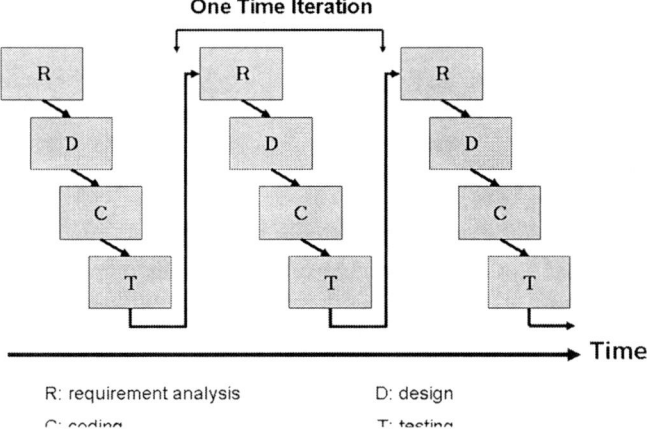

Fig. 1.58 Micro-Waterfall model applied with RUP for each iteration

- Construction incrementally fills-in the architecture with production-ready code produced from analysis, design, implementation, and testing of the functional requirements.
- Transition delivers the system into the production operating environment.

1.4.5 General Comments to All Process Models Existing with the Old-Established Software Engineering Paradigm

1.4.5.1 "All models Are Wrong, But Some Are Useful" [Box87]

We know that nothing in the world is completely perfect. Even the NSE paradigm to be introduced in details from Chaps. 3 to 24 in this book – for instance, the waterfall model can be applied with or without tool support, the RUP model can be applied with mainly static tool support, but NSE can only be applied with mainly dynamic tool support, such as those tools offered by Panorama++ to perform defect prevention and defect propagation prevention, and to establish bidirectional traceabilities. Of course, this disadvantage of NSE can also be considered as an advantage, because that in twenty-first century dynamic tools should be used in every software development company; otherwise, the company may lose its competition power.

In fact, each process model has its advantages and disadvantages – it mainly depends on the application environment, the project size, and the project complexity. For instance, when the requirements are fixed such as in the projects for scientific computing to solve some mathematical problems, the oldest waterfall model is still a considerable candidate.

1.4.5.2 The Common Limitations of the Existing Process Models

Unfortunately, the existing software engineering paradigm is established with linear thinking, reductionism, and the superposition principle that **the whole of a system is the sum of its parts,** so that it handles a nonlinear complex software product as a linear system, making all tasks/activities be performed linearly, partially, and locally.

The software engineering paradigm itself is a complex system consisting of many closely related parts including the software development methods, the process models, the software visualization paradigm, the software testing paradigm, the software quality assurance paradigm, the software documentation paradigm, the software maintenance paradigm, and the software project management paradigm. All the parts are connected and interactive. As a complex system, the overall behavior and characteristics of the software engineering paradigm cannot be inferred simply from the behavior of any individual part only but emerge from the interaction of all its parts – it means that only improving one or two of its parts such as the process improvement and management improvement without improving the other parts such as the software development methods, the software testing paradigm, the software quality assurance paradigm, and the software maintenance paradigm will not be able to make significant improvement to the entire software engineering paradigm.

In fact, the old-established software engineering paradigm based on linear thinking, reductionism, and the superposition principle limits the features and usability of all existing process models, no matter how great a process model itself may be – the common drawbacks of the existing software process models include the following:

(a) None of them is created to efficiently handle the essential issues existing with a software product – complexity, invisibility, changeability, and conformity, as defined by Brooks [Bro95-p182].
(b) None of them is able to efficiently solve the most critical problems with software products – low quality and productivity, and high cost and risk.
(c) None of them is able to make significant improvement to the software project success rate – it is still very low.
(d) All of them are linear process models with no upstream movement at all, making the defects introduced in upper phases easy to propagate to the lower phases, so that the defect removal cost will increase tenfold several times, and the maintenance of a software product is performed linearly, blindly, and locally with high risk.

So, generally speaking, today's software developed with the existing software engineering paradigm is not sufficiently engineered at this time to fulfill the role of "foundation."

1.5 Why the Current Software Is Not Sufficiently Engineered at This Time to Fulfill the Role of "Foundation"

Many critical problems exist with today's software products developed through the old-established software engineering paradigm: low productivity and quality, and high cost and risk. Many process models have been proposed for improving the software development processes. Those process models claim that they are based on "the best practices." But the question is: do we really have "the best practices" in software development with the existing software engineering paradigm? If so, what are "the best practices"? Some people believe they are as follows [Amb04]:

- Develop iteratively
- Manage requirements
- Use components
- Model visually
- Verify quality
- Control changes

It is agreeable to develop a software product iteratively, because in most cases, not all requirements are completely known by the customers in the beginning, so the customers need to review the implementation result as early as possible even it is not the complete product. But how is each iteration performed? It is still performed linearly using a micro-waterfall model with the existing software engineering paradigm, so that the defect introduced in the upper phases easily propagate to the lower phases to make the defect removal cost increase tenfold several times.

Managing requirements is important for software development, but can the requirements of a software product be best managed without various bidirectional traceabilities, particularly the automated traceability among the related documents and the test cases and the source code? The article written by Andrew Kannenberg and Hossein Saiedian and published in the Issue of Jul/Aug 2009 of CrossTalk argued that "Software Requirements Traceability Remains a Challenge" [Kan09].

About the use of the components, it is related to two different approaches, one is based on **Constructive Holism** applied with the existing software engineering paradigm, and another one is based on **Generative Holism** applied with NSE. According to the Constructive Holism principle, components are completed first, then the whole of the system is built with the completed components – it handles a software system like a machine. According to the Generative Holism, the whole of a system is developed first as an embryo, then the system grows up with its components. The benefits of the software development based on Generative Holism will be discussed in Chap. 10.

Visual tools used with the current models are based on linear thinking and the superposition principle so that they often generate many small pieces of charts/diagrams only, without the capability to show the entire system holistically. Even if an entire system chart/diagram can be generated/created, there are too many connection lines with the chart/diagram, making them hard to read and hard to understand without the capability to trace and highlight a module and all of the related modules calling and called by it.

How to verify the quality of a software produce developed with the existing software engineering paradigm? Current software quality assurance is mainly based on product review and testing after production – it is too late. There is no dynamic way to efficiently verify the quality in the requirement development phase and the product design phase without various automated and bidirectional traceabilities.

How should we control software changes? Some possible ways are as follows:

- Track changes
- Trace the changes
- Ensure quality
- Be sure changes are tested
- Inform users
- Update the related documents
- Perform system-level, file-level, module-level, and statement-level version comparison, etc.

With the existing software engineering paradigm, do we have the best practices in those fields?

When linear process models are used in the software development lifecycle, software change control is performed locally, and cannot trace the changes holistically and globally in system-level to identify how many requirements may be affected and how many other source modules may be affected – making the quality hard to ensure, and the related documents hard to update consistently, and the version comparison is often not performed in system-level.

Furthermore, with the existing software engineering paradigm, we do not have the best practices in software requirement development (see Chaps. 2 and 11); we do not have the best practices in software design (see Chaps. 2 and 12); we do not have the best practices in software coding (see Chaps. 2 and 13); we do not have the best practices in software testing (see Chaps. 2 and 16); we do not have the best practices in software quality assurance (see Chaps. 2 and 17); we do not have the best practices in software documentation (see Chaps. 2 and 19), we do not have the best practices in software maintenance (see Chaps. 2 and 18); we do not have the best practices in software project management too (see Chaps. 2 and 20) – it

Table 1.2 Software project success rates reported by Standish Group

	Date	Success rate (%)
First CHAOS report	1994	16
"Extreme CHAOS"	2001	28
Most recent CHAOS	2003	31

is why the software project success rate is so low as shown in Table 1.2 (For more information, see the Standish Group Website at http://www.standishgroup.com/).

In the article "Software development productivity and project success rates: Are we attacking the right problem?", the CEO of Ravenflow, Joe Marasco pointed out that "My conclusion is that we are making progress on the success-rate front, but slowly. The improvement is about 1.7 percentage points a year and appears to be linear based on this small sample of data. If the current improvement rate continues, we should achieve a 50 percent success rate in the year 2014." (http://www.ibm.com/developerworks/rational/library/feb06/marasco/).

How about the contribution of CMM/CMMI on the improvement of the software project success rate? As pointed by Ojelanki Ngwenyama and Peter Axel Nielsen that "Ever since its first presentation, CMM has been extremely influential on software engineering practices around the world. The model has served as a framework for software process and quality improvement efforts in thousands of software organizations and the resources expended on CMM-based SPI are in the billions of dollars. Despite the large investments of resources, the failure rate for SPI programs is high – too high many would say. The most recent report from the Software Engineering Institute puts the rate of failure at around 70%; a prior report showed equally dim results" [Ngw03].

Because today's software products are not sufficiently engineered, software disasters happen often (DevTopics Software Development Topics, http://www.devtopics.com/20-famous-software-disasters/).

1.6 What Does a Revolution Mean?

It means a drastic, complete, and fundamental change of paradigm to resolve some outstanding and generally recognized problem that can be met in no other way. According to *"The Structure of Scientific Revolutions"* [Kuh62], science does not progress continuously, by gradually extending an established paradigm. It proceeds as a series of revolutionary upheavals.

1.6.1 Three Phases of Scientific Revolutions

Kuhn described that there are three phases with Scientific Revolutions: the first phase, which exists only once, is the **preparadigm phase**, in which there is no consensus on any particular theory, though the research being carried out can be considered scientific in nature – this phase is characterized by several incompatible and incomplete theories; the second phase, is the **normal science** – if the actors in the preparadigm community eventually gravitate to one of these conceptual frameworks and ultimately to a widespread consensus on the appropriate choice of methods, terminology and on the kinds of experiments that are likely to contribute to increased insights, then the **normal science** begins, in which puzzles are solved within the context of the dominant paradigm. As long as there is general consensus within the discipline, normal science continues; the third phase is the **revolutionary science** phase – over time, progress in normal science may reveal anomalies, facts which are difficult to explain within the context of the existing paradigm. While usually these anomalies are resolved, in some cases they may accumulate to the point where normal science becomes difficult and where weaknesses in the old paradigm are revealed; Kuhn refers to this as a crisis. After significant efforts of normal science within a paradigm fail, science may enter the third phase, that of **revolutionary science**, in which the underlying assumptions of the field are reexamined and a new paradigm is established. After the new paradigm's dominance is established, scientists return to normal science, solving puzzles within the new paradigm. A science may go through these three phases cycles repeatedly, though Kuhn notes that it is a good thing for science that such paradigm shifts do not occur often or easily.

1.6.2 Progress Through Revolutions

The first edition of *The Structure of Scientific Revolutions*, ended with a chapter entitled "Progress Through Revolutions," in which Kuhn stated his views on the nature of scientific progress. Because Kuhn considered problem solving to be a central element of science, he saw that for a new paradigm candidate to be accepted by a scientific community, "First, the new candidate must seem to resolve some outstanding and generally recognized problem that can be met in no other way. Second, the new paradigm must promise to preserve a relatively large part of the concrete problem solving activity that has accrued to science through its predecessors."

1.7 What Is NSE?

NSE (Nonlinear Software Engineering paradigm) based on complexity science, is established with the objectives to revolutionarily solve the critical problems existing with the old-established software engineering paradigm. Those critical problems can be summarized as follows:

(a) **Incomplete** – there is no defined process model and support for software maintenance which takes 75% or more of the total effort and cost for a software product
(b) **Unreliable** – the quality of a software product mainly depends on software testing after production which has been proven impossible to ensure high quality

1.7 What Is NSE?

(c) **Invisible** – the existing visualization methods, techniques, and tools do not offer the capability to make the entire software development lifecycle visible, the generated charts and diagrams are not holistic and not traceable

(d) **Inconsistent** – the documents and the source code are not traceable to each other and not consistent after code modification again and again

(e) **Unchangeable** – the implementation of requirement change or code modification is performed locally and blindly with high risks

(f) **Not maintainable** – software maintenance is performed partially and locally without support for bidirectional traceability to prevent side effects, so that each code modification will have a 20–50% of chance to introduce new defects into the software product being maintained

(g) **Low productivity and quality** – most resources are spent in inefficient software maintenance, the quality cannot be ensured with the blind and local implementation of software changes

(h) **High cost and risk** – most cost is spent in blind and local maintenance of the software products, which makes a software product unstable day by day in responding to needed changes

(i) **Low project success rate** – it is still less than 30% for projects with budgets over $1 million

(j) **Often the software projects developed with the old-established software engineering paradigm are capable of becoming a monster of missed schedules, blown budgets, and flawed products** – because the old-established software engineering paradigm is based on linear thinking, reductionism, and superposition principle, so that almost all tasks/activities are performed linearly, partially, and locally

It is clear that those problems are related to the entire software engineering paradigm with all of its components, including the process models, the software development methodologies, the visualization paradigm, the software testing paradigm, the quality assurance paradigm, the documentation paradigm, the maintenance paradigm, the project management paradigm, and the related techniques and tools. It means that a local and partial solution will not work – **we need a holistic and global solution in almost all aspects of software engineering: a complete revolution.**

For solving those critical problems existing with today's software development efficiently, a new software engineering paradigm, NSE is established by me and implemented by me and my colleagues. The essential difference between the old-established software engineering paradigm and NSE is how to handle the relationship between the whole and its parts of a software system. **The former adheres to the reductionism principle and superposition principle that the whole is the sum of its parts,** so that nearly all software development tasks/activities are performed locally, such as the implementation of requirement changes. **The latter complies with the Holism principle of complexity science, that a software product is a Complex Adaptive System (CAS [Hol95]) having multiple interacting agents (components), of which the overall behavior and characteristics cannot**

be inferred simply from the behavior of its individual agents but emerge from the interaction of its parts, so that with NSE nearly all software development tasks/activities are performed globally and holistically to prevent defects in the entire software lifecycle [Xio09-1], [Xio09-2].

Some primary applications show that the NSE paradigm with its support platform, Panorama++, can make revolutionary changes to almost all aspects in software engineering to efficiently handle software complexity, invisibility, changeability, and conformity, and solve the critical problems (low productivity and quality, high cost and risk) existing with the old-established software engineering paradigm – NSE makes it possible to help software development organizations double their productivity, halve their cost, and remove 99–99.99% of the defects in their software products.

From Chaps. 3 to 24 in this book, I will introduce NSE in detail, including the foundation for establishing NSE, the framework for establishing NSE, and each of its components – NSE brings revolutionary changes to almost all aspects in software engineering, including the following:

- **The foundation** (see Chaps. 3 and 4)
  From: based on linear thinking and the reductionism principle and superposition principle that the whole is the sum of its parts, so that nearly all software development tasks/activities are performed linearly, partially, and locally, such as the implementation of requirement changes.
  To: based on nonlinear thinking and complexity science – to comply with the essential principles of complexity science, particularly the Nonlinear Principle and the Holism Principle that the whole of a complex system is greater than the sum of its parts – the characteristics and the behavior of a complex system is an emergent property of the interactions of its components (agents), so that with NSE nearly all software development tasks/activities are performed nonlinearly, holistically, and globally to prevent defects in the entire software lifecycle – for instance, if there is a need to change a requirement, with NSE and the support platform Panorama++ the implementation of the change will be performed nonlinearly, holistically, and globally through various bidirectional traceabilities: (1) Performs forward tracing for the requirement change (through the corresponding test cases) to determine what modules should be modified. (2) Performs backward tracing to check related requirements of the modules to be modified for preventing requirement conflicts. (3) Checks what other modules may also need to be changed with the modification by tracing the modules to find all related modules on the corresponding call graph shown in J-Chart innovated by me. (4) Checks where the global variables and static variables may be affected by the modification. (5) After modification, checks all related statements calling the modified module for preventing inconsistency defects between them. (6) Performs efficient regression testing through backward tracing from the modified module to find the related test cases. (7) Performs backward tracing to find and modify inconsistent documents after code modification.
- **The process model(s)** (see Chap. 8)

1.7 What Is NSE? 51

From: linear ones based on linear thinking and the reductionism principle and superposition principle, including the waterfall model, the incremental development models, the iterative development models, or the incremental and iterative development models with which there is only one track in one direction – no upstream movement at all, always going forward from the upper phases to the lower phases, so that defects introduced in the upper phases will easily propagate to the lower phases to make the defect removal cost greatly increase.

To: a nonlinear one (called the NSE process model, innovated by me) based on nonlinear thinking and complexity science with which there are multiple tracks in two directions through various traceabilities to prevent defects and defect propagation, so that experience and ideas from each downstream part of the construction process may leap upstream, sometimes more than one stage, and affect the upstream activity. With NSE, the software development process and software maintenance process are combined together closely; the software development process and the project management process are also combined together closely so that the project management documents are traceable with the implementations of software requirements and the source code. With the NSE process model, requirement validation and verification can be done easily through forward traceability in parallel, and code modification can be done with side-effect prevention through backward traceability in parallel too.

- **The software development methodologies** (see Chap. 10)
  From: the software development methods based on Constructive holism – **"building"** a software system with its components – the components are

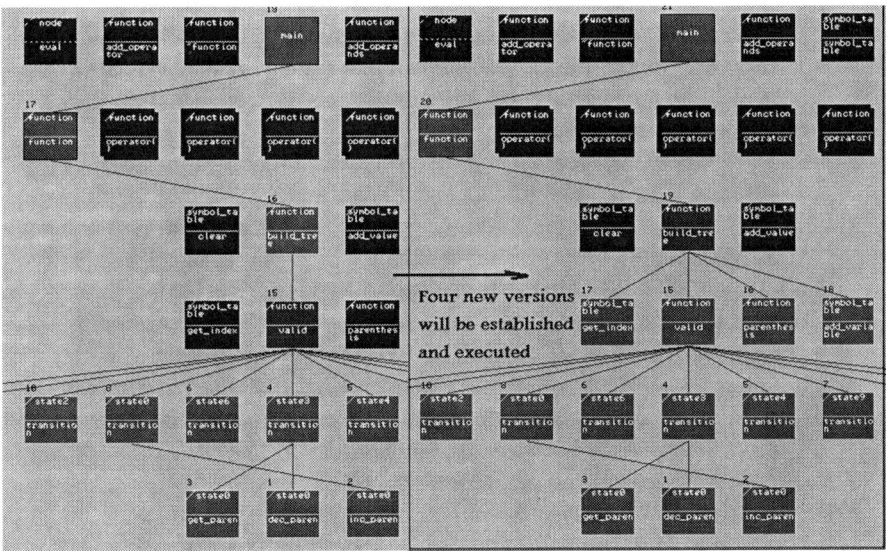

Fig. 1.59 An application example of incrementally growing up of a software system

developed first, then the system of a software product is built through the integration of the components developed. From the point of view of quality assurance, those methodologies are test-driven, but the functional testing is performed after coding; it is too late. These methodologies consider a software product as a machine rather than a logical product created by human beings. They all comply with the reductionism principle and superposition principle.

To: the software development method (NSE software development method, innovated by the me) based on generative Holism of complexity science – having the whole dummy system first, then "**growing up**" with its components as shown in Fig. 1.59.

The benefit by adding only one module each time is that if something unexpected happens, it is much easier to find and fix the problems.

From the point of view of quality assurance, the NSE software development method is defect prevention- and traceability-driven to assure the quality from the first step to the end.

- **The software testing paradigm** (see Chap. 16)

  From: mainly based on functional testing using the Black-box testing method being applied after the entire product is produced, structural testing using White-box testing method being applied after each software unit is coded for the incremental software development, and iterative software development [Coc08]. Both methods are applied separately without internal logic connections.

  To: mainly based on the Transparent-box method innovated by me to combine functional testing and structural testing seamlessly: to each set of inputs, it not only verifies whether the output (if any, can be none) is the same as the expected value, but also helps users to check whether the execution path covers the expected path with the capability to automatically establish bidirectional traceability among all of the related documents and the source code for inconsistency defect checking.

- **The quality assurance paradigm** (see Chap. 17)

  From: a test-driven approach, mainly using Black-box testing method plus structural testing method and code inspection after coding.

  To: NSE-SQA – defect prevention-driven approach innovated by me, mainly using the Transparent-box testing method in all phases of a software development lifecycle from the first step to the end because having an output is no longer a condition to use the Transparent-box testing method dynamically. The priority of NSE-SQA for assuring the quality of a software being developed is ordered as (1) defect prevention; (2) defect propagation prevention; (3) Refactoring applied to highly complex modules and module(s) that are performance bottlenecks; (4) Deep and broad testing.

- **The software diagramming paradigm** (see Chap. 7)

  From: drawing the diagrams manually or using graphic editors or using a tool to generate partial charts/diagrams which are neither interactive nor traceable in most cases. Even if some charts/diagrams for an entire software system can

1.7 What Is NSE?

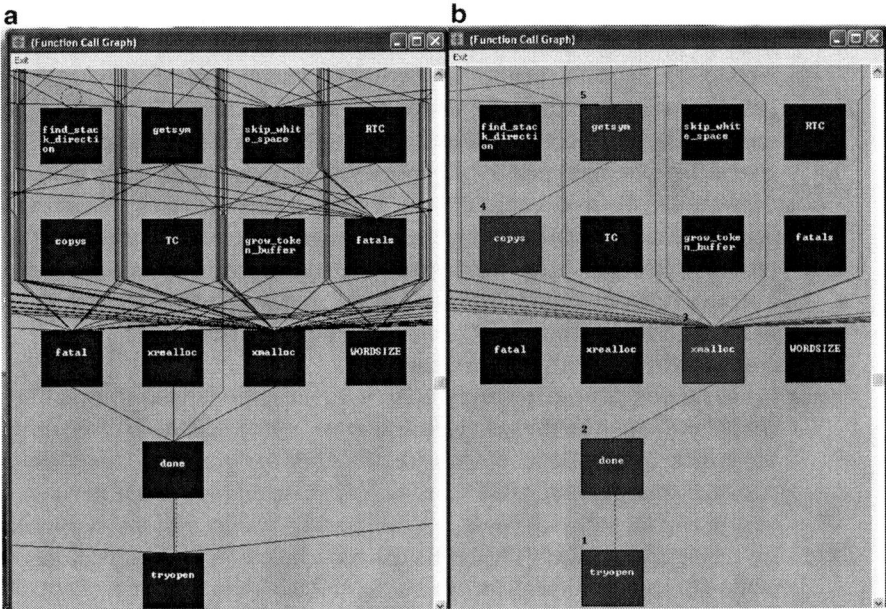

Fig. 1.60 A call graph shown in J-Chart notation defined by Jay Xiong. (**a**) A complex program structure. (**b**) A module and all of the related modules highlighted with the bottom-up orders for incremental coding and unit testing

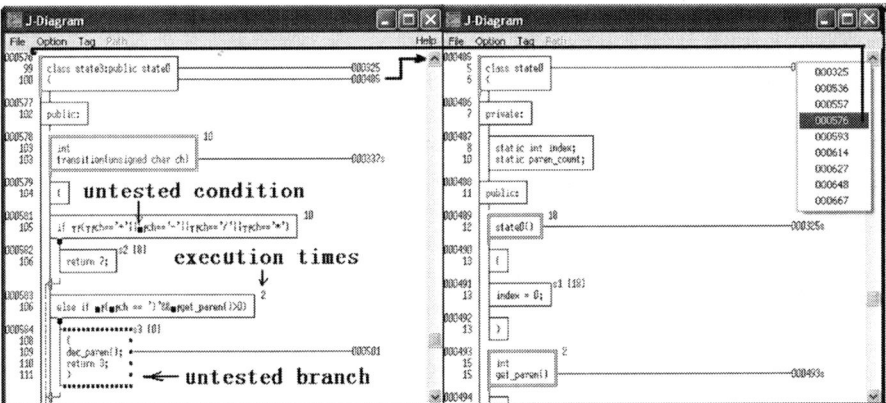

Fig. 1.61 A holistic and traceable logic diagram shown in J-Diagram notations defined by Jay Xiong with untested branches and conditions highlighted

be generated, they are still not useful because there are too many connection lines to make the charts/diagrams hard to view and hard to understand without the capability to trace an element with all the related elements.

To: holistic, interactive, traceable, and virtual software diagramming paradigm innovated by me to make an entire software development lifecycle visible. The charts/diagrams are dynamically generated from several Hash tables from the database and the source code through dummy programming or reverse engineering virtually without storing the hard copies in hard disk or memory to greatly reduce the space. The generated charts/diagrams are interactive and traceable between related elements – users can highlight an element with all of the related elements easily as shown in Figs. 1.60 and 1.61.

- **The documentation paradigm** (see Chap. 19)

  From: (a) separated from the source code without bidirectional traceability; (b) inconsistent with the source code after code modifications; (c) requiring huge disk space and memory space to store the graphical documents; (d) the display and operation speed is very slow; (e) hard to update; (f) not very useful for software product understanding, testing, and maintenance.

  To: (a) managed together with the source code based on bidirectional traceability; (b) consistent with the source code after code modification; (c) most documents are dynamically generated from several Hash tables and exist virtually without huge storage space; (d) the display and operation speed is very fast; (e) most documents can be updated automatically; (f) very useful for software product understanding, testing, and maintenance.

- **The software maintenance paradigm** (see Chap. 18)

  From: performed blindly, partially, and locally without the capability to prevent the side effects for the implementation of requirement changes or code modifications, takes about 70% of the total effort and cost in the software system development in most software organizations.

  To: performed visually, holistically, and globally using a systematic, disciplined, quantifiable approach innovated by me to prevent the side effects for the implementation of requirement changes or code modifications through various automated traceabilities; takes only about 25% of the total effort and cost in software system development, because with NSE there is no big difference between the software development process and the software maintenance process – both support requirement changes or code modification with side-effect prevention.

- **The software project management paradigm** (see Chap. 20)

  From: performed separately from the software product development process, often making the necessary actions being done too late.

  To: performed closely with the software development process, makes the project management documents such as the product development schedule, the cost reports, and the progress reports traceable with the requirement implementation or the corresponding test cases or the source code, making the necessary actions being done in time. Figure 1.62 shows a schedule chart traced and opened when a test case is selected for forward tracing.

1.7 What Is NSE?

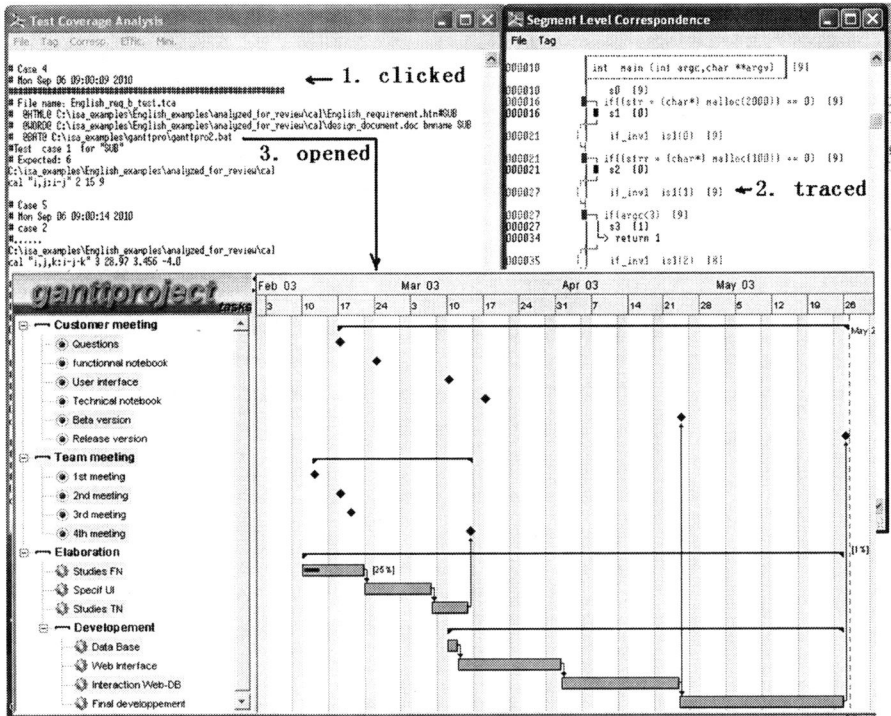

Fig. 1.62 An example of how management documents can be traced and automatically opened with bidirectional traceability from a requirement implementation, test case, or the source code

Why should NSE bring revolutionary changes to almost all aspects in software engineering? The answer is that

(a) According to complexity science, the characteristics and behaviors of the whole of a complex system emerge from the interaction of its components and the interaction between it and the environment, cannot be inherited from one or a few of its individual components, so that partial and local "**revolution**" for one or a few components of the entire software engineering paradigm will not work – for instance, focusing on software process improvement and management improvement only without changing the linear process models, the outdated software development methodologies based on reductionism and superposition principle, the inefficient software testing paradigm which cannot be dynamically used in upstream where most critical software defects are coming from, the inefficient software quality assurance paradigm based on software testing after production which violates Deming's product quality assurance principles, the inefficient software visualization paradigm by which the generated local and partial charts and diagrams are not interactive and not traceable, the inefficient software documentation paradigm by which the generated documents are not traceable to the source code, the blind software maintenance paradigm without support of automated and self-maintainable traceability, etc., is impossible to bring revolutionary changes to today's software development. It is also important to point out that even if all the components of the software engineering paradigm have been changed revolutionarily, it does not guarantee that the whole of the software engineering paradigm has been changed revolutionarily – it depends on

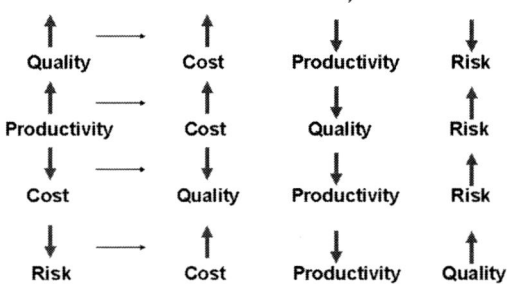

Fig. 1.63 The interactive effect among the critical problems existing with today's software development

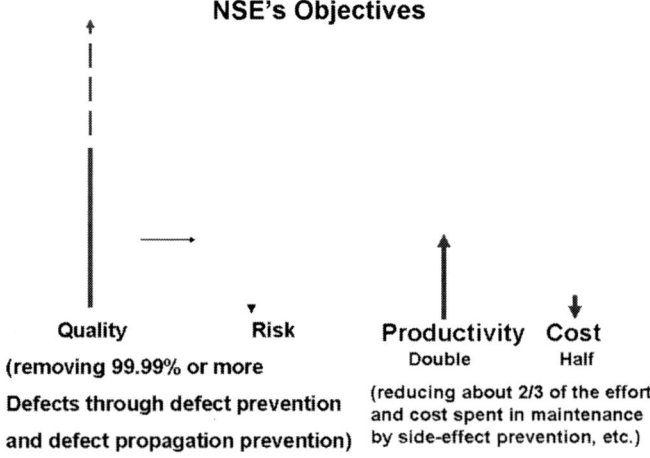

Fig. 1.64 NSE's objectives

the interaction between them: how they are working together, how they can share the resources such as the computer memory and the file system, how they can use the same database, how the obtained results from one component can be efficiently used by others, and so on to realize the whole of the software engineering paradigm greater (rather than less) than its components.
(b) It is also related to the objectives of NSE – to solve almost all of the critical problems existing with today's software development at the same time: low productivity and quality, high cost and risk. With the old-established software engineering paradigm, it is impossible to solve those critical problems together at the same time – see Fig. 1.63 about their limitation and effects brought by one to others, and Fig. 1.64 NSE's objectives.

1.8 Summary

Software is becoming the foundation of modern civilization – it affects almost all aspects of our lives and our everyday activities. With software engineering, many tasks/activities are defined, including requirement development, product design, coding, testing, deployment and support, maintenance, configuration, and project management. For supporting software development, many software process models are proposed and used in practice, including the Waterfall lifecycle model, the Prototype model, the Spiral Model, CMMI, Agile models, and RUP.

But unfortunately, today's software products are still not sufficiently engineered to fulfill the role of "foundation." There are many critical problems existing with today's software engineering paradigm: low productivity and quality, and high cost and risk.

The root cause of those critical problem comes from the fact that not only a software product but also the software engineering paradigm itself is a complex system consisting of many closely related parts, where the characteristics and behavior of the whole system emerge from the interaction of its parts – but the existing software engineering paradigm is established with linear thinking, reductionism, and the superposition principle that **the whole of a system is the sum of its parts,** so that it handles a nonlinear complex software product as a linear system, making all tasks/activities be performed linearly, partially, and locally.

For efficiently solving the critical problems existing with the old-established software engineering paradigm, a new revolutionary software engineering paradigm NSE (Nonlinear Software Engineering) paradigm has been established, which is based on nonlinear thinking and complexity science.

With NSE, "software" is redefined to include not only a computer program, the data used, and the documents traceable to the source code but also the database built through static and dynamic measurement of the program and a set of Assistant Online Agents to make the program adaptive and maintainable, and the acceptance testing can be performed in a fully automated way.

With NSE, software maintenance can be performed holistically and globally with side-effect prevention through various traceabilities.

With NSE, the quality of a software product is ensured from the first step down to maintenance through defect prevention and defect propagation prevention.

With NSE, the entire software development process is visible, and the documents are traceable to the source code.

The detailed descriptions on the all related topics will be introduced from Chaps. 3 to 24 of this book.

1.9 Points and Questions to Ponder

(a) What are the major differences between the traditional **software** definition and the new one defined with NSE? Do you think it is necessary to provide a software product to the customer (not the end user) with the database built through static and dynamic measurement of the product, and a set of Assisted Online Agents? Why?
(b) Are today's software products sufficiently engineered? Why?
(c) What are the common limitations existing with current software process models?
(d) For efficiently supporting software maintenance, what conditions do you think a process model or software development approach should satisfy?
(e) Although the software engineering paradigm itself is a complex system consisting of many related parts which are connected closely and interactively, some people still believe that only improving one or two parts of the software engineering paradigm without improving its other parts can still dramatically improve the overall characteristics, performance, behavior, and the problem-solving capability of the software engineering paradigm – do you agree with their conclusion? Why?

1.10 Further Reading and Information Source

(a) http://www.comdig.org/ complexity digest – subscribe to the newsletter
(b) http://www.brint.com/Systems.htm Complexity, Complex Systems & Chaos Theory Organizations as Self-Adaptive Complex Systems

References

[Amb04] Ambler SW, Nalbone J, Vizdos M (2004) Enterprise unified process: extending the rational unified process. Prentice Hall PTR, Upper Saddle River

[Amb05] Ambler S (2005) A manager's introduction to the Rational Unified Process (RUP). http://www.ambysoft.com/downloads/managersIntroToRUP.pdf. Accessed 20 Feb 2009

References

[Box87]	Box GEP, Draper NR (1987) Empirical model-building and response surfaces. Wiley, New York, p 424. ISBN 0471810339
[Bro95-p120]	Brooks FP Jr (1995) The mythical man-month. Addison-Wesley, Reading, P120
[Bro95-p122]	Brooks FP Jr (1995) The mythical man-month. Addison-Wesley, Reading, P122
[Bro95-p182]	Brooks FP Jr (1995) The mythical man-month. Addison-Wesley, Reading, P182
[Coc08]	Cockburn A (2008) Using both incremental and iterative development, CrossTalk, May Issue
[GSAM00]	USAF Software Technology Support Center (2000) Guidelines for the Successful Acquisition and Management of Software Intensive Systems (GSAM), version 3, chapter 5, USAF Software Technology Support Center, May
[GSAM03]	USAF Software Technology Support Center (2003) Condensed GSAM handbook, chapter 2, CrossTalk
[Hol95]	Holland JH (1995) Hidden order: how adaptation builds complexity. Addison-Wesley, Reading
[Kuh62]	Kuhn T (1962) The structure of scientific revolutions. The University of Chicago Press, Chicago
[Kan09]	Kannenberg A et al (2009) Why software requirements traceability remains a challenge, CrossTalk, Jul/Aug Issue
[Mck95]	McKenzie CA (1999) MIS327 – Systems analysis and design, course schedule
[Ngw03]	Ngwenyama O, Nielsen PA (2003) Competing values in software process improvement: an assumption analysis of CMM from an organizational culture perspective. IEEE Trans Eng Manag 50(1):100–112. doi:10.1109/TEM.2002.808267
[Ree95]	Sorensen R (1995) A comparison of software development methodologies, Crosstalk, Jan Issue
[Roy70]	Royce WW (1970) Managing the development of large software systems concepts and techniques. In: Proc. WESCON, August 1970
[Pre95-p4]	Pressman RS (2005) Software engineering: a practitioner's approach. McGraw-Hill, New York, Part 4
[Pre05-p78]	Pressman RS (2005) Software engineering: a practitioner's approach. McGraw-Hill, New York, p P78
[Xio09-1]	Xiong J, Xiong J (2009) A complete revolution in software engineering based on complexity science, WORLDCOMP'09 – SERP (Software Engineering Research and Practice 2009), Las Vegas, pp 109–115
[Xio09-2]	Xiong J (2009) Tutorial: a complete revolution in software engineering based on complexity science, WORLDCOMP'09, Las Vegas, 13–17 July 2009

Chapter 2
Is the Old-Established Software Engineering Paradigm Entirely Out of Date?

> Major software projects have been troubling business activities for more than 50 years. Of any known business activity, software projects have the highest probability of being cancelled or delayed. Once delivered, these projects display excessive error quantities and low levels of reliability.
>
> Capers Jones
>
> One of the primary reasons that many businesses fail is an attempt to solve a non-linear (or wicked) problem with a linear process. All people problems and issues are non-linear because they exist in a dynamic rather than a static environment.
>
> Cityzone, Process Versus Non-Linear Thinking
> http://www.city-zone.com/modules/publishing/item.aspx?iid=138

Software has become a driving force for the development of science, engineering, and business in the twenty-first century.

Since the term *software engineering* first appeared in the 1968 NATO Software Engineering Conference, it is more than 40 years past. Within that period of time, great progress in software engineering has been achieved, particularly the following people and their great contributions (without their contributions, it is impossible for me to write this book) listed by ***CompHist.org(http://comphist.org/)***:

Engineering:

1968:   **Peter Naur et al** coined the term "software engineering" at the NATO conference in Garmisch-Partenkirchen and pointed out that software should follow an engineering paradigm, it was the response to a software crisis where the quality was too low, the delivery was too late, and the costs went way over the budget.

1975:   **Frederick P. Brook, Jr.** book on "Software Engineering" which tackles the question of how to organize and manage large-scale programming projects.

Programming and Design Methodologies:

1972:   **E.W. Dijkstra** book on structured programming
1972:   **D.L. Parnas** "Parnas Module" which proposed information hiding.

1975: **M.A. Jackson** book on "Principles of Program Design," which model data and algorithms largely separated.
1978: **G.J. Myers** articles "Composite/Structured Design" for composite design.
1979: **Edward Yordon and L.L. Constantine** book on structured design.

They affected heavily how programming languages were being structured afterwards.

User's Requirements, Requirement Engineering and Description Technologies:

1977: **D. Teichrow and E. Hershey** paper on prototyping as a tool in the specification of user requirements.
1977: **D. Ross** paper on structured analysis.
1977: **M.W. Alford** paper on the use of lexical affinities in requirements extraction

Project Management Technologies:

1981: **Barry Boehm** book on "Software Engineering, Economics" which addresses cost estimation issues
1976: **T.J. MaCabe** paper on software complexity measurement and the detection of risky factors.
1977: **M.H. Halstead** book – "Elements of Software Science" which coined the term E measurement – efforts measurement.

...

At this phase, procedures started to be separated from the data; furthermore, related procedures and data were brought together into subsystems.

1980–1990 Prototyping technologies and formalization, partial automation in upstream, *includes analysis of dynamic, formal methods, and CASE tools.*

1986: **William. W. Agresti** paper on appearance of prototyping technologies, which discarded the waterfall model and shifted to prototyping.

Analysis of Dynamic Behavior of Specification:

1983: **M.A. Jackson** book on **JSP** (Jackson Structured Programming), a method for designing programs as compositions of sequential processes and **JSD** (Jackson System Development), a method for specifying and designing systems
1986: **Paul T. Ward** paper on real-time data flow
1986: **Pamela Zave and William Schell** paper on **PAISLey**, an executable specification language which is accompanied by a set of specification methods, analysis techniques, and software support tools.
1986: **Giorgio Bruno and Giuseppe Marchetto** paper on **PROTnet,** a Process-Translatable Petri Nets for the Rapid Prototyping of Process Control Systems

...

Formal Methods:

> **ISO standardization**, such as **GKS** (1985), the computer graphics standard, and **PREMO** (1998) the multimedia standard.
> SRI's PVS (Prototype Verification System) Theorem Prover
> **Bell Labs's SPIN** model checker

CASE (Computer Aided Software Engineering) Tools:

1988: Meilir Page-Jones book "The Practical Guide to Structured System Design," which features SA/SD – structured analysis/structured design with modularized view; a structure chart is used to show the programmers of a system how the system is partitioned into modules.

...

Around this time, subsystems began to be layered.

1985–1995 Software Process Model, *this includes process programming, CMM, integrated environment, and analyzing and supporting human factors.*

Software Process and SPI – Software Process Improvement:

1986: **Frederick P Brooks, Jr.** paper on information processing which address the essence and accidents in software development and the ratio between them, summarized as "No Silver Bullet"

...

1989: **Watts S. Humphrey** book "Managing the Software Process," featured **CMM** – Capability Maturity Model, which optimized the software process in five levels: initial, repeatable, defined, managed, optimizing.

Integrated Environments:

1993: **Lois Wakeman and Jonathan Jowett** book "PCTE – The Standard for Open Repositories" which discussed tool integration.

Analyzing and Supporting Human Factors:

1986: **Bill Curties** paper on protocol and human factors analysis
1988: **Colin Potts and Glenn Bruns** paper on design decision, which discussed communication support.

...

1985 to present – the Network Age, *this includes Object oriented technologies, distributed computing, open source software development and web engineering.*

Object Oriented Technologies:

Programming Language

1967: **O.J. Dahl** papers on **SIMULA**, a precursor to the OO language Simula, which featured class, instance and module.

1976: **Lampson et al.** introduced **EUCLID**, a related type systems Euclid, one of the first languages that considered the problem of aliasing, and included constructs to express it.

1976: **Niklaus Wirth** introduced **Modula**, a language derived from Pascal, which featured the module.

1977: **B. Liskov** paper on **CLU**, which was the first implemented programming language to provide direct linguistic support for data abstraction and featured clusters.

1979: **JD Ichbiah et al. Ada**, a programming language which featured packages

1981: **Alan kay and Dan Ingalls et al./Xerox** introduced **Smalltalk 80**, an object-oriented programming language.

1986: **Brad Cox** introduced the first **Objective-C** compiler

1986: **Bjarne Stroustrup** introduced **C++** Programming Language

1988: **Bertrand Meyer Eiffel,** an elegant object-oriented language, designed to support reuse, and including support for logical assertions.

1989: **David. A. Moon** introduced **CLOS** – Common Lisp Object System

1995: **James Gosling/Sun Microsystems** introduced **Java**, a simplified C++ like OOP which is expressly designed for use in the distributed environment of the Internet.

Object-Oriented Analysis and Design

1986: **G. Booch** introduced **OOD**(Object-Oriented Design)

1988: **Shlare-Mellor** papers on viewing systems as architecture, corresponding to breaking a large system up into components.

1991: **Peter Coad, Edward Yourdon** book on the principles of object-oriented technology

1991: **J. Rumbaugh** book on Object-Oriented Modeling and Design and introduced **OMT** (Object Modeling Technique).

1995: **Ivar Jacobson** paper on using case driven approach, which introduced **OOSE** (Object-Oriented Software Engineering).

...

1997: **Clemens Szypersky** book "Component Software – beyond object-oriented programming" introduced software components

1999: **Ivar Jacobson, James Rumbaugh, Brady Booch** books on the unified software development process, modeling and language, which introduced **UML**

Here, the big object orientation methodologies, layering, and OOP advancements quickly complemented each other.

Open Source Software Development

1997: **Eric S. Raymond** outlined the core principles of open source movement in a manifesto called "The Cathedral and the Bazaar."

Today many software products are about 10,000 times more complex than those written in 40 years ago. Unfortunately, the old-established software engineering paradigm is crisis-ridden and frequently disastrous, which is entirely outdated.

2.1 The *20 Famous Software Disasters* Reported

Software errors cost the US economy about $60 billion annually in rework, lost productivity, and actual damages.

DevTopics Software Development Topics listed the 20 Famous Software Disasters (http://www.devtopics.com/20-famous-software-disasters/), particularly these:

...

2. Hartford Coliseum Collapse (1978)
Cost: $70 million, plus another $20 million damage to the local economy
Disaster: Just hours after thousands of fans had left the Hartford Coliseum, the steel-latticed roof collapsed under the weight of wet snow.
Cause: The programmer of the CAD software used to design the coliseum incorrectly assumed the steel roof supports would only face pure compression. But when one of the supports unexpectedly buckled from the snow, it set off a chain reaction that brought down the other roof sections like dominoes.

...

4. World War III... Almost (1983)
Cost: Nearly all of humanity
Disaster: The Soviet early warning system falsely indicated the United States had launched five ballistic missiles. Fortunately the Soviet duty officer had a "funny feeling in my gut" and reasoned if the U.S. was really attacking they would launch more than five missiles, so he reported the apparent attack as a false alarm.
Cause: A bug in the Soviet software failed to filter out false missile detections caused by sunlight reflecting off cloud-tops.

...

5. Medical Machine Kills (1985)
Cost: Three people dead, three people critically injured
Disaster: Canada's Therac-25 radiation therapy machine malfunctioned and delivered lethal radiation doses to patients.
Cause: Because of a subtle bug called a race condition, a technician could accidentally configure Therac-25 so the electron beam would fire in high-power mode without the proper patient shielding.

...

6. Wall Street Crash (1987)
Cost: $500 billion in one day
Disaster: On "Black Monday" (October 19, 1987), the Dow Jones Industrial Average plummeted 508 points, losing 22.6% of its total value. The S&P 500 dropped 20.4%. This was the greatest loss Wall Street ever suffered in a single day.
Cause: A long bull market was halted by a rash of SEC investigations of insider trading and by other market forces. As investors fled stocks in a mass exodus, computer trading programs generated a flood of sell orders, overwhelming the market, crashing systems and leaving investors effectively blind.

...

8. Patriot Fails Soldiers (1991)
Cost: 28 soldiers dead, 100 injured
Disaster: During the first Gulf War, an American Patriot Missile system in Saudi Arabia failed to intercept an incoming Iraqi Scud missile. The missile destroyed an American Army barracks.
Cause: A software rounding error incorrectly calculated the time, causing the Patriot system to ignore the incoming Scud missile.

...

10. Ariane Rocket Goes Boom (1996)
Cost: $500 million
Disaster: Ariane 5, Europe's newest unmanned rocket, was intentionally destroyed seconds after launch on its maiden flight. Also destroyed was its cargo of four scientific satellites to study how the Earth's magnetic field interacts with solar winds.
Cause: Shutdown occurred when the guidance computer tried to convert the sideways rocket velocity from 64-bits to a 16-bit format. The number was too big, and an overflow error resulted. When the guidance system shut down, control passed to an identical redundant unit, which also failed because it was running the same algorithm.

...

15. Y2K (1999)
Cost: $500 billion
Disaster: One man's disaster is another man's fortune, as demonstrated by the infamous Y2K bug. Businesses spent billions on programmers to fix a glitch in legacy software. While no significant computer failures occurred, preparation for the Y2K bug had a significant cost and time impact on all industries that use computer technology.
Cause: To save computer storage space, legacy software often stored the year for dates as two digit numbers, such as "99″ for 1999. The software also interpreted "00″ to mean 1900 rather than 2000, so when the year 2000 came along, bugs would result.

...

18. Cancer Treatment to Die For (2000)
Cost: Eight people dead, 20 critically injured
Disaster: Radiation therapy software by Multidata Systems International miscalculated the proper dosage, exposing patients to harmful and in some cases fatal levels of radiation. The physicians, who were legally required to double-check the software's calculations, were indicted for murder.
Cause: The software calculated radiation dosage based on the order in which data was entered, sometimes delivering a double dose of radiation.

Why do software disasters happen so frequently? There are many reasons, but the root cause is that the current software engineering paradigm is entirely out of date; it does not meet the need for twenty-first century software development, because it is based on linear thinking and the superposition principle.

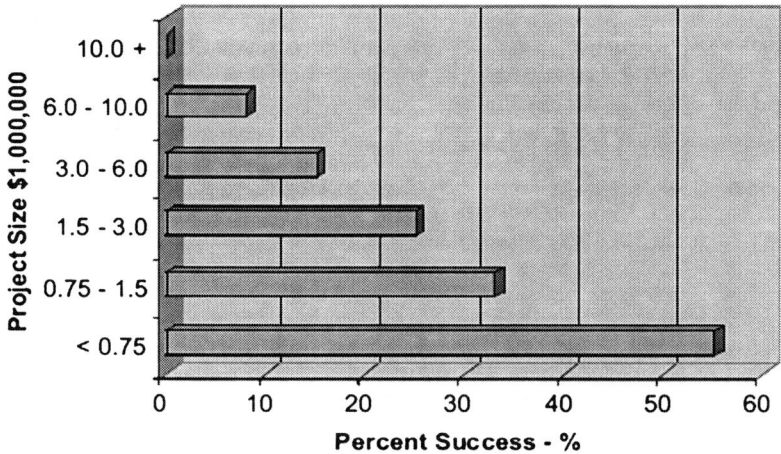

Fig. 2.1 Software project success rate based on size

2.1.1 Very High Project Failure Rate Reported

In the article of "Why Big Software Projects Fail: The 12 Key Questions," Watts S. Humphrey (the innovator of CMM/CMMI) reported that the software project success rate is still very low as shown in Fig. 2.1 [Hum05].

The definition of a successful project is one that completed within 10% or so of its committed cost and schedule and delivered all of its intended functions.

As shown in Fig. 2.1, the success rate for a software project with more than $1,000,000 is about 30% – it means about 70% of the projects have failed.

2.2 What Is the Root Cause for Software Disasters and Very High Software Project Failure Rate?

There are many different answers to this question:

Several researchers have suggested that "CMM does not effectively deal with the social aspects of organizations" [Ngw03].

Timothy K. Perkins believes as follows:

> the cause of project failures is knowledge: either managers do not have the necessary knowledge, or they do not properly apply the knowledge they have. [Per06]

Capers Jones concluded as follows:

> Both technical and social issues are associated with software project failures. Among the social issues that contribute to project failures are the rejections of accurate estimates and the forcing of projects to adhere to schedules that are essentially impossible. Among the technical issues that contribute to project failures are the lack of modern estimating

approaches and the failure to plan for requirements growth during development. However, it is not a law of nature that software projects will run late, be cancelled, or be unreliable after deployment. A careful program of risk analysis and risk abatement can lower the probability of a major software disaster. [Jon06]

Joe Marasco pointed out as follows:

All the effort has gone into two areas: managing requirements and something called "requirements traceability." Requirements management is the art of capturing requirements, cataloging them, and monitoring their evolution throughout the development cycle. Requirements are added, dropped, changed, and so on, and we now have requirements management systems that allow us to keep track of all this. That is a good thing. Traceability is a bit more ambitious. It attempts to link later-stage artifacts, such as pieces of a system and their test cases, back to the original requirements. That way, we can assess if we are actually meeting the requirements that were called out. This is a harder problem, but, once again, there has been substantial progress. To all this I say, wonderful, but not good enough.

For more information, see the Standish Group Web site at http://www.standishgroup.com/
Poor Estimation: Major Root Cause of Project Failure.

Galorath Incorporated, http://www.galorath.com/wp/poor-estimation-major-root-cause-of-project-failure.php

IT projects have been considered a tough undertaking and have certain characteristics that make them different from other engineering projects and increase the chances of their failure. Such characteristics are classified in seven categories (Peffers, Gengler & Tuunanen, 2003; Salmeron & Herrero, 2005): 1) abstract constraints which generate unrealistic expectations and overambitious projects; 2) difficulty of visualization, which has been attributed to senior management asking for over-ambitious or impossible functions, the IT project representation is not understandable for all stakeholders, and the late detection of problems (intangible product); 3) excessive perception of flexibility, which contributes to time and budget overrun and frequent requests of changes by the users; 4) hidden complexity, which involves difficulties to be estimated at the project's outset and interface with the reliability and efficiency of the system; 5) uncertainty, which causes difficulty in specifying requirements and problems in implementation of the specified system; 6) the tendency to software failure, which is due to assumptions that are not thought of during the development process and the difficulty of anticipating the effects of small changes in software; 7) the goal to change existing business processes, which requires IT practitioners' understanding of the business and processes concerned in the IT system and good processes to automate and make them quicker. Such automation is unlikely to make a bad process better.

International Management Review, 2009 by Al-Ahmad, Walid, et al., *A Taxonomy of an IT Project Failure: Root Causes,* Business Publications, http://findarticles.com/p/articles/mi_qa5439/is_200901/ai_n31965631/?tag=content;col1

In the article "Why Big Software Projects Fail: The 12 Key Questions" [Hum05], Watts S. Humphrey listed those questions as follows:

Question 1: Are All Large Software Projects Unmanageable?
Question 2: Why Are Large Software Projects Hard to Manage?
Question 3: Why Is Autocratic Management Ineffective for Software?
Question 4: Why Is Management Visibility a Problem for Software?
Question 5: Why Can't Managers Just Ask the Developers?
Question 6: Why Do Planned Projects Fail?
Question 7: Why Not Just Insist on Detailed Plans?

2.3 The "Software" Definition Is Outdated

Question 8: Why Not Tell the Developers to Plan Their Work?
Question 9: How Can We Get Developers to Make Good Plans?
Question 10: How Can Management Trust Developers to Make Plans?
Question 11: What Are the Risks of Changing?
Question 12: What Has Been the Experience So Far?

Root causes of project failure ...

- Ad hoc requirements management.
- Ambiguous and imprecise communication.
- Brittle architectures.
- Overwhelming complexity.
- Undetected inconsistencies in requirements, designs, and implementations.
- Insufficient testing.
- Subjective project status assessment.
- Failure to attack risk.
- Uncontrolled change propagation.
- Insufficient automation.

devdaily, http://www.devdaily.com/java/java_oo/node7.shtml

In my opinion, they are reasonable answers to the question, but not the fundamental reason for software project failure.

According to the essential principles of complexity science, particularly the Nonlinearity principle and the Holism principle, software is a nonlinear complex system where the whole is greater than the sum of its parts, the behaviors and characteristics of the whole emerge from the interaction of its parts and the interaction between the system and its environment, small differences in the initial condition or a small change to the system may produce large variations in the long-term behavior of the system – the "Butterfly-Effect."

But unfortunately, the existing software engineering paradigm is based on linear thinking, reductionism, and the superposition principle that the whole is the sum of its parts, so that almost all tasks/activities are performed linearly, partially, and locally. It means that the foundation of the existing software engineering paradigm is wrong. The wrong foundation makes almost all things wrong in software engineering, particularly the process models, the development methods, the visualization paradigm, the testing paradigm, the quality assurance paradigm, the documentation paradigm, the maintenance paradigm, and the project management paradigm – in fact the existing software engineering paradigm is entirely outdated.

2.3 The "Software" Definition Is Outdated

The current software is defined as (1) instructions (computer programs) that when executed provide desired features, function, and performance; (2) data structures that enable the programs to adequately manipulate information; and (3) documents that describe the operation and use of the programs [Pre05-p4]. The simplest definition of a software is: a program + data + documents.

This definition separates the documents and the source code without a facility to establish the traceability to represent the internal relationship among the documents, the test cases, and the source code, and gives up the development history and the database built through static and dynamic measurement, making a software product hard to understand, test, review, and maintain.

In fact, a software is working in a changing environment dynamically, so that it should be made adaptive and easy to maintain.

This old definition of software has been replaced by a new one with NSE (see Sect. 1.1 and Chap. 8).

2.4 The Current Software Development Process Models Are Out of Date

Current main software development process models are discussed in Sect. 1.4.

A process model recommended by Alistair Cockburn to combine both Incremental and Iterative development together [Coc08] is shown in Fig. 2.2.

These software engineering process models are out of date because they are linear models with only one track forward to unidirectional without upstream movement at all, complying with the superposition principle that the whole of a software system is the sum of its parts, so that all tasks are performed linearly, locally and partially, making the defects introduced into a software product at the upper phases easy to propagate to the lower phases and the defect removal cost increase tenfold several times as shown in Fig. 2.3.

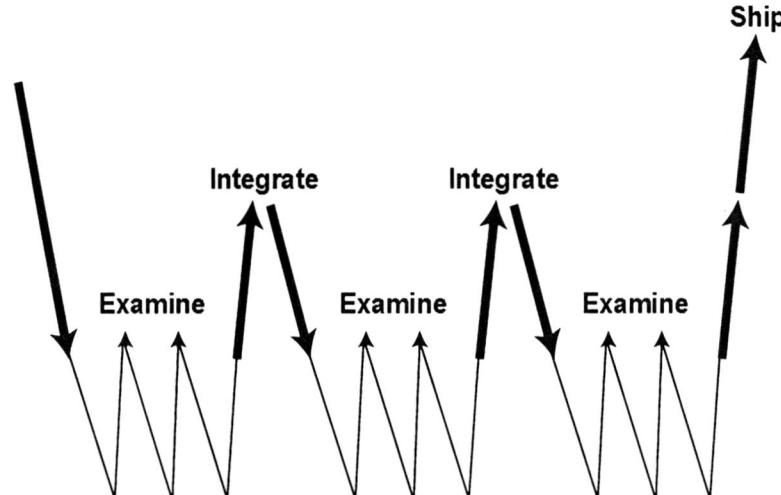

Fig. 2.2 Putting iterative and incremental development together

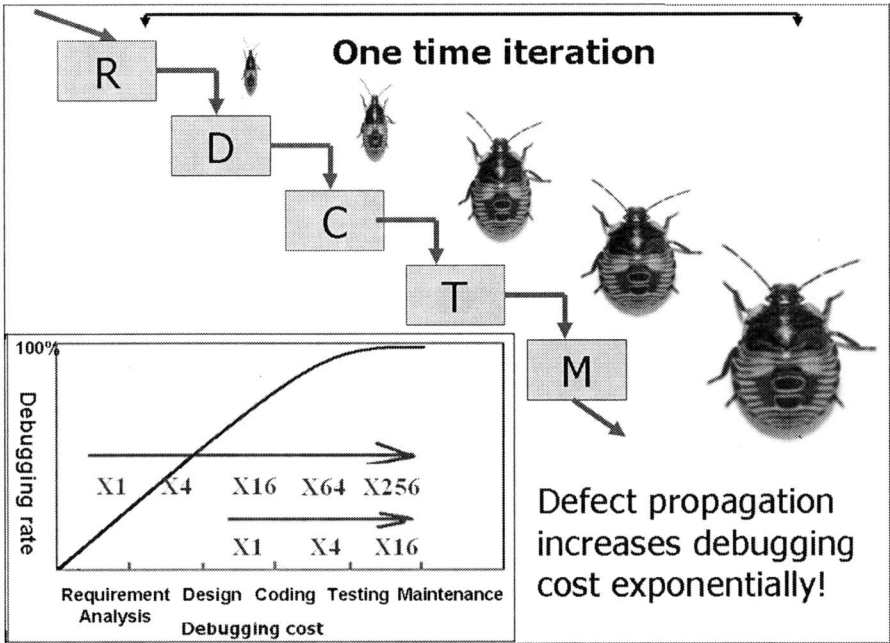

Fig. 2.3 The cost for removing a defect propagated from the requirement phase to the maintenance phase with linear process models

As shown in Fig. 2.2, a linear process model requires that people always do all things right without making any mistake, but can we drive a car from our home to another city always on an one-way with one track traffic only without U-Turns at all? No. For instance, sometimes we may forget something so that we should go back to do something – people are also nonlinear, often making wrong decisions which need to be corrected. Because there is only one track, when the engine of a car suddenly stops working, the entire traffic will be blocked.

With NSE these "one-way and one track" process models will be replaced by NSE process model with "two-way and multiple tracks." Chapter 8 will introduce the details.

2.5 Current Software Development Methodologies Are Out of Date

With current software development methodologies, software components are developed first, then the system of a software product is **built** through the integration of the components developed. From the point of view of quality assurance, those methodologies are test-driven, but the functional testing is performed after coding – it is too late. These methodologies handle a software product as a machine rather than a logical product created by human beings. They all comply with the superposition principle. With those methodologies, all tasks/activities are performed linearly, partially, and locally.

Is current CBSD (Component-Based Software Development) Out of Date Too?

The basis of CBSD is components which are developed with the old-established software engineering paradigm based on linear thinking and the superposition principle, so they are hard to ensure the quality and hard to maintain. From this point of view, the current CBSD is out of date too – it should be shifted to a new development methodology with the components developed using a novel software engineering platform based on complexity science.

2.6 The Existing Software Modeling Approaches Are Outdated

The existing software modeling approaches are outdated because they are outcomes of reductionism and superposition principle, using different sources for human understanding and computer understanding of a software system separately with a big gap between them. The obtained models are not traceable for static defect removal, not executable for debugging, not testable for dynamic defect removal, not consistent with the source code after code modification, and not qualified as the road map for software development.

2.7 Current Software Testing Paradigm Is Out of Date

Current software testing paradigm is mainly based on functional testing (plus structural testing, load testing, and stress testing) being performed after coding. It is too late, the functional testing cannot be performed in the requirement development phase and the design phase dynamically, so that it has no way to find defects introduced in the requirement development phase and the design phase dynamically using the existing software testing paradigm.

The current software testing paradigm separates functional testing and structural testing rather than combining them together seamlessly. To each set of inputs, the functional testing tools only check whether the output is the same as the expected value without checking whether the program execution path is the same as what is expected.

2.8 Current Software Quality Assurance Paradigm Is Out of Date

Current software quality assurance paradigm is mainly based on software testing and inspection using untraceable documents and untraceable source code, particularly the functional testing performed after coding.

NIST (National Institute of Standards and Technology) recommends that "Briefly, experience in testing software and systems has shown that testing to high degrees of security and reliability is from a practical perspective not possible. Thus, one needs to build security, reliability, and other aspects into the system design itself and perform a security fault analysis on the implementation of the design." ("Requiring Software Independence in VVSG 2007: STS Recommendations for the TGDC," November 2006, http://vote.nist.gov/DraftWhitePaperOnSIinVVSG2007-20061120.pdf).

With current process models and methodologies, the implementation of requirement changes and code modifications is performed locally rather than globally and holistically – without the capability to prevent the side effects, so that the quality of the modified product is hard to ensure.

2.9 Current Software Visualization Paradigm Is Out of Date

The current software visualization paradigm generates partial charts or diagrams rather than a complete chart or diagram for a software product. Most tools developed with the current software visualization paradigm are used for modeling only, rather than for the entire software development process.

Note: Even if a complete chart or diagram can be generated for an entire software product, it is still useless because there are too many connection lines, making the generated chart or diagram very hard to understand. Without traceability among related elements and the capability to highlight a module with all the related modules, a generated chart or diagram is not useful.

2.10 Current Software Documentation Paradigm Is out of Date

The current software documentation paradigm generates and manages documents separated from the source code – they are not traceable to each other.

Note: When the source code is modified the generated documents cannot be updated without bidirectional traceability, so the documents are often inconsistent with the source code as shown in Fig. 2.4, making them not very useful.

The visual documents generated with the current software visibility paradigm requires a huge amount of space to store, and the display speed is very slow.

2.11 Current Software Maintenance Paradigm Is Out of Date

The current software maintenance paradigm offers a blind, partial, and local approach for software maintenance, without support of various traceabilities. There is no way to prevent the side effects of the implementation of requirement changes or code modifications.

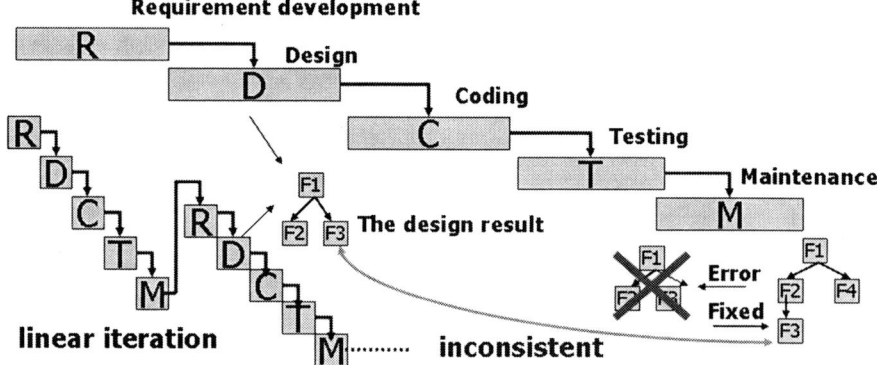

Fig. 2.4 The documents and the source code are inconsistent after code modification with the current software engineering paradigm

Note: Local and partial software maintenance is risky – each time when a bug is fixed, there is a 20–50% of chance of introducing another into the software product. It is why today software maintenance takes more than 75% of the total effort and total cost for software product development.

2.12 Current Software Project Management Paradigm Is Out of Date

According to the current software project management paradigm, software project management is separated from the software development process – the project development schedules and the cost reports are not traceable with the implementations of requirements and the source code.

Note: With the current software project management paradigm, often it is too late in finding and solving the problems.

2.13 "The Mythical Man-Month" Is an Outcome of Linear Thinking; The "No Silver Bullet" Conclusion Is Out of Date

"The Mythical Man-Month" written by Frederick P. Brooks, Jr. is a great book with many advanced concepts and ideas. I have learnt a lot from it, and will continue to learn more.

But unfortunately, because the old-established software engineering paradigm is based on linear thinking, reductionism, and superposition principle so that almost all tasks/activities are performed linearly, partially, and locally which limits all related process models, software development methods, software development

2.13 "The Mythical Man-Month" Is an Outcome of Linear Thinking

techniques and tools – it also affects all books in software engineering, including "The Mythical Man-Month."

In the 1995 edition of "The Mythical Man-Month," Frederick P. Brooks, Jr. criticized his 1975 edition of the book that "Don't Build One to Throw Away – The Waterfall Model Is Wrong! ...The biggest mistake in the 'Build one to throw away' concept is that it implicitly assumes the classical sequential or waterfall model of software construction. ...Chapter 11 is not the only one tainted by the sequential waterfall model; it runs through the book, beginning with the scheduling rule in Chapter 2. "

Unfortunately, in the 1995 edition of the book, it also assumes a sequential model – "An Incremental – Build Model" which is "a series of Waterfalls" [GSAM03] as shown in Fig. 2.5.

Comparing it with the one-time waterfall model, the Incremental – Build Model can help in reducing risk and waiting time, but it keeps all the major drawbacks of the one-time waterfall model. For instance, the defects introduced into a software product in the upper phases can easily propagate to the lower phases, making the final defect removal cost increase more than 100 times; the requirement changes and code modifications are implemented locally and blindly without support of bidirectional traceabilities, making software maintenance take more than 75% of the total effort and total cost in a software product development.

Brooks' law: "No Silver Bullet" – "There is no single development, in either technology or management technique, which by itself promises even one order-of-magnitude improvement within a decade in productivity, in reliability, in simplicity" is out of date – in fact only the bidirectional traceability technique by itself promises one order-of-magnitude improvement within a decade in productivity, in reliability, in simplicity.

> Software traceability can help bring software development into the 21st century. It reduces costs, gives better visibility and adequate test coverage, and helps software engineers meet customer needs. Changes can be implemented much faster and new projects can be estimated more accurately.
>
> Rick Coffey, Document Control Supervisor, Tyco Healthcare/Mallinckrodt

In Chap. 24 we will discuss three Candidates of "Silver Bullet."

After the establishment of NSE based on nonlinear thinking and complexity – complying with the essential principles, particularly the nonlinearity principle and the holism principle to perform almost all tasks/activities holistically and globally,

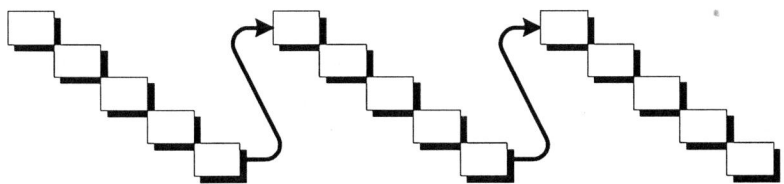

Fig. 2.5 Incremental Model [GSAM03]

there are many more conclusions stated in "The Mythical Man-Month" book that are outdated, such as these:

"*The fundamental problem with program maintenance is that fixing a defect has a substantial (20-50 percent) chance of introducing another. So the whole process is two steps forward and one step back*" – with NSE, this problem can be solved by performing software maintenance holistically and globally through side-effect prevention.

"*All repairs tend to destroy the structure, to increase the entropy and disorder of the system.*" – with NSE, repairs are performed with side-effect prevention.

"*Adding manpower to a late software project makes it later*" – with NSE a software system is diagrammed graphically with various traceabilities to make the product much easier to read and understand; the documents and the source code are managed together with bidirectional traceability which make the software product much easier to understand, test, and maintain; a project Web site and the technical forum will be set and the Web pages are traceable to the implementation of requirements and the source code to reduce the time and resources for communication; not only the program and the data used and the documents available, but the database built through static and dynamic measurement and a set of Assisted Online Agents are available to support visibility, testability, reliability, traceability, conformity, changeability, and maintainability – so that the new members of the development team can learn the system by themselves quickly, and begin to make contributions quickly. About the detailed discussion on this topic, please see Chap. 24.

"*Theoretically, after each fix one must run the entire bank of test cases previously run against the system to ensure that it has not been damaged in an obscure way.*" – No, it is time consuming, inefficient, and costly. With NSE, the regression testing after software modification is performed efficiently through test case efficiency analysis and test case minimization, plus intelligent test case selection through backward tracing from the modified modules or branches to find what test cases can be used to retest them. Sometimes, new test cases need to be designed and used.

2.14 Summary

The old-established software engineering paradigm, including the process models, the software development methods, the test paradigm, the quality assurance paradigm, the documentation paradigm, the maintenance paradigm, the project management paradigm, and the definition of software, is entirely out of date, because not only a software system but the software engineering paradigm itself is a nonlinear, dynamic, and complex system that cannot be handled as a linear one.

The old-established software engineering paradigm based on linear thinking and superposition principle should be replaced by a new revolutionary one based on nonlinear thinking and complexity science which should be able to remove the drawbacks of the old-established software engineering paradigm efficiently and bring revolutionary changes to all aspects in software engineering.

2.15 Points and Questions to Ponder

(a) How is a successful project defined?
(b) What is the root cause that about 70% of software projects are failures?
(c) Is the existing software engineering paradigm updated or outdated? Why?

2.16 Further Reading and Information Source

(a) Zambonelli F, Parunak HVD (2002) Signs of a revolution in computer science and software engineering, Madrid, Spain. http://citeseer.ist.psu.edu/zambonelli02signs.html
(b) Brooks FP Jr (1995) The mythical man-month. Addison-Wesley, Upper Saddle River
(c) Wikiversity. Unsolved problems in software engineering. http://en.wikiversity.org/wiki/Unsolved_problems_in_software_engineering

References

[Coc08]   Cockburn A (2008) Using both incremental and iterative development. CrossTalk, May Issue
[GSAM03]   Department of the Air Force Software Technology Support Center (2003) Condensed GSAM handbook, Chap 2, CrossTalk
[Hum05]   Humphrey WS (2005) The Software Engineering Institute, Why big software projects fail: the 12 key questions. CrossTalk, Mar Issue
[Jon06]   Capers J (2006) Social and technical reasons for software project failures. CrossTalk, Jun Issue
[Ngw03]   Ngwenyama O, Nielsen PA (2003) Competing values in software process improvement: an assumption analysis of CMM from an organizational culture perspective. IEEE Trans Eng Manag 50(1):100–112. doi:10.1109/TEM.2002.808267
[Per06]   Perkins TK (2006) Knowledge: the core problem of project failure. CrossTalk, Jun Issue
[Pre05-p4] Pressman RS (2005) Software engineering: a practitioner's approach. McGraw-Hill, New York, p 4
[Sei08]   What is CMMI? Software Engineering Institute. Accessed 30 October 2008, http://www.sei.cmu.edu/cmmi/general/index.html

Chapter 3
Foundation for Establishing NSE: Complexity Science

The next century will be the century of complexity

Stephen Hawking, January 2000

This chapter introduces the foundation for establishing NSE – complexity science. Complexity science is the scientific study of nonlinear, dynamic, complex systems and the process of self-organization. Complexity science is the driving force for the development of sciences, engineering, and business in the twenty-first century. Complexity science explains how holism emerges in the world, and more. It is the intellectual successor to systems theory and chaos theory. Complexity science is a field derived from multiple disciplines – physics, chemistry, biology, and mathematics. Definitions of complexity are often tied to the concept of a complex system – something with many parts that interact to produce results that cannot be explained by simply specifying the role of each part. This concept contrasts with traditional machine or Newtonian constructs, which assume that all parts of a system can be known, that detailed planning produces predictable results, and that information flows along a predetermined path. Elements of complexity theory have been incorporated into a number of fields including genetics, immunology, cognitive science, economics, computer science, and linguistics. Currently, the most robust research in complexity science involves the study of inanimate systems such as computer networks and hydrodynamic systems as well as certain cellular networks (Ashok M. Patel, M.D., Thoralf M. Sundt III, M.D., and Prathibha Varkey, M.D., *Complexity Science – Core Concepts and Applications for Medical Practice,* http://www.minnesotamedicine.com/PastIssues/February2008/ClinicalFebruary2008/tabid/2462/Default.aspx); [Ber76], [Sar06].

If you are familiar with complexity science, please skip this chapter.

3.1 The Basis of Complexity Science

The basis of complexity science is important to the establishment of NSE and the innovation of the paradigm-shift framework, FDS (Five-Dimensional Structure Synthesis Method) to be described in Chap. 4, and directly or indirectly related to a prediction that a new round of industry revolution in many kinds of businesses from the old-established one based on linear thinking and reductionism to a new one based on nonlinear thinking and complexity science (see Sect. 4.1).

3.1.1 Linear and Nonlinear

"Linear" and "nonlinear" are mathematical terms commonly used to distinguish the function $y = f(x)$. An equation whose graph is a straight line is called a linear function; other functions are nonlinear functions (see Fig. 3.1).

$$Y = f(x)$$

3.1.2 Reductionism

Reductionism is sometimes seen as the opposite of holism. Reductionism holds that a complex system can be explained by *reduction* to its fundamental parts – a complex system can always be understood by breaking them down into simpler or more fundamental components. The old-established software engineering paradigm is based on reductionism and superposition principle that the whole of a complex system is the sum of its parts, so that almost all tasks/activities are performed linearly, partially, and locally.

3.1.3 Chaos Theory

The first experimenter in chaos was a meteorologist, Edward Lorenz. In 1960, he was working on the problem of weather prediction. He had a computer to model the weather. One day, he entered the decimal 0.506 instead of entering the full 0.506127 as one of the required conditions to rerun the program. It was expected that the rounding off would have little or no effect on the final results. However, surprisingly, what Lorenz found was that the final results were dramatically different. It means that a small change made in a system can cause major changes in the final output (sensitivity to initial conditions). This process is popularly known as "the butterfly

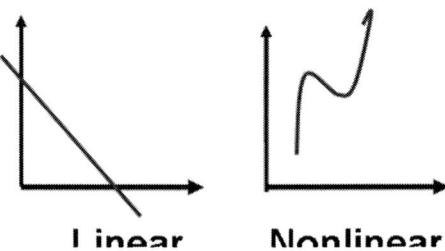

Fig. 3.1 Linear and nonlinear functions

effect," because it reflects the idea that a butterfly fluttering its wings in Taiwan could cause a hurricane in California. If small changes in the initial state of a complex system can drastically alter the final outcome, then long-term weather prediction is impossible as there is no way to perfectly measure and describe the weather at any one point in time. There is always a further level of accuracy to be measured. In other words, the deterministic nature of these systems does not make them predictable. This behavior is known as deterministic chaos, or simply *chaos*. Chaotic behavior can be observed in many natural systems, such as the weather [Sne97]. Chaos theory is a field of study in mathematics, physics, and philosophy studying the behavior of dynamical systems that are highly sensitive to initial conditions.

3.1.4 System

A system is a collection of interacting elements or components that are organized for a common purpose.

3.1.5 System Categories

Systems can be classified into natural systems, artificial systems, or a combination of both.

3.1.6 Linear System

A linear system is defined as that the whole of the system is the sum of its parts, complying with the superposition principle. As long as we know its initial conditions, we can understand its past and predict its future.

3.1.7 Nonlinear System and Complex System

A nonlinear system is a system not satisfying the superposition principle, or its output is not proportional to its input, small changes in its initial conditions may eventually cause the entire system to be changed greatly, and its long-term behavior is unpredictable.

A complex system is a system having multiple interacting components, of which the overall behavior cannot be inferred simply from the behavior of the components, but emerge from the interaction of its components and the interaction between it and its environment. Complex systems include IT networks, ecosystems, brains, markets, cities, and businesses. Of course, a complex system is also a nonlinear system.

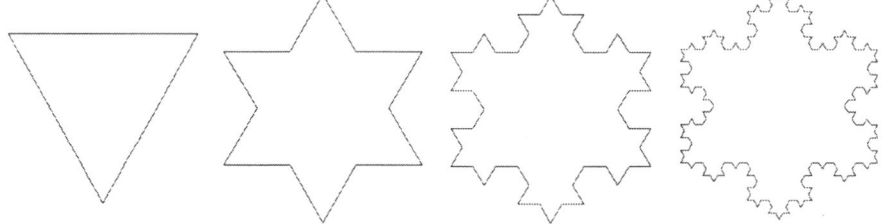

Fig. 3.2 An example of fractals: Koch island described by Helge von Koch in 1904

3.1.8 Feedback

Feedback refers to messages or information that are sent back to the source from the output.

3.1.9 Fractal

An irregular shape with self-similarity which can be split into parts, each of which is (at least approximately) a reduced-size copy of the whole (see Fig. 3.2) [Man82].

3.1.10 Fractal Dimension

A measure of a geometric object that can take on fractional values. At first used as a synonym to Hausdorff dimension, fractal dimension is currently used as a more general term for a measure of how fast length, area, or volume increases with decrease in scale.

3.1.11 Dynamical System

A dynamic system is a system that is constantly changing over time, like the human body system.

3.1.12 Dissipation Structure

According to the Belgian physicist and Nobel Prize winner Ilya Prigogine's proposed doctrine, an open system far from equilibrium, can form spatial and temporal structures (dissipative structures) that can exist as long as the system is held far from equilibrium due to a continual flow of energy or matter through the system.

3.1.13 Li–Yorke Theorem: Period Three Theorem

Li–Yorke Theorem holds that any one-dimensional system which exhibits a regular cycle of period three will also display regular cycles of every other length as well as completely chaotic cycles.

3.1.14 Self-Organization

The essence of self-organization is that system structure (at least in part) appears without explicit pressure or constraints from outside the system. In other words, the constraints on form are internal to the system and result from the interactions between the components, while being independent of the physical nature of those components. The organization can evolve either in time or space, can maintain a stable form or can show transient phenomena. General resource flows into or out of the system are permitted but are not critical to the concept.

The field of self-organization seeks to discover the general rules under which such structure appears, the forms which it can take, and methods of predicting the changes to the structure that will result from changes to the underlying system. The results are expected to be applicable to any system exhibiting the same network characteristics (Self-Organization FAQ, http://psoup.math.wisc.edu/archive/sosfaq.html).

3.1.15 Synergetics

Synergetics is an interdisciplinary field of research. It deals with open systems that are composed of many individual parts that interact with each other and that can form spatial, or functional structures by self-organization. Synergetics can refer to a school of thought on thinking and geometry developed by Buckminster Fuller or a school of thought on thermodynamics and other systems phenomena developed by Hermann Haken.

3.1.16 Catastrophe Theory

Originated by the winner of the highest award from the Mathematics – Fields Medal, the French mathematician Rene Thom in the 1960s, catastrophe theory is a special branch of dynamical systems theory. It studies and classifies phenomena that small changes in certain parameters of a nonlinear system can cause large and sudden changes of the behavior of the system.

3.1.17 Complex Adaptive System

The term *complex adaptive systems* (CAS) was coined at the interdisciplinary Santa Fe Institute (SFI), by John H. Holland, Murray Gell-Mann, and others. Complex Adaptive Systems involve many components (agents) that adapt or learn as they interact – are at the heart of important contemporary problems [Hol92]. Examples of complex adaptive systems include the stock market, social insect and ant colonies, the biosphere and the ecosystem, the brain and the immune system, the cell and the developing embryo, and manufacturing businesses.

3.1.18 Meta-Synthesis

The meta-synthesis approach is a method for solving the open giant complex systems problems proposed by Professor Qian Xuesen and his colleagues in China. The point of meta-synthesis is to unite organically the expert group, data, all sorts of information, and the computer technology, and to unite scientific theory of various disciplines and human experience and knowledge [Dai95]. The development phases of meta-synthetic social intelligence engineering are as follows:

1. From "qualitative and quantitative combined meta-synthesis" to "meta-synthesis from qualitative to quantitative"
2. From "meta-synthesis" to "hall for workshop of meta-synthetic engineering (HWME)"
3. Meta-synthesis of intelligent systems
4. From theoretical frameworks to operable platforms
5. From HWME to CWME – Cyberspace for Workshop of Meta-synthetic Engineering, a prototype of HWME
6. From methodology to applications

3.1.19 Cellular Automata

Cellular automata, also known as grid automata, were invented in the 1940s by the mathematicians John von Neuman [Neu66] and Stanislaw Ulam [Sip02]. Cellular automata (CA) are – by definition – dynamical systems which are discrete in space and time, operate on a uniform, regular lattice – and are characterised by "local" interactions. CAs are dynamical systems in which space and time are discrete. A cellular automaton consists of a regular grid of cells, each of which can be in one of a finite number of k possible states, updated synchronously in discrete time steps according to a local, identical interaction rule. The state of a cell is determined by the previous states of a surrounding neighborhood of cells [Wol84], [Tof87].

The infinite or finite cellular array (grid) is n-dimensional, where $n = 1, 2, 3$ is used in practice. The *identical* rule contained in each cell is essentially a finite state

3.1 The Basis of Complexity Science 85

machine, usually specified in the form of a *rule table* (also known as the *transition function*), with an entry for every possible neighborhood configuration of states. The *neighborhood* of a cell consists of the surrounding (adjacent) cells. For one-dimensional CAs, a cell is connected to r local neighbors (cells) on either side, where r is a parameter referred to as the *radius* (thus, each cell has $2r + 1$ neighbors, including itself). For two-dimensional CAs, two types of cellular neighborhoods are usually considered: five cells, consisting of the cell along with its four immediate nondiagonal neighbors, and nine cells, consisting of the cell along with its eight surrounding neighbors. The term *configuration* refers to an assignment of states to cells in the grid. When considering a finite-sized grid, spatially periodic boundary conditions are frequently applied, resulting in a circular grid for the one-dimensional case, and a toroidal one for the two-dimensional case. (Moshe Sipper, *A Brief Introduction To Cellular Automata*, http://www.cs.bgu.ac.il/~sipper/ca.html, http://www.moshesipper.com/).

3.1.20 Genetic Algorithm

Living organisms are consummate problem solvers. They exhibit a versatility that puts the best computer programs to shame. This observation is especially galling for computer scientists, who may spend months or years of intellectual effort on an algorithm, whereas organisms come by their abilities through the apparently undirected mechanism of evolution and natural selection. Pragmatic researchers see evolution's remarkable power as something to be emulated rather than envied. Natural selection eliminates one of the greatest hurdles in software design: specifying in advance all the features of a problem and the actions a program should take to deal with them. By harnessing the mechanisms of evolution, researchers may be able to "breed" programs that solve problems even when no person can fully understand their structure. Indeed, these so-called genetic algorithms (GA) have already demonstrated the ability to make breakthroughs in the design of such complex systems as jet engines. Genetic algorithms make it possible to explore a far greater range of potential solutions to a problem than do conventional programs. Furthermore, as researchers probe the natural selection of programs under controlled and well-understood conditions, the practical results they achieve may yield some insight into the details of how life and intelligence evolve in the natural world. (John H. Holland, *Genetic Algorithms*, http://econ2.econ.iastate.edu/tesfatsi/holland.gaintro.htm).

Genetic algorithms come from the classic evolutionary computation methods – stochastic global optimization algorithms, according to the "survival of the fittest" law of biological genetics and natural selection through computer simulation. The genetic algorithm can be applied with the following steps:

1. Define an objective function, for example, using the 26 English lower case letters plus a space character, to generate random 35 character strings and make it evolve into the "systems science is very interesting string."

2. A feasible solution of groups under certain constraints is initialized, for example, by randomly generating 500 35-character strings, each with a feasible solution to encode a vector x, called a chromosome with the representative weight vector gene, which corresponds to a particular decision variable feasible solution.
3. Calculate the groups for each chromosome x_i ($i = 1, 2, ..., n$) corresponding to the objective function value (n is an integer, such as the value of 500), and calculate the fitness value F_i; according to the size of the F_i, evaluate whether the feasible solution is good or bad – for example, in a chromosome where there are ten characters in the previous cases of the target line (that are correctly placed), its adaptive value is $10/35 = 0.2857$.
4. Using the mechanism of survival of the fittest, according to their fitness values, certain chromosomes will survive, whereas certain ones will be eliminated, and reproduction of randomly selected chromosomes will be carried out to form new groups.
5. Through hybridization and mutation operations to produce offspring, two randomly selected chromosomes (parents) will exchange genes and generate two new individuals (hybrids), with genetic mutations and variations at certain points (characters).
6. Repeat steps 3-5 for offspring groups, to generate a new round of genetic evolution, until the iterations converge (stable adaptation value) or to find the optimal or quasi-optimal solution. After 46 next-generation iterations, "systems science is very interesting" strings can be fully obtained for certain.

3.1.21 Soliton

Solitary waves and solitons in nonlinear science are important concepts.

In August of 1834, Bertrand Russell observed the solitary wave. In 1895, Korteweg and Defree proposed the KDV (Korteweg-De Vries) equation and its soliton solution. The soliton solution is a single peak traveling wave, where wave propagation is constant and the speed is also constant, where the shape and speed after any collisions remain unchanged.

3.2 Linear Thinking and Nonlinear Thinking

Linear Thinking: To continue to look at something from one point of view. To take information or observations from one situation, place this data in another situation (usually later), and make a conclusion in the later situation (see Fig. 3.3) (http://socialstudies.nelson.com/arnold/skimm/main/items/linearthinking.html).

Defined by Edward de Bono, nonlinear thinking is also called lateral thinking which can help us conjure creative solutions to emerge a winner in an increasingly

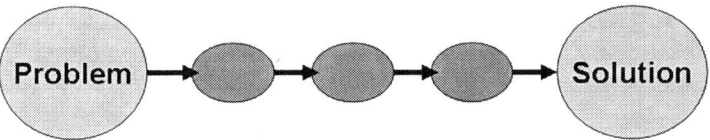

Fig. 3.3 Linear thinking

complex world. According to de Bono, intelligence is a potential and thinking is a skill to use that potential [Bon68]. De Bono has developed several techniques of lateral thinking under three broad categories: Challenge, Alternatives, and Provocation. The creative challenge is a challenge to exclusivity, which does not accept the status quo and is particularly relevant in those areas where ideas have become obsolete with time. Circumstances and situations often restrict the choice of alternatives and, therefore, it is better to assume a dynamic state of affairs. Limits and components are changed to enable new ways of doing things to emerge successful. Provocation is more in the nature of hypothesis where a situation is first conceived or imagined and then one proceeds to arrive at unique plausible conclusions.

3.3 The Essential Principles of Complexity Science

The essential principles of complexity science includes the following:

Nonlinearity principle – A complex system is not linear, that is, the system does not satisfy the superposition principle, or whose output is not proportional to its input. The behavior of a nonlinear system can change drastically in response to small changes in the system's initial conditions.

Holism principle – Holism is sometimes seen as the opposite of Reductionism. Holism holds that all the properties of a given complex system cannot be determined or explained by its components alone. Instead, the behaviors and characteristics of the whole of a complex system emerge from the interaction of its parts, and the interaction between it and the environment dynamically. The general principle of holism was concisely summarized by Aristotle in the Metaphysics: **"The whole is more than the sum of its parts."**

Complexity arises from simple rules principle – Complexity arises from the interaction of agents following simple rules; Complex systems are based on simple rules which feedback on itself, or iterate on themselves, and this can explain all phenomena everywhere.

Initial Condition Sensitivity principle – To a complex system, small causes may have large effects to the entire system – "Butterfly Effects."

Sensitivity to Change principle – To a complex system, small changes may have large effects to the entire system. It is similar to the **Initial Condition Sensitivity principle.**

Dynamics principle – A complex system is a dynamic one adaptive to its changing environment.

Openness principle – A complex system and its environment are inseparable, and it is constantly interacting with the environment.

Self-organization principle – see Sect. 3.1.14.

Self-adaptation principle – see Sect. 3.1.17.

3.4 Applications of Complexity Science

Many successful applications of nonlinear thinking and complexity science for various complex nonlinear systems were reported [Art99], [Nor08], [Den00], [Fan04], [Kim04].

Some published books and papers show that now more and more software scientists are applying complexity science to attack the problems facing software development/IT such as those with the following titles:

"Adaptive Software Development: A Collaborative Approach to Managing Complex Systems" [Hig00]
"Intelligent Agents: Software Technology for the new Millennium" [Fal00]
"Complexity Science and Software Development, An Introduction to Complexity Science and Its Applications in Agile Software Development" [Lam03]
"Agent-Oriented Software Engineering" [Jen00]

But unfortunately, the applications of complexity science has not reached the level expected by people – for instance, when complexity science was applied to solve the critical problems with an individual software system by us before, we did not get the expected result in productivity increase or quality improvement. Why? The main reason is that before making paradigm shift of the entire software engineering paradigm from the old one based on linear thinking, reductionism, and superposition principle to the new one based on nonlinear thinking and complexity science, it is almost impossible to directly apply complexity science to solve an individual software system problems because with the old-established software engineering paradigm, the process models, the development methodology, the testing paradigm, the quality assurance paradigm, the maintenance paradigm are based on linear thinking, reductionism, and superposition principle too. We finally realized that there should be two major steps: the first one is to complete the entire paradigm shift in the software engineering from the old one based on linear thinking, reductionism, and superposition principle to the new one based on nonlinear thinking and complexity science; then the second one is to apply complexity science to solve the problems of an individual software system after the completeness of the paradigm shift.

3.5 Complexity Science and NSE

As described in Chap. 4, NSE paradigm is established through FDS (Five-Dimensional Structure Synthesis Method), a paradigm-shift framework which requires the new revolutionary paradigm being established by complying with the essential principles of complexity science, particularly the nonlinearity principle and the holism principle, so that with NSE almost all tasks/activities are performed nonlinearly, globally, and holistically.

3.6 Summary

Complexity science is the driving force for the development of sciences, engineering, and business in the twenty-first century. Complexity science explains how holism emerges in the world, and more.

The foundation for establishing NSE nonlinear software engineering paradigm is complexity science. NSE complies with the essential principles of complexity science, particularly the Nonlinearity principle and the Holism principle that all the properties of a given complex system cannot be determined or explained by its components alone. Instead, the behaviors and characteristics of the whole of a complex system emerge from the interaction of its parts, and the interaction between it and the environment dynamically, so that with NSE almost all tasks/activities are performed nonlinearly, globally, and holistically.

3.7 Points and Questions to Ponder

(a) What is complexity science?
(b) What are the major differences between Reductionism and Holism?
(c) What are the essential principles of complexity science? How are they related to the establishment of NSE?

3.8 Further Reading and Information Source

(a) Waldrop MM (1992) Complexity: the emerging science at the edge of order and chaos. Viking, London
(b) Gleick J (1988) Chaos: making a new science. Cardinal, London
(c) Castellani B, Hafferty FW (2009) Sociology and complexity science: a new field of inquiry. Springer, Heidelberg

References

[Art99]	Arthur WB (1999) Complexity and the economy. Science 284:107–109
[Ber76]	Von Bertalanffy L (1976) General system theory. George Braziller, New York
[Bon68]	de Bono E (1968) New think: the use of lateral thinking in the generation of new ideas. Basic Books, New York
[Dai95]	Dai R (1995) Metasynthetic social intelligence engineering: a review. Institute of Automation, Chinese Academy of Sciences, Beijing
[Den00]	Dent EB (2000) Complexity science: a paradigm shift. Emergence 1(4):5–19
[Fal00]	Faltings B (2000) Intelligent agents: software technology for the new millennium. Informatik/Informatique 1:2–5
[Fan04]	Francis J (2004) Managing BPM, BPM and Nonlinear Thinking, June Issue
[Hol92]	Holland JH (1992) Complex adaptive systems. American Academy of Arts & Sciences, Cambridge
[Hig00]	James A. Highsmith III, Adaptive Software Development: A Collaborative Approach to Managing Complex Systems, DORSET HOUSE PUBLISHING CO., INC., 2000.
[Jen00]	Jennings NR, Wooldridge M (2000) Agent-oriented software engineering. Department of Electronic Engineering, Queen Mary & Westfield College, University of London, London
[Kim04]	Kimball L, Weinstein N, Silber T (2004) Maximizing facilitation skills using principles of complexity science. OD Network Conference, October 2004
[Lam03]	Lamoreux M (2003) Complexity science and software development, an introduction to complexity science and its applications in agile software development, http://comdig.unam.mx/article.php?id_article=13746&find=complexity
[Man82]	Mandelbrot BB (1982) The fractal geometry of nature. W.H. Freeman, San Francisco. ISBN 0-7167-1186-9
[Mer06]	Yasmin M, McKelvey B (2006) Using complexity science to effect a paradigm shift in information systems for the 21st century. J Inform Technol 21:211–215
[Neu66]	Von Neumann J (1966) Theory of self-reproducing automata. Edited and completed by A.W. Burks. University of Illinois Press, Urbana
[Nor08]	Norreys PA (2008) PHYSICS: complexity in fusion plasmas. Science 319:1193
[Sar06]	Sardar Z, Abrams I (2006) Caos Para Todos/Introducing chaos. Icon Books, Cambridge
[Sip02]	Sipper M (2002) Machine nature: the coming age of bio-inspired computing. McGraw-Hill, New York
[Sne97]	Raymond Sneyers (1997) Climate chaotic instability: statistical determination and theoretical background. Environmetrics 8(5):517–532
[Tof87]	Toffoli T, Margolus N (1987) Cellular automata machines. The MIT Press, Cambridge
[Wol84]	Wolfram S (1984) Universality and complexity in cellular automata. Physica D 10:1–35

Chapter 4
Prediction and Practices: A New Round of Industrial Revolution Driven by Complexity Science and a General Paradigm-Shift Framework

> *Framework is a set of ideas, principles, agreements, or rules that provides the basis or outline for something intended to be more fully developed at a later stage.*
>
> *Dictionary (http://encarta.msn.com/dictionary_1861613305/framework.html)*

This chapter describes a prediction – a new round of industrial revolution driven by Complexity Science, and a paradigm-shift framework, the Five-Dimensional Structure Synthesis method (FDS). Many businesses fail because of an attempt to solve nonlinear problems with linear processes. With FDS, the paradigm shift for an industry can be performed efficiently – from the old-established paradigm based on linear thinking, reductionism, and superposition principle to a new paradigm based on nonlinear thinking and complexity science in compliance with the common principles of complexity science. FDS has been successfully used in the paradigm shift of the software industry and could be successfully used for other industries too.

4.1 Prediction: A New Round of Industrial Revolution Driven by Complexity Science Is Coming

Today, more and more industries are becoming increasingly aware that traditional approaches to design and engineering are failing to keep up with the increasing scale of systems [Mck99]. The foundation of those traditional approaches is based on linear thinking and established science complying with the reductionism and superposition principle that **the whole of a system is the sum of its parts**. But in fact, all people problems and issues are nonlinear which do not comply with the superposition principle because they exist in a dynamic and changeable environment, rather than a static one [Lim05]. Complexity science tackles some of science and engineering's most challenging and fundamental questions [Mck99]. The FDS is innovated by me as a framework for making the **paradigm shift** (defined as "one conceptual world view is replaced by another" by Thomas Kuhn [Kuh62]) efficiently. Using FDS to perform the paradigm shift of an industry, it is required to comply with the essential principles (which are common to almost all theories of complexity science) to redefine the process models, redevelop the methodologies and technologies, redesign the productivity and the quality tools, reset the quality assurance standards, etc. FDS has been successfully used in the paradigm shift of the software industry (software engineering) with revolutionary changes made in almost all aspects of software engineering for efficiently handling almost all critical issues existing with the old-established paradigm, including the issues of complexity, changeability, invisibility, and conformity. It is possible to use FDS for making the paradigm shift efficiently in other industries to greatly improve the productivity and the product quality too.

A Prediction: a deeper and broader industry revolution driven by complexity science is coming because

(a) In various industries the old-established paradigms based on linear thinking and simplified science themselves have become obstacles to the system development rather than the driving forces in the twenty-first century. For instance, the computer software industry is a typical one. As pointed out by Capers Jones, "Major software projects have been troubling business activities for more than 50 years. Of any known business activity, software projects have the highest probability of being cancelled or delayed. Once delivered, these projects display excessive error quantities and low levels of reliability" [Jon06].
(b) Application results show that complexity science is the most powerful weapon for handling many critical issues in various complex systems.
(c) Now more and more people realize that nonlinear, complex adaptive systems are the best way to understand systems involving people [Gha04], so that it is the time to shift the old-established paradigm based on linear thinking and simplified science to a new one based on nonlinear thinking and complexity science for various industries.

4.2 The Contribution and Limitation of Hall's Systems Engineering Framework

In 1962 and 1969, A. D. Hall published his three-dimensional morphology for systems engineering (Hall's framework) [Hal62], [Hal69] as shown in Fig. 4.1.

Hall's framework has been used successfully in many industries in the late twentieth century. But unfortunately, his framework itself is a linear one. Looking at the "Phases" coordinate axis in Fig. 4.1, we can easily see that the process phases are done individually according to a sequence order. When applying Hall's framework to software engineering, a waterfall model (or a micro-waterfall model) would be logically established. There is also nothing related to the environment, which means that with Hall's framework, systems engineering can be isolated without considering the effects of the environment. It is also clear that Hall's framework is designed to be used for a detailed engineering project or for detailed systems design rather than that used for both the paradigm shift of an industry, and engineering for an individual project after the paradigm shift.

4.4 The Objectives of Innovating FDS

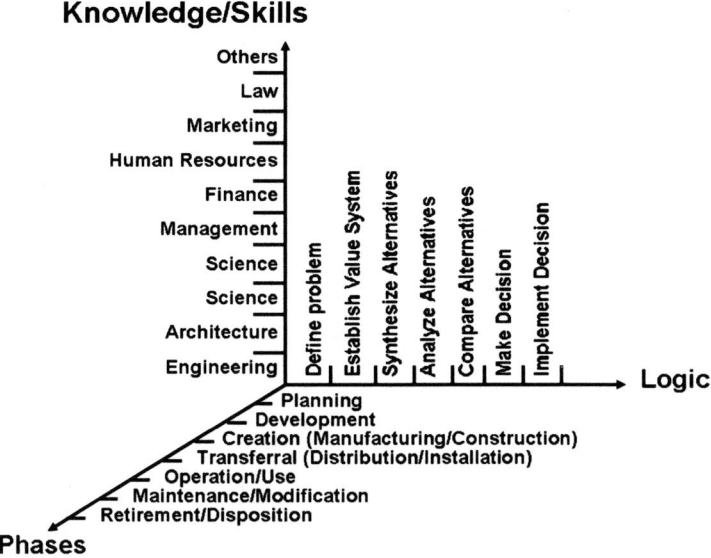

Fig. 4.1 Hall's systems engineering framework

4.3 The Background for the Innovation of FDS

Many businesses fail because of an attempt to use linear process models and methodologies to handle a complex nonlinear system, such as an EDA (Electronic Design Automation) system for VLSI (Very Large Scale Integration) chip design, where a linear order of processes is followed through chip partitioning, global placement, global routing, detailed placement, detailed routing, timing simulation, rule check, and verification, etc. The output obtained from an upper process becomes an input to a lower process. Often, the optimized result of an upper process (such as the detailed placement) does not satisfy the requirements of a lower process (such as the detailed routing), so that the upper process must be performed again and again, because the new optimized result of the upper process is obtained blindly and locally, which could be worse than the old one for the lower process. The same problems exist in many industries, such as the software industry, where the project success rate is as low as 30% today.

Complexity science can be used to efficiently solve those problems as introduced in Chap. 3.

It is the time to perform paradigm shifts for many industries, from the old-established paradigm based on linear thinking, reductionism, and superposition to a new revolutionary one based on nonlinear thinking and complexity science.

4.4 The Objectives of Innovating FDS

As pointed out by Warfield, J. N. that there are at least five schools of thought on complexity science [War96]. They are suitable for different applications, so that it will be better to combine all of the theories of complexity science together to form a powerful set of common principles of complexity science which should be complied with in performing paradigm shifts for various industries.

Since complexity science is still very young, it will not be easy for individual engineers to use it for solving a detailed problem, because using complexity science to perform a detailed task within the limitation of the old-established paradigm (without changing the entire old-established paradigm including the process models, the product development methods, the testing paradigm, and the quality assurance paradigm, etc.) will be very difficult to get the expected result. But performing an entire paradigm shift in an industry is also very hard to do for a small company – it should be done by a tool vendor, a research organization, or a company with a strong professional team. It means that there are some obstacles in applying complexity science to handle a real complex nonlinear system.

For applying complexity science deeply and broadly within an industry, it is needed to complete the entire paradigm shift first for that industry, from the old-established paradigm based on linear thinking and simplistic science to a new revolutionary one based on nonlinear thinking and complexity science, by complying with the common principles of complexity science to redefine the process models, redevelop the methodologies, redesign the productivity tools, reset the quality standards, and so on.

How can an old-established paradigm in an industry be efficiently replaced by a new revolutionary one? It needs at least two things as the primary prerequisites:

(a) A systematic paradigm-shift framework
(b) A successful application example of the paradigm-shift framework – people work well through examples

4.5 The Description of FDS

FDS is graphically shown in Fig. 4.2.

Based on the theories of complexity science and Hall's three-dimensional morphology for systems engineering, FDS is designed with changeability to meet and adapt to different applications in the paradigm shift of various industries.

There are five axes with FDS.

4.5.1 The "Principles of Complexity Science" Axis

Complexity science is still very young, where there are at least five existing schools of thought on complexity science [War96] which are suitable for different applications. It seems that it would be the best choice to combine all the theories of complexity science together to form a powerful synthesis with a set of common principles to drive the paradigm shift for various industries. As described in Chap. 3, those common principles include the following:

4.5 The Description of FDS

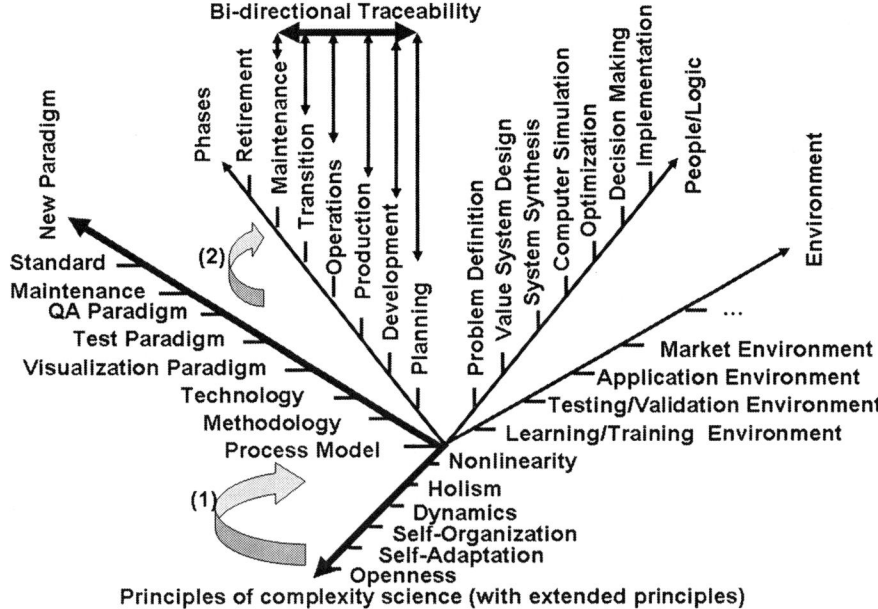

Fig. 4.2 The Five-Dimensional Structure Synthesis method (FDS)

- The **Nonlinearity** principle
- The **Holism** principle
- The **Dynamics** principle
- The **Self-organization** principle
- The **Self-adaptation** principle
- The **Openness** principle
- The **Initial Condition Sensitivity** principle
- The **Sensitivity to Change** principle
- The **Complexity Arises From Simple Rules** principle, and more

About the definition and meaning of each principle, please visit the Web site of *http://complexity.orconhosting.net.nz/fractal.html* (***Complexity Pages***) and the Web site of http://www.complexity.ecs.soton.ac.uk/index.php?page=q2 (**Complexity Science Focus**).

The essential one within those principles is the **Holism** principle, which states that a complex system is a system having multiple interacting agents (components), of which the overall behavior and characteristics cannot be inferred simply from the behavior of its individual agents. With FDS, it is required to comply with these common principles of the complexity science for performing a paradigm shift in an industry. For instance, when applying FDS for the paradigm shift of the software industry, the new redefined process model must comply with the Holism principle, the Nonlinearity principle, and other principles of complexity science, so that a waterfall-like process model will not satisfy the requirement. For meeting the Holism principle, any redefined candidate models must require each task to be done globally rather than locally. It means that there is a need for a revolutionary change in the design of the process models and methodologies, because there are no existing process models or methodologies meeting this requirement.

For some applications, there may be a need to establish some additional principles which may not be available in the existing theories. For instance, to establish a new paradigm for software engineering, the "Synthesis Design" and "Incremental Integration" principles are needed as pointed out by Brooks:

> "NSB ('No Silver Bullet') advocates a wholehearted attack on the problem of complexity, quite optimistic that progress can be made. It advocates adding necessary complexity to a software system:
>
> - **Hierarchically, by layered modules or objects**
> - **Incrementally, so that the system always works.**" [Bro95]

4.5.2 The "Environment" Axis

Based on the Openness principle of complexity science, a complex system is inseparable from its environment – the mutual interaction between the environment and the system will unceasingly influence the system's complexity. Openness means that the behavior of open (living) systems can be understood only in the context of their environment, so the environment is considered as an important element in FDS. In different applications, the items of "Environment" may be different. In most cases, the items on the "Environment" axis include the following:

- The **Learning/Training** environment.
- The **Testing/Validation** environment.
- The **Operation** environment.
- The **Application** environment.
- The **Market** environment – for instance, software requirements should be ordered according to their importance, so that the most important requirements can be implemented early to meet the market needs: if necessary, some optional requirements can be ignored to get the products ready on the market in time.

4.5.3 The "People/Logic" Axis

The items of this axis are almost the same as those in Hall's framework, except that the **Develop Requirement** is replaced with **Computer Simulation** because **Develop Requirement** may be combined in the **Development** part of the **"Phases,"** and **Computer Simulation** is a powerful tool for solving many complexity issues in a complex system.

4.5.4 The "New Paradigm" Axis Modified from the "Knowledge/Skills" Axis in Hall's Framework

With FDS, the Knowledge/Skills axis is considered as the essential condition for the people to perform the paradigm shift in an industry. The design purpose of FDS is mainly for the use in paradigm shifting, so the Knowledge/Skills axis is replaced with the "New Paradigm" axis.

The items in the "New Paradigm" axis may be different for different applications. In most cases, it could consist of "Process Model," "Methodology," "Technology," "Tool and Platform," "Quality Assurance," "Visual Technique," "Testing Method," "Maintenance Approach," "Quality Assurance Standard," "Project Management," "self-recovery," etc. Within them, the most important parts are the "Process Model," "Methodology," and "Technology" elements. It means that making revolutionary changes to the process model and the methodology and technology from the old-established paradigm based on linear thinking and simplistic science to the new revolutionary paradigm based on nonlinear thinking and complexity science is essential for establishing the new paradigm of an industry.

4.5.5 The "Phases" (Workflows) Axis

The items in this axis are the same as those specified in Hall's framework. But it is are recommended to perform those after the paradigm shift of the corresponding industry has been completed by a tool vendor or the organization itself. With FDS, the phases being performed do not follow a linear order. As Professor Brooks points out in his seminal work, *The Mythical Man-Month*: "**There has to be upstream movement**. Like the energetic salmon… experience and ideas from each downstream part of the construction process must leap upstream, sometimes more than one stage, and affect the upstream activity." [Bro95]. This idea is represented with a bidirectional traceability bar with this axis. Automated and self-maintainable traceability is crucial for handling changes globally to meet the Holism principle and the Self-adaptation principle.

FDS itself is designed as an adaptive framework – when FDS is used for the paradigm shift of an industry, the contents of each axis may represent different items.

The meanings of other items in FDS are the same as specified in Hall's framework. For detailed descriptions of those items and their meaning, please read A. D. Hall's original papers [Hal62], [Hal69].

The relationships among the five elements represented in the five axes of FDS are shown in Fig. 4.3.

As shown in Fig. 4.3, the principles of complexity science should be applied to all other items, not only those shown in the "New Paradigm" axis. For instance, when

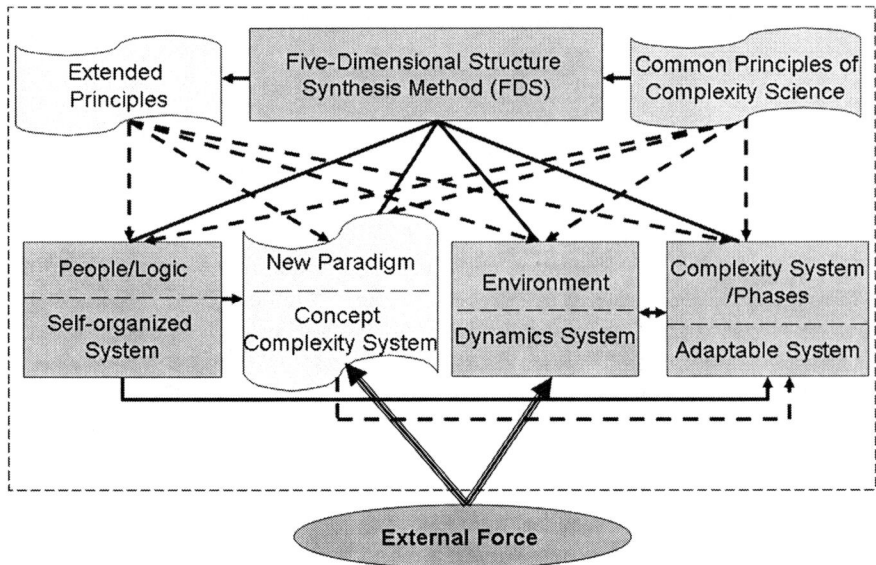

Fig. 4.3 The five elements of FDS and their relationships

FDS is used for software system development after the paradigm shift is completed, at least the extended "Synthesis Design" and the "Incremental Integration" principles should also be applied to the items shown on the "Phases" axis.

It is recommended to handle a complex system design or engineering in two major steps: the first one is to complete the paradigm shift by the organization performing the tasks or a tool vendor, then the second one is to handle the detailed tasks by applying the corresponding new paradigm established in the first step.

4.6 The Major Features of FDS

The major features of FDS are as follows:

1. **Based on complexity science** – The essential principles of complexity science become the requirements to be satisfied for the establishment of a new revolutionary paradigm in an industry.
2. **General** – It is innovated for the paradigm shift in many different industries where the existing paradigm is based on linear process, reductionism, and the superposition principle. FDS does not follow an individual school of thought but follows the common essential principles of complexity science.
3. **Operational** – FDS has been used to complete the paradigm shift of the software industry.
4. **Adaptive** – It is recommended to make necessary changes to FDS to meet the needs for different applications. For instance, the Environment part can be quite different in different applications.

5. **Useful for both** – The paradigm shift of an industry and the framework for solving an individual complex system after the completeness of the paradigm shift in the corresponding industry.

4.7 Applications of FDS

Sampling is a good approach for human beings to work with. As pointed out by Alistair A. R. Cockburn, *"Working from examples.* Some cognitive psychologists convincingly argue that our deductive mechanisms are built around constructing specific examples of problems. CRC cards and use cases are two software development mechanisms centered on examples, and are repeatedly cited by practitioners as effective. 'Instance diagrams' are often preferred by newcomers to object-oriented design, and still are used by experienced designers" [Coc99] – so that as an application example of the FDS method innovated, the paradigm shift for the software industry (engineering) is chosen.

The reason to choose the software industry as an application example of FDS is because

(a) Software has become an indispensable technology and a driving force for business, science, and engineering in the twenty-first century.
(b) Software affects almost every aspect of our lives and has become deeply pervasive in our commerce, our culture, and our everyday activities.
(c) But unfortunately, low productivity, high cost, and poor quality are the major problems existing with software industry for the past 50 years. Until today, the project success rate in software industry is only about 30%.
(d) Software product is a typical complex nonlinear system where the "Butterfly Effect" (a phrase which encapsulates the more technical notion of sensitive dependence on initial conditions in chaos theory) is a common occurrence.
(e) The paradigm shift for the software industry is very hard to perform.

As pointed out by Franco Zambonelli and H. Van Dyke Parunak in their paper titled "Signs of a Revolution in Computer Science and Software Engineering" [Zam03] that *"We are on the edge of a revolutionary shift of paradigm.* The change in the modeling and understanding of complex software systems will also impact how such systems are designed, maintained, and tested."

In the application example, the entire paradigm has been shifted in almost all major aspects in software engineering, including the following:

(a) The process models – from linear waterfall models or one-way iteration "micro-waterfall" models to a nonlinear two-way iteration model, the NSE model supported with facilities for automated and bidirectional traceability among all artifacts (including requirement specifications, design documents, project development plans, test cases, manuals, test results, QA reports, and the source code), so that the tasks can be done globally rather than locally to prevent the "Butterfly Effect" (side-effect propagation) in the implementation of software change.

(b) The software development methodology – from test-driven approaches to defect prevention and traceability driven approach, the NSE methodology, in the viewpoint of quality assurance. NSE methodology complies with the principles of complexity science, including the "Synthesis Design" and "Incremental Integration" principles with the "intention" to respond to requirement changes in real time. The new software development methodology is based on Generative holism rather than Constructive holism (see Chap. 10).

(c) The software testing system – from a Black-box approach (which can be used only in the case that the system has been completely coded so that to an input there is a corresponding output to be able to check whether the output is the same as what is expected) to a transparent-box approach, which can be dynamically used in the entire lifecycle of a software product development and maintenance, including the requirement development phase and the primary design phase, because to each input the NSE test system not only verifies whether the output (if any, can be none) is the same as what is expected but also verifies whether the specified execution path is covered with the execution of the corresponding test case, and whether some modules and/or branches that are prohibited to hit, have been hit with the execution of the test case, plus the capability to establish automated bidirectional traceability between the source code and the test case (can be expanded for all related documents) for identifying and removing inconsistent defects.

(d) The software maintenance process and system – from the old approach of blindly and locally responding to requirement change and code modification to a visible, systematic, disciplined, and quantifiable approach to respond to requirement changes and code modifications globally, with defect prevention capabilities through automated bidirectional traceability.

(e) The quality assurance system – from a testing and correction approach to defect prevention and defect propagation prevention ("An ounce of prevention is worth a pound of cure!" [Fra]) plus a deep and broad testing approach, the NSE SQA system.

(f) The visual technologies, tools, and their applications – from the technology and tools without traceability and used in modeling only to the interactive and traceable 3J graphics (J-Chart, J-Diagram, and J-Flow) defined and implemented by me which can be used in the entire lifecycle of a software development (see Chap. 6).

4.8 Bringing Feedback to the Research and Development of Complexity Science

How can we make more contributions to push the research and development of complexity science? The best way is to apply it in handling complex nonlinear systems in the real world, and then get feedback to drive the development of complexity science.

FDS is designed as a bridge between complexity science and its applications.

4.9 Summary

FDS is designed as a general framework for paradigm shift in an industry from the old-established paradigm based on linear thinking (linear process), reductionism, and superposition principle to a new one based on nonlinear thinking (nonlinear process) and complexity science by complying with the essential principles of complexity science, particularly the Nonlinearity principle and the Holism principle.

There are five axes with FDS: the "Principles of Complexity Science" axis, the "Environment" axis, the "People/Logic" axis, the "New Paradigm" axis, and the "Phases (Workflows)" axis.

As an application example, FDS has been successfully used to complete the paradigm shift of software engineering – the establishment of NSE nonlinear software engineering paradigm.

4.10 Points and Questions to Ponder

(a) What are the major differences between Hall's framework and FDS?
(b) Why is it recommended to apply complexity science to solve the problems of a complex system in an industry through two major steps (the first one is to complete the paradigm shift by the organization performing the tasks or a tool vendor, then the second one is to handle the detailed tasks by applying the corresponding new paradigm established in the first step)?

4.11 Further Reading and Information Source

(a) Abran A, Moore JW, Bourque P, Dupuis R (eds) (2004) Guide to the software engineering body of knowledge – 2004 Version. IEEE Computer Society. p. 1–1. ISBN 0-7695-2330-7.
(b) Bolton D. About.com Guide, Definition of Framework. http://cplus.about.com/od/glossar1/g/frameworkdefn.htm

References

[Bro95] Brooks FP Jr (1995) The mythical man-month. Addison-Wesley, Reading
[Coc99] Cockburn AAR (1999) Characterizing people as non-linear, first-order components in software development. HaT Technical Report 1999.03, Oct 21
[Fra] Franklin B (1736) An ounce of prevention is worth a pound of cure. Philadelphia's 1706–1790
[Gha04] Gharajedaghi J (2004) Systems methodology: a holistic language of interaction and design seeing through chaos and understanding complexities, http://www.acasa.upenn.edu/JGsystems.pdf

[Hal62]	Hall AD (1962) A methodology for systems engineering. Van Nostrand, Princeton
[Hal69]	Hall AD (1969) Three-dimensional morphology of systems engineering. IEEE Trans Syst Sci Cybern SSC-5(2):156–160
[Jon06]	Jones C (2006) Social and technical reasons for software project failures. CrossTalk, June Issue
[Kuh62]	Kuhn T (1962) The structure of scientific revolutions. University of Chicago press, Chicago
[Lim05]	Lindberg C (2005) Complexity, the science of relationships. Nursing, the profession of relationships. Plexus Institute, Allentown
[Mck99]	McKelvey B (1999) Complexity theory in organization science: seizing the promise or becoming a fad? Emergence 1(1):5–32
[War96]	Warfield JN (1996) Five schools of thought about complexity. In: Proceedings of the Society for Design and Process Science: integrated design and process technology, vol 2. SDPS, Austin
[Zam03]	Zambonelli F, Van Dyke Parunak H (2003) Signs of a revolution in computer science and software engineering. Springer, Berlin, http://www.newvectors.net/staff/parunakv/ZambonelliParunakAOSE02.pdf

Chapter 5
Outline of the NSE Paradigm

The whole is more than the sum of its parts.

A.1, Aristotle

This chapter will briefly introduce the NSE paradigm, including the development objectives, the basic idea, the technical route to achieve its development objectives, the structure, the components, and the major features and characteristics of the NSE paradigm.

5.1 A Tree Will Not Fall at One Blow: The Difficulty in Software Engineering Revolution

The software engineering paradigm itself is a very complex system consisting of many parts including the engineering process models, the software development methodology, the software testing paradigm, the quality assurance paradigm, the software visualization paradigm, the software documentation paradigm, the software maintenance paradigm, the software project management paradigm, the applied technologies, the used algorithms, the support tools, the support platforms, and more.

Unfortunately, as described in Chap. 2 of this book, almost all parts of the existing software engineering paradigm are established/created/designed based on linear thinking, reductionism, and the superposition principle – it means that almost all of the parts of the existing software engineering paradigm are outdated:

- The existing engineering process models are outdated because they are linear ones without upstream movement at all, where almost all software engineering tasks/activities are performed linearly, partially, and locally, making the defects introduced into a software product easy to propagate into the maintenance phase and making the defect removal cost increase tenfold many times.
- The existing software development methods are outdated because they are linear ones complying with the superposition principle to complete the components of a software product first, then *"Assemble the product from the product components,*

ensure the product, as integrated, functions properly and deliver the product." [CMMI1.1] It seems that those methods handles a software product as a linear system like a machine which can be *assembled*. But based on the **Generative Holism principle of complexity science,** the whole of a complex system should exist first as an embryo, then it "grows up" with its components. From the quality assurance view, the existing software development methods are test-driven, but the testing is performed after production which has been proven impossible to ensure high quality of a software product as reported by NIST (National Institute of Standards and Technology) ("Requiring Software Independence in VVSG 2007: STS Recommendations for the TGDC," November 2006, **http://vote.nist.gov/ DraftWhitePaperOnSIinVVSG2007-20061120.pdf**).

- The existing software testing paradigm is outdated – it separates functional testing and structural testing and can be dynamically used only after production.
- The existing software quality assurance paradigm is outdated – it depends on testing after coding and inspection rather than defect prevention.
- The existing software visualization paradigm is outdated – it offers partial, untraceable capability for software diagramming.
- The existing software documentation paradigm is outdated – it makes the software documents separate from the source code without bidirectional traceability.
- The existing software maintenance paradigm is outdated – it offers linear, blind, partial, and local approaches for software maintenance without the capability to prevent the side effects for the implementation of requirement changes or code modifications, making software maintenance take 75% or more of the total effort and total cost in software product development.
- The existing software project management paradigm is outdated – it separates the project management process and the software product development process. The documents for project management are not directly traceable to the requirement implementation and source code.
- Most of the existing software development techniques and tools are outdated – they are all working with the linear process models complying with reductionism and the superposition principle.
- The definition of "Software" is outdated too – it includes the program, the data used, and the documents separate from the source code, and ignores the history of the static and dynamic measurement of the program, with no efficient tools to help maintainers to handle the complexity, changeability, invisibility, conformity, traceability, and maintainability.

The improvement of only one or two parts of the existing software engineering paradigm will not work well for the whole.

Some models mainly focus on software process improvement (SPI) and project management improvement. Why is the success rate of the implementation of those models still as low as about 30% [Ngw03], [Nia06]? The root causes are as follows:

1. According to the Holism principle of complexity science, the behavior and characteristics of the whole of the software engineering paradigm cannot be inferred

5.2 The Objectives for Establishing NSE

from one or some of its parts, but emerge from the interaction of all of its parts – not only the process and the management. It means that without bringing revolutionary changes to the outdated software testing paradigm, the outdated software quality assurance paradigm, the outdated software documentation paradigm, the outdated software visualization paradigm, and the outdated software maintenance paradigm, it is impossible to efficiently improve the whole of the engineering paradigm to solve the critical issues existing with today's software development. For instance, no matter how good the process has been improved, if the outdated test paradigm has not been changed revolutionarily, the quality assurance still mainly depends on functional testing using the Black-box method after production; there is no efficient way to prevent and remove the defects introduced into a software product in the requirement development phase, the design phase, and the coding phase. Even if later on, a requirement defect or design defect is removed through testing, the cost for the removal of the defect will increase tenfold many times. Even if all the other parts of the existing software engineering paradigm have been changed revolutionarily except the software maintenance paradigm which is still performed blindly, partially, and locally so that each time when a defect is removed there is a chance of 20–50% to introduce a new defect into the software system, the software product will become unstable day by day.
2. Some popular models focus on SPI but ignore the most important process improvement – the life cycle models themselves: those popular models require software organizations to **select** one of the existing life cycle model such as the waterfall model, the iteration model, or the incremental model (**which is "a series of Waterfalls"**[GSAM03]) for their projects. It is questionable that if the life cycle model selected by a software organization for its projects is unsuitable, how can those popular models help the organization to improve the process? In this case it is possible that the better the process is improved, the worse the result obtained!

It is clear that the establishment of a new revolutionary software engineering paradigm based on complexity science itself is a big engineering project – only the support platform will consist of more than 10,000 function points with about one million lines of source code, and more than 100 new algorithms to be innovated for the establishment of the new revolutionary paradigm and the development of the support platforms.

It is why the establishment of NSE takes several years to complete.

5.2 The Objectives for Establishing NSE

The objectives for establishing NSE include

- Efficiently handling the essential issues with software and software engineering – the complexity, invisibility, changeability, and conformity, as defined by Brooks [Bro95-p182].

- Making it possible to help software development organizations double their productivity, halve their cost, while removing 99–99.99% of defects in their software products developed with NSE.
- Making it possible for software organizations to double their project success rate.
- Being candidates of the silver bullet to slay the software werewolf – missed schedules, blown budgets, and flawed products [Bro95-p181].
- Making a software product much easier to read, understand, test, and maintain in both the product development site and the product maintenance site.
- Helping software engineers relax from their daily hard work.

5.3 The Strategy to Achieve the Objectives of NSE

According to complexity science, the property of a complex system is determined by both the whole and its parts, so that the strategy to achieve the objectives of NSE is as follows:

1. The first thing is to bring revolutionary changes to all parts of software engineering by complying with the essential principles of complexity science, particularly the **Nonlinearity** principle and the **Holism** principle to innovate all required techniques and develop the related tools.
2. But only this is not good enough, so the second thing is to make all parts work together closely to change the behaviors and characteristics of the whole to what we desire, such as developing the support platform to integrate all the related tools together to share a tiny database using the unique data format, and making the documents produced by third party tools traceable with the implementation of requirements and the source code using batch files, etc.
3. The third thing is to apply the new software engineering paradigm in real software product development to see whether it works as what we expected, and then get the feedback from the users to improve the entire software engineering paradigm.

5.4 The Establishment of NSE

NSE is established as an application example of the FDS (Five-Dimension Synthesis Method) paradigm-shift framework as shown in Fig. 5.1.

As shown in Fig. 5.1, each part of the NSE paradigm is developed/created by complying with the essential principles of complexity science, particularly the Nonlinearity principle and the Holism principle, so that with NSE almost all software engineering tasks/activities are performed holistically and globally.

The FDS framework can also be used for an individual software product development after the establishment of NSE.

5.6 The Components of NSE

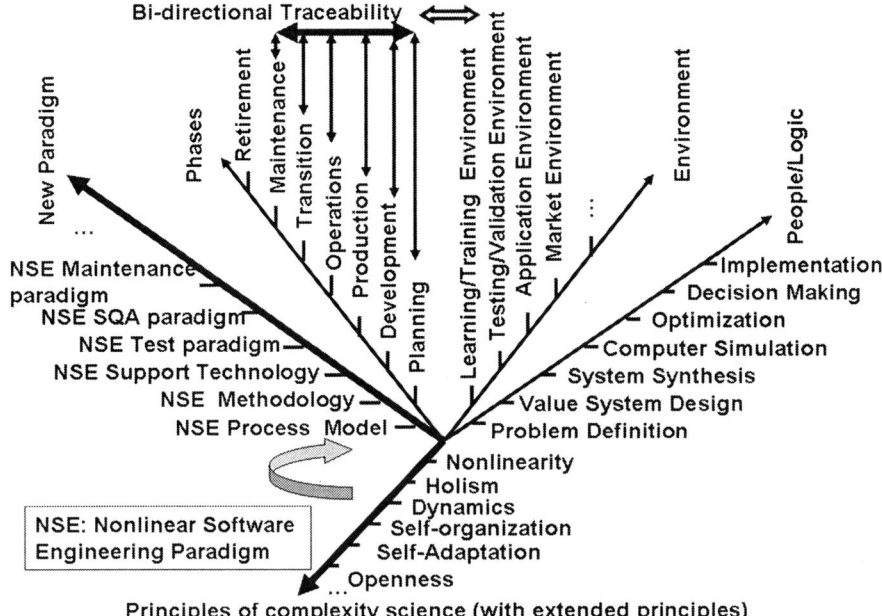

Fig. 5.1 The FDS (the Five-Dimensional Structure Synthesis) paradigm-shift framework

5.5 The Structure of NSE

The structure of NSE is shown in Fig. 5.2.

As shown in Fig. 5.1, NSE consists of ten parts, including (1) NSE process model, (2) NSE software development methodology, (3) NSE diagramming (visualization) paradigm, (4) NSE testing paradigm, (5) NSE quality assurance paradigm, (6) NSE documentation paradigm, (7) NSE maintenance paradigm, (8) NSE project management paradigm, (9) NSE support techniques, and (10) NSE support tools and platforms. They all work together closely.

The NSE paradigm has been implemented and supported by Panorama++ and SilverBullet platforms.

5.6 The Components of NSE

1. **The NSE process model** – It is the core part of NSE, a roadmap of the Nonlinear Software Engineering paradigm. The NSE process model is nonlinear, through two way iteration with multiple tracks (see Chap. 8) supported by automated and self-maintainable traceabilities (see Chap. 9).
2. **The NSE software development methodology** – It is based on **Generative Holism** and driven by defect prevention and traceability, different from the

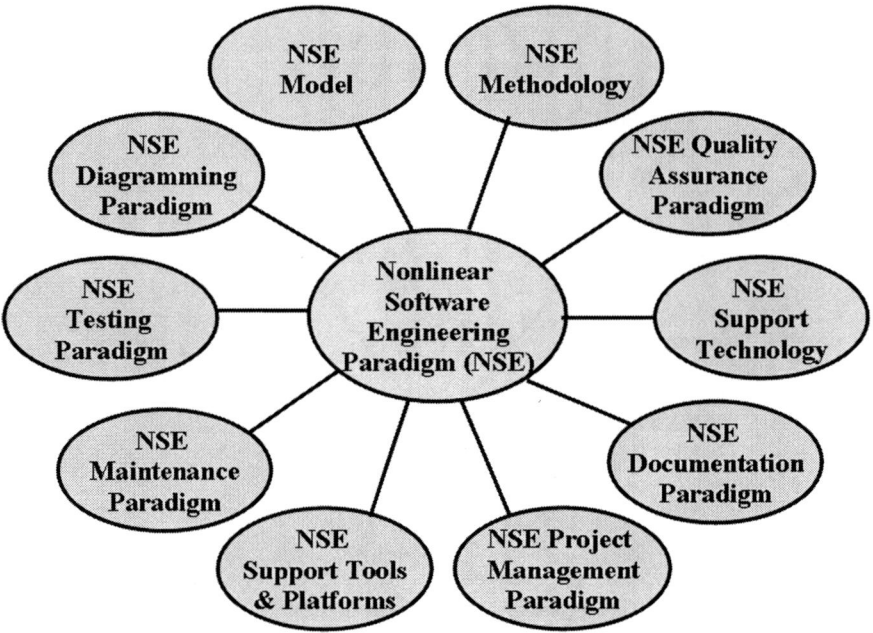

Fig. 5.2 The NSE structure

existing software development methodology based on **Constructive Holism** and driven by testing (see Chap. 10).

3. **The NSE diagramming (visualization) paradigm** – It makes the entire software engineering process visible from the first step down to the maintenance phase using interactive and traceable 3J graphics by generating the overall charts/diagrams for the entire software system and detailed logic diagrams and control flow diagrams for each file/class/function, with the capability to highlight untested conditions and branches when working with the MC/DC test coverage measurement tools integrated into the NSE support platforms (see Chap. 7).

4. **The NSE testing paradigm** – It is based on the Transparent-box testing method which combines functional testing and structural testing together seamlessly; to each test case it not only checks whether the output (if any, can be none) is the same as what is expected, but it also helps users to check whether the real execution path covers the expected one specified in control flow diagram, and then it automatically establishes bidirectional traceability among the related documents and test cases and the source code through the use of bookmarks and Time Tags inserted into both the test case description and the test coverage database for mapping the test cases and the tested source code together, so that it can be used dynamically in the entire software development and maintenance process, including the requirement development phase and the design phase, to greatly reduce the amount of defects introduced into a software product developed with NSE (see Chap. 16).

5. **The NSE quality Assurance paradigm** – It is based on defect prevention and defect propagation prevention from the first step down to the maintenance phase (see Chap. 17).
6. **The NSE documentation paradigm** – It makes the documents traceable to and from the source code to keep consistency with the source code at all times. The generated documents exist virtually to greatly reduce the required space and to speed up the display much faster (see Chap. 19).
7. **The NSE maintenance paradigm** – It helps users perform software maintenance holistically and globally with side-effect prevention for the implementation of requirement changes or code modification supported by various traceabilities to ensure the product quality and greatly reduce the cost through the use of test case minimization and intelligent test case selection in regression testing after code modification (see Chap. 18).
8. **The NSE project management paradigm** – It combines the software development process and project management process together, making the project management documents (such as the schedule chart, the project development plan, the cost estimation tables) traceable with the implementation of requirements and the source code for finding and fixing the management problems in time (see Chap. 20).
9. **The NSE support techniques** – They are the driving force for the establishment of NSE: 14 advanced techniques are innovated and applied into NSE and the support platforms (see Chap. 6).
10. **The NSE support tools and support platforms** – They help software organizations to apply NSE in their software product development easily, no matter if it is used for new software development, or to test or maintain an existing software product (see Chap. 22).

5.7 The Major Feature and Characteristics of NSE

- **It is based on a solid foundation – complexity science:** The entire NSE paradigm is established by complying with the essential principles of complexity science, particularly the Nonlinearity principle and the Holism principle.
- **It is complete** – NSE itself is complete, including its own process model, software development methodology, visualization paradigm, testing paradigm, QA paradigm, documentation paradigm, maintenance paradigm, management paradigm, etc.
- **It brings revolutionary changes to almost all aspects in software engineering** – It makes them change from the old one based on linear processes and the superposition principle, to the new one based on nonlinear processes and complexity science.
- **It offers both "what to do" and "how to do"** – Different from some popular models which only offer "what to do" but ignore "how to do," NSE offers both.
- **With it almost all software engineering tasks/activities are performed holistically and globally** – With NSE, from requirement development down to

maintenance, all tasks/activities are performed holistically and globally with defect prevention including side-effect prevention for the implementation of requirement changes and code modification.
- **It combines the software development process and software maintenance process together closely** – With NSE, requirement changes are welcome and implemented with side-effect prevention though various bidirectional traceabilities (see Chaps. 8 and 18).
- **It combines the software development process and software management process together closely** – It makes all documents including the management documents such the schedule chart and the cost reports traceable to the implementation of requirements and the source code to control a software project better and to find and fix the related issues in time (see Chaps. 8 and 20).
- **It ensues software product quality from the first step to the final step through defect prevention and dynamic testing using the Transparent-box testing method** – NSE offers many means to prevent defects introduced into a software product by people (the customers and the developers) with dynamic testing using the Transparent-box testing method which combines functional testing and structural testing seamlessly, can be dynamically used in the cases where there is no real output from the software system such as a dummy system with dummy modules only without detailed program logic (see Chaps. 11, 17, and 18).
- **With NSE, the design becomes precoding (top-down), and the coding becomes further design (bottom-up)** – With NSE, in most cases the design through dummy programming using dummy modules becomes precoding, and the coding becomes further design through reverse engineering (see Chaps. 12 and 13).
- **It makes software documents traceable to and from source code** – With NSE all related documents and test cases and the source code are traceable forwards or backwards though automated and self-maintainable traceabilities.
- **It supports real-time communication through traceable Web pages and traceable technical forum** – With NSE, the bidirectional traceability is extended to include Web pages and BBS for real-time communication.
- **It makes the entire software development process visible from the first step down to the final step** – The NSE visualization paradigm is capable of making the entire software development process visible through dummy programming and reverse engineering.
- **It makes a software product much easier to read, understand, test, and maintain** – With NSE, a software is represented graphically and shown in both the overall structure of the entire product and the detailed logic diagram and control flow diagram with various traceabilities and where the untested conditions and branches are highlighted.
- **It can be applied at any time in any stage for a software product development using any original method** – NSE can be added onto a software product being developed using any other approach by adding bookmarks in the related documents and modifying the test cases to use some key words to indicate the format of a document and the file path plus the bookmark, then the other work can be performed by the NSE support platform automatically.

5.7 The Major Feature and Characteristics of NSE

- **It requires much less time, resources, and manpower to apply compared with other existing approaches** – One just needs to reorganize the document hierarchy using bookmarks and modifying the test case description using some simple rules; all of the other work can be performed automatically by the NSE support platform with many automated and intelligent tools integrated together, including the creation of huge amounts of traceable and virtual documents based on static and dynamic measurement of the software, the diagramming of the entire software product to generate holistic and detailed system call graphs and class inheritance charts, the holistic and detailed test coverage measurement results shown in J-Chart and J-Diagram or J-Flow diagram with untested conditions and branches highlighted, the holistic and detailed quality measurement results shown in Kiviat diagram for the entire software product and each class or function, the holistic and detailed performance measurement results shown in J-Chart and bar chart with branch execution frequency measurement result shown in J-Diagram or J-Flow diagram to locate the performance bottleneck better, the software logic analysis results shown in J-Diagram with various kinds of traceability for semiautomated code inspection and walk through, the software control flow analysis results shown in J-Flow with untested conditions and branches highlighted, the GUI test operation capture and selective playback for regression testing after code modification, the test case efficiency analysis and test case minimization to form a minimized set of test cases to replace all the test cases to speed up the regression testing process and greatly save the required time and resources, the establishment of bidirectional traceability among all related documents and the test cases and the source code, the generation of more than 100 reports based on the static and dynamic measurement of the software which can be stored in HTML format for being used on the internet, the Cyclomatic complexity measurement results shown in J-Chart and J-Flow diagram for performing refactoring on the over complicated modules to reduce possible defects, and more.
- **It is possible for NSE to help software organizations double their productivity, halve their cost, and reduce 99–99.99% defects in their software products** – With NSE, the quality of a software product is ensured from the first step through defect prevention and defect propagation prevention rather than testing after coding, so that the amount of defects introduced into a software product is greatly reduced, and the defects propagating to the maintenance phase are also greatly reduced; software maintenance is performed holistically and globally with side-effect prevention; the regression testing after software modification is performed using a minimized test case set and some test cases selected through backward traceability from the modified modules and branches; software testing is performed in the entire software development process dynamically using the Transparent-method which combines functional testing and structural testing together seamlessly, and can be dynamically used in the case that there is no real output in running some test cases, when it is used in the requirement development phase and the software design phase.

5.8 Summary

The old-established software engineering paradigm is entirely outdated because it is based on linear thinking, reductionism, and the superposition principle that the **whole of a complex system is the sum of its parts,** so that almost all tasks/activities in software engineering are performed linearly, partially and locally – it is the root cause why many software projects fail.

The NSE paradigm is established with the FDS (Five-Dimension Synthesis Method) paradigm-shift framework by complying with the essential principles of complexity science, particularly the Nonlinearity principle and the Holism principle that **the behavior and characteristics of the whole of the software engineering paradigm cannot be inferred from its parts, but emerge from the interaction of all its parts,** so that with NSE almost all tasks/activities in software engineering are performed holistically and globally.

The NSE paradigm consists of ten major parts including the (1) NSE process model, (2) NSE software development methodology, (3) NSE diagramming (visualization) paradigm, (4) NSE testing paradigm, (5) NSE quality Assurance paradigm, (6) NSE documentation paradigm, (7) NSE maintenance paradigm, (8) NSE project management paradigm, (9) NSE support techniques, and (10) NSE support tools and platforms. They all work together closely.

5.9 Points and Questions to Ponder

(a) What are the major problems existing with today's software development? Why are those problems so hard to solve?
(b) Why does today's software maintenance take 75% or more of the total effort and total cost in software product development?
(c) What is NSE?

5.10 Further Reading and Information Source

(a) Brooks FP Jr (1995) The mythical man-month. Addison-Wesley, Reading
(b) DevTopics Software Development Topics, 20 Famous Software Disasters (http://www.devtopics.com/20-famous-software-disasters/)
(c) Xiong J, Xiong J (2009) A complete revolution in software engineering based on complexity science. In: WORLDCOMP'09 – SERP (Software Engineering Research and Practice 2009), pp 109–115

References

[Bro95-p181]	Brooks FP Jr (1995) The mythical man-month. Addison-Wesley, Reading, p 181
[Bro95-p182]	Brooks FP Jr (1995) The mythical man-month. Addison-Wesley, Reading, p 1282
[CMMI1.1]	Phillips M (2002) CMMI Program Manager, CMMI V1.1 and Appraisal Tutorial, http://www.sei.cmu.edu/cmmi/
[GSAM03]	USAF Software Technology Support Center (2003) Condensed GSAM Handbook, Chap 2, CrossTalk
[Ngw03]	Ngwenyama O, Nielsen PA (2003) Competing values in software process improvement: an assumption analysis of CMM from an organizational culture perspective. IEEE Trans Eng Manag 50(1):100–112. doi:10.1109/TEM.2002.808267
[Nia06]	Niazi M (2009) Keele University, Software process improvement implementation: avoiding critical barriers. CrossTalk, Jan Issue

Chapter 6
The Techniques Innovated to Support NSE

> *The road ahead for software engineering is driven by software technologies.*
>
> Roger S. Pressman, *"SOFTWARE ENGINEERING: A Practitioner's Approach"*

There are a set of unique techniques innovated to support the NSE process model and the entire NSE software engineering paradigm for efficiently solving the essential problems in software development: the complexity, changeability, conformity, and invisibility described by Frederick P. Brooks Jr. in his book, "The Mythical Man-Month" [Bro95-p182], plus testability, reliability, traceability, and maintainability. This set of unique techniques are innovated by me and implemented by me and my colleagues through the Paradigm-shift framework, FDS, also innovated by me (see Chap. 4), as shown in Fig. 6.1.

The related techniques innovated for the establishment of NSE is shown in Fig. 6.2.

6.1 Definitions

There are some definitions to be described first.

Dummy Module: a source code module having an empty body or a simple body with some function call statements only without real program logic.

Dummy System: a software system consisting of dummy modules, and can be compiled, executed, and tested without producing any real output.

Dummy Programming (Bone Programming): the process for designing and coding a dummy software system.

Time Tag: a time mark automatically inserted into a test case description in a test case script file and the corresponding test coverage database, to indicate the date and time when a test case is executed and where the corresponding test coverage result is located in the database. It is used for mapping a test case and the corresponding source code tested by the test case to establish bidirectional traceability. When a test case is

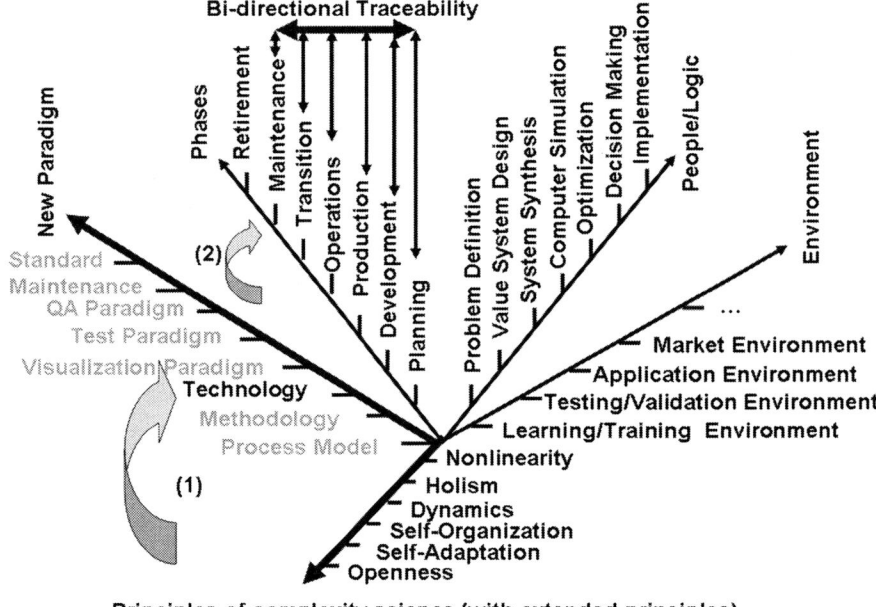

Fig. 6.1 Technology development complying with the principles of complexity science

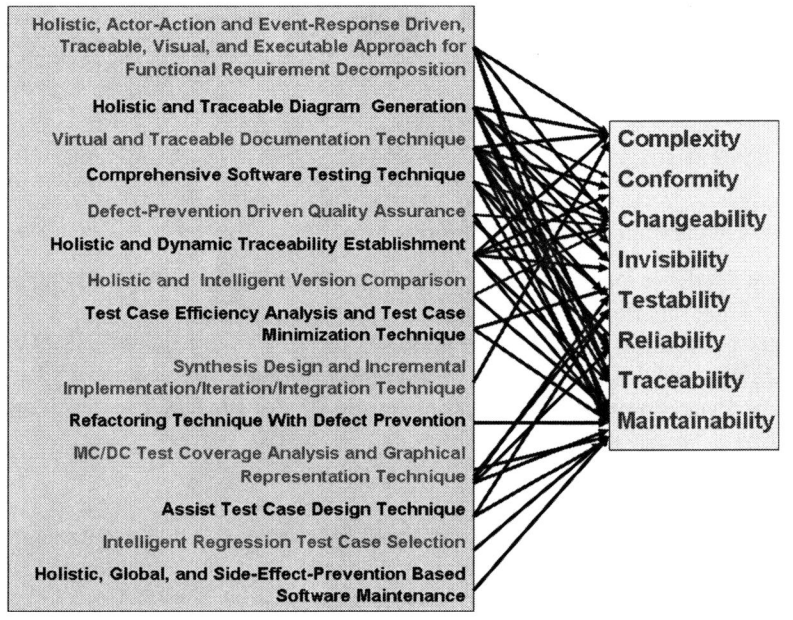

Fig. 6.2 Techniques innovated for the establishment

modified or the corresponding source code is modified, after rerunning the test case script, a new Time Tag will be inserted to replace the old one – so the bidirectional traceability can be automatically maintained without manual modification.

3J Graphics (J-Chart, J-Diagram, and J-Flow): new types of charts, logic diagrams, and control flow diagrams innovated by me which are interactive and traceable; used for making the entire software development lifecycle visible and the software product much easier to understand, test, and maintain.

Transparent-box Testing Method: a new software testing method innovated and implemented by me which combines functional testing and structural testing together seamlessly – to each test case a test tool developed with this method will check whether the output (if any, can be none – when applied in the requirement development phase and the preliminary design phase, there is no real output at all) is the same as what is expected, but also help users to check whether the real execution path covers the expected path specified using J-Flow diagram, and whether the execution hits some modules or some branches prohibited for the execution of the test case; after the execution of the test case, the tool will also build a bidirectional traceability facility to help users check the consistency among all of the related artifacts (including the source code too) through forward tracing and backward tracing. Different from traditional Black-box testing method which can be used after coding to find functional defects only, the Transparent-box testing method can be used to find functional defects, structural defects, and inconsistency defects among all of the artifacts in all phases of the software development life cycle, because having an output is no longer a condition for the use of the Transparent-box method dynamically. Chapter 16 will discuss this method in detail.

6.2 Holistic, Virtual, and Traceable Diagram Generation Technique

The Holistic, Virtual, and Traceable Diagram Generation technique is a new software diagram generation technique innovated by me and implemented by me and my colleagues which uses interactive and traceable 3J graphics to diagram the whole of a software system and its parts to solve the invisibility issue – making the entire software development process and the program structure/logic/control flow of an entire software product visible in all levels. The generated charts, logic diagrams, and control flows are traceable between a source file and the included files, a program tree and the related modules, a function definition body and the corresponding function call statements, a class and the inherited classes, an object and the class definition and the constructor, a module in a call graph and the corresponding logic diagram or control flow diagram of the module, or a module and all related modules calling and called by it, and so on. This technique provides an ideal solution to solve the critical problems faced with traditional software diagram generation techniques: software systems become more and more complex, so that using a big call graph to show the whole of a software system with many connection lines will make it very hard to view as shown in Fig. 6.3, but using many small call graphics to show the different parts of the entire

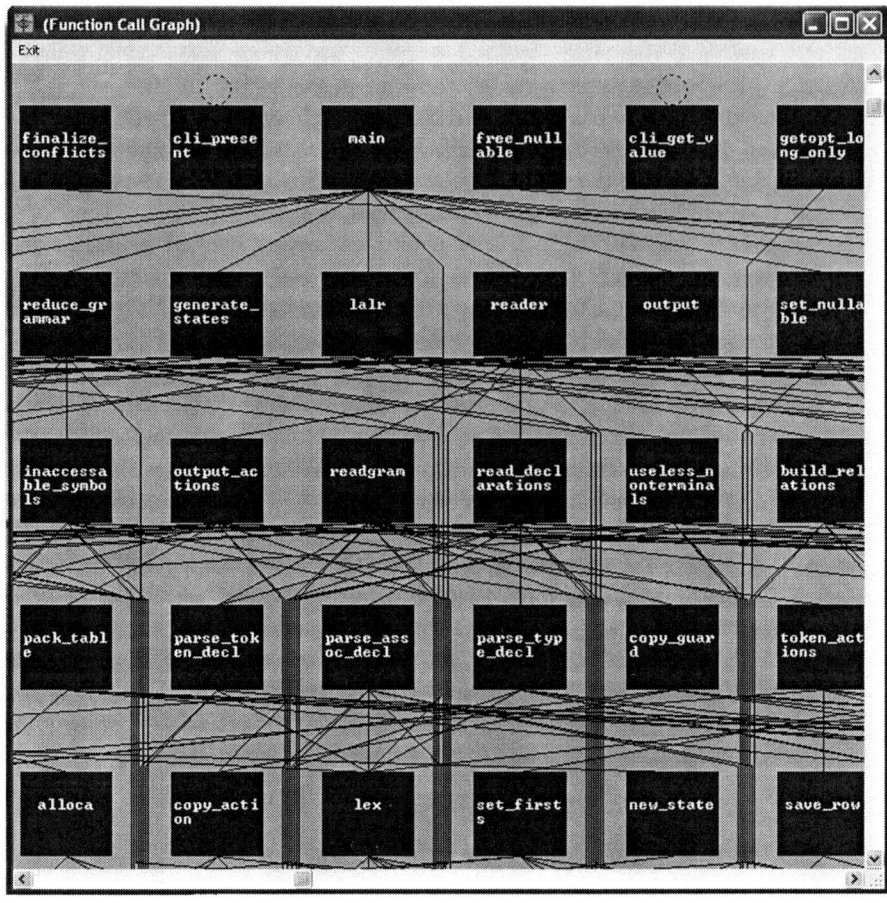

Fig. 6.3 An entire call graph of a complex software system generated with Virtual and Holistic Diagramming technique

software product separately with many connection arrows will not be able to show the big picture of the software system clearly, and often causes confusion. How can the Holistic, Virtual, and Traceable Diagram Generation technique and the corresponding tools solve this problem? The answer is that the generated charts/diagrams are traceable as shown in Fig. 6.4 when a user traces a module on the generated call graph shown in J-Chart notation, all the related modules calling and called by it are highlighted while the unrelated connection lines are invisible. A user can also select any module as a new root to let the corresponding tools to generate a subset of the system and show it in a new window. In addition, a bar chart can be attached on a module in the generated J-Chart to show the related information. All charts and diagrams are generated virtually from some simple hash tables without storing a hard copy in the hard disk or the memory of a computer to greatly save the required space and greatly reducing the time for displaying and monitoring the charts and diagram (see Sect. 6.3). Sample traceability of a J-Diagram is shown in Fig. 6.5.

6.3 Virtual and Traceable Documentation Technique

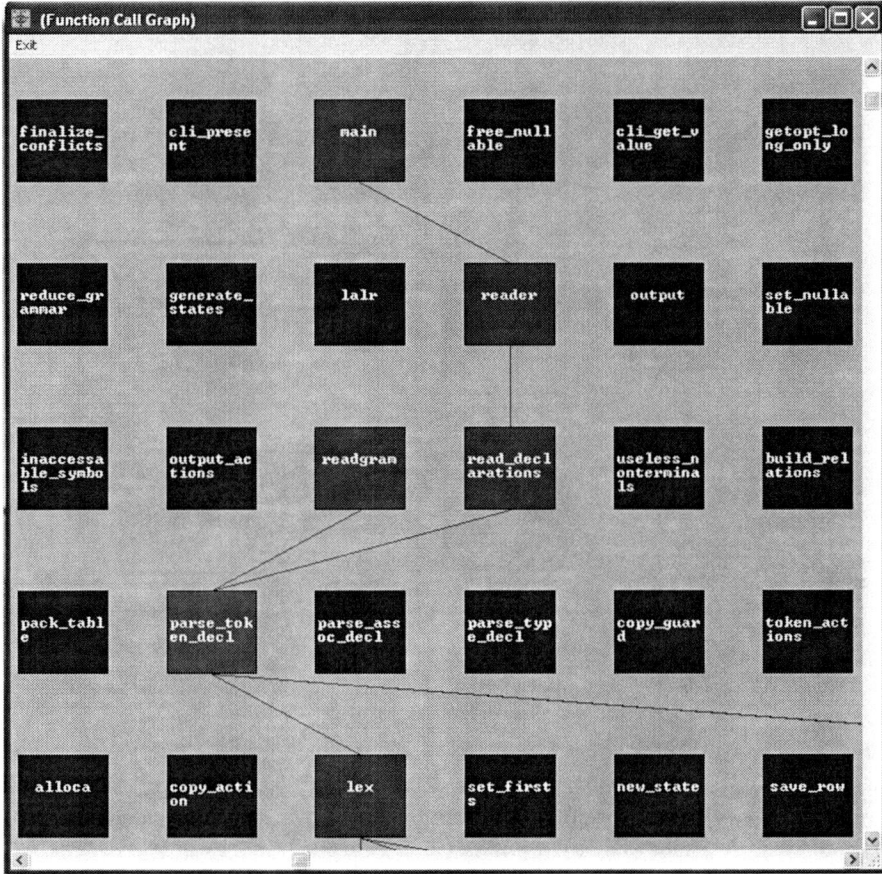

Fig. 6.4 A module and the related modules (calling and called by it) traced and highlighted

The generated charts and diagrams are virtually existing without storing any hard copy in disk or memory – they are generated dynamically from several hash tables from the database. This technique greatly reduces the requirement space (only needs about 1/100 of the space required by traditional approaches), and makes the display speed about 1,000 times faster (compared with the old versions of our own tools). Chapter 7 will discuss this technique in detail.

6.3 Virtual and Traceable Documentation Technique

The Virtual and Traceable Documentation technique is a new software documentation technique innovated by me and implemented by me and my colleagues. The corresponding tools developed from this technique automatically generate a great amount of documents/diagrams (the size is about 100 times bigger than the size of

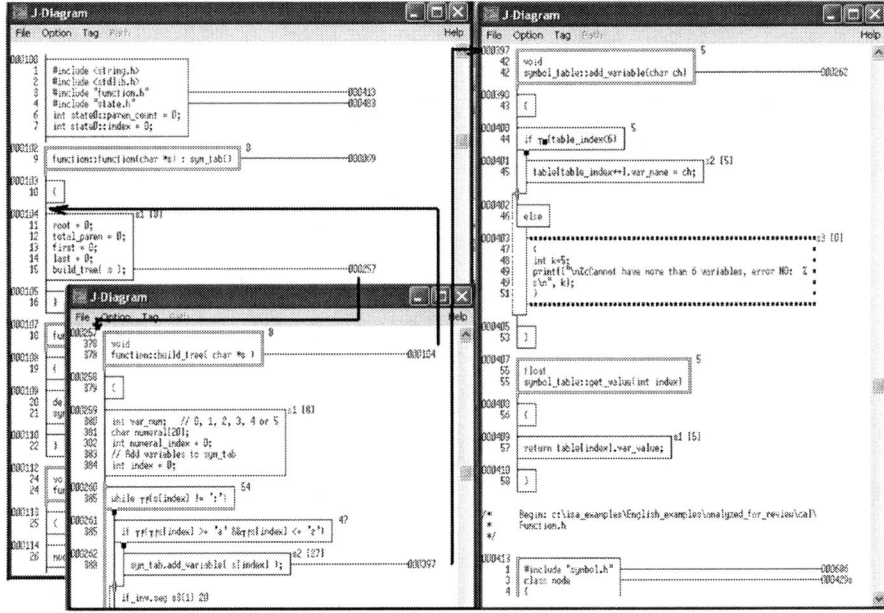

Fig. 6.5 Traceable J-Diagram (logic diagram) automatically generated

the source code) directly from the source code to document the overall measurement result for the whole system and the detailed result for each component of the software system through system structure analysis, program logic analysis, program control flow analysis, test coverage analysis, performance analysis, function cross reference analysis, global and static variable analysis, version comparison and quality measurement – with NSE, all graphics and diagrams are also virtually generated (including J-Charts, J-Diagrams, and J-Flow diagrams), but the size of the space needed for the documents/diagrams in memory and hard disk is about the same as the size of the source code, because the generated documents/diagrams are virtually existing without any hard copy to be stored in hard disks or the computer memory (unless the users require it) – each time, the document/diagram is generated from the corresponding database dynamically with the size being the same as the opened window for showing the document/diagram; when there is a need to trace an element to the related elements, for instance from a function call statement located in block 10 of the entire logic diagram to the called function body located in block 100,000 of the diagram of an entire complex software product, there is no real diagram movement performed from block 10 to block 100,000, but a new logic diagram is dynamically generated from the corresponding database and displayed from block 100,000. In this way, the users will see the graphical representation result as if all the documents/diagrams exist in the memory or hard disk, but the required space can be reduced to about 1/100 (compared with the traditional approaches), and the time required to display the generated documents/diagrams or operate them can be reduced to about 1/1,000. Chapter 19 will discuss this technique and its applications in details. A virtually generated quality measurement result for an entire software product and any individual class or function is shown in Fig. 6.6 – it is easy to imagine that if the graphical representation of the entire product and any class or function is not generated virtually, a huge amount of disk space and memory space will be required. But with NSE they are generated virtually from the corresponding database consisting of only several hash tables, and the total required space is almost the same as the size of storing the program source code.

6.4 Holistic and Intelligent Version Comparison Technique

The Holistic and Intelligent Version Comparison technique is a new software version comparison technique innovated by me and implemented by me and my colleagues which compares any two versions of an entire software system in system level, file level, and module/statement level using a virtual and holistic diagramming technique as shown in Fig. 6.7 in Black-white but the original one shown on the screen is colorful-blue is used for showing unchanged modules, red for changed ones, brown for deleted ones, and green for added ones. A new version of a module with more space characters in some lines will not be treated as modified – the tool developed using this technology is, in fact, an expert system which understands the grammar of the target language.

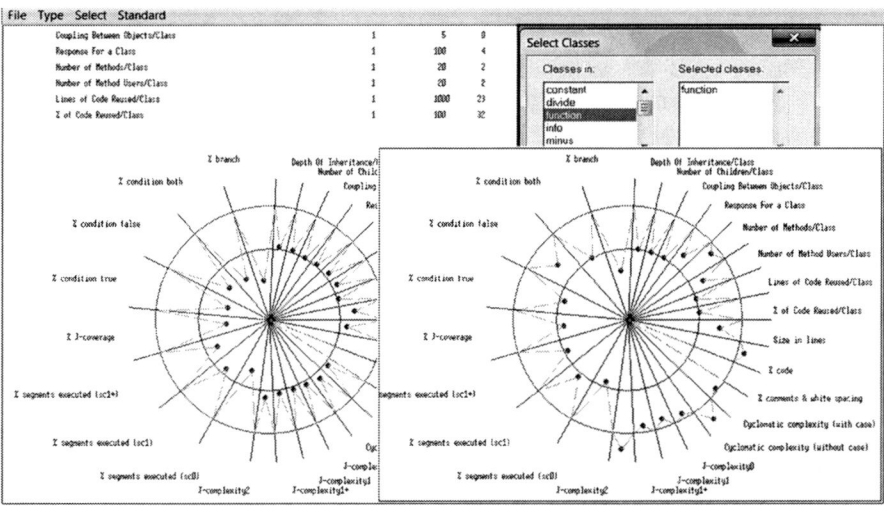

Fig. 6.6 A virtually generated quality measurement result for an entire software product and an individual class

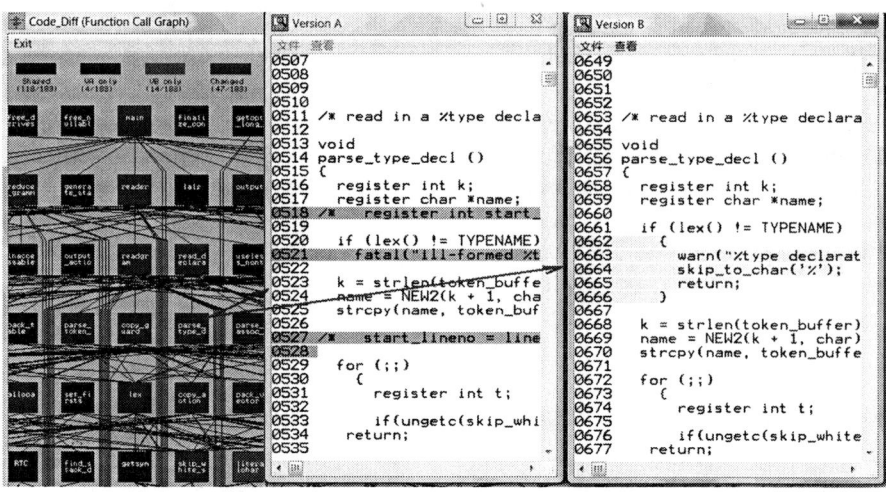

Fig. 6.7 An application example of the Holistic and Virtual Version Comparison Technique

6.5 Holistic and Dynamic Traceability Technique

The Holistic and Dynamic Traceability technique is a new software traceability establishment technique innovated by me and implemented by me and my colleagues, which is used to establish a self-maintainable facility offering automated and bidirectional traceability among all artifacts (including all related documents, test case scripts, test results, and the source code) of an entire software product (see Chap. 9), with the capability to selectively and dynamically play back the captured GUI operations of a traced or selected test case, and show the test coverage results graphically at the same time. This technique and the corresponding tools are particularly useful for defect prevention in the entire software development lifecycle (including the side-effect prevention in software maintenance for implementing software changes), and realizing full automation of software acceptance testing though mouse clicks only. Even if the GUI operation capture is performed for an entire test case script file with many test cases being executed together, a tool developed with this technique still can selectively play the captured GUI operations of only one test case back through Time Tags for data mapping. An application example is shown in Fig. 6.8.

6.6 Comprehensive Software Testing Technique Mainly Based on the Transparent-Box Method

The Comprehensive Software Testing technique is mainly based on the Transparent-box testing method to combine functional and structural testing together seamlessly, with the capability to establish automated and bidirectional traceability among all

6.7 Defect Prevention Driven Quality Assurance Technique

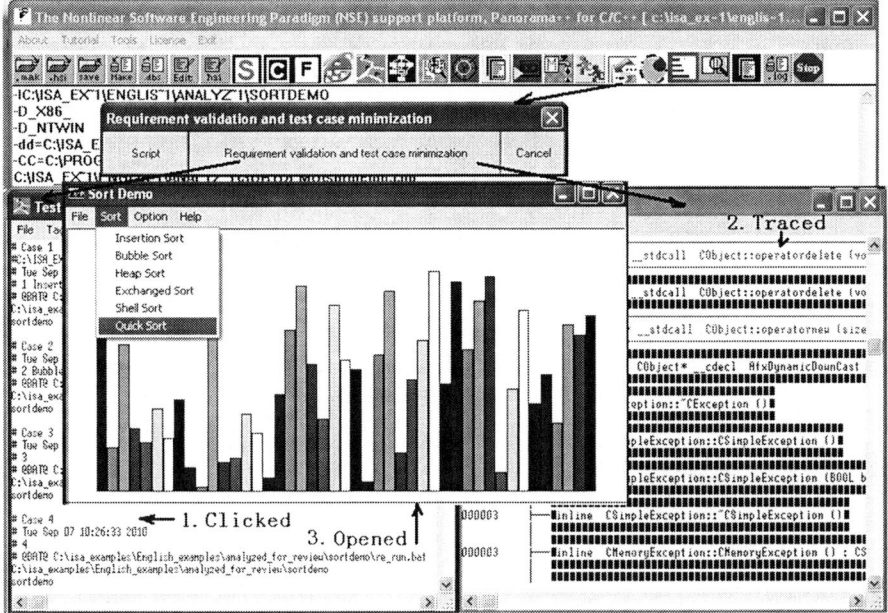

Fig. 6.8 Selective playback: click on a test case to automatically play the captured GUI operations back with the source code tested being highlighted

related documents and the source code. To each set of inputs, the tools developed with this kind of testing approach not only check whether the output (if any, can be none) is the same as what is expected, but also check whether the execution path covers the expected path.

Beside the Transparent-box testing, the Comprehensive Software Testing Technique also offers the capability for memory leak and memory usage violation analysis, performance analysis with program branch execution frequency measurement (for locating performance bottlenecks better), run-time error execution path analysis, incremental unit testing and integration testing, embedded software system testing, and GUI operation capture and playback.

6.7 Defect Prevention Driven Quality Assurance Technique

Traditional software quality assurance techniques are mainly based on software testing using the Black-box method and the structural testing method. But both methods are used after production (coding) – it violates Dr. W. Edwards Deming's principles for product quality control, *"Cease dependence on inspection to achieve quality. Eliminate the need for inspection on a mass basis by building quality into the product in the first place."* The proposed Defect Prevention Based Quality Assurance technique is used in the entire software engineering process to prevent defects through automated traceabilities and many other ways, including dynamic testing using the Transparent-box testing approach. Chapter 17 will describe the details. An application example of preventing inconsistency defects in module interfaces (between a module definition and the corresponding function call statements) in the coding process is shown in Fig. 6.9. It is done through incremental coding based on an assigned coding and unit testing order on the call graph obtained in the design phase. When writing a function call statement, the engineer can view the source code of the called function to prevent inconsistency defects because according to the assigned coding order, the called function must be completed and tested already. This Defect Prevention Driven Quality Assurance Technique is particularly

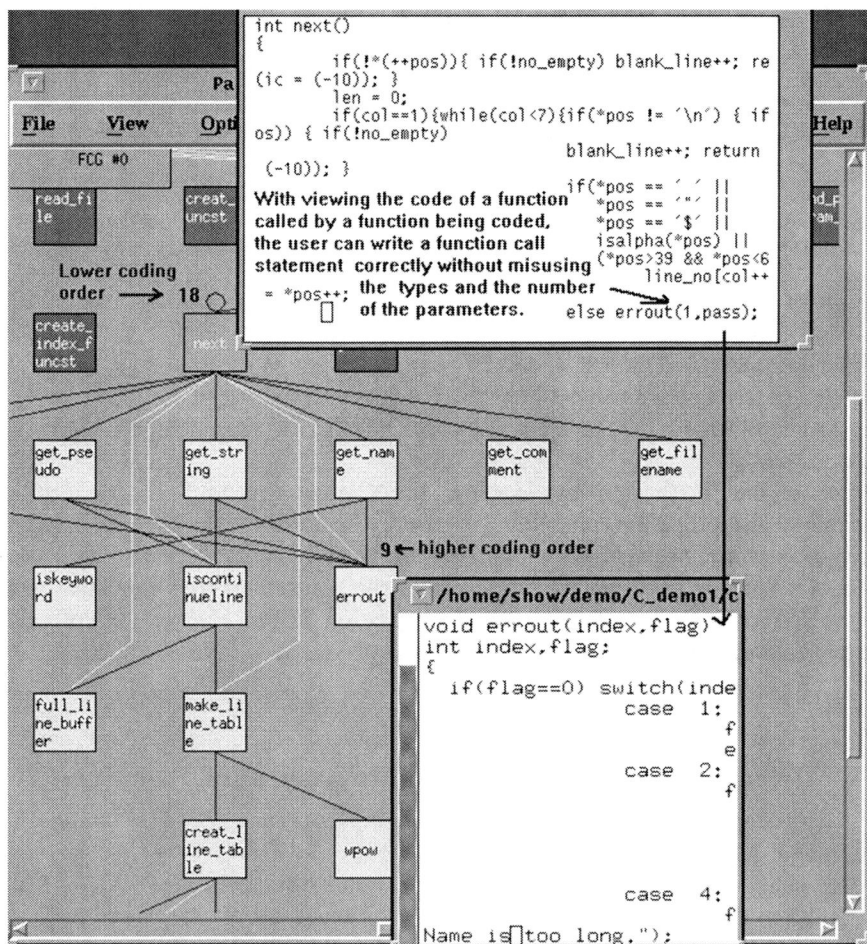

Fig. 6.9 Coding with defect prevention

useful in software maintenance when a requirement is modified, a forward tracing from the requirement to its implementation is performed to determine what modules need to be modified, and then a backward tracing is performed from the modules to the related requirements (which may be more than one) and documents as well as the modules calling and called by a module to be modified to prevent side effects – see Chap. 18 for the detailed description.

6.8 Test Case Efficiency Analysis and Test Case Minimization Technique

For testing a large and complex software product, a huge amount of test cases are designed and used. But within them the major parts are often useless which just repeat what have been tested by other test cases. With the old-established software engineering paradigm the functional testing and structural testing are performed separately, so that it is difficult to know what test cases are useful or useless. With NSE, the Transparent-box testing approach is used which combines functional testing and structural testing together seamlessly with the capability to measure the source code test coverage, so that it is easy to obtain the test efficiency for each test case by measuring the test coverage contribution as shown in the left side of Fig. 6.10. With the information of the test efficiency for each test case, the corresponding tool developed with NSE can

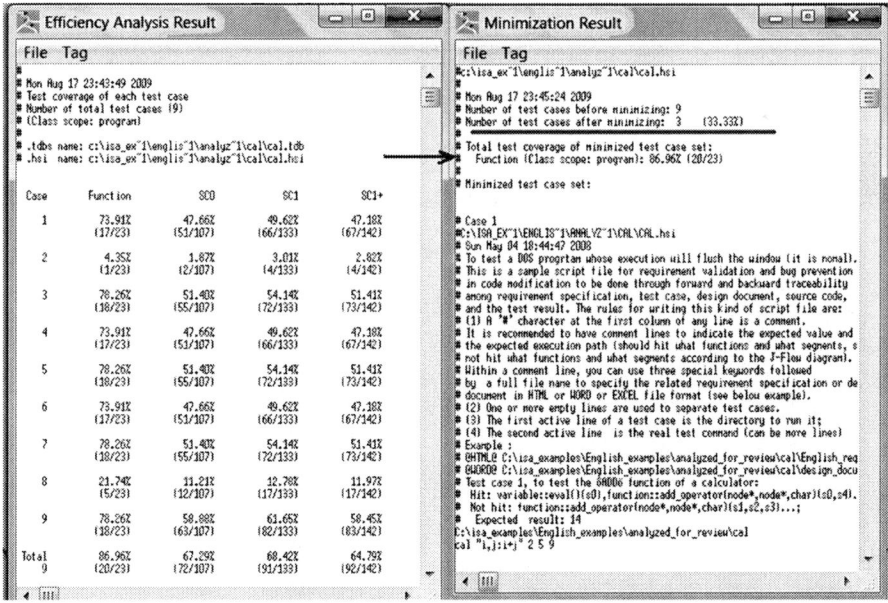

Fig. 6.10 An example of test case efficiency analysis and test case minimization

further perform test case minimization to select a minimized set of test cases which can be used to get the same test coverage result in regression testing after code modification. The corresponding algorithm for test case minimization will be introduced in Chap. 21. The key point is that whether a test case will be selected or not is not dependent on its single contribution in code test coverage, but the accumulated contribution with all the selected test cases. Usually a test case which has found a defect will be selected into the minimized set of test cases because the execution path of that test case is different from other test cases. If a test case covered one branch which has not been covered by all of other test cases, it will be selected into the minimized set of test cases. Usually the more test cases are used for testing a software product, the less the percentage the minimized set of test cases occupy. An application example is shown in the right of Fig. 6.10.

6.9 Refactoring Technique with Defect Prevention

The Refactoring technique with Defect Prevention is a program improvement approach by restructuring the program to remove duplication, improve communication, simplify the program, or add flexibility to the program without changing its behavior, which is performed with defect prevention to avoid the side effects of the modification. With NSE this approach is mainly used in those modules where: (1) the Cyclomatic complexity (the number of branch statements such as "if," "for," etc.) of the module is too big (over 30, for instance, as shown in Fig. 6.11), because often 80% of the defects exist in about 20% of the more complex modules; (2) it is a

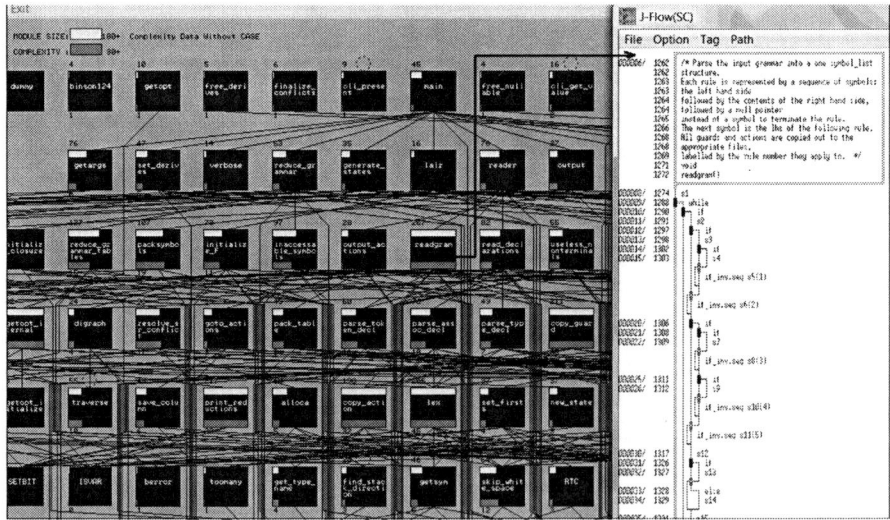

Fig. 6.11 An example of Cyclomatic complexity analysis and the control flow diagram of a complex program module

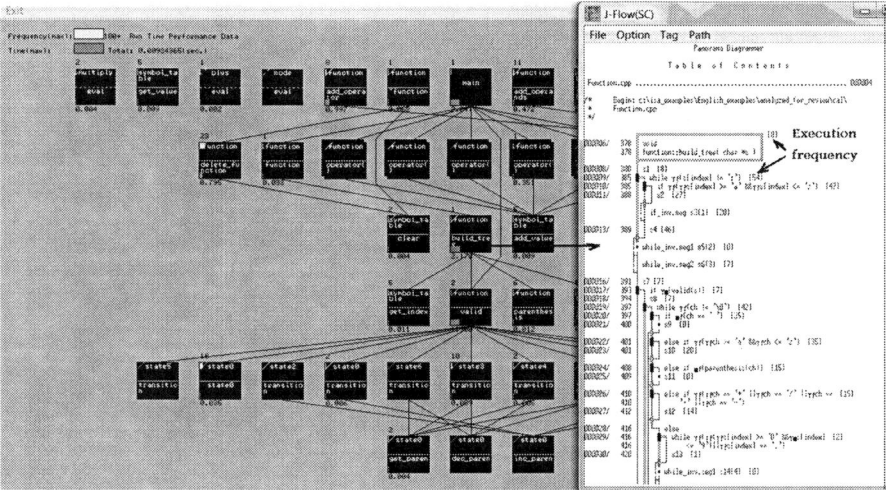

Fig. 6.12 An example of a performance bottleneck and the branch execution frequency analysis

performance bottleneck as shown in Fig. 6.12. Somehow, refactoring can be considered as a backward iteration. After refactoring, there is a need to modify the related design documents in the upper phases. Of course, refactoring can also be done through a forward approach by modifying the design and the corresponding documents first before the code modification. After refactoring, the program should be fully retested, including functional testing and structural testing with performance analysis, test coverage analysis, memory leak and usage violation analysis, variable analysis, and more, to ensure the quality of the modified program.

6.10 Holistic MC/DC Test Coverage Analysis and Graphical Representation Technique

According to the RTCA/DO-178B standards (Joseph Wlad, Product Marketing Manager Wind River, Alameda, CA, *DO-178B and Safety-Critical Software Technical Overview,* http://www.opengroup.org/rtforum/jul2001/slides/wlad.pdf), MC/DC (Modified Condition/Decision Coverage) is required for top quality software testing. The difficulty includes not only how to perform the MC/DC test coverage analysis but also how to visually show the test coverage results. The Holistic MC/DC Test Coverage Analysis and Graphical Representation technique innovated by me can be used to not only perform MC/DC test coverage for an entire software product and its parts but can also be used to show the test coverage results in interactive and traceable J-Chart, J-Diagram, and J-Flow diagrams with the capability to clearly highlight the untested branches and conditions using small black boxes in the generated J-Diagram and J-Flow diagrams as shown in Fig. 6.13.

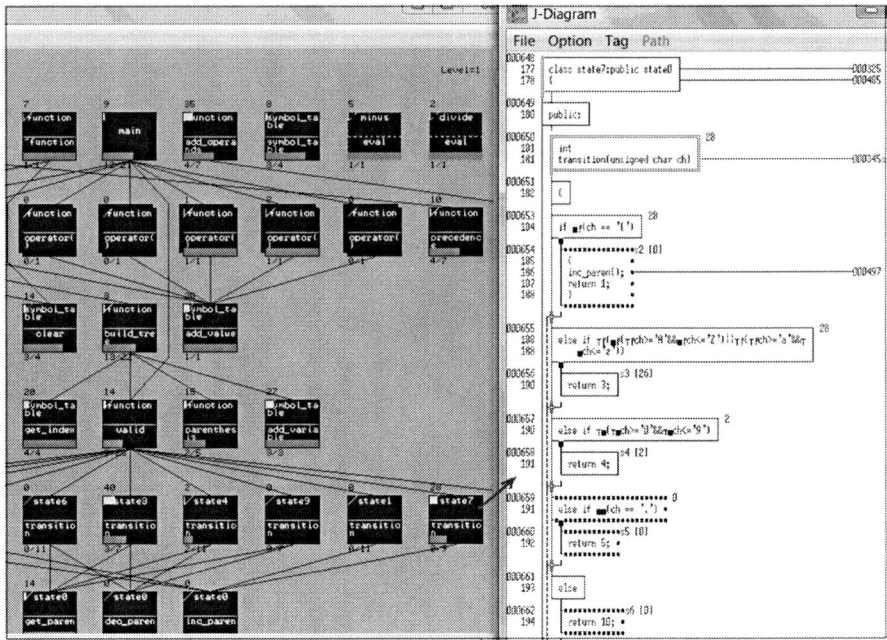

Fig. 6.13 Call graph shown in J-Chart notations with the MC/DC test coverage analysis result for the whole and its parts of a software system (here, untested conditions and branches are highlighted in *small black boxes*)

6.11 Assisted Test Case Design Technique

At the beginning of software testing for a software product, it seems easy to design test cases, but later on when the test coverage result reaches 50% or more, the test case design will become more and more difficult if we want to design a new test case which will cover code branches or conditions which have not been covered by the previous test case execution. The proposed Assisted Test Case Design technique works with the path and test coverage analysis to automatically compare all untested paths and select one with the most untested branches, then it automatically extracts the test conditions of the selected path to help users design the corresponding test case which could be better than ten test cases randomly designed. An application example is shown in Fig. 6.14.

6.12 Intelligent Regression Test Case Selection Technique

With the old-established software engineering paradigm, even if only one source module or only a few branches of the module have been modified, all test cases should be used for regression testing because without automatic and bidirectional

6.12 Intelligent Regression Test Case Selection Technique

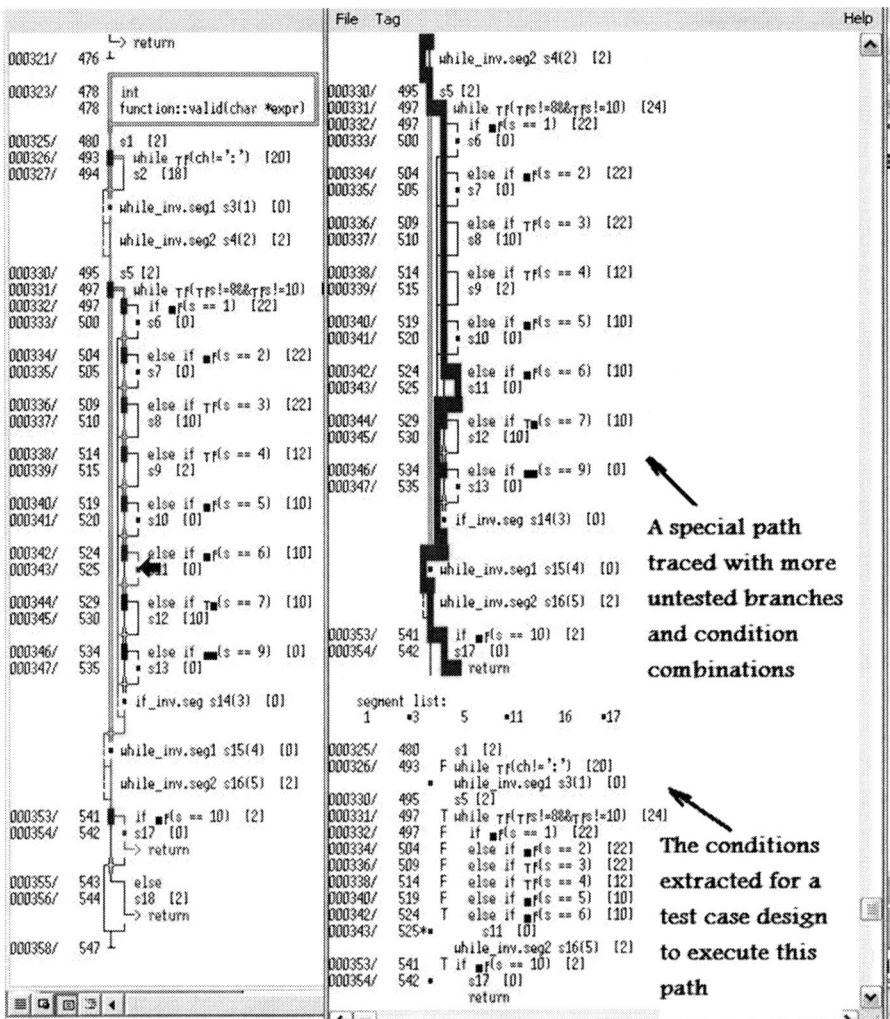

Fig. 6.14 An application example of Assisted Test Case Design

traceability, it is almost impossible to know what test cases can be used to retest the modified product. It is very clear that a complex software product is a nonlinear system where a small change may ultimately cause great changes in the entire system – the Butterfly Effect. The proposed Intelligent Regression Test Case Selection technique is mainly used to solve this kind of problem – when only a few modules or only a few source program branches are modified, the maintainers can easily perform backward tracing from a modified module or a modified branch to find out all related test cases which can be used to retest the modified product, using the NSE support platform. An application example is shown in Fig. 6.15 where for the modified branch S3 (Segment 3), only one test case was found which can be

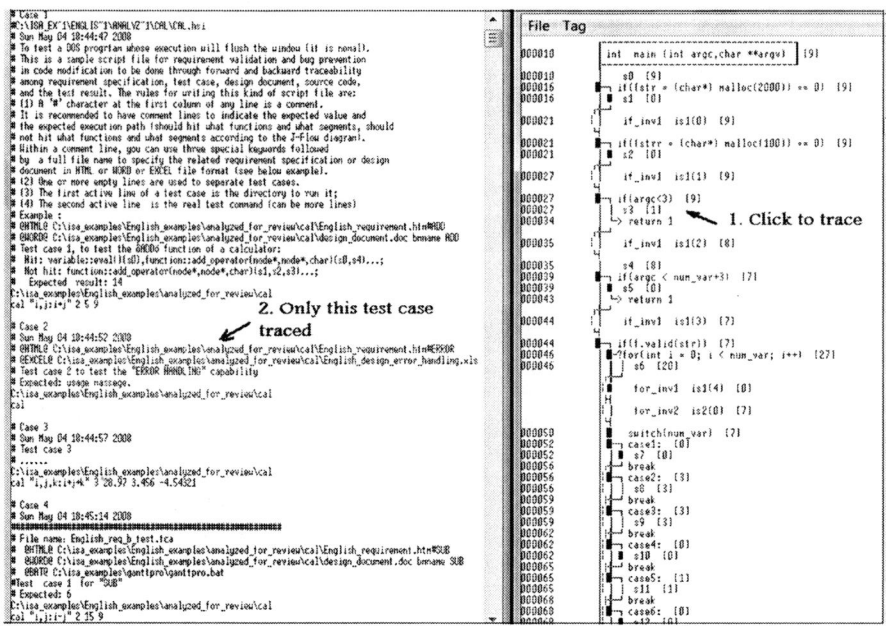

Fig. 6.15 An application example of the Intelligent Regression Test Case Selection

used to retest branch S3; the other test cases do not go through S3, so they are useless in retesting the branch S3. Of course, sometimes we need to add some new test cases to retest the modified product better.

6.13 Holistic, Actor–Action and Event–Response Driven, Traceable, Visual, and Executable Technique for Requirement Development

The Holistic, Actor-Action and Event-Response Driven, Traceable, Visual, and Executable technique is a new requirement development technique innovated and implemented by me for capturing customers' requirements through top-down dummy programming for making the requirements much easier to review and understand. The technique supports **Holistic, Actor–Action and Event–Response driven, Traceable, Visual, and Executable** requirement development, including the decomposition of functional requirements, and the nonfunctional requirements. About the applications of this technique, see step 1 of the main process of the NSE process model to be described in Chap. 8. It is innovated mainly for improving the Use Case approach which is not holistic, not suitable for event–response applications, and the obtained result is not really traceable for inspection and review, not easy to map to the product design, and not directly executable for finding and removing defects.

6.14 Synthesis Design and Incremental Growing Up (Implementation and Integration) Technique

The Synthesis Design and Incremental Growing Up (Implementation and Integration) technique is a new technique for software development innovated and implemented by me to efficiently develop a software product and solve the issue of software complexity. As pointed out by Frederick P. Brooks Jr., "The complexity of software is an essential property, not an accidental one. ... many classical problems of developing software products derive from this essential complexity and its non-linearity increases with size.... Much of the complexity in a software construct is, however, not due to conformity to the external world but rather to the implementation itself – its data structures, its algorithms, its connectivity." He then suggested an approach to efficiently handle the complexity issue:

- "Hierarchically, by layered modules or objects
- Incrementally, so that the system always works" [Bro95-p212]

But, it is not good enough for solving the complexity issues, because there are several different complexities to be handled, including the following:

1. Apparent complexity: appears complicated but simple patterns lie underneath the surface
2. Detail complexity: has great number of different parts
3. Dynamic complexity: has a great number of possible interconnections between parts
4. Inherent complexity: extremely complex with great number of different parts that have a great number of possible interconnections and feedback loops (http://www.businessdictionary.com/definition/types-of-complexity.html), or

Formulaic complexity:
    Description complexity
    Generative complexity
    Computational complexity

Compositional complexity
    Constitutional complexity
    Taxonomical complexity

Structural complexity
    Organizational complexity
    Hierarchical complexity

Functional complexity
    Operational complexity
    Normic complexity

So we need much more advanced techniques and tools to solve the complexity issue. Here, "**Synthesis Design**" means the following activities:

1. Collect the information and data related to the requirements, including the solution method comparison reports, prototype design and risk analysis reports,

test results, customer evaluation results, and the documents of the algorithms used, etc.
2. Perform functional requirement decomposition and defect removal through dynamic testing using the Transparent-box approach.
3. According to the functional requirement decomposition results plus nonfunctional requirements, design an executable dummy system (the preliminary architecture) through dummy programming.
4. Remove the defects introduced into the designed dummy system through visual diagramming and inspection, particularly dynamic testing using the Transparent-box approach.
5. Perform optimization of the designed dummy system to reduce the coupling degree.
6. Design the preliminary data structures (class structures) according to the collected information and data.
7. Compile and execute the designed dummy system mapping to the functional requirement decomposition plus the nonfunctional requirements.
8. Further decompose the system, as detailed as possible.
9. Work with the Incremental Implementation and Integration of requirements to make the system grow up with new versions of the system executable.

Here, **Incremental growing up** means the following activities:

1. Select one or a set of requirements according to the requirement priority assigned.
2. From the corresponding call graph (shown in J-Chart notation) of the designed system, highlight the critical module with all modules calling and called by modules for the selected requirement(s), assign a bottom-up design and coding order on the automatically generated system hierarchy.
3. Perform incremental unit coding according to the assigned order to prevent inconsistency defects between the interfaces of the calling modules and the called modules (see Fig. 6.9).
4. Carry out unit testing and integration testing to remove possible defects through comprehensive testing (including functional testing, structural testing, memory leak and usage violation checking, quality measurement, and performance analysis, etc.).
5. Recompile the entire program to establish a new version of the program, and then run the program again dynamically.
6. Different from traditional incremental iteration approaches which complete the subsystem design and coding for the selected requirement(s) first then carry out integration, with NSE the incremental implementation and iteration is done with integration at the same time – each time only one module of the subsystem for the selected requirements will be coded, tested, and integrated to establish a new version of the executable program, so that if something wrong is found, the problems often come from the one added module only rather than the entire subsystem implemented for the selected requirement(s). An application example of the **Incremental Implementation, Iteration, and Integration technique** is shown with step 1 of the main process described in Chap. 8.

7. Combine the processes of software development, testing, and maintenance together closely through many automated and bidirectional traceabilities for defect prevention in the entire software product development lifecycle.
8. If some critical problems are found in any phase, go back to the upper phases to solve the problem – it is possible to give up the previously selected solution method such as in the case where the performance is very bad because of the misuse of the virtual memory – it is a nonlinear way for requirement implementation.

6.15 Holistic, Global, and Side-Effect-Prevention Based Software Maintenance Technique

The Holistic, Global, and Side-Effect-Prevention Based Software Maintenance technique is a new software maintenance technique innovated by me and implemented by me and my colleagues which is also a Systematic, Disciplined, and Quantifiable approach for software maintenance. The key part of this technique is the various traceabilities that are established through **Transparent-box testing** and the **Holistic and Traceable Diagram Generation technique.** It is the most important technique of NSE for greatly reducing the cost and the effort spent in software system development. This technique and its applications will be described in detail in Chap. 18.

6.16 Summary

Fourteen unique software engineering techniques are innovated to support NSE. Those techniques are developed through the FDS framework by complying with the essential principles of complexity science, particularly the nonlinearity principle and the holism principle.

Those techniques include the **HAETVE** for Requirement Development used to replace the Use Case approach which is not holistic and the results obtained are not traceable and not directly executable for removing defects; the **Holistic, Virtual and Traceable Diagram Generation technique for** generating interactive and traceable charts/diagrams to make an entire software development process visible; the **Holistic and Dynamic Traceability technique** to establish automated and self-maintainable traceability among documents and test cases and source code for defect prevention and defect propagation prevention; the **Comprehensive Software Testing technique mainly based on the Transparent-box method** combining functional testing and structural testing together seamlessly with the capability to establish bidirectional traceability to find functional defects, logic defects, and inconsistency defects dynamically in the entire software development lifecycle; the **MC/DC Test Coverage Analysis and Graphical Representation technique** for highest quality software product development; the **Defect Prevention Driven Quality Assurance technique**, which is the key technique to ensure the quality of a software product; the **Refactoring technique with Defect Prevention** for further reducing the defects introduced to a software product because often the 20% most complex modules will have about 80% of the total defects; the **Intelligent Regression Test Case Selection technique** and the **Test Case Efficiency Analysis and Test Case Minimization technique** to greatly reduce the cost for regression testing after software modification; the **Virtual and Traceable Documentation technique** which makes the documents traceable to and from the source code to keep consistency; the **Holistic, Global, and Side-Effect-Prevention Based Software Maintenance technique** which is the key technique making it possible to help software organizations double their productivity and halve their software development cost; and the **Holistic and Intelligent Version Comparison technique** for version control and quick location of new defects after code modification. With those techniques and the corresponding tools, NSE with its support platform can efficiently handle the issue of software complexity, changeability, invisibility, and conformity, and efficiently solve the critical problems with today's software development: the low productivity and quality, and high cost and risk.

6.17 Points and Questions to Ponder

(a) What are the driving forces for the establishment of NSE (Nonlinear Software Engineering paradigm)? Describe them in as much detail as possible.
(b) Which principles do the techniques introduced in this chapter comply with? Why?

6.18 Further Reading and Information Source

Pressman RS (2005) Software engineering: a practitioner's approach. McGraw-Hill, New York

Jones C (2002) Software quality in 2002: a survey of the state of the art, Six Lincoln Knoll Lane, Burlington, MA. http://www.SPR.com July 23, 2002

Kannenberg A et al (2009) Why Software Requirements Traceability Remains a Challenge. CrossTalk, Jul/Aug Issue

Xiong J, Xiong J, A complete revolution in software engineering based on complexity science. In: WORLDCOMP'09 – SERP (Software Engineering Research and Practice 2009), p 109–115

References

[Bro95-p182] Brooks FP Jr (1995) The mythical man-month. Addison-Wesley, Reading, p 182
[Bro95-p212] Brooks FP Jr (1995) The mythical man-month. Addison-Wesley, Reading, p 212

Chapter 7
NSE Software Engineering Visualization Paradigm

"One Picture is Worth Ten Thousand Words."

Chinese Idioms

This chapter introduces the NSE software visualization paradigm which is holistic, and the generated charts and diagrams are interactive, colorful, and traceable; it is to be used in the entire software development lifecycle to make all the processes and the obtained work products visible.

"One Picture Is Worth Ten Thousand Words" – a holistic, interactive, colorful, and traceable chart/diagram will be more useful in the description of a complex software product.

The NSE software visualization paradigm is established through the FDS paradigm-shift framework by complying with the essential principles of complexity science as shown in Fig. 7.1, particularly the Nonlinearity principle and the Holism principle.

7.1 The Old-Established Software Engineering Visualization Paradigm Is Outdated

The old-established software engineering visualization paradigm is outdated because it is

1. **Based on linear thinking, reductionism, and the superposition principle**
   The traditional software engineering visualization techniques and tools are based on linear thinking, reductionism, and the superposition principle that **the whole of a system is the sum of its parts**, so that almost all diagramming tasks are performed locally and partially.
2. **Not Holistic**
   They are not holistic and global diagramming techniques and tools. The application results obtained consist of many small pieces without a complete chart/diagram to show an entire software product.

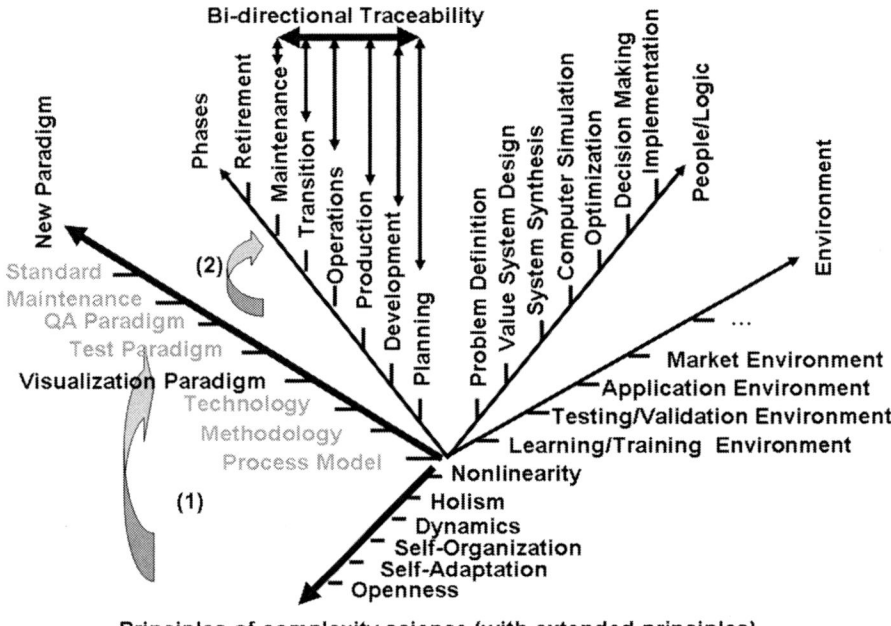

Fig. 7.1 The essential principles applied to the innovation of the NSE visualization (diagramming) paradigm through the paradigm-shift framework FDS

3. **Not automatable in most cases**
   Most charts/diagrams are created using graphic editors, and are not automatically generated.
4. **Not interactive**
   Most charts/diagrams generated using the old-established software visualization paradigm are not interactive, making it hard to manipulate.
5. **Not traceable**
   Even if a complete chart/diagram for an entire software product can be obtained by using a few diagramming tools, it is still useless because without traceability and the capability to highlight an element with all of the related elements, there are too many connection lines, making the chart/diagram hard to view and hard to understand.
6. **Not accurate**
   Often when the source code is modified, the generated charts and diagrams cannot be automatically updated to keep consistency with the source code.
7. **Not precise**
   For instance, when a logic diagram is used to show the result of program test coverage measurement, it cannot show whether an invisible "else" part (an "if" statement without an explicit "else" part) is tested or not. Almost all existing visualization tools cannot graphically show whether a condition in a decision statement is tested or not when applied to show the result of MC/DC (Modified Condition/Decision Coverage) test coverage measurement.

7.2 The Revolutionary Solution Offered by NSE

8. **Often not consistent with the source code**
   The charts/diagrams are often not consistent with the source code after software modification.
9. **Not consistent among all related charts and diagrams**
   Often they are generated with different formats using different information, making it hard to keep consistency among them.
10. **Not virtual**
    The generated charts and diagrams are stored in hard copies or XML or postscript format in the memory or hard disk, requiring much more disk space and long loading times.
11. **Not complete**
    The traditional software engineering visualization techniques and tools are not integrated together to completely support the following areas:
    (a) **Visualization of the entire software engineering lifecycle**
    (b) **Visualization for requirements engineering**
    (c) **Visualization for design engineering**
    (d) **Visualization for coding engineering**
    (e) **Visualization for software inspection**
    (f) **Visualization for software testing**
    (g) **Visualization for software maintenance**
    (h) **Visualization for software architectures**
    (i) **Visualization for the source code of an entire software product**
    (j) **Visualization for software debugging**
    (k) **Visualization for reverse engineering**
    (l) **Dynamic program behavior visualization**
    (m) **Integration of visualization tools in the software engineering tool chain**
    (n) **Visualization for software debugging**

7.2 The Revolutionary Solution Offered by NSE

The revolutionary solution offered by NSE will be described in detail in this chapter, including Sect. 7.8 about the major features of the NSE visualization paradigm. Here is the outline of the solution:

1. **Based on nonlinear thinking and complexity science**
2. **Holistic**
3. **Automatic**
4. **Interactive**
5. **Traceable**
6. **Accurate**
7. **Precise**
8. **Consistent among all related charts and diagrams**
9. **Linkable automatically between different charts and diagrams**
10. **Virtual**

11. **UML are supported indirectly**
    When there is a need to generate some UML charts or diagrams using graphic editors, a freeware product, Fujaba (http://www.fujaba.de/), is used with the NSE visualization paradigm.
12. **Complete in software engineering visualization, including**
    (a) **Visualization of the entire software engineering lifecycle**
    (b) **Visualization for requirements engineering**
    (c) **Visualization for design engineering**
    (d) **Visualization for coding engineering**
    (e) **Visualization for software inspection**
    (f) **Visualization for software testing**
    (g) **Visualization for software maintenance**
    (h) **Visualization for software verification/validation**
    (i) **Visualization for software architectures**
    (j) **Visualization for the source code of an entire software product**
    (k) **Visualization for reverse engineering**
    (l) **Dynamic program behavior visualization**
    (m) **Integration of visualization tools in the software engineering tool chain**
    (n) **Visualization for software debugging**

7.3 The 3J graphics (J-Chart, J-Diagram, and J-Flow)

The 3J graphics (J-Chart – a new type call graph, J-Diagram – a new type of logic diagram, and J-Flow – a new type of control flow diagram) are innovated by me and implemented by me and my colleagues. J-Chart/J-Diagram/J-Flow is a trinity: an Object-Oriented and structured chart/logic diagram/control flow diagram, the chart/diagram generator which is always running when the chart/diagram is shown, and the interface (using the chart/diagram itself) between the generator and the user for controlling the chart/diagram dynamically with a multiway online traceability/cross reference facility through which the users can view the related objects easily.

7.4 J-Chart

J-Chart not only can be used to represent the class inheritance relationship, the function call graph, and the class–function coupling structure graphically but can also be used to display incremental unit test order or the related test coverage and quality data in bar graphics overlaid on each object-box to help users view the overall results of testing and quality measurement. J-Chart is useful in system understanding, inspection, test planning, test result display, and reengineering. The J-Chart notations are shown in Fig. 7.2.

7.4 J-Chart

A comparison between J-Chart and the most traditional call graphs

	J-Chart	Traditional call graph
Is it holistic for directly showing a very complex software product?	Yes	No
Is it interactive for highlighting a path or getting related information?	Yes	No
Is it traceable such as to highlight a module with all the related modules?	Yes	No[a]
Is it supported to use a module as the root to generate a subchart?	Yes	No
Can a bar chart be added to a module box to show related information?	Yes	No
Can the source code be directly edited from a module box?	Yes	No
Can the logic diagram be linked from a module box?	Yes	No
Can the control flow diagram be linked from a module box?	Yes	No
Can bottom-up coding orders be assigned to the modules?	Yes	No
When used for software version comparison, can different colors be used to show "unchanged modules," "changed modules," "deleted modules," and "added new modules" separately?	Yes	No

[a] Some tools claim that they can provide dynamic function call graphs, but I have not seen their application examples.

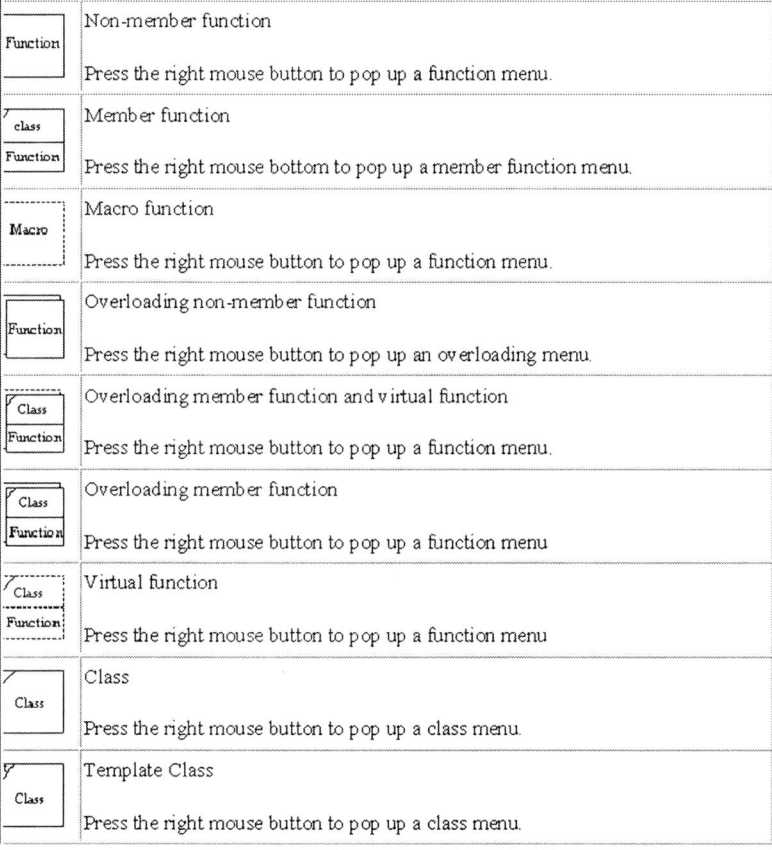

Fig. 7.2 J-Chart notations

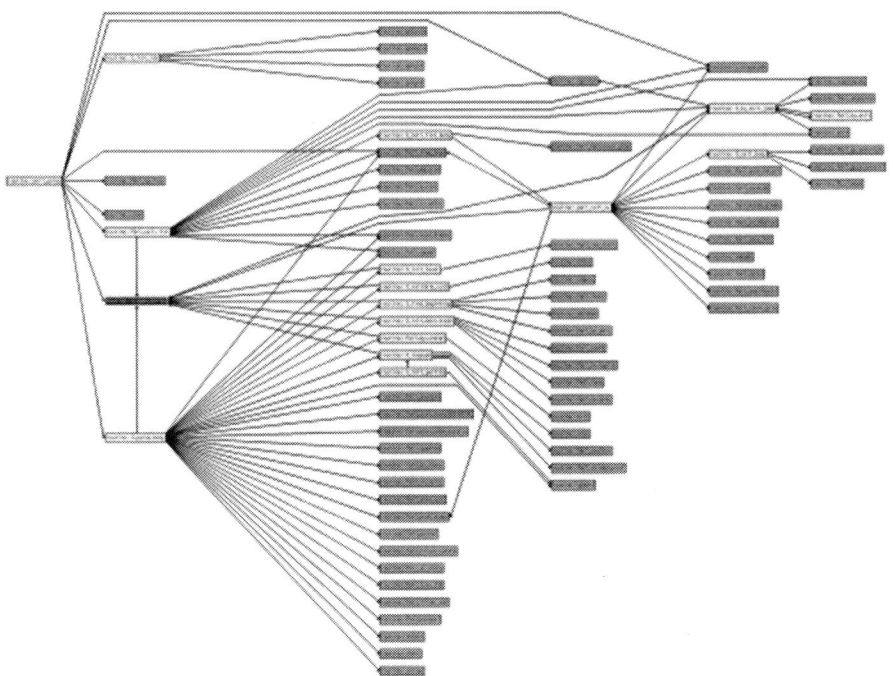

Fig. 7.3 An example of the traditional call graph. (http://www.aisee.com/graph_of_the_month/perlsm.gif)

For easy comparison, two traditional call graphs are shown in Figs. 7.3 and 7.4.

An example of J-Chart is shown in Fig. 7.5 – a call graph showing the result of the Cyclomatic complexity measurement (the number of branch statements such as "if," "for," etc.) with automated and self-maintainable traceability to highlight a module and all of the related modules calling and called by it. When the source code is modified, after rebuilding the database automatically, all the related traceabilities will be automatically updated. The automated and self-maintainable traceability is an important feature to make J-Chart much more useful than the old-established software visualization techniques and tools. Particularly, when a module needs to be modified, the traceability can be used to highlight all the related modules which may also need to be modified to keep consistency.

7.5 J-Diagram

J-Diagram not only can be automatically generated from source code in all levels including the class hierarchy tree, class structure diagram, and the class member function logic diagram with unexecuted class/function/segments/condition outcomes highlighted but also can be automatically linked together for an entire software

7.5 J-Diagram

Fig. 7.4 Another example of the traditional call graph. (http://gilliganscorner.files.wordpress.com/2009/07/healthcare_flow_chart.jpg)

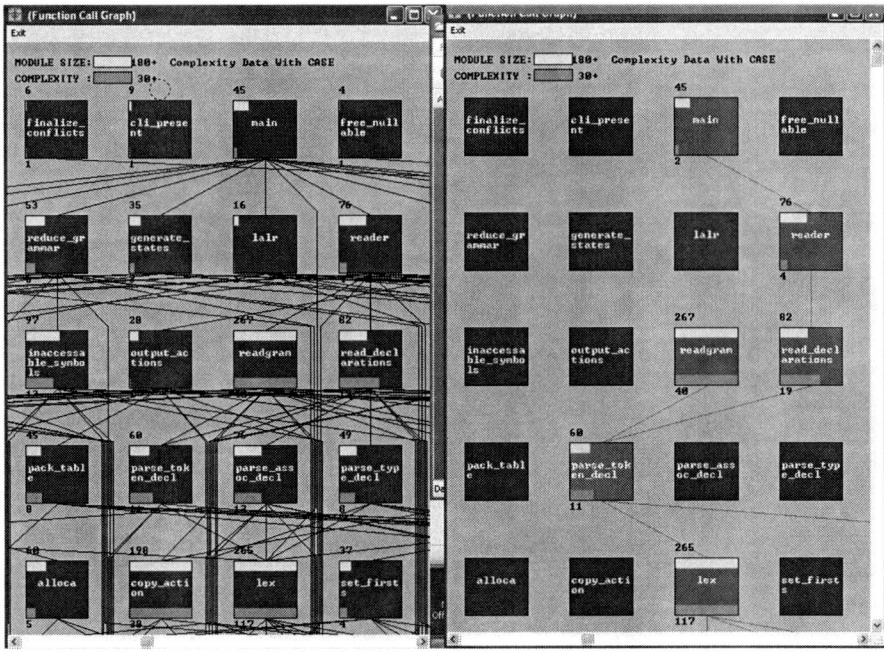

Fig. 7.5 A complex call graph shown with Cyclomatic complexity in J-Chart with the capability to highlight a module with all the modules calling and called by it

The major differences between J-Diagram and most Flow Charts

	J-Diagram	Flow Charts
Is it structured?	Yes	No
Can it show a very complex entire software product?	Yes	No
Is it unique?	Yes	No (arbitrary)
Is the location of the program logic indicated?	Yes	No
Can it show the result of test coverage measurement?	Yes	No
Can it show the branch execution frequency?	Yes	No
Does it offer traceability between related elements?	Yes	No
Can it be converted to a control flow diagram?	Yes	No
Does it exist virtually without huge storage space?	Yes	No

7.5 J-Diagram

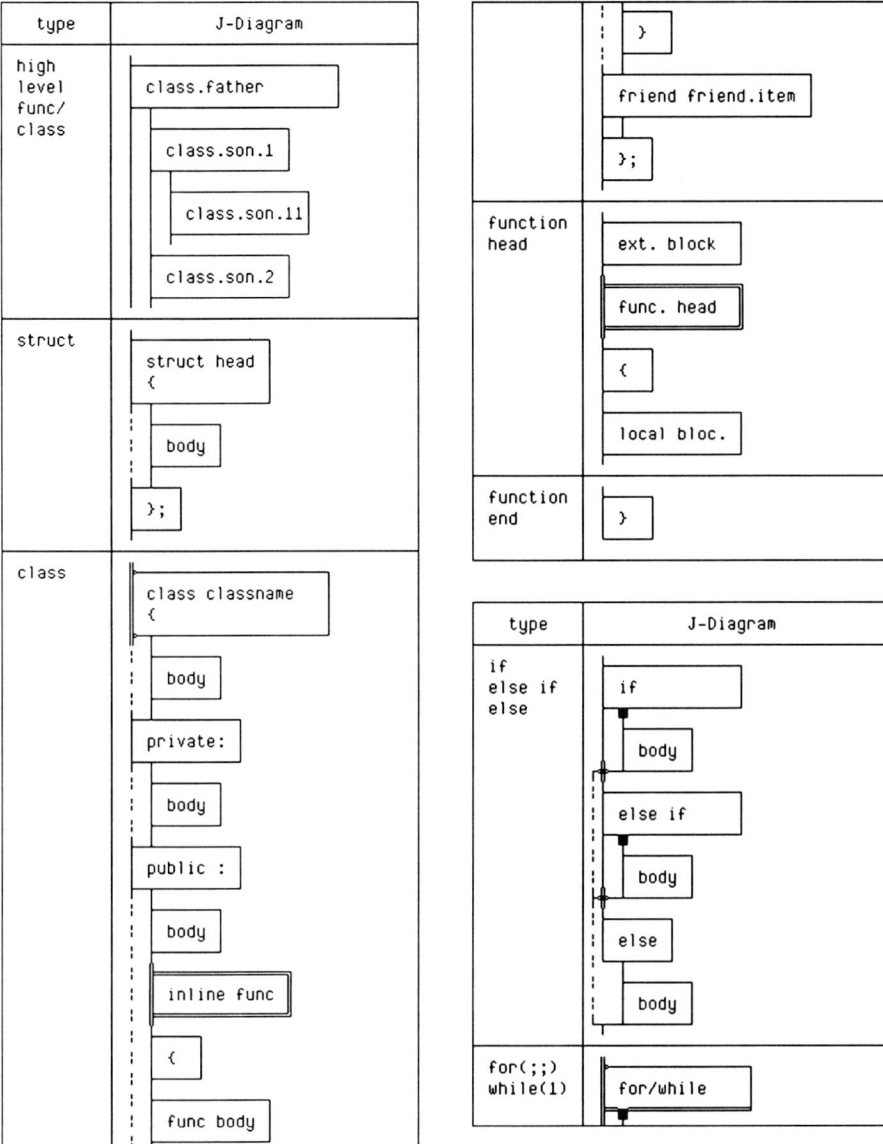

Fig. 7.6 J-Diagram notations

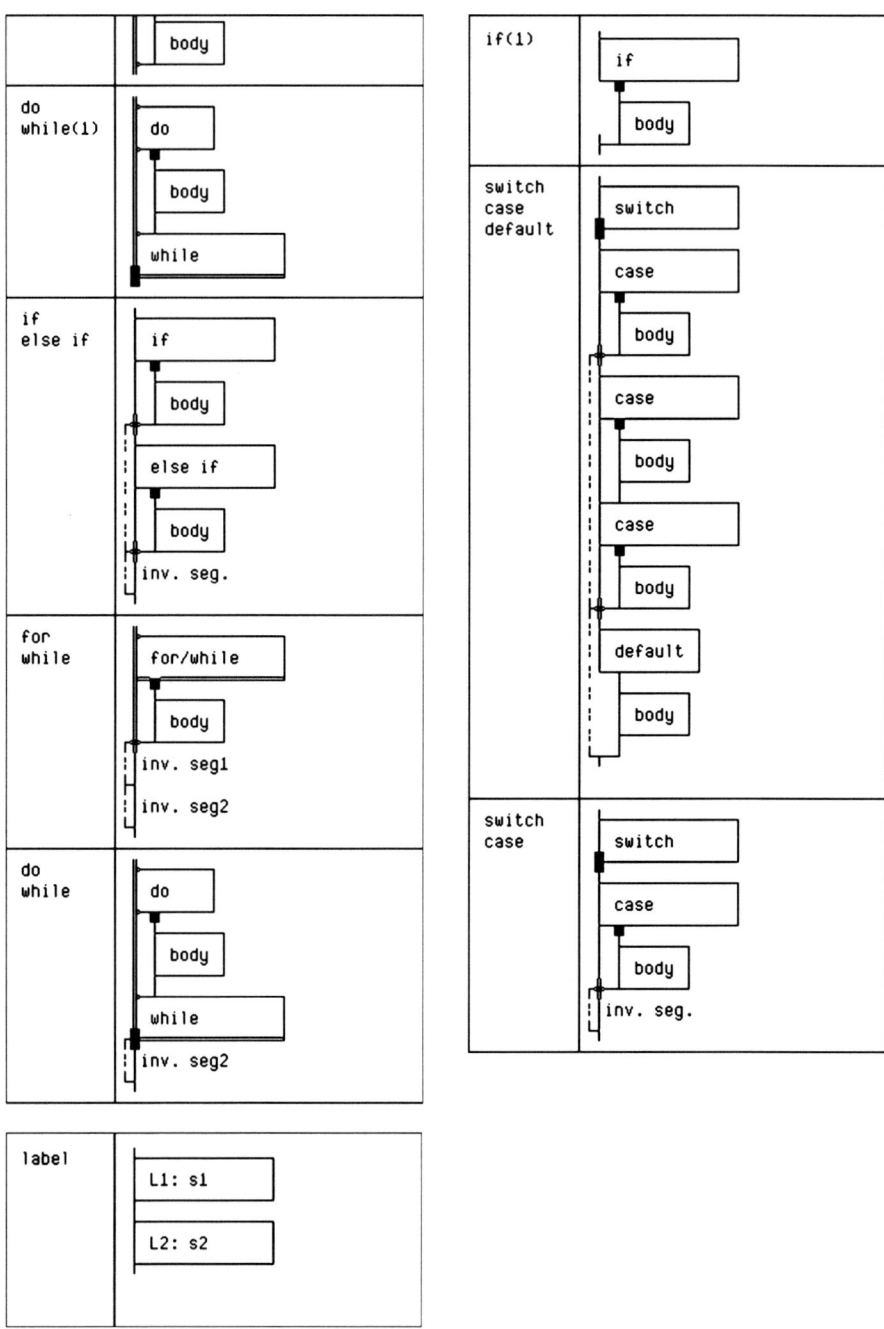

Fig. 7.6 (continued)

7.5 J-Diagram

The major differences between J-Flow and traditional control flow diagram

	J-Flow	Traditional control flow
Is it structured?	Yes	No
Can it show a very complex entire software product?	Yes	No
Is it unique?	Yes	No (arbitrary)
Is the source code locations of the control flow indicated?	Yes	No
Can it show the result of test coverage measurement?	Yes	No
Can it show the branch execution frequency?	Yes	No
Can it be automatically converted to a logic diagram?	Yes	No
Can it highlight a path with most untested elements?	Yes	No
Does it exist virtually without huge storage space?	Yes	No

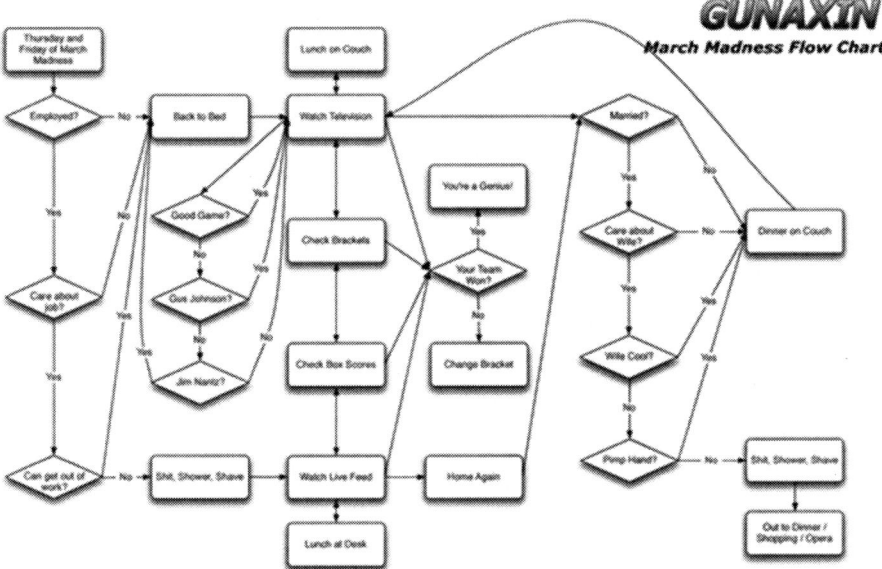

Fig. 7.7 A sample Flow Chart. (http://images.dailyradar.com/media/uploads/ballhype/story_large/2009/03/19/march_madness_flow_chart.png)

product to make the diagrammed code traceable in all levels. J-Diagram can be automatically converted into J-Flow diagram. J-Diagram is particularly useful in Object-Oriented software understanding, inspections, walkthroughs, and testing.

J-Diagram notations are shown in Fig. 7.6.

For easy comparison, two flow charts are shown in Figs. 7.7 and 7.8 separately.

Two sample J-Diagrams are shown in Figs. 7.9 and 7.10.

Another application example of J-Diagram representing a complex source module is shown in Fig. 7.11.

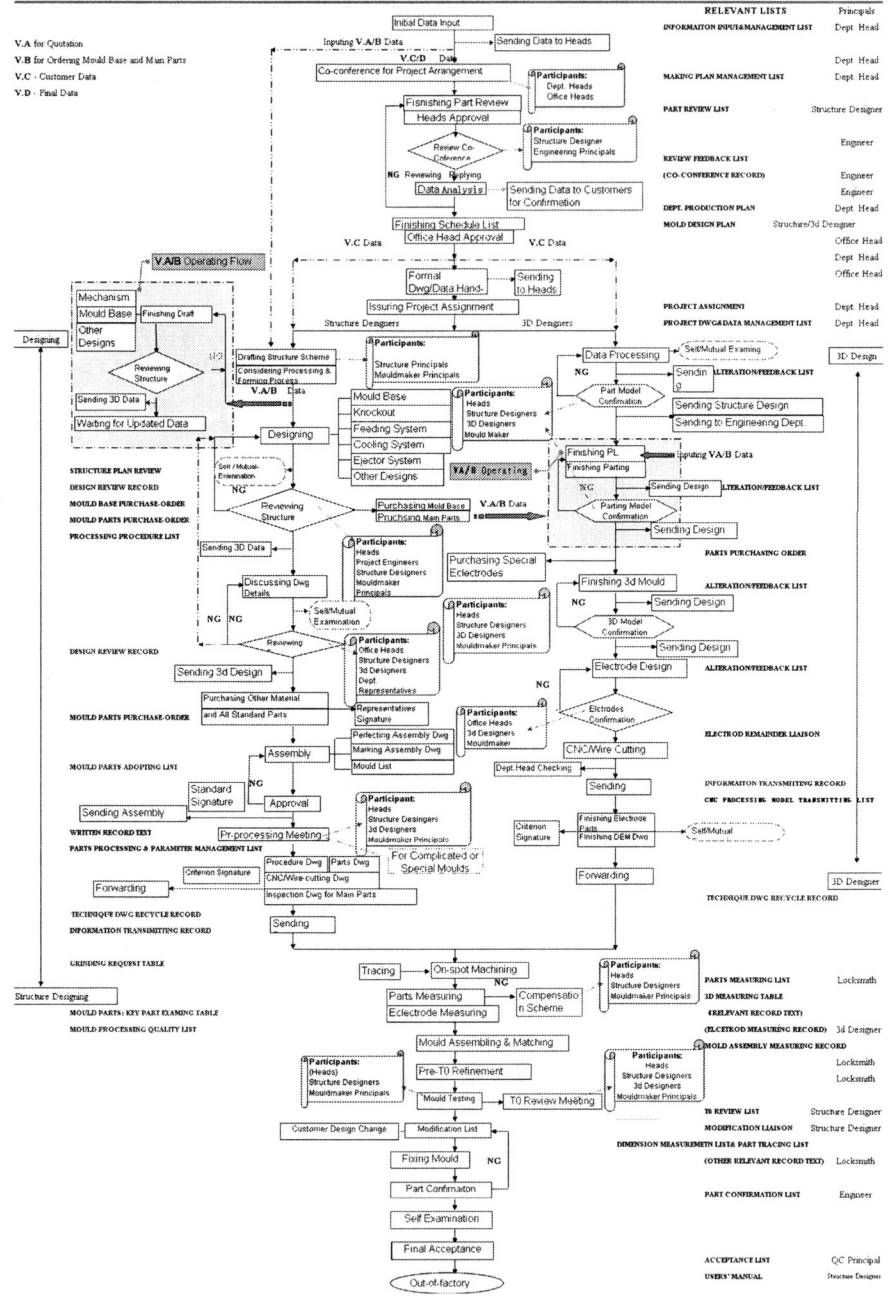

Fig. 7.8 Another sample Flow Chart. (http://www.mdmould.com/images/flow%20chart-english.jpg)

7.5 J-Diagram

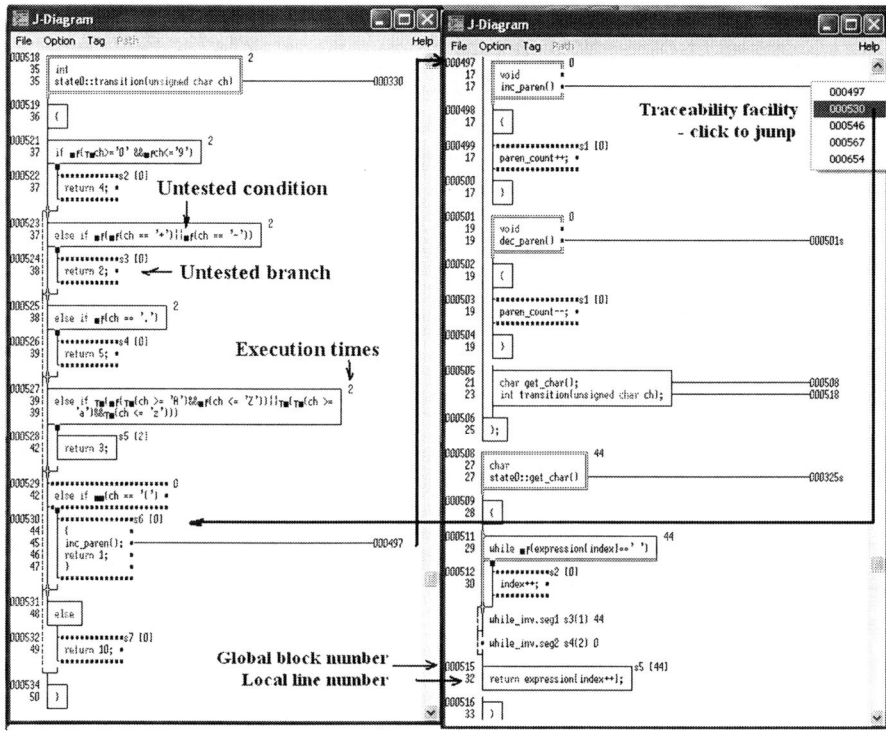

Fig. 7.9 A J-Diagram shown with the detailed information

148   7 NSE Software Engineering Visualization Paradigm

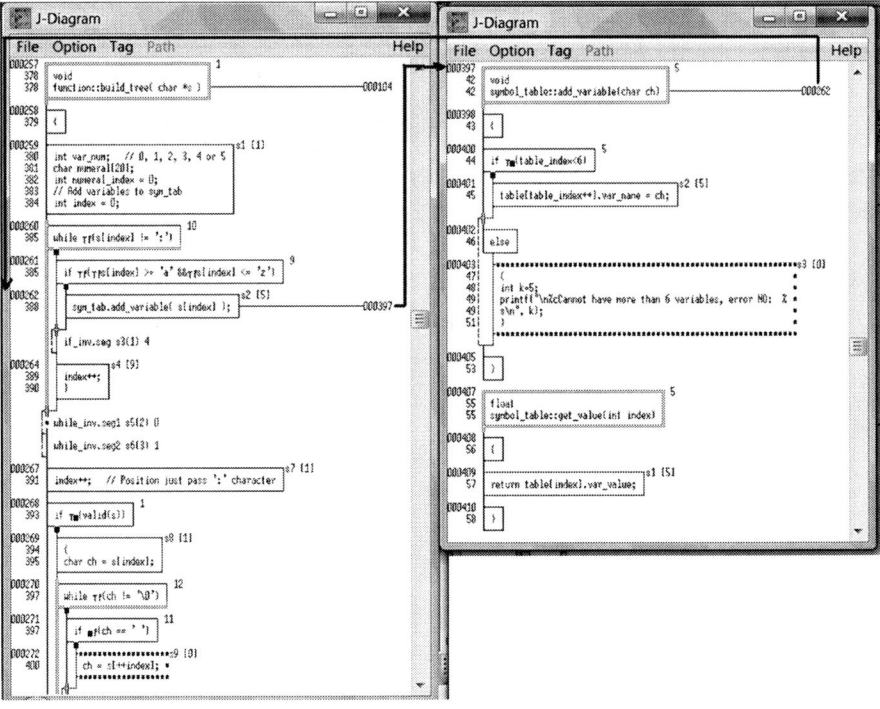

Fig. 7.10 A sample J-Diagram shown with the traceability

Interactive and traceable J-Diagram not only makes a software product much easier to read, understand, test, and maintain but also makes the code inspection and walk through much easier to perform in a semiautomated way.

7.6 J-Flow

Most traditional control flow diagrams are unstructured. They often use the same notation to represent different program logic and cannot display the logic conditions and the source code locations. The J-Flow diagram, on the other hand, is Object-Oriented and structured, uses different notations to represent different logic with the capability to show logic execution conditions and the corresponding source code locations. J-Flow is particularly useful in logic debugging, path analysis, test case and code correspondence analysis, and class/function-level test coverage result display with unexecuted elements (path, segments, and unexecuted condition outcomes) highlighted.

The notations of J-Flow diagram are shown in Fig. 7.12.

7.6 J-Flow

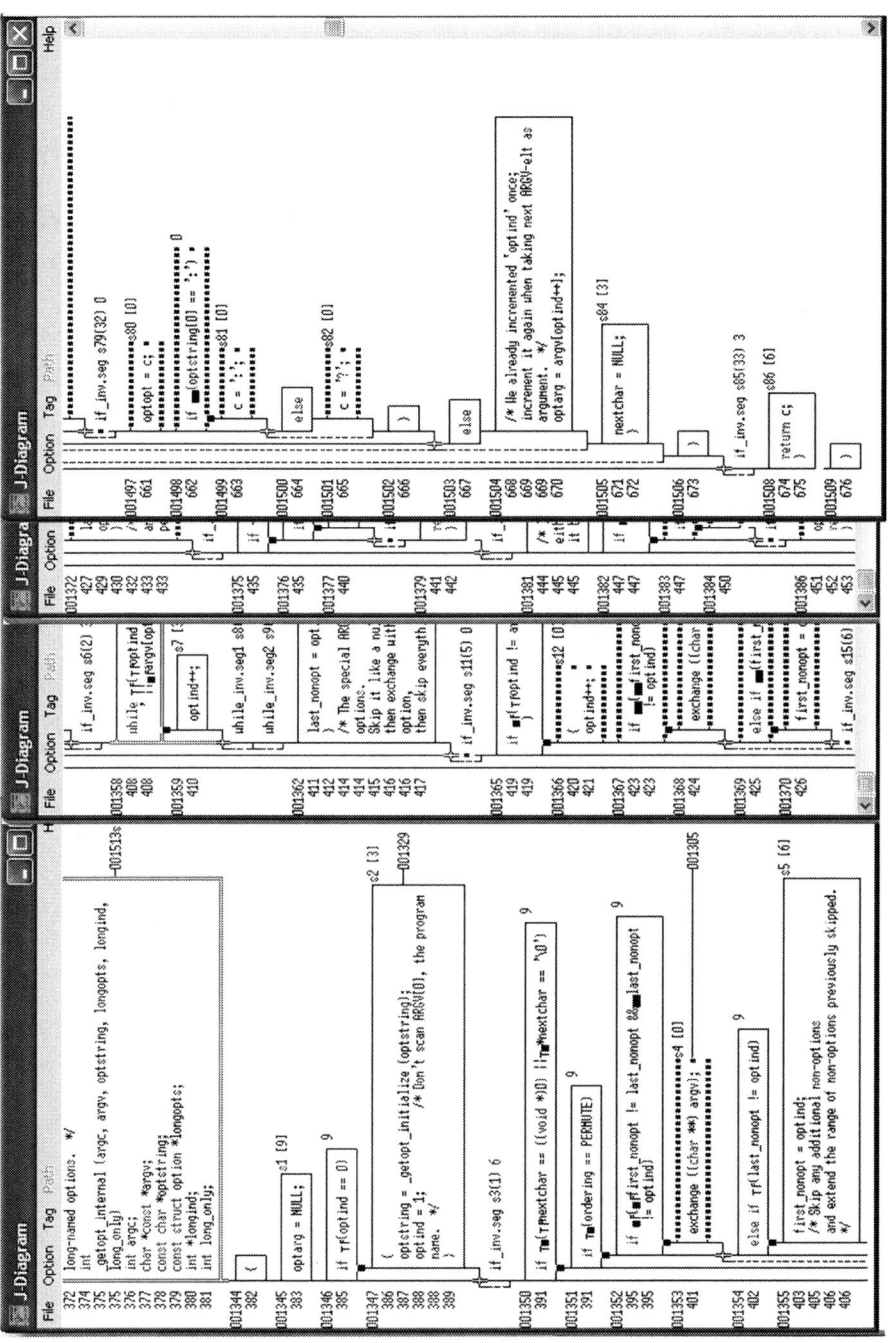

Fig. 7.11 The logic diagram of a complex program module shown with MC/DC test coverage analysis result

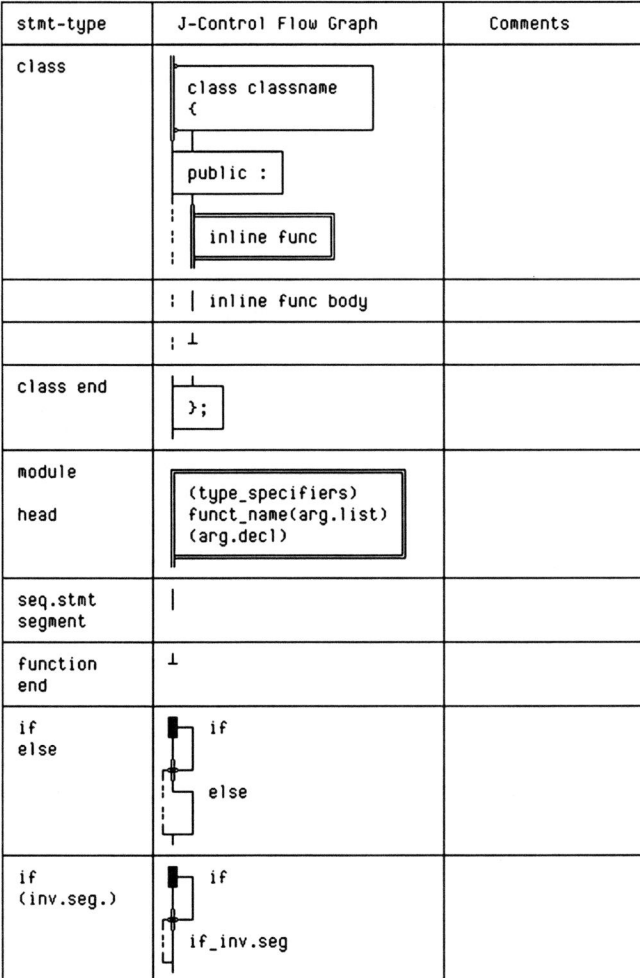

Fig. 7.12 J-Flow notations

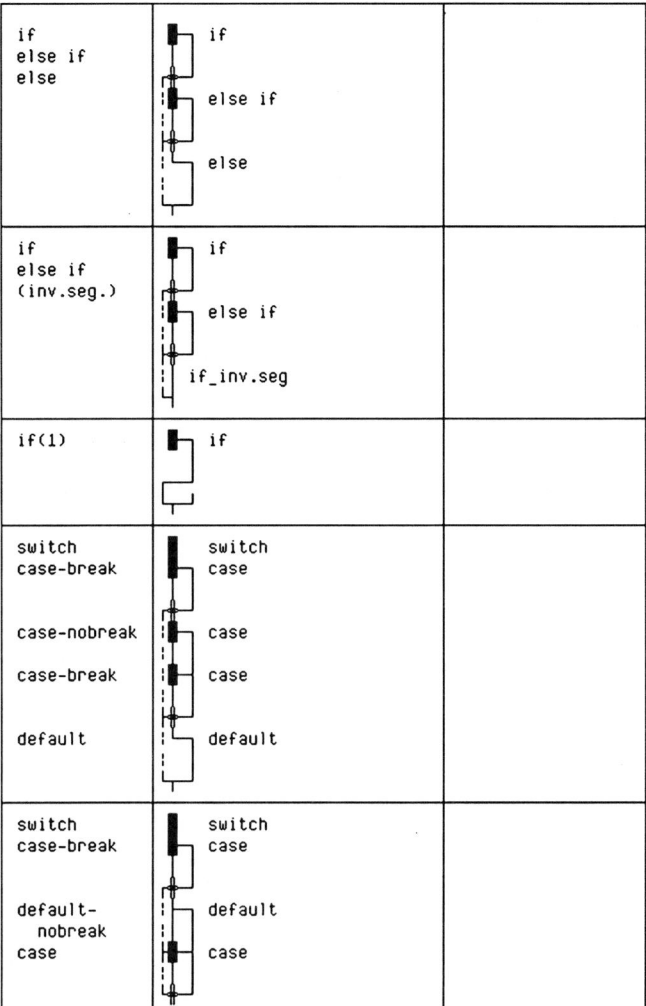

Fig. 7.12 (continued)

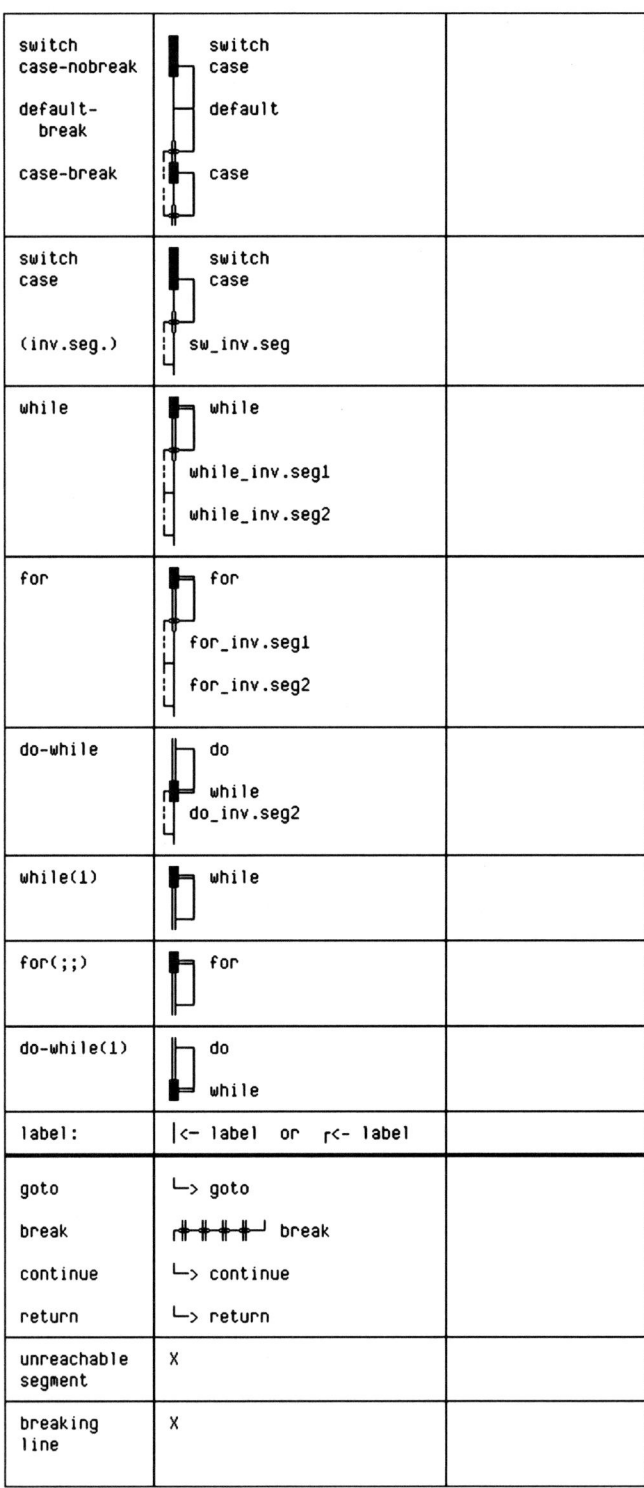

Fig. 7.12 (continued)

7.7 Entire Software Life Cycle Visualization with NSE

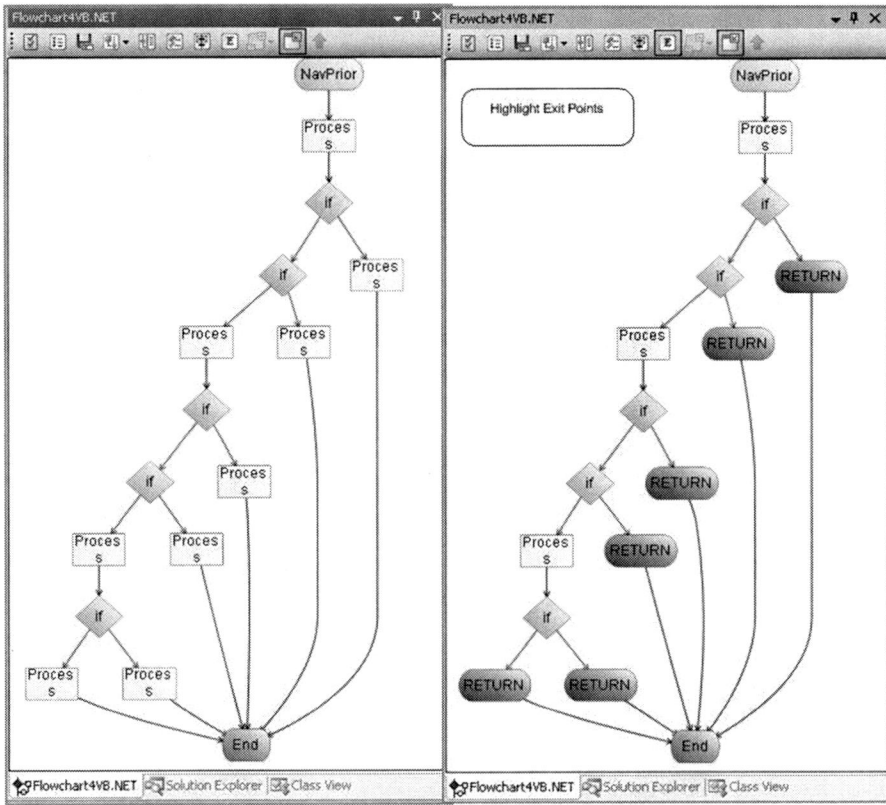

Fig. 7.13 Sample traditional control flow diagrams (http://www.codeswat.com/cswat/fc4vb/images/highexit.jpg)

For easy comparison, some traditional control flow diagrams are shown in Figs. 7.13 and 7.14.

Two sample J-Flow diagrams are shown in Figs. 7.15 and 7.16.

Another application example representing the control flow of a complex program model is shown in Fig. 7.17.

Interactive and traceable are the important features of the J-Flow diagram that are particularly useful for software testing.

7.7 Entire Software Life Cycle Visualization with NSE

With NSE the entire development process of a software product is visible from the first step in the software requirement development phase down to the final step in the software maintenance phase as shown in Fig. 7.18.

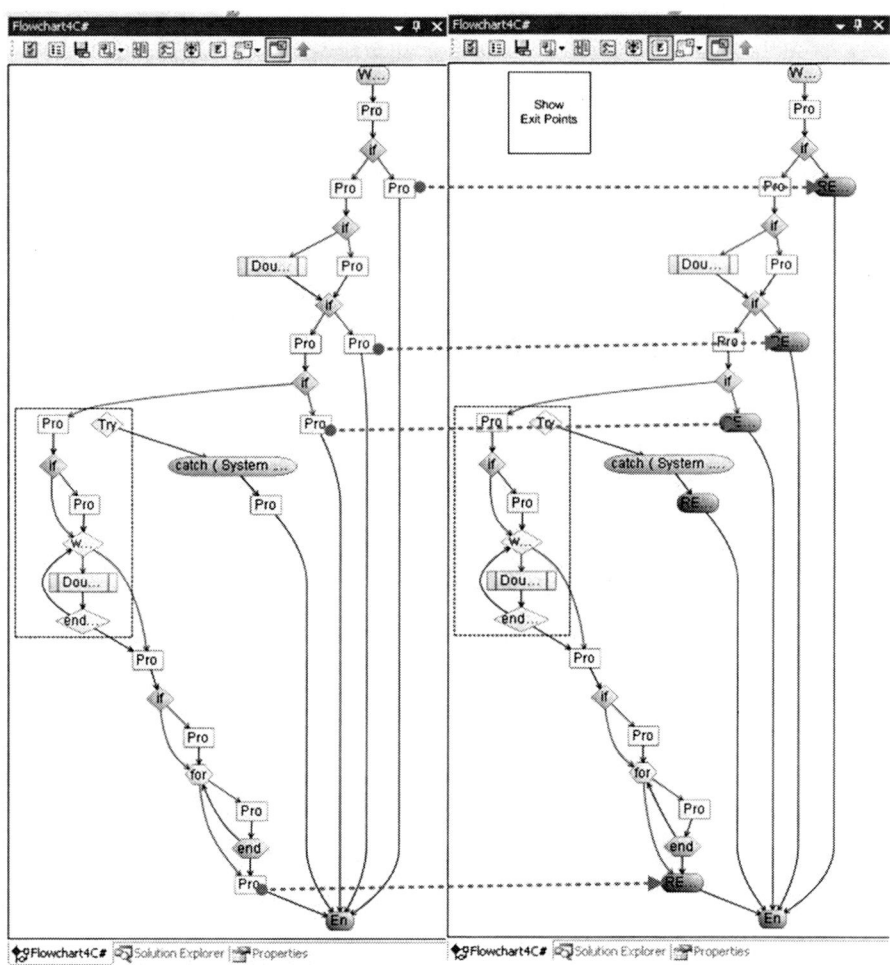

Fig. 7.14 Another traditional control flow diagram. (http://www.codeswat.com/cswat/fc4cs/images/highexit.jpg)

7.8 Rich Options for Generating 3J Graphics   155

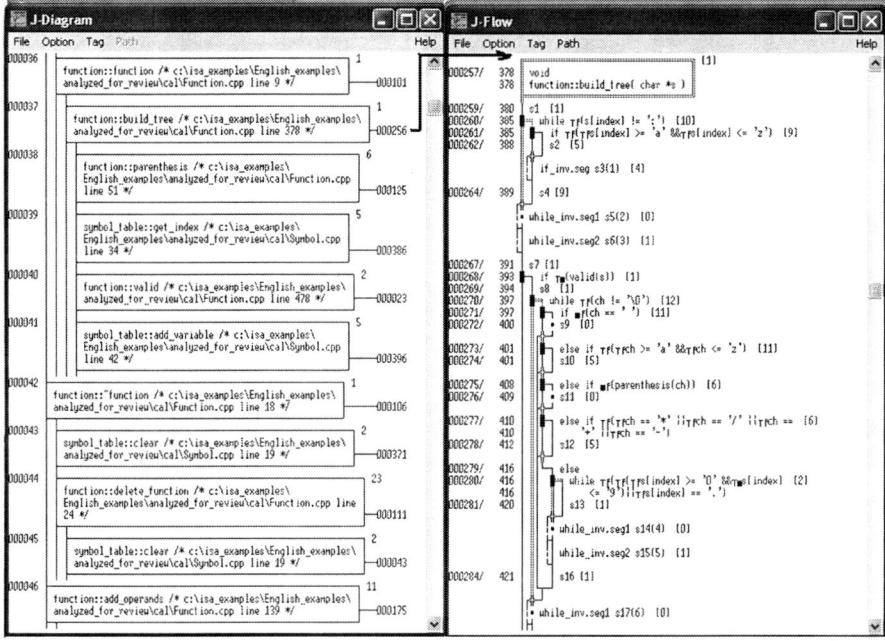

Fig. 7.15 A sample J-Flow diagram shown with the program tree and the test coverage analysis result

The charts/diagrams are generated in two ways: (a) in the requirement development phase and the design phase through dummy programming using dummy modules, each one of which may have an empty body or only some function call statements without detailed logic – so that the dummy programs are very easy to write for any programmer without extra training; (b) in lower software development phases, the charts/diagrams are generated from the source code through forward or reverse engineering.

7.8 Rich Options for Generating 3J Graphics

There are rich options for generating 3J graphics.

7.8.1 For J-Chart Generation

The interface of a J-Chart generator (Panorama++ OO-Browser) is shown in Fig. 7.19 with options for selecting the type of J-Chart and the related information to be shown together such as the Cyclomatic complexity with or without counting

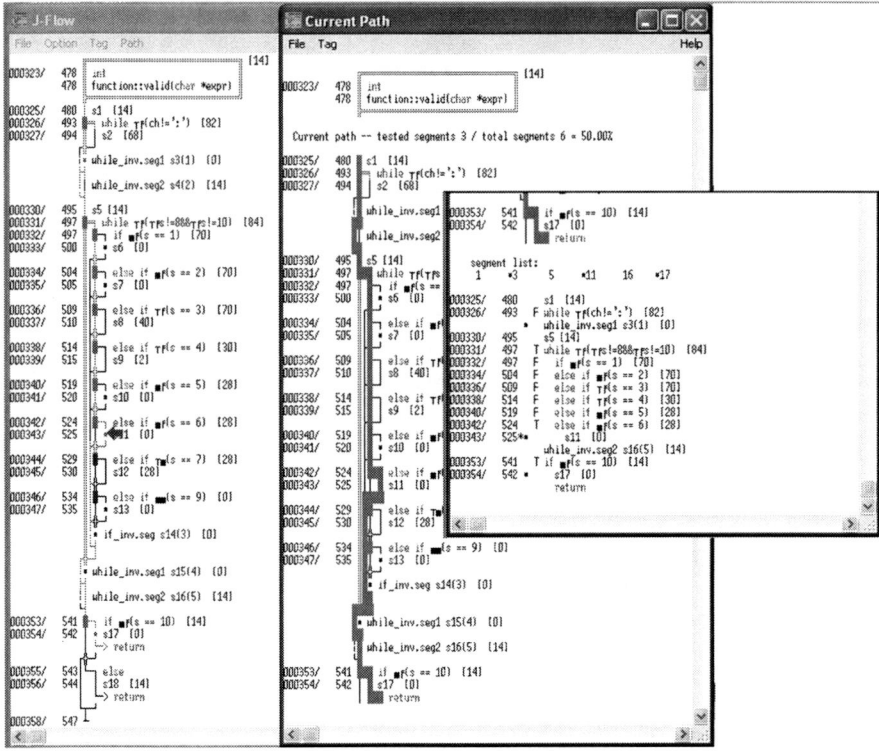

Fig. 7.16 A sample J-Flow diagram showing a special path with the most untested elements being highlighted and its execution conditions being extracted for semiautomated test case design

the "case" statement, the accumulated or last-run test coverage measurement result, the incremental coding/unit-testing order, etc.

Within a J-Chart, there is a detailed menu to provide more options for users to select as shown in Fig. 7.20.

From each module box, there is a pull-down menu for choosing the related operations as shown in Fig. 7.21.

Figure 7.22 shows the chart printing options – in general, there is no hard copy of a J-Chart being stored in the hard disk or the computer memory, because it virtually exists for greatly saving space, unless users want to save it or print it out for documentation, Web page design, or project presentation without using the NSE support platforms.

A sample output of a J-Chart showing the call graph of the cal example with test coverage measurement result is shown in Fig. 7.23 (where the size of the original output cal.ps in postscript format is about 19.9 KB, the size of the cps.pdf transferred by Adobe tool is about 13.1 KB with two pages – a big software may consists of hundred or more pages for making the output visible, the size of Fig. 7.23 after merging the two pages to one TIFF file is about 99.9 KB).

7.8 Rich Options for Generating 3J Graphics 157

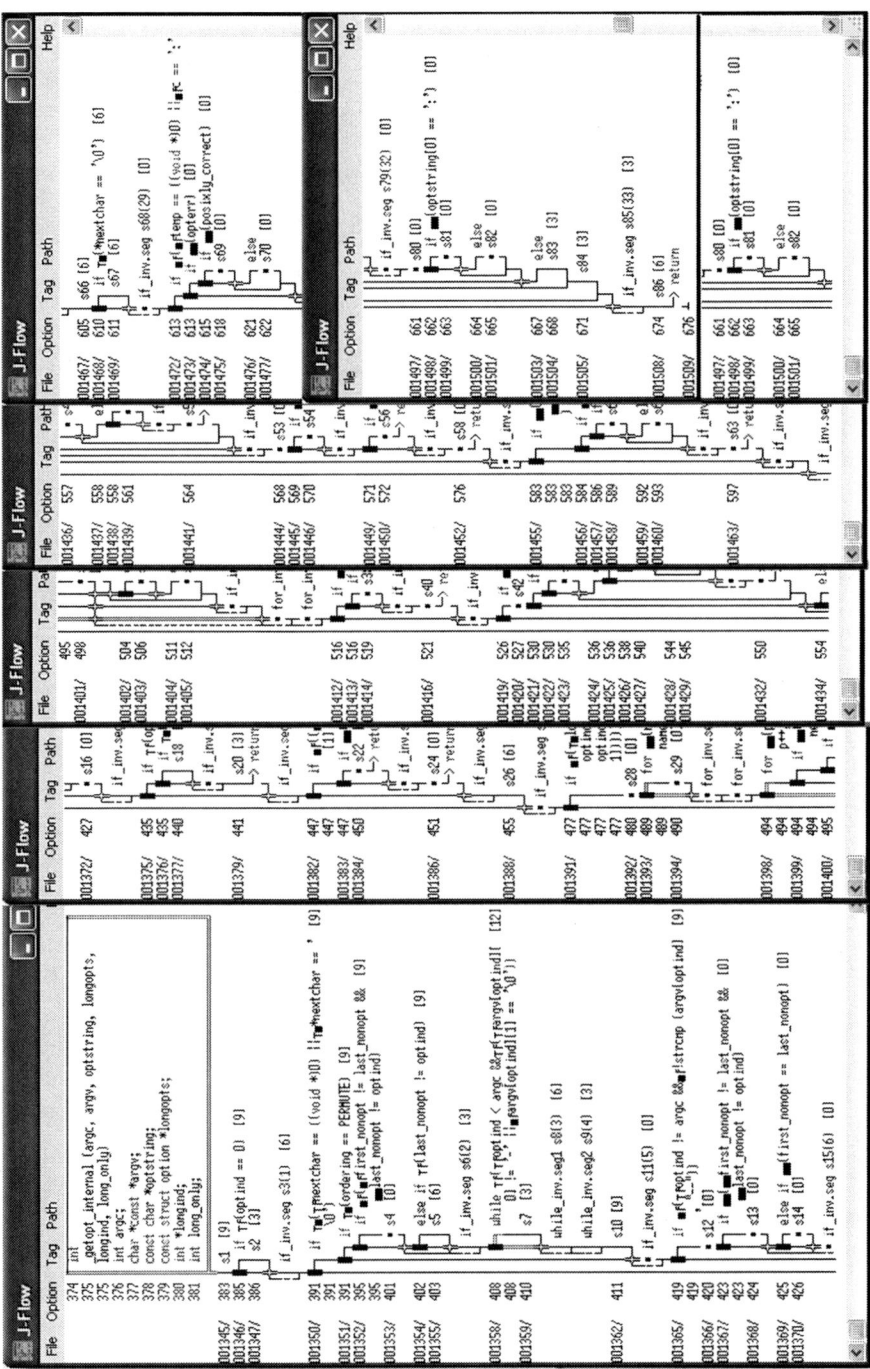

Fig. 7.17 The control flow of a complex program module shown in J-Flow diagram

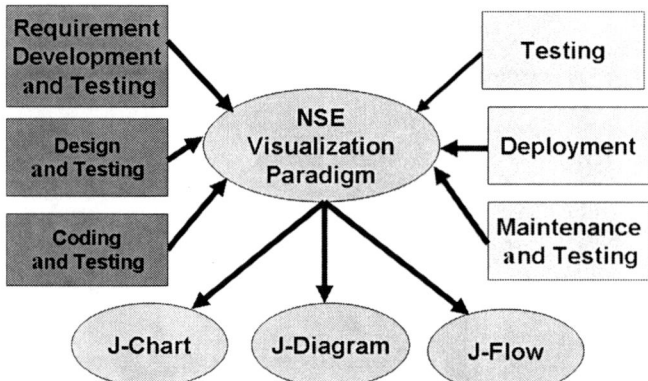

Fig. 7.18 Entire Software Life Cycle Visualization with NSE

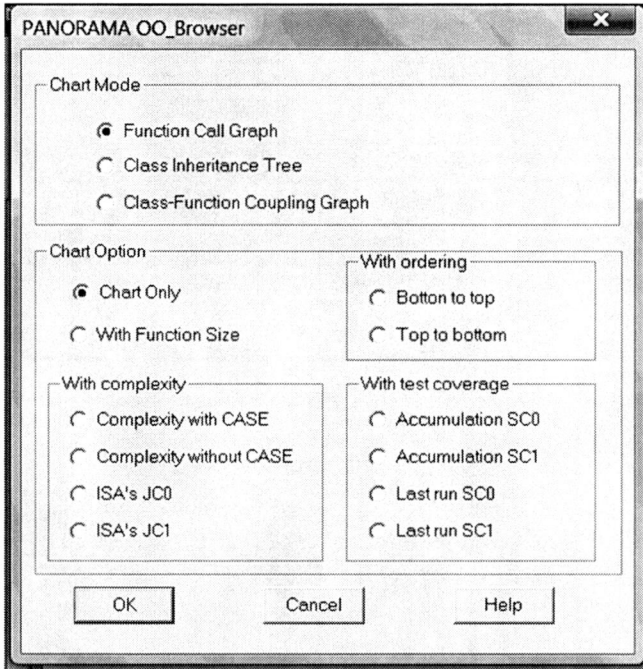

Fig. 7.19 The interface of J-Chart generator

7.8 Rich Options for Generating 3J Graphics

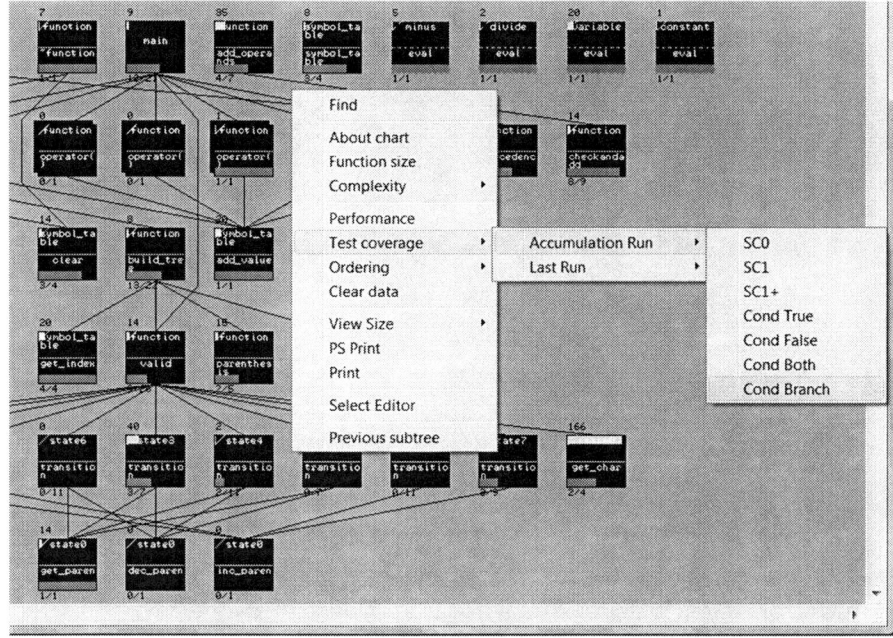

Fig. 7.20 Rich options for J-Chart to show the related information

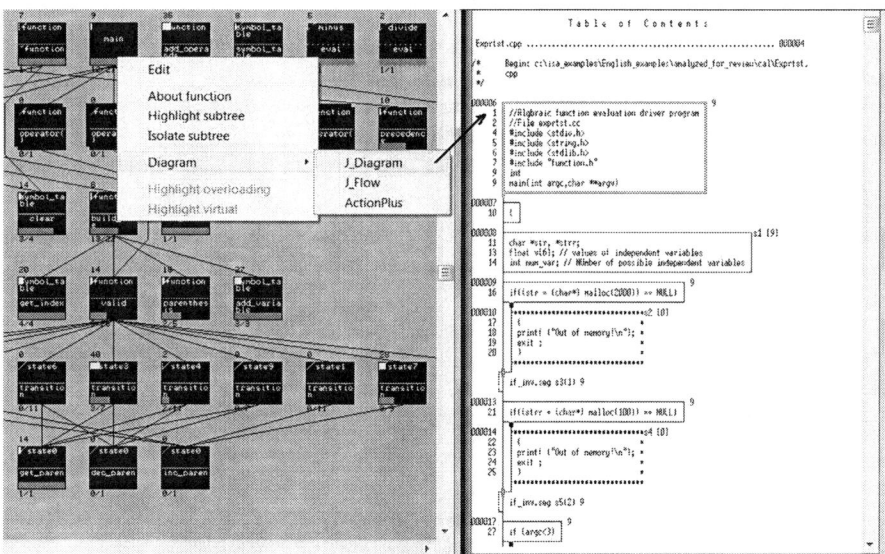

Fig. 7.21 A pull-down menu with each module box and the usability

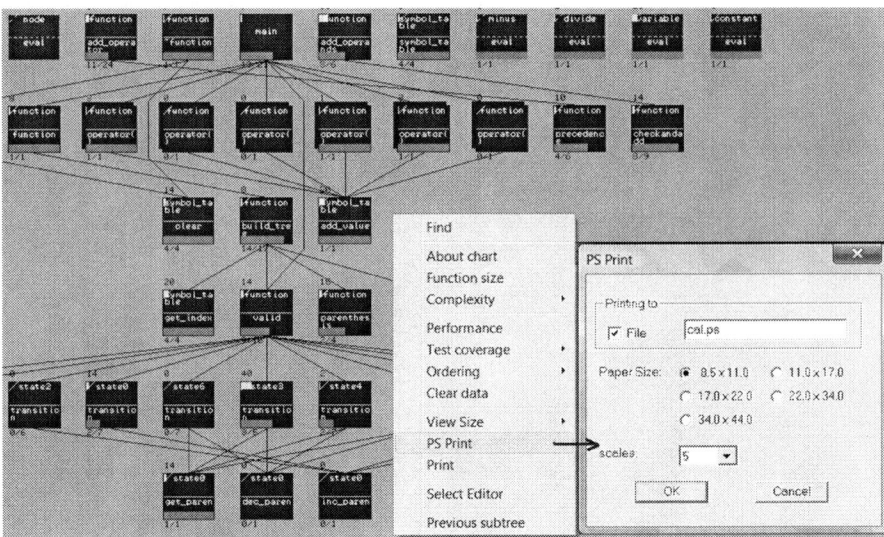

Fig. 7.22 The chart printing options

7.8.2 For J-Diagram and J-Flow Generation

The interface of a J-Diagram and J-Flow generator (Panorama++ OO-Diagrammer) is shown in Fig. 7.24 with options for selecting the type of diagram and the related information to be shown together such as the accumulated or last-run test coverage measurement result, the holistic program tree for the entire software product (function cross references), class cross references, system-level and module-level test coverage summary, Cyclomatic complexity summary, etc.

There are more pull-down menus for selecting related information to show with the generated diagrams (see Fig. 7.25).

Figure 7.26 shows the "file" part for object search and diagram printing options (users may select to print the entire diagram or only a part of the diagram highlighted by the users).

Figure 7.27 shows the associated click-to-jump facility.

Figure 7.28 shows the associated facility for manually setting the locations for jumping.

Figure 7.29 shows the associated facility for semiautomatic test case design.

7.9 The Major Features of NSE Software Visualization Paradigm

The major features of 3J graphics and NSE software visualization paradigm (which not only can generate 3J graphics but also can generate other software graphics such as bar charts and ActionPlus diagrams) include

7.9 The Major Features of NSE Software Visualization Paradigm 161

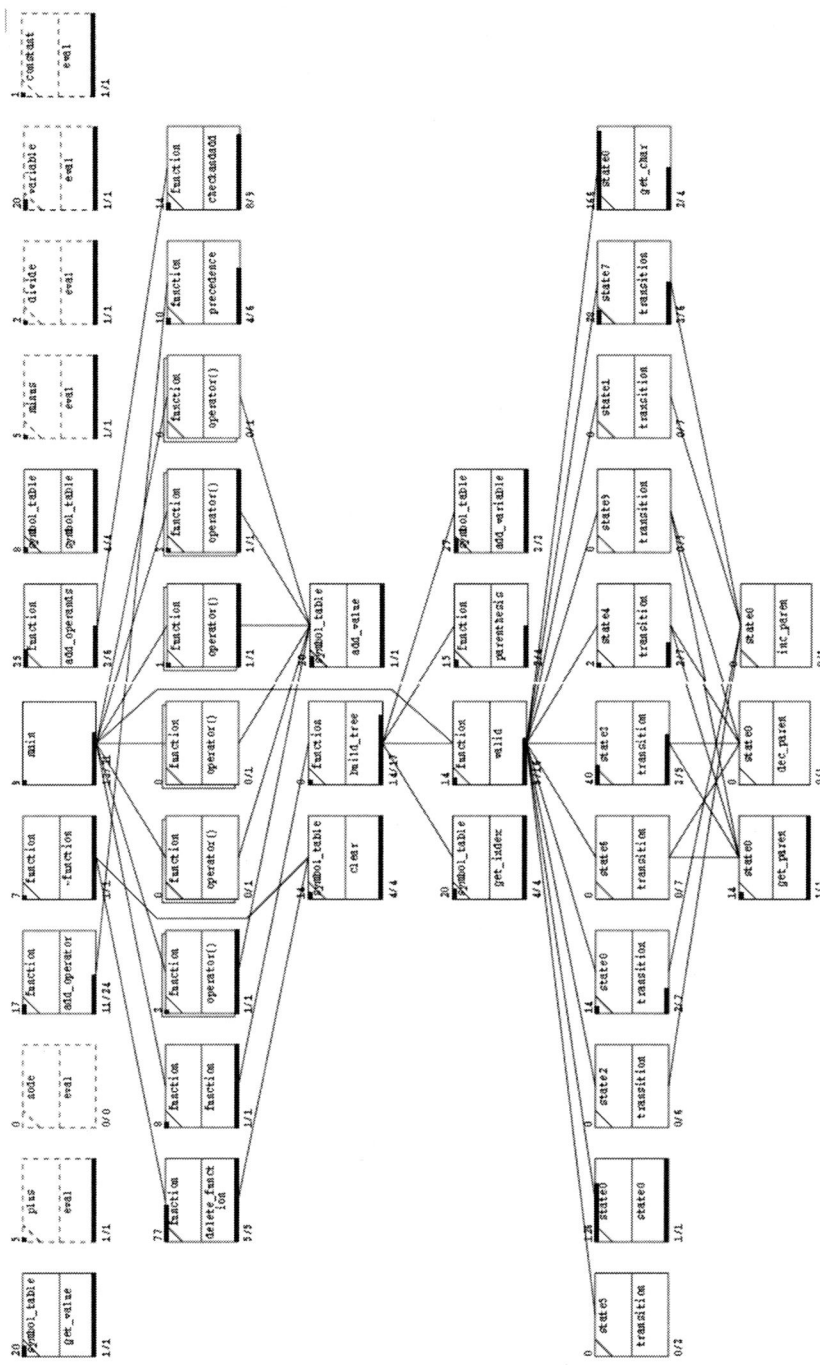

Fig. 7.23 A sample output of J-Chart showing the call graph of the cal example with test coverage data

1. **Based on nonlinear thinking and complexity science**
   The NSE software visualization paradigm complies with the essential principles of complexity science, particularly the nonlinearity principle and the holism principle.
2. **Holistic**
   The NSE software visualization paradigm generates entire charts/diagrams of a software product to show both the overview of the structure of the product and the detailed logic or control flow for an entire product and each file/class/function, including

   (a) The function call graph of the entire software system
   (b) The class inheritance chart of the entire software system
   (c) The class and independent function relation chart of the entire software system
   (d) The program tree of the entire software system
   (e) The overall MC/DC test coverage measurement result of the entire software system
   (f) The overall quality measurement result shown in Kiviat diagram
   (g) The overall performance measurement result of the entire software system
   (h) The overall Cyclomatic complexity measurement result of the entire software system
   (i) The logic diagram of the entire software system
   (j) The control flow diagram of the entire software system
   (k) The overall version comparison result shown in J-Chart with unchanged modules shown in blue, changed modules in red, deleted modules in brown, and added modules in green (originally on screen in color and not black and white).

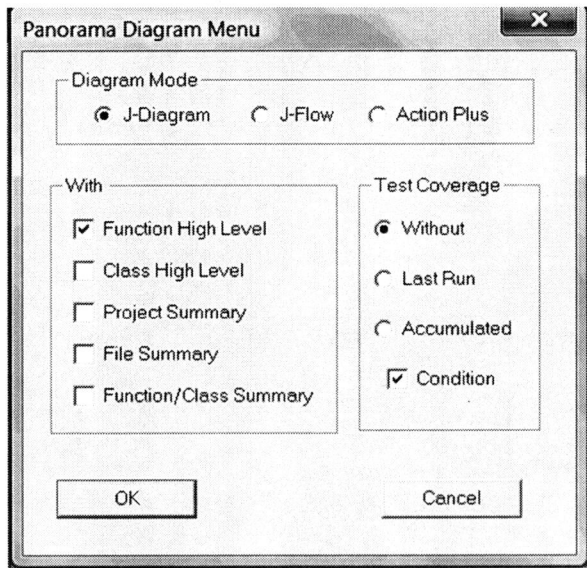

Fig. 7.24 The interface of a J-Diagram generator

7.9 The Major Features of NSE Software Visualization Paradigm 163

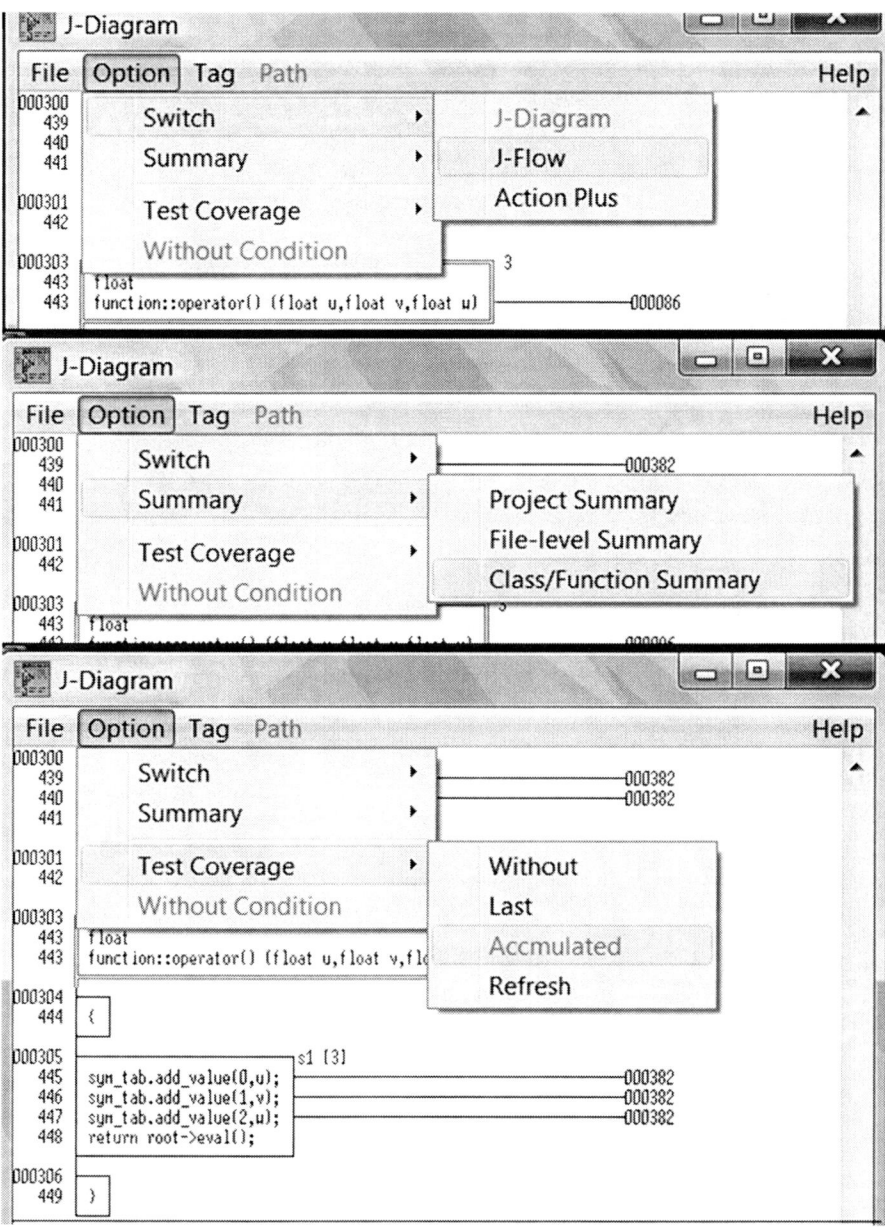

Fig. 7.25 More options to be chosen for showing the related information with the diagram

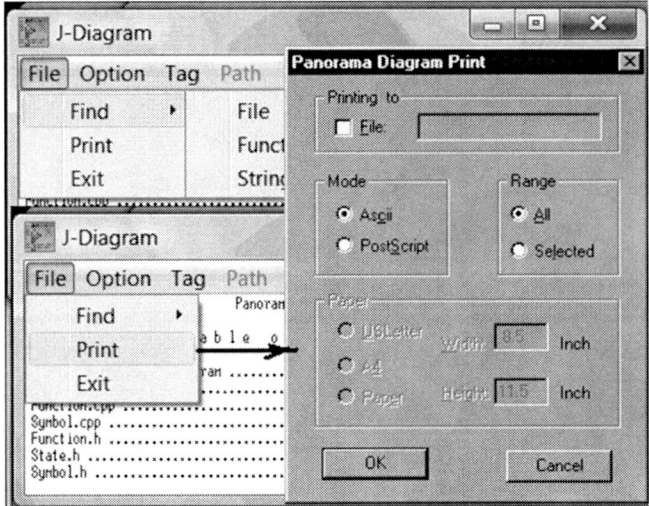

Fig. 7.26 Options for object search and printing

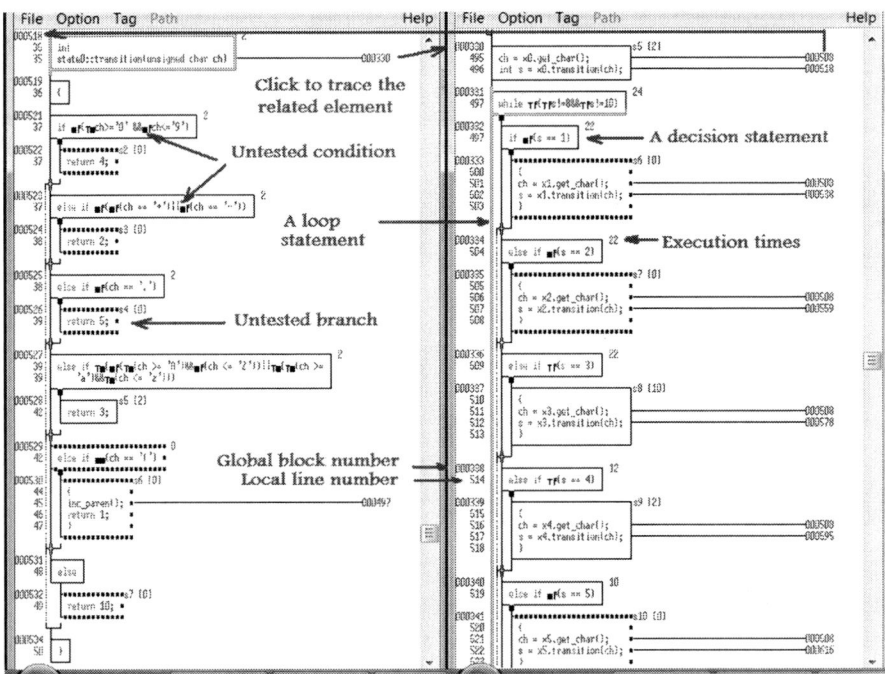

Fig. 7.27 Associated click-to-jump facility

7.9 The Major Features of NSE Software Visualization Paradigm

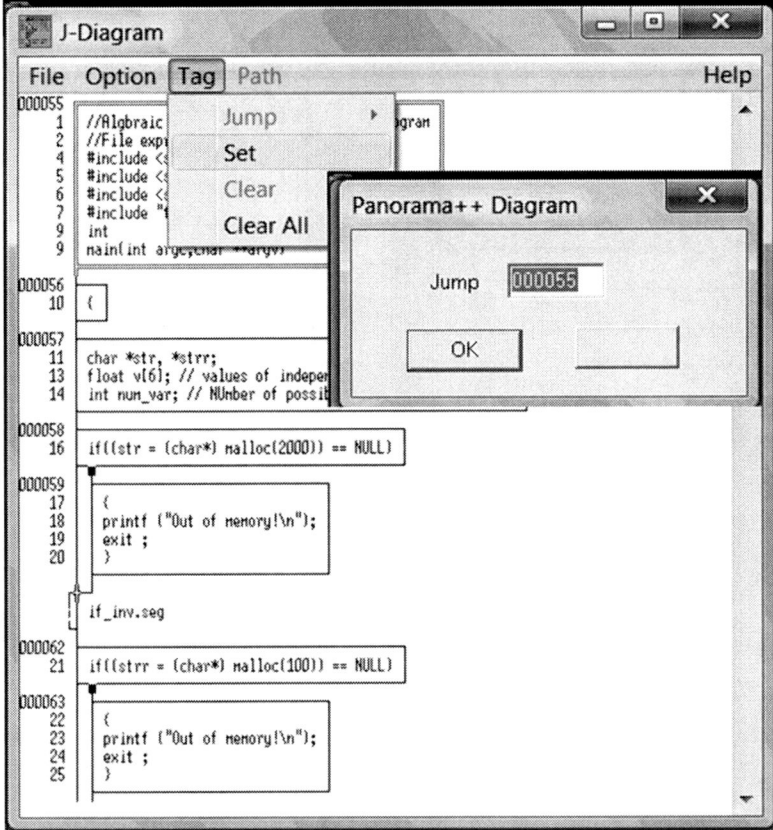

Fig. 7.28 Associated facility for manually setting the locations for jumping

3. **Automatic**

   With the NSE software visualization paradigm all system-level, file-level, and module-level charts and diagrams are generated automatically from the dummy programming source code (using dummy modules with empty bodies or only some function call statements) or the regular program source code.

4. **Structured without size limitation**

   All the 3J graphics generated are structured without size limitation, and can be used to graphically represent very big software products.

5. **Easy to update**

   After the dummy programs or the regular programs are modified, the NSE visualization paradigm will rebuild the database to automatically update the 3J graphics.

6. **Shown with detailed information**

   The generated 3J graphics can be shown with detailed information such as the code test coverage analysis result.

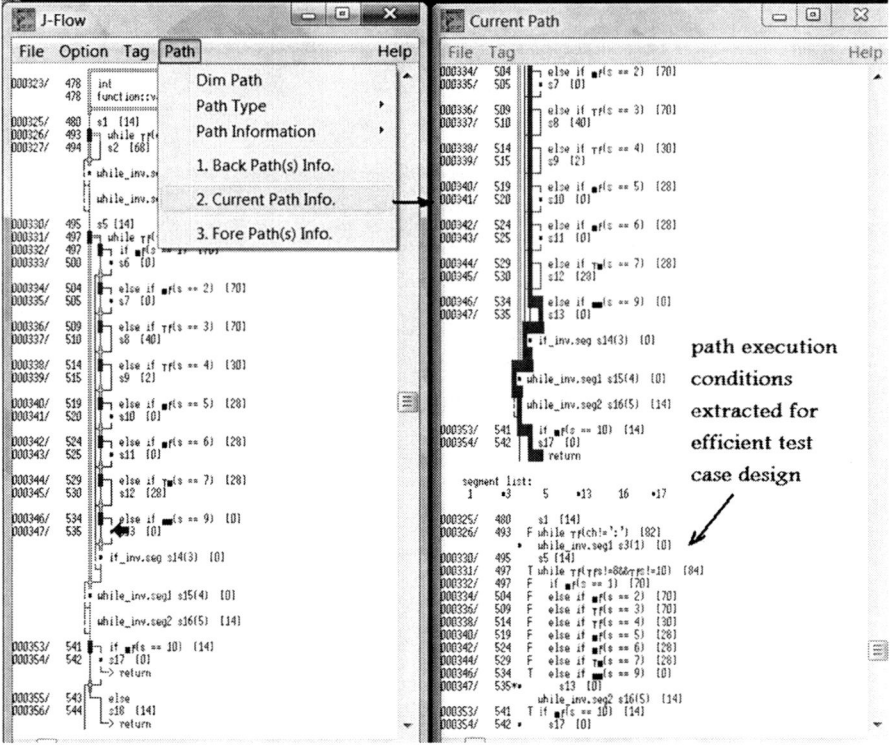

Fig. 7.29 The associated facility for semiautomatic test case design

7. **Independent from the source code writing style**
   Let us consider the following two different writing styles a and style b:
   The corresponding J-Diagrams show the same logic for writing style a and writing style b (see Fig. 7.30).
8. **Interactive**
   The generated charts and diagrams are interactive – the charts and diagrams themselves become the interfaces to accept users' commands. Figure 7.31 shows how a user can select a module as the new root to get the sub call graph.
9. **Traceable**
   The generated charts and diagrams are traceable from a module to trace the related modules calling and called by it, or from a module box to trace the detailed logic diagram or control flow diagram, or from a function call statement to the called function body, or from a class to the base classes, or from a #include statement to the included source file, etc., to support semiautomated software inspection and review. An application example is shown in Fig. 7.32.

7.9 The Major Features of NSE Software Visualization Paradigm

```
Style a:

 void function::add_operator( node *p, node *n, char ch )
 {
    if(n == 0)
    {
       switch(ch)
              {
              case '+':
                     root = new plus;
                     root->info = '+';
              break;
              case '*' :
 ...

 }
```

Style b:
void function::add_operator(node *p, node *n, char ch)
{
   if(n == 0){switch(ch){case '+':root = new plus;root->info = '+'; break;
   case '*' :
   ...
}

10. **Accurate – consistent with the source code**
    "To keep documentation maintained, it is crucial that it be incorporated in the source program, rather than kept as a separate document" [Bro95-p249]. The generated charts and diagrams are accurate to the source code – when the source code is modified, all charts and diagrams can be automatically updated after rebuilding the database automatically.
11. **Precise**
    The NSE software visualization paradigm generates charts and diagrams precisely, such that when a logic diagram is used to show the result of test coverage measurement of the source code, the generated logic diagram can highlight each untested branch and each untested condition combination precisely as shown in Fig. 7.33.
12. **Consistent among all related charts and diagrams**
    All charts and diagrams are generated from the same simple databases with several Hash tables only to keep consistency among them even if the formats are different.
13. **Linkable automatically**
    A module box in a generated call graph or a node of the generated program tree can be automatically linked to the detailed logic diagram or control flow diagram.

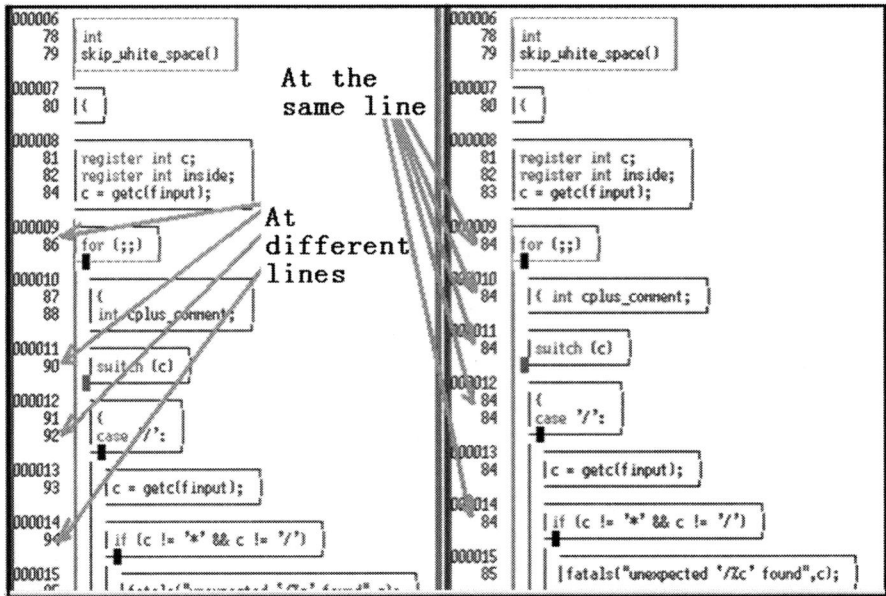

Fig. 7.30 Source code writing style-independent logic diagramming

An application example linking a module box in a call graph to the logic diagram is shown in Fig. 7.34 with the test coverage analysis result.

14. **Convertible** – See Fig. 7.35 to convert a J-Diagram to a J-Flow or ActionPlus diagram.
15. **Useful for software visualization and the support of incremental software development** – see Fig. 7.36.
16. **Virtual**
    The holistic charts and diagrams are generated dynamically from the database and shown within a Window no more or less, when a chart or diagram is needed to move, a new one will be regenerated dynamically without really moving the chart or diagram, so that the required time for tracing a diagram from block 10 to block 10,000 will be the same as that for tracing it from block 10 to block 20, but the user will still feel that the entire chart/diagram exists. In fact, with NSE there is no real holistic chart or diagram stored in the memory or the hard disk – they are dynamically generated virtually to greatly reduce the required space (only needing about 1/100 of the space required by a traditional approach with hard copies stored in memory or hard disk), and speed up the graphic display in about 1,000 times faster compared with the old version of the Panorama product with which hard copies of the charts or diagrams are stored in a disk or the memory of a computer. As shown in

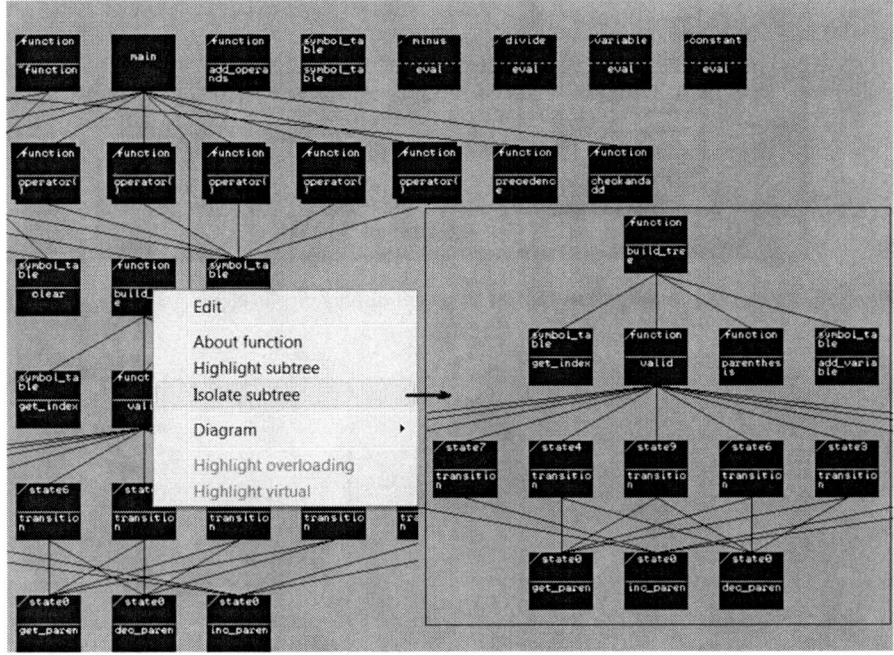

Fig. 7.31 An interaction example for getting a sub call graph

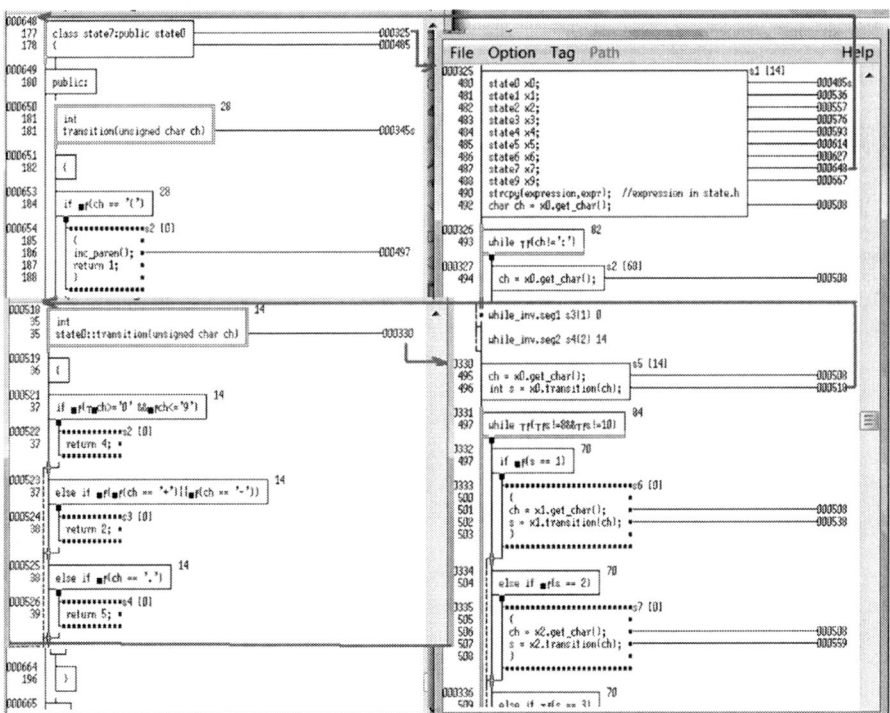

Fig. 7.32 Various traceabilities established with J-Diagram for semiautomated inspection

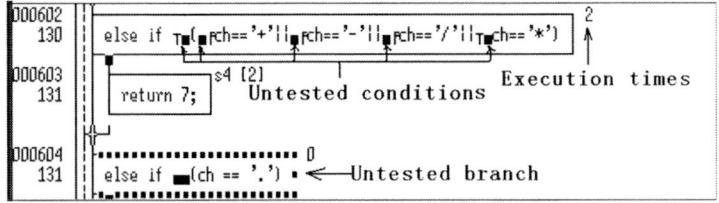

Fig. 7.33 Precise test coverage analysis and the result display graphically

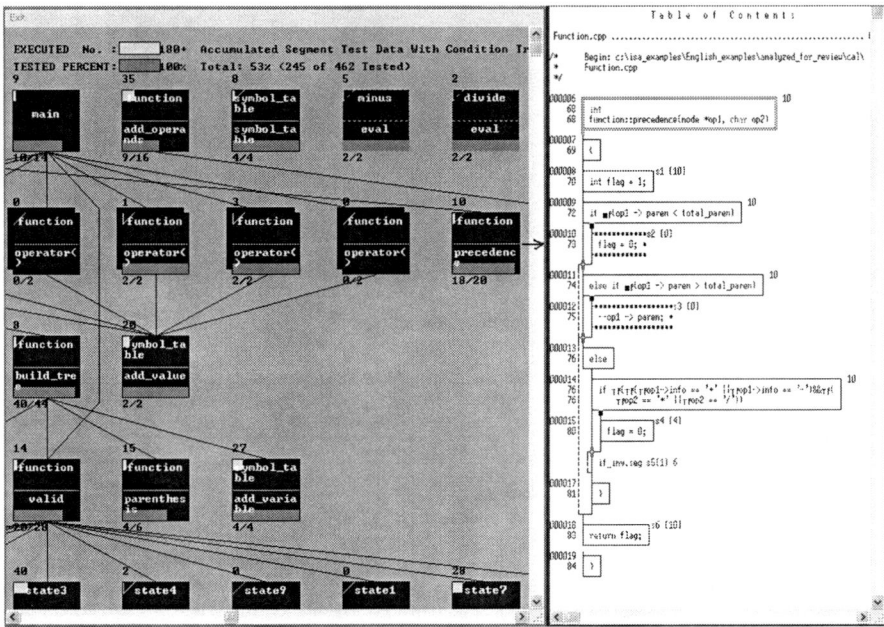

Fig. 7.34 An application example linking a module box in a call graph to the logic diagram shown with MC/DC test coverage analysis result.

Fig. 7.37, a GNU bison program version 1.24 is used for comparing the space needed in regular approach and the virtual approach: there are 10,932 lines of the source code with 34 files, and the size of the source code is 349 KB, but the size of the built database is only 143 KB, less than ½ of the size of the source code.

But as shown in Fig. 7.38, with the small database, a call graph showing the Cyclomatic complexity measurement result with 169 functions can be dynamically generated virtually.

For storing the call graph in postscript format using traditional approach, it requires 795 KB space as shown in Fig. 7.39.

7.9 The Major Features of NSE Software Visualization Paradigm

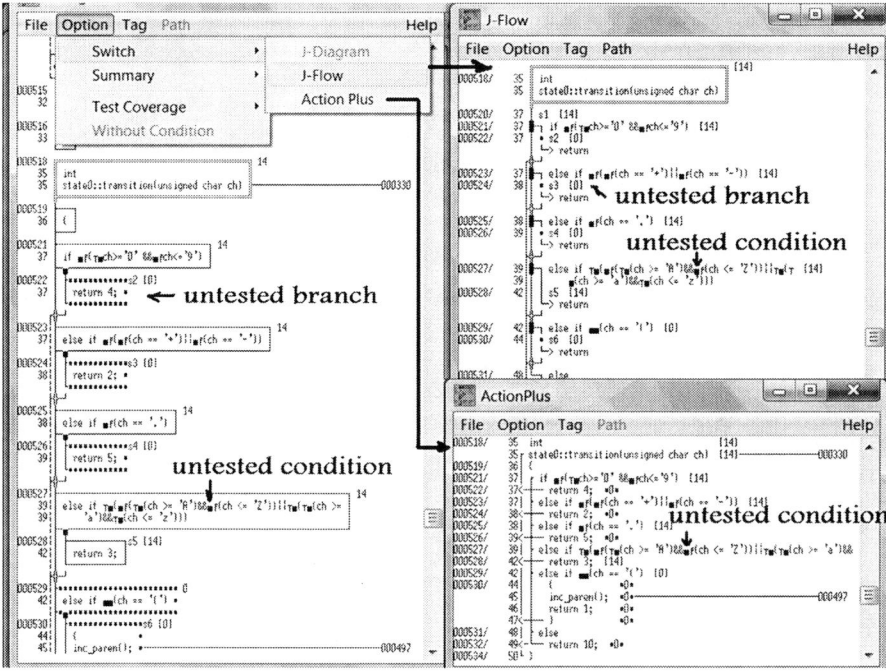

Fig. 7.35 An example of converting a J-Diagram to a J-Flow diagram

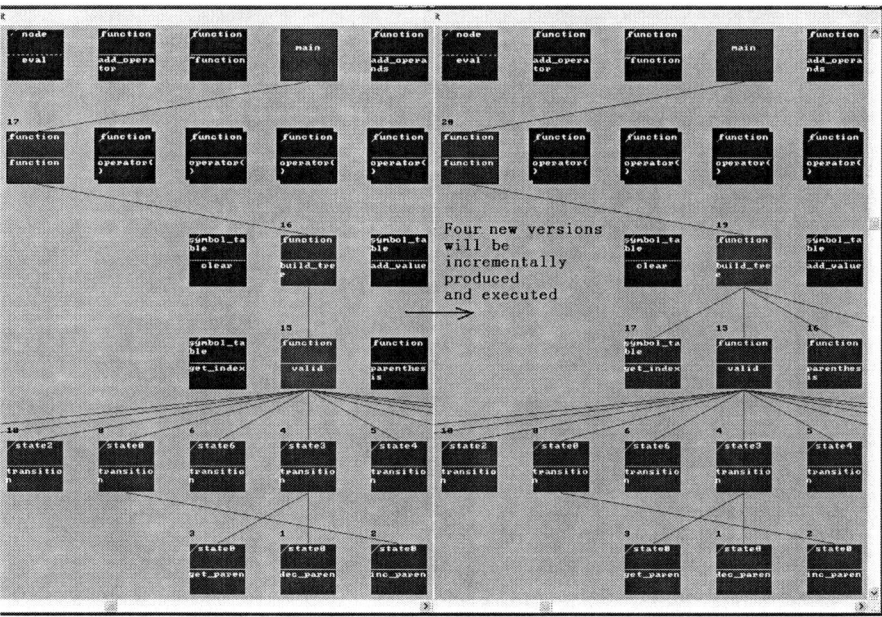

Fig. 7.36 Assigning bottom-up order for incremental software development, including incremental unit coding and unit testing

So storing the following listed charts and diagrams in postscript format will need the space more than 100 times of the size of the source code:

The system-level charts/diagrams

(a) The function call graph of the entire software system
(b) The class inheritance chart of the entire software system
(c) The class and independent function relation chart of the entire software system
(d) The program tree of the entire software system

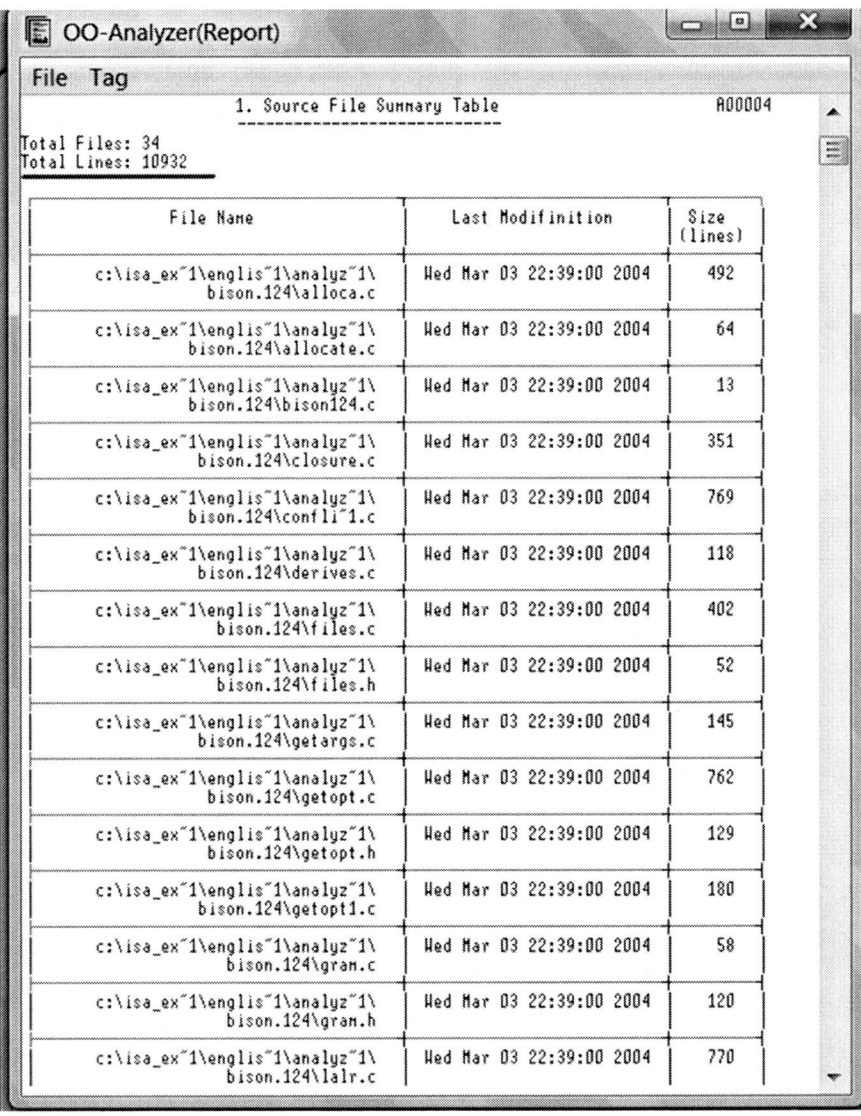

Fig. 7.37 The size comparison between the source code and the built database

7.9 The Major Features of NSE Software Visualization Paradigm 173

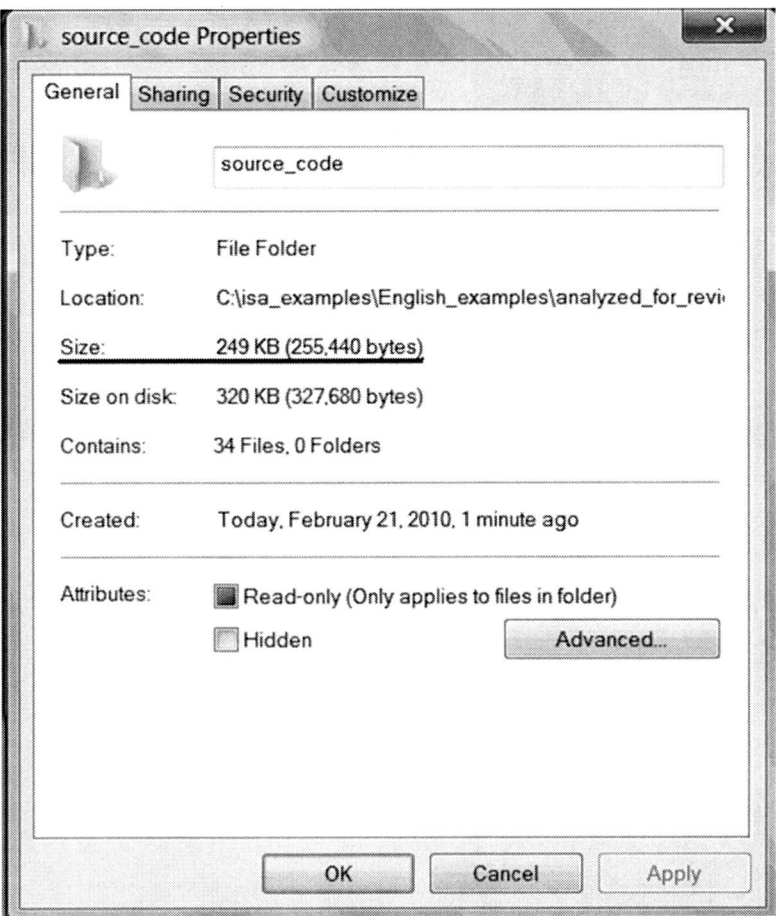

Fig. 7.37 (continued)

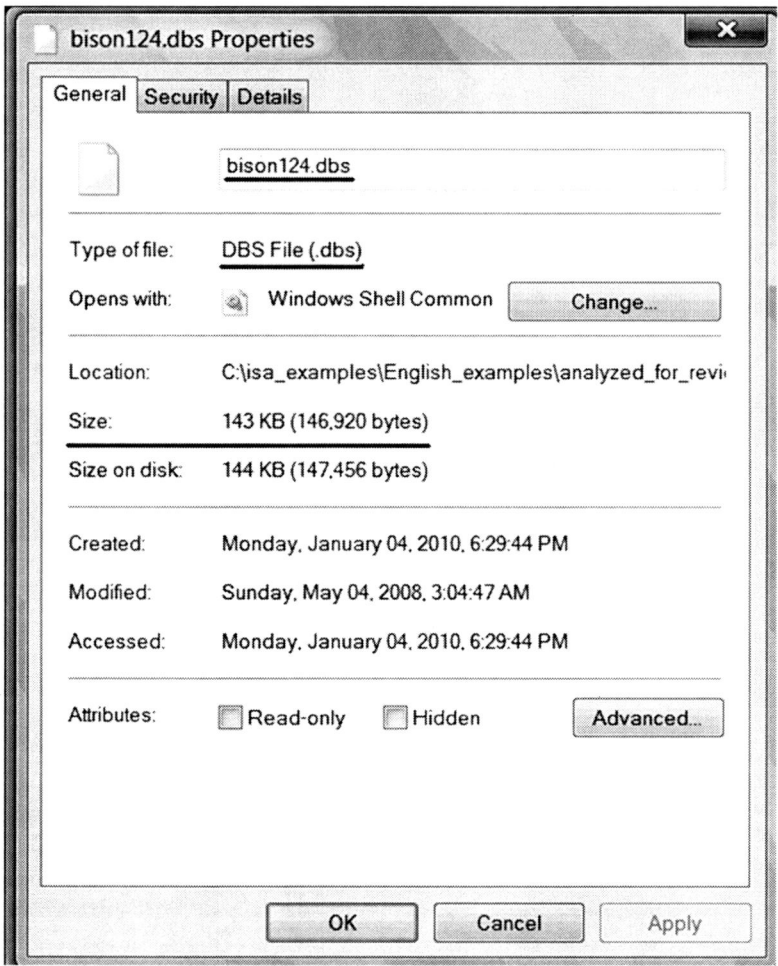

Fig. 7.37 (continued)

7.9 The Major Features of NSE Software Visualization Paradigm

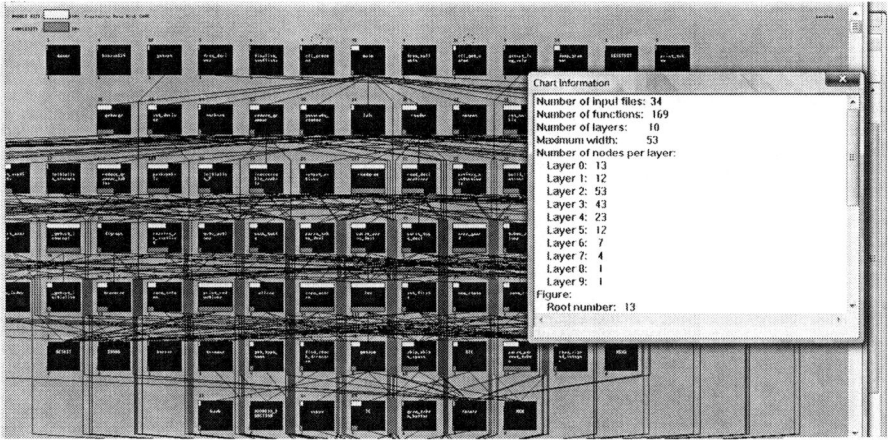

Fig. 7.38 The call graph of bison V1.24 with Cyclomatic complexity measurement result

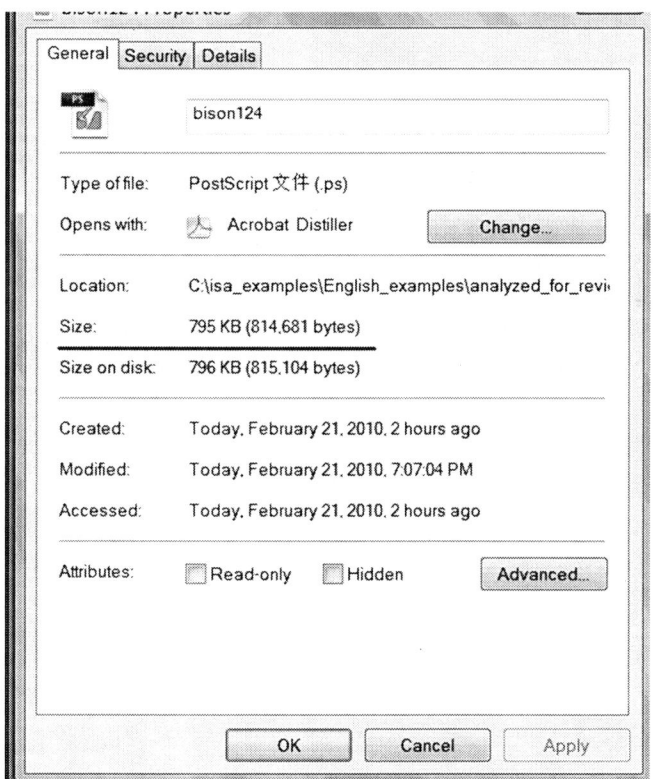

Fig. 7.39 The size of the chart stored in postscript format

```
112  trouble (x)
113  int x;
114  {
115  int i, t=1;
116  char c,*pc=NULL,ch[10],*p=NULL,*e=NULL;
117  if((e=malloc(4))==NULL)printf("Out of memory,x=%s",x), exit(-1);
118  for(i = x; i <= 8 && t; p=&ch[i++])
119    if(i % 2 ==1) {
120      p=&c; t=0; }
121  ch[0] = *p;     /* seg. fault when x > 8 */
122  i = x ;
123  while (i > -2 && i<=7 )(/*dead loop if x=7 or x=3*/
124    switch ( x + z ) {
125      case 0: case 1: x = z = 1; break;
126      case 2: y = 1; break; }
127    if ( i < 7 )
128      i += 4; }
129  if ( x < 5 )
130    pc = ch;
131  if( x < 6 )
132    fd=fopen("trouble.c", "r");
133  c = getc (fd);  /* seg. fault when x = 6 */
134  strcpy (pc, "ab"); /* seg. fault if x = 5 */
135  c = ch[y]; /* seg. fault when x = 4 */
136  z = x / z; /* Arith. excep. when x = 2 */
137  if((p=malloc(3))!=NULL) strcpy(p,"OK");
138  }
```

Fig. 7.40 A source code module with defects

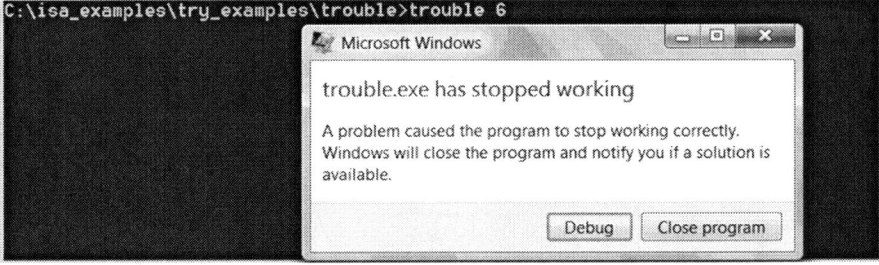

Fig. 7.41 An error message given by the system without showing the error location

(e) The overall MC/DC test coverage measurement result of the entire software system
(f) The overall quality measurement result shown in Kiviat diagram
(g) The overall performance measurement result of the entire software system
(h) The overall Cyclomatic complexity measurement result of the entire software system
(i) The logic diagram of the entire software system
(j) The control flow diagram of the entire software system

7.9 The Major Features of NSE Software Visualization Paradigm

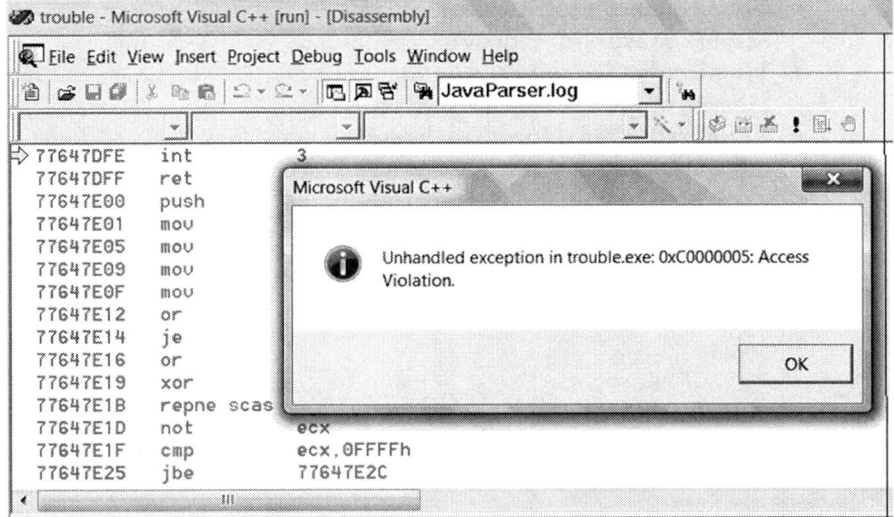

Fig. 7.42 The system debugger can only show the location of the object code which is not very useful

Fig. 7.43 When it is executed under NSE, an error message is given with the detailed source code location (line 133)

(k) The static and dynamic analysis result of the entire software system
(l) The overall version comparison result shown in J-Chart with unchanged modules shown in blue, changed modules in red, deleted modules in brown, and added modules in green.

Plus the file-level charts/diagrams, and the module-level charts/diagrams.

17. **Complete**
    NSE software engineering visualization paradigm completely support:

    (a) **Visualization of the entire software engineering lifecycle**

    - **Visualization for requirements engineering** – See Fig. 7.43 for an application example of functional decomposition of functional requirements in the first step.
    - **Visualization for design engineering** – See Fig. 7.44 for an application example of a top-down software system design.
    - **Visualization for coding engineering** – See Fig. 7.34 for an application example to assign bottom-up incremental coding orders.

- **Visualization for software inspection** – See Fig. 7.30 for an application example to establish various traceabilities for code inspection.
- **Visualization for software testing** – See Fig. 7.47 for an application example of MC/DC test coverage analysis.
- **Visualization for software maintenance** – See Figs. 7.47 and 7.48 for safe implementation of requirement changes or code modifications.

(b) **Visualization of software architectures** – See Fig. 7.5 for a program structure (function call graph), Fig. 7.15 for the program tree, and Fig. 7.41 for the data (class) structure of a program.

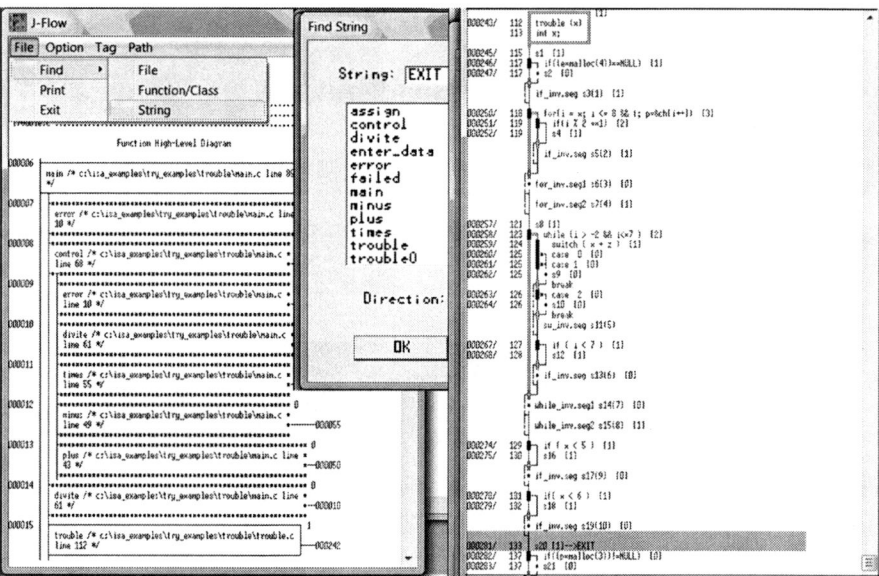

Fig. 7.44 Visually locating the error location in the control flow of the source code module where an "EXIT" string has been added to indicate the unexpected program termination location in the source code

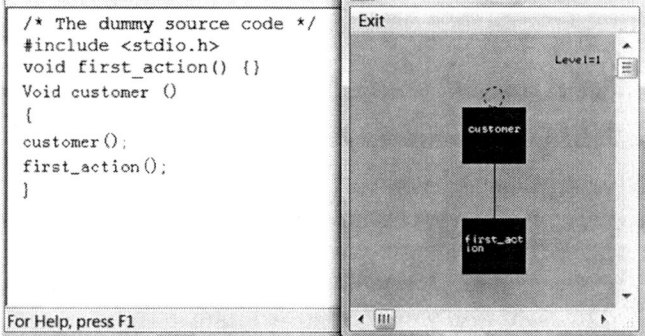

Fig. 7.45 An application example for requirement elicitation/gathering

7.9 The Major Features of NSE Software Visualization Paradigm

(c) **Visualization of source code** – See Fig. 7.11 for the detailed logic diagram of a source code module (shown with untested branches and conditions highlighted), and Fig. 7.15 for the control flow diagram of a source code module (shown with the untested branches and conditions highlighted).

(d) **Visualization in reverse engineering** – See Fig. 7.29 for the call graph and sub call graph in reverse engineering, and Fig. 7.28 for the logic diagram of a source code module shown with a source code writing style-independent way.

(e) **Dynamic program behavior visualization** – See Fig. 7.32 for overall MC/DC test coverage measurement result and the detailed test coverage result of a source code module with the untested branches and conditions highlighted, and Fig. 7.45 in Sect. 7.9 for the overall performance measurement result with the branch execution frequency of a module indicated in J-Flow diagram.

(f) **Integration of visualization tools in the software engineering tool chain** The NSE software engineering paradigm is integrated into the NSE software engineering paradigm and the support platform, Panorama++.

(g) **Visualization for software debugging** – see Figs. 7.40–7.44:
After compilation and execution of the program directly without using NSE tools, the system shows an error message without detailed information (see Fig. 7.41):

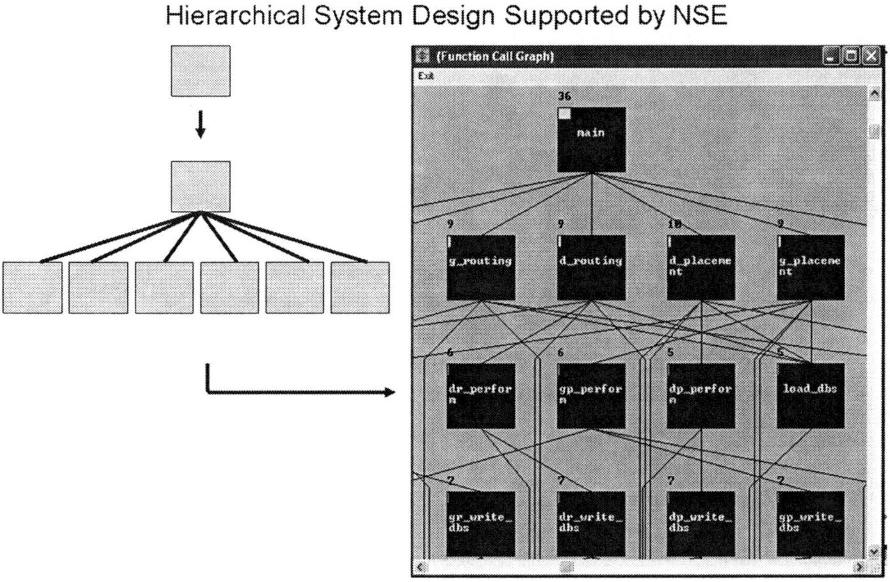

Fig. 7.46 A top-down system design process shown graphically through dummy programming and virtual diagramming

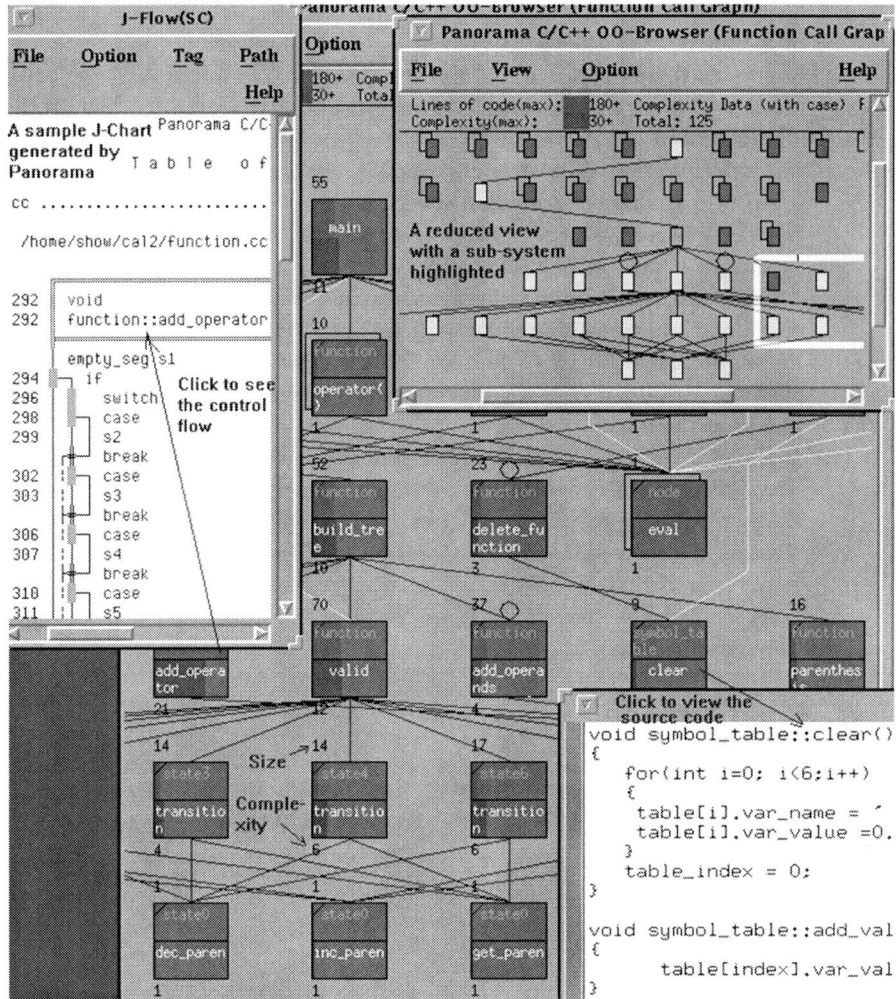

Fig. 7.47 An example of J-Chart shown with some related information

Debugging can also be performed visually with the NSE software engineering paradigm as shown in Fig. 7.44.

7.10 Applications

The NSE Software Visualization Paradigm can be applied in the entire software development process to make the software product much easier to understand, test, and maintain:

7.10 Applications

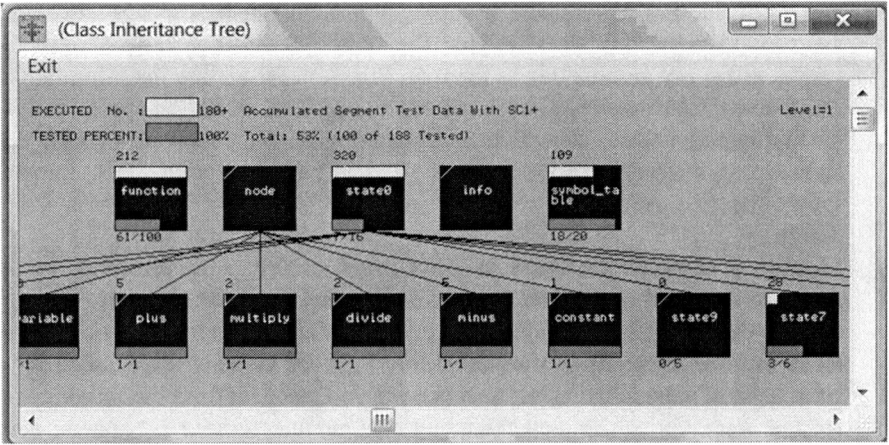

Fig. 7.48 Class test coverage analysis result

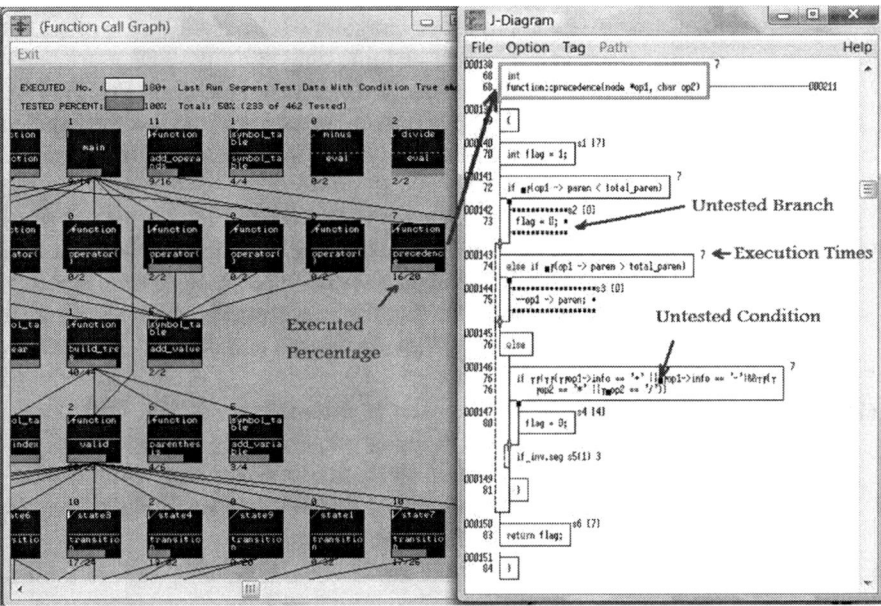

Fig. 7.49 The MC/DC test coverage measurement result shown in J-Chart and J-Diagram

(a) **Making the entire software development process visible** – See Fig. 7.5 for an overview of the structure of a complex software product shown with the Cyclomatic complexity measurement result, Fig. 7.9 for viewing the detailed program logic of a complex module and the related information, Fig. 7.20 for getting many overall program measurement results including the performance measurement, the Cyclomatic complexity measurement, the test coverage measurement, the module size, etc., Fig. 7.31 for viewing various sub call graphs using any module box as the root, Fig. 7.36 for getting the information about how the software product is organized, Fig. 7.45 shows the application in the first step for requirement elicitation/gathering using the innovated HAETVE technique through dummy programming and J-Chart generation – this example shows the first Actor's first action, and Fig. 7.46 for a top-down system design.

(b) **Making a complex software product much easier to understand** – See Fig. 7.15 for viewing the overall program tree and the detailed control flow of each module, Fig. 7.21 for viewing a call graph and the associated pull-down menu to view interesting information, Fig 7.30 for viewing the detailed program logic of a module diagrammed in a way independent from the source code writing style for easily understanding the module written by others, and Fig. 7.47 for getting more information from a call graph.

(c) **Making the diagrammed source code traceable** for semiautomated code inspection, review, and walk though – See Fig. 7.10 for tracing a function call statement to the called function body, Fig. 7.15 for tracing a module from the related program tree to the control flow diagram of the module, Fig. 7.21 for tracing a module from a call graph to its logic diagram, Fig. 7.27 to tracing a function with all the locations of the function call statements, and Fig. 7.34 for tracing a module with the test coverage measurement result to view the detailed logic diagram where untested branches and conditions are highlighted in small black boxes.

(d) **Making a software product much easier to test** – See Fig. 7.5 for test planning, Fig. 7.15 for test coverage analysis result, Fig. 7.16 for efficient test case design, and Fig. 7.48 for getting the class test coverage analysis result (note: a class cannot be directly executed, so that the test coverage analysis result is obtained from its instances), and Fig. 7.49 for overall and detailed MC/DC test coverage analysis result.

(e) **Making a software product much easier to maintain** – See Fig. 7.50 to trace a module to be modified to find how many requirements are related (in this example, two requirements are related so that the modification should satisfy both), and Fig. 7.51 to trace the module to find what other modules may be affected to prevent the side effects for the modification.

(f) **Locating the performance bottleneck easily** – see Fig. 7.52.

(g) **Finding logic defects better** – Programs written in textual format are hard to read and understand. A logic defect is not easy to find because a program with logic defects may run without providing an error message, but the results are often incorrect. With the NSE visualization paradigm, logic defects can be found through program logic analysis and diagramming to compare to the program algorithm. An application example is shown from Figs. 7.53–7.55.

(h) **Determining runtime error locations visually** – see Figs. 7.40–7.44.

(i) **Used for multilevel version comparison** – The NSE visualization paradigm can be used to compare two versions of an entire program holistically in system level, file level, module level, and statement level as shown from Figs. 7.56–7.60.

7.10 Applications

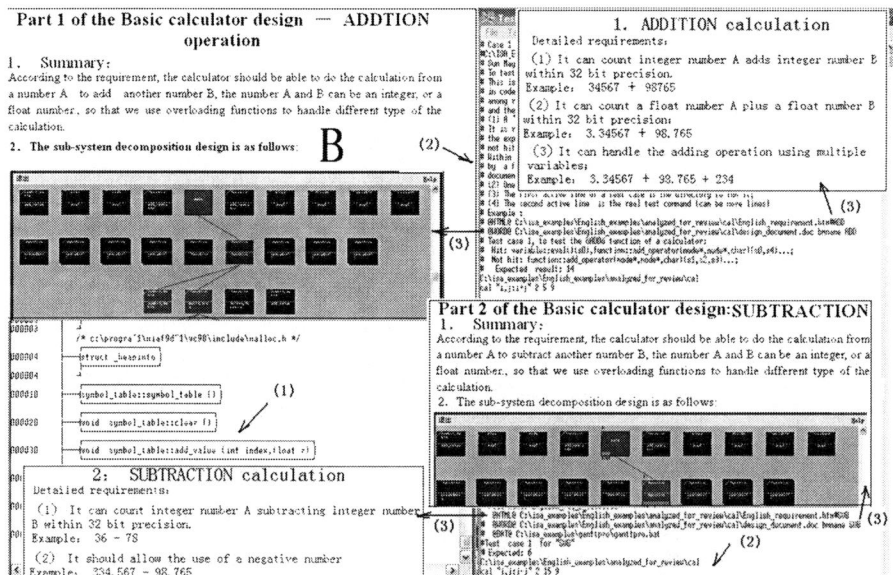

Fig. 7.50 Tracing a module to be modified to see how many requirements are related (in this example, two requirements are related which should all be satisfied in the module modification)

(j) **Used to efficiently handle the issues of complexity** – NSE software visualization paradigm can be used to efficiently handle the complexity issue because the "Complexity is levels" [Bro95-p211].

According to complexity science, a software system complexity includes:

- Formulaic complexity:
  - Description complexity – The NSE software visualization paradigm makes it possible to graphically describe a software system with the source code, including the program structure (see Fig. 7.5), the program logic of an entire software product (see Figs. 7.9–7.11), and the control flow of an entire software product (see Figs. 7.15 and 7.17) with various traceabilities established automatically.
  - Generative complexity – working with the NSE HAETVE requirement development support technique, the NSE software visualization paradigm supports the NSE software development methodology based on **Generative Holism** to form and display the whole of a software system first through dummy programming, then assigns incremental coding and unit-testing order to support the system **growing up** incrementally (see Figs. 7.36, 7.38, 7.45, and 7.46).
  - Computational complexity – The NSE software visualization paradigm makes the program algorithms much easier to understand through path analysis and logic diagram and control flow diagram generation as shown in Figs. 7.53–7.55.

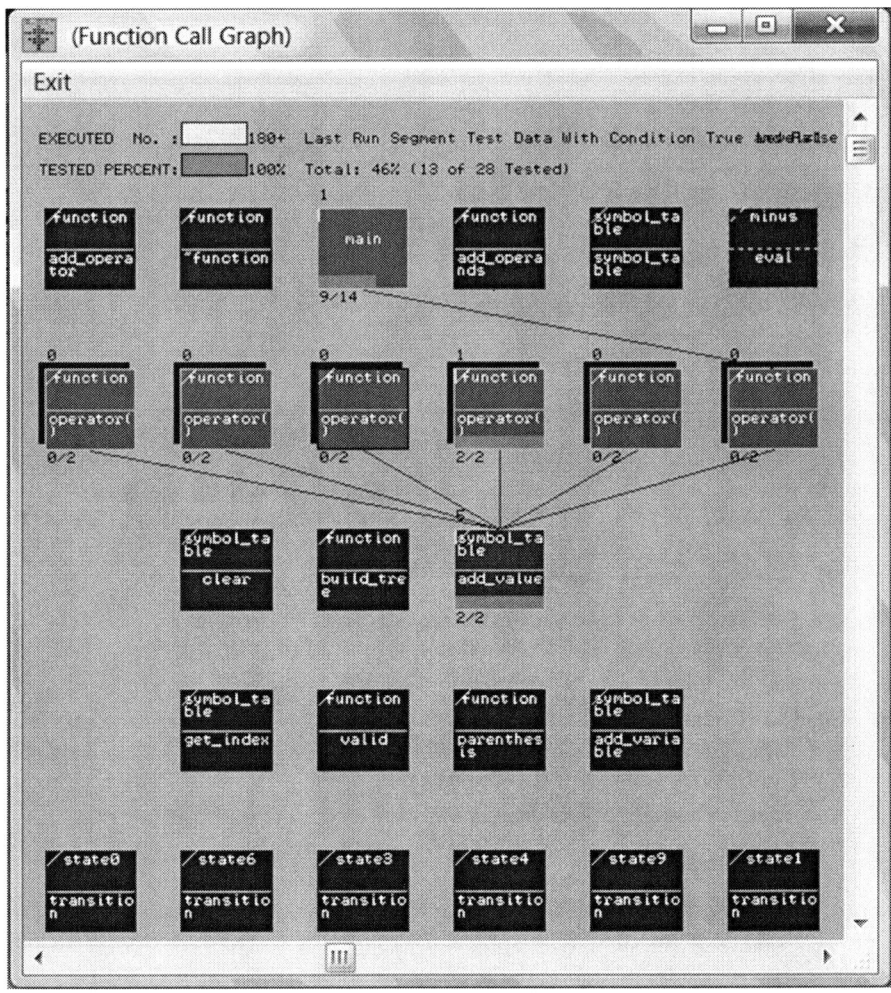

Fig. 7.51 Tracing a module to be modified to the related modules

7.10 Applications

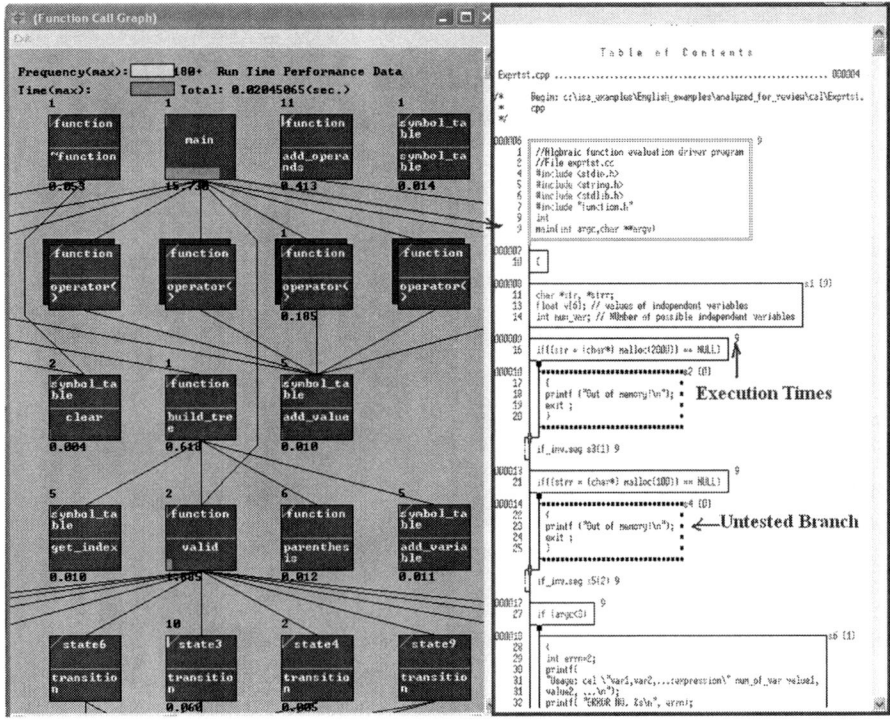

Fig. 7.52 A J-Chart showing the performance analysis result with a J-Flow diagram showing the branch execution frequency for locating the performance bottleneck easily

Fig. 7.53 Two similar versions of a program module with one having a logic defect

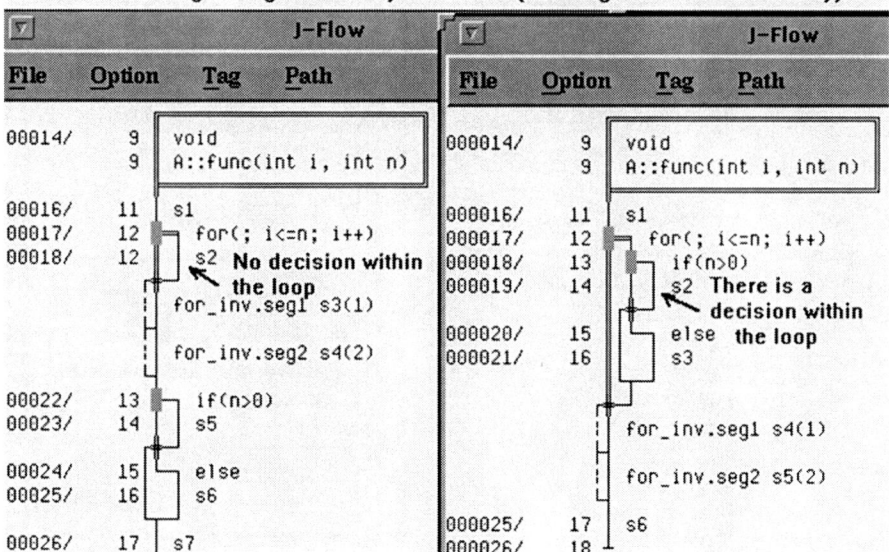

Fig. 7.54 The control flow diagrams shown in J-Flow are different clearly

7.10 Applications

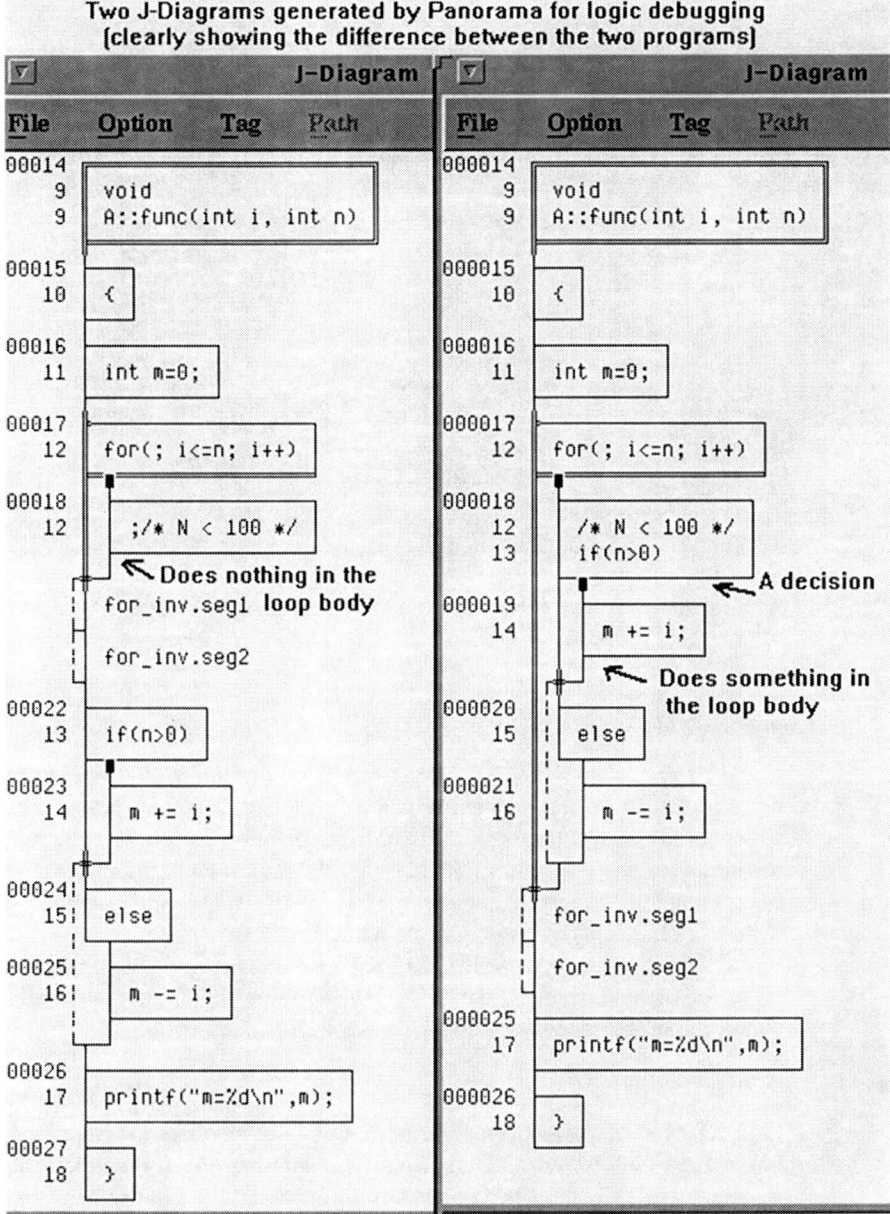

Fig. 7.55 The logic diagrams shown in J-Diagram can be used to find the logic defect easily

Fig. 7.56 The system-level version comparison result for a GNU bison program (V1.24 and V1.25)

- Compositional complexity:
  - Constitutional complexity – The NSE software visualization paradigm helps users to handle constitutional complexity in many ways. For instance, to a class, the NSE software visualization paradigm performs the structure analysis, the logic analysis, the data member analysis, the function member analysis, the control flow analysis, etc., and graphically represents the analysis results as shown in Fig. 7.61.
  - Taxonomical complexity – working with NSE static and dynamic program measurement tools, the NSE visualization paradigm graphically shows the measurement results as shown in Figs 7.62–7.64.
- Structural complexity:
  - Organizational complexity – The NSE software visualization paradigm helps users understand how a software product organized including the structure analysis, the file system analysis, the data (variable) analysis, the program logic analysis, and the control flow analysis.
  - Hierarchical complexity – The NSE software visualization paradigm helps users to understand the system hierarchy by generating the system interactive and traceable call graph, class inheritance chart, program tree, etc.

7.10 Applications

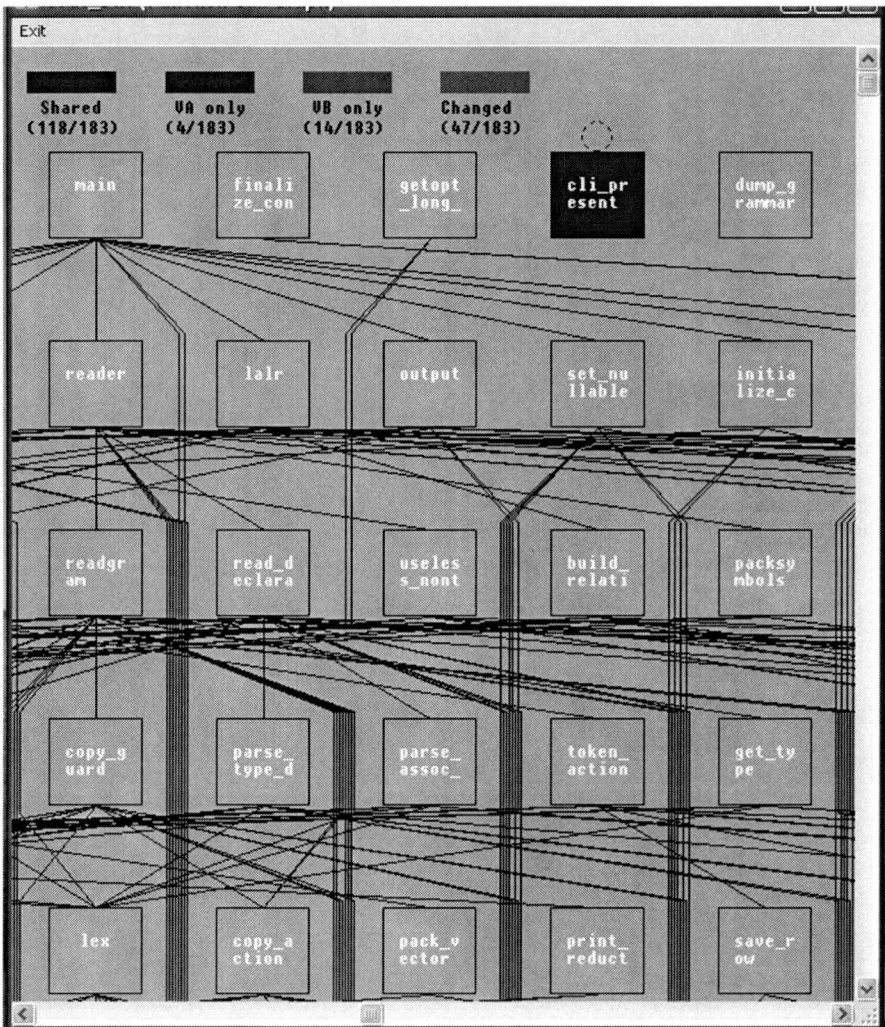

Fig. 7.57 The Modules deleted from version A to version B

- Functional complexity:
  - Operational complexity – NSE software visualization paradigm makes the Operation process visible, recordable, and easy to playback through backward traceability from the system control diagram shown in J-Flow diagram notation, including dynamically running a third-party complicated program using a batch file as shown in Fig. 7.63.
  - Function and rule complexity – Working with HAETVE requirement development technique, the NSE software visualization paradigm helps users to decompose the function of the functional requirements and make

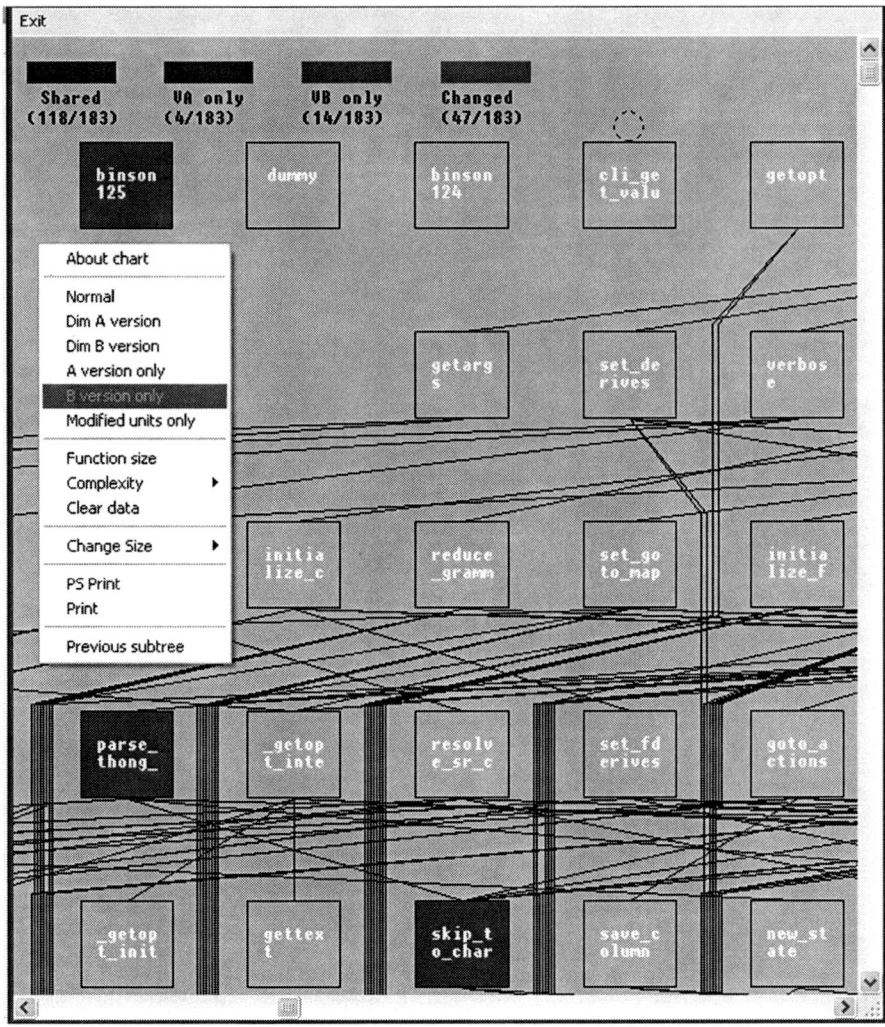

Fig. 7.58 The new modules added to version B

the decomposition visible as shown in Figs. 7.45 and 7.46 and represents function cross relationships graphically by generating the function call graph shown in J-Chart notations.

(k) **Used to efficiently handle the issue of software invisibility** – The NSE software visualization paradigm makes the entire software development process and the entire system visible from the first step (as shown in Figs. 7.45 and 7.46) down to the maintenance process (as shown in Figs. 7.50, 7.51, 7.56–7.60). The NSE software visualization paradigm represents software information graphically in many ways as shown in Fig. 7.64.

7.11 Self-Documenting

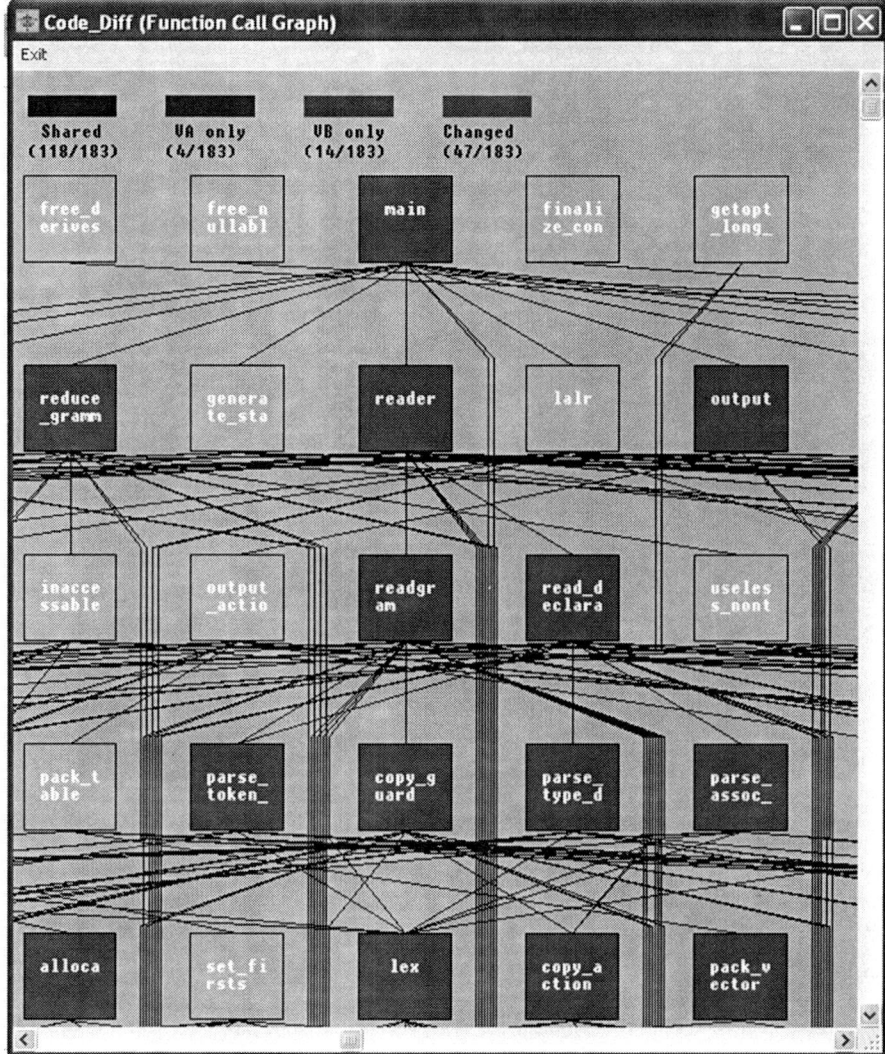

Fig. 7.59 The modules modified in version B

7.11 Self-Documenting

For easy maintenance, many kinds of documents can be merged into the source code such as the cross references. Sometimes, when there is a need to use something like the Sequence Diagram to expose time ordering of events/messages, we can describe the same thing within a program comment such as the use of a formatted table shown in C/C++ as follows:

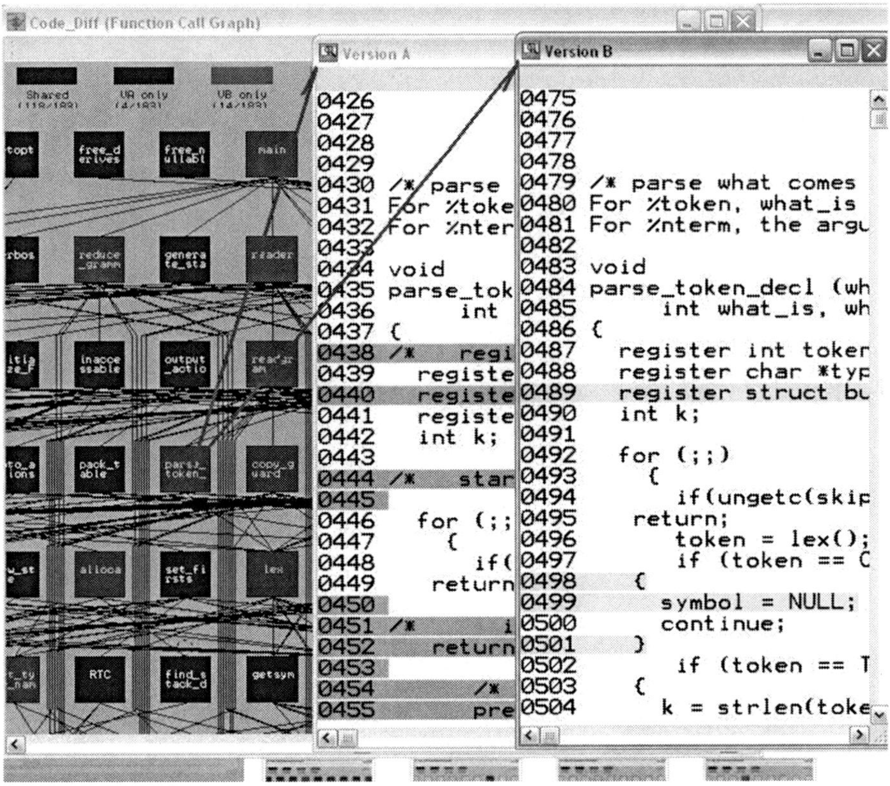

Fig. 7.60 The detailed difference between a modified module

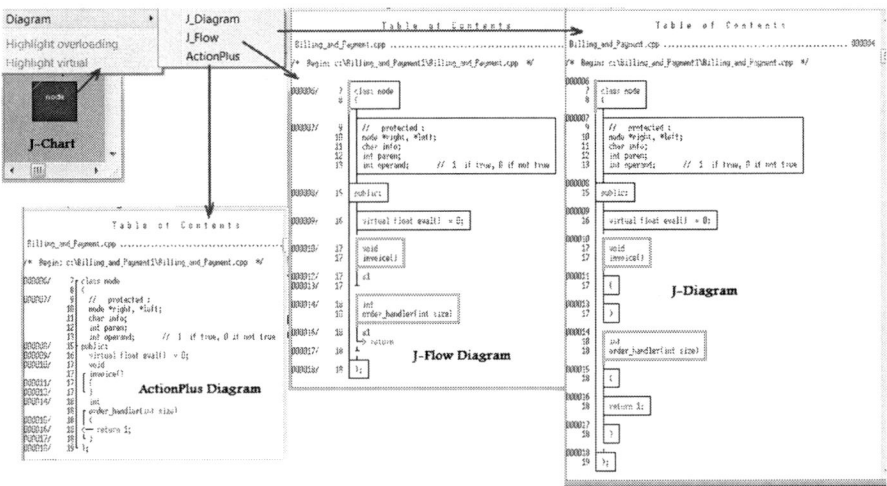

Fig. 7.61 Class analysis and the analysis result display shown in several ways

7.11 Self-Documenting

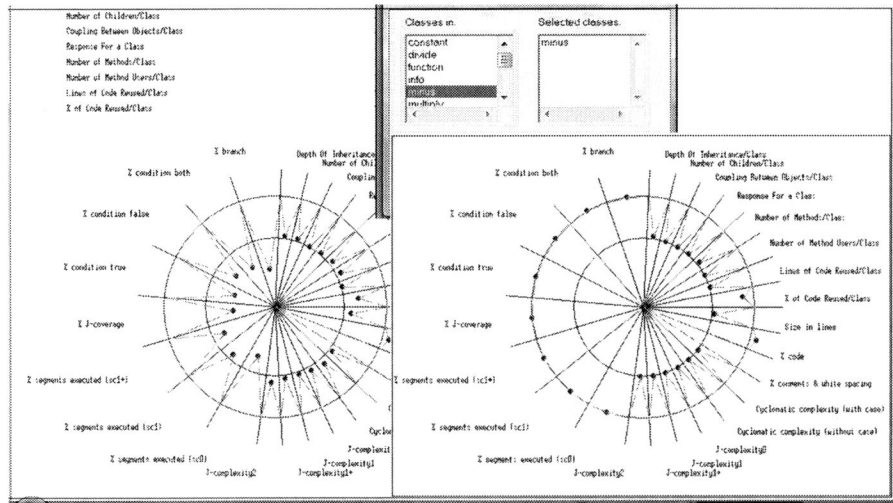

Fig. 7.62 The results of program static and dynamic measurement

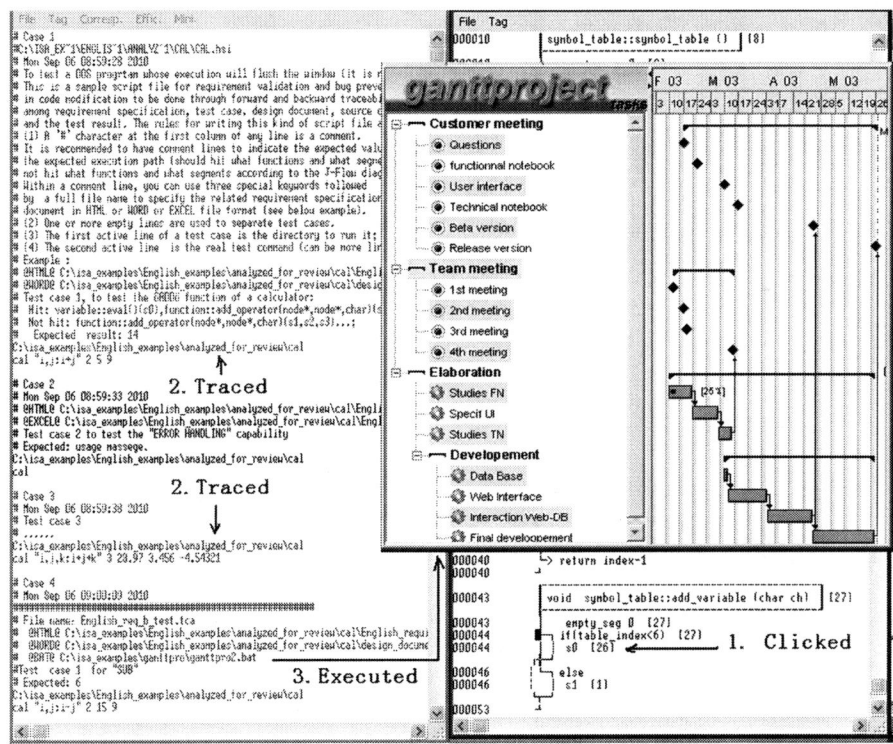

Fig. 7.63 Directly running a third-party program through backward traceability from a code branch shown in J-Flow diagram

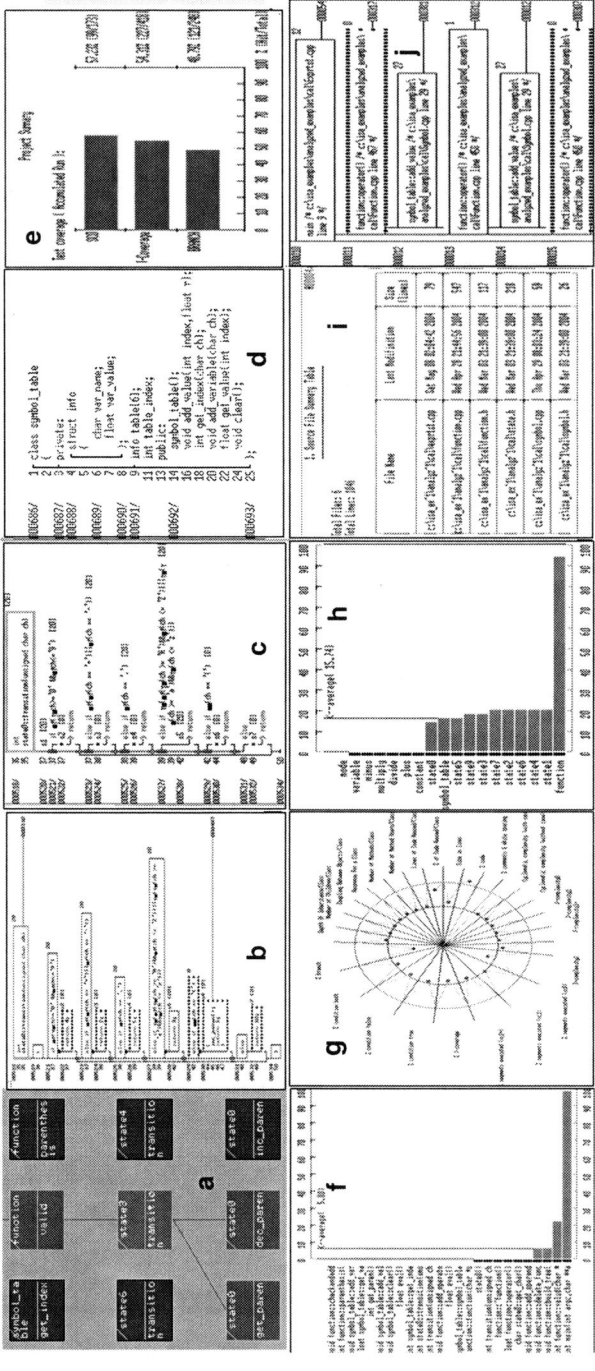

Fig. 7.64 The NSE visualization paradigm represents almost all information of a software product in graphics – (**a**) module-level relationship chart, (**b**) logic diagram with statement-level cross reference (called by and calling to), (**c**) control flow diagram (used for complexity analysis and control flow analysis), (**d**) ActionPlus diagram (used for class attribute analysis, etc.), (**e**) bar chart used for showing test coverage analysis, etc.), (**f**) a untested path highlighted with the test conditions, (**g**) Kiviat diagram for quality measurement of an entire product or each individual module, (**h**) complexity analysis result chart of classes or functions, (**i**) the size measurement report table of source files, (**j**) a program tree (untested modules are highlighted in small black boxes)

7.12 Summary

"One Picture Is Worth Ten Thousand Words." – a holistic, interactive, colorful, and traceable chart/diagram is more useful in the description of a complex software product. But unfortunately, the traditional software visualization paradigm works with linear process models complying with the superposition principle that the whole of a system is the sum of its parts, so that almost all visualization tasks/

```
/* Time-Event table:
 _____
|    Timing    |    t1    |    t2    |    t3    |    t4    |
|_____|_____|_____|_____|_____|
|    Events    |  Event1  |          |          |          |
|_____|_____|_____|_____|_____|
|              |          |  event2  |          |          |
|_____|_____|_____|_____|_____|
|              |          |          |  event3  |          |
}_____|_____|_____|_____|_____|
|              |          |          |          |  event4  |
|_____|_____|_____|_____|_____|

 _____
|    Timing    |    t5    |    t6    |    t7    |    t8    |
|_____|_____|_____|_____|_____|
|    Events    |  Event5  |          |          |          |
|_____|_____|_____|_____|_____|
|              |          |  event6  |          |          |
|_____|_____|_____|_____|_____|
|              |          |          |  event7  |          |
}_____|_____|_____|_____|_____|
|              |          |          |          |  event8  |
|_____|_____|_____|_____|_____|
...
*/
```

activities are performed linearly, partially, and locally, mainly only making the modeling process visible with graphic editors to produce many small pieces of charts or diagrams which are not interactive and not traceable in most cases, rather than complete, interactive, and traceable ones for graphically representing an entire software product. Even if a complete chart/diagram can be obtained by using a few diagramming tools, it is still useless because without automated and self-maintainable traceability and the capability to highlight an element with all of the related elements, there are too many connection lines making the chart/diagram hard to view and hard to understand.

The NSE software visualization paradigm is based on complexity science, complying with the Nonlinearity principle and the Holism principle, so that almost all visualization tasks/activities are performed holistically and globally to automatically generate virtual, interactive, and traceable 3J graphics (J-Chart, J-Diagram, and J-Flow) innovated to make the entire software development process visible. The NSE software visualization paradigm makes a software product much easier to understand, test, and maintain.

7.13 Points and Questions to Ponder

(a) What are the major differences between the NSE software visualization paradigm and the traditional software visualization paradigm?
(b) What are the major benefits of virtually existing charts and diagrams without storing hard copies in the hard disk and the memory of a computer?
(c) Point out the reasons why a system-level call graph or diagram should be made interactive and traceable.
(d) Write three small programs for generating the following three charts separately through dummy programming, then compile them and run the executable programs to correct possible defects.

7.14 Further Reading and Information Source

Brooks FP Jr (1995) The mythical man-month. Addison-Wesley, Reading, Chap 10.
Xiong J (2009) Tutorial. A complete revolution in software engineering based on complexity science. In: WORLDCOMP'09, Las Vegas, July 13–17, 2009.

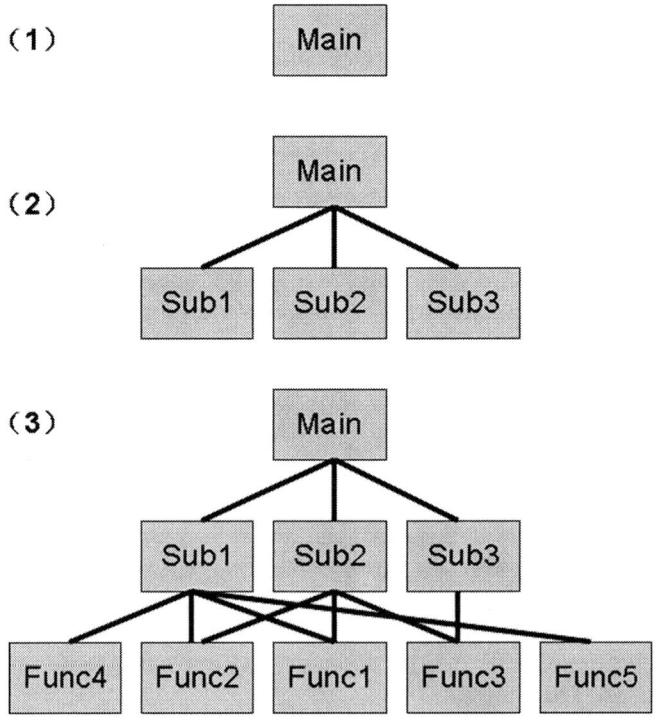

Xiong J, Xiong J (2009) A complete revolution in software engineering based on complexity science. In: WORLDCOMP'09 – SERP (Software Engineering Research and Practice 2009), pp 109–115.

References

[Bro95-p211]   Brooks FP Jr (1995) The mythical man-month. Addison Wesley, Reading, p 211
[Bro95-p249]   Brooks FP Jr (1995) The mythical man-month. Addison Wesley, Reading, p 249

Chapter 8
NSE Process Model

> *There has to be upstream movement... experience and ideas from each downstream part of the construction process must leap upstream, sometimes more than one stage, and affect the upstream activity.*
>
> Frederick P. Brooks, Jr.

This chapter describes an important component of the NSE (Nonlinear Software Engineering) paradigm – the NSE process model.

Software process is a road map for software managers and engineers to follow. A software process model defines a distinct set of activities, actions, tasks, milestones, and work products for developing and maintaining a software product.

The NSE Process Model is different from the old-established ones based on linear thinking and simplistic science. It is nonlinear, created through a paradigm-shift framework, the Five-Dimension Synthesis Method (FDS) proposed by me as shown in Fig. 8.1.

As shown in Fig. 8.1, the new process model is created by complying with the essential principles of complexity science, particularly the Nonlinearity principle and the Holism principle. Of course, a waterfall-like process model will not be created because it does not comply with the Nonlinearity principle and the Holism principle of complexity science.

8.1 Some Experts' Expectations

Many software engineering experts not only point out the problems existing with the old-established software engineering paradigm but also clearly express their expectations in software engineering innovation.

> **Professor Roger S. Pressman,** the author of the book, **"Software Engineering A Practitioner's Approach":**
> Originally... software engineering was approached as a linear activity in which a series of sequential steps were applied in order to solve problems. Yet, linear approaches to software development run counter to the way in which most systems are actually built. In reality,

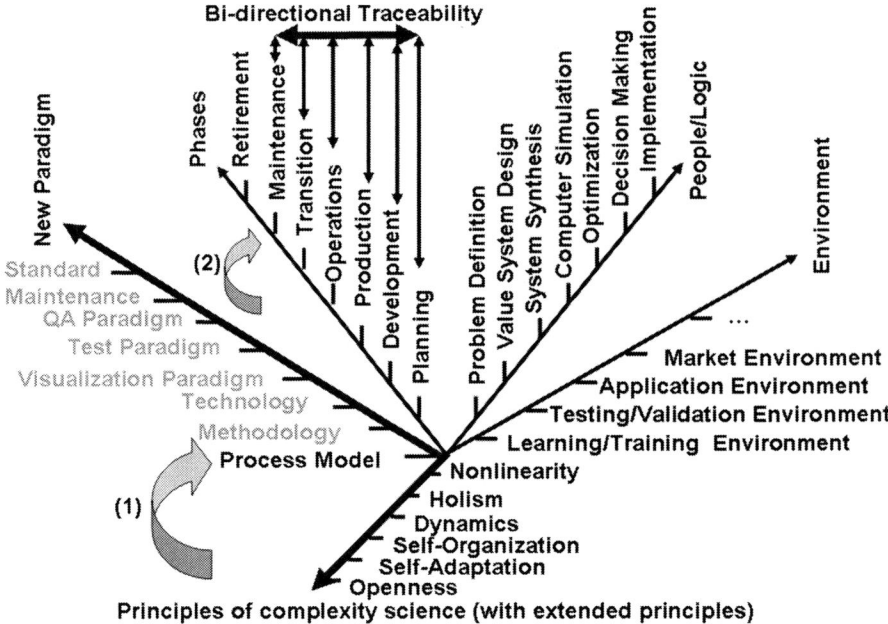

Fig. 8.1 Redefining the software process model by complying with the essential principles of complexity science through the proposed paradigm-shift framework, FDS

complex systems evolve iteratively, even incrementally. It is for this reason that a large segment of the software engineering community is moving toward evolutionary models of software development. [Pre05-p864]

Frederick P. Brooks Jr., the author of the book, "The Mythical Man-Month":
There has to be upstream movement... experience and ideas from each downstream part of the construction process must leap upstream, sometimes more than one stage, and affect the upstream activity. Designing the implementation will show that some architectures cripple performance; so the architecture has to be reworked. Coding the realization will show some functions to balloon space requirements; so there may have to be changes to architecture and implementation. One may well, therefore, iterate through two or more architecture-implementation design cycles before realizing anything as code. [Bro95-p122]

After all, software engineering, like chemical engineering, is concerned with the nonlinear problems of scaling up into industrial-scale process, and like industrial engineering, it is permanently confounded by the complexities of human behavior. [Bro95-p288]

The fundamental problem with program maintenance is that fixing a defect has a substantial (20-50 percent) chance of introducing another. So the whole process is two steps forward and one step back... Clearly, methods of designing programs so as to eliminate or at least illuminate side effects can have an immense payoff in maintenance costs. So can methods of implementing designs with fewer people, fewer interface, and hence fewer bugs. [Bro95-p122]

Franco Zambonelli, H. Van Dyke Parunak, **the authors of the paper "Signs of a Revolution in Computer Science and Software Engineering":**
We are on the edge of a revolutionary shift of paradigm, pioneered by the multiagent systems community, and likely to change our very attitudes in software systems modeling and engineering. [Zam08]

8.2 All of the Existing Software Engineering Process Models Are Outdated

As described in Sect. 2.5, the existing software engineering process models are out of date, no matter if they are waterfall style models, incremental development models, iterative development models, or a new one recommended by Alistair Cockburn to combine **both incremental and iterative development together [Coc08]**, because they are linear models with only one track forward in one direction without upstream movement at all, like one way traffic with only one track as shown in Fig. 8.2, but what we really need is a process model incrementally supporting bidirectional iteration with multiple tracks through various traceabilities, like two-way traffic with multiple tracks as shown in Fig. 8.3.

In fact, those existing process models themselves are outcomes of linear thinking, reductionism, and the superposition principle. It is clear that those process models handle a software product as a linear system like a machine which can be *assembled* to comply with the superposition principle that **the whole of a software system is the sum of its parts**. But it violates the Holism principle of complexity science that the **whole of a complex system is greater than the sum of its parts – the characteristics and behavior emerge from the interaction of its parts**. Based on the **Generative Holism** principle of

Fig. 8.2 One Way Traffic with only one track

Fig. 8.3 Two Way Traffic with multiple tracks

complexity science, the whole of a complex system should exist first as an embryo, then it "**grows up**" with its components as shown in Fig. 1.63 and Fig. 1.64 shown in Chap. 1.

8.3 Outline of the Revolutionary Solution Offered with NSE

With NSE, a revolutionary solution is offered – the NSE process model:

(a) The NSE process model is based on complexity science, complying with the essential principles of complexity science, particularly the Nonlinearity principle and the Holism principle.
(b) The NSE process model is supported by many new software engineering techniques.
(c) The NSE software process model has been commercially implemented with the support platform Panorama++ – it not only indicates what needs to be done but also provides models/techniques and tools to help users solve the issue of how to do it better.
(d) The NSE process model not only supports new software product development but also supports software product maintenance which often takes 75% of the total effort and total cost for product development with the old-established software engineering paradigm.
(e) Almost all of the tools developed to support the NSE process model are dummy ones, easy to understand and use.
(f) The NSE process model is established with the goal to solve all essential problems (complexity, changeability, invisibility, and conformity) and all critical problems (low quality and productivity, and high cost and risk) existing with today's software development.

8.4 The Driving Forces and The Support Techniques

The driving force for NSE and its process model is complexity science, applied to solve the essential software engineering difficulties defined by Brooks – complexity, conformity, changeability, and invisibility, plus testability, reliability, traceability, and maintainability which we added.

It is established by complying with the essential principles of complexity science described in Chap. 4, including the **Nonlinearity** principle, the **Holism** principle (that **a whole is greater than the** sum **of its** parts), the **Dynamics** principle, the **Self-organization** principle, the **Self-adaptation** principle, the **Openness** principle, the **Initial Condition Sensitivity** principle, the **Sensitivity to Change** principle, the **Complexity Arises From Simple Rules** principle, etc. to develop the required new techniques and tools to efficiently slay Fred Brooks' software engineering werewolf as shown in Table 8.1. A corresponding mapping between the innovated techniques described in Chap. 6 and the targeted Issues is shown in Fig. 8.4.

Table 8.1 Issues and the solution technique mapping

Issue	Solution Techniques
Complexity	1. Synthesis Design and Incremental Implementation/Iteration/Integration
	2. Holistic, Actor–Action and Event-Response driven, Traceable, Visual, and Executable Approach for Functional Requirement Decomposition
	3. Holistic and Dynamic Traceability Technique
	4. Holistic and Traceable Diagram Generation Technique
	5. Virtual and Traceable Documentation Technique
	6. Refactoring Technique with Defect Prevention
Conformity	1. Holistic and Dynamic Traceability Technique
	2. Virtual and Traceable Documentation Technique
	3. Holistic and Traceable Diagram Generation Technique
Changeability	1. Holistic, Actor–Action and Event-Response driven, Traceable, Visual, and Executable Approach for Functional Requirement Decomposition
	2. Holistic and Dynamic Traceability Technique
	3. Comprehensive Software Testing Technique
	4. Defect Prevention-Based Quality Assurance Technique
	5. Holistic and Traceable Diagram Generation Technique
	6. Virtual and Traceable Documentation Technique
	7. Holistic and Virtual Version Comparison Technique
Invisibility	1. Holistic and Traceable Diagram Generation Technique
	2. Virtual and Traceable Documentation Technique
	3. Holistic and Dynamic Traceability Technique
Testability	1. Comprehensive Software Testing Technique
	2. Holistic and Traceable Diagram Generation Technique
	3. Virtual and Traceable Documentation Technique
	4. MC/DC Test Coverage Analysis and Graphical Representation
	5. Assisted Test Case Design
	6. Intelligent Regression Test Case Selection Technique
Reliability	1. Comprehensive Software Testing Technique
	2. Defect Prevention-Based Quality Assurance Technique
	3. MC/DC Test Coverage Analysis and Graphical Representation

(continued)

Table 8.1 (continued)

Issue	Solution Techniques
Traceability	1. Holistic and Dynamic Traceability Technique 2. Holistic and Traceable Diagram Generation Technique 3. Virtual and Traceable documentation technique
Maintainability	1. Holistic and Dynamic Traceability Establishment Technique 2. Holistic and Traceable Diagram Generation Technique 3. Virtual and Traceable Documentation Technique 4. Holistic, Actor–Action and Event-Response driven, Traceable, Visual, and Executable Approach for Functional Requirement Decomposition 5. Comprehensive Software Testing Technique 6. Holistic and Dynamic Traceability Technique 7. Defect Prevention-Based Quality Assurance Technique 8. MC/DC Test Coverage Analysis and Graphical Representation 9. Refactoring Technique with Defect Prevention 10. Assisted Test Case Design 11. Test Case Efficiency Analysis and Test Case Minimization 12. Intelligent Regression Test Case Selection Technique 13. Holistic, Global, and Side-Effect-Prevention Based Software Maintenance Technique

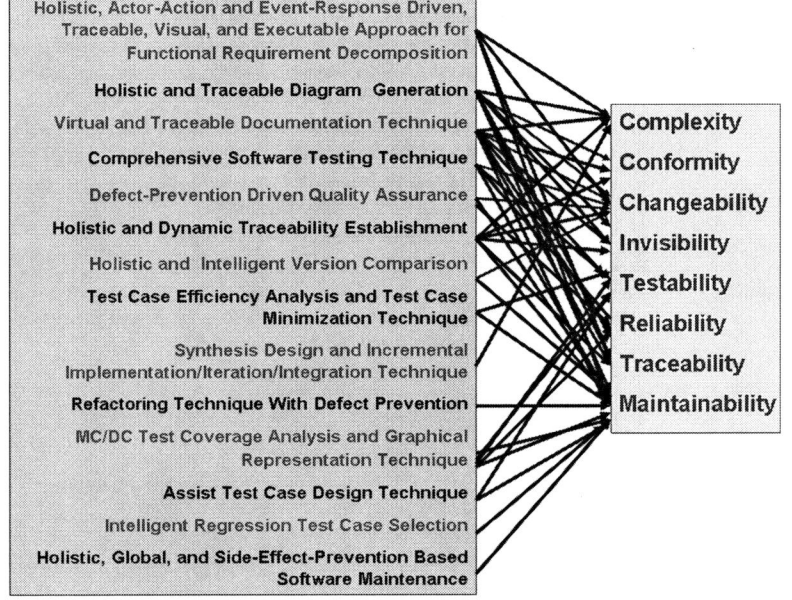

Fig. 8.4 Techniques and the targeted issues

8.5 The Graphical Representation of the NSE Process Model

The proposed NSE process model (Fig. 8.5) consists of the preprocess part and the main process part which is supported by a facility for automated and bidirectional traceability (see Fig. 8.6). Both parts are not really separated but combined together

8.5 The Graphical Representation of the NSE Process Model

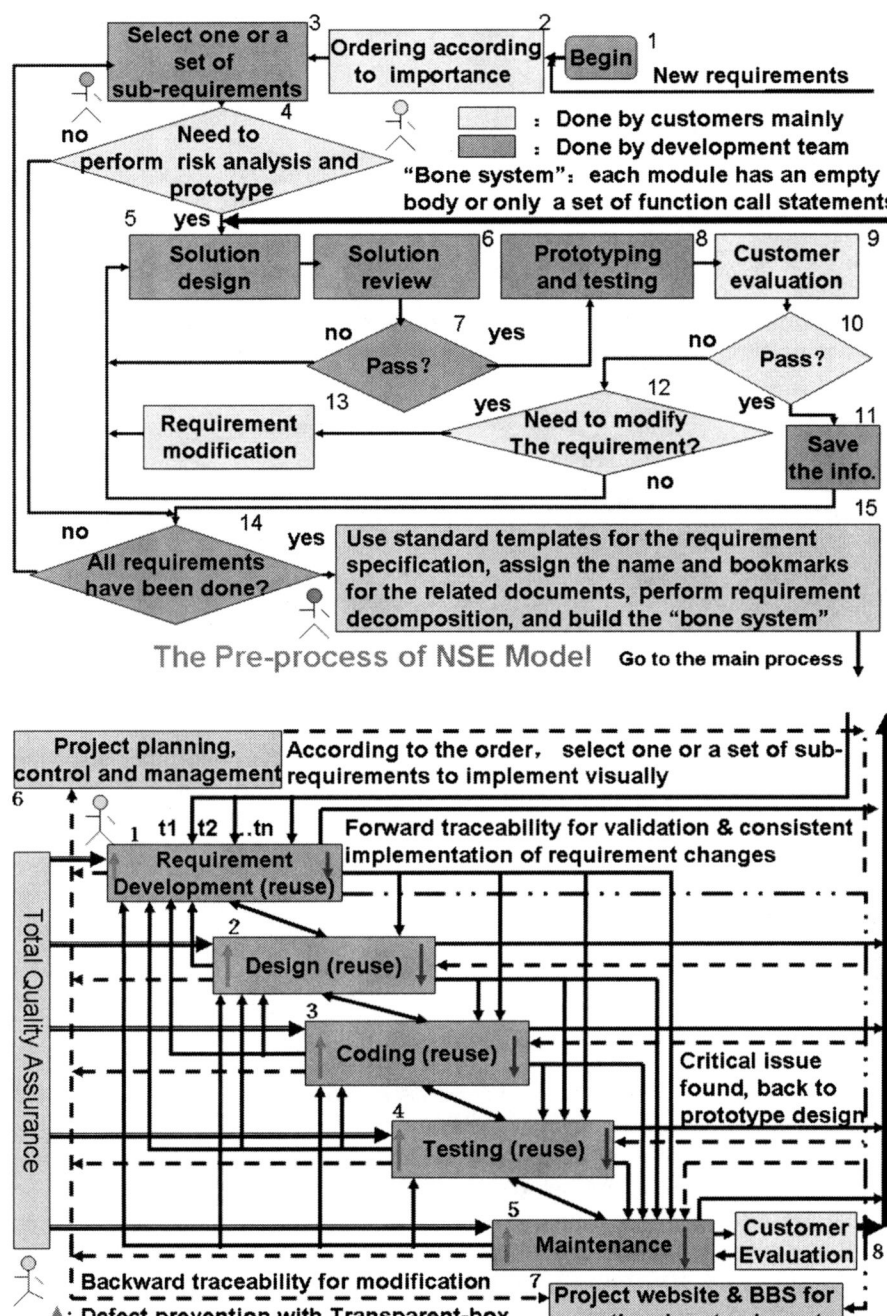

Fig. 8.5 NSE Process Model

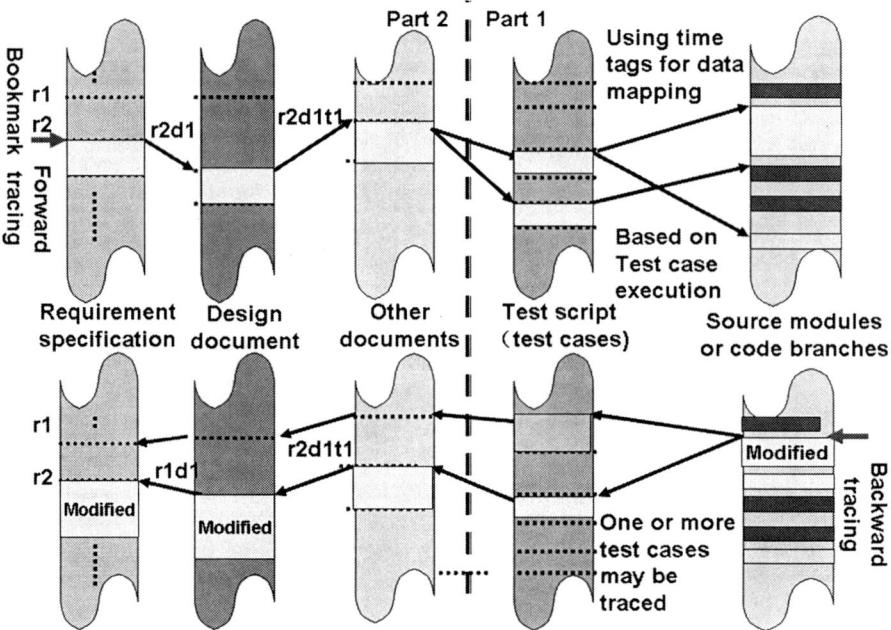

Fig. 8.6 The self-maintainable facility for bidirectional traceability

as shown in Fig. 8.5. If a critical problem is found in the main process for the implementation of a requirement using the solution method selected in the preprocess, the work flow may go back to the preprocess for the prototyping design and testing of a new solution method, and so on.

8.5.1 The Objectives of the Preprocess

The objectives of the preprocess are as follows:

(a) Working closely with the customer to assign priority to the requirements for better control of the development schedule and the budget.
(b) Performing prototyping design and evaluation for some unfamiliar requirements to reduce project development risk.
(c) Performing functional decomposition of the functional requirements using the Holistic, Actor–Action and Event–Response driven, Traceable, Visual, and Executable technique described in Chap. 7 (see the step 1 of the main process model for the application examples).
(d) Working closely with the customer to make a primary version of the requirement specification document using standard templates or the NSE requirement

specification template (see Appendix A) provided to prevent defects of missing something.
(e) Carrying out Synthesis Design of the system using the "Dummy Programming" technique (through the use of dummy modules) or reverse engineering from an old software system to complete a dummy system. According to the **Generative Holism Theory** of complexity science, the whole of a complex system may not be "**built**" from its components but exists (like a human embryo) earlier than its parts, then "**grows up**" with its parts (like human eyes). Some real application examples will be provided in Chap. 10.
(f) Performing cost estimation based on the necessary prototype design and test and review, and the functional decomposition of the functional requirements, and the dummy "whole" system designed.

8.5.2 The Objectives of the Main Process

The **objectives of the main process** are as follows:

(a) Implementing the requirements incrementally to make the software system "**grow up**" gradually.
(b) Combining the product development process and the maintenance process together through bidirectional traceabilities and defect prevention to greatly reduce the cost and effort spent in software maintenance through side-effect prevention.
(c) Combining project management and product development together by making the project plan and schedule charts and cost charts traceable with the requirement implementation to avoid budget overuse and schedule delay.
(d) Responding to requirement changes in real time with defect prevention through traceabilities among all artifacts.
(e) Supporting real-time communication among team members through traceable project Web sites for distributing development and speeding up the problem-solving process.
(f) Making the design documents and the source code traceable to each other.
(g) Performing dynamic testing in the entire software development lifecycle (including the requirement development phase and the design phase too) to prevent defects using the proposed Transparent-box method, which seamlessly combines functional testing and structural testing together.
(h) Assure the quality of the product being developed from the first step to the end through defect prevention and defect propagation prevention with various bidirectional traceabilities.
(i) Making it possible to help software organizations double their productivity, halve their cost, and remove 99–99.99% of the defects in their products.

8.5.3 The Objective of the Support Facility for Automated and Bidirectional Traceability

The objectives of the support facility for automated and bidirectional traceability are as follows:

(a) Helping software developers to prevent side effects in the implementation of software changes.
(b) Solving the conformity issue to make the documents and the source code traceable to each other.
(c) Removing the problems existing with a man-made Requirement-Traceability Matrix, which is inaccurate, time consuming, and almost unmaintainable.

This automated traceability facility is self-maintainable: no matter whether the contents of the documents are changed, or the test cases are changed, or the source code is changed, after regression testing, the bidirectional traceability will be automatically updated without manual work. For instance, when the source code is modified, after rerunning the test cases, new Time Tags will be inserted into the test case scripts and the test coverage database to map them together correspondingly.

8.6 The Major Steps of the Preprocess

The Major Steps of the preprocess are as follows:

Step 1. Start.
Step 2. Work with the customers to sort the initial requirements into several different classes such as "Critical," "Essential," "Needed," "Better to have," "Optional," and so on, and assign them corresponding priorities to control the product development plan and the schedule as well as the budget better. Usually the number of the initial requirements is about half of the final number of the requirements. With NSE, requirement changes or new requirements coming from the customers are welcome and responded to in real time to enhance the customers' market competition power and catch the best time for the customers' product to be available on the market. If necessary, some noncritical and nonessential functions may be temporarily given up.
Step 3. According to the assigned priorities, take one or a set of requirements to perform the preprocess (see the following steps).
Step 4. Check whether the requirement(s) are new to the development team to determine whether risk analysis and prototype design, testing, and evaluation for the requirement(s) are needed. If there is no need to do so, go to step 14; otherwise go to step 5.
Step 5. Compare different solution methods, then select the best one according to the development team's knowledge.
Step 6. Perform technology review and risk analysis for the selected solution method.
Step 7. If the selected solution method passes the technology review and risk analysis, go to step 8; otherwise return to step 5.

8.6 The Major Steps of the Preprocess

Step 8. Perform the prototype design and testing or reuse a suitable prototype and the test cases for the selected solution method.

Step 9. Provide all of the related material including the prototype design documents, the source code, and the test cases, as well as the test result to the customers for them to review.

Step 10. If the customers are satisfied with the prototype and the test result for the selected solution method, go to step 11; otherwise go to step 12.

Step 11. Save all of the information and make them ready to use for the implementation of the requirement(s) in the main process; then go to step 14.

Step 12. Get the customers' decision whether they want to modify the requirement(s) – if they want to modify the requirement(s), go to step 13; otherwise return to step 5.

Step 13. Perform the requirement modification by the customers, then go to step 5.

Step 14. Check whether all of the requirements have been handled – if so, go to step 15; otherwise go to step 3.

Step 15. After all of the requirements have been handled for the preprocess treatment, work closely with the customer to complete a preliminary version of the requirement specification using a standard template provided internally or the NSE requirement specification template (see Appendix A) to prevent something missing (more detailed requirement specifications should be completed incrementally in the main process); organize the requirement specifications and the related documents hierarchically (even if some documents have not been really designed) with inherited bookmarks as shown in Table 8.2 or meaningful bookmarks (so that when a

Table 8.2 Sample document hierarchy with inherited bookmarks

Document Hierarchy					
Project Name			Project Code		
Project Description					
The full path name of the Project feasibility report			Version number		
The full path name of the requirement specification			Version number		
Requirement 1	Bookmark	r1			
Description					
	The full path name of the related design document				
	Description		Bookmark	r1d1	
	The full path name of the related test specification				
	Description		Bookmark	r1d1t1	
	...				
Requirement 2	Bookmark	r2			
...					

document is traced, the document will be shown from the position indicated by the corresponding bookmark), or organize the document directories as shown in Fig. 8.7 with a "Bookmark Information List" file in each directory to indicate three elements in each line, including (1) the bookmark name; (2) the corresponding document name; and (3) the file type (the source

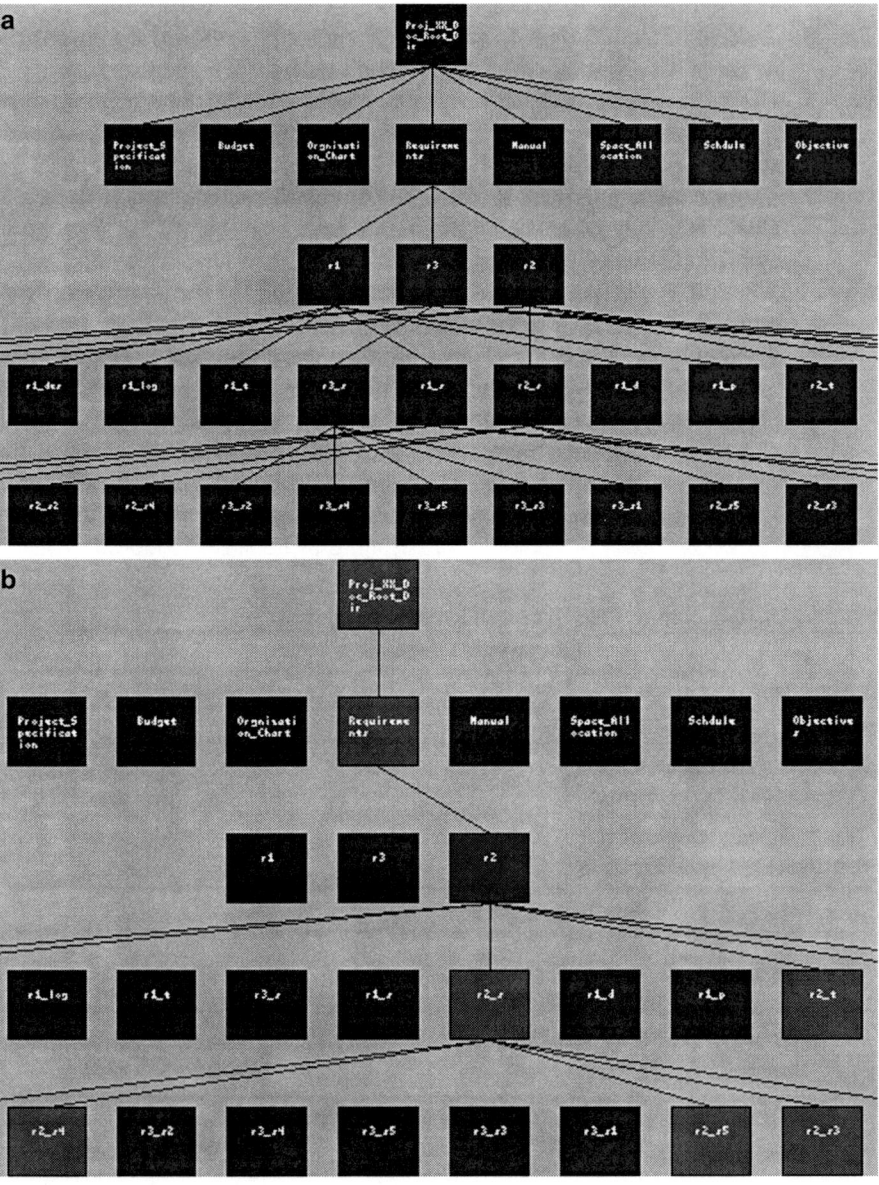

Fig. 8.7 Document directory hierarchy design through dummy programming: (**a**) An application example; (**b**) the document subdirectories of requirement r3 highlighted

8.6 The Major Steps of the Preprocess

code of the dummy program is listed in Appendix A), then perform the decomposition of the functional requirements and the required functions for nonfunctional requirements and defect removal through dummy programming (see Fig. 8.8) using the HAETVE technique introduced in Chap. 11 and the tools; perform top-down system decomposition as shown in Fig. 8.9, and complete the dummy system hierarchy and defect removal (through dummy programming too) according to the prototype design

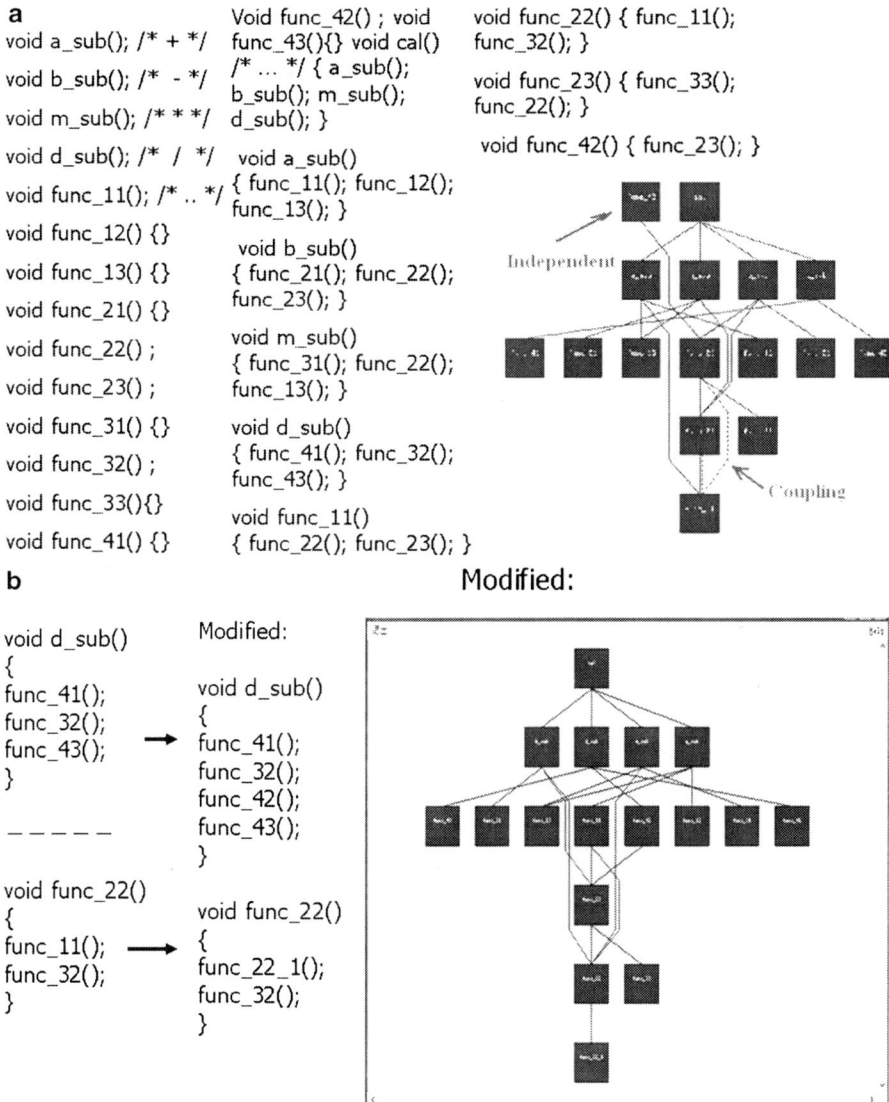

a

```
void a_sub(); /* + */
void b_sub(); /* - */
void m_sub(); /* * */
void d_sub(); /* / */
void func_11(); /* .. */
void func_12() {}
void func_13() {}
void func_21() {}
void func_22() ;
void func_23() ;
void func_31() {}
void func_32() ;
void func_33(){}
void func_41() {}
```

```
Void func_42() ; void
func_43(){} void cal()
/* ... */ { a_sub();
b_sub(); m_sub();
d_sub(); }

void a_sub()
{ func_11(); func_12();
func_13(); }

void b_sub()
{ func_21(); func_22();
func_23(); }

void m_sub()
{ func_31(); func_22();
func_13(); }

void d_sub()
{ func_41(); func_32();
func_43(); }

void func_11()
{ func_22(); func_23(); }
```

```
void func_22() { func_11();
func_32(); }

void func_23() { func_33();
func_22(); }

void func_42() { func_23(); }
```

b

```
void d_sub()
{
func_41();
func_32();
func_43();
}
-----
void func_22()
{
func_11();
func_32();
}
```

→

```
Modified:

void d_sub()
{
func_41();
func_32();
func_42();
func_43();
}

void func_22()
{
func_22_1();
func_32();
}
```

Modified:

Fig. 8.8 Decomposition of functional requirements of a sample project (**a**) and defect removal through dummy programming (**b**)

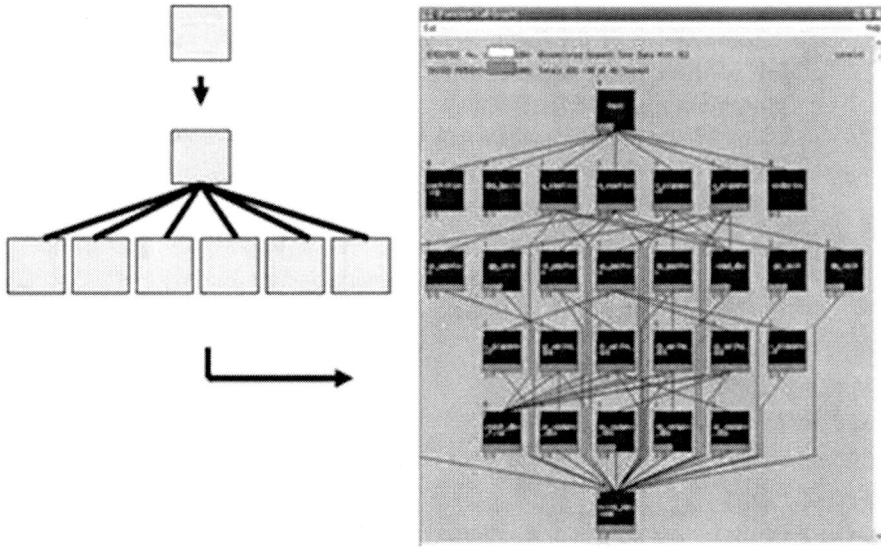

Fig. 8.9 Software system hierarchy design through dummy programming

System decomposition design (using "Bone" programming too) and error correction

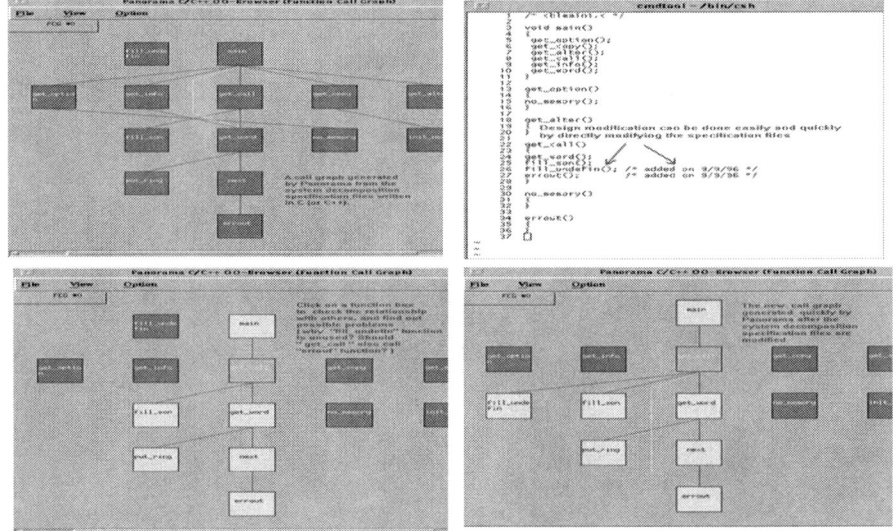

Fig. 8.10 Sample dummy system design and defect removal through dummy programming according to the result of prototyping and the result of the functional decomposition of the requirements

result and the result of the functional decomposition of the requirements as shown in Fig. 8.10; finally make a corresponding project development plan and cost estimation table, and complete the Project Feasibility Report.

8.7 The Major Steps of the Main Process

The major steps of the main process are as follows:

Step 1: According to the project development plan, the priority assigned to the requirements, and the "whole" dummy system designed in the preprocess, take one or a set of requirements to implement visually. It is recommended to select the critical and essential requirements (about 20% of the initial requirements) first to implement and form an essential version of the software product (which should be executable) through **incremental integration development** which is different from traditional **incremental development** as shown in Figs. 1.2 and 1.3. **Incremental integration development** means **making a software system grow up incrementally** – each time only a new module will be coded and tested to form a new executable version as shown in Fig. 1.4.

The NSE process model supports the *defect prevention and traceability* driven software development method (see Chap. 10 for details) through the following activities to be performed by the development team working closely with the customers:

1. Update the requirement specification **to prevent the defects of something missing.**
2. Design the document hierarchy (as shown in Table 8.2) including the test specification documents and the test case scripts **to prevent the defects of untestable requirements**.
3. Check and improve the result of the functional decomposition of the functional requirements performed initially in the preprocess **to further remove the defects in the functional decomposition of the functional requirements** using the **Holistic, Actor–Action and Event–Response driven, Traceable, Visual, and Executable approach** (see Chap. 11).
4. Use all of the documents related to the prototype design and testing performed in the preprocess phase to implement the corresponding requirement(s) according to the approved solution method; if there is a need to use a new solution method, the new solution method must pass the preprocess treatment with prototype design and testing and evaluation **to prevent the defects coming from unrealizable requirements.**
5. If it is possible, reuse approved documents and test cases (test script files) suitable for the corresponding requirements **to reduce the defect rate**.
6. If there is a need for the customers to add new requirements after some partially completed working versions have been delivered to the customers for review, respond in real time by going back to the preprocess through

the new solution method selection and inspection, prototype design and testing, customers' review, and so on if necessary, **to prevent the defects of unrealizable new requirements**.
7. If there is a need for the customers to modify some requirements in the main process phase, respond to it in real time too by implementing the modified requirements through bidirectional traceability **to prevent the defects coming from the side effects of the modification**. If it is necessary to use a new solution method, go back to the preprocess phase.
8. Realize visual development in the entire software development lifecycle (not only in modeling) **to greatly increase the defect removal rate**: according to complexity science, the characteristics and behavior of a complex system are determined by both the whole and its parts, so it is needed to use Holistic and Traceable Diagram Generation technique (see Chap. 7) and mainly the interactive and traceable 3J graphics (J-Chart, J-Diagram, and J-Flow) proposed and implemented by me to make the entire software development process visible.
9. Perform dynamic testing plus formal inspection and review in the entire software development lifecycle using traceable documents and traceable source code **to prevent various kinds of defects**; even if only the first one of the requirements is being handled before the beginning of the corresponding program design and coding, we should already have a set of related documents to be checked for consistency, including the objectives document, the project development plan/schedule, the requirement specification, the test requirement specification, the prototype design and test result and the inspection and review reports, and so on, so that we should design a virtual "main" program and the corresponding test script files first, then dynamically execute the program with the test scripts using the Transparent-box testing method (proposed and implemented by me, see Chap. 16) – It is a very important feature of NSE for ensuring the quality of a software product in the requirement development phase before the corresponding program design and coding using Transparent-box testing tools dynamically.

Step 2: Apply the **Synthesis Design and Incremental Growing up (Implementation, Iteration, and Integration) Technique** with the **Holistic and Traceable Diagram Generation Technique** to further perform preliminary design for the selected requirement(s) according to the detailed requirement specification to improve the corresponding part of the dummy system obtained in the preprocess phase, then perform formal inspection and review using traceable documents, and design the corresponding test cases to dynamically test the result of the preliminary design using the Transparent-box method to prevent inconsistency defects through bidirectional traceability that is established automatically. After that, perform detailed design for the selected requirement(s) according to the result of the preliminary design with formal inspection and review using traceable documents, and dynamic testing like what was done in the

8.7 The Major Steps of the Main Process

preliminary design process. For a detailed description on software design engineering under NSE, see Chap. 12.

Step 3: Apply the **Synthesis Design and Incremental Growing up (Implementation, Iteration, and Integration) Technique** to perform incremental coding: on the generated system decomposition chart (a call graph), highlight the corresponding key module(s) and the related modules for the selected requirement(s), then assign an incremental bottom-up coding order to the modules as shown in Fig. 8.11.

As shown in Fig. 8.12, when we are writing a function call statement to a called module which has been coded, we can read the diagrammed source code in another window to know how many parameters are needed, their types, and their sequence to prevent inconsistency defects between the module interfaces.

Usually, a logic defect is hard to detect because the program source code is written in text format, and a program with a logic defect can be executed without providing error messages but the result is incorrect. For solving this kind of problem, users may use Panorama++ to generate the control flow diagram in J-Flow notation, or the logic diagram in J-Diagram

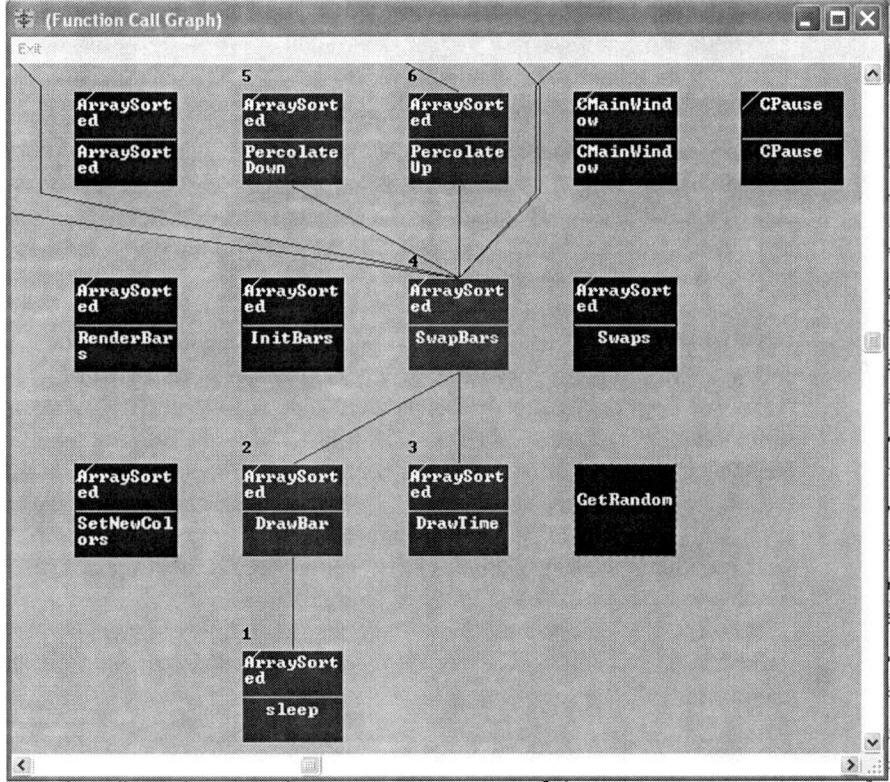

Fig. 8.11 An example of bottom-up ordering for incremental coding

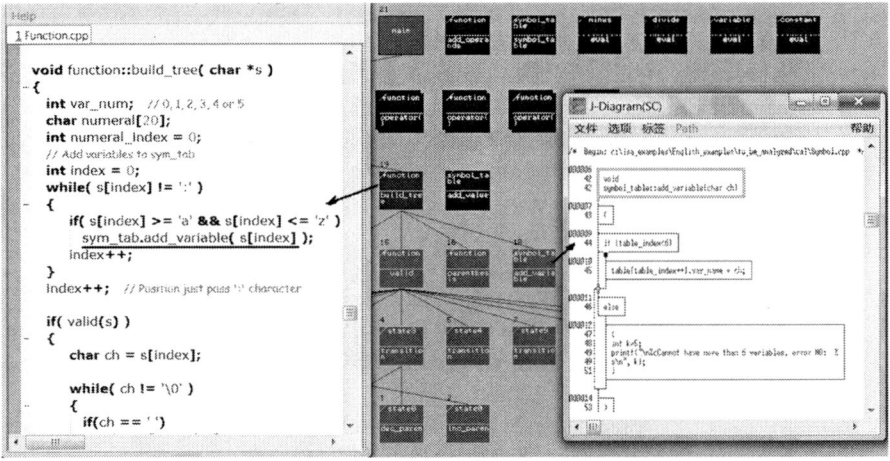

Fig. 8.12 Incremental coding with defect prevention

notation, to graphically represent the program for finding the logic defects better. An application example is shown in Fig. 8.13.

If something critical is found in the coding process, go to the upper phases through backward tracing, or if the solution method does not satisfy the requirement(s), go back to the preprocess again.

Step 4: Perform incremental unit testing with integration testing, and finally system testing, mainly using the Transparent-box approach to combine functional and structural testing together with the capability to establish automated and bidirectional traceability among all documents and the source code for helping users to remove the inconsistency defects. At the same time, perform MC/DC test coverage analysis, performance analysis, memory leak analysis, and memory usage violation check. According to the incremental coding and testing order, when we code a module, all modules called by it must have been coded already so that there is no need to design and use a stub module to replace a called module – in this way the unit testing also becomes integration testing with all modules being called together. When a module being called needs to return some special values, two applicable approaches are provided in the Appendix C.

If something critical is found in the testing process, treat the situation as some critical issues found in the coding process.

With the NSE support platform, Panorama++, unit testing can be performed in a semiautomated way through a tool called Panounit whose features include the following:

1. Semiautomatically designs the corresponding driver program – the main() function.
2. Automatically put the driver, the program unit being tested, and all modules called by the program unit together.

8.7 The Major Steps of the Main Process

a

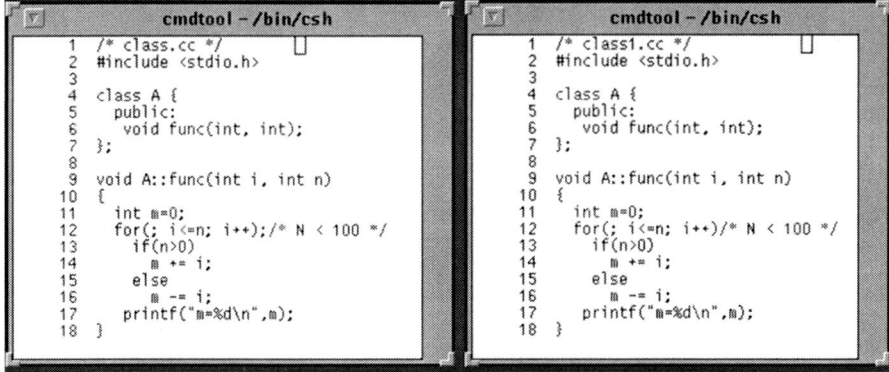

b

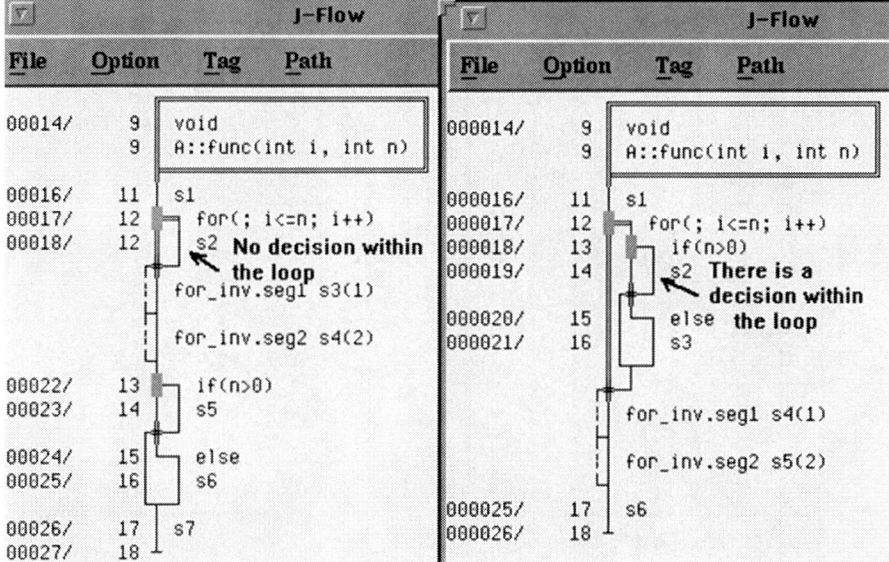

Fig. 8.13 Finding out logic defects through graphical representation of the source code

3. If it is retesting for an existing product, Panounit will search all locations where a value is assigned to a global variable or a static variable, and lists those values for users to choose.
4. Supports assertion setting and verifying the value in any valid location.
5. Supports semiautomated test case design.
6. Performs MC/DC test coverage analysis and test result display using J-Chart, J-Diagram, and J-Flow with untested branches and conditions highlighted.
7. Automatically determines the test result – pass or fail.

c Two J-Diagrams generated by Panorama for logic debugging (clearly showing the difference between the two programs)

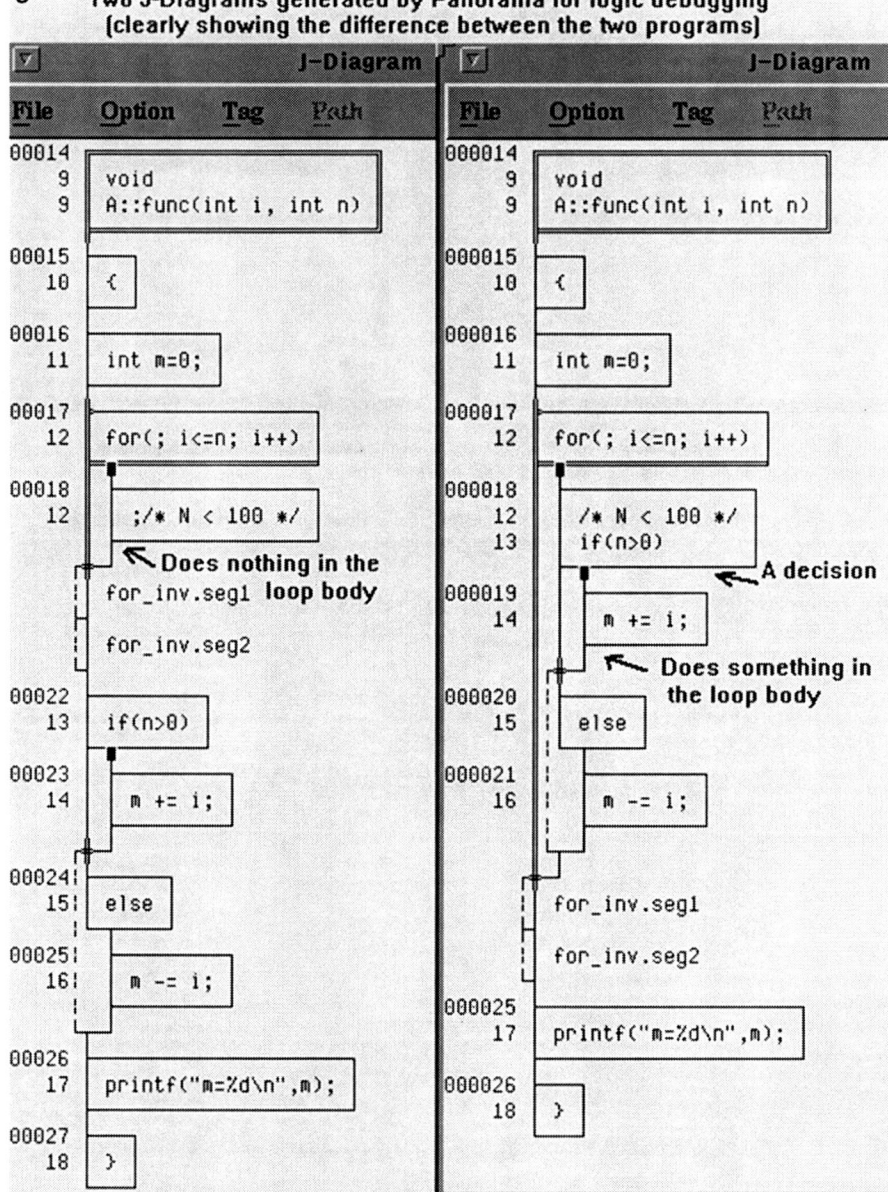

Fig. 8.13 (continued)

The GUI of Panounit is shown in Fig. 8.14.

In the system testing process, Panorama++ also offers the capability to capture users' GUI operations, and plays them back automatically for regression testing, and the capability for MC/DC test coverage analysis for the entire product, plus performance analysis, test case efficiency analysis

8.7 The Major Steps of the Main Process

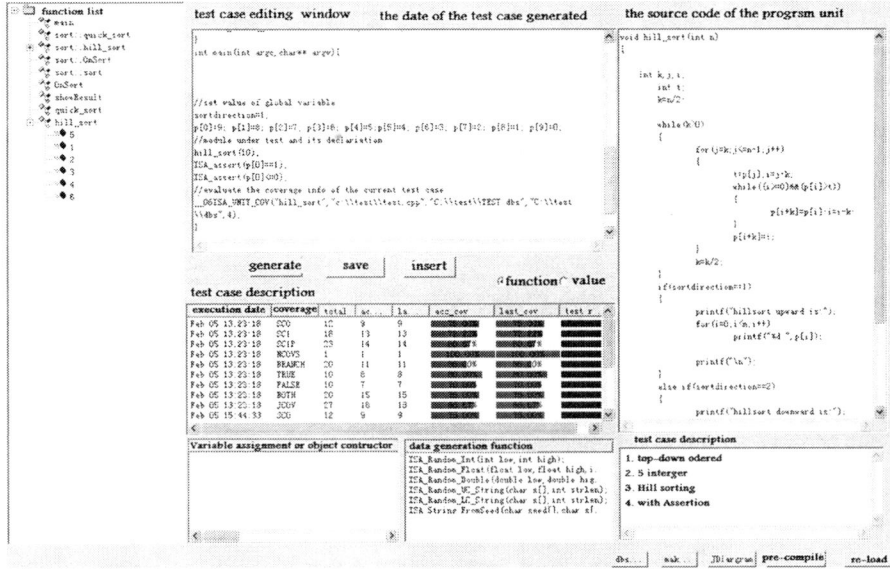

Fig. 8.14 The GUI of Panounit tool for unit and integration testing

and test case minimization for efficient regression testing after code modification. With system testing, an automated and bidirectional traceability among all artifacts including the source code will be established for defect prevention.

Chapters 14 and 15 will discuss the software testing support in detail.

Step 5: Perform systematic, disciplined, and quantifiable software maintenance using the **Holistic, Global, and Side-Effect-Prevention Based Software Maintenance** technique:

1. Respond to requirement changes and new requirements or code modifications in real-time to implement them holistically and globally with side-effect prevention.
2. Bring great savings to regression testing after requirement changes or code modification through test case efficiency analysis and test case minimization, plus intelligent test case selection through backward traceability between test cases and the source code.
3. Make it possible to reduce the cost and effort spent in software maintenance from more than 75% of the total with the old-established paradigm to about 25% of the total with NSE, so that it is possible for NSE to help software organizations to double their productivity and halve their cost – with NSE there is no essential difference between the software development process and the software maintenance process – in both processes, software changes are supported in real time with side effects prevented through various kinds of bidirectional traceabilities (see Chap. 18). An application example of software maintenance

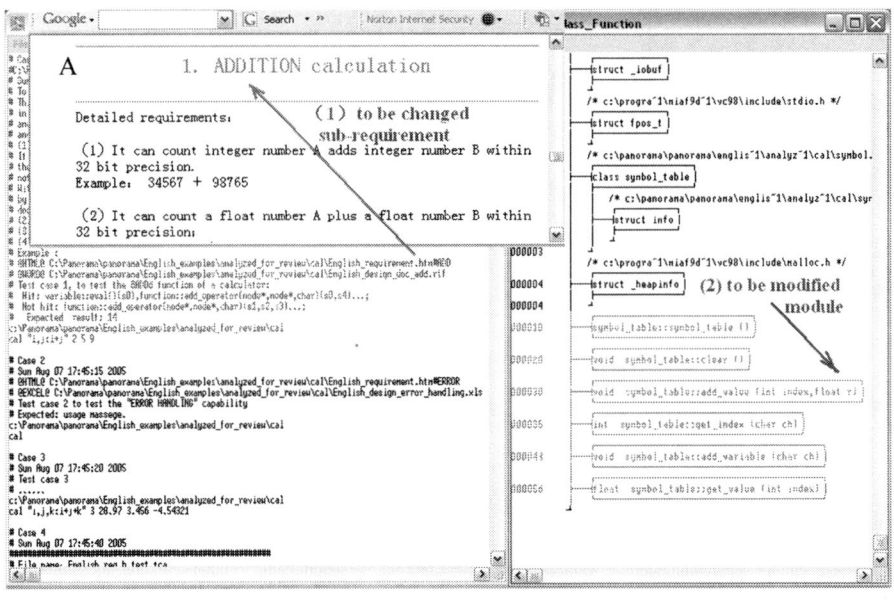
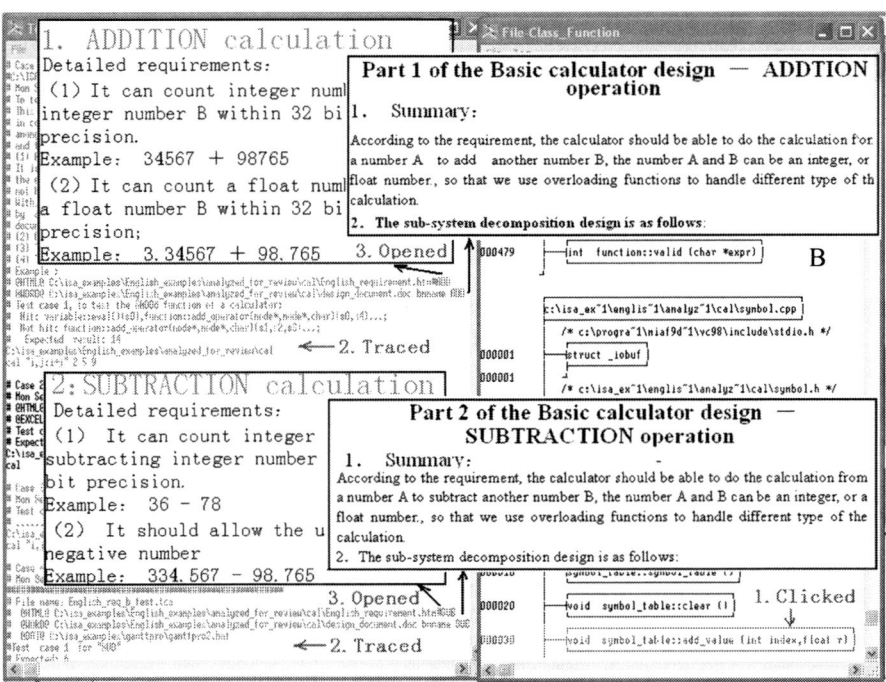

Fig. 8.15 Defect prevention for requirement changes performed by the NSE support platform, Panorama++: (**a**) Performs forward tracing for a requirement change (through the corresponding test cases) to determine what modules should be modified. (**b**) Performs backward tracing to check related requirements of the modules to be modified for preventing requirement conflicts (in this example, two requirements are related). (**c**) Checks what other modules may also need to be changed with the modification. (**d**) After modification, check all related call statements for defect

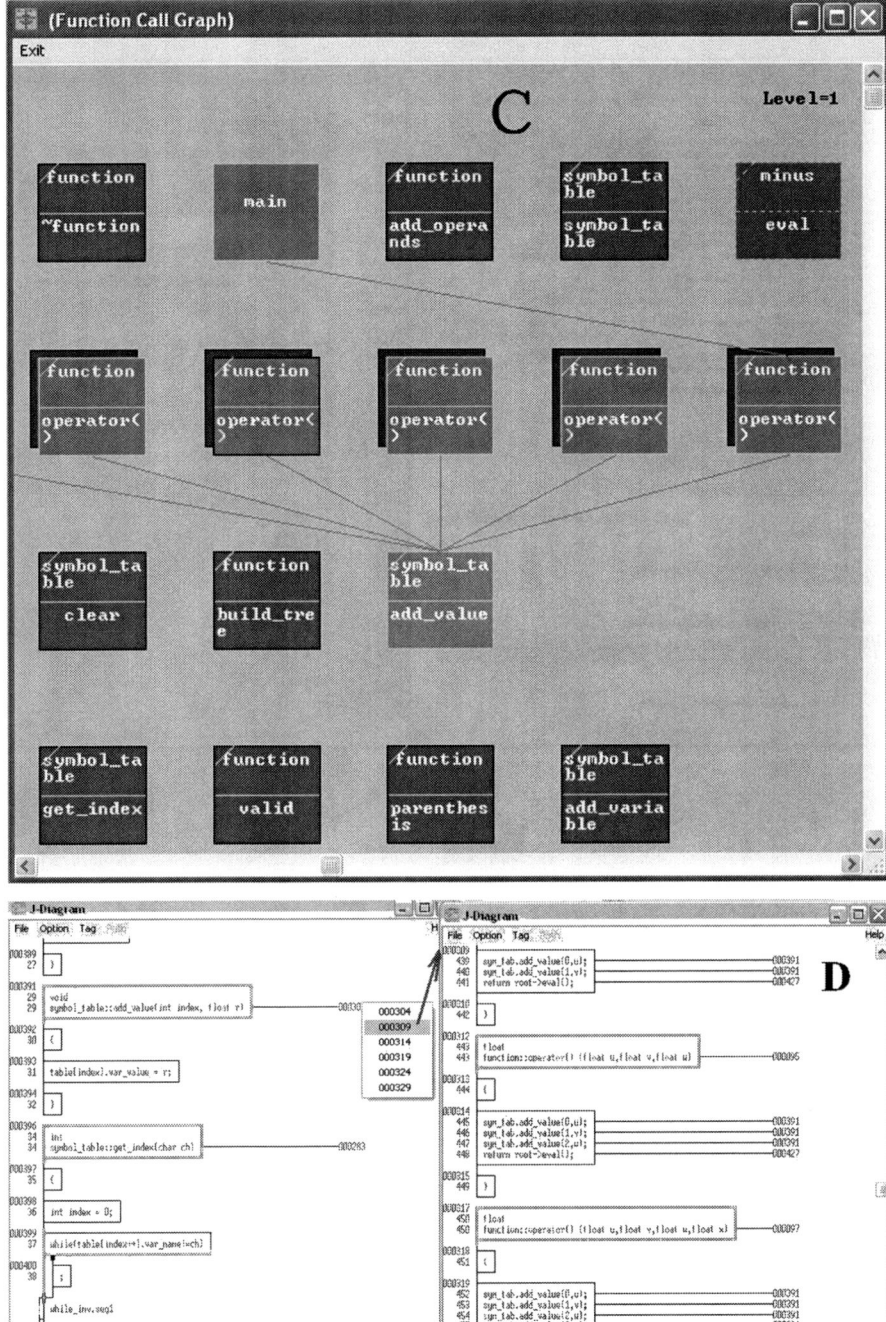

Fig. 8.15 (continued) prevention. (**e**) Efficient regression testing through related test case selection based on backward traceability. (**f**) Performs backward tracing to find and modify inconsistent documents after code modification

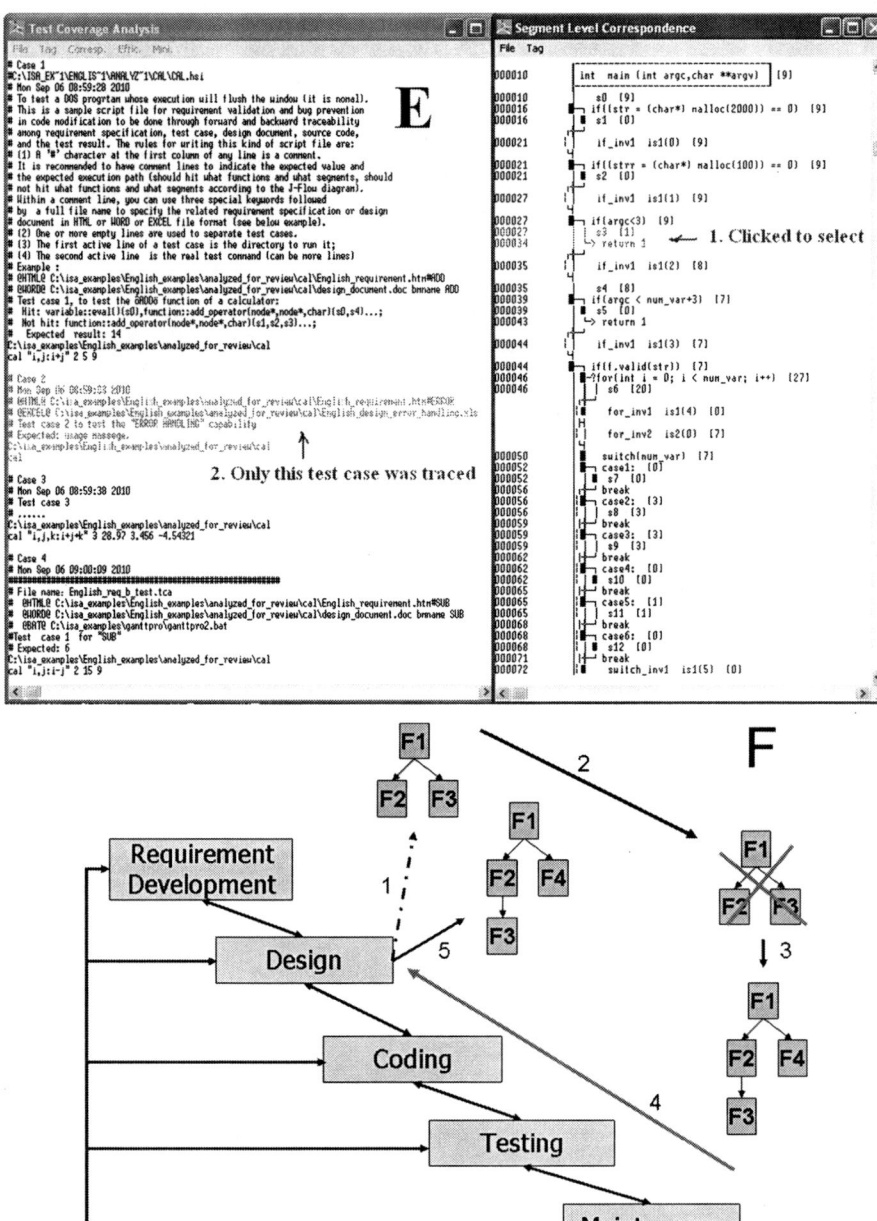

Fig. 8.15 (continued)

support is shown in Fig. 8.15 – defect prevention for requirement changes performed by the NSE support platform, Panorama++.

4. If there is still something wrong after the implementation of a requirement change or code modification, perform intelligent version comparison to locate the defects.

8.7 The Major Steps of the Main Process

Step 6: Closely combine the project management process and the product development process together, making the project plan, schedule charts, and cost estimation reports traceable with the requirement implementation and the source code, for better control of the cost and project development schedule. An application example to trace the project plan/schedule with the requirement implementation is shown in Fig. 8.16.

Step 7: Establish a project Web site and the related technical forum for real-time communication and technical discussion among team members to report the progress of the project, and to open technical discussions for brainstorming and report a variety of related events, error handling processes and results, and especially unexpected events in order to discuss the response, which can all be traced back through the bidirectional and automatic traceability mechanism to update them in real time. It goes without saying, setting up the project Web site and the project forums is not difficult. It allows the content of Web sites and forums or topics to be traceable with the corresponding requirements, design documents, test cases, and source code automatically for opening them and achieving real-time updates, which is its real value. An application example to trace a corresponding Web page from a test case is shown in Fig. 8.17.

Step 8: Frequently deliver working products to the customer for review and evaluation from the beginning to the end of the software development lifecycle, even if there is no real output for a dummy system designed in

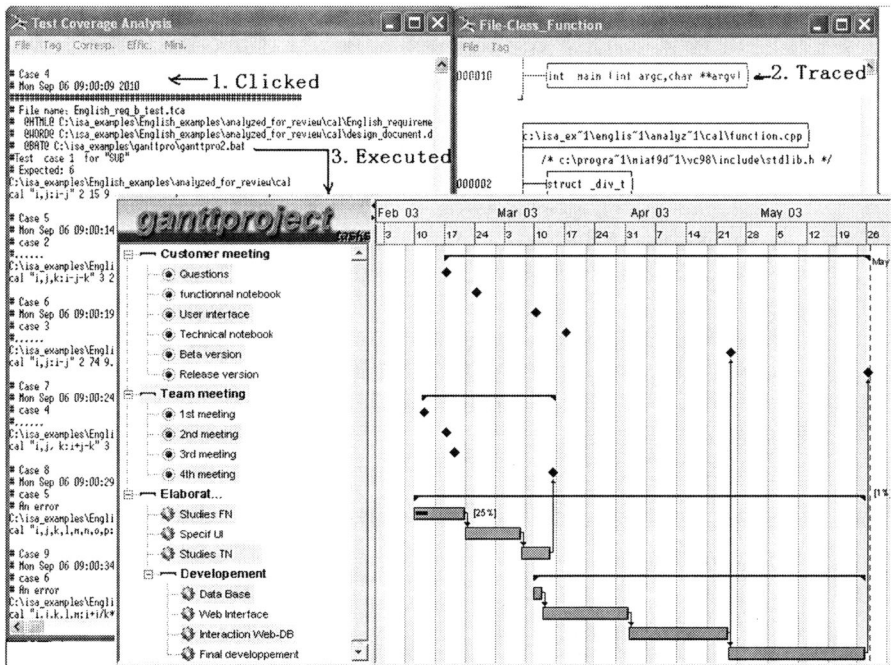

Fig. 8.16 An example of tracing a requirement to the project development schedule

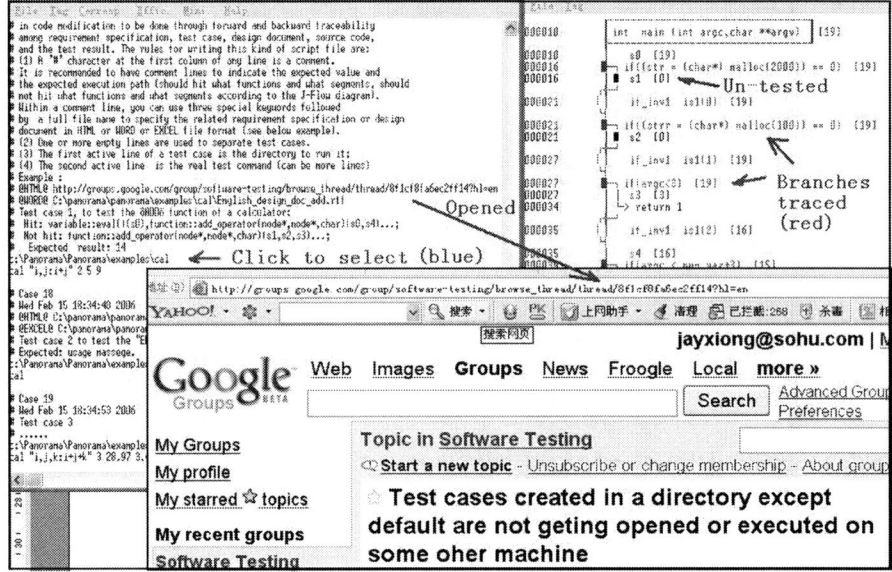

Fig. 8.17 An application example to trace a test case to a related Web page

the requirement development phase. Get the customer's feedback to improve the product development process and the corresponding result. Each time when a working version of the product is delivered to the customer, the related test case scripts should also be delivered so that the customer can easily duplicate the process and view the results directly. Finally, when the product is completed and delivered to the customer, not only should the entire product with the program, data and documents be delivered, but also the database built through static and dynamic program measurement plus a set of Assisted Online Agents (automated and intelligent tools – if the database is built using the NSE platform, those Assisted Online Agents can be distributed without charge) should be delivered to the customer to efficiently handle the issues of complexity, invisibility, conformity, changeability, reliability, and traceability, so that the acceptance testing can be done in a fully automated way and the delivered product can be easily maintained on the customer side.

8.8 The Support Facility for Automated and Bidirectional Traceability

As shown in Fig. 8.6, the main facility for bidirectional traceability consists of two parts:

1. Part I.
   Part I of the facility is related to the traceability between test cases and the corresponding source code executed by running the test cases. It is done with the use of Time Tags which are automatically inserted into both the test case descriptions and the corresponding test coverage database. For instance, if test case 1 is executed at 09:00 AM on September 2, 2009, and test case 2 is executed at 10:00 AM on the same day, and test case 3 is executed at 11:00 AM on the same day, then the three different Time Tags will be inserted into the three test cases and the corresponding test coverage database separately. So, when test case 2 is selected for forward tracing, the Time tag of 10:00 AM on September 2, 2009 will be taken from the test case description to search the test coverage data with the same time tag, so the corresponding test coverage data will be read and displayed on the corresponding control flow diagram shown in J-Flow notation. On the other hand, when a module or code segment shown on the J-Flow diagram is selected, the related time tags (which can be more than one) used to indicate what time the module or segment was executed will be taken to search the test case descriptions to see how many test cases there are with the mapping time tags through backward tracing, then it will highlight all test cases mapped on the window showing the test case script.
2. Part II.
   Part II of the facility is to extend the bidirectional traceability from test cases and the source code to include all related documents, the test cases, and the source code. It is done using some key words (written into the comment part of the description of the test case) such as @WORD@, @HTML@, @BAT@, @PDF@, and @EXCEL@ followed with the corresponding file path and a bookmark to indicate the format of the document, the full path name of the file, and the corresponding bookmark, so that when a test case is selected for forward tracing, or a module or segment is selected for backward tracing, the corresponding document will be opened and shown from the location indicated by the bookmark.

   This facility is self-maintainable without manual rework – when the document is modified or the test case parameter is modified or the source code is modified, after rerunning the test case script, new Time Tags will be inserted into both the test case description and the test coverage data to update the facility automatically.

   Chapter 9 will discuss this facility in more detail.

8.9 The Manifestation of the Essential Principles of Complexity Science in the NSE Process Model

The major essential principles of system science and complexity science are applied within the creation of the NSE process model as shown in Fig. 8.18, particularly the Holism principle which is not only applied in the preprocess but also

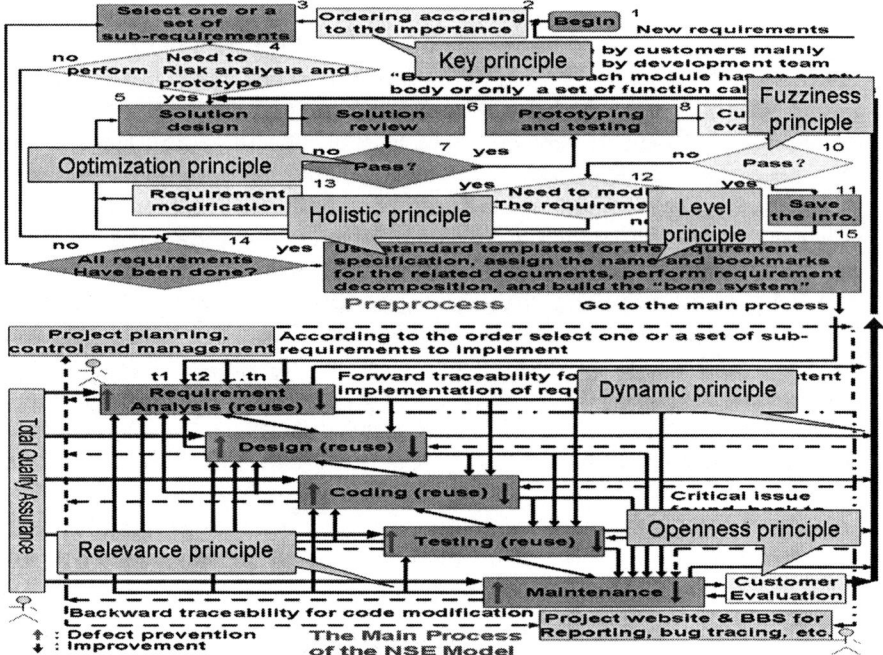

Fig. 8.18 The manifestation of the essential principles of complexity science

applied in the main process of the NSE process model, including holistic requirement development, holistic system design, holistic diagramming, holistic documentation, holistic testing, holistic quality assurance, holistic maintenance, and holistic version comparison.

Why is the Holism principle applied into all phases and so many activities? Because software is not a linear system but a nonlinear complex system, where small changes made locally will affect the entire system through the "Butterfly Effect." For instance, when the implementation of a requirement change or code modification is performed locally and blindly with the old-established software engineering paradigm without bidirectional traceabilities to prevent side effects, the entire system may be affected with inconsistency defects, so that the quality of the product will become unreliable. But with the NSE process model, the modification is performed holistically and globally with side effects prevented to avoid inconsistency defects.

8.10 The Major Features and Characteristics of the NSE Process Model

The major features and characteristics of the NSE process model include the following:

8.10 The Major Features and Characteristics of the NSE Process Model

1. **Dual-process:** NSE model consists of the preprocess and the main process. They are different but also closely linked together. The objectives of the preprocess and the main process are different as described in Sect. 8.3.
2. **Nonlinear:** The NSE model is established on complexity science and supported by facilities for two-way multilevel automated traceabilities to avoid a series of shortcomings existing with the linear process models under the old-established software development paradigm. Unlike the linear model which assumes that the upper processes are correct so that the only need is to continue to carry out the lower-level processes – it makes the existing defects easy to propagate from the upper phases to the lower phases and the cost for removing the defects increase 10 to 100 times or more – the NSE process model always assumes that there may be defects introduced in the upper phases so that there is a need to check and remove the defects in the upper phases through dynamic testing using the Transparent-box method and backward traceability that is established automatically. Similarly, changes made in the upper phases may affect the work products obtained in lower-level phases, so that there is also a need to check and remove the inconsistency defects in lower-level phases through forward traceability.
3. **Parallel with Multiple tracks:** "Much of software architecture, implementation, and realization can proceed in parallel" [Bro95-p233]. For reducing waiting time and speeding up software development processes, the NSE process model supports tasks being performed in parallel with multiple tracks through bidirectional traceability. Some application examples are shown in Figs. 8.19–8.21.
4. **Real time:** "Timely updating is of critical importance" [Bro95-p235]. The NSE process model supports real-time updating of the system – even if only one new module is completed and integrated, a new version of the entire executable

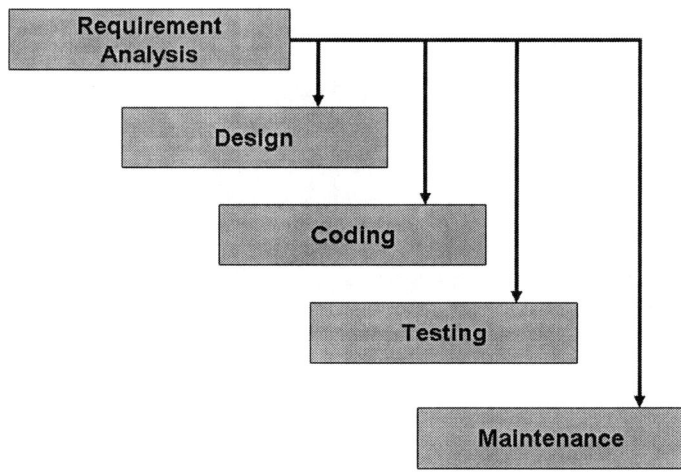

Fig. 8.19 Supporting parallel work for requirement validation through forward traceability

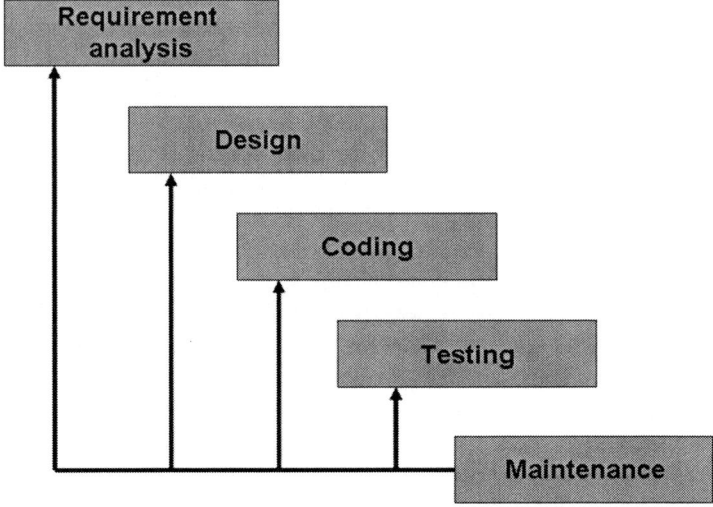

Fig. 8.20 Supporting parallel work for consistent code modification through backward traceability

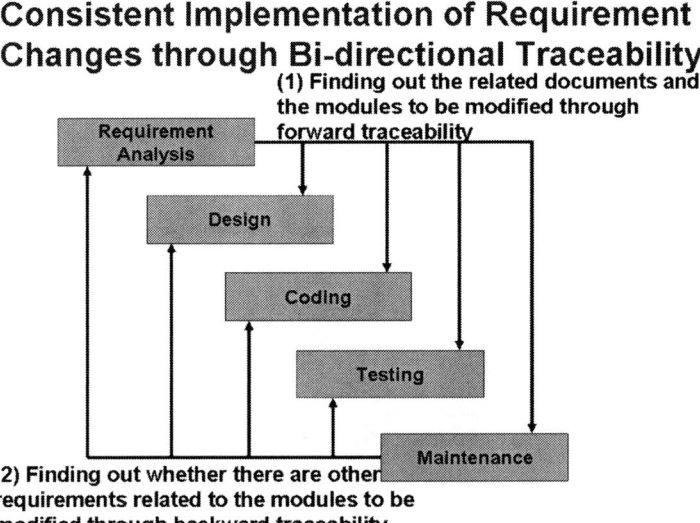

Fig. 8.21 Supporting parallel work for consistent implementation of requirement changes

system will be updated to check the progress and effects. The NSE process model also supports requirement changes in real time to implement the changes with defect prevention through bidirectional traceabilities for increasing the customer's market competition power.
5. **Incremental development with two-way iteration:** The NSE process model supports incremental development with two-way iteration, including refactoring to handle highly complex modules and performance bottlenecks with side-effect prevention through various traceabilities. When a critical issue is found in the main process, the work flow may go back to the preprocess for selecting a better solution method, and so on.
6. **The software development process and software maintenance process are combined together seamlessly:** With the NSE process model, there is no big difference between the software development process and software maintenance process – according to SPR's report[Jon02], "Requirements sometimes grow at >2% per month," so a 2-year product may double the requirements at the end – it means that the development process also needs to handle requirement changes. The NSE process model support safe software changes through various automated traceabilities to prevent side effects in the implementation of the changes, whether in the software development process or the maintenance process. Particularly, with the NSE paradigm, a software product will be delivered to the customer with the computer program, the data used, all related traceable documents, plus the database built through static and dynamic measurement of the product, and a set of Assisted Online Agents (automated and intelligent tools) to support testability, reliability, and efficiently handle the issues of complexity, changeability, conformity, and invisibility to make the product maintainable at the customer site with the same conditions as in the development site.
7. **The software development process and the project management process are combined together closely:** With the NSE process model, all documents including the project management documents such as the project development plan, the schedule chart, the cost estimation report are traceable with the requirement implementation and the source code for better control of the product development. NSE process model also supports the critical requirements and most important requirements being implemented early with the assigned priority to avoid budget overuse – if necessary, some optional requirements and not so important requirements can be ignored temporarily.
8. **Reusable Component-Based Software Development support:** The NSE process model supports component reuse in all phases if the reusable components are qualified as "Broken Limbs" rather than "Artificial Limbs" – based on complexity science, a complexity system is not **built** from its parts but is **growing up** with its parts, so that a reusable component must be qualified as a Broken limb with self-adaptive capability – at least no negative effects on the system quality, no overuse of the system memory, no memory leaks, no negative effects on the performance, fully tested with test cases for verification, and fully fulfills the functionality required.
9. **Adaptation focused rather than predictability focused:** The entire world is always changing, so the NSE process model is adaptation focused rather than

predictability focused – it supports requirement changes, code modifications, data modifications, and document modifications to make them consistent and updated with side-effect prevention in the implementation of the changes.

10. **Defect prevention driven**: The NSE process model is defect prevention driven in the entire software development lifecycle through various kinds of traceabilities and the use of Transparent-box testing, plus inspection and review using traceable documents and traceable and diagrammed source code.

11. **People are considered as the first-order driver for software development** – One of Manifestos for Agile Software Development is "Individual and interaction over processes and tools." In the paper, "**Characterizing people as non-linear, first-order components in software development,**" Alistair A.R. Cockburn stated that "I now consider the characteristics of people as 'the dominant, first-order' project driver," and "**People tend to inconsistency.**" When people like Alistair A.R. Cockburn consider "people as the first-order" to software development, they focus on how to trust and support people better for their jobs but ignore the other side of people's effect on software development – almost all defects introduced into software products are made by people, the developers, and the customers. So NSE supports people in two ways: one is to support them with better methodology, technology, and tools; another one is to prevent the possible defects to be introduced into the software products by people – it is done mainly through various automated and bidirectional traceabilities.

12. **Better support for people:** The NSE process model with its support platform Panorama++ provides better support for the software development team members and the customers:

    (a) **Empowered customers:** With the NSE process model, customers assign priority to the requirements, review the solution methods and the prototype design as well as the test results, have all working versions delivered to them for review from the dummy system to the final products, make requirement changes or add new requirements without worrying about the side effects because the implementation of requirement changes is done with defect prevention through various automated traceabilities, particularly the outsourcing products developed with the use of the NSE paradigm are now truly maintainable because the products are delivered to them with the programs, the data used, the documents, and the database built in the static and dynamic measurement of the product, plus a set of Assisted Online Agents to make the product visible, testable, reliable, and maintainable.

    (b) **Confident Project Manager:** Most software projects fail because of missed schedules, blown budgets, and flawed products. But with NSE, the requirements are assigned priorities according to their importance for better control of the development schedule and budget. With the application of the Holistic, Global, and Side-Effect-Prevention Based Software Maintenance technique, the effort and cost spent in software maintenance can be greatly reduced. The product quality is assured through defect

8.10 The Major Features and Characteristics of the NSE Process Model

prevention performed in the entire software development lifecycle though dynamic testing using the Transparent-box approach, and inspection using traceable documents and traceable source code. Now the project managers do not need to worry about requirement changes because the implementation of requirement changes will be done with side-effect prevention through various traceabilities. The NSE process model also supports rapid prototyping and customer reviews, frequent delivery of working products to the customers, incremental integration, and traceabilities among documents and project management materials, plus traceable project Web sites and technical discussion forums for efficient problem solving. The project managers do not need to worry about whether the original designers of the project have left the development team, because the related documents and the source code are linked together and traceable with each other, the static and dynamic measurement results can be duplicated with the corresponding database and a set of Assisted Online Agents is used to make the project much easier to maintain and improve.

(c) **Equipped Business Analyst:** The NSE process model and the support platform support Business Analysts with

- The **Holistic, Actor–Action Driven, Traceable, Visual, and Executable approach (HAETVE) for Functional Requirement Decomposition** through dummy programming without drawing diagrams by hand in most product development.
- The Transparent-box testing approach which can be dynamically applied in the requirement development phase to establish bidirectional traceability for removing inconsistency defects even if the product version is a dummy system without real outputs.
- The **Holistic and Traceable Diagram Generation technique** and tools.
- The **Virtual and Traceable Documentation technique** and tools.

(d) **High Efficiency Designer:** With the NSE process model and the support platform Panorama++, software designers have several automated diagramming tools in their hands to generate required charts and diagrams through dummy programming efficiently without manually drawing them, so that they can spend more time on design optimization. The defects introduced in the requirement development phase and software design phases can be removed efficiently through dynamic testing using the Transparent-box approach without waiting for running Black-box testing after coding. With the automated traceability facility established by running dynamic testing using the Transparent-box approach, the designers do not need to worry about the inconsistency issue among the documents and the source code. They can respond to the requirement changes in real time through various traceabilities to prevent the side effects in the implementation of changes, and more.

(e) **Programming engineer:** With the NSE process model and the support platform Panorama ++, coding can be easily done incrementally through a bottom-up coding order automatically assigned on the call graph shown in J-Chart notation, so that the programming engineers do not need to worry about the inconsistency issue in the interfaces between the modules, because according to the bottom-up coding order, the modules called by a module being coded must have been coded already. The programming engineers can open a window to view the source code of the called module to know the needed parameters when writing a module call statement for the module being coded. Usually a logic defect is hard to detect because the source code is written in textual format and a program with some logic defects is still executable without producing error messages, but the result is wrong; however, with the NSE process model and the support platform, logic defects can be easily found through logic diagram generation and control flow generation. Usually different programming engineers will have different coding styles so that it is not easy to understand the programs written by other engineers, but with NSE the logic diagram generated in J-Diagram notations is not writing-style dependent – this makes the programs written by other engineers much easier to read and understand. With the diagramming support tools, the source code of a product can be entirely diagrammed with traceability between a module call statement and the called module body, between a class and all inherited classes, an instance and the corresponding class, a #include statement and the included file, and so on – it makes the coded programs much easier to inspect and review for defect removal.

(f) **Fully Armed Software Testing Engineer:** With the NSE process model and the support platform, software testing engineers obtain almost all of the tools and support they need, including

- Test planning support through Cyclomatic complexity measurement.
- Semiautomated test case generation support through path analysis.
- Incremental unit testing support according to a bottom-up coding and testing order without designing and using stub modules to replace the modules called by the unit being tested. In this way the unit testing also becomes integration testing.
- Automated driver design for unit testing.
- Modified Condition/Decision Coverage (MC/DC) software test coverage analysis, and the graphical display of the test result with the capability to highlight untested branches and conditions.
- Performance measurement.
- Memory leak check and memory usage violation check.
- GUI test operation capture and automatic playback.
- The capability to trace the execution path for a runtime error.
- Test case efficiency analysis and test case minimization for efficient regression testing after code modification.
- Transparent-box testing approach which combines functional and structural testing together seamlessly and can be dynamically used in the

8.10 The Major Features and Characteristics of the NSE Process Model

entire software development and maintenance process with the capability to establish bidirectional traceability for defect prevention and inconsistent defect removal.

(g) **Happiest Software Maintainer:** A software product developed with linear process models used in the old-established software engineering paradigm based on linear thinking and the superposition principle is almost not maintainable, because

- With linear process models, defects will easily propagate from the upper phases down to the maintenance phase, making the software maintenance job very hard to perform.
- The documents and the source code are separated and often inconsistent after code modification.
- The implementation of requirement changes or code modifications is done locally and blindly without facilities for bidirectional traceabilities, so that each time when a bug is fixed, there is a 20–50% chance of introducing a new bug into the software system.
- The regression testing reuses all the test cases – it is time consuming and costly.
- But with the NSE process and the support platform, software maintenance is much easier to do because
- With defect prevention performed in the entire software development lifecycle, only a few defects will exist and propagate down to the maintenance phase.
- The documents and the source code of a software product are linked together and traceable with each other.
- The implementation of requirement changes or code modifications can be done holistically and globally with side-effect prevention through various automated and bidirectional traceabilities.
- The regression testing can be done efficiently using a minimized set of test cases obtained through test case efficiency analysis.
- If there is still something unexpected after the requirement changes or code modifications, the maintainers can compare the new version and the previous version in system-level, file-level, module-level, and statement-level to locate the problems using the Holistic and Intelligent Version Comparison technique and tools offered.

(h) **Relaxed Software Development Team:** With the NSE process model and the support platform, a software development team can greatly increase the productivity, reduce the costs, improve the product quality, realize complete information sharing, make the documents and the source code traceable to each other, perform software maintenance with side-effect prevention through various traceabilities, and more, so that the team can achieve sustainable development, working just 40 hours a week.

8.11 Summary

In this chapter, a core part of the NSE paradigm, the NSE process model, was described which (with the support techniques and platforms) brings revolutionary changes to almost all aspects in software engineering, including the following:

- **The Foundation of Software Engineering – from** linear thinking and the superposition principle **to** nonlinear thinking and complexity science.
- **The Definition of Software – from** program + data + documents **to** program + data + documents **+ the database built with the program development lifecycle through static and dynamic measurement + a set of Associated Online Agents for supporting testability and reliability, and efficiently solving the issues of complexity, changeability, invisibility, and conformity – to make the program adaptive and truly maintainable.**
- **The Software Development Methodology – from** "building" the software system with its components **to** having the whole dummy system first then "**growing up**" with its components.
- **The Software Diagramming Paradigm – from** drawing the diagrams manually or using graphic editors **to** automatically generating them virtually through dummy programming or real source code to efficiently solve the drawbacks that manually drawn diagrams or diagrams using editors have. These drawbacks include being hard to draw, being not holistic, requiring much more space to store (about 100 times more than the diagrams that exist virtually), taking much more time to display and operate (about 1,000 times longer than the diagrams that exist virtually), being hard to check whether they are correct, being hard to change, and being hard to use without traceabilities.
- **The Software Documentation Paradigm – from** the produced documents are separated from the source code and not traceable with the source code **to** the produced documents are linked with the source code and traceable with the source code.
- **The Software Testing Paradigm – from** mainly using the Black-box testing method for functional testing plus structural testing to be performed separately after coding (it is too late) **to** mainly using the Transparent-box method to combine functional testing and structural testing seamlessly: to each set of inputs, it not only verifies whether the output (if any, can be none) is the same as the expected value, but also helps users to check whether the execution path covers the expected path with capability to automatically establish bidirectional traceability among all of the related documents and the source code for inconsistent defect checking.
- **The Software Quality Assurance Paradigm – from** test-driven, mainly using Black-box testing method after coding (it violates Dr. W. Edwards Deming's principles for product quality control – *"Cease dependence on inspection to achieve quality. Eliminate the need for inspection on a mass basis by building quality into the product in the first place."*) **to** defect prevention driven, mainly using the Transparent-box testing method in all phases of the software development lifecycle.

- **The Software Maintenance Paradigm – from** a local, blind and nonengineering maintenance approach (each time when a bug is fixed, there is a 20–50% chance to introduce a new one into the system, so that more than 75% of the effort and cost are spent in software maintenance in most software organizations) **to** a holistic, global, and engineering maintenance approach to perform systematic, quantifiable, and disciplined software maintenance with side-effect prevention, so that it is possible to help software organizations to reduce two-third of the effort and cost spent in software maintenance – almost the same as what is spent in the new product development process. In fact, with the NSE process model, there is no big difference between the software development process and the maintenance process; both support requirement changes and code modifications in real time with side-effect prevention.
- **The Software Project Management Paradigm – from** the management process being separated from the software development process **to** the management process being combined with the software development process – for instance, the project schedule and progress chart and the cost report are traceable with the requirement implementation and the source code for better control. In the case that there are two projects, project A and project B that are related, the project plan, schedule, cost, and the progress of project A can be traced to and from project B, also the project plan, schedule, cost, and the progress of project B can be traced to and from project A for balancing the two projects.

8.12 Points and Questions to Ponder

(a) About the software process model, "There has to be upstream movement" – why?
(b) Why is there no upstream movement at all in all the existing software process models (excluding the NSE process model)?
(c) Why should software maintenance be performed globally and holistically? How can software maintenance be performed globally and holistically?
(d) Is a modified waterfall model with feedback as shown in the following figure a linear model or not? Why?

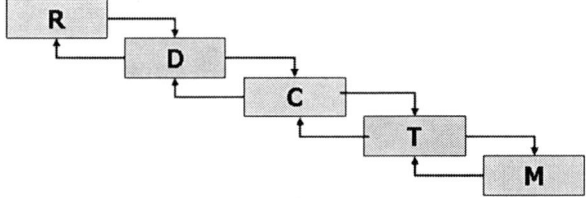

(e) List the drawbacks of a linear life cycle model without upstream movement.
(f) What are the major differences between the NSE process model and the existing process models?

8.13 Further Reading and Information Source

- Xiong J (2009) Tutorial, A complete revolution in software engineering based on complexity science. In: WORLDCOMP'09, Las Vegas, July 13–17, 2009
- Xiong J, Xiong J (2009) A complete revolution in software engineering based on complexity science. In: WORLDCOMP'09 – SERP (Software Engineering Research and Practice 2009), pp 109–115
- Jennings NR and Wooldridge M (2000) Agent-oriented software engineering. Department of Electronic Engineering, Queen Mary & Westfield College, University of London
- Merali Y, McKelvey B (2006) Using complexity science to effect a paradigm shift in information systems for the 21st century. J Inform Tech 21:211–215

References

[Bro95-p122]	Brooks FP Jr (1995) The mythical man-month. Addison-Wesley, Reading, p 122
[Bro95-p233]	Brooks FP Jr (1995) The mythical man-month. Addison-Wesley, Reading, p 233
[Bro95-p235]	Brooks FP Jr (1995) The mythical man-month. Addison-Wesley, Reading, p 235
[Bro95-p288]	Brooks FP Jr (1995) The mythical man-month. Addison-Wesley, Reading, p 288
[Jon02]	Jones C (2002) Software quality in 2002: a survey of the state of the art. Six Lincoln Knoll Lane, Burlington, MA. http://www.SPR.com July 23, 2002
[Pre05-p864]	Pressman RS (2005) Software engineering: a practitioner's approach. McGraw-Hill, New York, p 864
[Zam08]	Zambonelli F, Van Dyke PH (2003) Signs of a revolution in computer science and software engineering. Springer, Berlin

Chapter 9
The Facility for Automated and Self-Maintainable Traceability

> *Requirements traceability has been demonstrated to provide many benefits to organizations that make proper use of traceability techniques. This is why traceability is an important component of many standards for software development, such as the CMMI and ISO 9001:2000. Important benefits from traceability can be realized in the following areas: project management, process visibility, verification and validation (V&V), and maintenance ...In spite of the benefits that traceability offers to the software engineering industry, its practice faces many challenges. These challenges can be identified under the areas of cost in terms of time and effort, the difficulty of maintaining traceability through change, different viewpoints on traceability held by various project stakeholders, organizational problems and politics, and poor tool support.*
>
> Andrew Kannenberg, Garmin International
> Dr. Hossein Saiedian, The University of Kansas

This chapter introduces the facility for automated, dynamic, accurate, precise, and self-maintainable traceabilities among related software documents and test cases and source code established through test case execution and some keywords used within the test case descriptions to indicate the format of the documents as well as the file paths and the bookmarks for automatically opening the documents from the corresponding positions when the related test case is selected for forward tracing or traced backwards from the corresponding source code. When a test case is executed, a Time Tag will be automatically inserted into both the test case description and the database of the test coverage measurement results for mapping them together. No matter if the contents of a document are modified, or the parameters of a test case are changed, or the corresponding source code is modified, after rerunning the test case the traceabilities will be updated automatically without any manual rework. Here a "document" means a regular file for requirement specification, design description, test requirement specification, user manual, project development plan, cost report, or a Web page as well as a batch file for dynamically running a related program such as a tool for selectively playing back the GUI operations captured with the test case execution, and displaying the test coverage measurement result shown in a new type of control flow diagram which is interactive and traceable with untested source modules and branches highlighted at the same time for automated software acceptance testing.

This traceability facility is used to support the NSE process model described in Chap. 8.

The contents of a software development process model determines its graphical representation shape. In understanding the differences between the different process models, only comparing their graphical representation makes no sense – what really needs to be compared are the contents of the models: the features and characteristics.

9.1 The Importance of Requirement Traceability

Software is a nonlinear complex system where a small change can ripple through the entire system to cause major unintended impacts – "Butterfly-Effects," so that prior to performing the actual change, maintainers need facilities in order to understand and estimate how a change will affect the rest of the system. Traceability offers benefits to organizations in the areas of project management, process visibility, requirement validation and verification, and software maintenance. Traceability needs to be hardcoded into a process to be replicated iteratively on each and every project [Kan09]. Without bidirectional traceabilities, software maintenance cannot be performed globally and holistically to prevent side effects. Local and blind software changes will make the software product unstable and unreliable.

9.2 The Problems Addressed

The lack of traceability among software documents, test cases, test results, and source code is caused by several factors, including (1) the fact that these artifacts are written in different languages (natural language vs. programming language); (2) they describe a software system at various abstraction levels (requirements vs. implementation); (3) processes applied within an organization do not enforce maintenance of existing traceability links; (4) a lack of adequate tool support to create and maintain traceability [Ril07], (5) there are many different types of documents, some of which are created manually, some of which are generated automatically by internal tools, some of which are generated automatically by third party tools, some of which are designed using graphic editors; (6) some documents are stored locally, some documents are stored in other places through a network; (7) some related documents are Web pages, which can be read through the internet only; (8) some documents are related to the software development, while some documents are related to the project management which should also be traceable; and (9) some documents are not static materials, and must be viewed dynamically through a program execution. Unfortunately, neither manual traceability methods nor existing COTS traceability tools available on the market are adequate for the current needs of the software engineering industry. Poor methods and tool support are perhaps the biggest challenge to the implementation of traceability – when those tools are used, the traceability information is not always maintained, nor can it always be trusted to be up to date and accurate [Kan09]. Studies have shown that existing commercial traceability tools provide only simplistic support for traceability [Ram01]. Why does software maintenance take 75% or more of the total effort and total budget [Amb05] in most software projects? One of the critical issues is the lack of bidirectional traceabilities among the requirement specifications, the design documents, the test cases, the test results, and the source code of a software product.

9.3 The Solution Offered with NSE

The new requirement traceability approach proposed by me and implemented by me and my colleagues is graphically shown in Fig. 9.1.

The objectives of this traceability facility are as follows:

a) Helping software developers to prevent side effects in the implementation of software changes

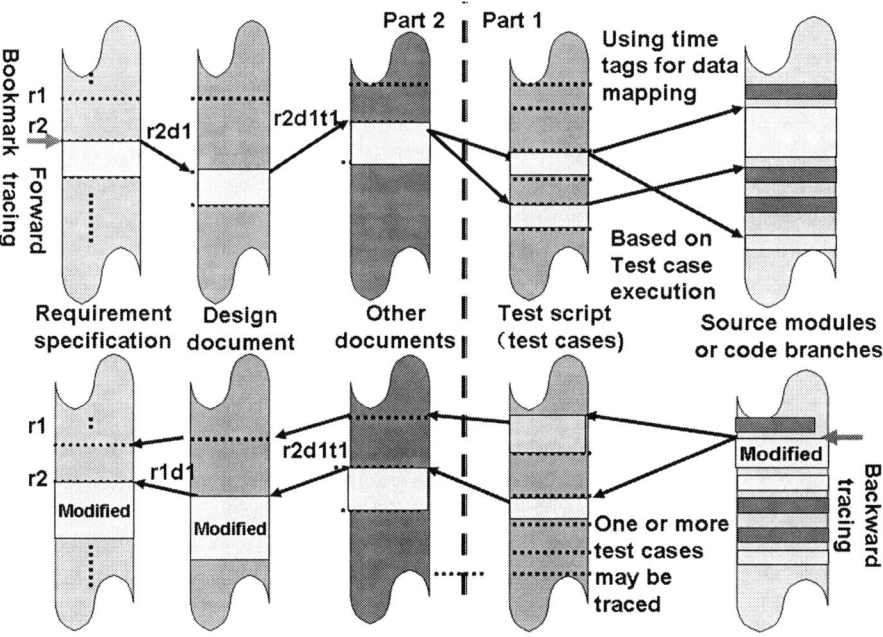

Fig. 9.1 The facility for automated, bidirectional, and self-maintainable traceability among the documents and the test cases and the source code of a software product

b) Solving the inconsistency issue to make the documents and the source code traceable with each other to keep consistency
c) Removing the problems existing with a man-made Requirement-Traceability Matrix which is inaccurate, time consuming, and almost unmaintainable
d) Making the software development process visible
e) Making the requirement validation and verification much easier to perform
f) Making the software product much easier to understand, test, and maintain

As shown in Fig. 9.1, this facility for bidirectional traceability consists of two parts.

9.3.1 Part 1

Part 1 of the facility is related to the traceability between test cases and the corresponding source code executed by running the test cases. It is done with the use of Time Tags which are automatically inserted into both the test case descriptions and the corresponding test coverage database. For instance, if test case 1 is executed at 09:00 AM on September 2, 2009, and test case 2 is executed at 10:00 AM on the same day, and test case 3 is executed at 11:00 AM on the same day, then the three different Time Tags will be inserted into the three test cases and the corresponding test coverage database separately. So, when test case 2 is selected for forward tracing, the Time tag of 10:00 AM on September 2, 2009 will be taken from the test case description to search the test coverage data with the same time tag, so the corresponding test coverage data will be read and the corresponding modules and branches will be highlighted on a control flow diagram. On the other hand, when a module or code segment shown on a control flow diagram is selected, the related time tags (which can be more than one) used to indicate what time the module or segment was executed will be taken to search the test case descriptions to see how many test cases have the mapping time tags through backward tracing, then highlights all test cases mapped on the window showing the test case script.

9.3.2 Part 2

Part 2 of the facility is to extend the bidirectional traceability from test cases and the source code to include all related documents, the test cases, and the source code. It is done using some key words (written into the comment part of the description of a test case) such as @WORD@, @HTML@, @BAT@, @PDF@, and @EXCEL@ followed with the corresponding file path and a bookmark to indicate the format of the document, the full path name of the file, and the corresponding location in the document, so that when a test case is selected for forward tracing, or traced backwards from a module or segment, the corresponding document will be opened and shown from the location indicated by the bookmark.

9.3 The Solution Offered with NSE

It is recommended to organize the requirement specifications and the related documents hierarchically (even if some documents have not been really designed) with inherited (or meaningful) bookmarks as shown in Table 9.1.

Table 9.1 Document Hierarchy

Document Hierarchy					
Project Name			Project Code		
Project Description					
The full path name of the Project feasibility report				Version number	
The full path name of the requirement specification				Version number	
Requirement 1	Bookmark	r1			
Description					
	The full path name of the related design document				
	Description			Bookmark	r1d1
	The full path name of the related test specification				
	Description			Bookmark	r1d1t1
	...				
Requirement 2	Bookmark	r2			
...					

It is important to make the document hierarchy include the test case scripts (test case numbers) so that when a requirement needs to be changed or selected for validation, it is easy to find what test cases need to be used.

The major steps for establishing and applying the bidirectional traceability are as follows:

Step 1. Organize the requirement specification and the related documents hierarchically with the bookmarks, clearly indicate each requirement and the corresponding test scripts and the test case numbers.

Step 2. Design the test case scripts with the corresponding keywords to indicate the formats and the file paths and the bookmarks for the related documents.

Step 3. Perform code instrumentation for test coverage analysis to the entire program.

Step 4. Compile the instrumented program.

Step 5. Execute the test case scripts with the corresponding tool.

Step 6. Show the modified test case script files with inserted time tags in a window.

Step 7. Show the program test coverage measurement result using a control flow diagram in another window.

Step 8. Perform forward tracing from a test case with a tool to map and highlight the corresponding modules and code branches tested by the test case through the inserted time tag – at the same time, open the related documents according to the document formats, file paths, as well as the bookmarks (or run the corresponding batch file if a @BAT@ keyword is used).

Step 9. Perform backward tracing from a program module or code branch with a tool to map and highlight the related test cases through the inserted time tags – at the same time, open the related documents according to the document formats, file paths, as well as the bookmarks (or run the corresponding batch file if a @BAT@ keyword is used).

Step 10. After the implementation of code modifications, go to step 3.

Step 11. If a related document is modified in the contents only without changing the bookmarks, there is nothing to do; but if the bookmarks are modified (such as the name of a bookmark is changed), modify the corresponding test case scripts according to the new bookmarks, then go to step 5.

Step 12. If only the test cases are modified, go to step 5.

Step 13. If the source code is modified, go to step 3.

Step 14. If it is the time to perform requirement validation and verification (V&V), use the document hierarchy information organized in step 1 to get each requirement and the corresponding test cases to perform forward tracing one by one to see whether the requirement is completely implemented.

Step 15. If a requirement needs to be modified: (1) get the test cases related to this requirement to perform forward tracing to locate the documents that need to be updated, and the source modules or branches that need to be modified; (2) perform backward tracing from those modules or branches to see whether more requirements are related – if it is related to more requirements, the implementation of the code modification must satisfy all of the related requirements to avoid requirement conflicts.

Step 16. If it is time to perform regression testing after modification, get the modified modules or branches to perform backward tracing to collect the corresponding test cases which can be used to retest the modified program efficiently. Sometimes, there may be a need to add new test cases.

9.4 How It Works

The facility for automated and self-maintainable traceability is based on source code test coverage analysis. The first step is to perform code instrumentation for recording the test coverage data.

9.4 How It Works

To a statement in C/C++ programming language:

        if (condition) printf("\OK\n");
Note: there is no "else" part in this example – it exists but it is invisible.
After code instrumentation, the statements are changed to
        if (condition? ++cov[i][0],1 : ++cov[i][1],0)

or   if(condition) {++cov[i][0]; printf("\OK\n");}
        else ++ cov[i][1];

Here, "i" is the record order number.
Let's consider a small program listed as follows:

```
#include <stdio.h>
static char *tp=NULL;
int r=1, x=0, y=1000000, z=0;
FILE *fd=NULL;
void trouble();
void error (message)
char *message;
{
printf("\n ERROR! %s", message);
}

main(argc, argv)
int argc;
char **argv;
{
int k=0;
if(argc == 1)
  enter_data();
else if(argc==2)
  trouble(atoi(argv[1]));
else if (argc==3)
  divite(atoi(argv[1]), atoi(argv[2]));
else if (argc == 4)
  control(atoi(argv[1]), atoi(argv[2]), atoi(argv[3]));
else
  error(" Too many arguments!");
}
```

```c
/*##############################################*/
/* trouble.c */
#include <stdio.h>
#include <malloc.h>
#ifdef ERROR_SIMULATION
#include "ISA_simu.h"
#endif

failed(message)
char *message;
{
printf("Failed! %d\n",message);
}

extern int x,y,z;
extern FILE *fd;
FILE *fi, *fo;
trouble (x)
int x;
{
int i, t=1;
char c,*pc=NULL,ch[10],*p=NULL,*e=NULL;
if((e=malloc(4))==NULL)printf("Out of memory,x=%s",x),
exit(-1);
for(i = x; i <= 8 && t; p=&ch[i++])
  if(i % 2 ==1) {
    p=&c; t=0; }
ch[0] = *p; /* seg. fault when x > 8 */
i = x ;
while (i > -2 && i<=7 ){/*dead loop if x=7 or x=3*/
  switch ( x + z ) {
  case  0: case 1: x = z = 1; break;
  case  2: y = 1; break; }
if ( i < 7 )
  i += 4; }
if ( x < 5 )
  pc = ch;
if( x < 6 )
  fd=fopen("trouble.c", "r");
c = getc (fd); /* seg. fault when x = 6 */
strcpy (pc, "ab"); /* seg. fault if x = 5 */
c = ch[y]; /* seg. fault when x = 4 */
z = x / z; /* Arith. excep. when x = 2 */
if((p=malloc(3))!=NULL) strcpy(p,"OK");
}
```

9.4 How It Works

The Makefile for controlling the builds of this program is as follows:

```
# Some macros for building Win32 applications

CPU=i386

cc = cl

link = link

cflags = -c

all: trouble.exe

OBJS = main.obj trouble.obj

trouble.obj: trouble.c
    $(cc) $(cflags) $*.c

main.obj: main.c
    $(cc) $(cflags) $*.c

trouble.exe: $(OBJS)
    $(link) -out:trouble.exe -subsystem:console main.obj trouble.obj  libc.lib kernel32.lib
```

For testing this small program, three test cases are used. After running the test cases under the NSE support platform, three different Time Tags were automatically inserted into the test coverage database and the test script as shown in Fig. 9.2, which are used for mapping the test cases and the corresponding source code tested.

9.4.1 Bidirectional Traceability Between the Test Cases and the Source Code Modules or Branches

For realizing bidirectional traceability between the test cases and the source code, two windows are opened for displaying the accumulated Test Case file

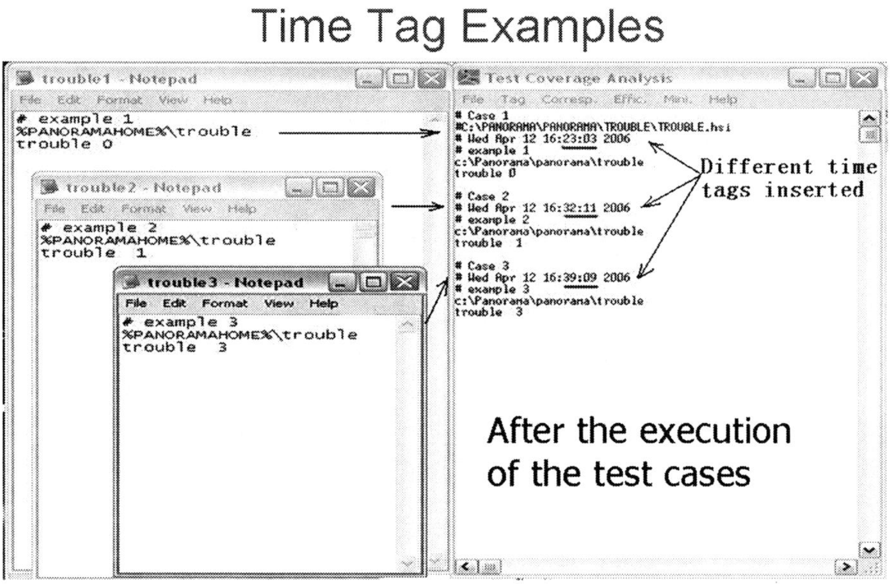

Fig. 9.2 Time tags automatically inserted into the test script

(for all test cases together) and the Source Code shown in J-Flow notation with untested segments highlighted in small black boxes separately.

The operations for forward tracing – click a test case on the Test Case window to select it, then the corresponding tool will highlight the selected test case description part in the Test Case window, while the source code modules and segments (a segment is a group of statements with the same execution conditions) will be highlighted in the Source Code window through the mapping of Time Tags – see Figs. 9.3 and 9.4.

The operations for backward tracing – click a segment (or module) on the Source Code window to select it, then the corresponding tool will highlight the selected segment or module in the Source Code window, while the test cases will be highlighted in the Test Case window through the mapping of the Time Tags – see Fig. 9.5.

9.4.2 Extending the Bidirectional Traceability to Include All Related Documents

It is done using some keywords such as @WORD@, @HTML@, @PDF@, @EXCEL@, and @BAT@ written within the comment part of a test case to indicate the format of the document, followed by the file path of the document and the bookmark used to open the document from the corresponding position rather than from the beginning of the document by the corresponding tool.
An application example for forward tracing is shown in Fig. 9.6.

9.4 How It Works

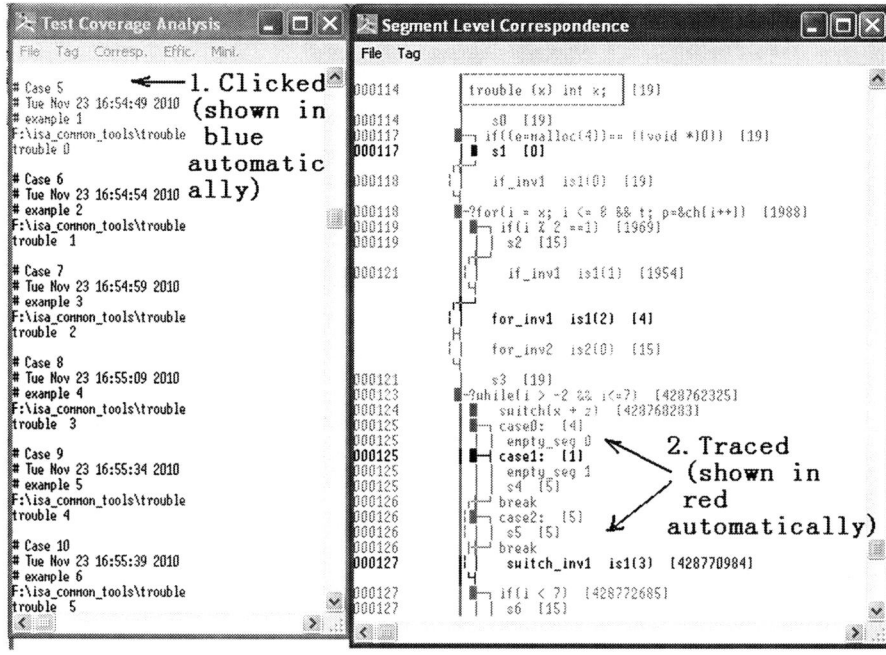

Fig. 9.3 An application example of forward traceability established

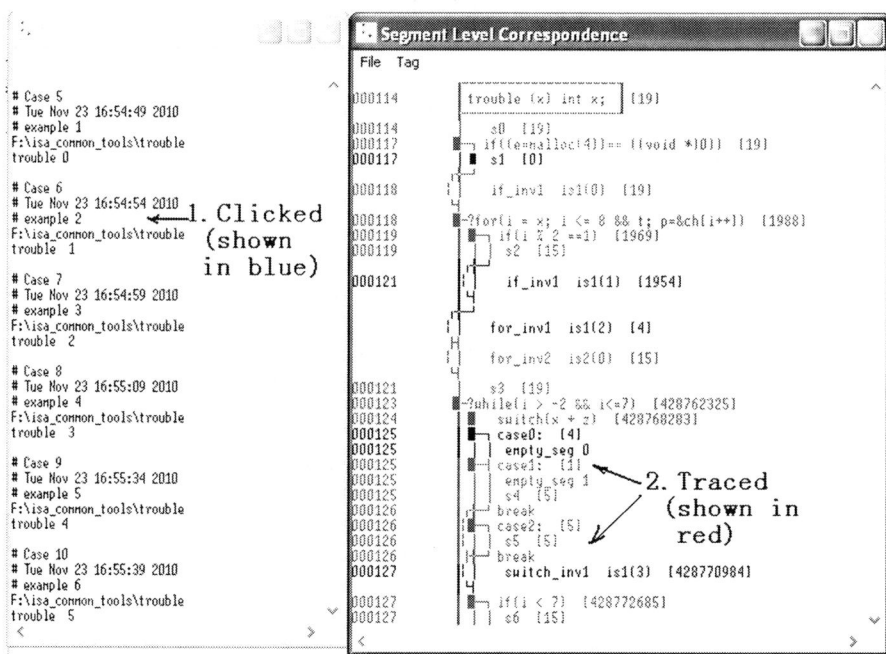

Fig. 9.4 Another application example of forward traceability established

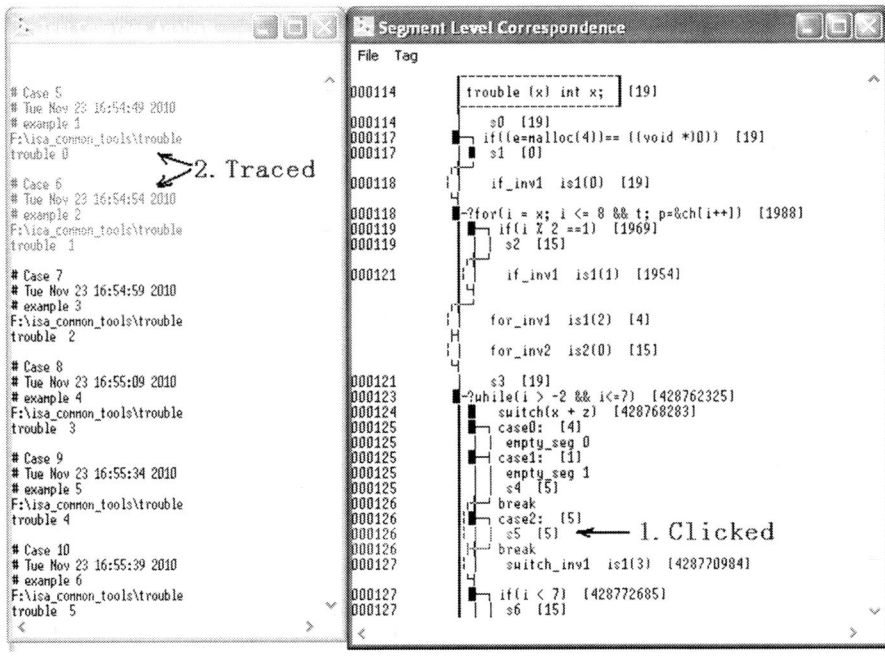

Fig. 9.5 An application example of backward traceability established

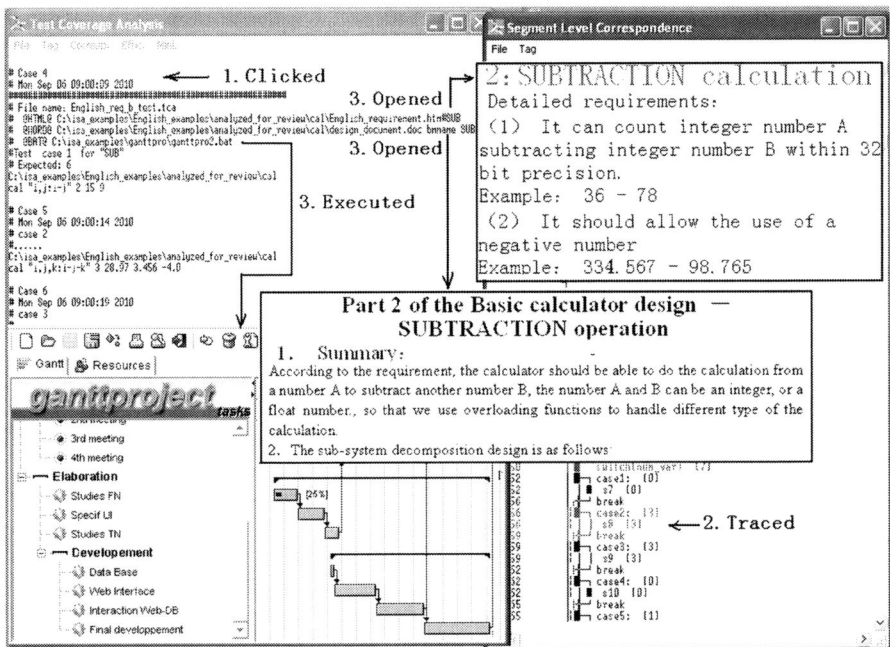

Fig. 9.6 Forward tracing from a test case to trace the source code with the related requirement specification, design document, and schedule chart opened

9.5 The Major Features

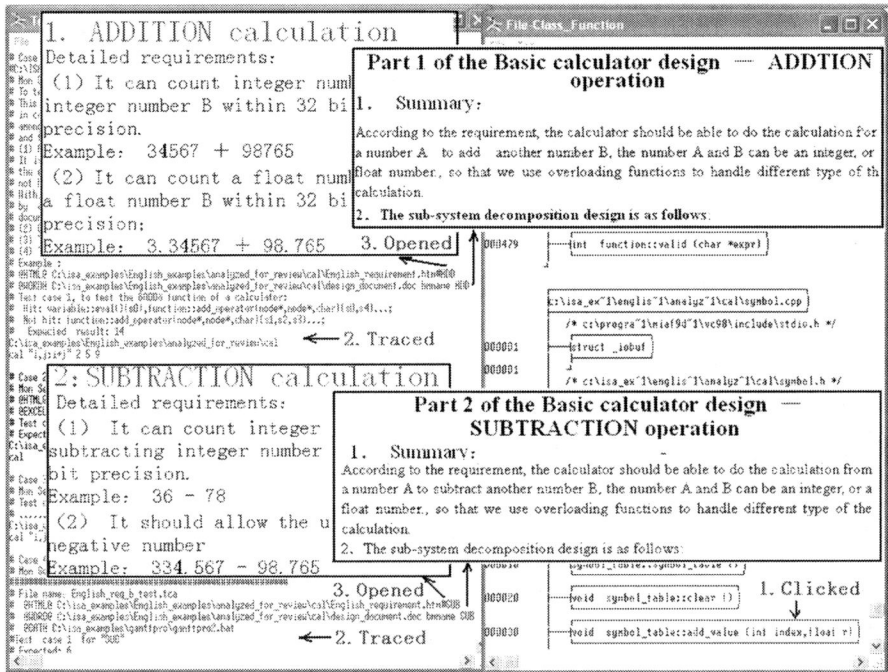

Fig. 9.7 Backward tracing from a module to trace the test cases (which can be used to test that module) and all of the related documents

An application example for backward tracing is shown in Fig. 9.7 – in this example, two requirements are traced from a module, and it means that the modification of that module must be done carefully to satisfy the two requirements at the same time.

9.5 The Major Features

9.5.1 Automated

This facility works automatically with the capability to insert the Time Tags into both the test case description part (see Fig. 9.2) and the database of the program test coverage measurement result, and highlight the test cases selected on the corresponding test script window, and the source code modules/branches shown in a control flow diagram in the corresponding source code window, or vice versa, as well as open the related documents traced from the locations pointed by the bookmarks.

9.5.2 Self-Maintainable

This facility is self-maintainable no matter if the contents of a document are modified, the parameters of a test case are modified, or the source code is modified – after rerunning the test case scripts, the traceability will be automatically updated without manual rework.

9.5.3 Methodology-Independent

This facility is methodology-independent, no matter which methodology or process models are used to develop the product.

9.5.4 Nonlinear, Bidirectional, and Parallel

This facility works in a nonlinear, bidirectional, and parallel style as shown in Fig. 9.8 – when a design defect is found after the product delivery, the developers can perform backward tracing to check the related requirement, and forward tracing to find and fix the related source code.

9.5.5 Accurate

This facility is based on the dynamic execution of the test cases and test coverage measurement and the time tags to map the test cases and the source code tested, so that it is accurate. After code modification or parameter changes of the test cases, we can rerun the test cases to automatically update the facility.

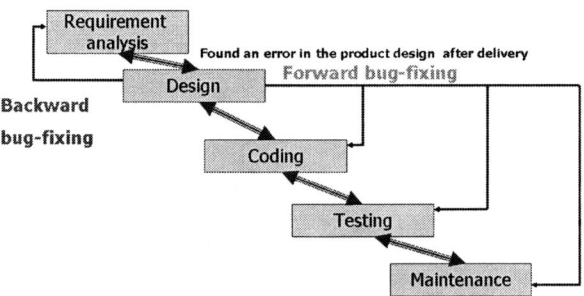

Fig. 9.8 Fixing a design defect through forward and backward traceability

9.5.6 Precise

This facility is precise to the highest level – up to the code statement/segment (a set of statements to be executed with the same conditions) level, bidirectionally. It is particularly useful for side-effect prevention in software maintenance.

9.5.7 Extended to Include Software Project Management Documents

This facility is extended to include not only the software development documents, but also include the project management documents such as the product development schedule charts, the cost estimation reports, and so on, to combine the software development process and the software management process together. If a project management document (such as a Gantt chart) is designed using a third party tool, a corresponding batch file should be designed and used with the @BAT@ keyword to indicate the location of the batch file in the test case description part such as the following example shown in Fig. 9.6, step (5).

9.5.8 Extended to Include Web Pages

For supporting Web-based software development and applications, this facility is extended to include Web pages to be traced and automatically opened through the use of @HTML@ keyword to indicate the URL address and the bookmark (#NAME) such as the following example:

@HTML@ http://www.stsc.hill.af.mil/CrossTalk/2010/01/index.aspx

When the corresponding test case is selected for forward tracing or backward tracing from a source code module or a source code branch mapped to the test case, the corresponding CrossTalk Web page will be opened automatically if the internet is connected – see Fig. 9.9.

9.5.9 Extended for Multiproject Support

This facility is extended to support multiproject development by making the related project progress reports, special event reports, schedules, budget control documents, and cost reports traceable between two related projects (or among more related projects) as shown in Fig. 9.10.

Fig. 9.9 A Web page opened automatically through the bidirectional traceability

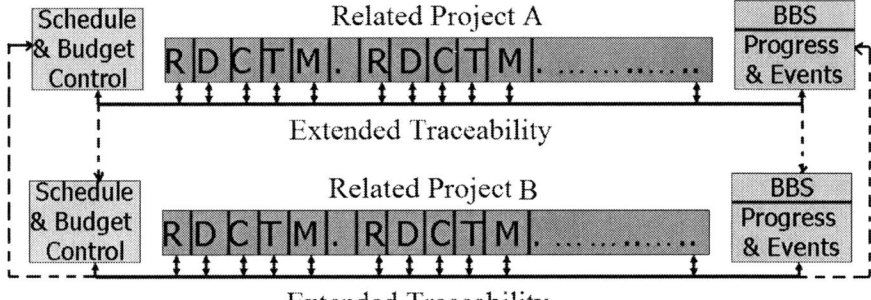

Fig. 9.10 Multiproject development support

9.5.10 *Dynamic*

This facility is extended to have the capability to trace a batch file and dynamically execute the batch file for many kinds of applications such as playing back the captured GUI operations selectively through the time tags in automated acceptance testing, or running a third party tool to handle the corresponding documents generated by that tool, or dynamically execute a related program for other purposes.

@BAT@ C:\isa_examples\analyzed_examples\sortdemo\re_run.bat. It will be used to automatically execute the batch file re_run.bat which includes the following statement:

9.5 The Major Features

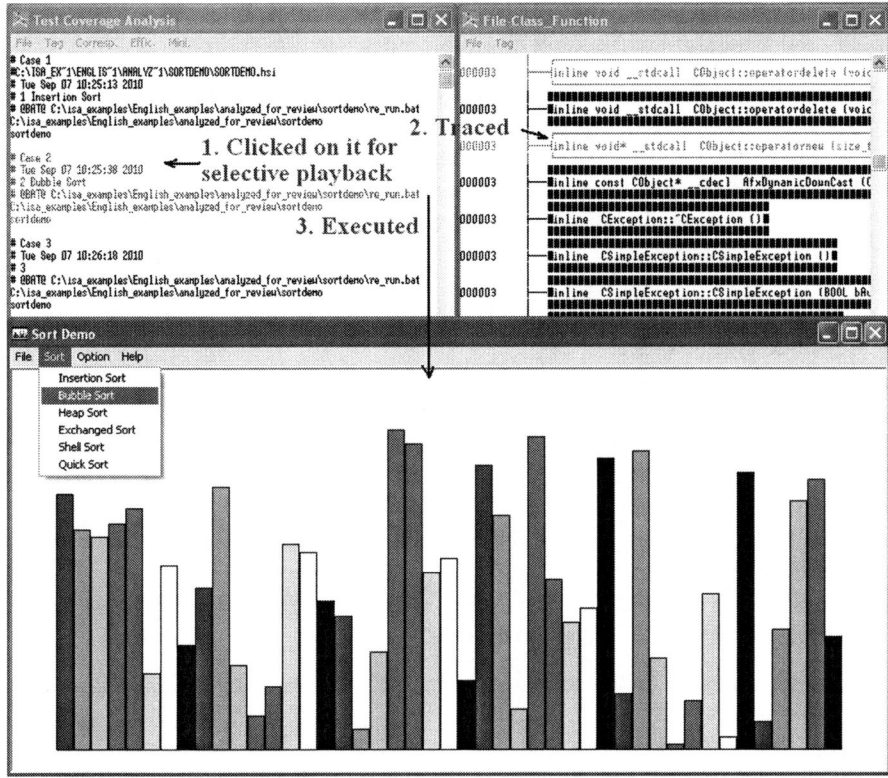

Fig. 9.11 An application example of Dynamic Traceability to play back the GUI operations captured

C:\isa_examples\\play -hsi=C:\isa_examples\analyzed_examples\sortdemo\sortdemo.hsi -dbspath=C:\isa_examples\analyzed_examples\sortdemo\dbs\ -tdb=C:\isa_examples\analyzed_examples\sortdemo\sortdemo.tdb -tdb=C:\isa_examples\analyzed_examples\sortdemo\playout.tdb

and the corresponding GUI operation capture records will be automatically played back as shown in Fig. 9.11.

9.5.11 *Easy to Add on at Any Time, In Any Status*

This facility can be added on at any time and at any status of a software product development project, even if it is in the requirement development phase where the product design and coding have not started yet – in this case, we can design a dummy main program without a real output which can be executed for checking the consistency between requirement specifications, prototype design documents, test requirements, and test scripts – it is recommended to design the test scripts with the requirement specifications at the same time before the product design. In the case, this facility is used for a product developed or being developed using other methodologies, the users only need to set bookmarks to the related documents and modify the test case description with simple rules listed as follows:

(a) An empty line means a separator between different test cases.
(b) A '#' character at the beginning position of a line means a comment.
(c) Within comments, users can use some keywords such as @WORD@, @HTML@, @PDF@, and @BAT@ to indicate the format of a document, followed by the full path name of the document, and a bookmark – for finding inconsistent defects.
(d) Within comments, users can use [path] and [/path] pair to indicate the expected execution path using control flow notation (segment numbers) for a test case – for finding logic defect.
(e) Within comments, users can use Expected Output to indicate the expected value to be produced – for finding functional defects.
(f) Within comments, users can also use [Not_Hit] and [/Not_Hit] marks to indicate modules or branches (segments) which are prohibited for the related test case execution to enter.
(g) After the comment part, there is a line to indicate the directory for running the corresponding program.
(h) The final line in a test case description is the command line (which may start a program with the GUI) and the options.

Other work can be done automatically by the corresponding tools.

9.6 Application

This automated and self-maintainable traceability technique has been successfully applied in requirement validation and verification, side-effect prevention for the implementation of requirement changes and code modifications, inconsistency checking among documents and test cases and source code, efficient regression testing through backward tracing from a modified module or branch to select the corresponding test cases, and quality assurance in the entire software development lifecycle through defect prevention and defect propagation prevention.

Some application areas of this automated and self-maintainable traceability techniques provided with NSE are listed in Fig. 9.12.

9.7 Summary

This chapter presented automated and self-maintainable traceability based on test case execution with Time Tags automatically inserted into both the test cases and the database of the corresponding program test coverage measurement result, plus the use of some keywords to indicate the format of the related documents,

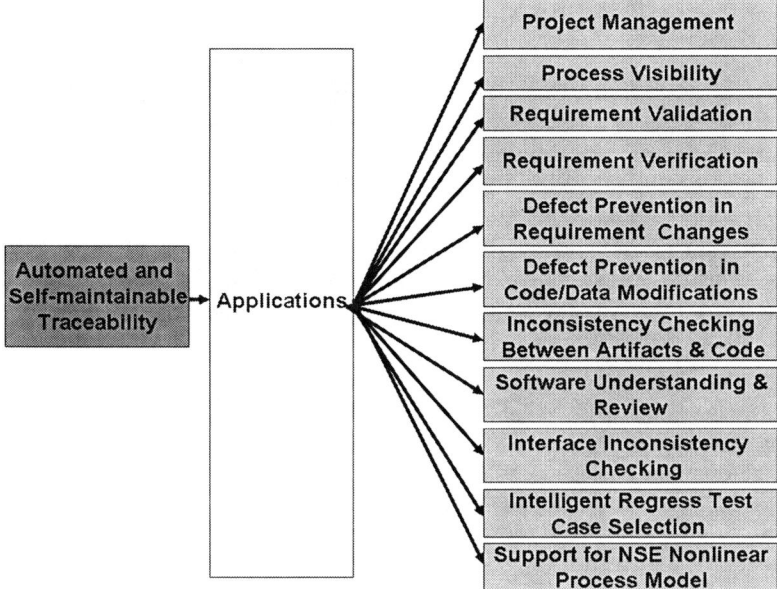

Fig. 9.12 Application area of the automated and self-maintainable traceabilities provided

the file paths, and bookmarks for automatically opening the documents from the positions pointed by the bookmarks. It is useful for requirement validation and verification, side-effect prevention in the implementation of requirement changes and code modifications, automated acceptance testing, conformity check of artifacts, efficient regression testing, and defect prevention for software quality assurance in the entire software development lifecycle of a software product.

9.8 Points and Questions to Ponder

(a) Why is software traceability, particularly requirement traceability, so important?
(b) Why should a bookmark be used to open a related document that is traced automatically?
(c) What are the benefits to use Time Tags for implementing the bidirectional traceability between the test cases and the source code?
(d) What are the major features of this automated and self-maintainable traceability?
(e) Where do you think this automated and self-maintainable traceability can be efficiently used in software engineering?
(f) How can this automated and self-maintainable traceability be used to make a document produced by a third party tool traceable with the requirements of a project being developed with this technique and tools?

9.9 Further Reading and Information Source

1. Gotel O, Finkelstein A (1994) An analysis of the requirements traceability problem. In: Proceedings of the first international conference on requirements engineering, Colorado Springs, 1994, pp 94–101
2. Dömges R, Pohl K (2008) Adapting traceability environments to project specific needs. Commun ACM 12:55–62
3. Palmer JD (1997) Traceability. In: Thayer RH, Dorfman M (eds) Software requirements engineering. IEEE Computer Society, New York
4. Wiegers K (2003) Software requirements, 2nd edn. Microsoft, Redmond
5. Boehm B (2003) Value based software engineering. ACM SIGSOFT Software Engineering Notes 2
6. Clarke S et al (1999) Subject-oriented design: towards improved alignment of requirements, design, and code. Proceedings of the 1999 ACM SIGPLAN Conference on object-oriented programming, systems, languages, and applications, Dallas, TX, pp 325–329

References

[Amb05] Ambler SW (2005) A manager's introduction to the rational unified process (RUP). Ambysoft. http://www.ambysoft.com/unifiedprocess/rupIntroduction.html

[Kan09] Kannenberg A et al (2009) Why software requirements traceability remains a challenge. CrossTalk, Jul/Aug 2009 Issue

[Ram01] Ramesh B, Jarke M (2001) Toward reference models for requirements traceability. IEEE Trans Software Eng 1:58–93

[Ril07] Rilling J et al (2007) CASCON 2007 workshop report, traceability in software engineering – past, present and future. IBM Technical Report: TR-74-211, October 25, 2007

Chapter 10
NSE Software Development Methodology Driven by Defect Prevention and Traceability

> *DEFECT PREVENTION: Technologies that minimize the risk of making errors in software deliverables.*
>
> Capers Jones

This chapter describes another important component of NSE – the NSE software development methodology.

Software development methodology is a framework used to structure, plan, and control the process of software product development and maintenance.

The NSE software development methodology is innovated by me through the FDS paradigm-shift framework described in Chap. 4 (see Fig. 10.1), so that it complies with the essential principles of complexity science, particularly the Nonlinearity principle and the Holism principle – with the NSE software development methodology, almost all software engineering tasks are performed holistically and globally.

10.1 Almost All Existing Software Development Methodologies Are Outdated

Almost all existing software development methodologies are outdated because

1. They are based on reductionism and the superposition principle that the whole of a system is the sum of its parts, so that almost all software development tasks and activities are performed linearly, partially, and locally.
2. They comply with the Constructive Holism principle that software components are developed first, then, "**Assemble** *the product from the product components, ensure the product, as integrated, functions properly and deliver the product.*" [CMMI1.1]
3. From the point of view of quality assurance, those methodologies are test-driven – through testing after coding and inspection. As pointed by NIST (National Institute of Standards and Technology) that "Briefly, experience in testing software and systems has shown that testing to high degrees of security and reliability is from a practical perspective not possible. Thus, one needs to build

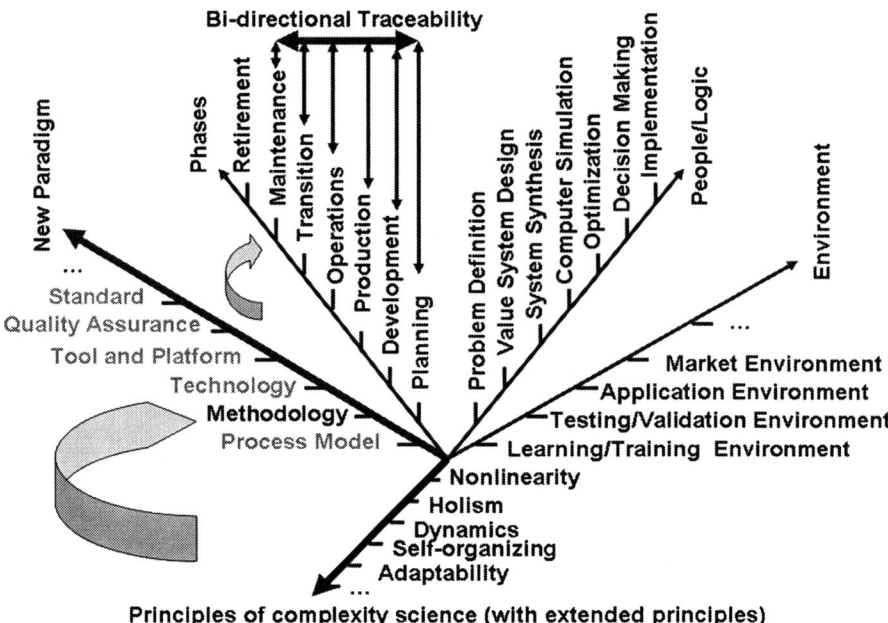

Fig. 10.1 Innovating the NSE software development methodology by complying with the essential principles of complexity science through the proposed paradigm-shift framework, FDS

security, reliability, and other aspects into the system design itself and perform a security fault analysis on the implementation of the design." ("Requiring Software Independence in VVSG 2007: STS Recommendations for the TGDC," November 2006, http://vote.nist.gov/DraftWhitePaperOnSIinVVSG2007-20061120.pdf).

4. The software development process and the obtained results are almost invisible – UML is applied mainly in the modeling process and the obtained graphics often consist of many small pieces without a holistic whole.
5. The application results of those methodologies show that today the software project success rate is still very low – only about 30%. It is not acceptable in any other industry.
6. Although the CBSD (Component-Based Software Development) method has been used successfully in some applications, the CBSD concept is not new but as old as the "Software Engineering" concept – the idea that software should be componentized – built from prefabricated *components* first became prominent with Douglas McIlroy's address at the NATO conference on software engineering in Garmisch, Germany, 1968, titled *Mass Produced Software Components*. The basic idea about CBSD is that software should be componentized – built from prefabricated *components* which are software packages or modules that encapsulate a set of related functions and data. But there are some fundamental issues with CBSD:
   (a) It is an outcome of linear thinking, reductionism, and the superposition principle that the whole of a complex system is the sum of its parts.

(b) It violates the Nonlinearity principle and the Holism principle of complexity science that the whole of a complex system is greater than the sum of its parts, the behavior and characteristics of the whole emerge from the interaction of its parts and the interaction between the system and its environment.
(c) According to the Generative Holism principle of complexity science, the whole of a complex system may exist earlier than its components, as an embryo, then grows up with its components.
(d) When a reusable component is integrated into the system, not only the functions and the interface of the component should be considered, but the effects to the whole of the system should also be considered such as the effects to the performance, the effects to the file system, and the effects to the product quality. For instance, if there is a memory leak with the new component, the performance of the entire product may become a critical problem; if there is a logic error in the new component, the entire system may become unstable.

In fact, today, a component is still designed using the old-established software engineering paradigm, and its quality is still hard to ensure and its maintenance is still hard to perform. From these points of view, the current CBSD is out of date too – it should be shifted to a new development methodology with the components being developed as, for instance, "Broken Arms" rather than "Artificial Arms" using a novel software engineering platform based on complexity science.

10.2 Outline of the Revolutionary Solution Offered by NSE

The revolutionary solution offered by NSE in software development methodology will be described in detail in this chapter later. Here is the outline of the solution:

1. It is based on complexity science by complying with the essential principles of complexity science, particularly the Nonlinearity principle and the Holism principle that **the whole of a complex system is greater than the sum of its components – the characteristics and behaviors of the whole emerge from the interaction of its components,** so that with NSE almost all tasks and activities in software development are performed holistically and globally. For instance, requirement changes are implemented holistically and globally with side-effect prevention through various traceabilities to avoid "Butterfly Effects" (see Chap. 3).
2. It complies with the Generative Holism principle of complexity science that the whole of a complex system exists (as an embryo) earlier than its components, then grows up with its components. As pointed by Frederick P. Books Jr. that "Incremental development – grow, not build software … that the system should first be made to run, even though it does nothing useful except call the proper set of dummy subprograms. Then, bit by bit it is fleshed out, with the subprograms in turn being developed into actions or calls to empty stubs in the level below." [Bro95-P200] "An Incremental-Build Model Is Better – Progressive Refinement … we should build the basic polling loop of a real-time system, with subroutine calls (stubs) for all the functions, but only null subroutines. Compile it; test it. …

After every function works at a primitive level, we refine or rewrite first one module and then another, incrementally growing the system. Sometimes, to be sure, we have to change the original driving loop, and or even its module interface. Since we have a working system at all times

(a) we can begin user testing very early, and
(b) we can adopt a build-to-budget strategy that protects absolutely against schedule or budget overrun (at the cost of possible functional shortfall)." [Bro95-P267]

3. From the point of view of quality assurance, the NSE software development methodology is driven by *defect prevention, defect propagation prevention, and traceability* where software quality is ensured from the first step down to the final step through defect prevention and defect propagation prevention supported by various automated and self-maintainable traceability and software visualization. With NSE, software development methodology software testing is performed dynamically in the entire software development lifecycle (including the requirement development phase, the product design phase, the coding phase, the testing phase, and the maintenance phase) using the innovated Transparent-box testing method (see Fig. 10.2B2 and Chap. 16 and the related reference [Xio09]) which combines functional testing and structural testing together seamlessly – to each test case, it not only checks whether the output (if any, can be none – having a real output is no longer a condition to use this software testing method dynamically) is the same as what is expected, but also checks whether the real program execution path covers the expected one specified in J-Flow (see Chap. 7), and then establishes the automated and self-maintainable traceability among the related documents, the test cases, and the source code to help users find and remove the inconsistency defects. It means that the NSE software development methodology complies with **W. Edwards Deming's product quality principle that *"Cease dependence on inspection to achieve quality. Eliminate the need for inspection on a mass basis by building quality into the product in the first place."*** [Dem86].

The defect prevention and defect propagation prevention is also performed through software visualization in the entire software development process.

Figure 10.2 shows a comparison of the software design strategy and the quality assurance strategy between the existing software development methodologies (part A) and the NSE software development methodology (part B).

4. It is visual – With NSE, the entire software development process and the obtained results are visible, supported by the NSE Software Visualization Paradigm.
5. It works with the NSE documentation paradigm to generate huge amounts of documents automatically. The generated documents are always updated, precise,

---

Fig. 10.2 A comparison of the software design strategy and the quality assurance strategy between the existing software development methodologies (**a**) and the NSE software development methodology (**b**) – (A1): the software product development strategy based on the Constructive Holism principle that the components of a software product are developed first, then the whole system is built

10.2 Outline of the Revolutionary Solution Offered by NSE

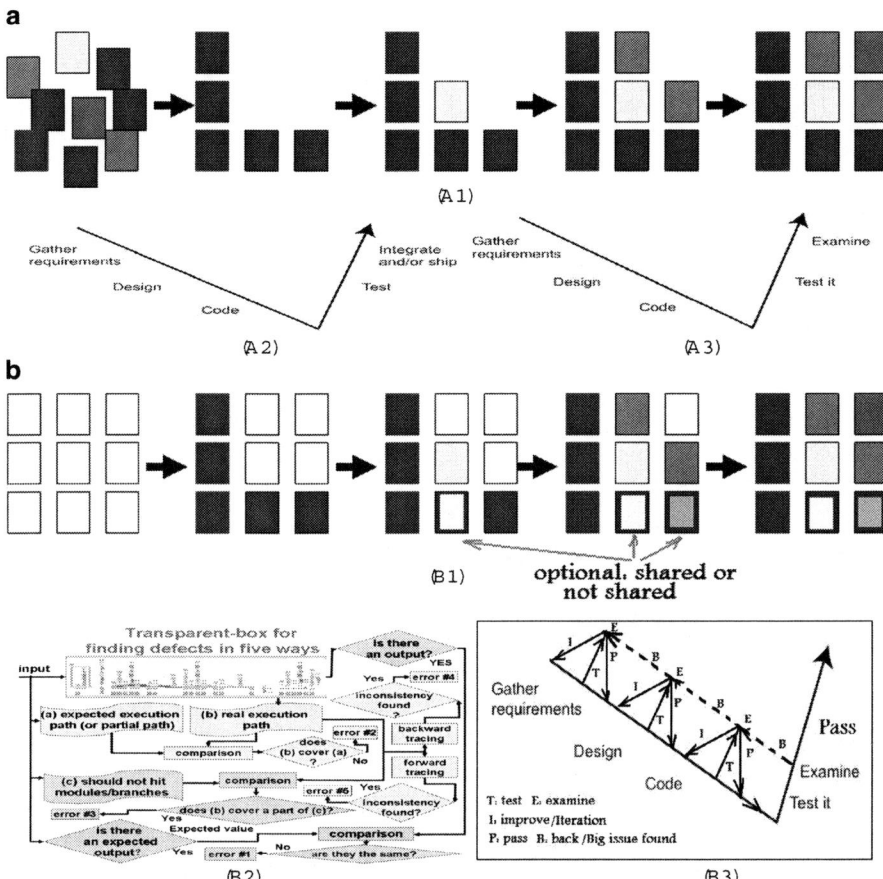

Fig. 10.2 (continued) with the components; (A2): the quality assurance strategy for the incremental software development method – mainly depends on testing after coding using the Black-box functional testing approach and structural testing approach [Coc08]; (A3): the quality assurance strategy for the iterative software development method – also mainly depends on testing after coding using the Black-box functional testing approach and structural testing approach [Coc08]; (B1): the software product development strategy based on the Generative Holism principle that the whole of a software product exists first (as an embryo), then grows up with its components; (B2): the quality assurance strategy for the NSE software development methodology – mainly depends on defect prevention and defect propagation prevention through dynamic testing using the Transparent-box testing method (see Chap. 16) combining functional testing and structural testing together seamlessly with the capability to establish automated and self-maintainable traceability among all related documents and the test cases and the source code through Time Tags automatically inserted into both the test case description part and the product test coverage measurement database for mapping them together, and some keywords (such as @WORD@, @HTML@, @PDF@, @EXCEL@, @BAT@) written in the test case description part to indicate the types of the related documents, the file locations, and the bookmarks for opening the traced documents from the corresponding locations, so that it can be used to find functional defects, structural defects, and inconsistency defects in the entire software development lifecycle. (B3): The defect prevention, mainly through dynamic testing using the Transparent-box testing method, is performed in all phases of a software development lifecycle

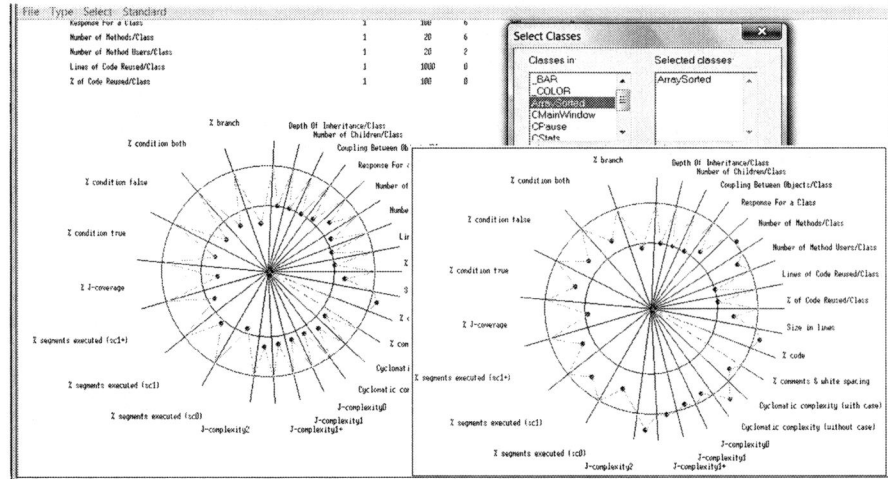

Fig. 10.3 Quality measurement results for an entire software product and each class/function generated by the NSE software documentation paradigm automatically, dynamically, and virtually

and exist virtually, thus greatly saving the space needed to store them and also speeding up the display time by about a thousand times (see Chap. 19). Figure 10.3 shows the quality measurement results using Kiviat diagrams for an entire software product and each class/function generated through only several hash tables dynamically, without storing any hard copies in the memory or hard disk (unless the user wants to store them).

6. The preliminary applications show that compared with the old-established software development methodologies based on reductionism and the superposition principle, it is possible for the NSE software development methodology (working with the NSE software development process model) to help software development organizations double their productivity, halve their cost, improve their product quality tenfold several times, and double their project success rate.
7. It brings revolutionary changes to the CBSD approach by shifting the software component development foundation from that based on reductionism and the superposition principle to that based on complexity science to greatly ensure the quality of the components themselves, and make the components adaptive and maintainable. According to the principle of complexity science that the behavior and characteristics of a complex system is determined by both the whole of the system and its components, with NSE a software component used for CBSD should at least satisfy the following listed conditions:

(a) Being 100% tested using the MC/DC (Modified Condition/Decision Coverage) test coverage metric, no matter whether it is provided as a class (a class cannot be directly executed, so the test coverage data should be collected through its instances) or a regular function (see Appendix B for an example showing how to realize 100% MC/DC test coverage with NSE)
(b) Being verified that there is no memory leak or memory usage violation found

(c) Being verified that it will not become a performance bottleneck to the application system
(d) Being verified that it will not bring bad effects to the file system and the I/O system for the applications
(e) Being verified that it satisfies the quality standard in the corresponding applications
(f) Being verified that it is provided with the related documents, the test cases, and (if possible) the source code traceable to and from the documents

10.3 The Driving Forces for the Innovation of the NSE Software Development Methodology

The NSE Software Development Methodology is driven by defect prevention and various automated and self-maintainable traceabilities:

1. Defect prevention:

   (a) Repeatable Defect Prevention through

   - Causal analysis
   - Preventive actions
   - Increase awareness of quality issues
   - Data collection
   - Improvement of the Defect Prevention Plan

   (b) New Defect Prevention (more useful) through bidirectional traceability to prevent

   - Inconsistent or changed requirement definitions that may contain conflicts
   - Inconsistent designs or design changes
   - Inconsistent coding (such as inconsistencies between function definitions and calling statements)
   - Inconsistent source code modification, etc.

   Some kinds of defects can be prevented are shown in Fig. 10.4.

2. Traceabilities, including the following:

   (a) Automated and self-maintainable traceability among documents and test cases and source code, including the documents obtained from project planning, requirement development, product design, coding, testing, and maintenance. This type of traceability is essential to software validation, verification, debugging, and the identification of unimplemented requirements and useless source code modules, requirements that are related to a module to be changed (for consistent modification), test cases that can be used for regression testing (whereby the efficiency of regression testing can be improved tenfold!), etc. This kind of traceability is established through dynamic testing using the Transparent-box testing method. The traceability between test cases

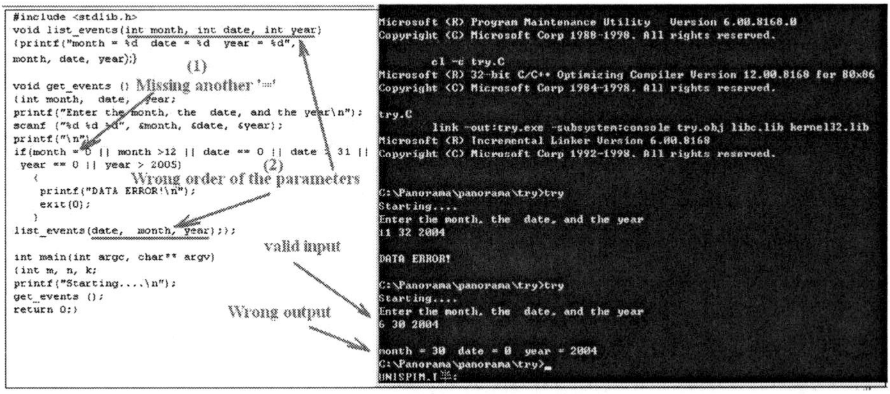

Fig. 10.4 Sample defects can be prevented

and the source code is established using Time Tags automatically inserted in the test cases description and the test coverage database.

The traceability between test cases and the source code has been extended to include all related documents using some keywords written in the description part of a test case to indicate the document formats, the file paths, and the corresponding bookmarks for showing the documents from the corresponding locations.

(b) Automated and bidirectional traceability within the source code, among source files, classes, functions, and detailed statements. It is established by diagramming the entire program, then creating the traceability automatically between header file and "#include"statement, program tree and function body, function definition and function call statement, class instance and class definition, goto statement and label, etc. This type of traceability is essential for efficient source code inspection and walkthrough, testing, bug checking, consistent source code modification, etc. See Fig. 10.5 for some examples of established automated traceability.

(c) Capability to trace a runtime error to the execution path and the related functions. This type of traceability is useful for debugging with testing. An example of this kind of traceability is shown in Fig. 10.6.

(d) Automated traceability in a systematic analysis of software changes to graphically show version comparison results at the system level, source file level, class and function level, and statement level. It includes identifying which modules are deleted (shown in brown), added (shown in green), changed (shown in red), and unchanged (shown in blue); these colors are shown on the original screen, and not black and white in the book. To a changed module, we can further trace the detailed source code to find which statements are deleted, added, and modified. This type of traceability is very useful for version comparison and debugging, particularly in the maintenance phase when some bugs have been removed but new bugs are found. An example of this kind of traceability is shown in Fig. 10.7.

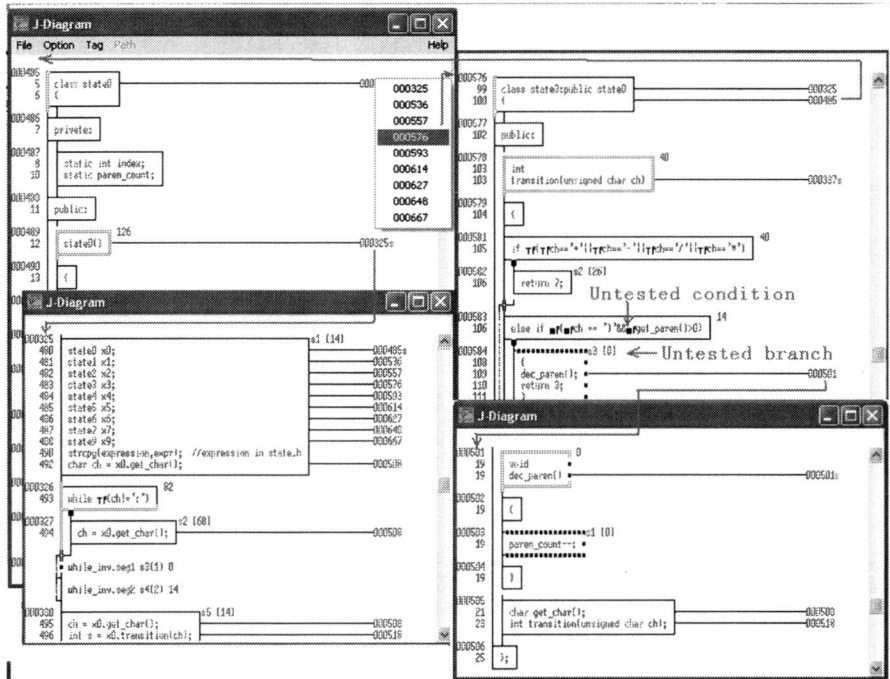

Fig. 10.5 Sample traceabilities automatically established with a J-Diagram

(e) Automated traceability among documents such as those for requirement management. requirement specifications, requirement changes, etc. To realize this type of automatic traceability, we use a set of predesigned templates in HTML/XML format. These templates will link together by themselves.

(f) Automatic traceability through all possible execution paths for each module from a call graph. This kind of traceability is useful in identifying which other modules may be affected by a change made within a module. An example of this kind of traceability is shown in Fig. 10.8.

10.4 The Related NSE Software Engineering Process Model

The NSE software development methodology works seamlessly with the NSE software engineering process model shown in Chap. 8.

The NSE software engineering process model consists of three parts – the preprocess, the main process, and the support facility for automated and self-maintainable traceability among the related documents such as the requirement specifications, the test cases, and the source code (see Chap. 9 for the detailed description).

The main purpose of the preprocess is to assign priority to the requirements according to its importance, perform prototyping for the important and unfamiliar

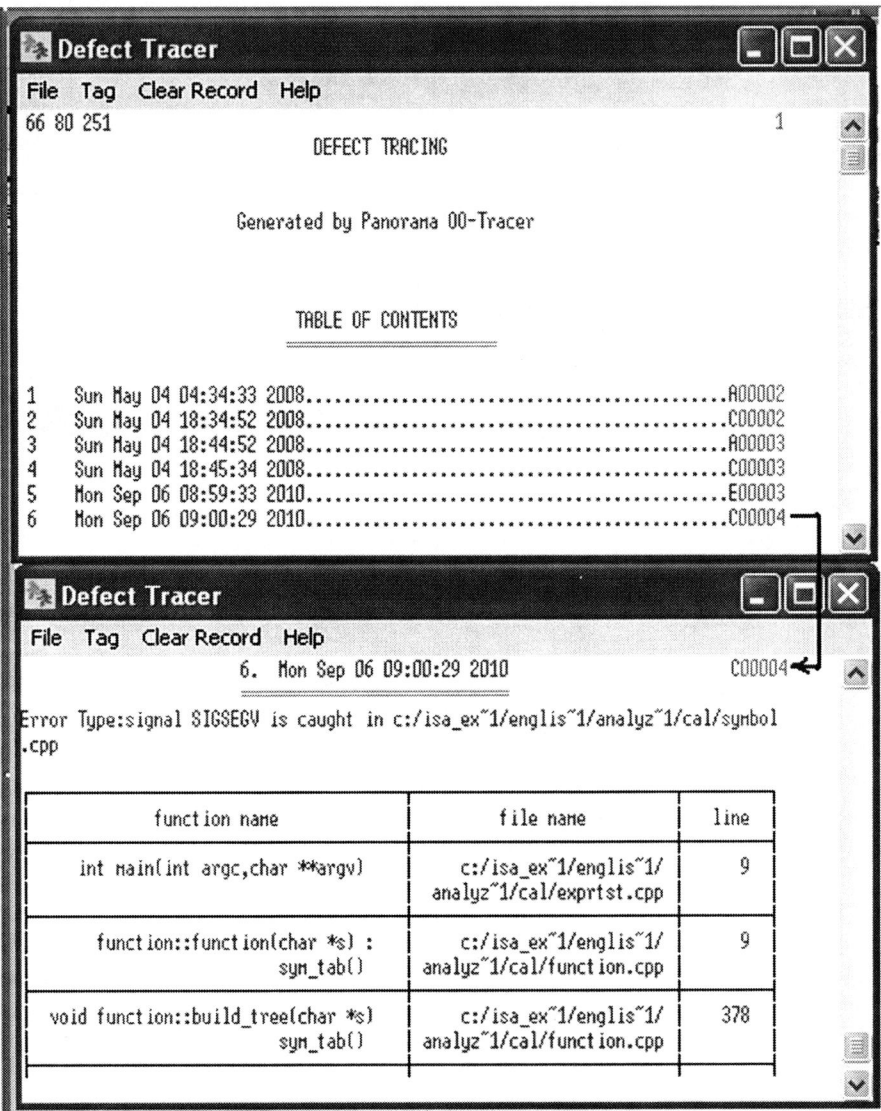

Fig. 10.6 Tracing the execution path for a runtime error

requirements to reduce risk, perform the functional decomposition for functional requirements and system preliminary design through dummy programming to form the whole of a software system as an embryo using dummy modules having an empty body or only some calling statements to call other low-layered modules without detailed program logic – see the Completeness Percentage axis of the graphical description of the NSE software development methodology shown in Fig. 10.9, the "Bone" system (about 5% of the product effort, the first milestone) is obtained in the preprocess.

10.5 Graphical Presentation of the NSE Software Development Methodology

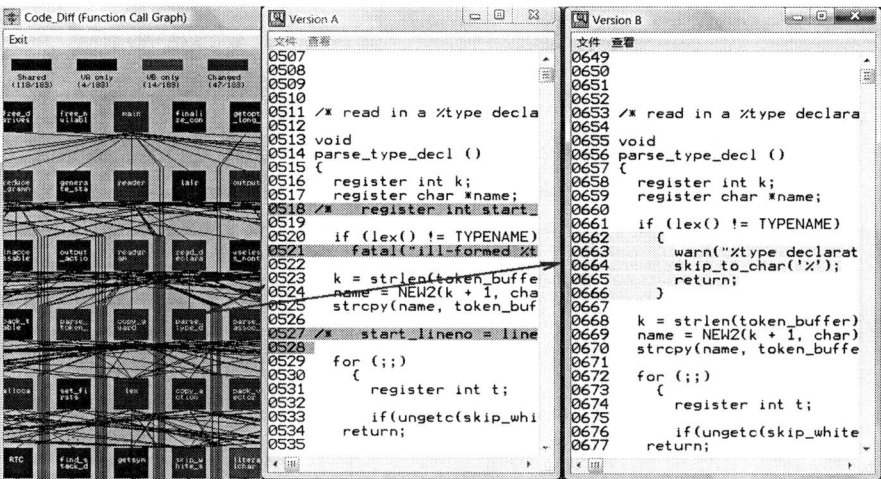

Fig. 10.7 Tracing an entire software product to a previous version

The implementation of requirements are performed with the main process incrementally through two-way iteration supported by automated and self-maintainable traceability – it is recommended to complete the implementation of about 20% of the most important requirements to form an essential version of the product – see the Completeness Percentage axis of the graphical description of the NSE software development methodology shown in Fig. 10.9; it corresponds to the "Essential" version (second milestone) of the product completeness.

After that, the whole system grows up with more incremental implementations of the requirements, until the final product is completed. With the NSE software development methodology, all versions including the "Bone" system are executable (even if there is no real output provided), and delivered to the customer for review and then the customer's feedback will be used to improve the product.

The NSE software engineering process model is a nonlinear one which assumes that the upper phases may have defects or something wrong, so there is a need to check the inconsistency with the upper phases to improve the product – when a critical issue is found, there may be a need to go back to the preprocess to design a better solution method for the corresponding requirement(s), and perform the prototyping again.

10.5 Graphical Presentation of the NSE Software Development Methodology

The graphical description of the NSE software development methodology is shown in Fig. 10.9.

As shown in Fig. 10.9, there are three axes representing the Work Flow, the Time, and the Completeness Percentage separately.

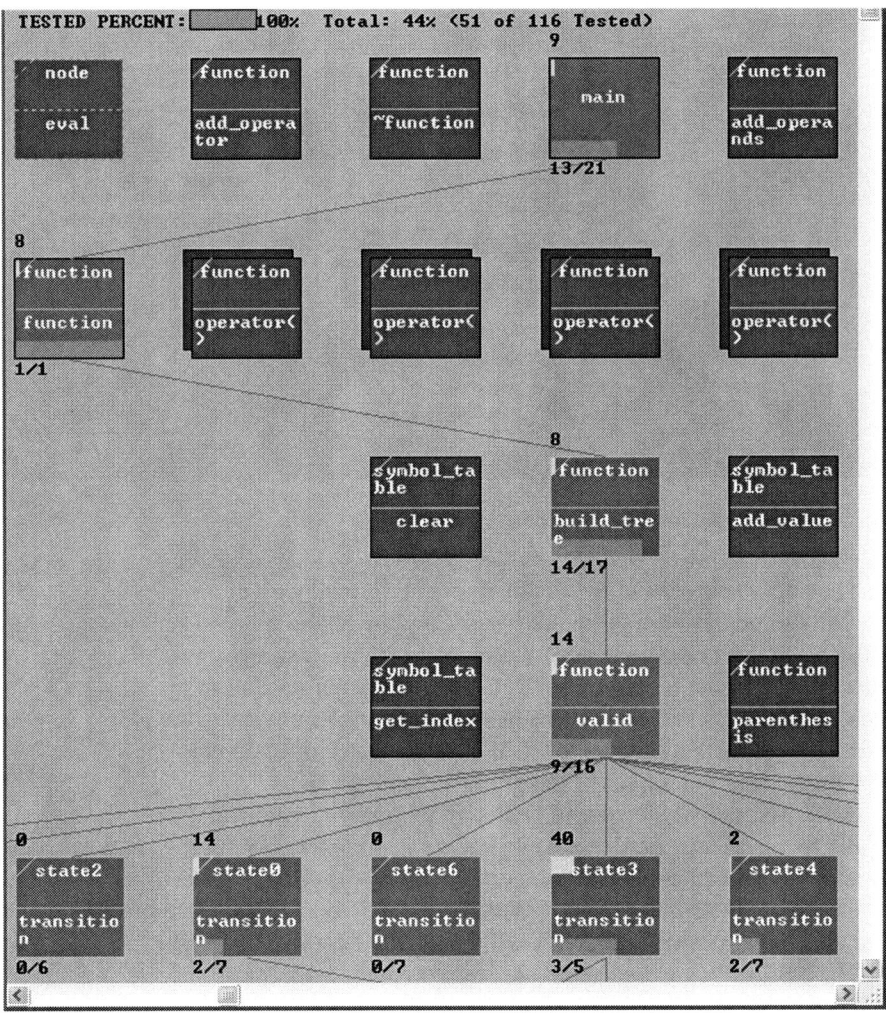

Fig. 10.8 Sample traceability between related modules

In the Work Flow axis, it not only includes the phases of requirement development, design, coding, testing, and maintenance but also includes the project management, the product delivery, and the support for the product Web site and BBS for communication – it combines the product development and maintenance together, and also combines software development and project management together. No matter in what phase, defect prevention and defect propagation prevention is performed to ensure the quality of the product being developed. It does not always follow a linear order – as shown on the right side, upstream movement is supported through traceability for two-way iteration, if necessary.

The Time axis represents the progress.

10.5 Graphical Presentation of the NSE Software Development Methodology

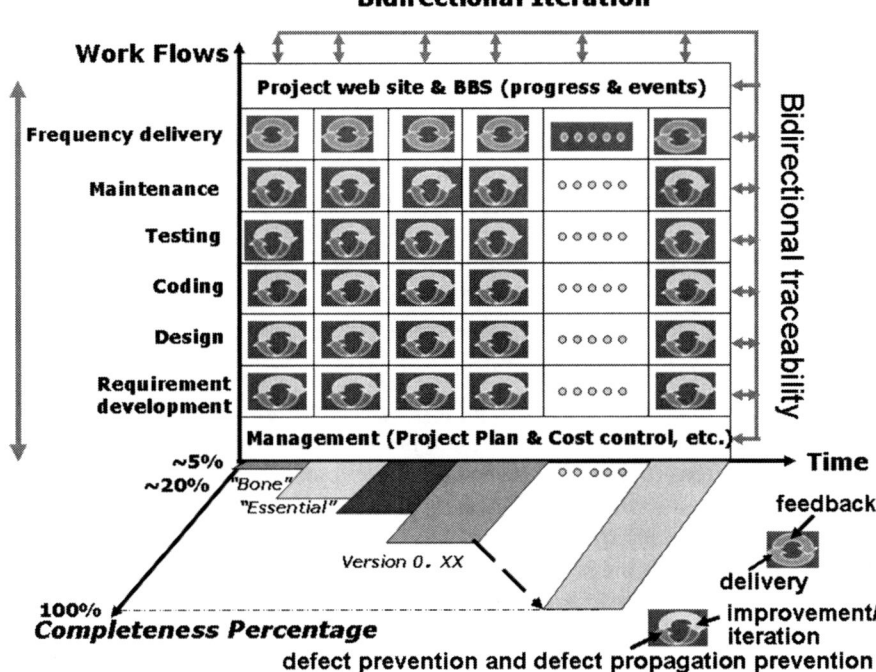

Fig. 10.9 NSE software development methodology

The Completeness Percentage axis represents how much percent of the product is completed – there are three milestones:

1. The first one is the "Bone" system completed through dummy programming. The "Bone" version is completed through the preprocess – after prototyping and risk analysis, functional decomposition of the functional requirements will be performed, then the requirement functional decomposition result will be used for the preliminary design to establish the "Bone" system through dummy programming (each dummy module has an empty body or a list of function call statements without detailed program logic).
2. The second one is the "Essential" version of the product – about 20% of the most important requirements have been implemented. It is completed in the main process incrementally for the most important requirements.
3. The third one is the final product. Often the number of the requirements will be doubled, compared with the initial number of the requirements – NSE supports requirement changes in both the software development process, and the software maintenance process with side-effect prevention through various traceabilities.

With NSE, when a product is delivered to the customer, not only are the computer program, the data used, and the documents traceable to and from the source code provided but also the database built through static and dynamic measurement of the program, and a set of Assisted Online Agents (automated and intelligent tools) to support the program's testability, visibility, reliability, conformity, changeability, and traceability, to make the program truly adaptive and truly maintainable.

10.6 Application

The NSE software development methodology has been implemented commercially and is supported by NSE application platform Panorama++ and SilverBullet++.

The following chapters will describe the detailed applications in the software development lifecycle:

(a) The applications in the requirement development phase – see Chap. 11
(b) The applications in the software design phase – see Chap. 12
(c) The applications in the software coding phase – see Chap. 13
(d) The applications in the software testing phase – see Chap. 16
(e) The applications in software quality assurance – see Chap. 17
(f) The applications in the software maintenance phase – see Chap. 18
(g) The applications in software documentation – see Chap. 19
(h) The applications in software project management – see Chap. 20

As described above, the NSE software development methodology is driven by defect prevention, defect propagation prevention, and traceability mainly through dynamic testing using the Transparent-box testing method, and software visualization in the entire software development process.

About the application examples of the NSE software development methodology using the Transparent-box testing method for defect prevention and defect propagation prevention, please read Chaps. 16 and 17.

10.6.1 Some Suggestions About the Applications of the NSE Software Development Methodology

(a) It is recommended to use NSE from the beginning for a software project, but in fact NSE can be applied at any time in any stage using any methodology originally, to update the product development approach – with NSE, it only requires users to rewrite the test cases with some simple rules, and set the corresponding bookmarks for the related documents; almost all the other work can be done by the NSE support platform automatically.
(b) Moving the quality assurance strategy from focusing on downstream to focusing on upstream, not only through static review but also through dynamic testing using the NSE software testing paradigm based on the Transparent-box method, because dynamic testing can be used to detect more defects – many defects cannot be found without really running the software system, including the "Bone" system without real output. In the meantime, dynamic testing using

the Transparent-box method can also automatically establish the traceability among the related documents and test cases and the source code (including the dummy programming source code where each dummy module has an empty body or only some function call statements without detailed program logic) to assist formal review in a semiautomatic way. Upstream defect prevention and defect propagation prevention (removing defects at their source) is the key to ensure the product quality and greatly reduce the software development cost.
(c) It is recommended to reach 100% MC/DC test coverage for each program unit for any commercial software product to ensure the quality of the product. See Appendix B for an application example – it shows that with NSE in the unit testing process, it is not difficult and not expensive to reach a 100% MC/DC test coverage analysis result.

10.7 The Major Features of the NSE Software Development Methodology

The major features of the NSE software development methodology are briefly listed as follows:

(a) It is based on complexity science.
(b) It complies with the Constructive Holism principle of complexity science.
(c) It supports incremental development with two-way iteration through various traceabilities.
(d) It combines software development and software maintenance together closely.
(e) It combines software development and software project management together closely.
(f) It makes the entire software development process visible.
(g) It enhances the communication among developers through traceable project Web sites and BBS.
(h) It supports multiple project development – two or more related projects' documents, including the management documents and the progress reports, can be traced to each other as shown in Fig. 9.10 in Chapter 9.
(i) It supports parallel development.
(j) It supports refactoring for the highly complex modules through defect prevention.
(k) It makes the design become pre-coding (through dummy programming), and the coding become further design (through reverse engineering and backward traceability to update the design documents).

10.8 Summary

This chapter presented the NSE software development methodology based on complexity science. It is driven by traceability, defect prevention, and defect propagation through dynamic testing using the Transparent-box testing method and software visualization. Preliminary applications show that compared with the existing software development methodologies it is possible for the NSE software development methodology (with the NSE process model and the support platform) to help software development organizations double their productivity and project success rate, halve their cost, and increase their product quality in tenfold many times.

10.9 Points and Questions to Ponder

(a) What are the differences in software development methodology between that based on Constructive Holism and that based on Generative Holism?
(b) What are the major differences between RUP (Rational Unified Process) and the NSE software development methodology?
(c) How can the NSE software testing paradigm be dynamically used in upstream quality assurance for defect prevention and defect propagation prevention?
(d) How can the NSE software visualization paradigm be used in software defect prevention, defect propagation prevention, software understanding, testing, and maintenance?

10.10 Further Reading and Information Source

(a) Kay A, Hewlett Packard (2004) "The computer revolution", "computer science", and "software engineering" haven't happened yet. http://portal.acm.org/citation.cfm?id=1017758. ISBN:1-58113-860-1
(b) Zambonelli F, VanDyke Parunak H (2003) Signs of a revolution in computer science and software engineering. Springer, Berlin. ISSN 0302-9743 (Print) 1611-3349 (Online)

References

[CMMI1.1] Phillips M (2002) CMMI Program Manager. CMMI V1.1 and Appraisal Tutorial. http://www.sei.cmu.edu/cmmi/, slide 118, titled "Product Integration"
[Bro95-P200] Brooks FP Jr (1995) The mythical man-month. Addison-Wesley, Reading, p 200
[Bro95-P267] Brooks FP Jr (1995) The mythical man-month. Addison-Wesley, Reading, p 267
[Xio09] Xiong J, Xiong J (2009) A complete revolution in software engineering based on complexity science. In: WORLDCOMP'09 – SERP (Software Engineering Research and Practice 2009), pp 109–115
[Dem86] Deming WE (1986) Out of the crisis. MIT Press, Cambridge. ISBN 0-911379-01-0. OCLC 13126265
[Coc08] Cockburn A (2008) Using both incremental and iterative development. CrossTalk, May Issue

Chapter 11
Requirement Engineering Under NSE: Source Code Driven Dynamic Software Modeling

> *The important thing is that one model is enough – either the code or the diagrams. They should be reproducible from one another.*
>
> Harry M. Sneed

This chapter introduces the detailed applications of the NSE process model and the NSE software development methodology in software requirement development.

As described in Chap. 1, software requirement engineering includes requirement elicitation/gathering, functional decomposition of functional requirements and the description of the nonfunctional requirements, requirement analysis, requirement specification, requirement validation, etc.

With NSE, several templates are provided for helping users to prevent possible defects such as missing something particularly in the requirement specification – a requirement specification template is attached in Appendix A for readers to review and use.

Here, in this chapter, we will focus on "Source Code Driven Dynamic Software Modeling and Engineering" using the innovated HAETVE (Holistic, Actor–Action and Event–Response driven, Traceable, Visual, and Executable) technique.

11.1 Are All the Existing Software Modeling Approaches Outdated?

Based on the results of a survey of 113 software practitioners conducted between April and December 2007, Andrew Forward reported that *"Problems and opportunities for model-centric versus code-centric software development ... Programmers that model extensively (versus those that do not model much) are more likely to agree that models become out of date and inconsistent with code"* [For08].

State Information Technology Consortium (http://www.state-itc.org/) listed the benefits and possible drawbacks of the Use Case approach as follows:

"Benefits of Use Cases

- Help define user requirements
- Identify and document current methods, systems, and stakeholders
- Drive detailed application analysis and design
- Develop scripts for testing
- Suggest prototyping activity
- Clarify architecture requirements
- Highlight risks and needs for risk management

Potential Use Case drawbacks

- Might be incomplete.
- Each case may not describe enough detail of use.
- Not enough Use Cases; may miss entire areas of functionality.
- Might be inaccurate.
- Might not have been reviewed.
- Might not have been updated when requirements changed.
- Might be ambiguous/unclear.
- Will not find many bugs."

How about the future of UML?

As indicated by Jim Arlow and Ila Neustadt in their book, "UML 2 and the Unified Process: Practical Object-Oriented Analysis and Design (Second Edition)" [Arl06] that

MDA – the future of UML
The future of UML may be a recent OMG initiative called Model Driven Architecture (MDA)... MDA defines a vision for how software can be developed based on models... In MDA software is produced through a series of model transformations aided by an MDA modeling tool. An abstract computer-independent model (CIM) is used as basis for a platform-independent model (PIM). The PIM is transformed into a platform-specific model (PSM) that is transformed into code.

But about MDA, Harry Sneed pointed that

Model driven considered harmful

- Model-driven tools magnify the mistakes made in the problem definition.
- Model-driven tools create an additional semantic level to be maintained.
- Model-driven tools distort the image of what the program is really like.
- The model cannot be directly executed. It must first be transformed into code which may behave other than expected.
- Model-driven tools complicate the maintenance process by creating redundant descriptions which have to be maintained in parallel.
- Model-driven tools are designed for top-down development.
- Top-down functional decomposition creates maintenance problems.

11.1 Are All the Existing Software Modeling Approaches Outdated?

"Summary:

- If a UML design can really replace the programming code as envisioned by Jacobson in his paper, UML all the way down, then it becomes just another programming language.
- The question then comes up as to what is easier to change
  - The design documents or
  - The programming language
- This depends on the nature of the problem and the people trying to solve it. If they are more comfortable with diagrams, they can use diagrams. If they are more comfortable with text, they should write text.
- Diagrams are not always the best means of modeling a solution. A solution can also be described in words. The important thing is that one model is enough – either the code or the diagrams. They should be reproducible from one another" [Sne07].

In my opinion, the major drawbacks of the existing software modeling approaches are as follows:

1. **They are outcomes of reductionism and the superposition principle that the whole of a complex system is the sum of its parts**, so that with them almost all software modeling activities are performed partially and locally – it is the root cause for their many drawbacks.
2. **They are designed to work with the existing linear process models with no upstream movement at all**, making the defects introduced in the requirement development phase and software design phase easily propagate down to the maintenance phase, and the defect removal cost increase tenfold several times.
3. **They comply with the Constructive Holism principle** that software components are developed first, then, "***Assemble** the product from the product components, ensure the product, as integrated, functions properly and deliver the product*." [CMMI1.1] – it makes the quality of a software product much difficult to ensure – for instance, when a runtime error happens in the product integration, it is hard to know where the error comes from.
4. **They use two kinds of sources for software engineering** with one in diagrams for human understanding of a software system and the another one in textual format for computer understanding of the product – there is a big gap between the two sources. Here I call them "Two Sources Approach" (TSA) as shown in Fig. 11.1.
5. **They miss the Big Picture** – the obtained results consist of many small pieces without a holistic whole for a complex software system.
6. **The obtained results are not traceable** – hard to review for static defect removal, and even if a holistic result can be obtained, without traceability to highlight the related elements it is still useless because there will be too many connection lines, making the result hard to view and hard to understand.
7. **The results obtained are hard to update and maintain** – the models are created in inefficient ways, not automatically generated.

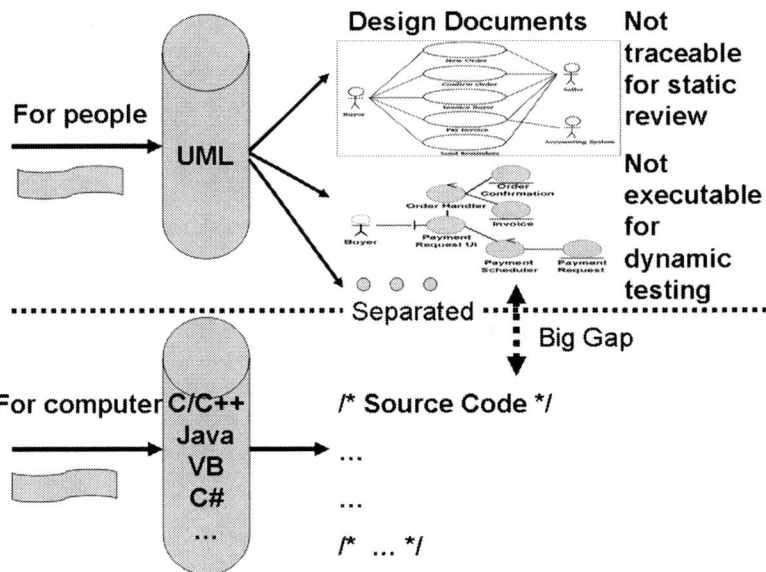

Fig. 11.1 The two sources approach for software modeling (TSA)

8. **The obtained results are not consistent with the source code** after code modification.
9. **They are static modeling approaches – the obtained results are not directly executable**, so that there is no debugging at all.
10. **There is no dynamic testing at all** – no way to ensure the quality of the models.
11. **No right to be wrong** – they offer "one time only" approaches – once the models are obtained, they become the road map for the product development following a linear process model with no upstream movement at all, so that the creators of the models have no right to be wrong – but it violates the nature of human beings: people are nonlinear and it is easy to make mistakes in thinking, working, reading, writing, hearing – everyone makes mistakes and wrong decisions, including in software modeling.
12. **They could be high risk approaches** – they force the obtained unqualified models (without being debugged and tested) to become the road map for software product development. I think it is one of the major reasons why about 70% of software projects are failures.

Conclusion: **the existing software modeling approaches are outdated**.

11.2 Outline of the Revolutionary Solution Offered by NSE

The revolutionary solution offered by NSE in software modeling and requirement engineering is described in detail in this chapter later. Here is the outline of the solution called "Source Code Driven Dynamic Software Modeling and Engineering"

11.2 Outline of the Revolutionary Solution Offered by NSE

using the HAETVE (a Holistic, Actor–Action and Event–Response driven, Traceable, Visual, and Executable) requirement development technique:

1. **It is based on complexity science** by complying with the essential principles of complexity science, particularly the Nonlinearity principle and the Holism principle that **the whole of a complex system is greater than the sum of its components – the characteristics and behaviors of the whole emerge from the interaction of its components,** so that with NSE almost all tasks and activities in software modeling and requirement engineering are performed holistically and globally.
2. **It works with the NSE process model which is a nonlinear one with two-way iteration** (see Chap. 8) supported by automated and self-maintainable traceabilities (see Chap. 9) to prevent defects brought into software products by the product developers and the customers.
3. **It complies with the Generative Holism principle** of complexity science that the whole of a complex system exists (as an embryo) earlier than its components, then grows up with its components. As pointed by Frederick P. Books Jr. that "Incremental development – grow, not build software ... that the system should first be made to run, even though it does nothing useful except call the proper set of dummy subprograms. Then, bit by bit it is fleshed out, with the subprograms in turn being developed into actions or calls to empty stubs in the level below." [Bro95-p200], we will have a working system at all times to begin user testing very early and adopt a build-to-budget strategy that protects absolutely against schedule or budget overrun (at the cost of possible functional shortfall).
4. **It uses one kind of source for both** human understanding and computer understanding of a software system. Here I call it SDM (Source Code Driven Dynamic Software Modeling) as shown in Fig. 11.2.

    It is the key to bring revolutionary changes to software modeling by making the models obtained traceable for static defect removal, executable for debugging, testable for dynamic defect removal using Transparent-box testing method (see Chap. 16), and always consistent with the source code. With it, the diagrams are generated automatically through forward engineering using dummy programs (a dummy module will have an empty body or only a list of function call statements) and reverse engineering using regular programs.

    Why are the diagrams generated from code rather than the code generated from diagrams? The reason is simple:

    (a) There is not enough detailed information from the diagrams to generate code.
    (b) If there is enough detailed information from the diagrams to generate code, then the diagrams are useless – there will be too many elements and too many connection lines to make the diagrams very hard to view and very hard to understand.

The difficulty for generating diagrams from the code is due to its big size – for instance, an EDA program may include 100,000 function points with more than ten million lines of source code. With NSE, there is no size limitation for generating diagrams from big programs, because NSE visualization paradigm offers features to generate virtual, holistic, interactive, and traceable diagrams virtually (see Chap. 7).

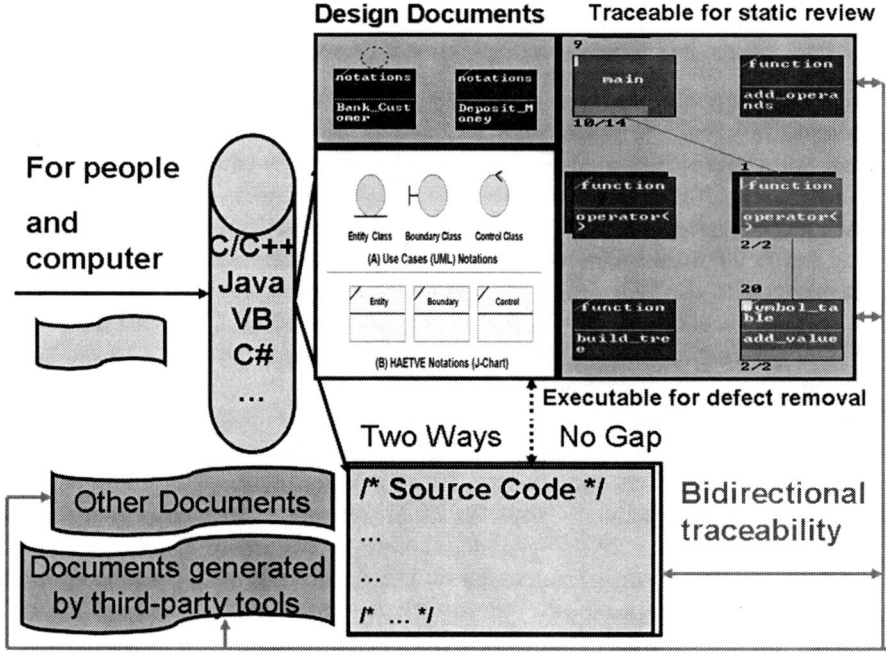

Fig. 11.2 Source Code Driven Dynamic Software Modeling (SDM)

5. **It offers holistic modeling results for an entire software system** to show the Big Picture of a software product automatically.
6. **The obtained results are traceable** – easy to review for static defect removal. For instance, users can trace a module in the functional decomposition chart of the functional requirements to highlight all the related functions.
7. **The results obtained are easy to update and maintain** – the models are automatically generated from the dummy programs or the regular source code.
8. **The obtained results are always consistent with the source code** after code modification – the models can be updated automatically.
9. **It offers dynamic modeling capability – the obtained results are executable**, so that a debugger can be used to find and fix some defects.
10. **With it, dynamic testing is performed from the first step to ensure the model quality** – it works closely with the NSE software testing paradigm using the innovated Transparent-box testing method which combines functional testing and structural testing together seamlessly: to each set of inputs to any working version of a software product (including the dummy whole system without providing any real output), it not only checks whether the output (if any, can be none) is the same as what expected, but also checks whether the real execution path covers a expected one specified in control flow, and then automatically establishes bidirectional traceability to help users remove inconsistent defects among the related documents and test cases and source code.

11. **We have the right to be wrong, but we also have the right to be right** – we are human beings, and it is easy to make mistakes in thinking, working, reading, writing, and hearing, so that with NSE modeling becomes preimplementation of requirements, and implementation of requirements becomes further modeling – when something wrong is found with the models in coding, we can fix the problem by modifying the source code, and then re-generate the models with the self-adaptation and self-maintenance capability.
12. **It is a high-quality approach** – with NSE, the dynamic modeling results are traceable for static review and defect removal, executable for debugging, and testable for defect removal performed dynamically to ensure the model quality. This is one of the major reasons why it is possible for NSE to help users double their project success rate.

11.3 Description of the HAETVE Technique

As the name suggests, HAETVE is a Holistic, Actor–Action and Event–Response driven, Traceable, Visual, and Executable technique for software requirement development. It is innovated by me according to the NSE process model and NSE software development methodology based on complexity science by complying with the essential principles of complexity science, particularly the Nonlinearity principle and the Holism principle, so that with HAETVE almost all tasks/activities in software requirement engineering are performed holistically and globally.

HAETVE is one of the most important means to the implementation of the **Upstream Quality Assurance** strategy and **Total Quality Management** strategy to ensure the quality of a software product through defect prevention and defect propagation prevention from the first step to the final step in the entire software development and maintenance process, to follow Deming's Product Quality Assurance Principles that *"Cease dependence on inspection to achieve quality. Eliminate the need for inspection on a mass basis by building quality into the product in the first place"* [Dem82].

With **HAETVE**, the graphical notations for representing an actor and an action for C/C++ programs are shown in Fig. 11.3 where the notation used for representing an actor is originally designed for representing a recursive program module.

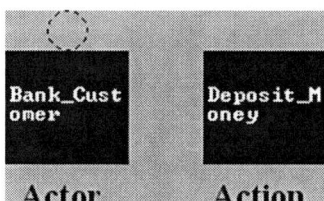

Fig. 11.3 The notations for representing an actor and the action for C/C++

Fig. 11.4 The notations for representing an actor and the action for Java

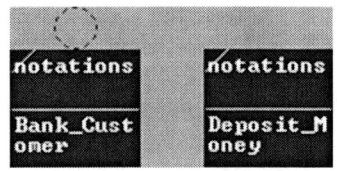

The corresponding dummy source code written in C/C++ is listed separately as follows:

```
Bank_Customer ()
{
Bank_Customer ();
}
Void Deposit_Money ()
{
}
```

The corresponding notations for Java are shown in Fig. 11.4.

```
public class notations {

    public static void Bank_Customer ()
    {
       Bank_Customer () ;
    }
    public static void Deposit_Money ()
    {
    }
}
```

The corresponding source code in Java dummy programming is listed as follows:

Why use Java? It is because Java is a platform-independent programming language, so that the results obtained in modeling are also independent from the target languages and platforms. If there is a need, the dummy java source code can be transformed to a target language.

Here a special Actor – **SuperActor** is defined and used to request nonfunctional requirements. With NSE and the support platform, Panorama++, a SuperActor can request the following nonfunctional requirement as shown in Table 11.1.

Similarly, a sample of Event–Response relationship notations are shown in Fig. 11.5.

The documents for actor–action are similar to Use Cases. But the event–response documents are different from those with Use Cases. An example of the event–response document is shown in Table 11.2.

Table 11.1 Nonfunctional requirements which can be required/specified by the SuperActor

Item of nonfunctional requirements	What can be required/specified
Interface design	(1) Commend-line (2) Graphic user interface (GUI) (3) Both commend-line and GUI (it needs to further describe the details)
Performance	1. Performance measurement (method and tool) 2. Code branch execution frequency measurement for locating performance bottlenecks better 3. Memory leak and usage violation check
Quality level related to structural testing	1. MC/DC (modified condition/decision coverage) test coverage measurement, or 2. Branch test coverage measurement, or 3. Statement test coverage (low quality level, not recommended to use) With all untested conditions and branches highlighted graphically

Quality measurement metrics	1. Cyclomatic complexity (the number of decision statements such as "if" and "for") for each module (less than 30 is recommended)
	2. Module size (less than 200 lines for instance)
	3. Class-related metrics:
	  (a) Lines of code per class (LOC)
	  (b) Number of methods per class (NOM)
	  (c) Number of method users per class (NMU)
	  (d) Weighted methods per class (WMC) in multiple complexity metrics
	  (e) Depth of inheritance tree (DIT)
	  (f) Number of children per class (NOC)
	  (g) Coupling between objects (CBO)
	  (h) Response for a class (RFC)
	  (i) Lines of code reused per class (LCR)
	  (j) Ratio of code reused per class (RCR)
	  (k) Test coverage per class (TCC) in multiple test coverage metrics
	4. Other metrics
	They are flexible, can be selected, and assigned standard values from the corresponding menu

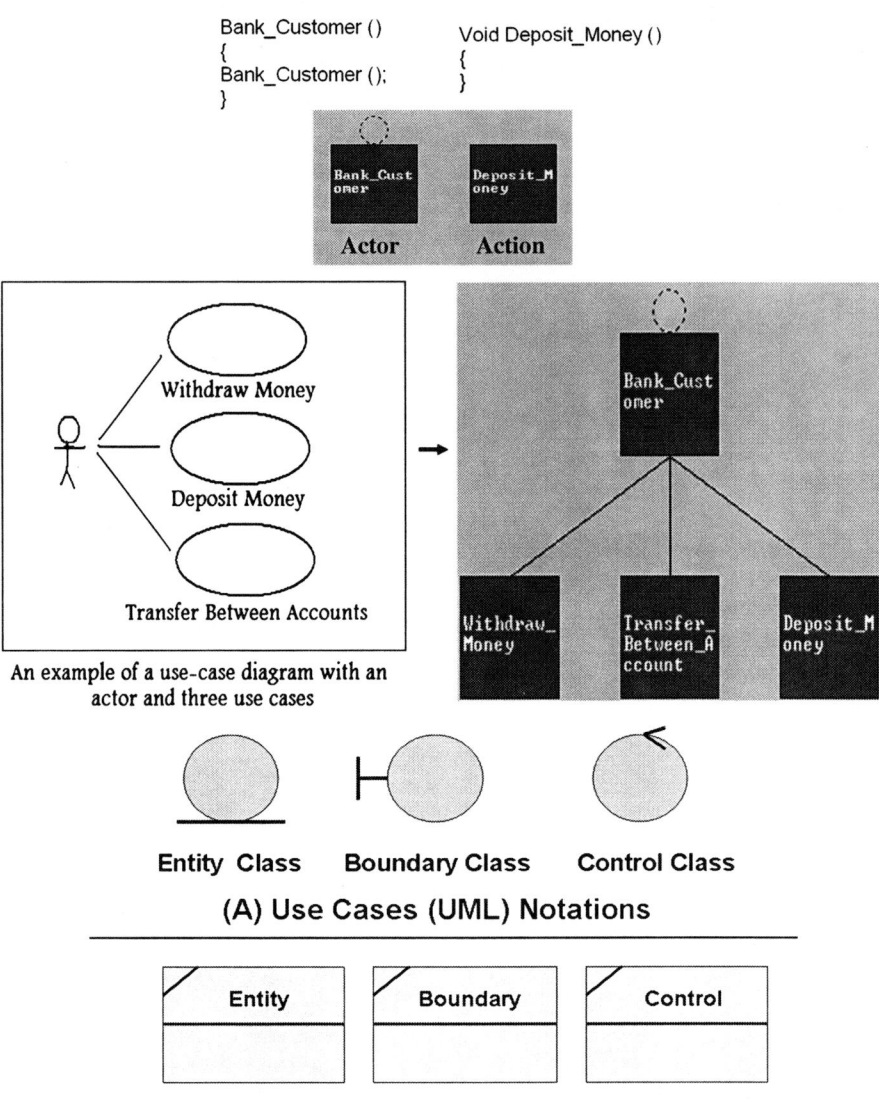

Fig. 11.5 A sample event–response relationship

An application example provided by Panorama++ with the unexpected termination event–response treatment is shown in Fig. 11.6.

Correspondingly, the unexpected termination location is also indicated by the error message and shown on a window after the program termination:

SIGSEGV is caught in trouble.c line 133

For the Actor–Action type applications, HAETVE is similar to the Use Case approach [Jac92], and is easy to map to Use Case notations as shown in Figs. 11.7–11.9.

Table 11.2 Sample event–response table for Panorama++ design

Event	System state	Response	Implementation hint
Unexpected termination of the program	Running user's program with test coverage analysis through the use of Panorama++	(1) Analyze the error type (2) Get the termination location and map it to the source code (3) Record the execution path (4) Close the opened files	Replace the on_exit() system function with isa_exit() to get the related messages earlier
User's action to terminate Panorama++	Running Panorama++ to handle a user's program	1. Close the files opened 2. Save related data 3. Close all tools being used 4. Exit Panorama++	Set a program termination button on Panorama++ interface for users to use
…	…	…	…
…	…	…	…
…	…	…	…

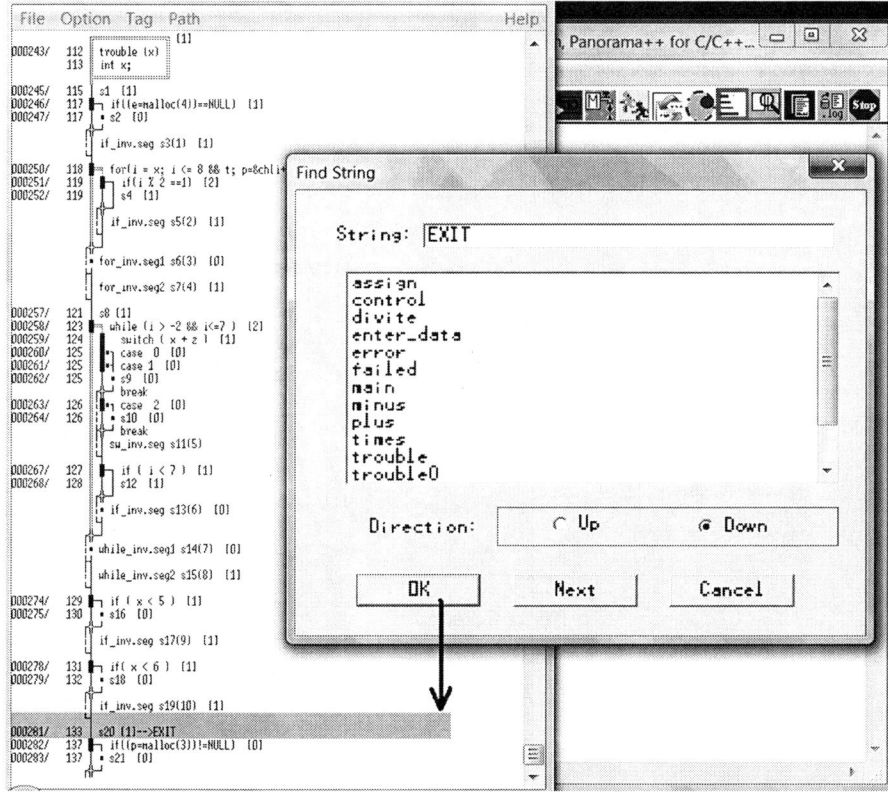

Fig. 11.6 An event–response treatment result provided by Panorama++ with the unexpected termination location (shown with an EXIT word inserted into the J-Flow diagram) mapped to the source code of the program being tested

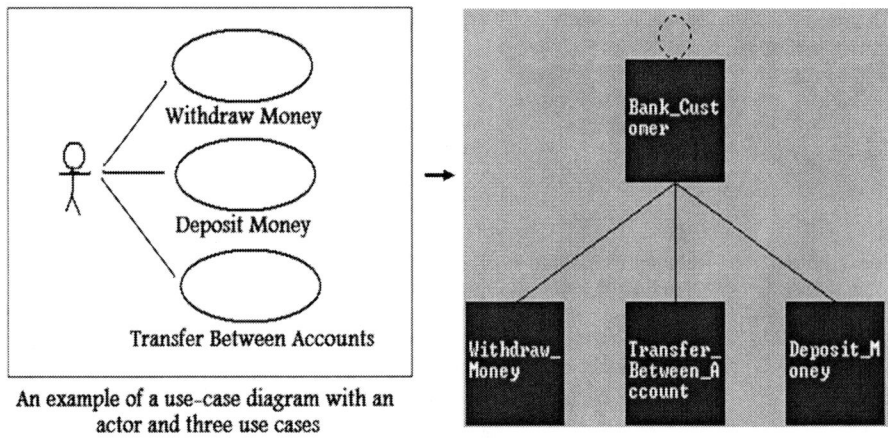

Fig. 11.7 Notation mapping between Use Case and HAETVE

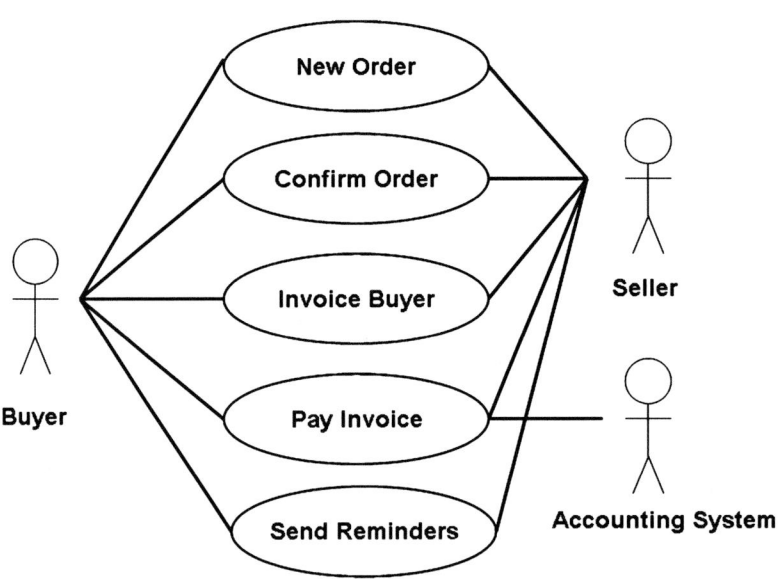

Fig. 11.8 An application example of Use Cases

11.3 Description of the HAETVE Technique

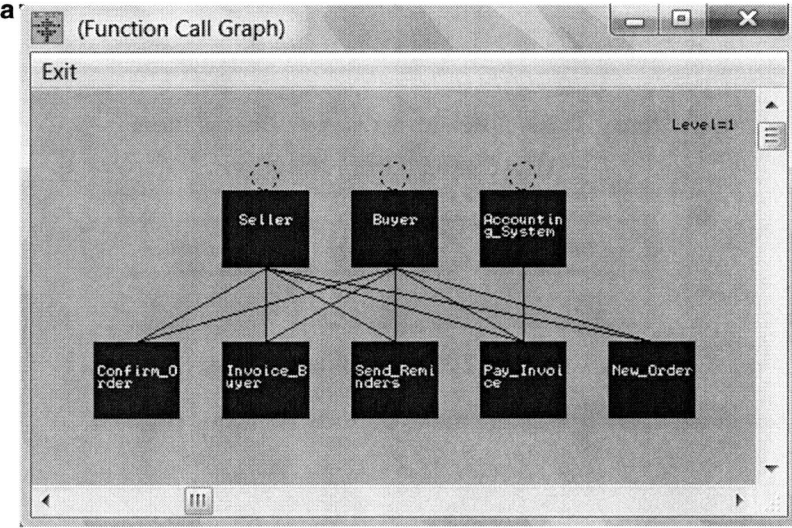

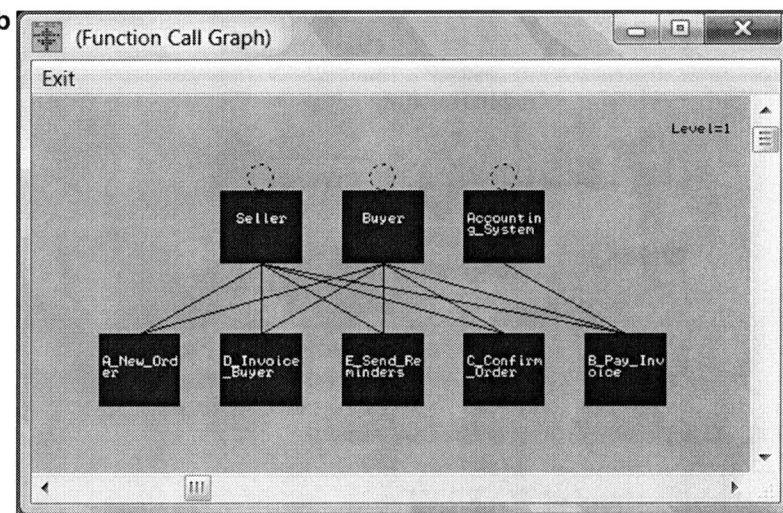

Fig. 11.9 Mapping from HAETVE to Use Cases: (**a**) mapping to Use Cases shown in Fig. 11.6. (**b**) A modified version of (**a**) with assigned priority (using A, B, C... a, b, c...to order them)

The analysis result of Use Cases can also be mapped to HAETVE as shown in Fig. 11.10.

Working with the NSE software visualization paradigm (see Chap. 7) and the NSE software testing paradigm (see Chap. 16) based on the Transparent-box testing method, the obtained results using the HAETVE technique are traceable for static defect removal and executable for dynamic defect prevention and defect propagation prevention (see next section).

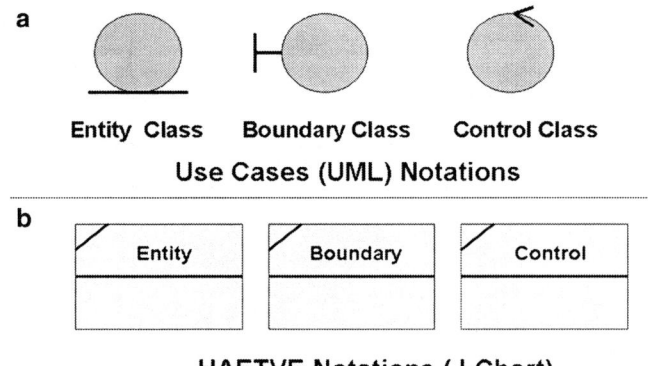

Fig. 11.10 Analysis notation mapping between Use Cases (UML) and HAETVE

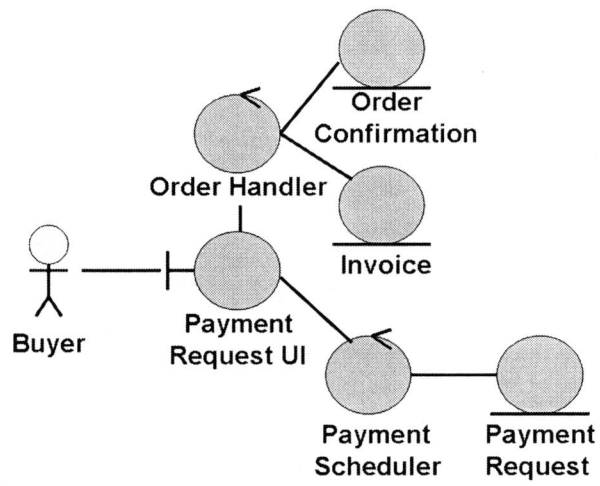

Fig. 11.11 An application example of Use Case Analysis

11.4 Applications of HAETVE

An application example in functional requirement decomposition using Use Cases is shown in Fig. 11.11. The mapping result using the HAETVE approach is shown in Fig. 11.12.

But there are some special things with HAETVE:

(a) It supports holistic functional requirement decomposition as shown in Fig. 11.13:
(b) The obtained result is traceable for static defect prevention and defect propagation prevention, see Fig. 11.14 – found a defect through traceability: the Order_ Handler should handle Order_Confirmation too:

The modified version with the defect removed is shown in Fig. 11.15.

11.4 Applications of HAETVE

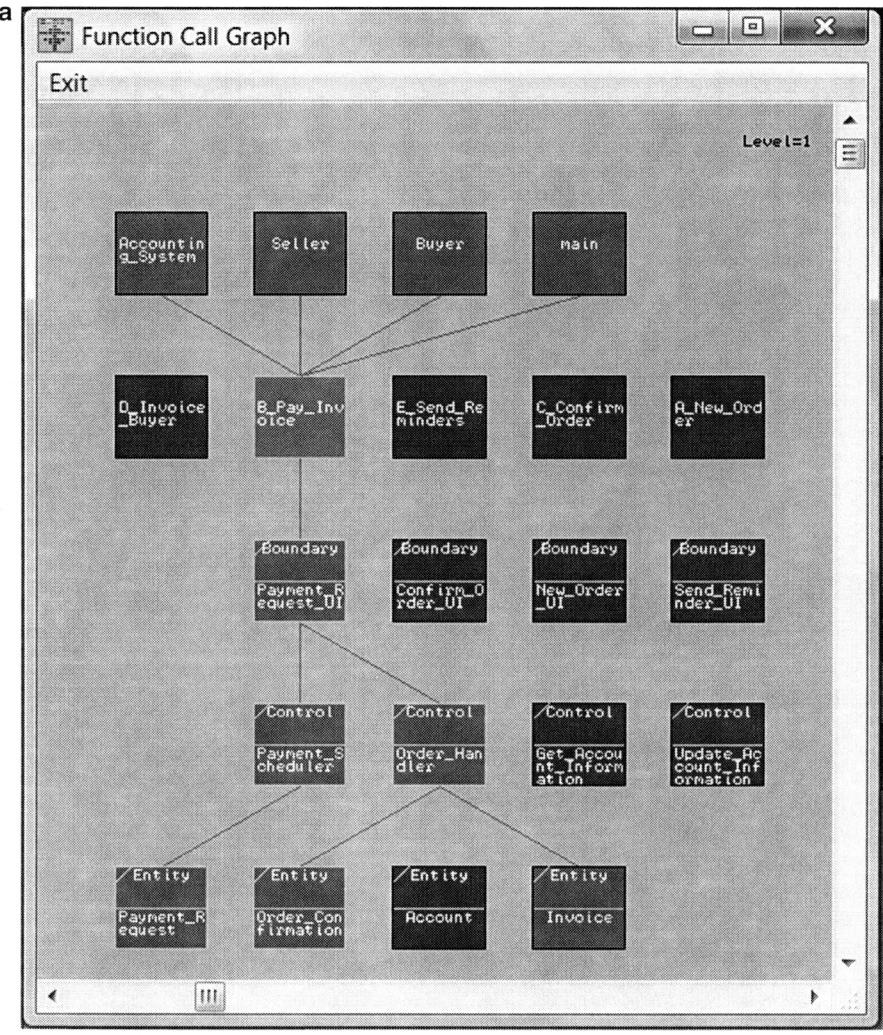

Fig. 11.12 The result mapping to the Use Case Analysis shown in Fig. 11.9: (**a**) using class notations and (**b**) using regular function notations

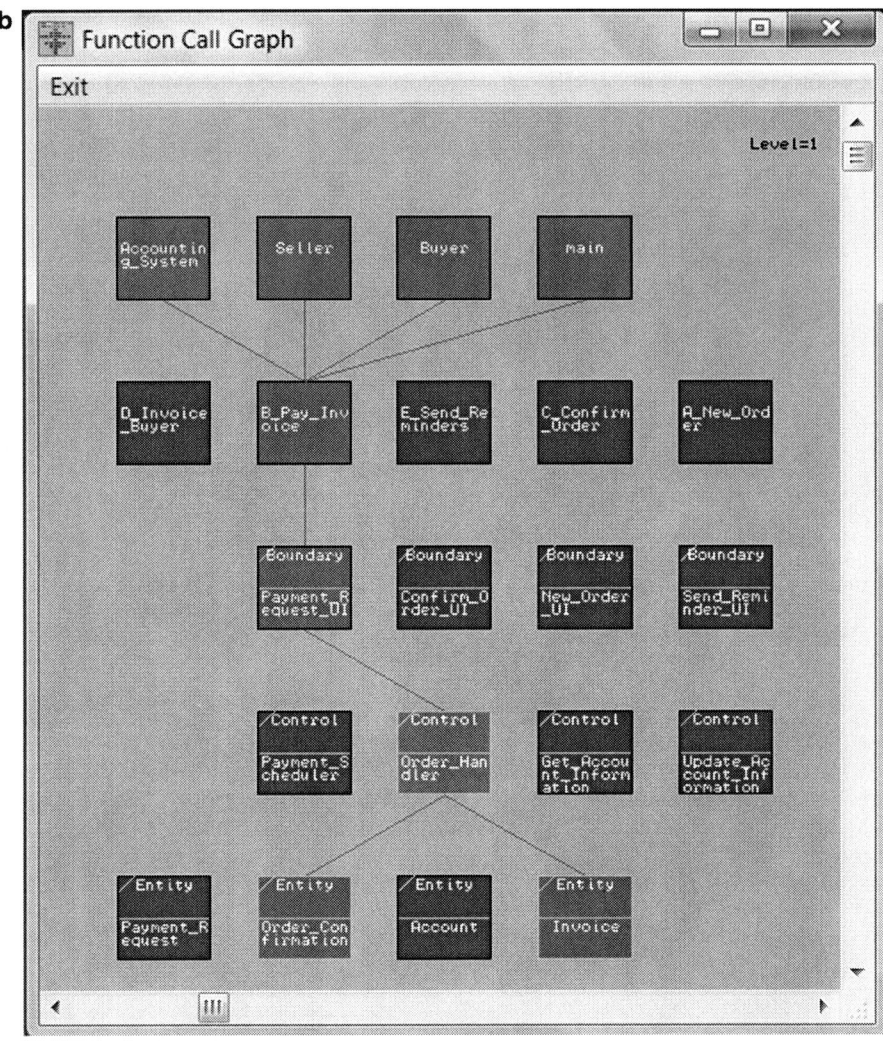

Fig. 11.12 (continued)

11.4 Applications of HAETVE

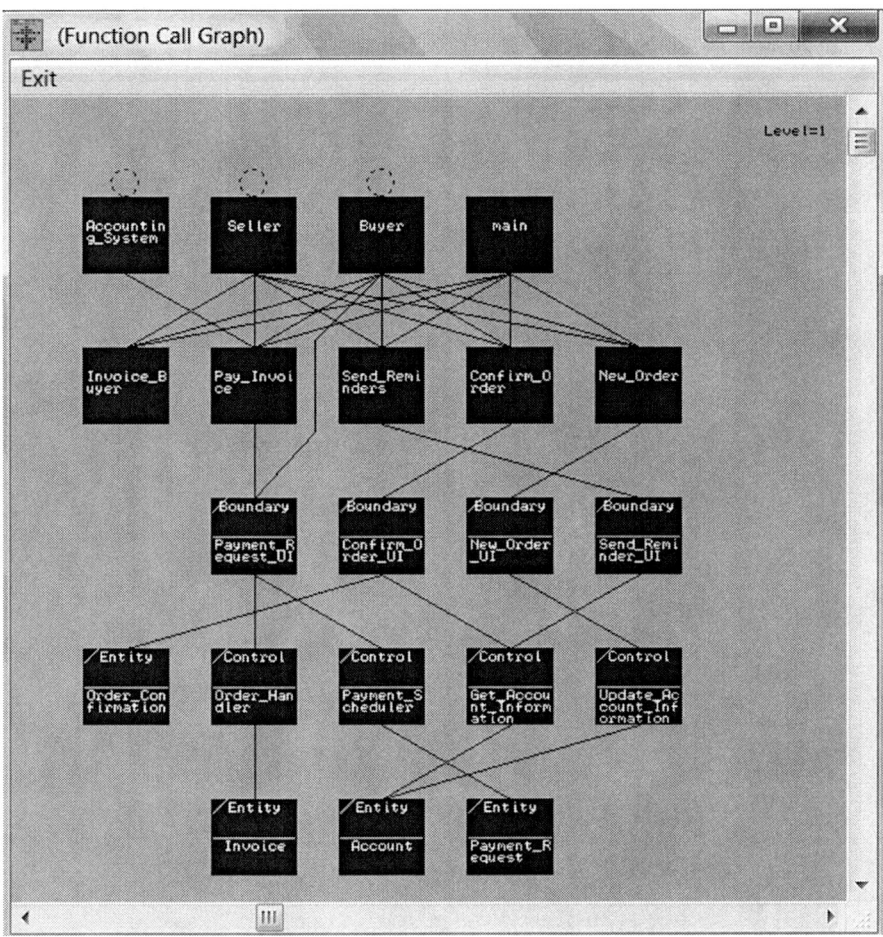

Fig. 11.13 An example of Holistic requirement functional decomposition

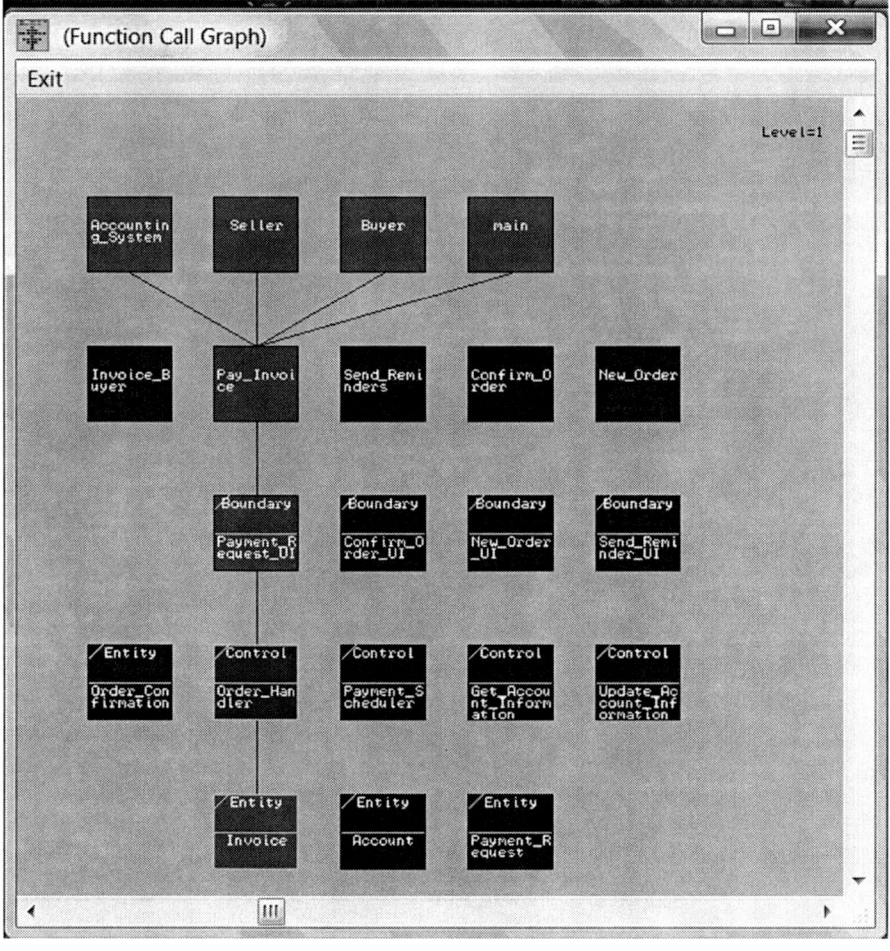

Fig. 11.14 Traceability used for static defect prevention and defect propagation prevention

11.4 Applications of HAETVE

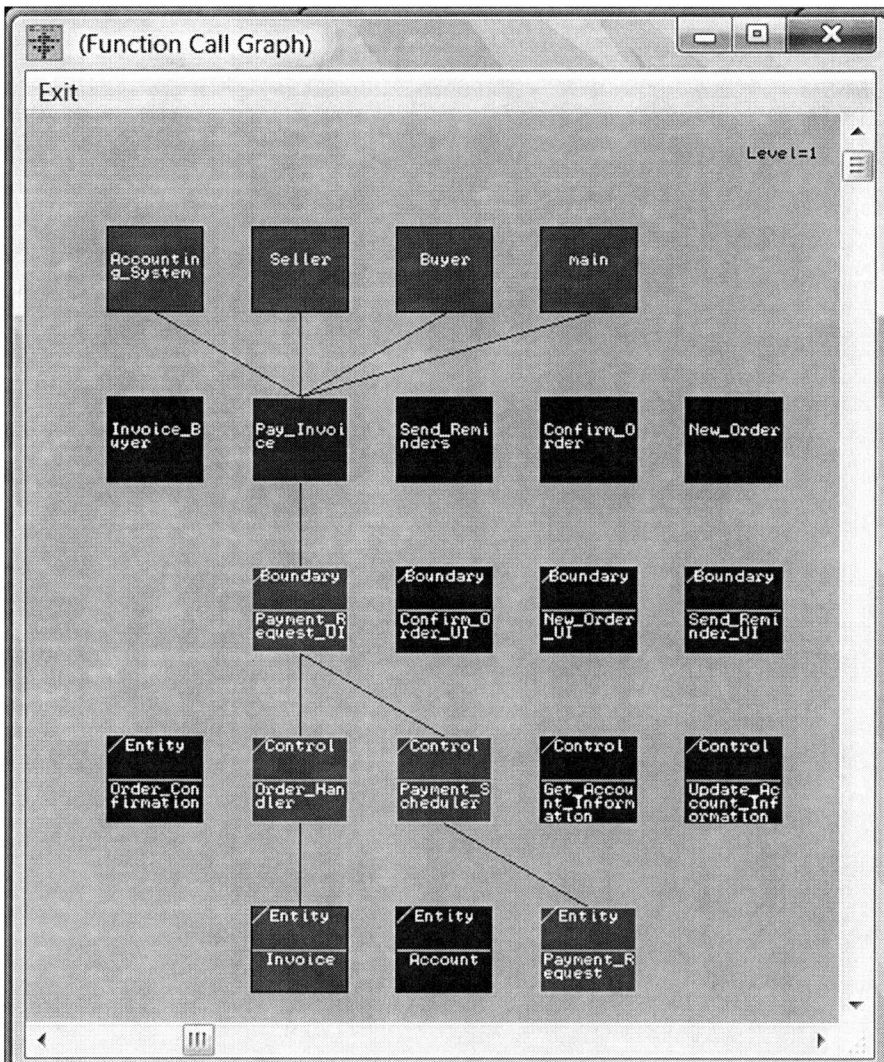

Fig. 11.14 (continued)

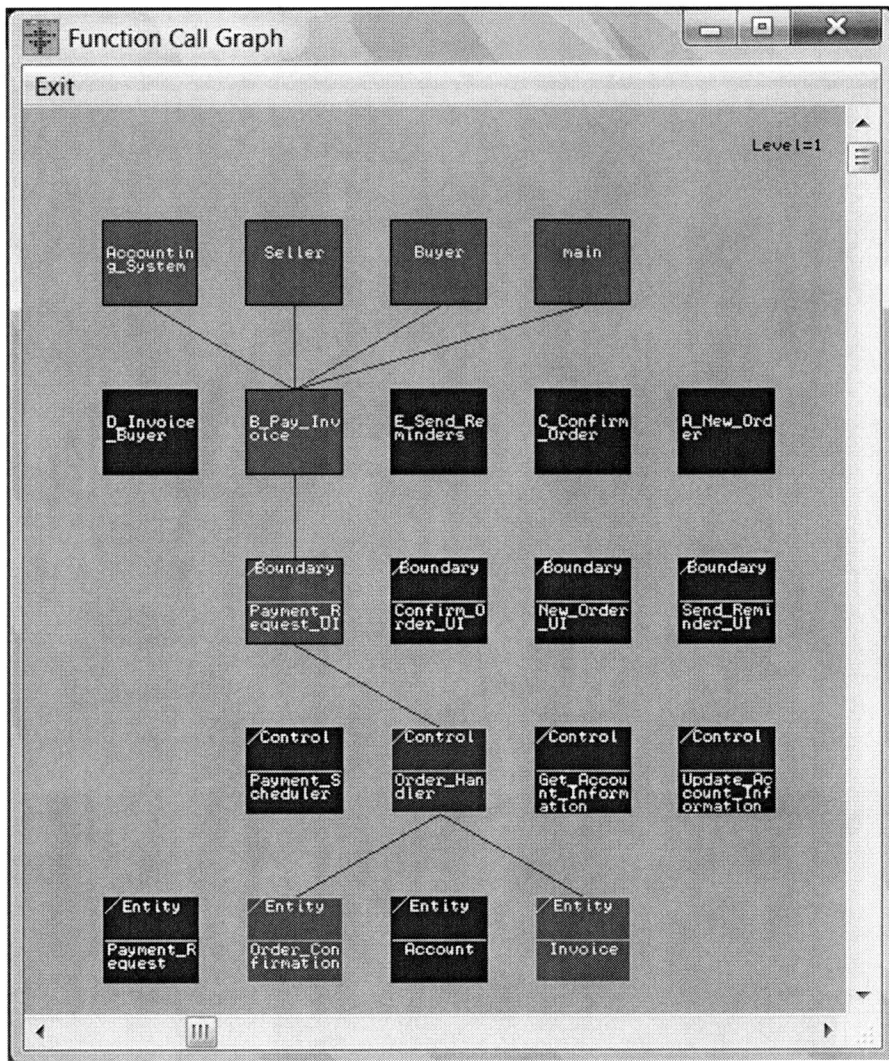

Fig. 11.15 The modified version of the Holistic requirement functional decomposition with the defect removed

11.4 Applications of HAETVE

The corresponding source code of dummy programming is listed as follows:

```
#include <stdio.h>
#include <string.h>
#include <stdlib.h>

class Entity
{
  friend class Control;
  friend class Boundary;

  protected :
           int operand;  // 1 if true, 0 if
  false

  public:
           void Payment_Request () { }
           void Invoice () { }
           void Order_Confirmation ()
  { }
           void Account () { }
};

Entity cp;

class Control
{
  friend class Entity;
  friend class Boundary;

  protected :
           int operand;    // 1 if true, 0
  if false

  public:
           void Order_Handler () {
              cp.Order_Confirmation();
              cp.Invoice();
           }
           void Payment_Scheduler () {
              cp.Payment_Request();
           }
           void
Get_Account_Information ()
           {
              cp.Account();
           }
           void
Update_Account_Information ()
           {
              cp.Account();
           }
};

Control mp;

class Boundary
{
  friend class Entity;
  friend class Control;

  protected :
           int operand;  // 1 if true, 0 if
  false

  public:
           void Payment_Request_UI ()
  {
  //        Control cp;
           mp.Payment_Scheduler();
           mp.Order_Handler();
           }
           void New_Order_UI () {
  //        Control np;
           mp.Update_Account_Informa
tion();
           }
           void Send_Reminder_UI(){
  //        Control sp;
           mp.Get_Account_Informatio
n();
           }
           void Confirm_Order_UI () {
  //        Control cp;
  //        Entity ep;
           mp.Get_Account_Informatio
n();
           cp.Order_Confirmation();
           }
};

static int buyer_n = 0;
static int seller_n = 0;
static int accounting_system_n = 0;

void New_Order(){
  Boundary op;
  op.New_Order_UI();
}
void Send_Reminders () {
  Boundary sp;
  sp.Send_Reminder_UI ();
}
void Pay_Invoice()
{
  Boundary up;
  up.Payment_Request_UI ();
}
void Confirm_Order() {
  Boundary bp;
  bp.Confirm_Order_UI ();
}

void Invoice_Buyer(){ }

void Buyer ()
{
  Boundary bp;
  New_Order();
  Confirm_Order();
  Invoice_Buyer();
  Pay_Invoice();
  Send_Reminders ();
  bp.Payment_Request_UI ();
  if(buyer_n == 0)
  {
     ++ buyer_n;
     Buyer ();
  }
}

void Seller ()
{
  New_Order();
  Confirm_Order();
  Invoice_Buyer();
  Pay_Invoice();
  Send_Reminders ();
  if (seller_n == 0) {++seller_n; Seller(); }
  return;
}

void Accounting_System ()
{
  Pay_Invoice() ;
  if(accounting_system_n==0)
           {++ accounting_system_n;
           Accounting_System ();
}
}

void main(int argc,char** argv)
{
  New_Order();
  Confirm_Order();
  Invoice_Buyer();
  Pay_Invoice();
  Send_Reminders ();
  printf("\n *** Executed. ***\n");
}
```

(c) The functional decomposition result is represented graphically in any level as shown in Fig. 11.16.

A functional decomposition result for functional requirements with actor–actions and event–response treatment is shown in Fig. 11.17.

Figure 11.18 shows an event and the response highlighted.
For comparing HAETVE with Use Case more easily, in the following examples, we only consider the actor–action driven applications (the main() function is also modified to accept different command line options – see Fig. 11.18) without considering the event–response applications.

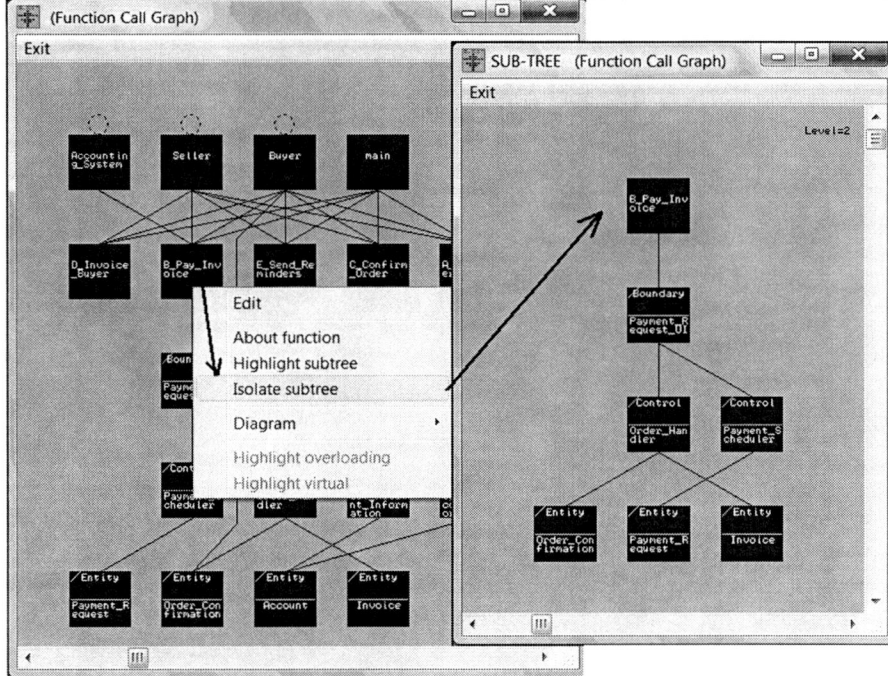

Fig. 11.16 An example of requirement functional decomposition in any level

(d) The program of functional requirement decomposition is executable dynamically for easily finding and removing defects as shown in Fig. 11.19: the module test coverage analysis process and result.

Using the Transparent-box software testing approach (see Chap. 16), we can further design many test cases to test the requirement functional decomposition result according to the different execution paths, and automatically establish the bidirectional traceability for removing inconsistency defects (see Chap. 9). An example of the corresponding test cases and the execution results are shown in Figs. 11.20–11.25.

When running the command, Billing_and_Payment.exe Invoice_Buyer, an error was found:

> C:\Billing_and_Payment9>Billing_and_Payment.exe Invoice_Buyer
> Invalid Commands:
> Billing_and_Payment.exe Invoice_Buyer
> *** Executed. ***

After checking the source code, it is clear that the problem comes from a typing error:

> …
> else if (strcmp(argv[1],"INvoice_Buyer")==0 ||

11.4 Applications of HAETVE

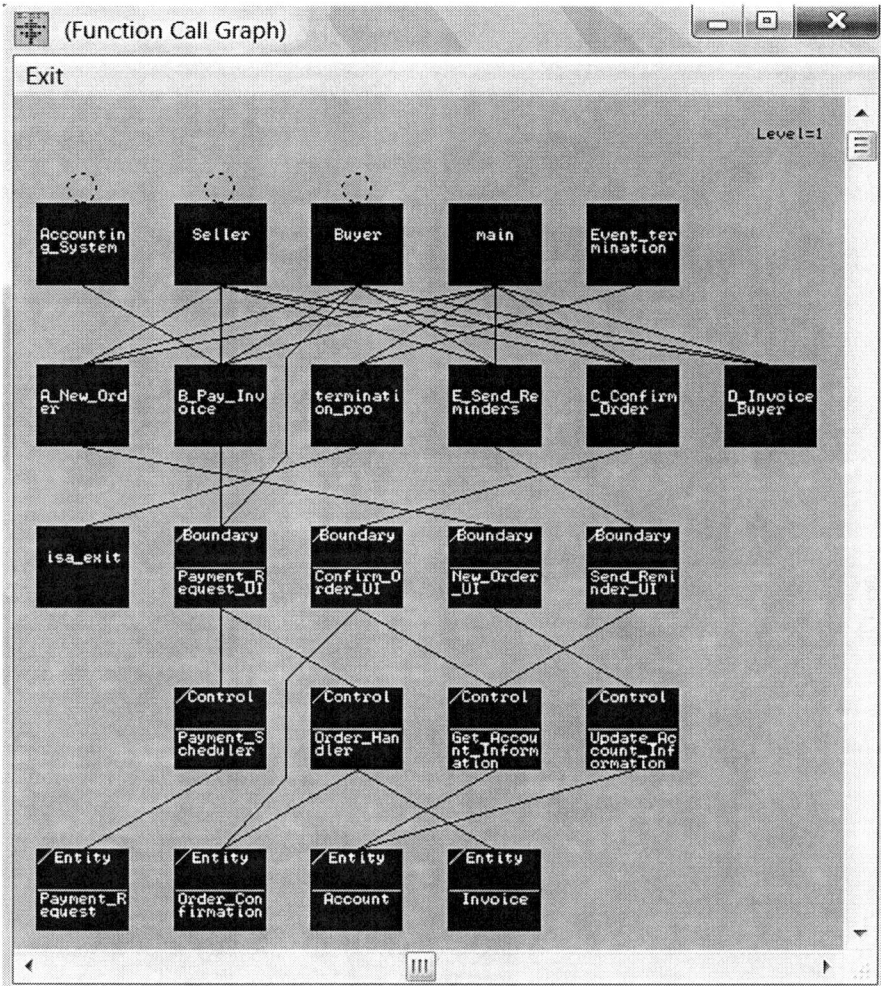

Fig. 11.17 An example of functional decomposition with actor–action and event–response treatments

strcmp(argv[1],"Invoice_buyer")==0
|| strcmp(argv[1],"invoice_buyer")==0)

After removing the error (changing "INvoice_Buyer" to "Invoice_Buyer") the program executed correctly:

    C:\Billing_and_Payment10>Billing_and_Payment.exe Invoice_Buyer
    *** D_Invoice_Buyer() called. ***
    *** Executed. ***

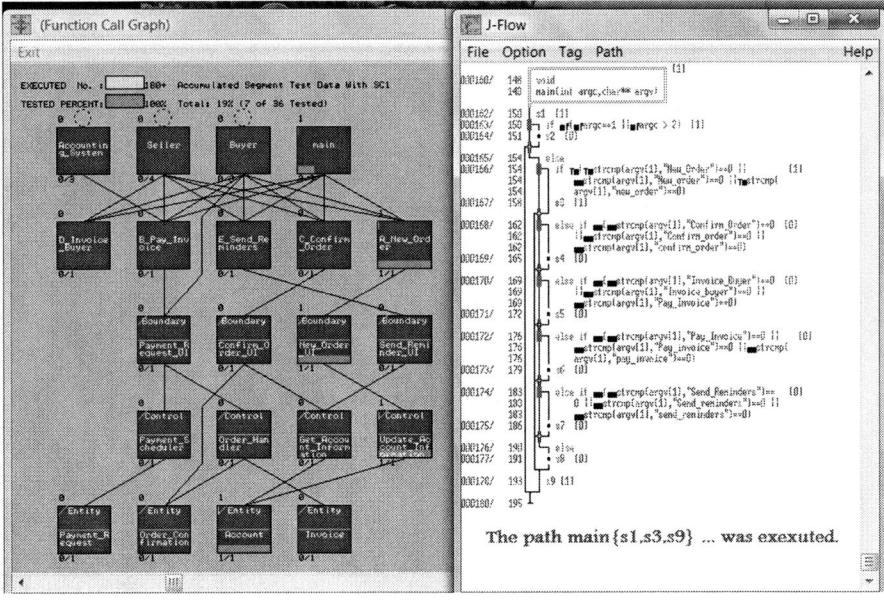
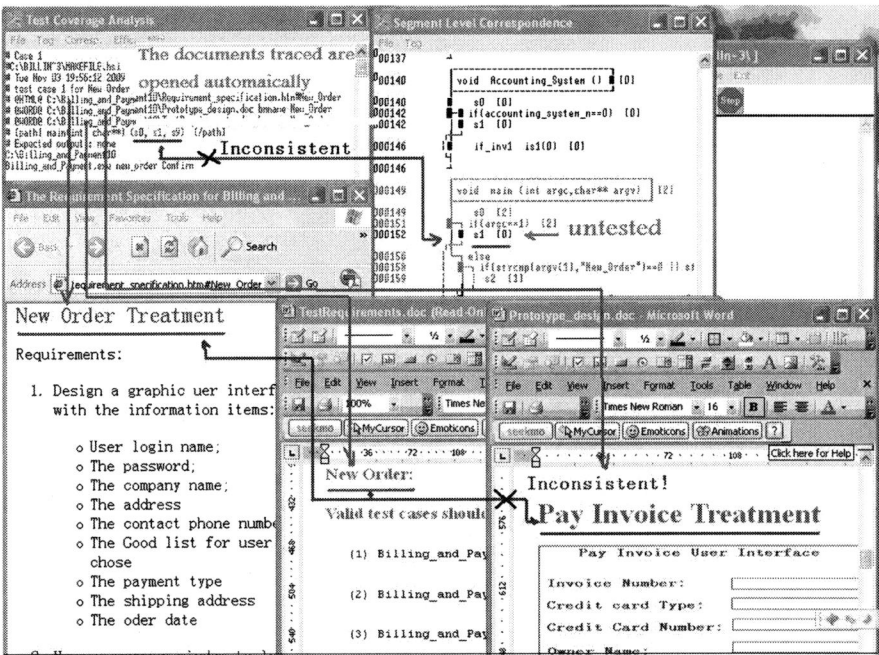

Fig. 11.18 An event and the response highlighted

11.4 Applications of HAETVE

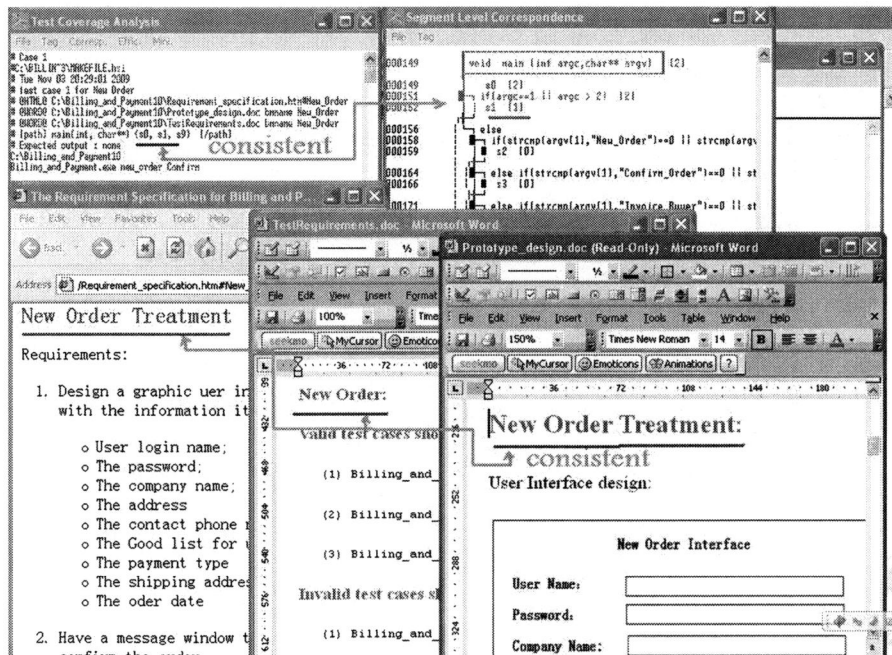

Fig. 11.18 (continued)

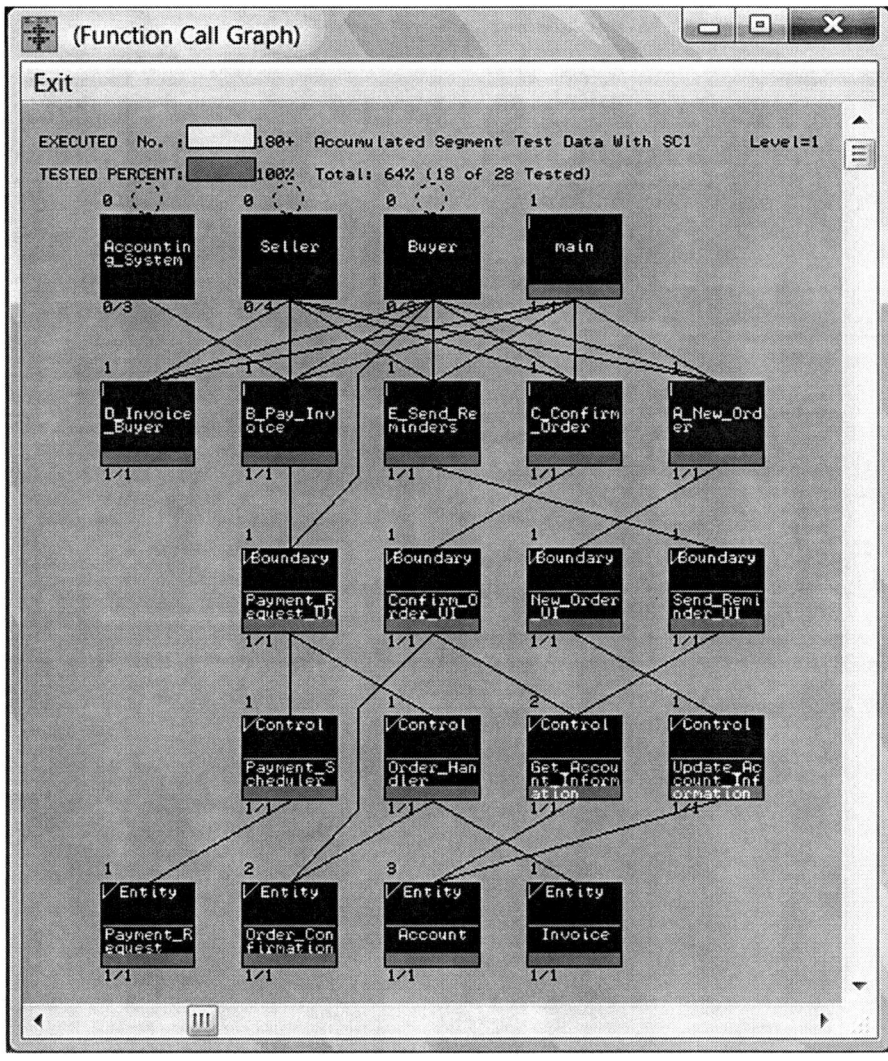

Fig. 11.19 The test coverage analysis result after dynamic execution

11.4 Applications of HAETVE

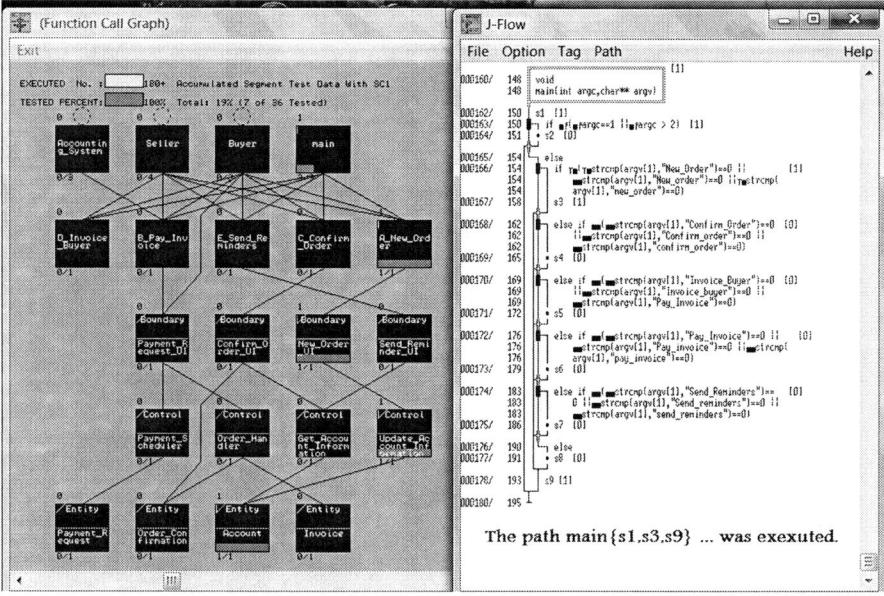

Fig. 11.20 The test coverage measurement result of the example after running the first test case (Billing_and_Payment.exe New_Order)

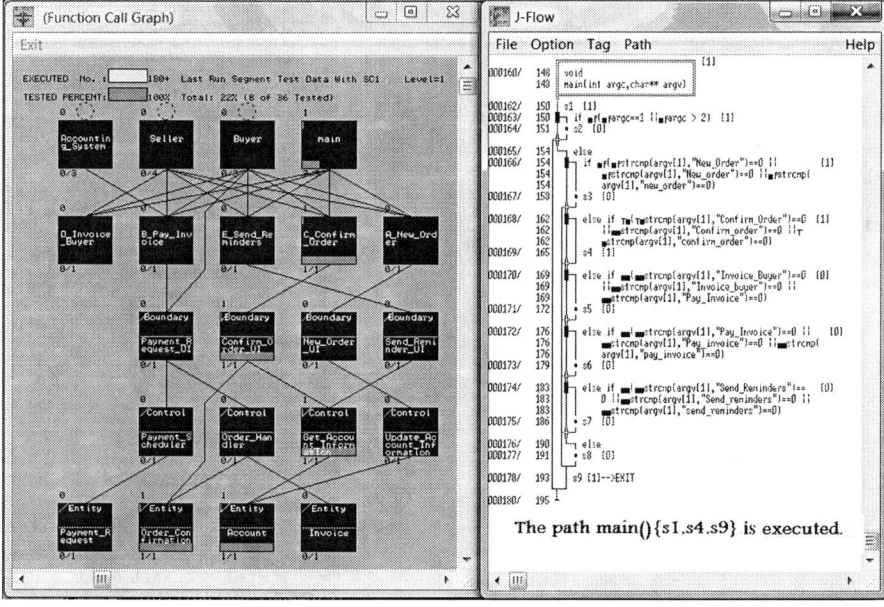

Fig. 11.21 The test coverage measurement result after running the second test case (Billing_and_Payment.exe Confirm_Order)

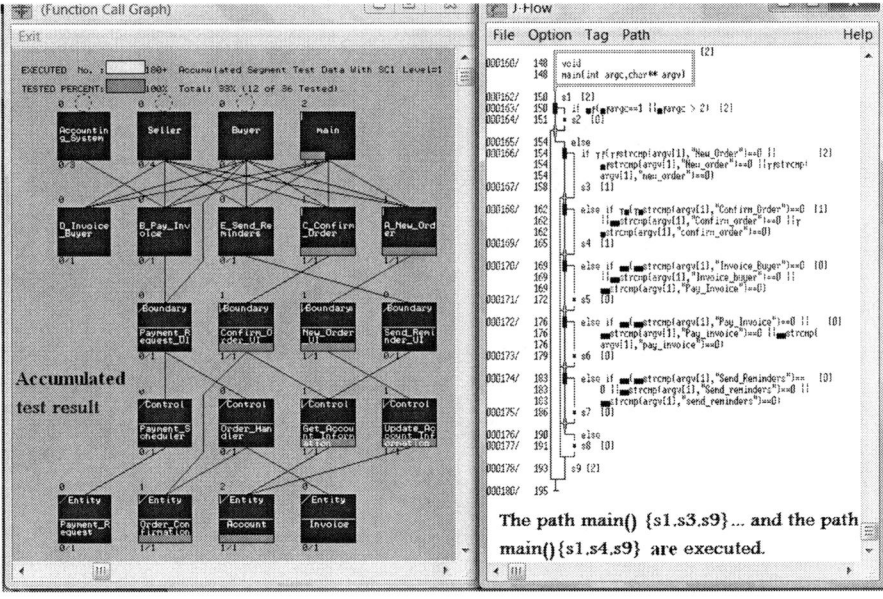

Fig. 11.22 The test coverage measurement result after running the first and second test cases

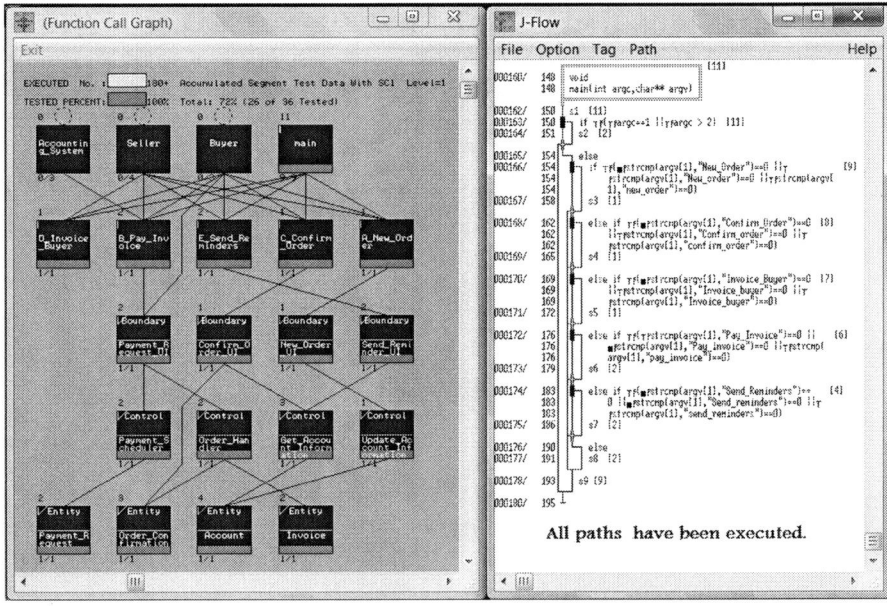

Fig. 11.23 The test coverage measurement after running four test cases

11.4 Applications of HAETVE

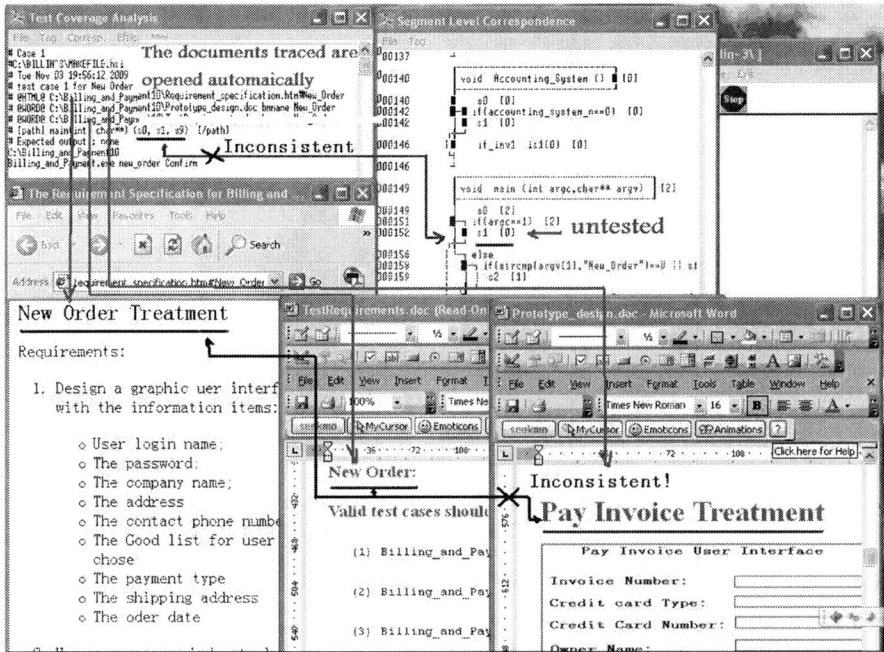

Fig. 11.24 Two defects are found through forward tracing from a test case to the source code

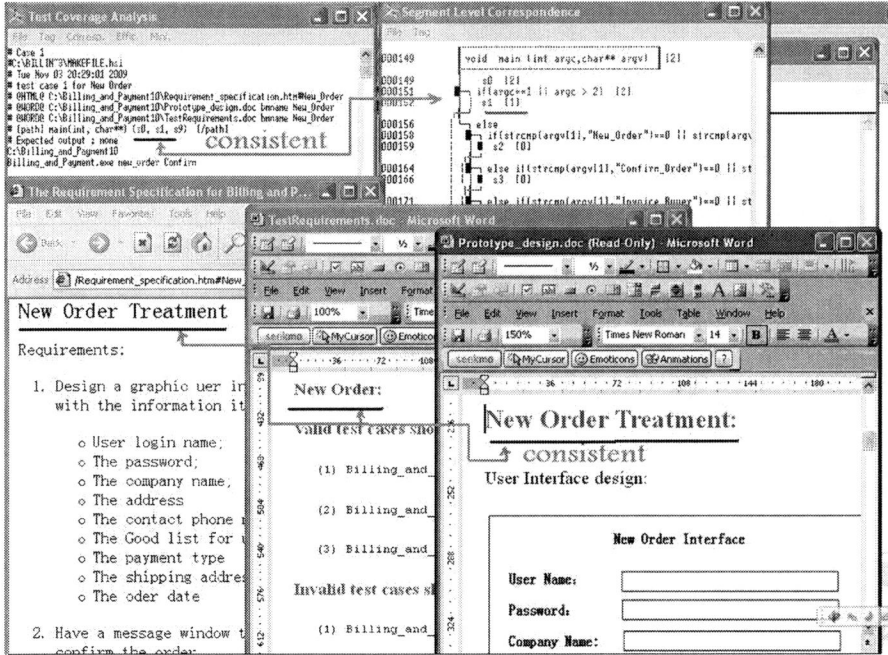

Fig. 11.25 The two defects have been removed

The following is a test case script written by complying with the very simple NSE test case design rules (see Chap. 9):

```
# test case 1 for New Order
# @HTML@ C:\Billing_and_Payment10\Requirement_specification.htm#New_Order
# @WORD@ C:\Billing_and_Payment10\Prototype_design.doc bmname New_Order
# @WORD@ C:\Billing_and_Payment10\TestRequirements.doc bmname New_Order
# [path] main(int, char**) {s0, s1, s9}   [/path]
# Expected output : none
C:\Billing_and_Payment10
Billing_and_Payment.exe new_order Confirm

# test case 2 for Pay Invoice
#@HTML@ C:\Billing_and_Payment10\Requirement_specification.htm#Pay_Invoice
#@WORD@ C:\Billing_and_Payment10\Protorype_design.doc Pay_Invoice
# @BAT@ C:\isa_examples\ganttpro\ganttpr9.bat
#[path] main(int, char**) {s1, s6, s9, }B-Pay_Invoice(void)  [/path]
# Expected output : none
C:\Billing_and_Payment10
Billing_and_Payment.exe Pay_Invoice
```

......

After running the test script, two defects are found as shown in Fig. 11.24

1. After checking the source code, we can easily find that there is a defect coming from an extra space character:

```
                                        |  an extra space character
                                        V
if(argc==1 /* Missing a parameter * /
     || argc > 2 /* Having an extra parameter */)
     {
     cout << "Invalid Commands: \n" << argv;
     }
else
{
if(strcmp(argv[1],"New_Order")==0  || strcmp(argv[1],"New_order")==0
     || strcmp(argv[1],"new_order")==0 )
     {
     A_New_Order();
     cout << "*** A_New_Order () called. ***\n";
     }
```

After code modification, the defect is removed:

```
...
if(argc==1 /* Missing a parameter */
     || argc > 2 /* Having an extra parameter */)
     {
     cout << "Invalid Commands: \n" << argv;
     }
else
{
if(strcmp(argv[1],"New_Order")==0 || strcmp(argv[1],"New_order")==0
|| strcmp(argv[1],"new_order")==0 )
{
A_New_Order();
cout << "*** A_New_Order () called. ***\n";
}
```

2. Another defect is found where two bookmarks (New_Order and Pay_Invoice) are pointing to the same location that is used for Pay Invoice Treatment part. This defect is corrected by changing the New_Order bookmark to point to the New_Order Treatment section in the prototyping document.

After fixing the problems, we can get the correct result shown in Fig. 11.25.

Of course, the functional requirement decomposition result is not the requirement implementation result, but it will become a basis for the requirement implementation.

Besides the functional requirements, there are some other requirements to be specified, such as the performance requirement and the UI (user interface) requirement which can be specified by a SuperActor.

11.5 How to Make a Hard Copy of a Graphical Requirement Document

Usually, there is no need to print out a graphical document because with NSE the graphical documents are all generated dynamically from several hash tables and exist virtually to greatly save space, unless users want to make them for documentation, Web page design, or project presentations without using the NSE support platforms. All graphical documents can be printed out on paper or to files (it is recommended to print out to a file in Postscript format, then use Adobe tools to transfer it to PDF format for easy viewing and saving the required space). Figure 11.26 shows an application example.

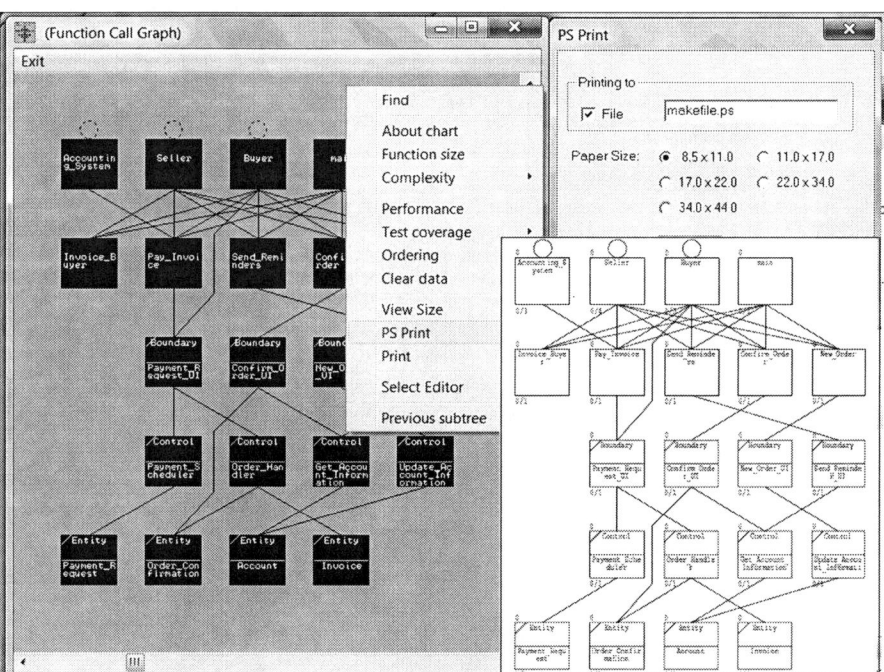

Fig. 11.26 An example about how to make a hard copy of a graphical document

11.6 Suggestions for the Requirement Documentation Design

How to design the requirement documents with NSE? There are some suggestions:

(a) Complete the requirement specification using the NSE requirement specification template (see Appendix A) to avoid missing something.
(b) Complete other related documents such as the Test Requirement Specification and the Test Script files according to the requirement specification to avoid something untested.
(c) Set the bookmarks to all the related requirement documents using inherited bookmarks or meaningful bookmarks (even if some related documents such as the Project Design Document have not been designed in detail yet – just a Table of Contents and the Section Headers). An application example for setting bookmarks in a word file is shown in Fig. 11.27.
(d) Complete an initial design of the Document Hierarchy Description table used for establishing automated and self-maintainable traceability among the documents and the test cases and the source code (see Chap. 9) as shown in the following template (see Table 11.3).

Also do not forget to list the corresponding test case script files and the corresponding test case numbers to be used to perform requirement validation and verification later through forward traceability – see Chap. 18 about the NSE software maintenance paradigm.

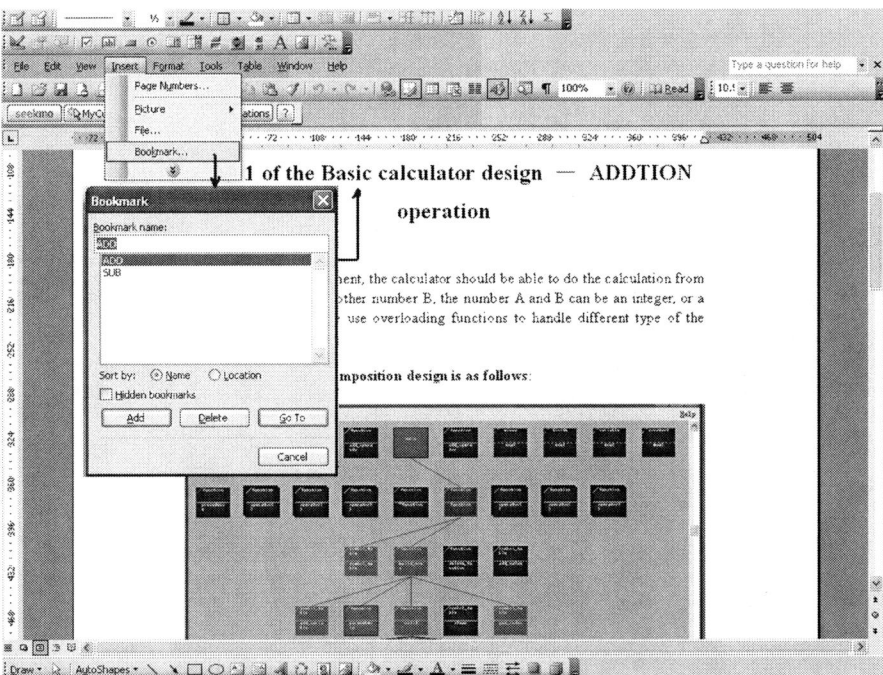

Fig. 11.27 Bookmark setting example

11.6 Suggestions for the Requirement Documentation Design

Table 11.3 Document Hierarchy Description template

<table>
<tr><td colspan="6" align="center">**Document Hierarchy**</td></tr>
<tr><td colspan="3">Project Name</td><td colspan="2">Project Code</td><td></td></tr>
<tr><td colspan="6">Project Description</td></tr>
<tr><td colspan="3">The full path name of the Project feasibility report</td><td colspan="2"></td><td>Version number</td></tr>
<tr><td colspan="3">The full path name of the requirement specification</td><td colspan="2"></td><td>Version number</td></tr>
<tr><td colspan="2">Requirement 1</td><td>Bookmark</td><td colspan="3">r1</td></tr>
<tr><td colspan="2">Description</td><td colspan="4"></td></tr>
<tr><td></td><td colspan="3">The full path name of the related design document</td><td></td><td></td></tr>
<tr><td></td><td colspan="3">Description</td><td>Bookmark</td><td>r1d1</td></tr>
<tr><td></td><td colspan="3">The full path name of the related test specification</td><td></td><td></td></tr>
<tr><td></td><td colspan="3">Description</td><td>Bookmark</td><td>r1d1t1</td></tr>
<tr><td></td><td colspan="3">...</td><td></td><td></td></tr>
<tr><td colspan="2">Requirement 2</td><td>Bookmark</td><td colspan="3">r2</td></tr>
<tr><td colspan="6">...</td></tr>
</table>

(e) Ignore the static "Requirement Traceability Matrix" which is time consuming to make, incomplete, not accurate, not precise, not bidirectional, hard to use, and hard to maintain.

(f) If you have some related documents made by third party tools, you can make them traceable too – design a batch file, then use the @BAT@ keyword in the corresponding test cases description, so that when the test case is selected for forward tracing or traced from the corresponding source code, the batch file will be automatically executed to use the third-party tools to open the related document(s). A sample batch file is as follows:

%PANORAMAHOME%\tool_j\bin\java -jar %PANORAMAHOME%\ganttpro\build\ganttproject-1.9.11.jar %PANORAMAHOME%\ganttpro\ganttproject-example3.xml.

The application example is shown in Fig. 11.28.

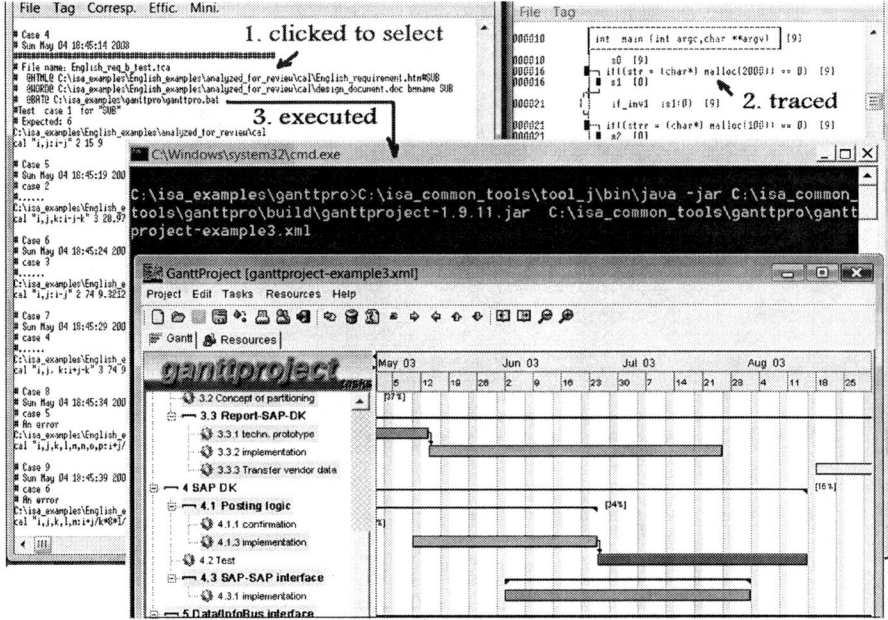

Fig. 11.28 A dynamic traceability example

11.7 The Major Features of HAETVE

The major features of the HAETVE technique include:

(a) It is an engineering approach for software requirement development.
(b) It is holistic – with HAETVE almost all tasks/activities of requirement development are performed holistically and globally.
(c) It supports both the actor–action driven software development and the event–response driven software development.
(d) It is visual – the application process and the obtained results are visible.
(e) The obtained results are traceable for review and static defect prevention and defect propagation prevention – traceability is particularly useful for a complex software product as shown in Figs. 11.29 and 11.30.
(f) The obtained results are executable for dynamic defect prevention and defect propagation prevention through the NSE software testing paradigm based on the Transparent-box method (see Chap. 16).
(g) It mainly works through dummy programming using dummy modules having an empty body or only some function call statements without detailed program logic.
(h) It uses the notations of 3J graphics (see Chap. 7) to show the obtained results graphically.

11.7 The Major Features of HAETVE 307

Fig. 11.29 The functional decomposition result of a complex software

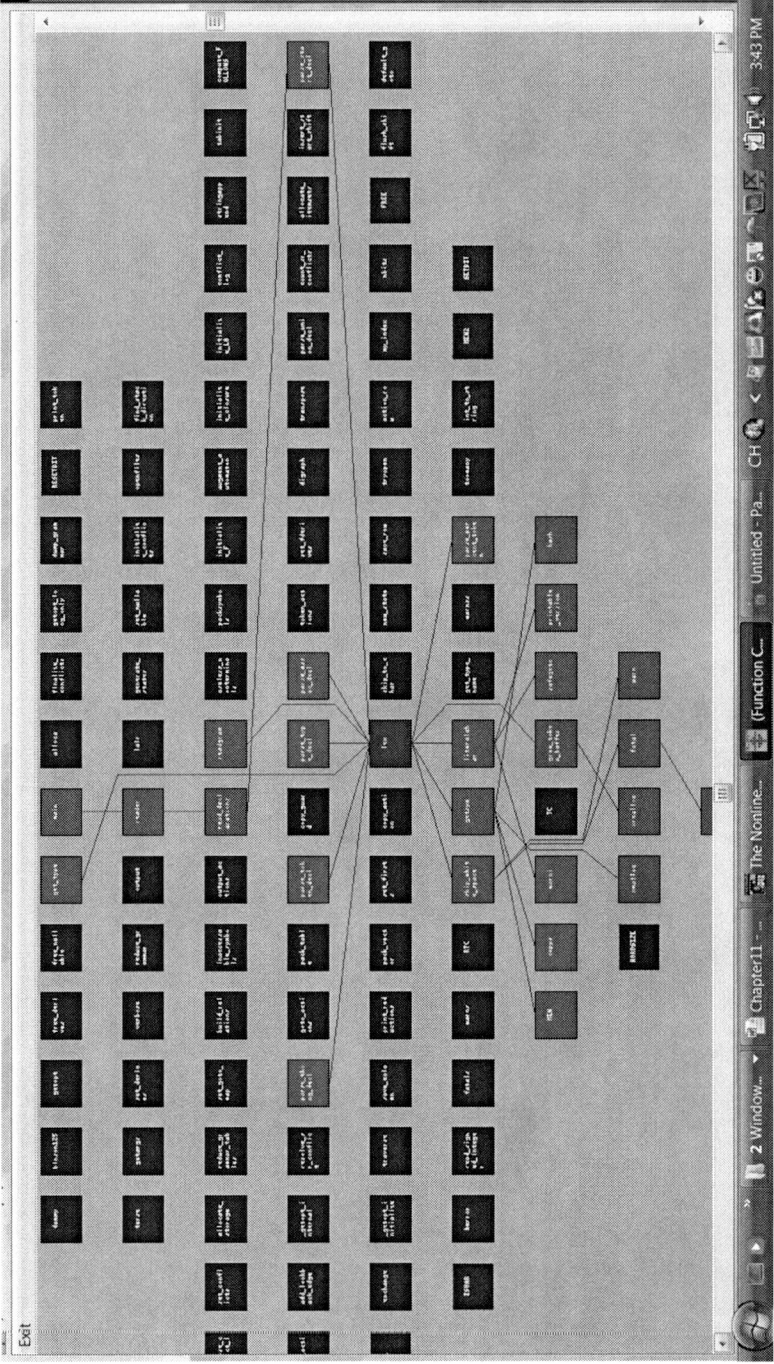

Fig 11.30 Tracing an element (lex()) from the chart shown in Fig. 11.29

(i) The graphics showing the obtained holistic results are generated automatically, dynamically, and virtually from the dummy source code without storing any hard copy in the memory and the hard disk to greatly save the needed space.
(j) UML is supported indirectly through the use of an open source product, Fujaba.
(k) It not only works for C, C++, and Visual Basic, but also works for Java to make the modeling results independent of the target programming languages and platforms.

11.8 More About Dynamic Modeling

Dynamic modeling means:

1. The generated diagrams/models are executable directly or indirectly through the corresponding code.
2. The generated diagrams/models not only can represent the static properties of a software product, but can also represent the dynamic properties of a software product, such as the code test coverage and the percentage of the execution time spent in each module.
3. The generated diagrams/models are interactive and traceable.
4. **The most important feature of Dynamic Modeling is that the generated diagrams/models no longer statically exist – they dynamically exist ("alive") – the generated diagrams/models, the generators of the diagrams/models, and the interfaces to accept users' commands (using the diagrams/models themselves) are three in one: when a diagram/model is shown in a Window, its generator is always working and waiting for a user's command through the diagram/model (acting as the interface) – after receiving a user's command, the generator will dynamically respond,** such as generating a subtree (see Fig. 7.11), printing out a chart (see Fig. 7.23), or performing untested path analysis and automatically highlighting a "best" one with most untested elements and automatically extracting the execution conditions to help users design the most efficient test case.
5. The generated diagrams/models and the corresponding source code are no longer separated, instead, they are combined together to form a powerful union to help users develop a software product better, understand a software product better, test a software product better, and maintain a software product better. For instance, clicking on a module box from the generated call graph to directly edit the source code of that module as shown in Fig. 11.31, or clicking on a module from the generated control flow diagram to trace the corresponding test cases and directly play the captured GUI test operations back dynamically as shown in Fig. 11.32 (*note*: during the playback process, we cannot directly make a screenshot).

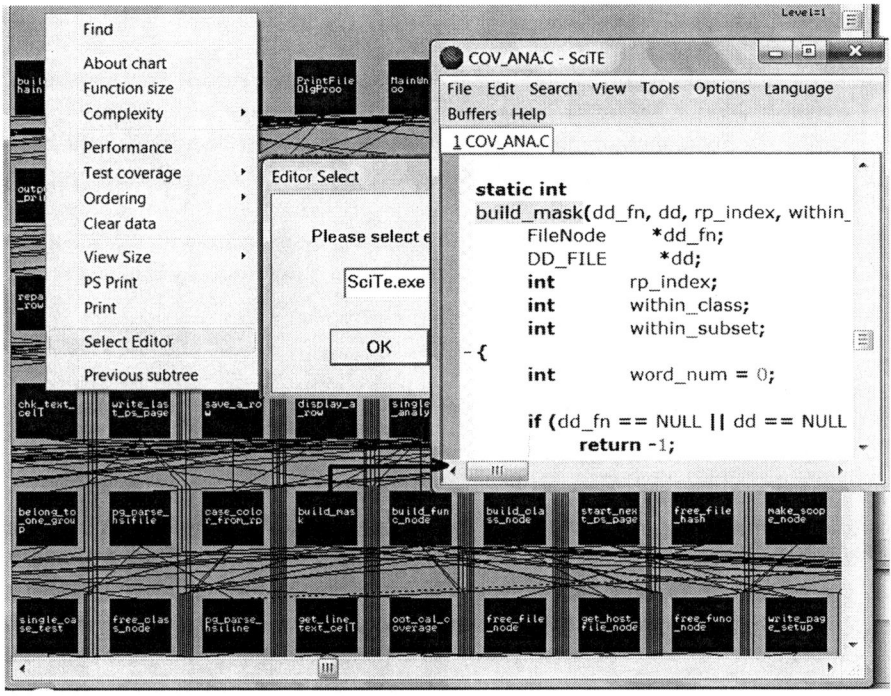

Fig. 11.31 Directly select an editor and then edit a module from a generated call graph

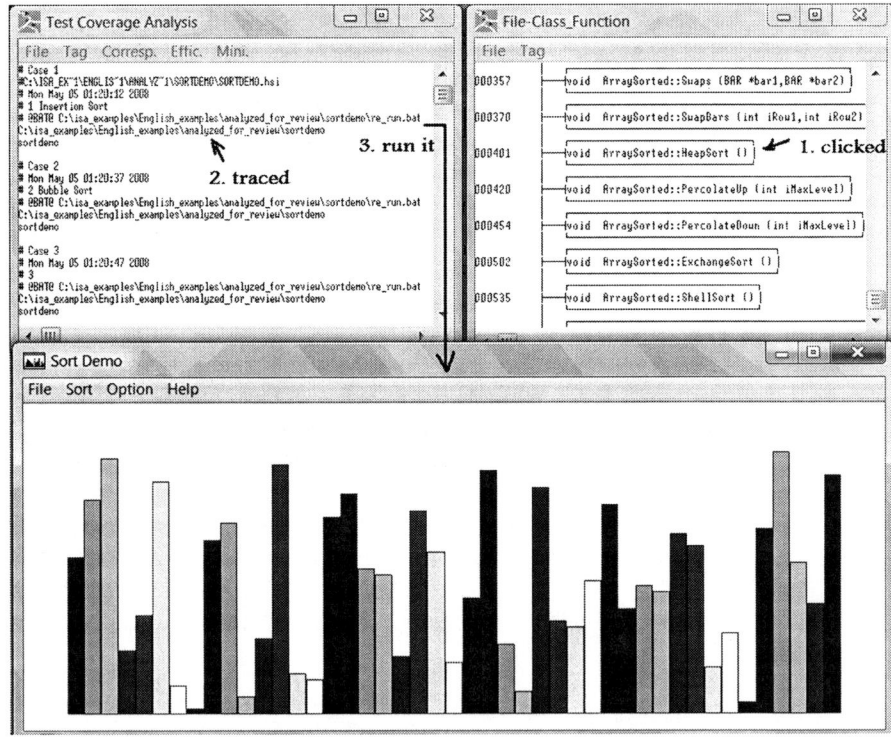

Fig. 11.32 Click on a module from a control flow to trace the test cases run on a program(s)

11.9 Summary

The existing software modeling approaches are outdated because they are outcomes of reductionism and the superposition principle, and use different sources for human understanding and computer understanding of a software system separately with a big gap between them. The obtained models are not traceable for static defect removal, not executable for debugging, not testable for dynamic defect removal, not consistent with the source code after code modification, and not qualified as the road map for software development.

This chapter introduced how to perform software requirement development with NSE through Source Code Driven Dynamic Software Modeling and Engineering using the innovated Holistic, Actor–Action and Event–Response driven, Traceable, Visual, and Executable (HAETVE) technique. HAETVE is one of the most important means to the implementation of the NSE's **Upstream Quality Assurance** strategy and **Total Quality Management** strategy to ensure the quality of a software product through defect prevention and defect propagation prevention from the first step to the final step in the entire software development and maintenance process, following Deming's Product Quality Assurance Principles that *"Cease dependence on inspection to achieve quality. Eliminate the need for inspection on a mass basis by building quality into the product in the first place."*

11.10 Points and Questions to Ponder

(a) What are the major differences between the Use Case approach and the HAETVE technique?
(b) Why should the graphical result of the function decomposition of the functional requirements of a product be made traceable?
(c) Why do we need not only static review, but also dynamic testing in the software requirement development phase?

11.11 Further Reading and Information Source

Resource of requirement specification templates:
http://www.systemsguild.com/pdfs/SpecTemplate6.1.pdf
http://www.docin.com/p-49779695.html
http://www.klariti.com/Software-Requirements-Specification-Template/
http://www.cs.iit.edu/~oaldawud/CS487/project/requirement_specification_document_template.htm
http://www.lcwu.edu.pk/etm/cs_projdoctemp/SRS.pdf
https://svn.origo.ethz.ch/jid08-team17/trunk/srs_team_view/Eloha%20view%20SRS.pdf

References

[Arl06] Arlow J, Neustadt I (2006) UML 2 and the unified process: practical object-oriented analysis and design, 2nd edn. Person Education, Inc., Boston

[Bro95-p200] Brooks FP Jr (1995) The mythical man-month. Addison-Wesley, Reading, p 200

[CMMI1.1] Phillips M (2002) CMMI Program Manager, CMMI V1.1 and Appraisal Tutorial, http://www.sei.cmu.edu/cmmi/

[Dem82] Deming WE (1982) Out of the crisis. MIT Press, Cambridge

[For08] Forward A (2008) Problems and opportunities for model-centric versus code-centric software development: a survey of software professionals. Proceedings of the 2008 international workshop on models in software engineering, Leipzig, Germany, pp 27–32

[Jac92] Jacobson J (1992) Object oriented software engineering: a use case driven approach. Addison-Wesley, Reading

[Sne07] Sneed H (2007) The drawbacks of model driven software evolution. IEEE CSMR 07 – workshop on model-driven software evolution, Amsterdam, 20 March 2007

Chapter 12
Design Engineering Under NSE

The whole is more than the sum of its parts.

Aristotle

In the software design phase, the major tasks include the planning of the solution according to the requirement specification, design of the software architecture, design of the data structure, design of the interfaces, design of the algorithms, design of the modules, and design of the documents.

This chapter introduces how software design engineering can be performed holistically, globally, virtually, visually, and efficiently using the innovated **Synthesis Design and Incremental growing up (Implementation and Integration)** technique in the software design phase.

12.1 The Major Problem Addressed

Although in software engineering many methods, techniques, and tools have been developed for software design and applied in practices, there are still many critical problems existing with the old-established software design paradigm because

(a) The old-established software design paradigm (including the methods, techniques, and tools) is based on linear thinking, reductionism, and the superposition principle that the whole of a complex system is the sum of its components, so that almost all software design tasks and activities are performed linearly, partially, and locally – through "**Analysis**."
(b) It follows the **Constructive Holism** principle that software components are developed first, then "*Assemble the product from the product components, ensure the product, as integrated, functions properly and deliver the product.*" [CMMI1.1]. It handles a software product as a machine which can be *assembled*, rather than a logic product created by people.
(c) Often the designed results consist of many small pieces without a holistic whole for a complex software system.

(d) The designed results are not traceable – hard to review and hard to understand.
(e) Even if a holistic result can be designed and shown graphically, without traceability it is still useless because there are too many connection lines, making the designed result hard to view and hard to understand.
(f) The designed results are not directly executable – an **Upstream Quality Assurance** strategy cannot be implemented dynamically through testing to prevent defects and prevent defect propagation early in the product design phase.
(g) The designed results are hard to update and maintain, no matter if they are represented in text or graphics.
(h) The designed results and the related documents are not traceable backwards to the requirements or forwards to the source code.
(i) The designed graphical documents are stored in hard copies or XML or PostScript format, requiring a huge amount of space.
(j) About the design documents, there are two extremes: one is requiring a huge amount of documents but most of which are useless because they are inconsistent with the source code after code modification performed again and again; another one is based on the concept that "Only the source code is the best document" so that only a few design documents will be provided – making the software product more difficult to maintain.
(k) Working with the linear process models, the defects introduced in the design phase easily propagate to the lower phases to make the defect removal cost increase tenfold several times.
(l) The application results of the old-established software design paradigm show that today the software project success rate is still very low – only about 30%.

12.2 Outline of the Solution for Software Design with NSE

The solution offered by NSE for software design using the innovated **Synthesis Design and Incremental growing up** technique is described in detail in this chapter later. Here is the outline of the solution:

(a) It is based on nonlinear thinking and complexity science by complying with the essential principles of complexity science, particularly the Nonlinearity principle and the Holism principle that the whole is greater than the sum of its parts, and the characteristics and behaviors of the whole emerge from the interaction of its parts and the interaction of it and the environment, so that with NSE almost all the software design tasks and activities are performed holistically and globally.
(b) It complies with the **Generative Holism** principle that the whole of a complex system may exist (as an embryo) earlier than its components, then grows up with its components incrementally.
(c) The designed results are holistic for the entire product.

(d) The designed results are traceable – easy to review and understand.
(e) With traceability, no matter how complex a software product is, we can easily highlight a module with all the related modules calling and called by it to make the designed results much easier to review and understand.
(f) The designed results are directly executable – an **Upstream Quality Assurance** strategy can be implemented dynamically through testing to prevent defects and prevent defect propagation early in the product design phase.
(g) The designed results are easy to update and maintain – after modifying the source code or the dummy programming source code, the design documents can be automatically updated.
(h) The designed results and the related documents are traceable backwards to the requirements or forwards to the source code (see Chap. 9).
(i) The designed graphical documents virtually exist without storing any hard copies in disk or the computer memory – they are automatically generated though several hash tables virtually.
(j) About the design documents, a huge amount of documents will be automatically generated which are always consistent with the source code– making the software product much easier to maintain.
(k) Working with the NSE nonlinear process models through defect prevention and defect propagation prevention using traceable documents and dynamic testing using the Transparent-box method (see Chap. 16), the defects introduced into a software product and the defects propagated to the maintenance phase will be greatly reduced.
(l) The application results show that working with the NSE process model and the NSE software development methodology, it is possible for the NSE software design paradigm to help software organizations double their product success rate.

12.3 Description of the Innovated "Synthesis Design and Incremental Growing Up" Technique

12.3.1 Basic Ideas

(a) As pointed out by Aristotle, "The whole is more than the sum of its parts."
(b) Software is people oriented – people are the first-order components in software development [Coc99].
(c) People are nonlinear.
(d) People make mistakes and wrong decisions often.
(e) So, design and coding should be a two-way process by combining design and coding together closely (top-down + bottom-up).
(f) With NSE **design is precoding, coding is further design.**

12.3.2 What is Synthesis? What is Analysis?

Synthesis means "to put together" and analysis means "to loosen up," respectively. Analysis is defined as the procedure by which we break down an intellectual or substantial whole into parts or components. Synthesis is defined as the opposite procedure: to combine separate elements or components together to form a coherent whole – "1 + 1 > 2."

According to the **Generative Holism** principle of complexity science, the whole of a complex system exists earlier than its components – as an embryo, then grows up with its components.

Here, "**Synthesis Design and Incremental growing up**" means:

(a) Combining all NSE components together and make them work together closely (such as sharing the unique database, using a common interface, etc.) to form the whole of NSE for the requirement implementation including software design and coding, including the NSE process model, NSE software development methodology, and particularly the NSE software visualization paradigm (see Chap. 7) and the NSE software testing paradigm (see Chap. 16).
(b) Combining software design and coding together, supported by the entire NSE paradigm as shown in Fig. 12.1.
(c) Combining the "top-down" design approach and "bottom-up" design approach together through two-way iteration.
(d) Combining human-intelligence and computer-computing power together to solve issues such as error simulation used for realizing 100% MC/DC test coverage, and getting the class test coverage results from their instances (a class cannot be directly executed).

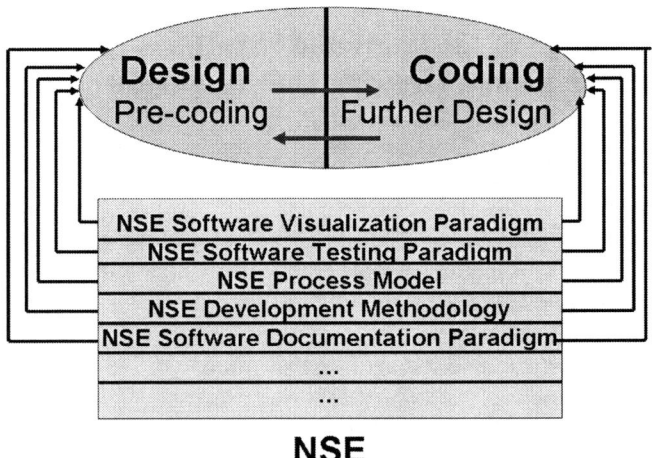

Fig. 12.1 Design and coding with NSE

12.3 Description of the Innovated "Synthesis Design and Incremental Growing Up" Technique

(e) Combining qualitative research and quantitative implementation together such as the test planning through cyclomatic complexity (the number of the decision statements) for the entire product and each individual module.
(f) Combining textual description and graphical representation together – generating the graphics directly from the dummy source code or the regular source code.
(g) Combining complexity science and reductionism together – "complexity is by levels" [Bro95-P211]. Sometimes we need to compare their application results as well.
(h) Collecting the information, documents, and data related to the requirements, including the solution method comparison reports, prototype design and risk estimation reports, test cases and the test results, customer evaluation results, the documents of the algorithms used, etc.
(i) According to the functional requirement decomposition results plus nonfunctional requirements, updating the executable dummy system (the preliminary architectures were designed in the preprocess) through dummy programming.
(j) Testing the designed results dynamically using the Transparent-box approach, and reviewing the result statically using traceable documents and test cases and the source code – even if there is only one top-level dummy module (main()) available and executable with different command-line options (see Section 12.4).
(k) Removing the defects introduced into the designed dummy system through software visualization and inspection, particularly dynamic testing using the Transparent-box approach.
(l) Performing optimization of the designed dummy system to reduce the coupling degree.
(m) Designing the preliminary data structures (class structures) according to the collected information and data.
(n) Compiling and executing the designed dummy system that maps to the functional requirement decomposition plus the nonfunctional requirements.
(o) Performing detailed design of the modules.
(p) Working with incremental coding to make the system grow up with new versions of the system executable.
(q) Updating the design results through "Design is precoding, and coding is further design." – for instance, the design shows function A calls function B only, but the coding engineers may find that the function A should call function B and function C – in this case, after coding, they can update the design documents by rebuilding the database to make the design result consistent to the code (they may select to modify the design documents first, then change the code).
(r) If critical issues found, going back to the preprocess to choose a suitable solution method and performing prototyping again.

With NSE, the preliminary design of the whole of the software system is performed in the preprocess (see Chap. 8) through dummy programming using dummy modules based on the result of the functional decomposition of the functional requirements and the description of the nonfunctional requirements. A dummy module has an empty body or only some function call statements without detailed program logic.

With NSE, defect prevention and defect propagation prevention should be performed in the entire software development process and the maintenance process using the Transparent-box testing method, plus formal inspection and review using traceable documents and test cases and the source code supported by various traceabilities, plus software visualization.

With NSE, the document hierarchy is specified with a table using bookmarks to indicate the relationship among the related documents and test cases (see Chap. 11), which will be used to establish the traceability among all the related documents and test cases and the source code through the execution of the test cases (see Chap. 9) using Time Tags that are automatically inserted into both the test cases and the test coverage database of the source code for data mapping between the test cases and the source code, and some keywords such as @WORD@, @HTML@, @EXCEL@, @PDF@, and @BAT@ written within the test case description to indicate the format of a document, the file path, and the bookmark to open the document from the corresponding location when the document is traced. The @BAT@ keyword is used for dynamic traceability to automatically execute a batch file.

With NSE, the design process and the designed results are visible for static defect prevention and defect propagation prevention.

With NSE, the design results are always executable.

12.3.3 Recommendation for Graphic Document Creation/Generation

It is recommended that, in most cases, one should not spend too much time in drawing design graphics manually or using a graphic editor (draft graphics are good enough to use for review only), because it is time-consuming, costly, not traceable, not executable, hard to change, and hard to maintain. I believe in most applications there is no need to draw a graphic manually or using a graphic editor – in most cases, all graphics can be generated automatically through dummy programming or regular source code.

Figure 12.2 shows a draft graphic drawn manually.

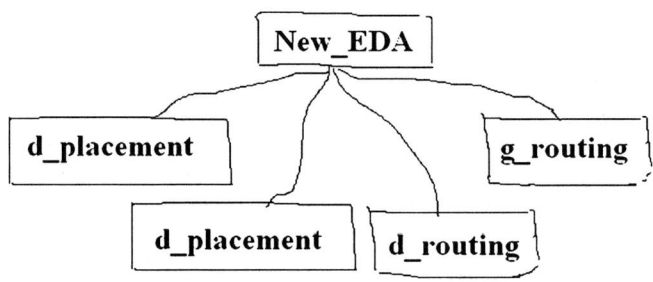

Fig. 12.2 A graphic made manually

12.3 Description of the Innovated "Synthesis Design and Incremental Growing Up" Technique

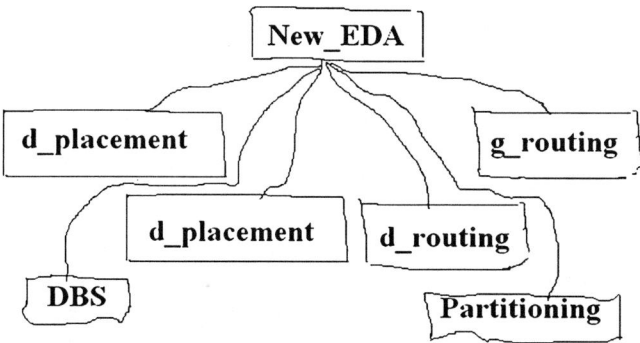

Fig. 12.3 A modified version of the graphic shown in Fig. 12.2

Figure 12.3 shows that a hand-made graphic is hard to change.

For obtaining the graphic shown in Fig. 12.2, the dummy program in C/C++ is very simple:

```
#include <stdio.h>
void   d_routing(){}
void   d_placement(){}
void   g_routing(){}
void   g_placement(){}
void NEW_EDA()
{
 g_placement();
 g_routing();
 d_placement();
 d_routing();
}
```

The corresponding J-Chart generated automatically is shown in Fig. 12.4.

Using dummy programming approach, the same modification is easy to perform:

```
void   d_routing(){}
void   d_placement(){}
void   g_routing(){}
void   g_placement(){}
void   dbs()  {}
void   paertitioning()  {}

void NEW_EDA()
{
 g_placement();
 g_routing();
 d_placement();
 d_routing();
 dbs();
 partitioning();
}
```

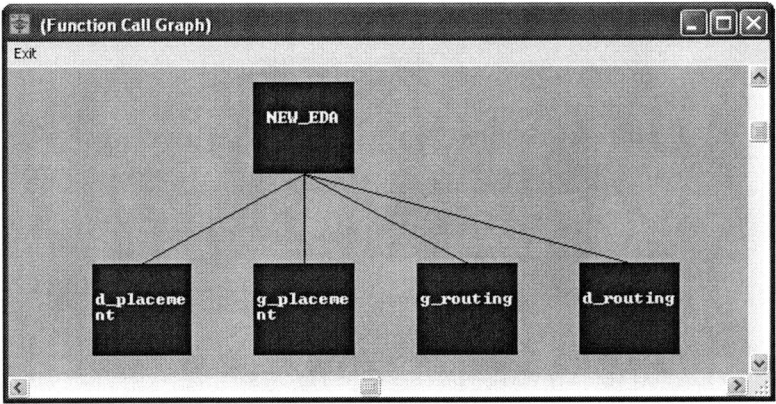

Fig. 12.4 The automatically generated call graph corresponding to Fig. 12.2

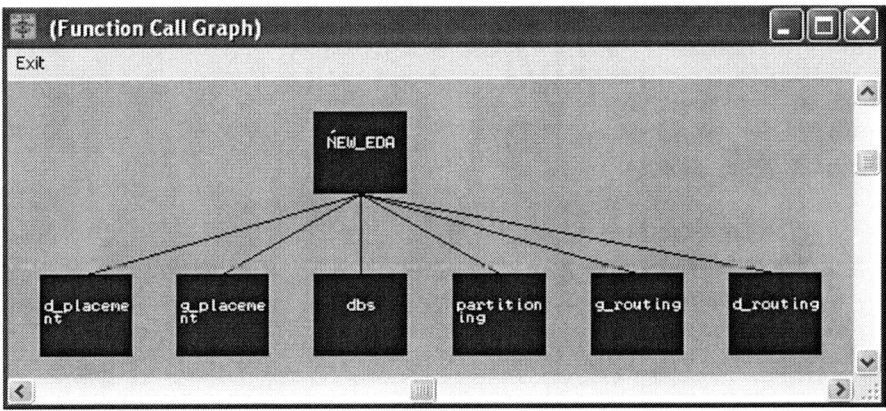

Fig. 12.5 The automatically generated call graph corresponding to that shown in Fig. 12.3

The modified call graph automatically generated from the modified dummy source code is shown in Fig. 12.5.

After changing the module NEW_EDA() to main(), the program can be compiled and executed.

12.3.4 Self-Documenting

As stated in Chap. 7, to easily maintain a software product, many kinds of documents can be merged into the source code such as the Sequence Diagram to expose time ordering of events/messages – we can describe the same thing within a program comment such as the use of a formatted table in C/C++ shown as follows:

```
/* Time-Event table:

    | Timing |   t1   |   t2   |   t3   |
    | Events | Event1 |        |        |
    |        |        | event2 |        |
    |        |        |        | event3 |
  }

    | Timing |   t4   |   t5   |   t6   |
    | Events | Event4 |        |        |
    |        |        | event5 |        |
    |        |        |        | event6 |
  }
...
*/
```

12.3.5 Detailed System Hierarchy Design

Through dummy programming, a detailed program hierarchy of a complex software product can be designed as shown in Fig. 12.6.

12.3.6 Static Defect Prevention and Defect Propagation Prevention Through Traceability

It is difficult to review a complex program hierarchy shown graphically with many modules connected to each other. With NSE, all generated charts and diagrams are traceable for helping users perform static defect prevention and defect propagation prevention as shown in Fig. 12.7.

12.3.7 Dynamic Defect Prevention and Defect Propagation Prevention

Even if only one top module (the main() function, for instance) is preliminarily designed with some command-line options, we can design a set of test cases to test the module dynamically through different command-line options, then the testing

Fig. 12.6 The call graph of a complex software product

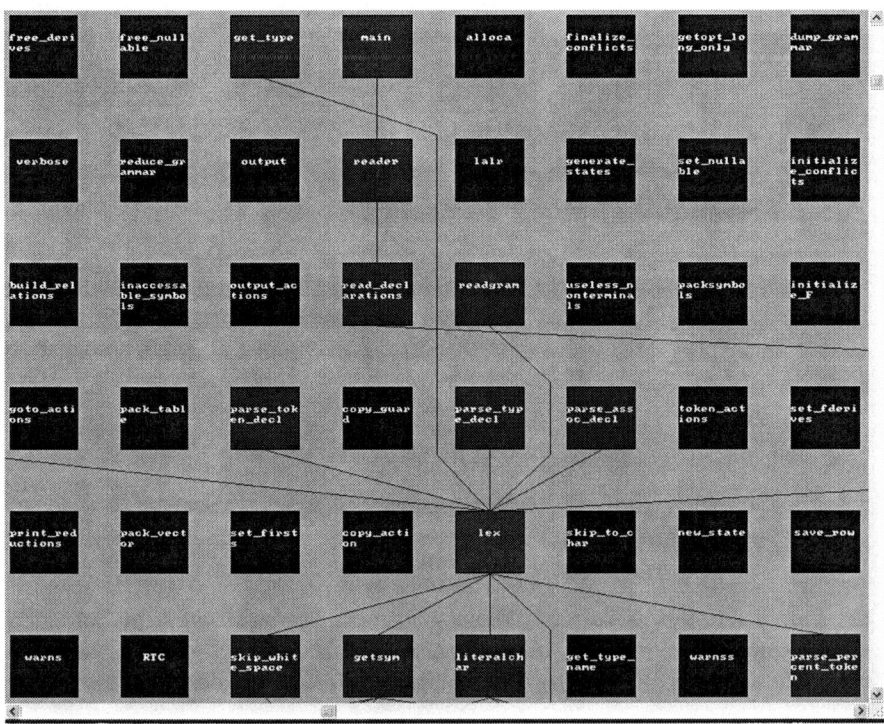

Fig. 12.7 A module and the related modules highlighted for static defect prevention and defect propagation prevention

12.3 Description of the Innovated "Synthesis Design and Incremental Growing Up" Technique

tool using the Transparent-box testing method will establish the automated and self-maintainable traceability among the related documents, the test cases, and the source code for preventing inconsistency defects – see Section 12.4.

12.3.8 Data Structure Design

Data structure design is one of the most important tasks in software design. With NSE, it is also being done through dummy programming. Figure 12.8 shows a class inheritance chart of a designed program.

12.3.9 Detailed Logic Design of the Modules

With NSE, it is recommended to perform detailed module design using J-Diagram. Figure 12.9 shows a program design example using the activity diagram of UML.

A sample programming source code used for representing the corresponding product design specified in the activity diagram is listed as follows:

The J-Diagram generated from the listed dummy programming source code is shown in Fig. 12.10.

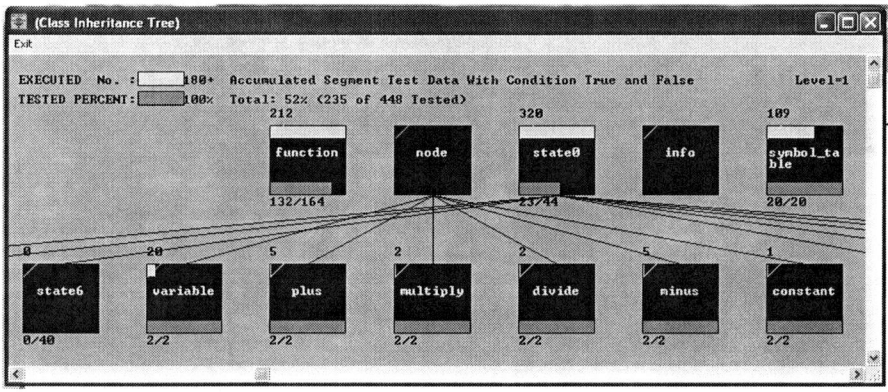

Fig. 12.8 A sample chart showing the class relationship of a program

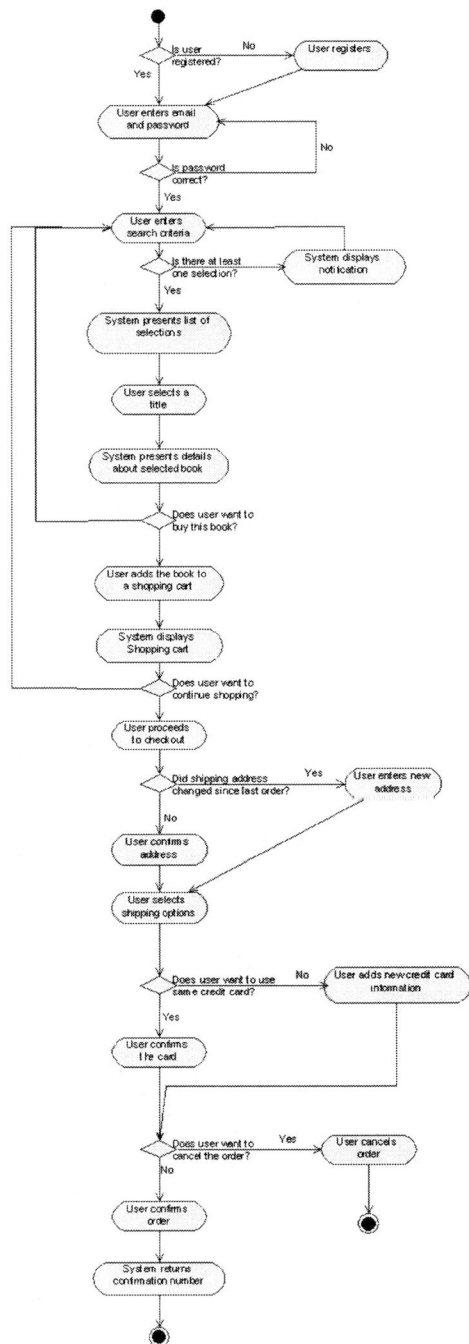

Fig. 12.9 A typical activity diagram. *Source*: Peter Zielczynski, Director of Technology Solutions, The A Consulting Team, Inc., "Traceability from use cases to test cases", 2004, http://www.ibm.com/developerworks/rational/library/04/r-3217/?S_TACT=105AGX52&S_CMP=cn-a-r#author1

12.4 Application

```
#include <stdlib.h>
#include <string.h>
#include <stdio.h>
#include <iomanip.h>
#include "my_include.h"

void main(int argc, char** argv)
{
user_login_name = get_user_login_name();
saved_user_password = get_saved_user_password();
if(!user_registred())
        /* handling user register. */ ;
  if(saved_user_password!=NULL && user_password!=NULL)
        {while(!strcmp(saved_user_password, user_password))
        /* user Enters Email and Password. */ ;
        }
  else /* user Enters Email and Password. */;
  is_true =1;
  while(is_true)
  {
  if(there_is_at_least_one_selection=user_enter_search_criteria())
              {
              /*
              system presents selection list;
              user selects a title;
              system presents details about selected book; */
              if(user_want_to_buy_this_book())
                      {
                      /*
                      user adds this book to shopping car; */;
                      if(user_wants_to_continue_shopping()) ;
                      else break;
                      }
              else
                      /*
                      system displays notification;
                      */;
        /* user processes to check out. */ ;
        if(the_shipping_address_changed_since_last_order())
              /* user enters new address. */ ;
        else
              /* user confirms address. */ ;
        /* user selects shipping optyions. */ ;
        if(user_uses_same_credit_card())
              /* user confirms  the card. */ ;
        else
              /* user adds new credit card information. */ ;
        if(user_wants_cancel_the_order())
              {
              /* user cancels order process. */ ;
              exit;
              }
        /* user confirms order ;
        system_returs_confirmation_number. */ ;
        break;
            }
  }
  cout << "\n This program has been executed successfully.\n";
}
```

The included file:

```
class Customer_passwd {
private:
  char login_name[80];
  char password[80];
public:
  int check_password(char* login_name){}
  int change_password(char* login_name, char* password) {}
};

class Customer_infor: Customer_passwd {
private:
  char name[80];
  char shipping_address[80];
public:
  int update_info(char* name, char* shipping_address){}
/* other menber function. */
};

char* user_password;
char* saved_user_password;
char* user_login_name;
int there_is_at_least_one_selection;
int is_true;

int m=1,n=3;
char* p="abc";

int user_registred(){return m;}
char* get_saved_user_password(){return p;}
char* get_user_login_name(){return p;}
int user_want_to_buy_this_book(){return m;}
int user_wants_to_continue_shopping(){return n;}
int the_shipping_address_changed_since_last_order(){return n;}
int user_uses_same_credit_card(){return n;}
int user_wants_cancel_the_order(){return n;}
int user_enter_search_criteria(){return n;}
```

The following shows an application example for a new EDA software product design.

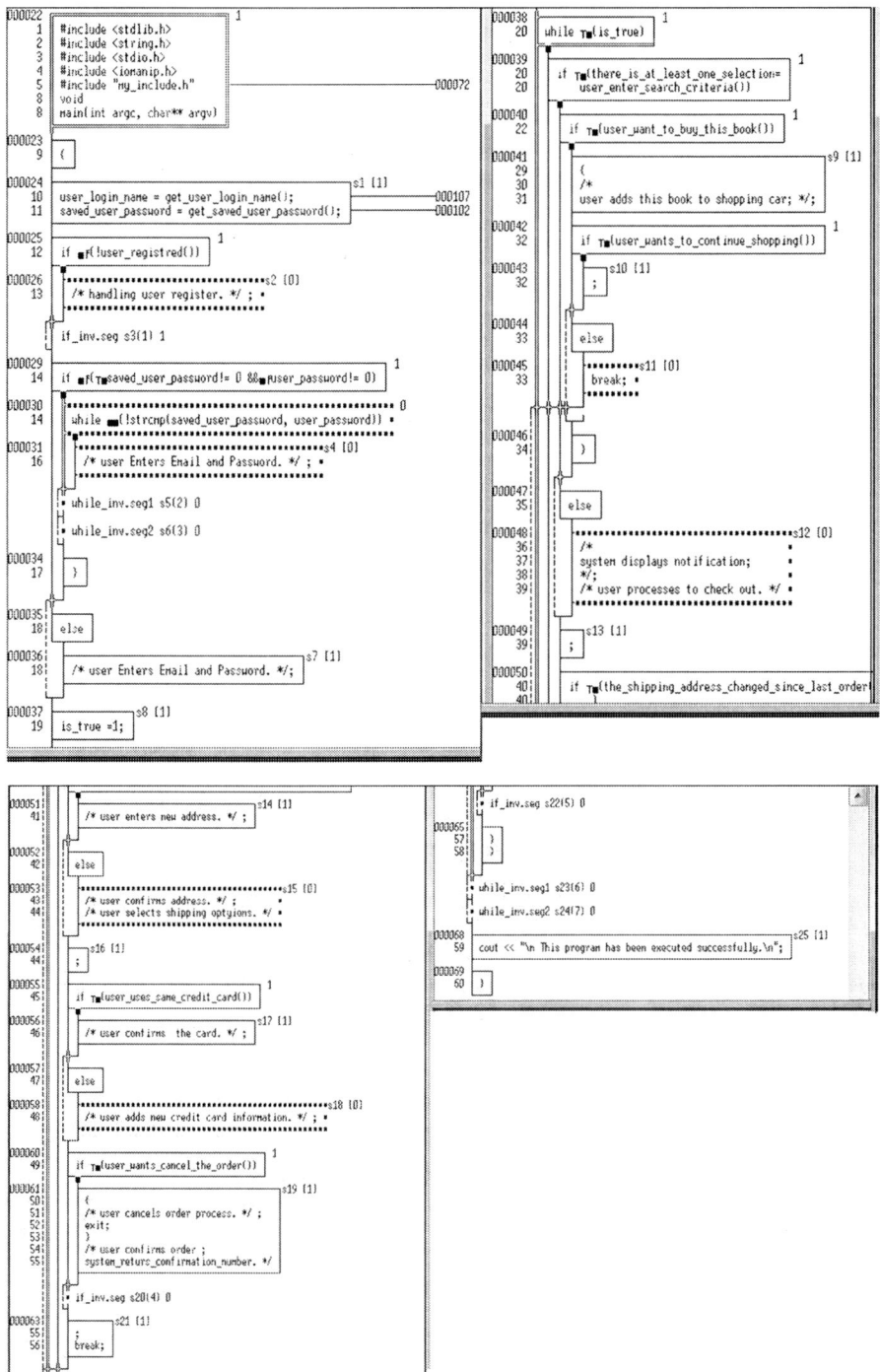

Fig. 12.10 The generated J-Diagram from the listed dummy source code, shown with the test coverage measurement result after dynamic testing

1. Complying with Deming's product quality assurance principles – *"Cease dependence on inspection to achieve quality. Eliminate the need for inspection on a mass basis by building quality into the product in the first place"*[Dem86] – with NSE, the quality of a software product being designed is ensured from the first step in top-down product hierarchy design with only the main() program for handling some command-line options – see the dummy source code listed below:

```
#include <stdio.h>
#include <string.h>

void main(int argc, char** argv)
{
int ERROR_CODE;
if(argc != 3 && argc != 4)
    printf("Error found in the command-line.\n");
else if (argc == 3){
if(strcmp(argv[1],"global_placement")==0)
    ; // calling  g_placement(argv[2]);
else if(strcmp(argv[1],"global_routine")==0)
    ; // calling   g_routing(argv[2]);
else if(strcmp(argv[1],"detailed_placement")==0)
    ; // calling   d_placement(argv[2]);
else if(strcmp(argv[1],"detailed_routing")==0)
    ; // calling                d_routing(argv[2]);
else if(strcmp(argv[1],"partititionning")==0)
    ; // calling              partitioning(argv[2]);
else if(strcmp(argv[1],"ordering")==0)
    ; // calling   ordering(argv[2]);
else
    ; // calling   printf("Invalid name: %s\n",argv[1]);
} else if (strcmp(argv[2],"dbs_build") == 0)
    ; // calling   dbs_build(argv[2],argv[3]);
else printf("Error! Invalid name: %s\n",argv[1]);
}
```

Usually, with NSE in the beginning of product design, some documents should be ready for use, including the corresponding requirement specification, the test requirement specification, the prototyping documents, the product development plan, etc., so that according to the test requirement specification and the command-line options (GUI operation options), we can design a corresponding test script file as follows:

12.4 Application

```
# An example of using transparent-box to prevent defects in requirement analysis phase
#                   and initial design phase of a software product development.
# test case 1
# test purpose : to find the inconsistency among all related documents for the global placement sub-system
# The related requirement specification  : @HTML@ C:\EN_transparent_box\new_EDA_specifications.htm#G_PLACEMENT
# The related prototype design document :  @WORD@  C:\EN_transparent_box\prototyping.doc  bmname g_placement
# The related development plan : @BAT@ C:\EN_transparent_box\ganttpro6.bat
# The expected execution path:
# [path] main (int, char**) {s0, s2}
# [NOT_HIT} !path [/NOT_HIT]
# Expected output : none
# The directory and the execution command:
C:\EN_transparent_box
new_EDA global_placement -dbs=my_dbs

# test case 2
# test purpose : to find the inconsistency among all related documents for the global routing sub-system
#              of a software product, NEW_EDA.
# The related requirement specification  : @HTML@ C:\EN_transparent_box\new_EDA_specifications.htm#G_ROUTING
# The related testmrequirement specification: @WORD@  C:\EN_transparent_box\Test_Requirement_Specification.doc
# The related prototype design document  :  @WORD@  C:\EN_transparent_box\prototyping.doc  bmname g_router
# The expected execution path:
# [path] main (int, char**) {s0, s3}
# [NOT_HIT} !path [/NOT_HIT]
# The expected output : none
# The directory and the execution command:
C:\EN_transparent_box
new_EDA global_routing -dbs=my_dbs

# test case 3
# test purpose : to find the inconsistency among all related documents for the detailed placement sub-system
#              of a software product, NEW_EDA.
# The related requirement specification  : @HTML@ C:\EN_transparent_box\new_EDA_specifications.htm#D_PLACEMENT
# The related prototype design document :  @WORD@  C:\EN_transparent_box\prototyping.doc  bmname d_placement
# The related development plan : @BAT@ C:\EN_transparent_box\ganttpro6.bat
# The expected execution path:
# [path] main (int, char**) {s0, s4}
# [NOT_HIT] !path [/NOT_HIT]
# Expected output : none
# The directory and the execution command:
C:\EN_transparent_box
new_EDA detailed_placement -dbs=my_dbs

# test case 4
# test purpose : to find the inconsistency among all related documents for the detailed routing sub-system
#              of a software product, NEW_EDA.
# The related requirement specification  : @HTML@ C:\EN_transparent_box\new_EDA_specifications.htm#D_ROUTING
# The related prototype design document:  @WORD@  C:\EN_transparent_box\prototyping.doc  bmname d_routing
# The related development plan  : @BAT@ C:\EN_transparent_box\ganttpro6.bat
# [path] main (int, char**) {s0, s5}
# [NOT_HIT] !path [/NOT_HIT]
# Expected output : none
C:\EN_transparent_box
new_EDA detailed_routing -dbs=my_dbs
```

After running the four test cases, the test coverage result is shown in Fig. 12.11.

Now it is the time to perform defect prevention and defect propagation prevention through dynamic testing and review using traceable documents and the test cases and the source code:

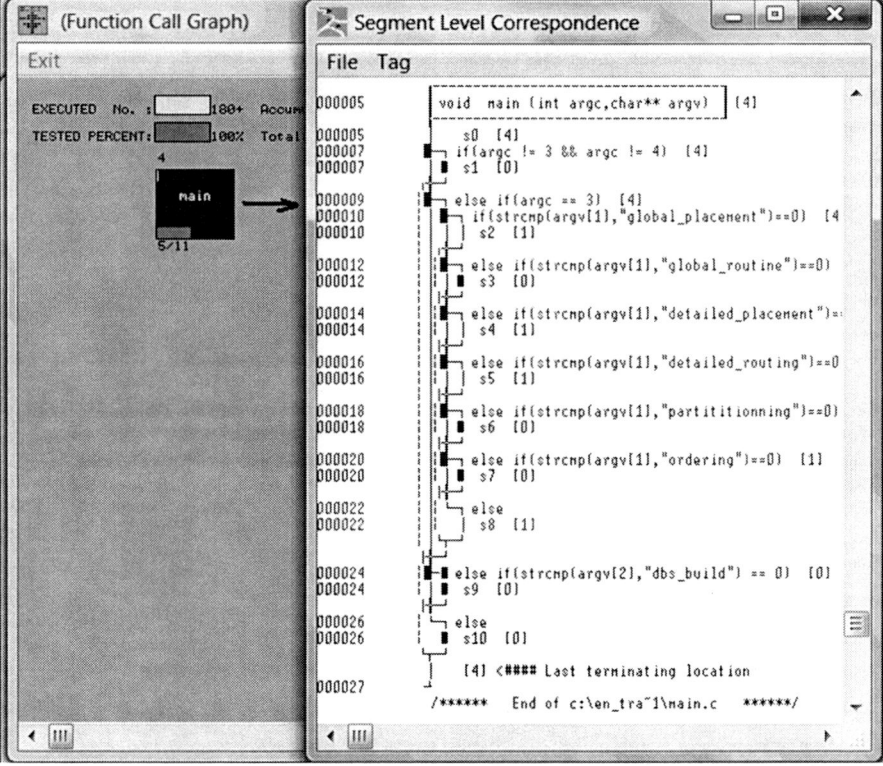

Fig. 12.11 The test coverage measurement result of the main() program

After running the test cases, we can perform forward tracing and backward tracing to find and remove defects – see Figs. 12.12–12.16.

Removing the defects:

(a) Find the location for the first defect and modify the main() program

12.4 Application

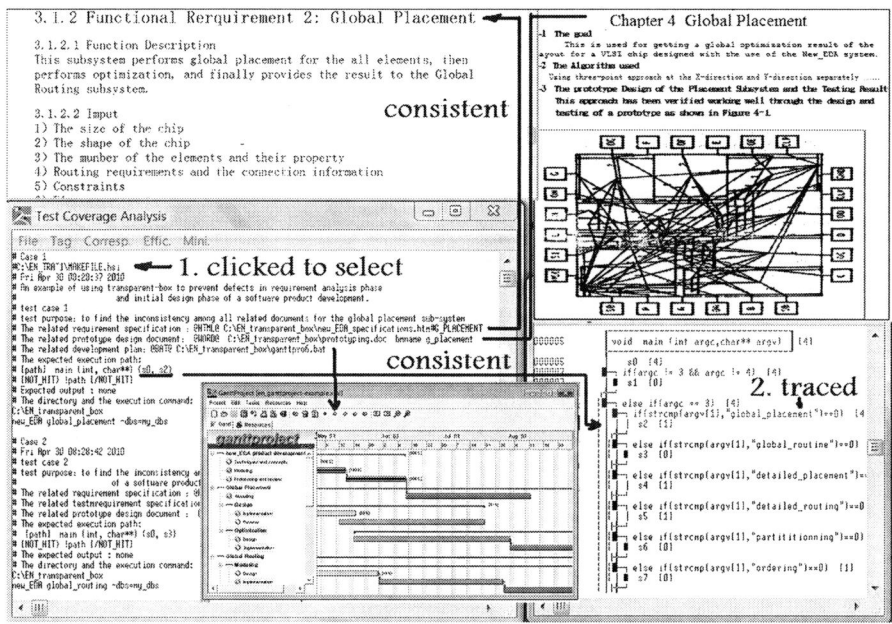

Fig. 12.12 Tracing the first test case to find the corresponding source code and automatically open the related documents (the requirement specification, the prototyping documents, and the project development plan) – in this operation, no defect was found

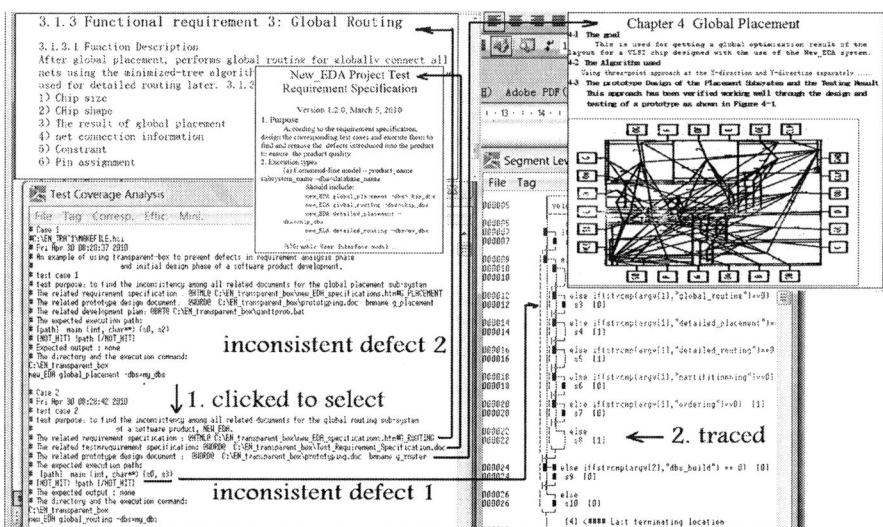

Fig. 12.13 Tracing the second test case with two inconsistency defects found: (1) The real execution path did not cover the expected execution path main(int, char**) {s0, s3} – segment s3 is highlighted as untested; (2) The bookmark for opening the global routing section of the prototyping document, g_router, pointed to the wrong section – the global placement section

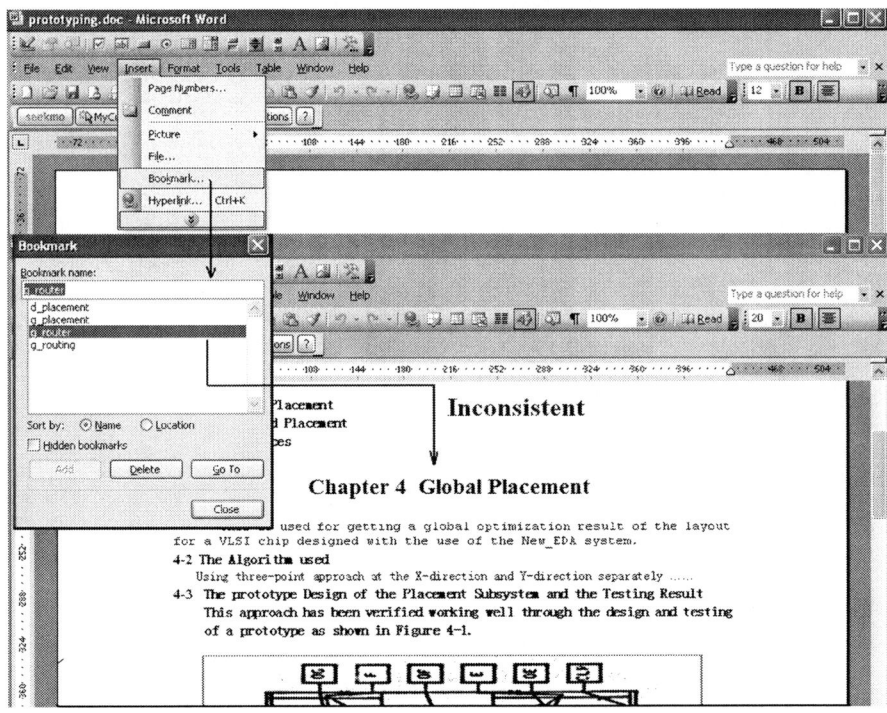

Fig. 12.14 Locating the mistake of the bookmark, g_router

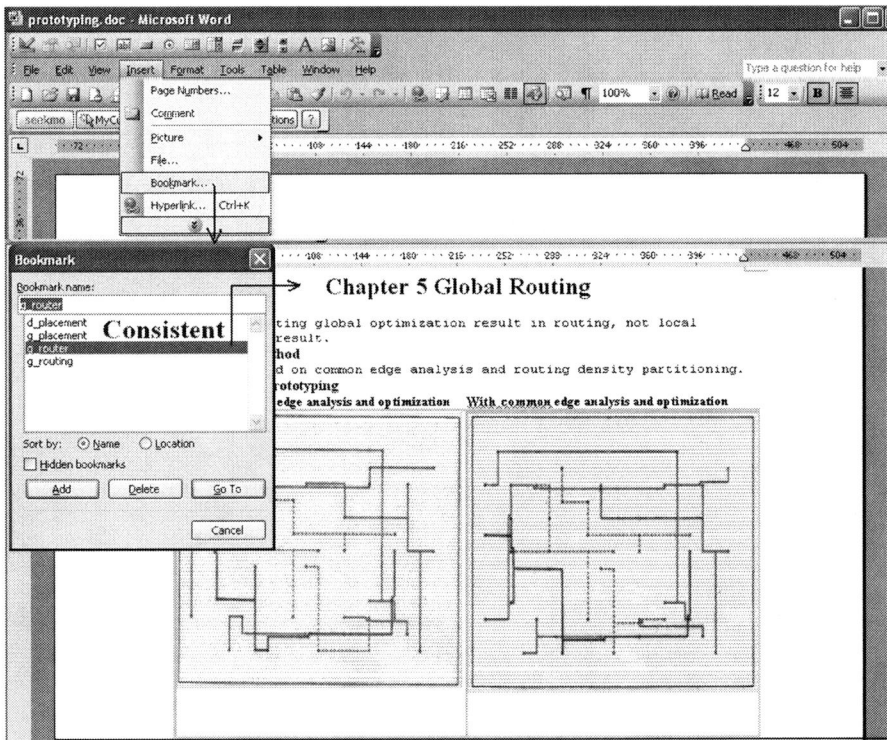

Fig. 12.15 Fixing the bookmark mistake

12.4 Application

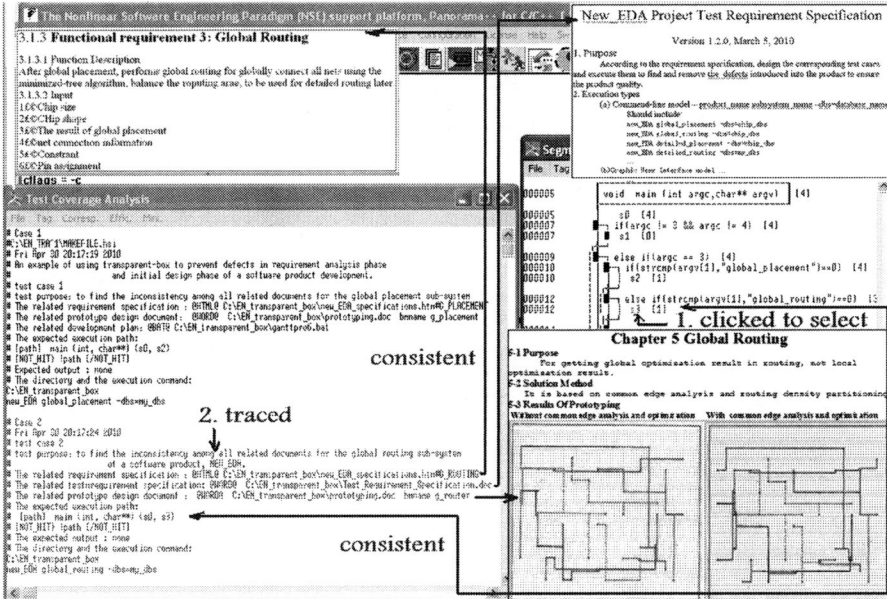

Fig. 12.16 Verifying that the two defects have been removed through backward tracing from the segment s3 (the test case 2 was traced and the related documents were opened without defects)

From:
```
...
else if(strcmp(argv[1],"global_routine")==0)
  ; // calling    g_routine(argv[2])
                  ^
```

TO:
```
...
else if(strcmp(argv[1],"global_routine")==0)
  ; // calling         g_routing(argv[2]);
                       ^
```

(b) Find the mistake related to the bookmark, g_router (Fig. 12.14), and fix it (Fig. 12.15):
2. The designed result after adding the second level is shown in Fig. 12.17.
3. The top-down design result after adding some more levels is shown in Fig. 12.18.

Figure 12.19 shows that the designed results are always traceable.

The corresponding dummy programming source code for the main() module is listed as follows:

```
#include <stdio.h>
#include <string.h>
```

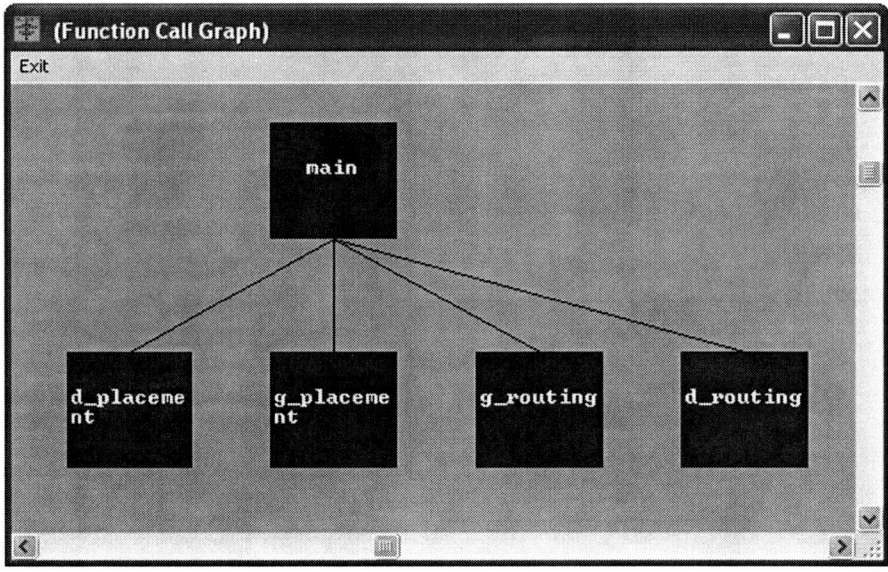

Fig. 12.17 The design result after adding four subsystems

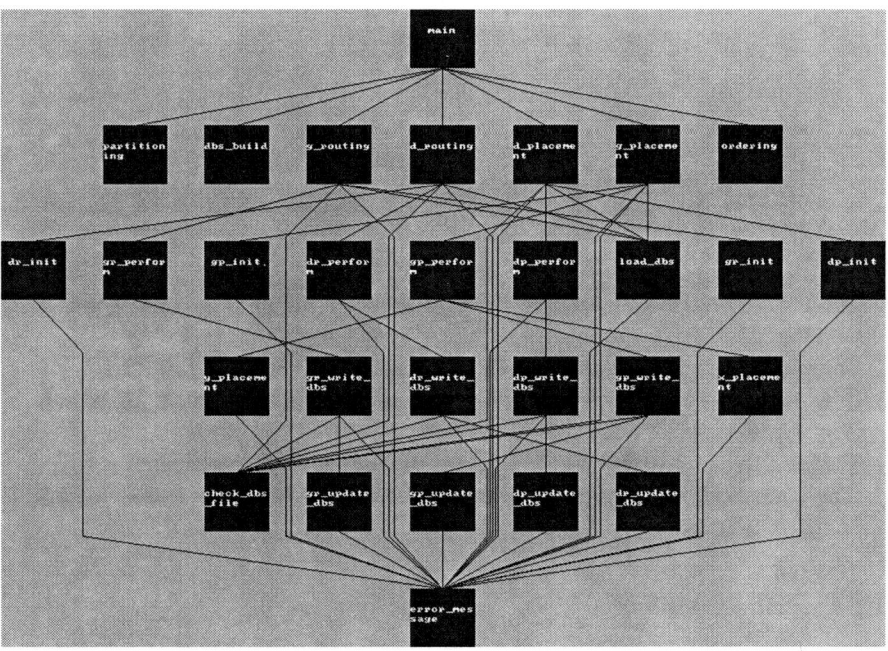

Fig. 12.18 The top-down design result after adding some more levels

12.4 Application

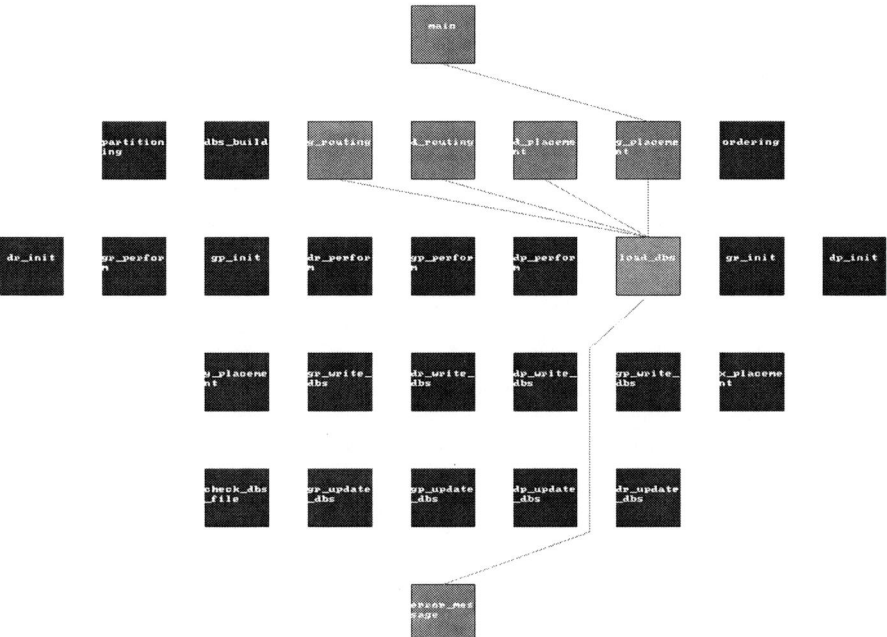

Fig. 12.19 Tracing a module (load_dbs()) to highlight all the related modules

```
extern int dbs_build(char*, char*);
extern int d_routing(char*);
extern int d_placement(char*);
extern int g_routing(char*);
extern int g_placement(char*);
extern int partitioning(char*);
extern int odering(char*);

void main(argc, argv)
int argc;
char **argv;
{
if(argc != 3 && argc != 4)
{
printf("Usage: \n");
printf("    new_EDA dbs_build -conditions=condition_file
-dbs=database_file \n");
printf("    new_EDA global_placement -dbs=database_file \n");
printf("    new_EDA global_routing -dbs=database_file \n");
printf("    new_EDA detailed_placement -dbs=database_file \n");
printf("    new_EDA detailed_routing -dbs=database_file \n");
printf("    new_EDA partitioning -dbs=database_file \n");
printf("    new_EDA ordering -dbs=database_file \n");
}
else if (argc == 3)
```

```
{
if(strcmp(argv[1],"global_placement")==0)
   g_placement(argv[2]);
else if(strcmp(argv[1],"global_routing")==0)
   g_routing(argv[2]);
else if(strcmp(argv[1],"detailed_placement")==0)
   d_placement(argv[2]);
else if(strcmp(argv[1],"detailed_routing")==0)
   d_routing(argv[2]);
else if(strcmp(argv[1],"partititionning")==0)
   partitioning(argv[2]);
else if(strcmp(argv[1],"ordering")==0)
   ordering(argv[2]);
else
   printf("Invalid name: %s\n",argv[1]);
}
  else if (strcmp(argv[2],"dbs_build") == 0)
    dbs_build(argv[2],argv[3]);
else printf("Error! Invalid name: %s\n",argv[1]);
}
```

4. About the final result of the top-down design of the NEW_EDA system.
   It is the project I may develop with my colleagues in the future based on complexity science – the existing EDA products for VLSI chip design are completely outdated, because they are outcomes of reductionism and the superposition principle. The NEW_EDA system may have more than 50,000 function points with more than ten million lines of source code.

12.5 The Major Features of the Software Synthesis Design Technique

The major features of the **Software Synthesis Design (and Incremental growing up) technique** include:

(a) It is an engineering approach for software design.
(b) It works with the NSE process model and the NSE software development methodology based on complexity science by complying with the essential principles of complexity science, particularly the Nonlinearity principle and the Holism principle.
(c) It complies with the Generative Holism principle of complexity science that the whole of a complex system exists earlier (as an embryo) than its components, and then grows up with its components.
(d) It follows the rule that people is the first-order element in software engineering, and the natural law about human beings that people are nonlinear and they easily make mistakes and wrong decisions, so that it combines software design and

software coding together to make design become precoding and coding (see Chap. 13) become further design.
(e) It meets the NSE Upstream Quality Assurance strategy from the first step to the end of the software development lifecycle through defect prevention and defect propagation prevention by dynamic testing using the Transparent-box method, review and inspection using the traceable documents and source code, and software visualization.
(f) The work products designed using this technique are holistic, visual, traceable, and always executable.
(g) It is a component of the entire NSE software engineering paradigm for efficiently handling the essential issues existing with today's software development: the complexity, changeability, invisibility, and conformity, defined by Brooks [Bro95-P182].

12.6 Summary

The old-established software design paradigm works with the linear process models based on reductionism and the superposition principle that the whole of a complex system is the sum of its components, so that with it almost all software design tasks and activities are performed linearly, partially, and locally – through "**Analysis**." The obtained work products using the old-established software design paradigm are not holistic, not traceable, not visible, and not directly executable – it means the quality of the product design is hard to ensure.

With NSE, software design engineering is performed using the **Software Synthesis Design (and Incremental growing up) Technique** working with the NSE process model and the NSE software development methodology based on complexity science by complying with the essential principles of complexity science, particularly the Nonlinearity principle and the Holism principle, so that with NSE almost all the software design tasks and activities are performed holistically and globally – through "**Synthesis**." The obtained work products are holistic, visible, traceable, and directly executable for defect prevention and defect propagation prevention mainly using the Transparent-box testing method – it means the quality of the product design is easy to ensure.

With NSE design becomes precoding, and coding becomes further design.

12.7 Points and Questions to Ponder

(a) What are the major problems with today's software design?
(b) What are the benefits to use the **Software Synthesis Design (and Incremental growing up) technique** for software design?

(c) Complete the dummy program for generating the top-down design result shown as follows:

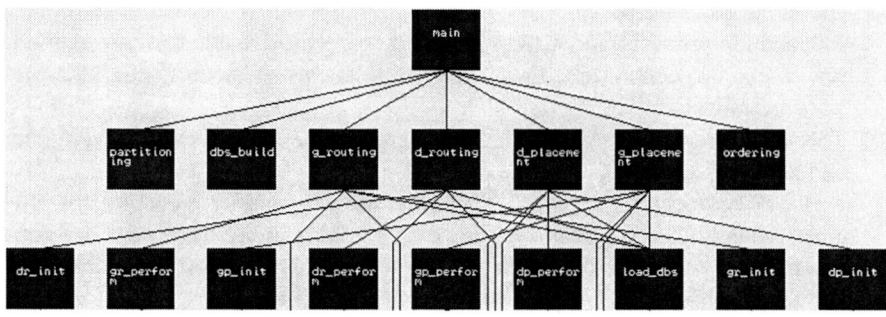

12.8 Further Reading and Information Source

Software design, From Wikipedia, the free encyclopedia. http://en.wikipedia.org/wiki/Software_design

Software design document template:

http://www.klariti.com/templates/Design-Document-Template.shtml
http://www.rspa.com/docs/Designspec.html
http://en.wikipedia.org/wiki/Software_Design_Description
http://ant.comm.ccu.edu.tw/course/97_Programming/7_SampleCode/Design%20Document%20Template%20-%20Chapters.pdf

References

[Bro95-P182] Brooks FP Jr (1995) The mythical man-month. Addison-Wesley, Reading, p 182
[Bro95-P211] Brooks FP Jr (1995) The mythical man-month. Addison-Wesley, Reading, p 211
[CMMI1.1] Phillips M (2002) CMMI program manager, CMMI V1.1 and appraisal tutorial. http://www.sei.cmu.edu/cmmi/
[Coc99] Cockburn AAR (1999) Characterizing people as non-linear, first-order components in software development, Humans and Technology, HaT Technical Report 1999.03, 21 Oct 1999
[Dem86] Deming WE (1986) Out of the crisis. MIT Press, Cambridge

Chapter 13
Coding Engineering with NSE

> *Add one component at a time.* This precept, too, is obvious, but optimism and laziness tempt us to violate it. To do it requires dummies and other scaffolding, and that takes work.... Note that one must have thorough test cases, testing the partial systems after each new piece is added.
>
> Frederick P. Brooks, Jr.

This chapter introduces software coding engineering with NSE on which software coding and software design are combined together closely using the innovated **Synthesis Design and Incremental Integration (growing up)** technique – with NSE, software design becomes precoding and software coding becomes further design. The quality of the work products (source code and the documents) is ensured by defect prevention and defect propagation prevention through dynamic testing using the innovated Transparent-box testing method, inspection using traceable documents and the source code, and software visualization.

13.1 The Problems Addressed

Many useful programming techniques have been proposed by software engineering experts and successfully applied in practices such as Object-Oriented programming technique [Coa93], the Pair Programming technique [Bec99], and the langrage-specified program editing technique.

But unfortunately, there are still many critical problems exist with today's software coding engineering paradigm:

(a) It complies with the linear process models and linear software development methodologies based on reductionism and the superposition principle, so that almost all software programming tasks are performed linearly, partially, and

locally, such as the implementation of program modification, program documentation, test planning, program refactoring, and program version comparison.
(b) It is performed after software design following a top-down order without upstream movement (bottom-up) at all.
(c) It follows the **Constructive Holism** principle that the components of a software product are completed first, then "Assemble" the whole of the entire product from the components or subsystems, so that system testing and user evaluation is done at or close to the end of the programming.
(d) The coding process and the work products (source code and the documents) are not visible.
(e) The work products are not traceable.
(f) Code inspection is performed with separated documents and source code.
(g) Modules are coded randomly without systematic ordering support.
(h) There is a need to design and use stub modules to replace the real modules called by the unit being coded in unit testing – but stubs will not return the real value so that it is different from the real application execution.
(i) The quality of the modules coded are low – besides unit testing using stubs, the code inspection is performed inefficiently without the support of various traceabilities, and often the modules do not satisfy 100% MC/DC (Modified Condition/Decision Coverage) test coverage [RTCA92]: most software testing tools used for structural testing only offer the capability for statement-level test coverage analysis or branch-level test coverage analysis; although some tools claim that they do support MC/DC test coverage analysis, the test results are shown in textual format without the capability to highlight untested condition graphically and directly, so that the test results are hard to review and hard to improve.
(j) It is not supported by a coding-style-independent graphical representation technique and tool, so that the source code written by others is much difficult to read and understand.
(k) There is no powerful technique and tool to automatically document a source program and make the documents (such as the call graph, the class inheritance chart, the logic diagram of the entire product, and the control flow diagram of the entire product) always consistent with the source code.
(l) The related documents are often inconsistent with the source code after code modification – there is no systematic way available to update them in time.
(m) Programming productivity is hard to calculate without a systematic technique and tool to count the total amount and the percentages of the comment lines, partial comment lines, empty lines, and active lines of the source code.
(n) There is a lack of systematic technologies and tools for the support of re-engineering and reverse engineering to automatically generate huge amount of graphic documents of the program architectures, the program logic of an entire software product and the entire control flow diagram of an entire software product with various traceabilities established, etc.

13.2 The Solution: Software Coding Engineering with NSE Using the Synthesis Design and Incremental Integration Technique

Here, the solution is called "NSE-Coding," which is the application of the innovated **Synthesis Design and Incremental Integration (growing up) technique** in coding engineering. This technique combines software design and coding engineering together to make design become precoding and coding become further design.

Here, the **Incremental Implementation and Integration** means the following activities:

1. Select one or a set of requirements according to the requirement priority assigned.
2. From the corresponding call graph (shown in J-Chart notation) of the designed system, highlight the critical module with all modules calling and called by the module for the selected requirement(s), assign a bottom-up coding order on them.
3. Perform incremental unit coding according to the assigned order to prevent inconsistency defects between the interfaces of the calling modules and the called modules (see Sect. 13.7).
4. Carry out unit testing and integration testing together to remove possible defects through comprehensive testing (including functional testing, structural testing, memory leak and usage violation checking, quality measurement, performance analysis, etc., see Chap. 16).
5. Recompile the entire program to establish a new version of the program, and then run the program again dynamically.
6. Different from traditional incremental integration approaches which complete the subsystem design and coding first then carry out integration for the whole system, with NSE the incremental implementation and integration is done at the same time – each time only one module of the subsystem for the selected requirements will be integrated to establish a new version of the executable program (although different critical paths or different subsystems can be coded in parallel, integration should be done by adding one module at a time), so that if something is found wrong, the problems often come from the one added module only rather than the entire subsystem implemented for the selected requirement(s). An application example of the **Incremental Implementation, Iteration, and Integration** is shown with the step 1 of the main process described in Chap. 8.
7. Combine the processes of software development, testing, and maintenance together closely through many automated and bidirectional traceabilities for defect prevention in the entire software product development lifecycle.
8. If some critical problems are found in coding phase, go back to the upper phases to solve the problem – it is possible to give up the previous selected solution method such as in the case that the performance is very bad because of the misuse of virtual memory – in this case, go back to the preprocess to design new solution methods.

The major offerings of NSE-Coding are as follows:

(a) It complies with the NSE nonlinear process model and the NSE software development methodologies based on complexity science, so that almost all software programming tasks are performed nonlinearly, holistically, and globally, such as the implementation of program modification, program documentation, test planning, program refactoring, and program version comparison – see Fig. 13.1, an application example of system-level test planning support through Cyclomatic complexity (the number of decision statements) measurement of an entire software product being coded.

(b) With it, software design and coding are combined together closely, making design become precoding, and coding becomes further design to improve the design – no matter in software design or coding, people are nonlinear, and it is easy to make mistakes and wrong decisions (When a critical issue is found in coding process, the work flow should even go back to the requirement development phase or the preprocess phase) – see Fig. 13.2, an application example of design becoming precoding – to code a module by editing the source code directly from a call graph (shown in J-Chart) generated in the design process.

For instance, in the case that the design shows function A calling function B, but the coding engineers find that function A should call function C and function C should call function B – after coding they can update the design documents by rebuilding the database to make the design result consistent with the code (in this case, they may choose to modify the design first, then edit and change the code).

An example of coding becoming further design is shown in Figs. 13.3 and 13.4.

(c) It follows the **Generative Holism** principle that the whole of a software product exists (as an embryo, but executable) earlier than its components, and then grows up with its components to continuously form new executable versions, so that system testing and user evaluation is done at or close to the beginning of product development.

(d) The coding process and the work products (source code and the documents) are always visible as shown in Figs. 13.2 and 13.5.

(e) The work products are traceable internally and externally or even dynamically – see Sect. 13.5.

(f) Code inspection is performed with traceable documents and traceable source code – see Sect. 13.5 too.

(g) Modules are coded incrementally with systematic ordering support – see Fig. 13.6 (ordering for all modules) and Fig. 13.7 (ordering for a critical path).

(h) In unit testing with the capability to control the return value (see Sect. 13.3 and Appendix C about how to control the return values for a function call statement), there is no need to design and use stub modules to replace the real modules called by the unit being coded.

13.2 The Solution: Software Coding Engineering with NSE 343

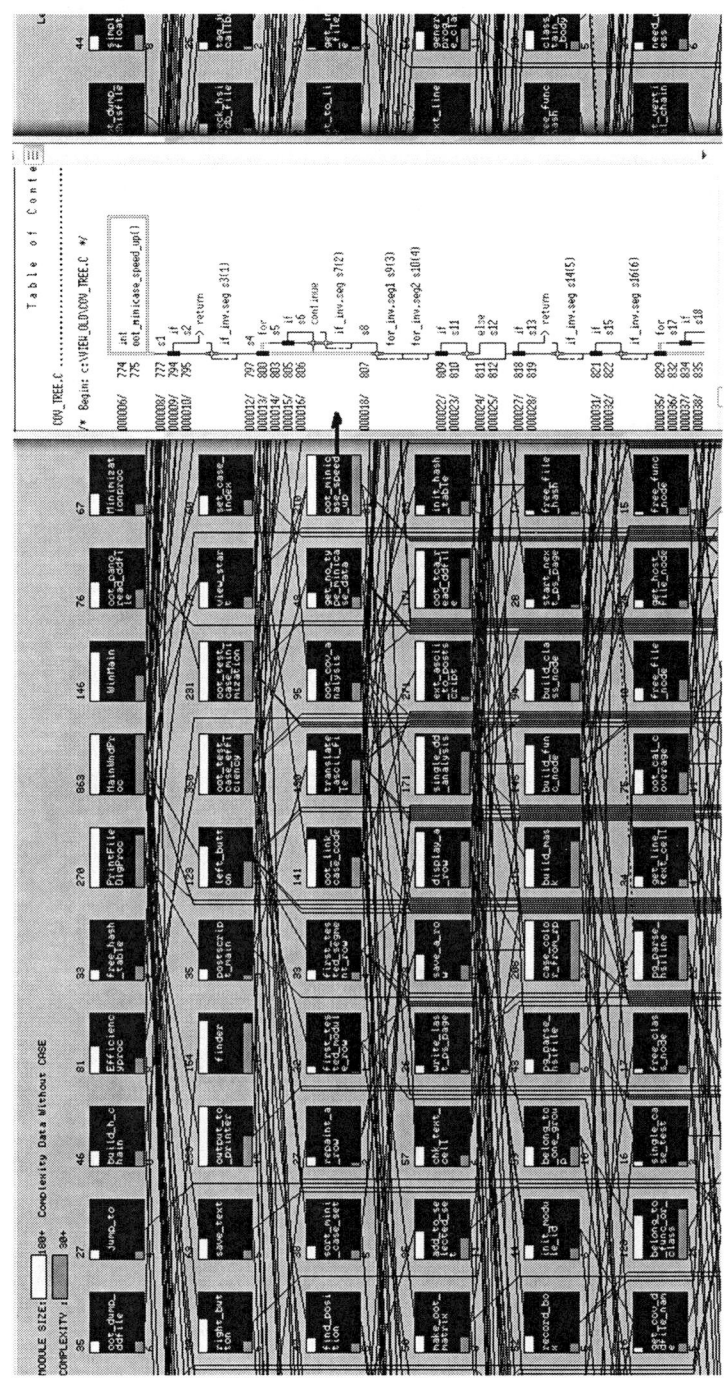

Fig. 13.1 System-level test planning through Cyclomatic complexity measurement

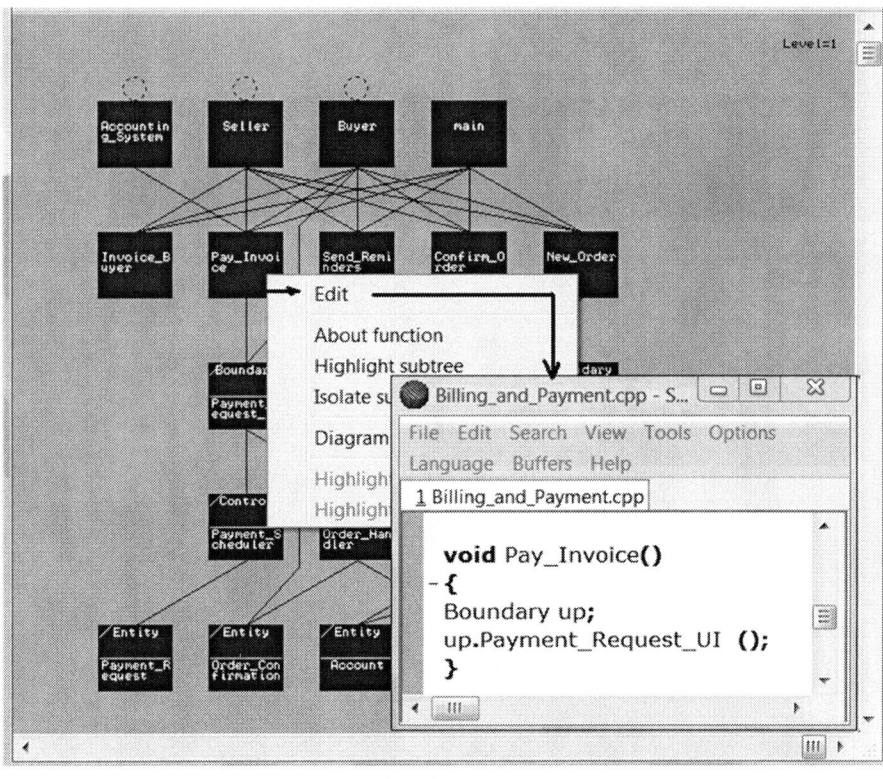

Fig. 13.2 Directly coding from a call graph generated in the design process

Fig. 13.3 Two function call statements are added in the coding process of the state4::transition (unsigned char) module designed without using them

13.2 The Solution: Software Coding Engineering with NSE 345

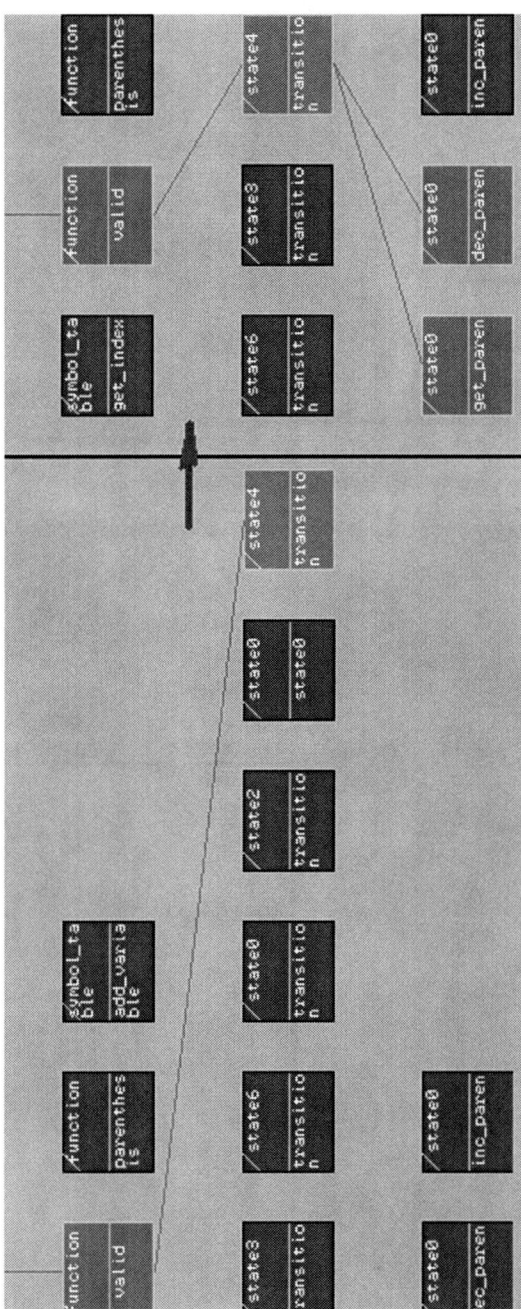

Fig. 13.4 After rebuilding the database, the corresponding design documents are updated

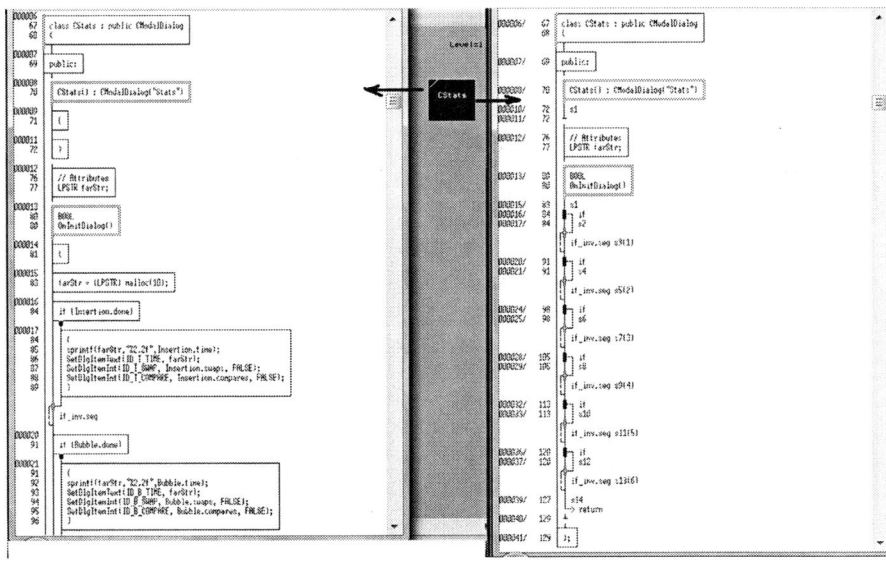

Fig. 13.5 A class shown visually with its logic and control flow

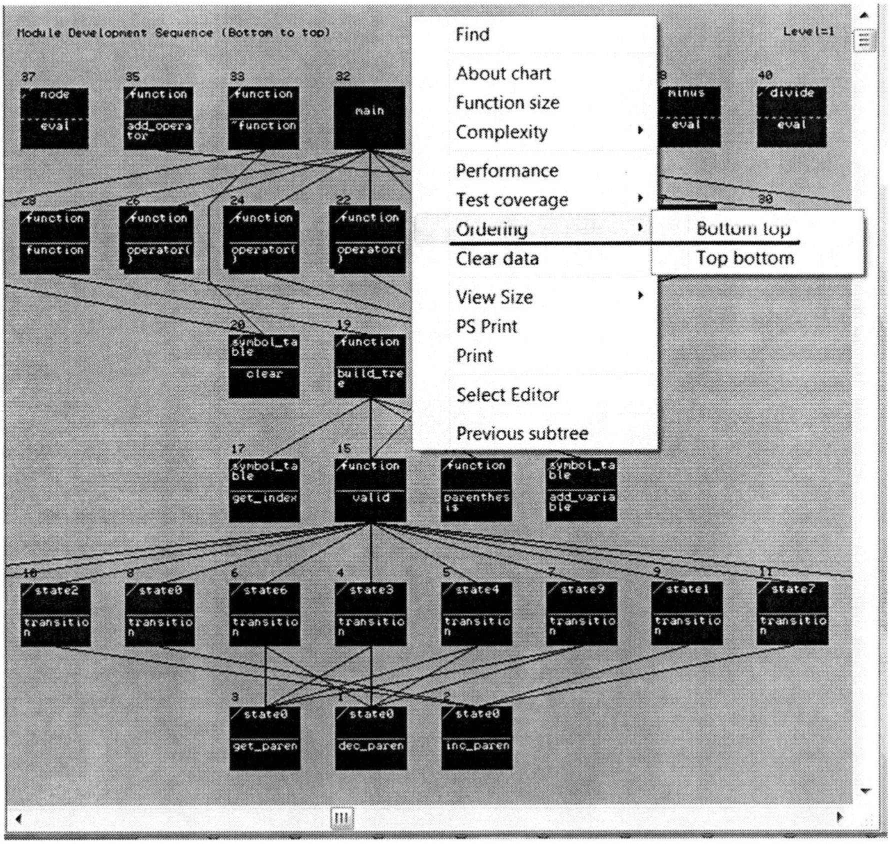

Fig. 13.6 Module coding ordering support for an entire product

13.2 The Solution: Software Coding Engineering with NSE

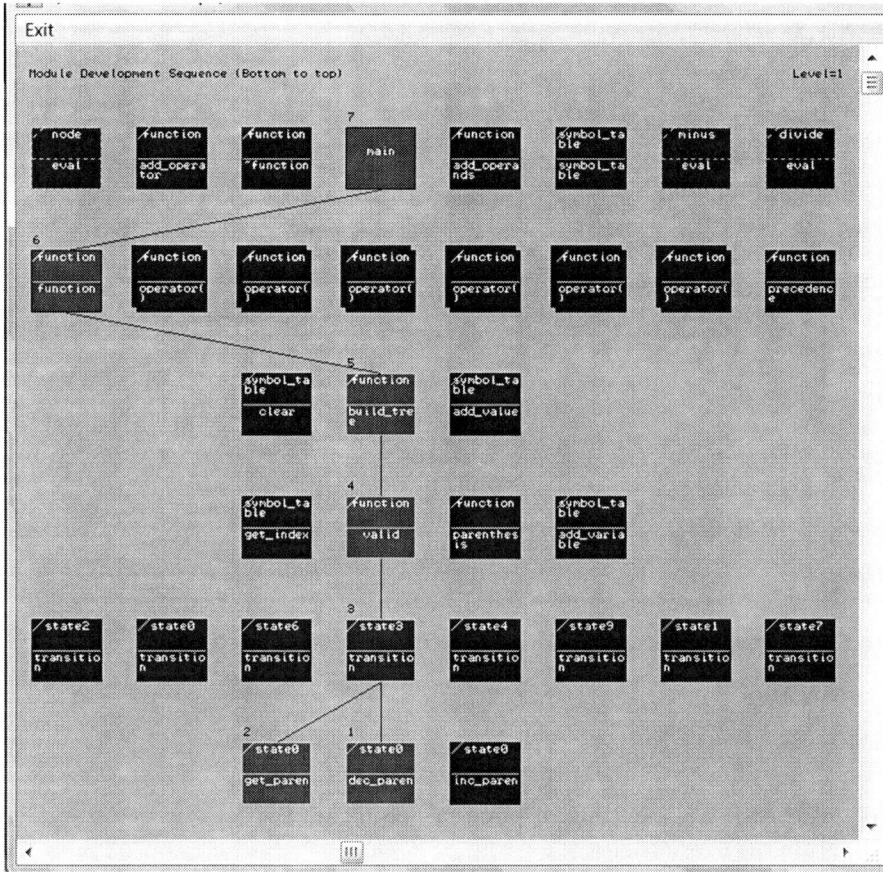

Fig. 13.7 Incremental coding ordering support for a critical path

(i) The quality of the modules coded will be high – besides unit testing in real conditions without using stubs, the quality is ensured through defect prevention and defect propagation prevention supported by dynamic testing using the Transparent-box testing method, semiautomated inspection using traceable source code and documents (the code inspection is performed efficiently with the support of various traceabilities – see Sect. 13.5), software visualization (see Sect. 13.6), and 100% MC/DC (Modified Condition/Decision Coverage) test coverage result support (see Sect. 13.4).

(j) It is supported by a coding-style-independent graphical representation technique and tools, so that the source code written by others is also easy to read and understand – see Fig. 13.8.

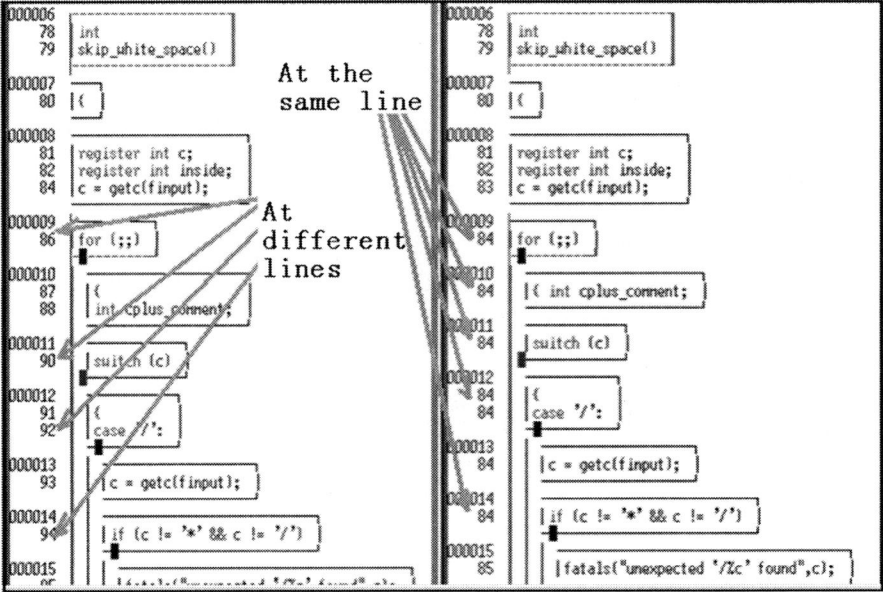

Fig. 13.8 Coding-style-independent program representation shown in J-Diagram

(k) With NSE, there is a set of powerful techniques and tools to automatically document a source program – with NSE, the source code is also the source for generating many graphical documents – see Chap. 11 and Figs. 13.1–13.8.

(l) The source code is consistent with the documents after code modification through bidirectional traceability and rebuilding the corresponding database.

(m) Programming productivity is easy to calculate with a systematic technique and tool to count the total amount and the percentages of the comment lines, partial comment lines, empty lines, and active lines of the source code of an entire software product as shown in Fig. 13.9.

(n) There are a set of systematic technologies and tools for the support of re-engineering and reverse engineering – to automatically generate huge amounts of graphical documents of the program architectures, the program logic of an entire software product, the control flow diagram of an entire software product with various traceabilities established, etc. (see Chap. 19).

(o) It supports parallel coding performed in different subsystems (see Fig. 13.10) or different critical paths (see Fig. 13.11), but the integration should still be done by **adding one component at a time** [Bro95-p149] for easily locating the possible defects. At any time, the updated whole product should be executable.

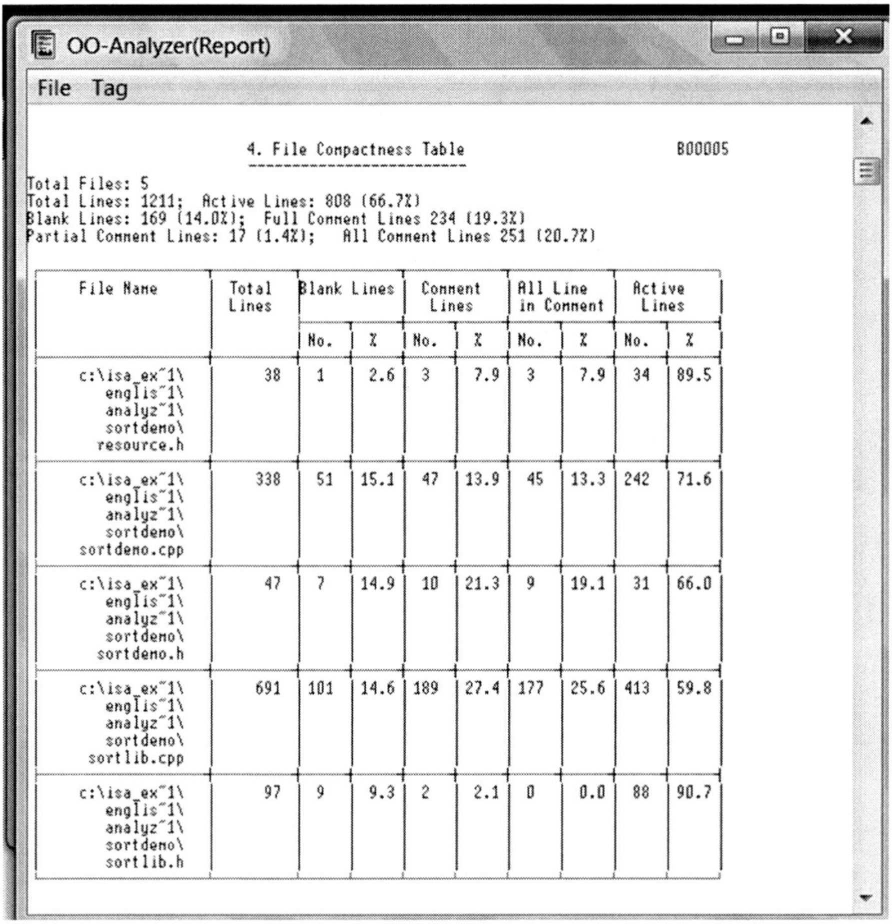
OO-Analyzer(Report)
File Tag
4. File Compactness Table
B00005
Total Files: 5
Total Lines: 1211; Active Lines: 808 (66.7%)
Blank Lines: 169 (14.0%); Full Comment Lines 234 (19.3%)
Partial Comment Lines: 17 (1.4%); All Comment Lines 251 (20.7%)

File Name	Total Lines	Blank Lines		Comment Lines		All Line in Comment		Active Lines	
		No.	%	No.	%	No.	%	No.	%
c:\isa_ex~1\englis~1\analyz~1\sortdemo\resource.h	38	1	2.6	3	7.9	3	7.9	34	89.5
c:\isa_ex~1\englis~1\analyz~1\sortdemo\sortdemo.cpp	338	51	15.1	47	13.9	45	13.3	242	71.6
c:\isa_ex~1\englis~1\analyz~1\sortdemo\sortdemo.h	47	7	14.9	10	21.3	9	19.1	31	66.0
c:\isa_ex~1\englis~1\analyz~1\sortdemo\sortlib.cpp	691	101	14.6	189	27.4	177	25.6	413	59.8
c:\isa_ex~1\englis~1\analyz~1\sortdemo\sortlib.h	97	9	9.3	2	2.1	0	0.0	88	90.7

Fig. 13.9 Productivity measurement support

13.3 Unit Testing and Integration Testing Support

With NSE-Coding, unit testing and integration testing are combined together closely using the PanoUnit toolset. The interface of PanoUnit is shown in Fig. 13.12.

Figure 13.13 shows the test case generation options.

As described above, with NSE the incremental unit testing is performed without designing and using stubs to replace other units called by the unit being tested, because according to the incremental coding and testing order, those units called by the unit being tested must have been coded and tested already. So, it is real product testing – when a stub unit is used for the traditional unit testing, it is not real product testing because the stub unit will not return the real value needed.

350   13 Coding Engineering with NSE

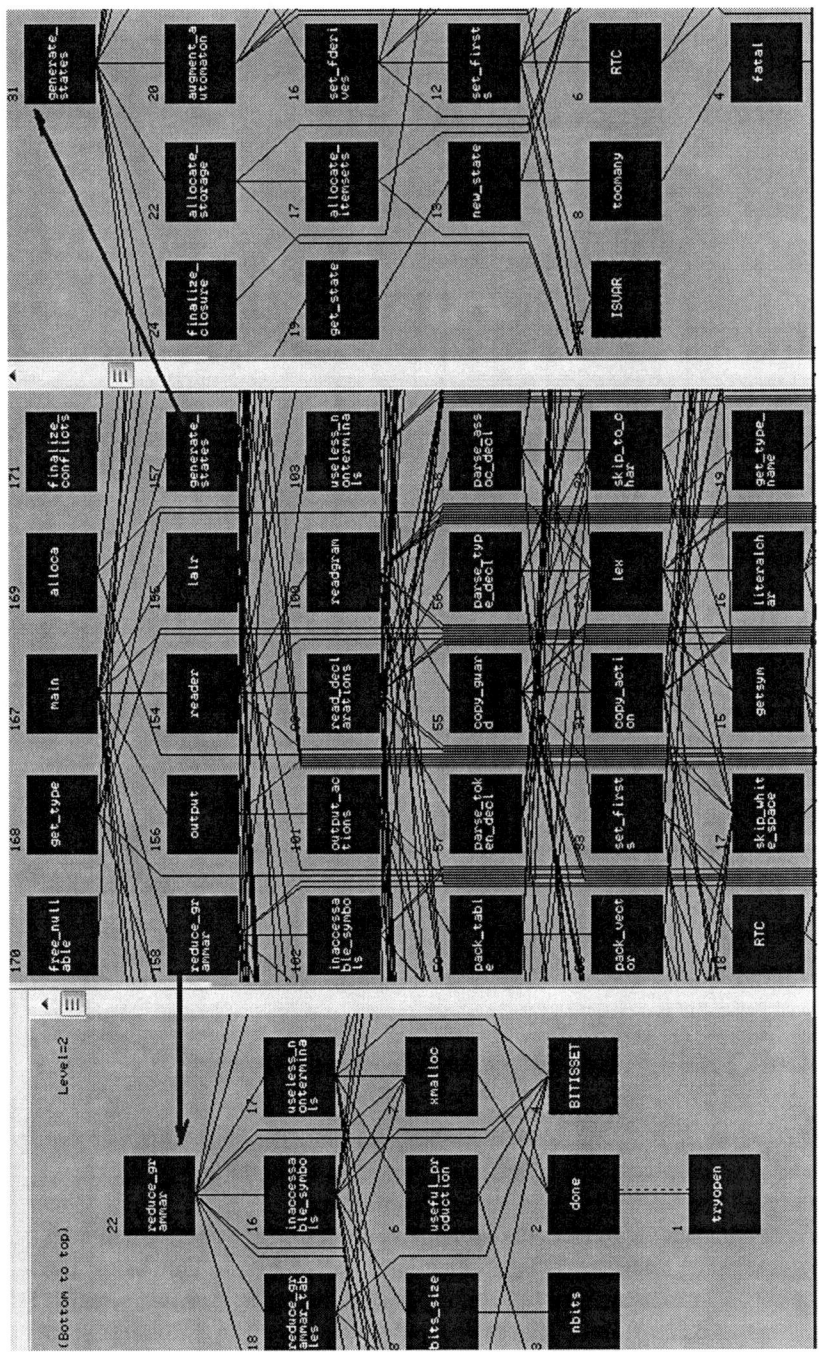

Fig. 13.10 Parallel coding for different subsystems

13.3 Unit Testing and Integration Testing Support

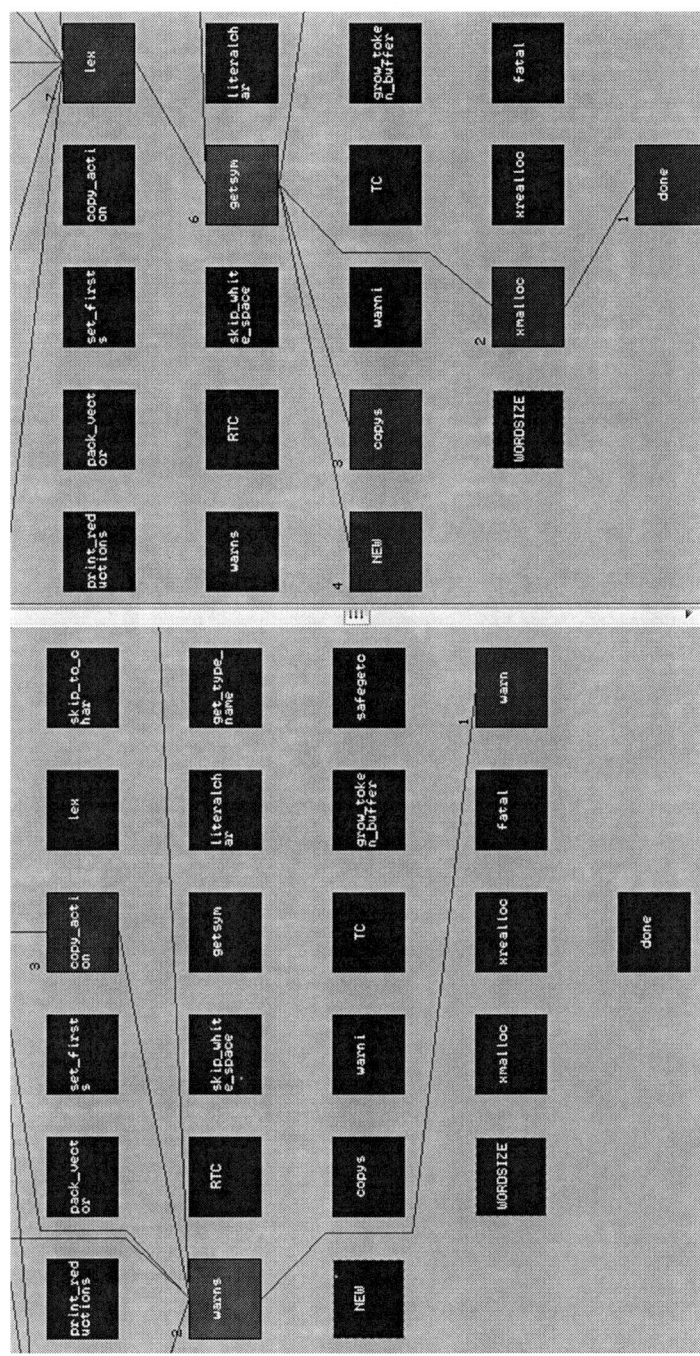

Fig. 13.11 Parallel coding for different paths

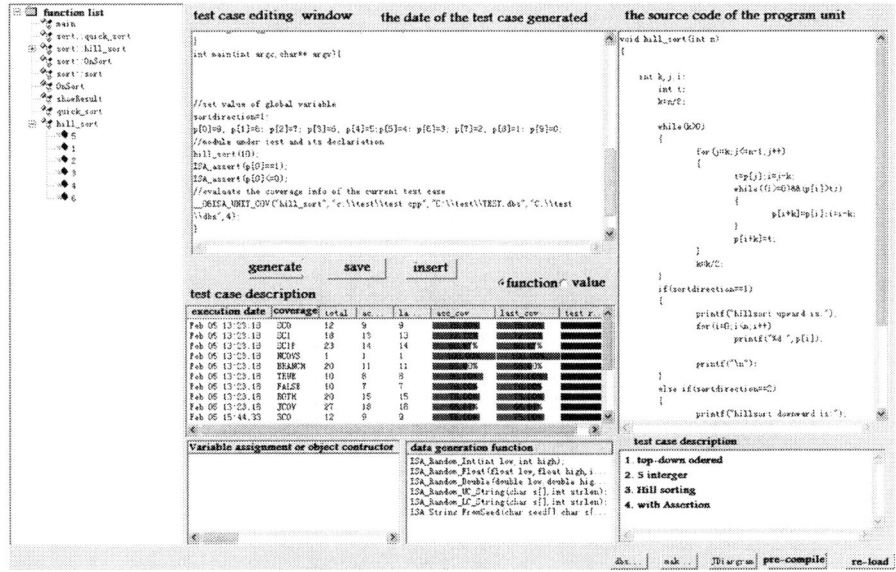

Fig. 13.12 The interface of PanoUnit for unit testing and integration testing

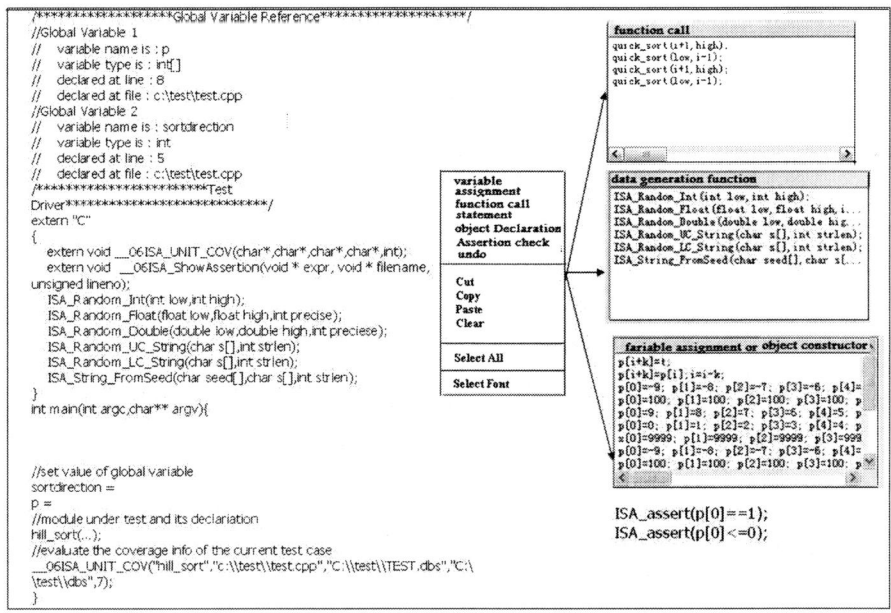

Fig. 13.13 The options for test case generation

Sometimes, we may want a called unit to return some special values for error simulation. It is supported in two automatic ways – see Appendix C.

The major features of PanoUnit include:

1. Automatically collects all the units called by the unit being tested together.
2. Semiautomatically creates the test driver to help users complete the driver design.
3. A set of data generation functions are provided for users to choose.
4. It helps users to easily insert assertions for checking the test result.
5. If it is re-testing an existing software product, PanoUnit can collect all the possible values assigned to a global variable or static variable in different locations for users to choose, can also collect the values used to meet the requirement of the constructor of a class object for users to choose.
6. It can compile the program with the driver and the unit as well as all the units called together, and execute the test cases automatically.
7. It can also perform MC/DC test coverage measurement, memory leak checking, etc.
8. It can show the test results in graphics with the untested branches and conditions highlighted.
9. It can automatically identify whether a test passed or not.

About system testing support, please read Chap. 16.

13.4 MC/DC Test Coverage Measurement Support

With NSE-Coding, it is strongly recommended to realize 100% MC/DC (Modified Condition/Decision Coverage) test coverage for any module in any commercial application and any engineering project, not only for meeting the RTCA/DO-178B level A requirements. Why?

Often people believe that statement-level test coverage is not good enough for the quality assurance of commercial software, but branch-level test coverage may meet the quality assurance requirements. Is it true?

Before answering the question, let us see some examples.

Func1 is a C program module with the source code as follows:

```
int func1 (int a, int b, int c)
{
    if(a && b && (c==1 || c==11 ||
       c==111 || c==1111 || c==11111))
            return c + c/10 + c/100 + c/1000 + c/10000;
    else
            return 0;
}
```

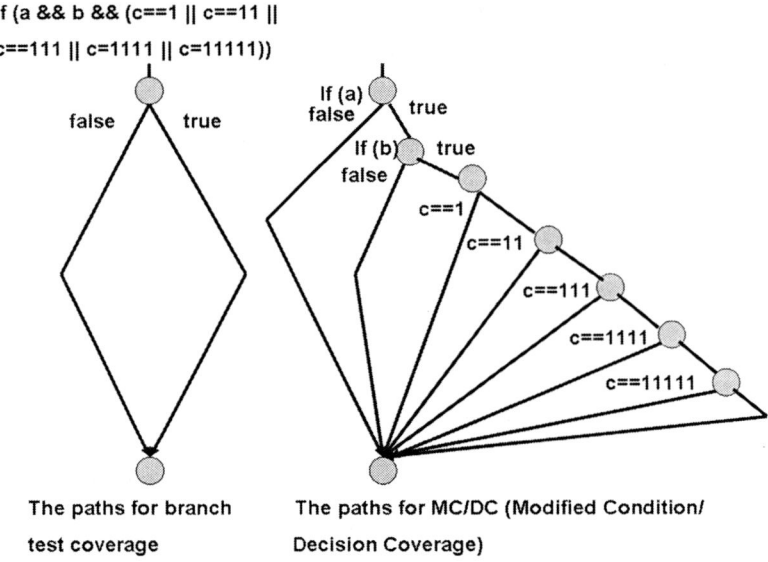

Fig. 13.14 The logic paths of the func1 program module

If we consider branch-level test coverage only, then there are two logic paths, but if we consider MC/DC test coverage, there are eight logic paths as shown in Fig. 13.14.

Func2 is another C program module with the same functionary as func1 but written in different style without using multiple conditions in a decision statement:

```
int func2 (int a, int b, int c)
{
if (a)
   {
    if (b)
      {
      switch (c)
        {
        case 1:
                return 1;
        case 11:
                return 12;
        case 111:
                return 123;
        case 1111:
                return 1234;
        case 11111:
                return 12345;
        default:
                return 0;
        }
      }
    }
   return 0;
}
```

13.4 MC/DC Test Coverage Measurement Support

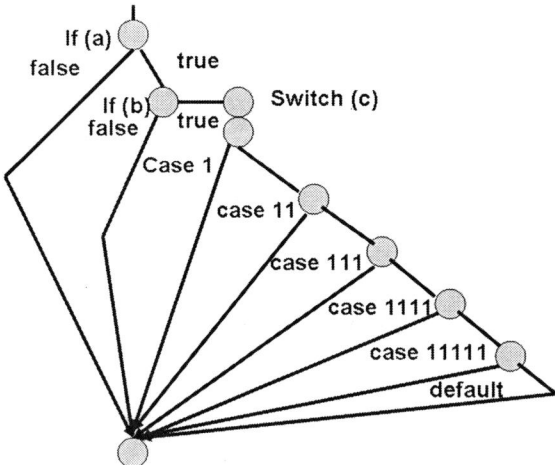

The paths for branch or MC/DC (Modified Condition/Decision Coverage)
test coverage measurement for func2 module

Fig. 13.15 The number of logic paths of func2 program module

The number of source lines of func2 is 25 while those of func1 is 8.
The number of logic paths for func2 is eight as shown in Fig. 13.15.
The source code of a corresponding main() program used to test func1 module and func2 modules is listed as follows:

```
void main(int argc, char** argv)
{
printf(" c == %d\n",
  func1(atoi(argv[1]),atoi(argv[2]),atoi(argv[3])));
printf(" c == %d\n",
  func2(atoi(argv[1]),atoi(argv[2]),atoi(argv[3])));
}
```

A simple "Makefile" for running this program is listed as follows:

```
# Makefile

LINK = link
CC = cl

main.exe:    main.C
      $(CC) -c main.C
      $(LINK) -out:main.exe -subsystem:console main.obj libc.lib kernel32.lib

clean:
      -erase main.exe
      -erase main.obj
```

After compilation, it is easy to verify that func1 and func2 have the same functionary – the execution command lines and the obtained results are as follows:

C:\Analyzer1>main 1 1 1
c==1
c==1

C:\Analyzer1>main 1 1 11
c==12
c==12

C:\Analyzer1>main 1 1 111
c==123
c==123

C:\Analyzer1>main 1 1 1111
c==1234
c==1234

C:\Analyzer1>main 1 1 1111
c==12345
c==12345

C:\Analyzer1>main 0 1 1
c==0
c==0

C:\Analyzer1>main 1 0 1
c==0
c==0

To achieve 100% branch-level test coverage result for func1 module, only two test cases are needed:

C:\Analyzer1>main 0 1 1
c==0
c==0

C:\Analyzer1>main 1 1 1
c==1
c==1

The corresponding branch-level test coverage result for func1 is 100% tested as shown in Fig. 13.16.

But if we consider the MC/DC test coverage result, we will find that there are many conditions (and six paths) untested as shown in Figs. 13.17 and 13.18.

13.4 MC/DC Test Coverage Measurement Support

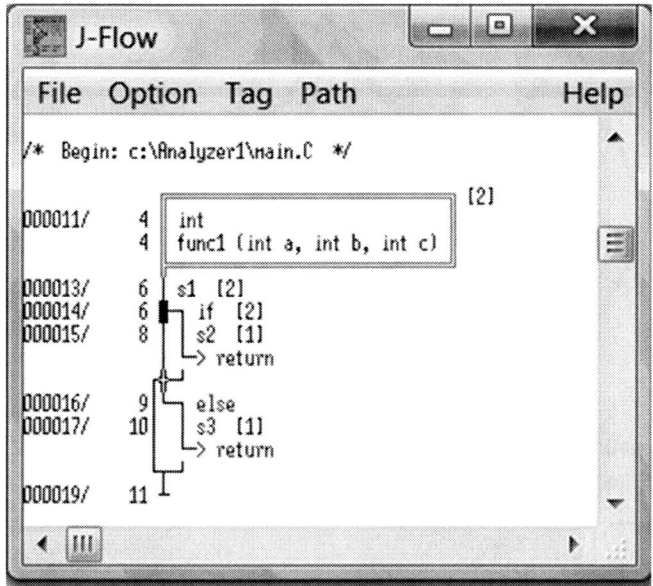

Fig. 13.16 The branch-level test coverage measurement result for func1

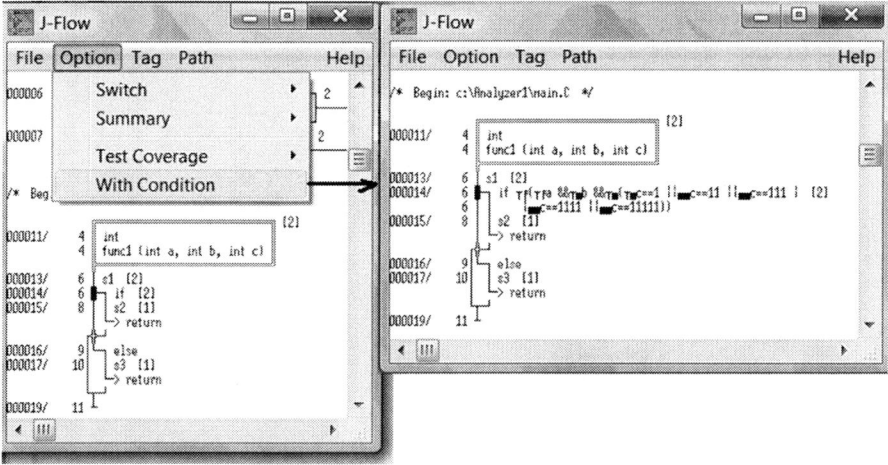

Fig. 13.17 The MC/DC test coverage measurement result (untested branches/segments/conditions are highlighted in *small black box*)

This result is also shown clearly in Fig. 13.18.

The corresponding branch test coverage measurement result of func2 is shown in Fig. 13.19.

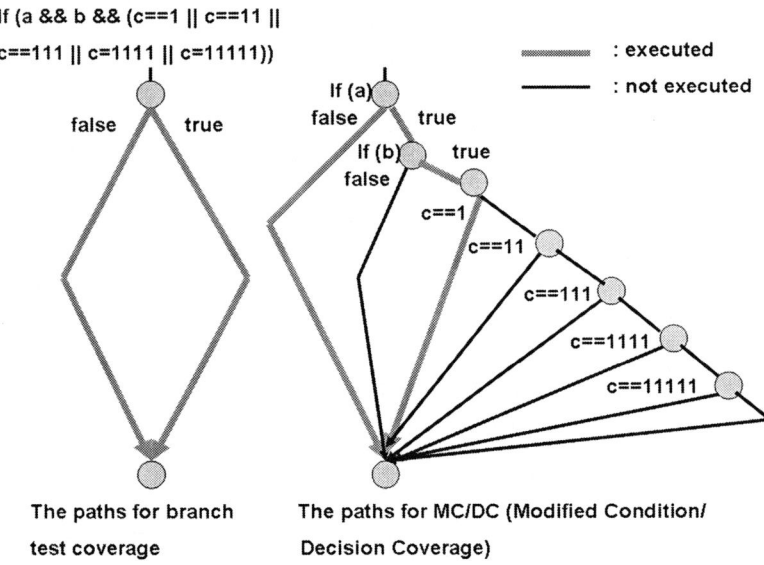

Fig. 13.18 The corresponding logic paths executed for func1

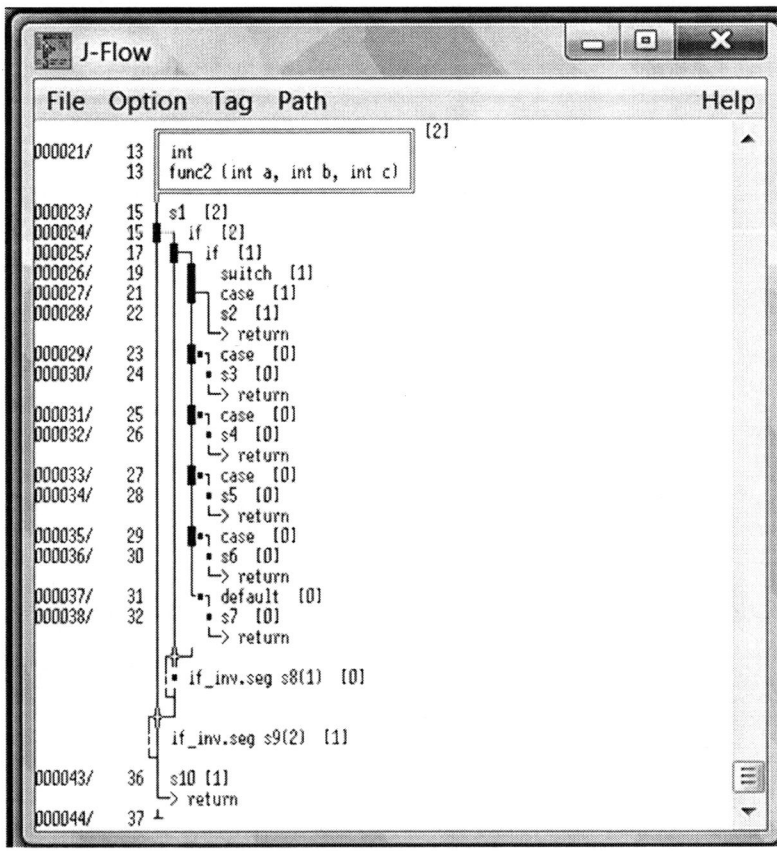

Fig. 13.19 The branch-level test coverage measurement result for func2

13.4 MC/DC Test Coverage Measurement Support

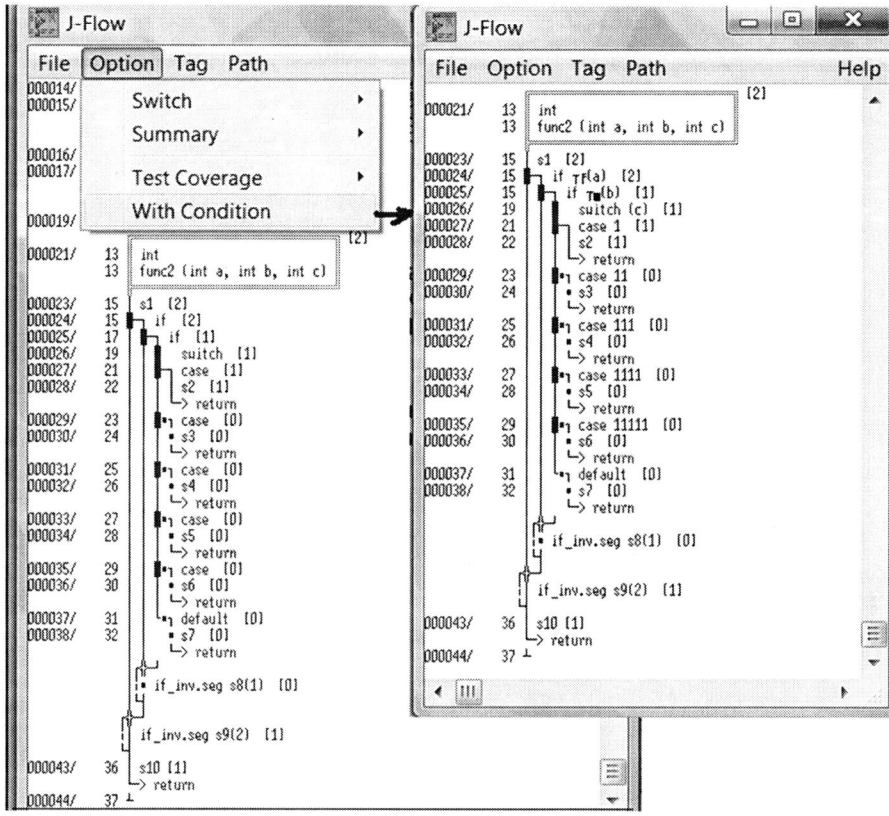

Fig. 13.20 The untested paths for func2 in branch-level test coverage measurement and MC/DC test coverage measurement are the same (six paths are untested)

The corresponding MC/DC test coverage measurement result for func2 is shown in Fig. 13.20 – the untested paths are the same as that for branch-level test coverage measurement.

This result is also represented clearly in Fig. 13.21.

For getting 100% MC/DC test coverage result for func1, at least six more test cases are needed as shown in the following list:

main 1 0 1
main 1 1 0
main 1 1 11
main 1 1 111
main 1 1 1111
main 1 1 11111

After running those test cases, the MC/DC test coverage result for func1 and func2 are shown in Figs. 13.22 and 13.23, respectively.

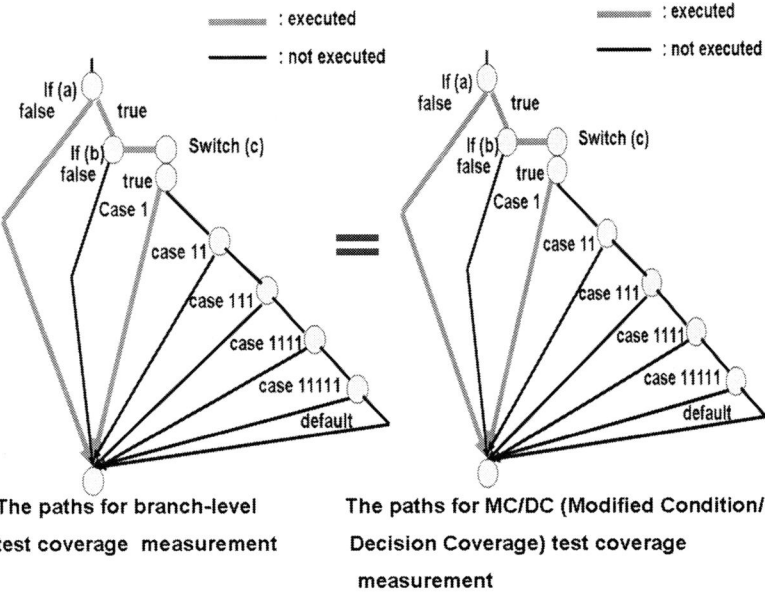

Fig. 13.21 The corresponding paths executed for func2

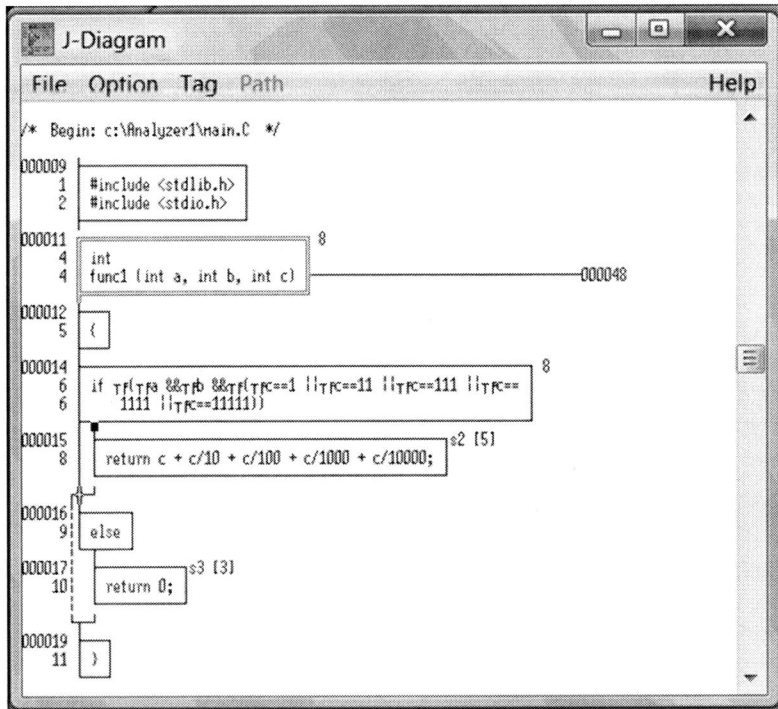

Fig. 13.22 All paths in func1 have been tested

13.4 MC/DC Test Coverage Measurement Support

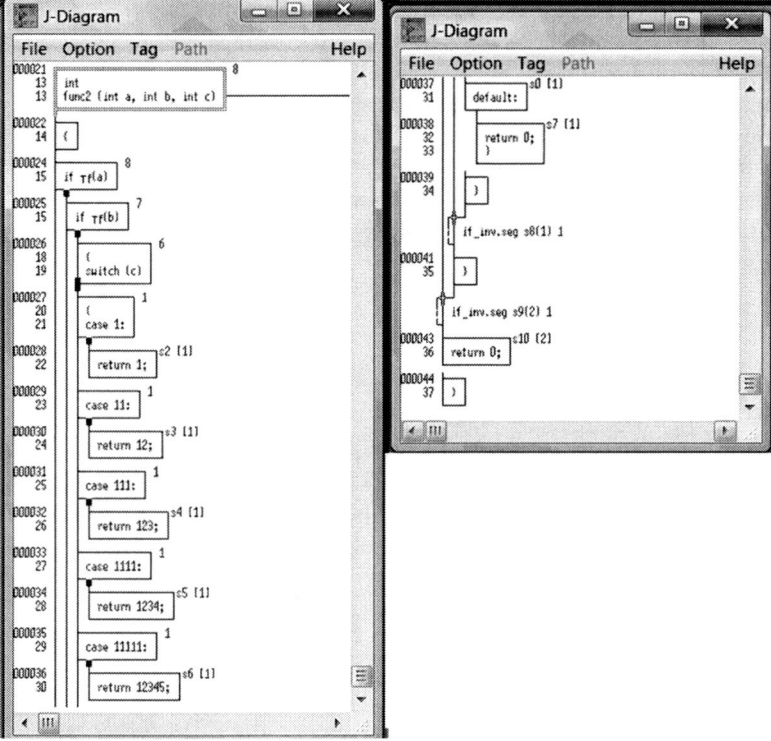

Fig. 13.23 All paths in func2 have been tested

How about statement-level test coverage measurement? It is much worse than the branch-level test coverage measurement, so that it should not be used for any commercial software testing at all – see Chap. 16.

With NSE and the support platform, Panorama++/Silver Bullet, it is not difficult and not expensive to achieve 100% MC/DC test coverage measurement result in unit testing process – see a real application example shown in Appendix B.

13.4.1 Conclusion

(a) A program module written with multiple conditions in a decision statement is much easier to read and understand than without multiple conditions written in a decision statement.
(b) The size of a program module written without multiple conditions in a decision statement is much bigger than that of a program module written with multiple conditions in a decision statement (in this example, the size of func2 is about three times bigger than that of func1).

(c) As shown in this example, to a program written with multiple conditions in a decision statement, "100% branch-level test coverage" result may be equal to only 20% of the MC/DC test coverage result – it will not be accepted for any commercial or engineering software product: the risk is too high! – in many cases, the execution part of a decision statement with multiple conditions will be much more complicated and dependent on the conditions, so that if the untested paths in the product development site are executed in the customer site in the real applications, something unexpected may happen to harm the customer's business.

13.5 Semiautomated Inspection Support

Inspection has been proven a useful technique for finding defects. But traditional inspection is performed using separated documents and source code without automated and self-maintainable traceability. Alternatively, with NSE, software inspection can be done in a semiautomated way supported by various traceabilities – see Figs. 13.24–13.26.

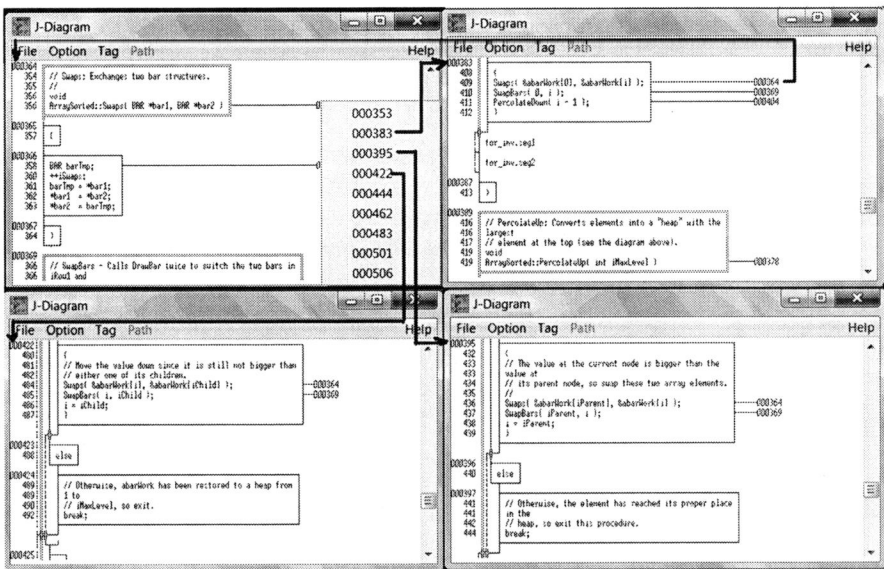

Fig. 13.24 Internal traceability within the source code

13.5 Semiautomated Inspection Support

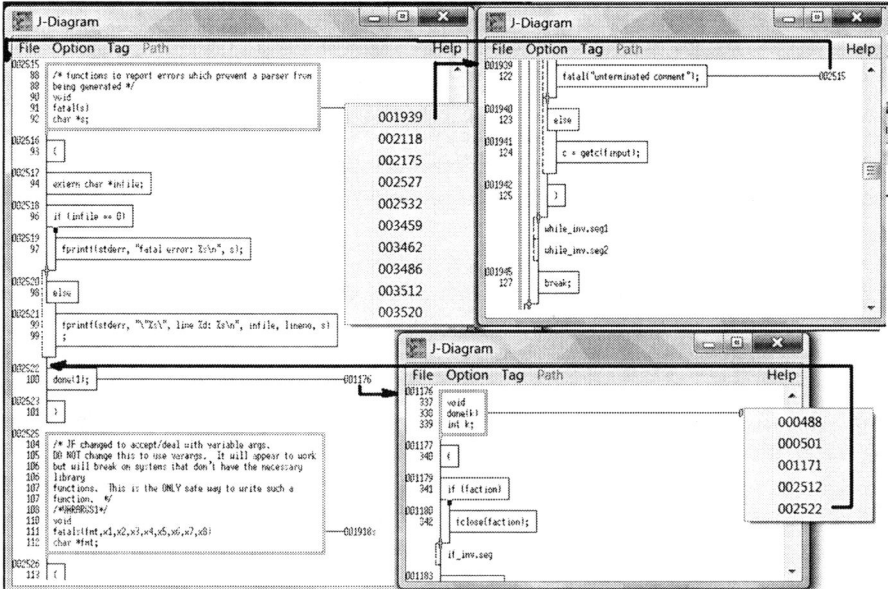

Fig. 13.25 Various traceabilities for supporting code inspection

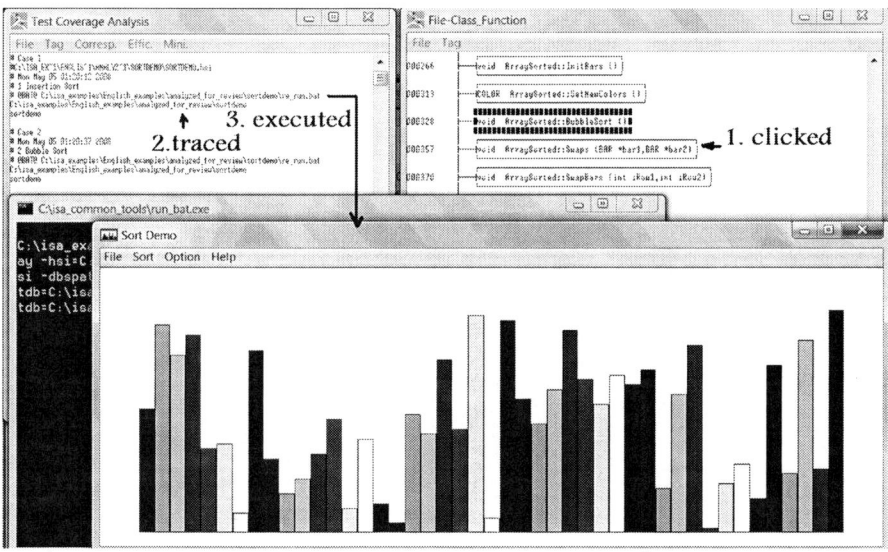

Fig. 13.26 An example of external traceability and dynamic traceability among source code and test cases and the related document or program execution

13.6 Defect Prevention Driven Quality Assurance in Programming

As described before, with NSE the quality of a software product being developed is ensured with defect prevention and defect propagation prevention through software testing dynamically using the Transparent-box testing method, inspection, and software visualization.

How to prevent defects and defect propagation through dynamic testing using the Transparent-box method has been introduced with many examples in Chaps. 11 and 12; here, let us discuss how to prevent defects and defect propagation with software visualization.

Figure 13.27 shows defect prevention through incremental ordering and software visualization: when writing a function call statement for a module being coded (in this example, the module with order number 6) to call a module coded (in this example, the module with order number 4), we can see the diagrammed source code of the called module in a new window to know how many parameters are needed and their order to avoid inconsistent defects in writing the calling statement.

In the following sample program, there is a logic defect that is not easy to find because a program is represented in textual format, and a program with logic defects may execute normally without providing error messages but the result could be wrong:

```
#include <stdlib.h>
#include <stdio.h>
int func1 (char *s, int m, int n)
{int value;
switch (s[0])
  {
  case '+':
  value = m + n;
  case '*':
  value = m * n;
  break;
  case '-':
  value = m - n;
  break;
  case '/':
  value = m / n;
  break;
  default:
  value = -1;
  }
return value;
}
```

13.6 Defect Prevention Driven Quality Assurance in Programming

```
void main(int argc, char** argv)
{
printf(" The value == %d\n",
func1(argv[1],atoi(argv[2]),atoi(arg
v[3])));
}
```

Some sample test cases may not be able to find the errors such as the following test cases and the results:

```
C:\tem_dir>main+0 0
The value==0

C:\tem_dir>main+2 2
The value==4
```

But through program visualization, the logic error is much easier to find as shown in Fig. 13.28.

Of course, to this very simple program, we can use more test cases to find it such as the following test cases:

```
C:\tem_dir>main+2 3
The value==6

C:\tem_dir>main*2 3
The value==6
```

But in the real application programs, the execution part for each "case" statement may be complicated, and it is hard to find the defect that a "break" statement is missing.

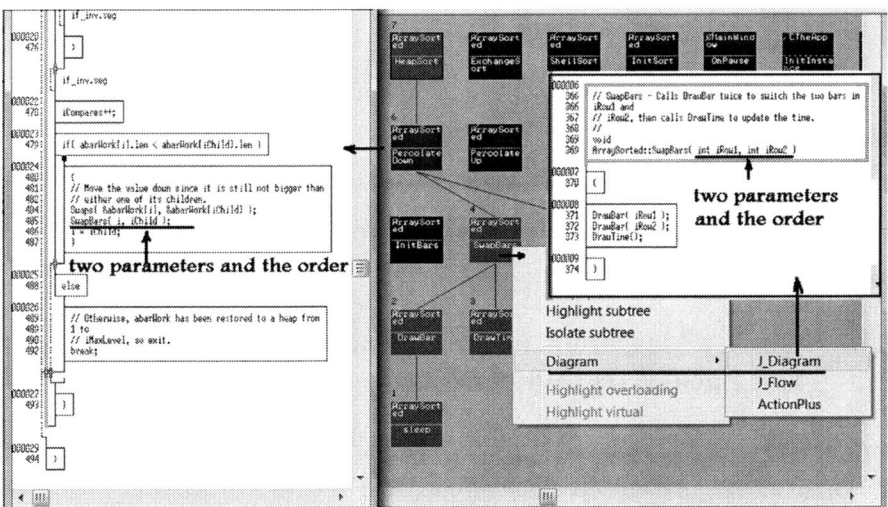

Fig. 13.27 Defect prevention through incremental ordering and visualization

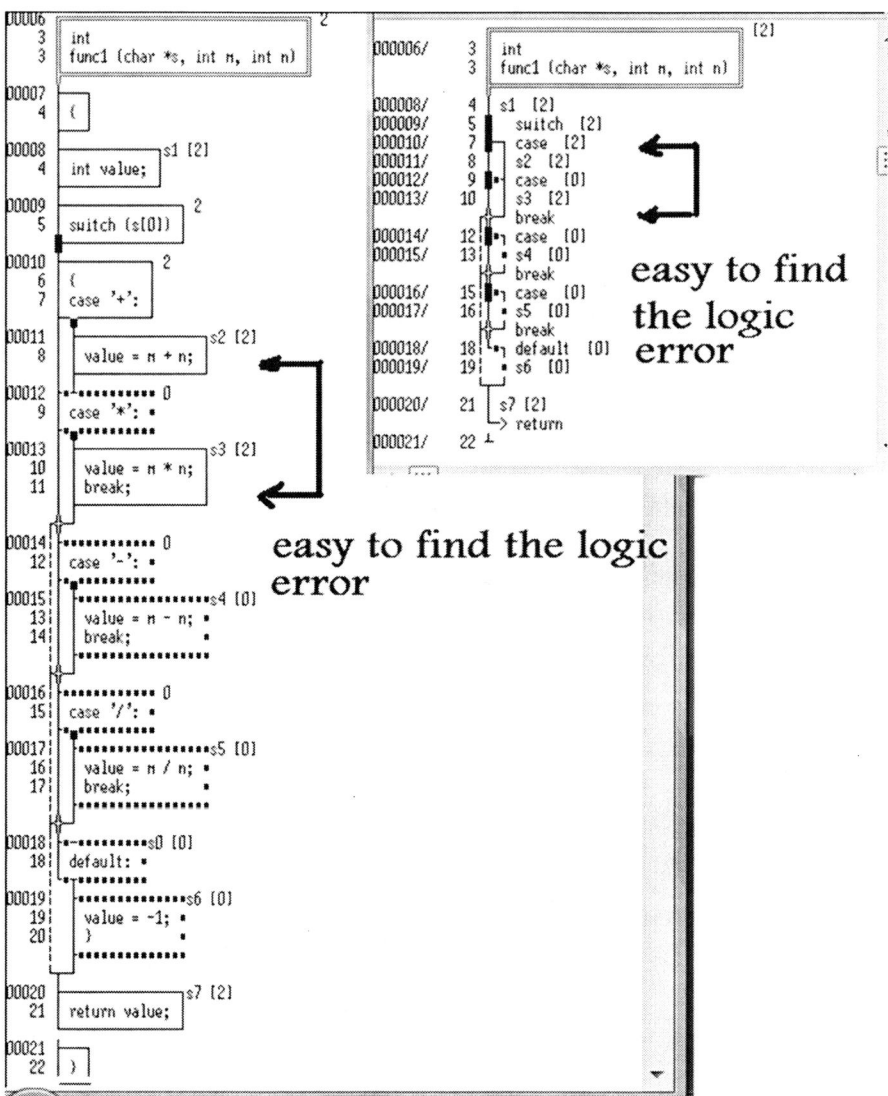

Fig. 13.28 Finding logic defects through software visualization

13.7 Quality Measurement for an Entire Software Product and Each of Its Components

With NSE-Coding, not only the quality of an entire software product will be measured, but also any individual modules (units) will be measured and shown in Kiviat diagram – see Fig. 13.29, an application example.

13.8 Application

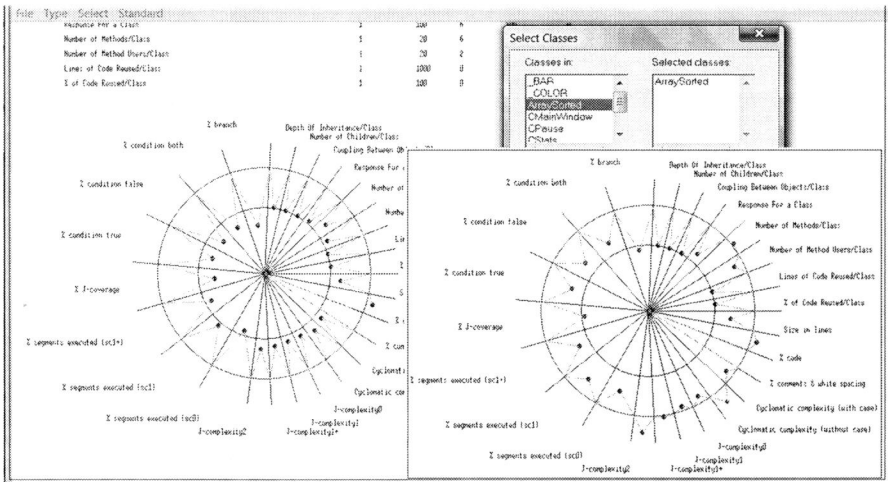

Fig. 13.29 Quality measurement example (the result out of the big circle and inside the small circle means not satisfying the required quality standard)

For measuring the quality of classes, some special metrics are used, including

- Lines of code per class (LOC)
- Number of methods per class (NOM)
- Number of method users per class (NMU)
- Weighted methods per class (WMC) in multiple complexity metrics
- Depth of inheritance tree (DIT)
- Number of children per class (NOC)
- Coupling between objects (CBO)
- Response for a class (RFC)
- Lines of code reused per class (LCR)
- Ratio of code reused per class (RCR)
- Test coverage per class (TCC) in multiple test coverage metrics

Users can set the standard value through the OO-SQA toolkit of Panorama++.

13.8 Application

NSE-Coding using the **Synthesis Design and Incremental Integration** Technique has been successfully applied in practices, including the improvement of the NSE support platform, Panorama++. All screenshots shown in this chapter are from the application examples of NSE-Coding.

13.9 The Major Features

The major features of NSE-Coding are briefly summarized as follows:

(a) Holistic – it is based on **Generative Holism** principle.
(b) Incremental – coding is done incrementally.
(c) Parallel – parallel coding is supported to avoid waiting for something.
(d) Visual – the coding process and the work products (the source code and the documents) are visible.
(e) Traceable – the work products are traceable internally and traceable to the test cases and the documents.
(f) Consistent – the design documents and the source code are consistent.
(g) Combined with software design together closely.
(h) Driven by defect prevention and defect propagation prevention.
(i) The source code is always entirely executable.

13.10 Summary

This chapter introduced NSE-Coding which applies the innovated **Synthesis Design and Incremental Integration** technique. With NSE, coding is performed holistically and incrementally by complying with the NSE nonlinear process model and the NSE software development methodology based on complexity science. The coding process and the work products (source code and the related documents) are visible. The quality of the coded programs is ensured through defect prevention and defect propagation prevention performed with dynamic testing using the Transparent-box testing method and software visualization plus inspection using traceable documents and traceable source code.

With NSE, product design becomes precoding and the coding becomes further design – **source code is the source for generating most graphical software documents – to keep the documents consistent with the program, and traceable to and from the source code.**

13.11 Points and Questions to Ponder

(a) What are the major problems existing with today's software programming?
(b) Why, with NSE, does software design become precoding and coding become further design?
(c) Why should all commercial software products satisfy 100% MC/DC test coverage?

13.12 Further Reading and Information Source

(a) Pressman RS (2005) Software engineering: a practitioner's approach. McGraw-Hill, New York
(b) Sources for coding standards:
http://drupal.org/coding-standards
http://drupal.org/node/302199
http://www.amazon.com/Coding-Standards-Rules-Guidelines-Practices/dp/0321113586

References

[Bec99]   Bech K (1999) Extreme programming explained: embrace change. Addison-Wesley, Boston
[Bro95-p149] Brooks FP Jr (1995) The mythical man-month. Addison-Wesley, Reading, p 149
[Coa93]   Coad P, Nicola J (1993) Object-oriented programming. Prentice Hall, Englewood Cliffs
[RTCA92]   RTCA/DO-178B (1992) Software considerations in airborne systems and equipment certification. RTCA, Washington, DC

Chapter 14
The Basis of Software Testing

> *Practical wisdom is only to be learned in the school of experience.*
>
> Samuel Smiles (1812–1904)

This chapter introduces the basic concepts and knowledge of software testing, including the purpose of software testing, the basic test methods and technology, and their characteristics. The complexity and size of today's software makes writing bug-free code extremely difficult, even for highly experienced programmers [Pat00], so that a software should be tested.

14.1 The Purpose of Software Testing

About software testing purposes, although there are different perspectives, in general it includes the following points:

1. **Validating whether it is the right software product** – meeting the applicable standards and customer needs, whether it works as expected.
   After a software product is built, it should be validated whether the product meets customer needs, particularly the functionary, through a large number of test cases to test in order to make accurate judgments. For example, if it is a sorting program to be used to sort many different types of data, we should use a variety of test cases to check whether the product is working properly, including the input of a group of integers, real number, strings, etc.
2. **Verifying whether the product is right** – matching the requirement specification.
   Besides the functionary, a software product should be developed as specified in the requirement specification, such as meeting the internal product development standard, and other requirements not directly related to customer's needs, such as the coupling-degree of the program modules.
3. **Finding defects/bugs** introduced into the software product to help the product development team improve the quality of the product.

A software product not only should meet customer's needs and the requirement specification, but also should be stable and reliable, so that we must test the product through a variety of testing methods and test cases, to find the defects/bugs as much as possible. For this purpose, we should not only use legitimate data, but also illegal input data to check an error handling capability.

For instance, to the sorting software, in order to identify possible errors, for any data type (such as integers), we should at least design a series of test cases to be used to test the program with

(a) No input data.
(b) Only one input data.
(c) Two input data, in the order from small to big and from big to small.
(d) Three input data in the order of (1) large, medium, small; (2) large, small, medium; (3) medium, large, small; (4) medium, small, large; (5) small, large, medium; and (6) small, medium, large.
(e) Groups of randomly selected input data, more or less.
(f) In the case that the software is designed with maximum processing number MAX (if the software is used to sort the name of the students of a school up to 1,000 students, the program may set the maximum value for treatment MAX = 1,000), then enter the three sets of data, the number was MAX − 1, MAX, and MAX + 1.
(g) Enter the number of illegal data, such as punctuation and negative numbers.

After running these test cases, some possible errors may be found. But some other types of defects/bugs need different test methods and tools.

Software defect/bug types

(a) From the nature of the defects/bugs
 The defects/bugs can be divided into function, structure, performance, reliability, service, user support (user manuals and product brochures), and other properties of the errors.
(b) From the process phases where the defects/bugs introduced
 The defects/bugs can be divided into requirements, design, coding, integration, interoperability, system, operation, and version control errors.
 Therefore, to address the nature and classification of various errors, we should design many different test cases to find them.

4. **Collecting required data/information for helping the development team improve the product**
 Consider the sorting programs – some may run very fast and some may run very slowly. A program may run fast at the beginning, but later on it runs very slowly. Why? The testing may find that there are memory leaks to reduce the amount of available memory, so that the system must swap something often between the memory and the hard disk.
5. **Providing test documents as an important part of the entire product documentation**

14.3 Structural Testing and the White-Box Method

The test documentation includes:

(a) The testing requirements – can be partially provided by the customer.
(b) The testing can be divided into unit test plans, integration test plans, and system test plans, including test purpose, test content, the required resources (software and hardware equipment), manpower allocation, and so on.
(c) The test scripts and test cases – better to have a test case manager (tool) to perform test case efficiency analysis and test case minimization to get the minimum equivalent set of test cases for efficient regression testing after code modification.
(d) The online overall test coverage analysis report: including project level, file level, and block level, with the module call graph (Call Graph), bar (Bar Chart), and the statements given.
(e) The online detailed test coverage analysis report: the logic diagram and the control flow diagram shown with the untested branch and conditions highlighted.
(f) That software errors have been resolved and recorded.
(g) Already known but unsolved software error list, a description of the impact level, and how to "bypass" the errors, so the software can continue to work.
(h) And the summary.

14.2 Functional Testing and the Black-Box Method

Functional testing is to test the functionary of a software product according to the customer's needs. It verifies that the system behaves correctly from the user's/business's perspective and functions according to the requirements, models, storyboards, or any other design paradigm used to specify the application. Functional testing is performed using the Black-box testing method (proposed by Myers [Mye79]) which handles the software as a "black box," no matter how the internal structure is and what algorithm is used. The Test Designer/Tester designs and implements the test cases to validate that the product performs in accordance to the requirements. To each test case, the tester needs to check whether the output is the same as what is expected.

14.3 Structural Testing and the White-Box Method

Structural testing uses an internal perspective of the product to design test cases based on internal structure. It requires programming skills to identify all logic paths through the software. The tester chooses test case inputs to exercise paths through the code to determine whether there is something wrong. For instance, there may be enough memory to be used in the development site, but in the customer site, it is not always guaranteed – when other programs are running, there may not be enough memory to be assigned to the corresponding program. In this case, if the code branch used to handle the event of running out of memory has never been tested in the product development process, then something unexpected may happen in the customer site. In this case, there is a need to use error simulation techniques to complete the test.

It is clear that testing all possible paths is almost impossible, particularly in the case that there are many loop statements in the program, so we mainly consider the logic paths rather the absolute paths.

Often structural testing is performed in the unit testing stage.

14.3.1 Test Coverage Metrics

There are three major metrics used in most software development organization: the **statement test coverage, branch (segment) test coverage, and MC/DC (Modified Condition/Decision Coverage) test coverage.**

Considering the fact that software disasters happens often, with NSE, it is strongly recommended for any commercial software product to realize 100% MC/DC test coverage result for all program modules. With NSE and the support platform, Panorama++, it is not difficult to do so (see Appendix B).

High percentage statement test coverage result may mislead developers and customers into thinking that there should be no problem with the program structure. An application example shows that there are seven defects/bugs, after reaching 100% statement test coverage analysis result, none of the defects/bugs are found – see Sect. 16.4.

How about the branch test coverage metric? As shown in Sect. 13.4, 100% branch test coverage result may be equal to only 20% of the MC/DC test coverage result – about 80% of the logic paths are untested!

14.3.2 Instrumentation Methods

Mainly there are two different instrumentation methods:

1. Instrumentation is performed into the object code of the software product.

    (a) The main advantage

    - It does not require source code, only to have the object code (target program).
    - It can be compiled in a very short time.
    - It is easier to handle a variety of computer software written in many different programming languages.

    (b) The main drawback

    - The program execution is much slower (based on my own tests, it requires about five times longer to run, compared with the instrumentation method performed into the source code).

- In the absence of source code, it can only tell how much the test coverage is, but cannot show the corresponding locations where the untested elements are, so that it cannot help users to effectively improve the test coverage result.
- It cannot provide the test coverage results of classes and inline functions for C++ programs, because the header files are expanded and compiled with the program.
- It cannot update the test coverage database incrementally, each time when one module or even only one statement is modified, the test coverage data obtained before will be cleaned up.

2. Instrumentation is performed into the source code of the software product.

   (a) The main advantage

   - The additional overhead is small in program execution, usually only about 20% (but if the original program is very complicated, the overhead may increase).
   - It not only can provide the test coverage results of the entire program and each program module, but can also indicate where the branches are not tested with the line numbers. With NSE, the support platform can further highlight untested branches and conditions in small black boxes on the generated logic diagram or control flow diagram.
   - With NSE, it can provide the test coverage results of classes (a class cannot directly execute, so that the test coverage analysis result of a class is collected from its instances) and inline functions as shown in Fig. 14.1.
   - It can easily update the test coverage database incrementally if only a few source files of a product are modified.

   (b) The main disadvantages

   - Need to have the source code.
   - The compilling process takes a longer time.
   - For dealing with programs written in different computer languages, it requires different tools.

14.4 Gray-Box Testing

There are two different descriptions about Gray-box testing:

(a) Gray-box testing = Black-box testing + White-box testing.
(b) Gray-box testing is a testing method that tests a software while already having some knowledge of its underlying code or logic. It implies more understanding of the internals of the program than Black-box testing, but less than White-box testing.

In Chap. 16, a new software testing method (Transparent-box) truly combining functional testing and structural testing together with internal connection is introduced.

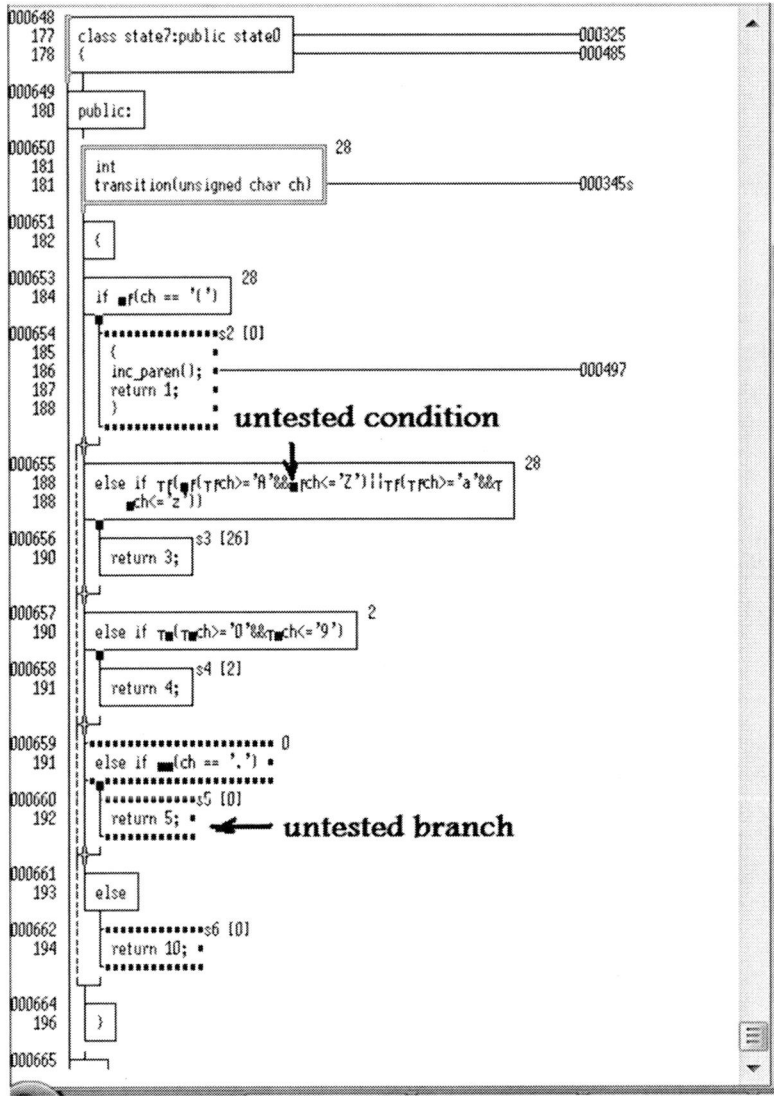

Fig. 14.1 Sample class test coverage measurement result provided by Panorama++

14.5 Performance Testing and the Testing Method

Software performance testing is used to determine the speed or effectiveness of a software program, how many percent of the total execution time is spent in each program unit, and where is the performance bottleneck. An application example of performance testing is shown in Fig. 14.2.

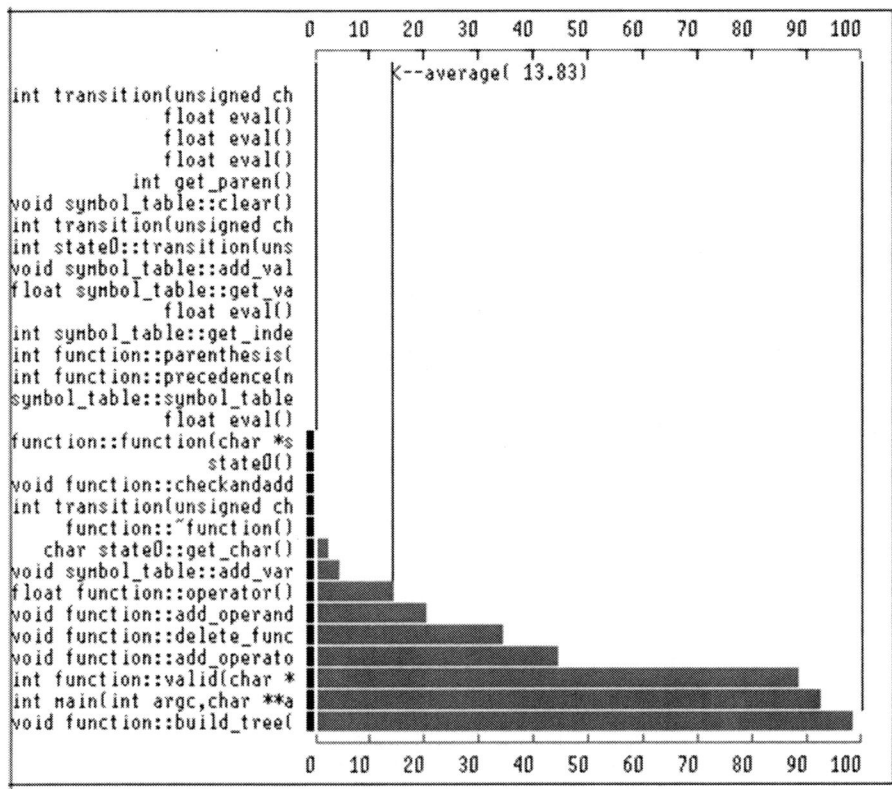

Fig. 14.2 A performance measurement report provided by Panorama++

14.6 Other Nonfunctional Testing

Other nonfunctional testing includes

1. The load testing – simulating multiple users to run the software under testing to check its processing and response capabilities.
2. The stress testing – to simulate a heavy load of work to view the application at its peak. The idea of stress testing is to stress a system to the breaking point in order to find bugs that will make that break potentially harmful.
3. The reliability testing – reliability refers to the consistency of the measurement. A test is considered reliable if we get the same result repeatedly. It also checks for how long the product can work correctly.
4. Compatibility testing – is the testing conducted on the application to evaluate the application's compatibility with the computing environment such as the computing capacity of the hardware platform the bandwidth handling capacity of networking hardware, the peripherals, the operating systems, the database, and other system software (Web server, networking/messaging tool, etc.)

14.7 Unit Testing, Integration Testing, and System Testing

Software systems are hierarchically structured, usually designed from top to bottom. The smallest unit is called a module. It goes without saying that the entire software system functionality, performance, and quality are highly dependent on the characteristics of each unit. Therefore, each unit must be first tested. Integration testing is performed for a related group or subsystem. Finally, when the entire product is ready, a system-level testing should be performed to see whether the product meets the customer's needs and whether the product works as expected. Usually, unit testing focuses on structural testing, whereas system testing focuses on functional testing.

It is recommended to test software units incrementally, and combine unit testing and integration testing together seamlessly without designing and using stubs. The NSE support platform can help users perform unit testing in this way.

14.8 Regression Test After Code Modification

Regression testing makes sure that the previous functionality still works after code modifications or new functionality is added – the intent of regression testing is to provide a general assurance that no additional errors were introduced in the process of fixing other problems or implementing a new/changed requirement.

With NSE, regression testing is performed using the minimized set of test cases selected through test case efficiency analysis and test case minimization. If only a few braches or modules are modified, regression testing can be performed with the corresponding test cases only, chosen or directly executed through backward traceability.

14.9 Object-Oriented Software Testing

Object-Oriented software testing and Process-Oriented software testing are basically the same, but the test plans and strategies need to be appropriately changed. First, the basic unit of object-oriented software is a class (Class) with member functions and independent functions. (For the Java programming language, there are only classes.) Class can be inherited. Thus, for test planning, we need to analyze not only the Cyclomatic complexity of the classes, but also the complexity of the parent classes in order to more accurately complete the testing plan. Figure 14.3 shows the complexity analysis result of each class itself, and the class with its parent class provided by NSE support platform, Panorama++ products.

The most difficult part in Object-Oriented software testing is the class which is the most important unit in, for instance, C++ program testing, and cannot execute directly, so that the test coverage results must be obtained through the instance objects of a class, then the test coverage results must be mapped to the original class source code.

14.9 Object-Oriented Software Testing

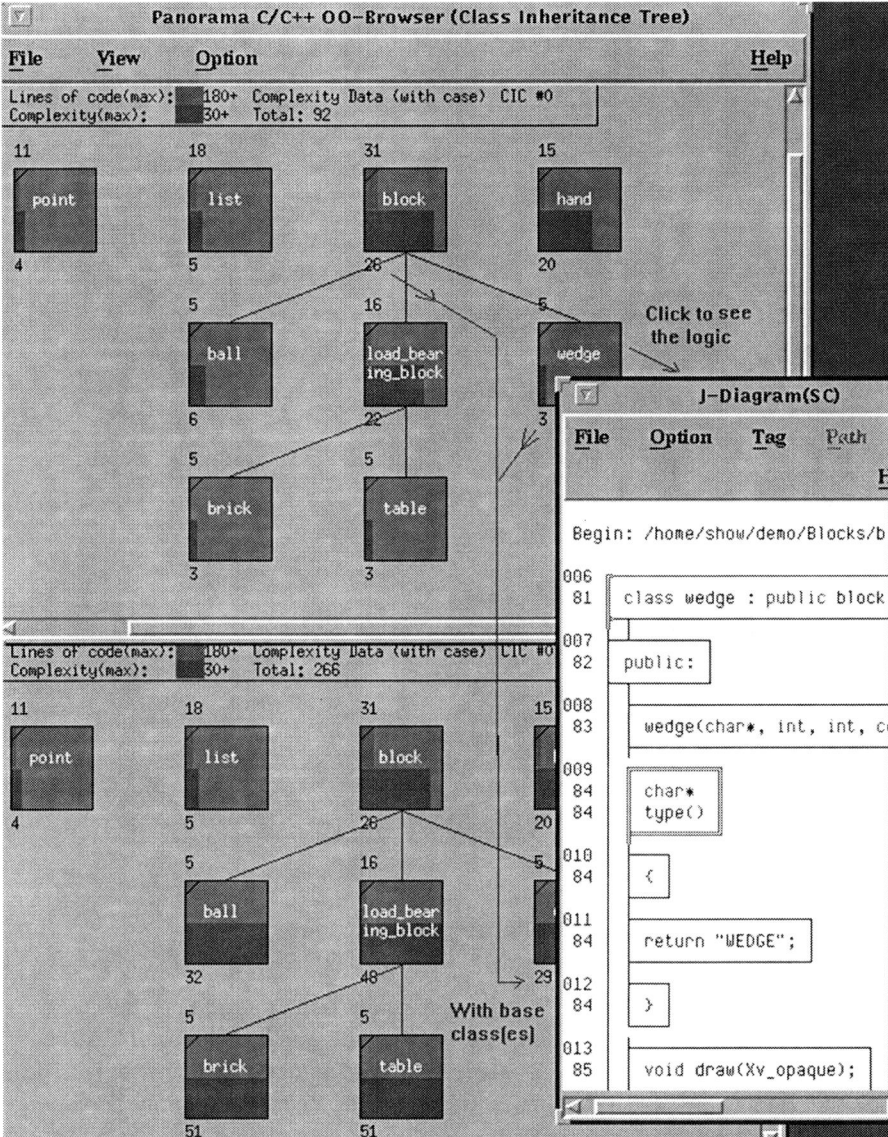

Fig. 14.3 The Cyclomatic complexity of a class with and without the parent classes

With special header files and the interpreter, the NSE support platform, Panorama++, not only can provide the overall test coverage analysis results of classes, but also can provide the MC/DC test coverage analysis results shown in J-Diagram with untested branches and conditions highlighted as shown in Fig. 14.4.

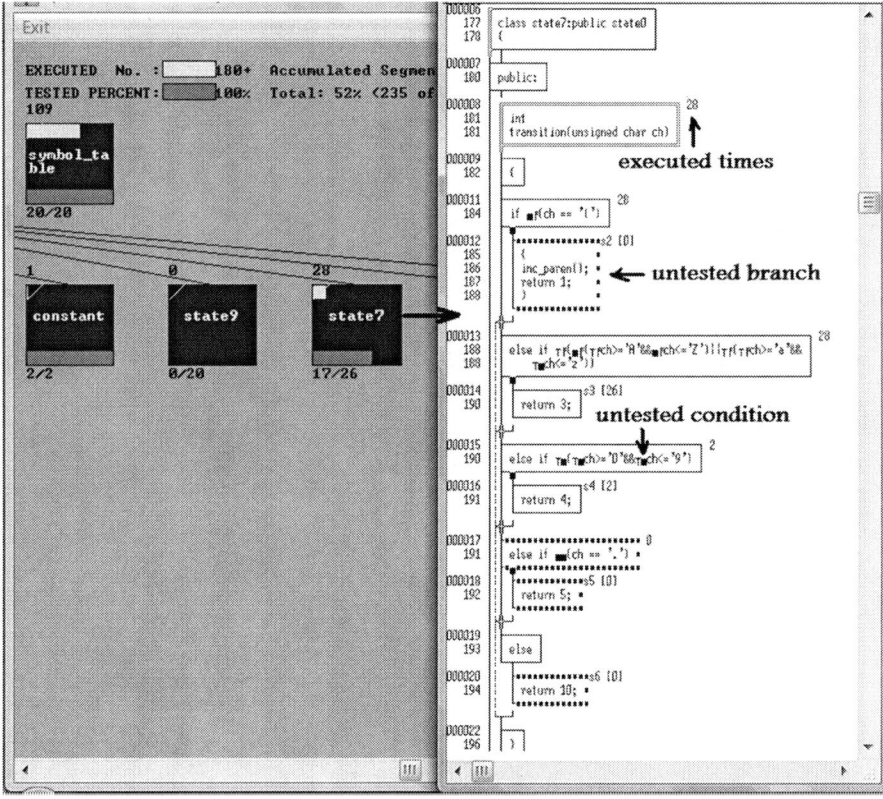

Fig. 14.4 An example of class test coverage analysis provided by Panorama++

14.10 Web Application Testing

The main purpose of Web (Internet) application testing is almost the same as traditional software testing – validating whether the applications meet customers' needs, verifying whether it works as expected according to the requirement specification, and finding possible defects/bugs to help the developers to improve the product quality. The major difference between Web application testing and the traditional program testing is that Web application is running on the Internet with a variety of operating systems, browsers, communication protocols, hardware environments for interaction, etc., so that the test engineers need to learn and acquire more knowledge – this is certainly a new challenge. On specific test content, in addition to functional testing, performance testing and structural testing, there should also be particular emphasis on safety testing, compatibility testing, load testing, interoperability testing, and product navigation. Web application testing requires specialized testing tools to simulate the test environment – for example, simulate a large number of users to simultaneously use the software to be tested to check its capacity and response time.

Fortunately, many Web application software testing tools are freeware or open source products. The following example shows a simple load testing result provided by a freeware tool, OpenLoad:

URL: http://www.sohu.com:80/
Clients: 10
MaTps 0.02, Tps 0.02, Resp Time 47.319, Err 0%, Count 1
MaTps 0.03, Tps 0.07, Resp Time 61.332, Err 0%, Count 2
MaTps 0.06, Tps 0.35, Resp Time 64.211, Err 0%, Count 3
MaTps 0.06, Tps 0.07, Resp Time 78.310, Err 0%, Count 4
MaTps 0.16, Tps 1.06, Resp Time 79.531, Err 0%, Count 6
MaTps 0.15, Tps 0.09, Resp Time 91.179, Err 0%, Count 7
MaTps 0.20, Tps 0.63, Resp Time 92.756, Err 0%, Count 8
MaTps 0.24, Tps 0.56, Resp Time 94.812, Err 0%, Count 10
MaTps 0.22, Tps 0.03, Resp Time 68.576, Err 0%, Count 11
MaTps 0.27, Tps 0.73, Resp Time 86.867, Err 0%, Count 12
MaTps 0.25, Tps 0.07, Resp Time 67.334, Err 0%, Count 13
MaTps 0.24, Tps 0.16, Resp Time 92.344, Err 0%, Count 14
MaTps 0.24, Tps 0.23, Resp Time 61.737, Err 0%, Count 15
MaTps 0.23, Tps 0.14, Resp Time 86.724, Err 0%, Count 16
MaTps 0.22, Tps 0.11, Resp Time 95.176, Err 0%, Count 17
MaTps 0.20, Tps 0.08, Resp Time 95.895, Err 0%, Count 18
MaTps 0.19, Tps 0.10, Resp Time 103.881, Err 0%, Count 19
MaTps 0.24, Tps 0.66, Resp Time 105.927, Err 0%, Count 20
Total TPS: 0.10
Average Response Time: 82.413 sec.
Maximum Response Time: 105.927 sec
Simulated User Number: 20
Total Error Number : 0

14.11 Embedded Software Testing

Embedded systems are becoming larger and more complex with an increasing amount of software, leading to a growing need for testing methods which help to tackle the typical problems in embedded software testing.

The difficulty for testing embedded systems is from the following listed factors:

1. It highly depends on the hardware systems.
2. There are many different application software embedded in many different types of systems.
3. Many embedded systems are real-time systems.
4. Often there is no file system to be used.

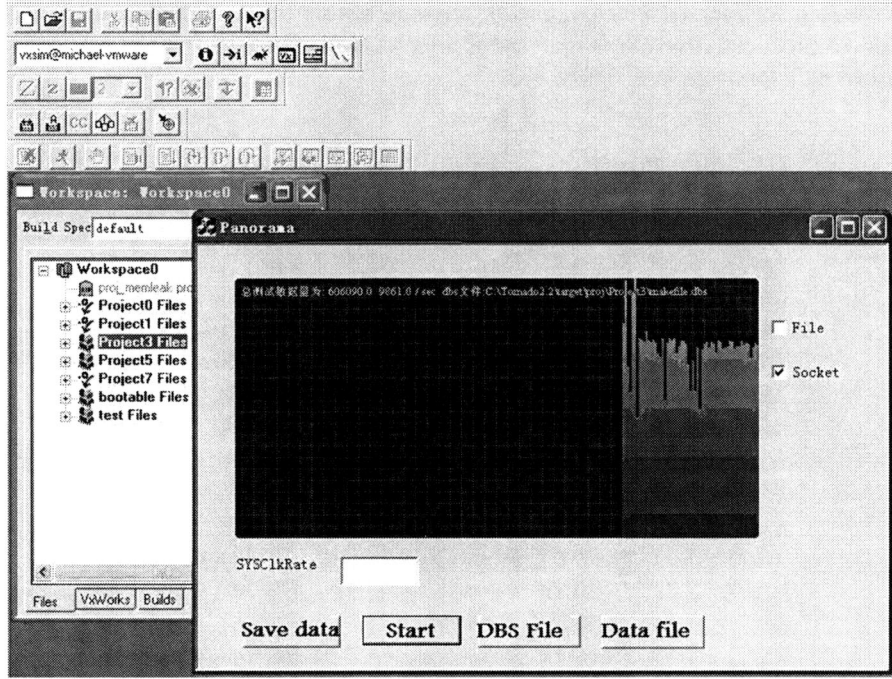

Fig. 14.5 The data transfer process between the target system and the host system for MC/DC test coverage analysis, provided by Panorama++

5. The amount of available memory is small.
6. Often the software compilation environment is different from the execution environment.

It means that the test technology, the test process, and the test tools are quite different from regular software testing.

Figure 14.5 shows an application example of the NSE support platform, Panorama++, used to test an embedded software running on VxWorks environment.

14.12 GUI Operation Capture and Playback

Often the graphical user interface (GUI) of a software product has many operations that need to be tested – a very small program such as Microsoft WordPad has more than 300 possible GUI operations. In a large program, the number of operations can easily be an order of magnitude larger, so that the GUI testing is a time-consuming process, and that after program modification the testing process should be repeated again and again.

Sometimes a tester may find that after a complex combination of the GUI operations, an unexpected error appeared, but the tester cannot show the error to people because he/she did not remember the procedure of the test operations.

GUI operation capture and playback technique and tools can be used to solve those problems – they can capture the GUI operations and then play them back automatically.

14.13 Acceptance Testing

Acceptance testing is performed on a system by the developer prior to its delivery or by the customer prior to accepting the transfer of the ownership of a software product.

As described in Chap. 1, with NSE acceptance testing can be performed in a fully automated way with mouse clicks only, including rerun the test cases dynamically.

14.14 Why Should Software Testing Tools Be Used

Any commercial software product and engineering software product should be tested manually and using tools, because many kinds of software testing tasks cannot be performed manually, such as the structural testing and memory usage violation testing.

14.15 The Major Drawback of the Major Existing Software Testing Paradigm and the Solution

The existing major software testing methods, technologies, and tools are working with the old-established software engineering paradigm based on linear process, reductionism, and the superposition principle that the whole of a complex software system is the sum of its components, so that almost all software development tasks and activities are performed linearly, partially, and locally. With those methods, technologies, and tools, dynamic software testing is performed after coding. But most software defects are introduced upstream rather than downstream.

The concept of the "black box" is questionable – for instance, for a mathematical problem, should the teacher only check the answer given by a student without checking the student's problem-solving process? No!

In 2002, the National Institute of Standards and Technology calculated the annual cost of these operational test failures in the US public and private sectors as $59.5 billion [RTI02]. An independent study found that more than half of IT acquisitions doubled their initial budget and schedule projections, the average acquisition provided only 61% of the desired functionality, and one-third of software-intensive projects were ultimately canceled ([May03], [Fak05]).

For removing those drawbacks of the existing software testing paradigm, a new revolutionary software testing paradigm based on the innovated Transparent-box method has been established – the NSE software testing paradigm (see Chap. 16).

14.16 Summary

Software testing is important to ensure the quality of a software product. There are many types of testing, including functional testing, structural testing, performance testing, load and stress testing, unit and integration and system testing, etc.

Many types of software testing use automated tools.

The major drawback of the existing software testing paradigm is that most defects are introduced into a software product in the upstream phases, but the existing testing is dynamically performed in the downstream phases of a software product development lifecycle – too late.

14.17 Points and Questions to Ponder

(a) Why should a software product be tested before its application?
(b) How many kinds of tests are needed?
(c) Can a software product be tested manually only, without using tools?
(d) Who should test a software product – the product developers, other teams or groups but not the developers, or both? Why?

14.18 Further Reading and Information Source

(a) Requiring software independence in VVSG 2007: STS recommendations for the TGDC. November 2006. http://vote.nist.gov/DraftWhitePaperOnSIinVVSG2007-20061120.pdf
(b) Software QA and testing frequently-asked-questions. http://www.softwareqatest.com/qatfaq1.html

References

[Fak05] Falcone S (2005) A correlated strategic guide for software testing. CrossTalk, July Issue

[May03] Maybury M, King A, Brooks J (2003) Software intensive system acquisition – best practices. 2003 Acquisition Conference, Arlington, VA, 28–30 January 2003. http://www.sei.cmu.edu/products/events/acquisition/2003-presentations/maybury.pdf

References

[Mye79] Myers GJ (1979) The art of software testing. John Wiley and Sons, New York. ISBN 0-471-04328-1

[Pat00] Patton R (2000) Software testing. SAMS, Indianapolis

[RTI02] National Institute of Standards and Technology, Planning report 02-3: the economic impacts of inadequate infrastructure for software testing. National Institute of Standards and Technology, Washington, DC, May 2002. http://www.nist.gov/director/prog-ofc/report02-3.pdf

Chapter 15
Software Test Case Design

> *He who would search for pearls must dive below.*
> John Dryden (1631–1700)

This chapter introduces how to design test cases for efficiently testing a software product.

15.1 What Is a Test Case?

1. The traditional definition of test case is [Che06]

    (a) A set of test inputs, execution conditions, and expected results developed for a particular objective such as to execute a particular program path **or** to verify compliance with a specific requirement.
    (b) Documentation specifying inputs, predicted results, and a set of execution conditions for the test item.

2. The definition of test case with NSE:

    (a) A set of test inputs, execution conditions, expected functional results to verify compliance with a specific requirement, **the expected program execution path specified in control flow and used to verify whether it is covered by the real execution path, a list of modules and branches which should not be hit by the test case execution** (see Sect. 15.6 and Chap. 16 about the Transparent-box testing method), and a list of related document information specified by some keywords (@WORD@, @HTML@, @PDF@, @EXCEL@, and @BAT@) to indicate the formats of the documents, followed by the file paths, and the bookmarks used to open the traced documents from the corresponding locations (see Chap. 9).
    (b) Documentation specifying inputs, predicted results, expected execution path, a list of modules and branches which are not allowed to be entered for the execution of the specified test case, and a set of execution conditions for the test item.

15.2 The Basis of Test Case Design

If you open any professional book on software testing, it is easy to find a chapter describing how to design test cases – the contents in the different books are similar, coming from previous practice and experience.

15.2.1 Equivalence Class Partition and Boundary Value Analysis

1. Equivalence class partition [Bei95] is used to identify whether two (or more) tests are equivalent. When the two inputs are equivalent, you can expect them to encounter the same sequence of operations or they will follow the same path in the source code. Thus, when two or more test cases are equivalent, only one is usually needed to be implemented in order to save the testing time.
   Equivalent class examples:

   (a) Digital scope (e.g., all the figures 10 to 99)
   (b) Group members (date, time, and name of country)
   (c) Illegal input (e.g., input symbols used to calculate numbers)
   (d) Produce equivalent output of the events (all produce equivalent output of the input)
   (e) Equivalent operating environment
   (f) Duplication of activities

2. The boundary value is equivalent to the change point. It might be some limit that defines the boundary between supported inputs and unsupported inputs. Therefore, the boundary-conditions test for finding undiscovered errors is often more effective [Mye79].
   Typically, each equivalent class is divided by its boundary values. Of course, not all equivalence classes have a border.

   However, each equivalent class also represents a potential risk. Therefore, in the application of a method to design the equivalent class test cases, it is best to design nine test cases for each partition, including:

   (a) Values in the legal division of the region
   (b) Values in the low end of the legal division of the border area
   (c) Values in the low end of the legal division of district boundaries +1
   (d) Values in the low end of the legal division of district boundaries −1
   (e) Values in the high end of the legal division of the border area
   (f) Values in the high end of the legal division of district boundaries +1
   (g) Values in the high end of the legal division of district boundaries −1
   (h) Values in much less than its low-end boundary
   (i) Values in much larger than its high-end boundary

   For example, if an integer is equivalent to the scope of class 1 and 100, then we can design the nine test cases – the corresponding distribution of the input integer values: 50, 1, 2, 0, 100, 101, 99, −32,769, 123,456,789. Among them, the "basic boundary test set" of values is: 50, 0, 101.

15.2 The Basis of Test Case Design

3. The design of test cases and equivalent class analysis includes the following main steps:

   (a) Determine the equivalent class.
   (b) Determine the boundary.
   (c) Determine all outputs for the legal inputs.
   (d) Determine the error handling part for illegal inputs.
   (e) For each equivalent class, finish the test cases table (up to nine test cases).

4. The serious shortcoming of white-box testing tools which only offer capability for statement coverage analysis is that the error detecting ability of these kinds of tools is very poor and not suitable for testing commercial software products – they often report "100% of statements have been tested" without checking the boundary values.

15.2.2 State Transition Analysis

State Transition Analysis is the analysis of the status of an application conversion, conversion trigger events, and conversion results.

To design test cases using this method, the following four steps should be followed:

1. Determine all the states supported by an application.
2. For each test case, define the following:

   (a) Initial state
   (b) Input events causing state transitions
   (c) The output or event for each state transition
   (d) End state

3. Draw a diagram to describe the state, event, and application response relationship.
4. Make a test case table for each state transition.

15.2.3 Conditions Combination Method

The challenge in software testing is the long time it takes to implement all possible test cases. However, from the cost and practical considerations, we should try to compress the required number of test cases, while achieving the required quality requirements.

There are many ways to reduce the number of test cases.

The Conditions Combination Method includes the variable conditions of portfolio analysis. Each represents a combination of conditions using the same test scripts and test processing sequence to test a condition.

The main steps of this method are as follows:

1. Identify the variables used.
2. For each variable, assign a set of unique values.
3. Create a table for various variables and the values assigned.

For example, suppose a total of three variables A, B, C to be assigned, each with three unique values, respectively, 1, 2, 3; 4, 5, 6; and 7, 8, 9, then the total combined number of test cases conditions of $3 \times 3 \times 3 = 27$ is:

Test case number	(A, B, C) value
1.	(1, 4, 7)
2.	(1, 4, 8)
3.	(1, 4, 9)
4.	(1, 5, 7)
5.	(1, 5, 8)
6.	(1, 5, 9)
7.	(1, 6, 7)
8.	(1, 6, 8)
9.	(1, 6, 9)
10.	(2, 4, 7)
11.	(2, 4, 8)
12.	(2, 4, 9)
13.	(2, 5, 7)
14.	(2, 5, 8)
15.	(2, 5, 9)
16.	(2, 6, 7)
17.	(2, 6, 8)
18.	(2, 6, 9)
19.	(3, 4, 7)
20.	(3, 4, 8)
21.	(3, 4, 9)
22.	(3, 5, 7)
23.	(3, 5, 8)
24.	(3, 5, 9)
25.	(3, 6, 7)
26.	(3, 6, 8)
27.	(3, 6, 9)

15.3 Semiautomated Test Case Design

In unit testing, it is clear that different test cases will execute in the same or different paths consisting of some branches. The more branches in a test path, the higher the efficiency of the test case. But to a complex unit with many decision statements, it is difficult to know which path will include more untested branches.

A tool provided with the NSE software visualization paradigm can be used to automatically perform untested path analysis and then automatically select the "longest" one with more untested branches, highlight it with red color in the generated control flow diagram, and then automatically extract the execution conditions to help the users design the corresponding test cases easier. It can be used from the beginning of unit testing with no path tested or after some paths have been tested. A sample application for a very complex program unit is shown in Figs. 15.1–15.3.

15.4 Test Case Efficiency Measurement

For testing a complex software product deeply, a huge amount of test cases will be designed. But within those test cases, the test case efficiency is quite different. But it is hard to know the efficiency of the test cases without using the corresponding tools. The result of test case efficiency measurement can be used to realize test case minimization to automatically choose a minimum set of test cases which can be used to get the same test coverage result as that obtained by all the test cases used.

With NSE, test case efficiency can be automatically measured. An application example is shown in Fig. 15.4.

In Fig. 15.4, SC0 means the test coverage of visible segments (a segment is similar to a branch, but more accurate – a segment is a set of statements with the same execution conditions), SC1 means the test coverage of visible and invisible segments (a "if" statement without the corresponding "else" part means there is an invisible segment), SC1+ means SC1 plus loop boundary test coverage – with NSE, a loop statement will be handled as three segments. For more detailed information about SC0, SC1, SC1+, please see "Glossary" of this book.

15.5 Test Case Minimization

The test case minimization technique is used to automatically select a minimum set of test cases from all the test cases used before for testing a complex software product. It is required to greatly reduce the running time and resources used in regression testing after program changes. The minimized set of test cases can be used to get the same test coverage result as that using all the test cases. The algorithm for test case minimization with NSE is described in Chap. 21. The key point is that the test case having the biggest test coverage contribution will be selected first, whereas the selection of the other test cases does not depend on test coverage contribution, but on the accumulated test coverage contribution which covers more untested elements which have not been covered by all selected test cases.

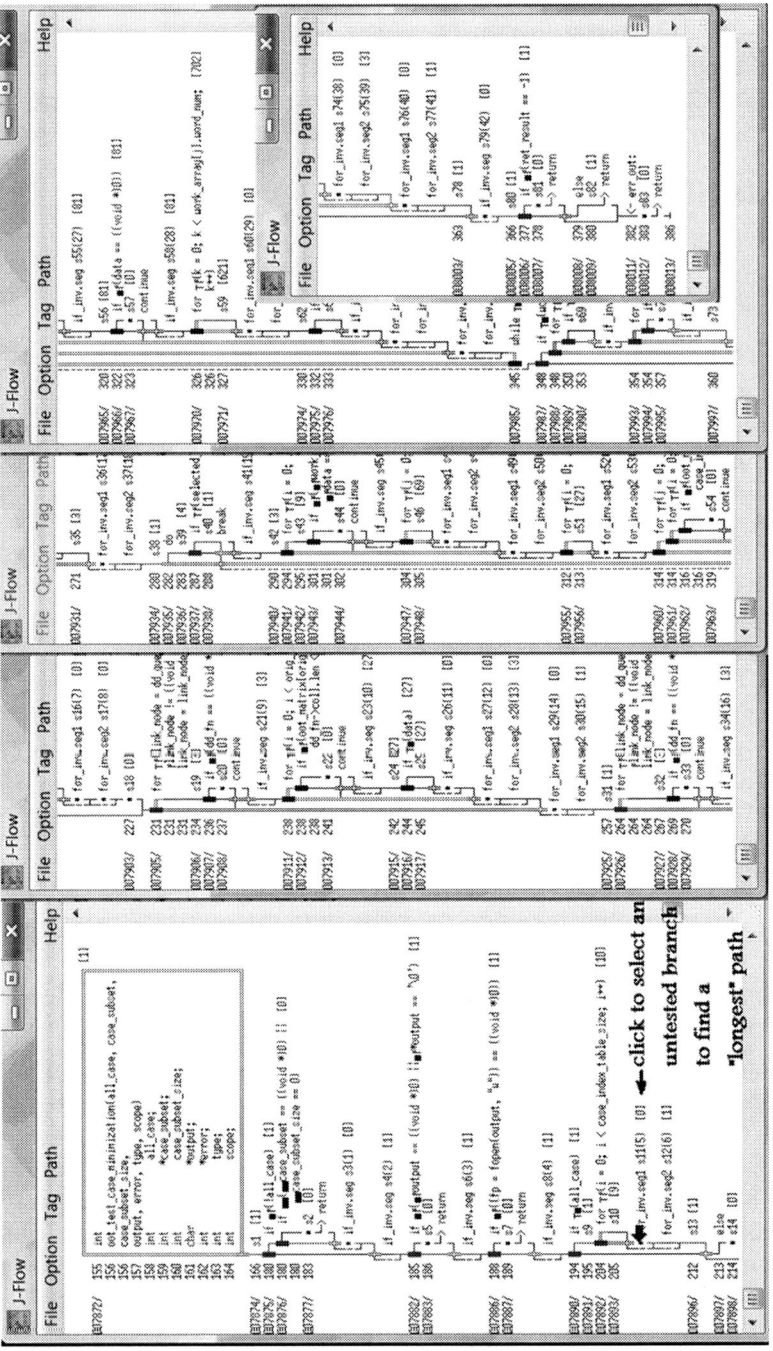

Fig. 15.1 After some paths have been tested, click an untested branch to let the tool to find a "longest" path with most untested branches

15.5 Test Case Minimization 393

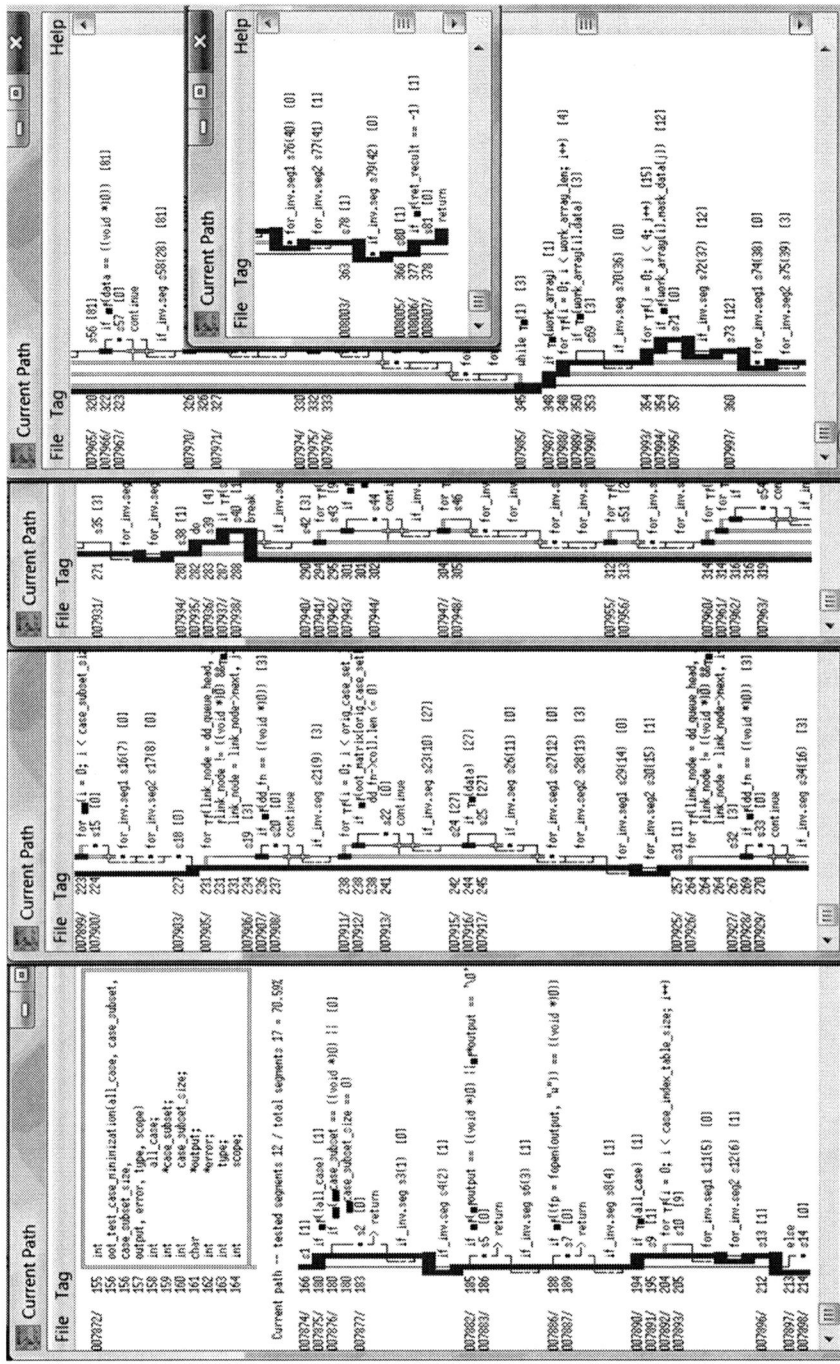

Fig. 15.2 The "longest path" highlighted correspondingly

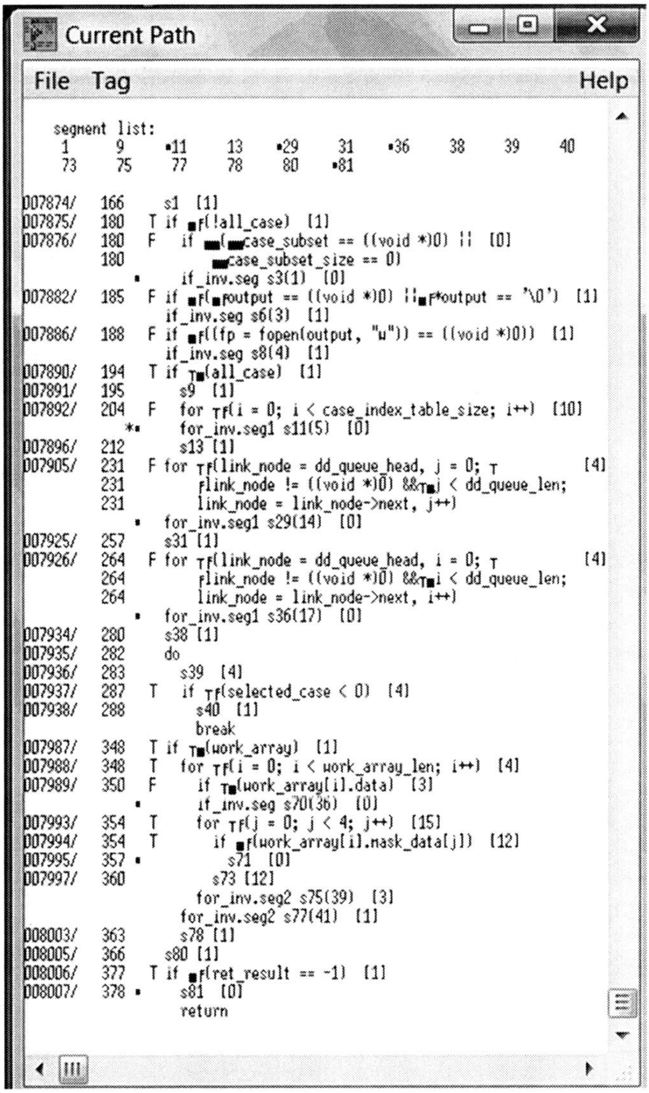

Fig. 15.3 The extracted conditions for helping users to design the corresponding test case to test the "longest path" (In this figure, a small "T" character means the condition is True, "F" is false)

15.6 Test Case Minimization

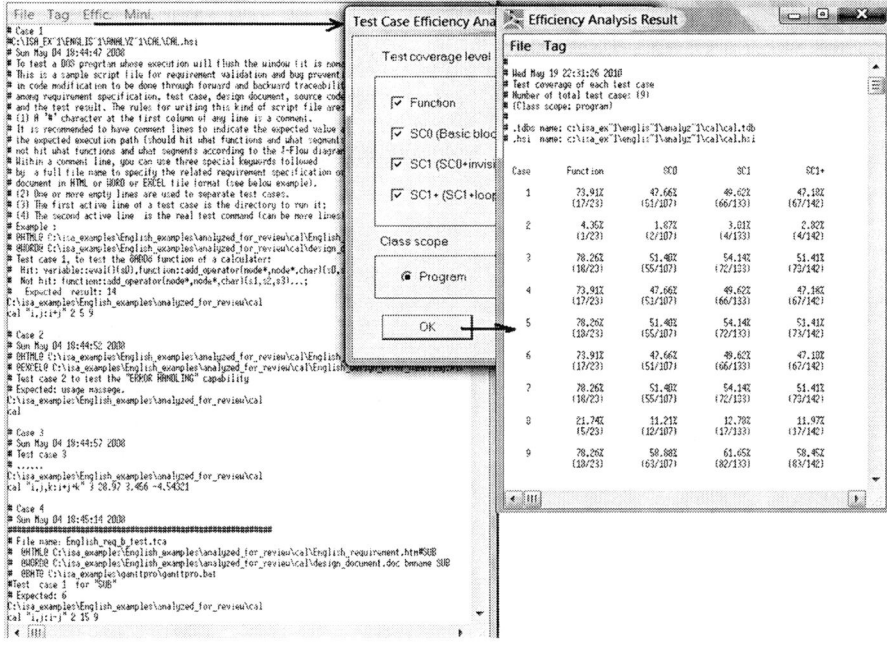

Fig. 15.4 An application example of test case efficiency measurement

Usually, a test case that can be used to find a defect/bug will be selected into the minimized set of test cases, because its execution path will be different from that of other test cases which have not been able to find a defect/bug.

An application example of test case minimization is shown in Fig. 15.5.

About the algorithm for test case minimization, please see Chap. 21.

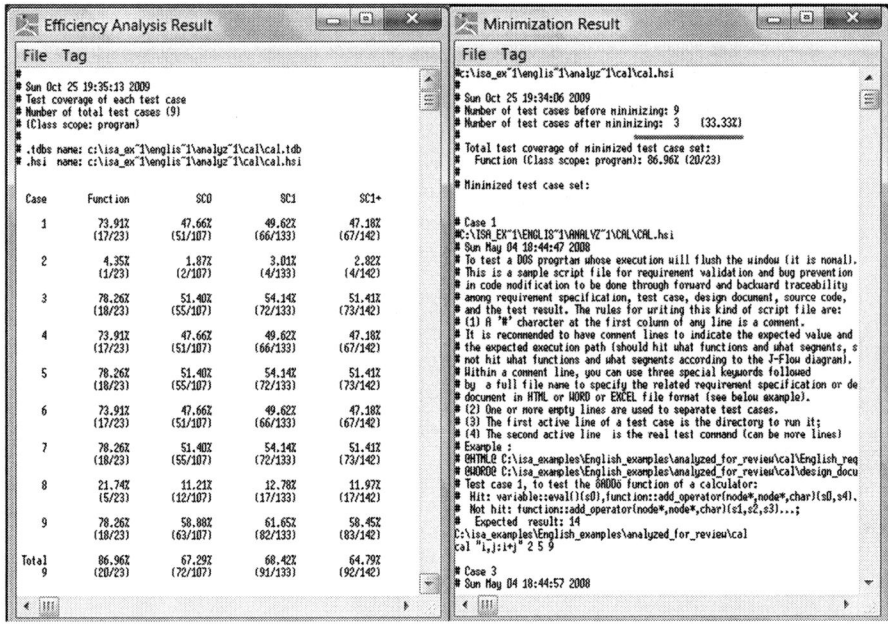

Fig. 15.5 An application example of test case minimization

15.6 NSE Test Case Design with HAETVE Technique for Both Functional Testing and Structural Testing

Functional test cases should be designed according to the requirement specification and test requirement specification. With NSE, test cases can be designed and dynamically used from the beginning of requirement development using the HAETVE technique as described in Chap. 11.

HAETVE means Holistic, Actor–Action and Event–Response driven Traceable, Visual, and Executable requirement development technique working with the dummy programming technique using dummy modules having an empty body or only some function call statements.

An example of the dummy source code written in C/C++ for representing an actor is listed as follows:

Bank_Customer ()
{
Bank_Customer ();
}

15.6 NSE Test Case Design with HAETVE Technique

An example of the dummy source code for representing an action is as follows:

Void Deposit_Money ()
{
}

The corresponding notations are shown in Fig. 15.6.

It is easy to map the notations to Use Cases as shown in Figs. 15.7 and 15.8.

Now we can add a main() module to the dummy program with the source as follows:

The modified call graph is shown in Fig. 15.9.

Different from Use Case, the dummy program is executable after adding the main() module.

Now we can further design the corresponding test cases to dynamically test this very simple dummy program for defect prevention and defect propagation prevention.

```
void main(int argc,char** argv)
{
int key;
if(strcmp(argv[1],"New_Order")==0)
  New_Order();
else if (strcmp(argv[1],"Confirm_Order")==0)
  Confirm_Order();
else if (strcmp(argv[1],"Invoice_Buyer")==0)
  Invoice_Buyer();
else if (strcmp(argv[1],"Pay_Invoice")==0)
  Pay_Invoice();
else
  Send_Reminders ();
}
```

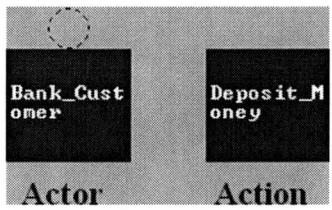

Fig. 15.6 Notations for representing an actor and an action

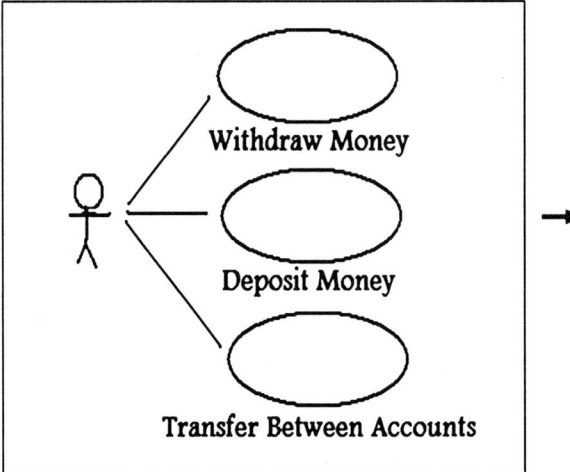

An example of a use-case diagram with an actor and three use cases

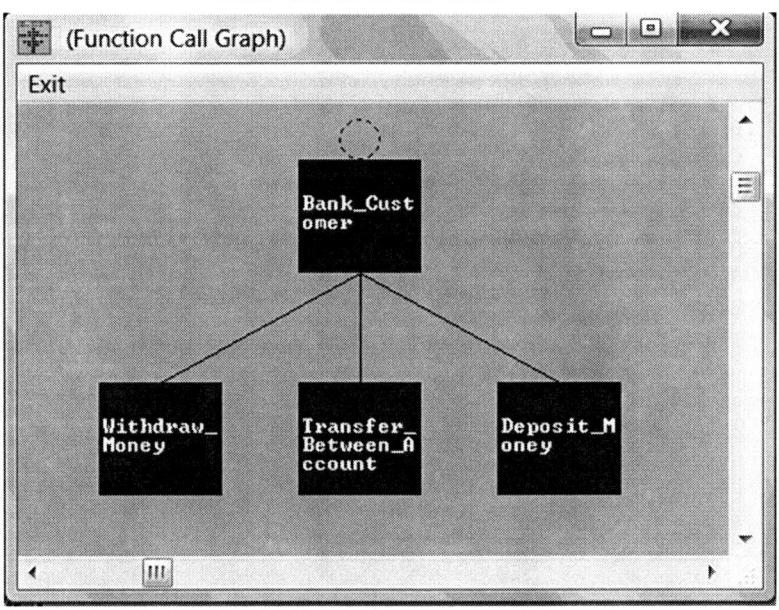

Fig. 15.7 Notation mapping between Use Cases and HAETVE

The simple rules for designing a test case are listed as follows:

(a) An empty line means a separator between different test cases.
(b) A '#' character at the beginning position of a line means a comment.

15.6 NSE Test Case Design with HAETVE Technique

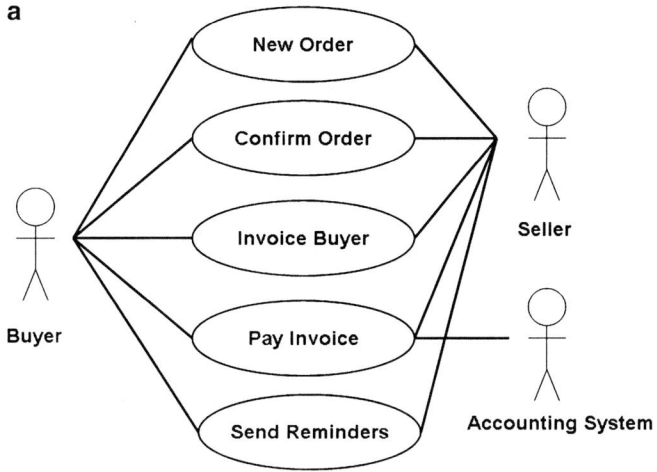

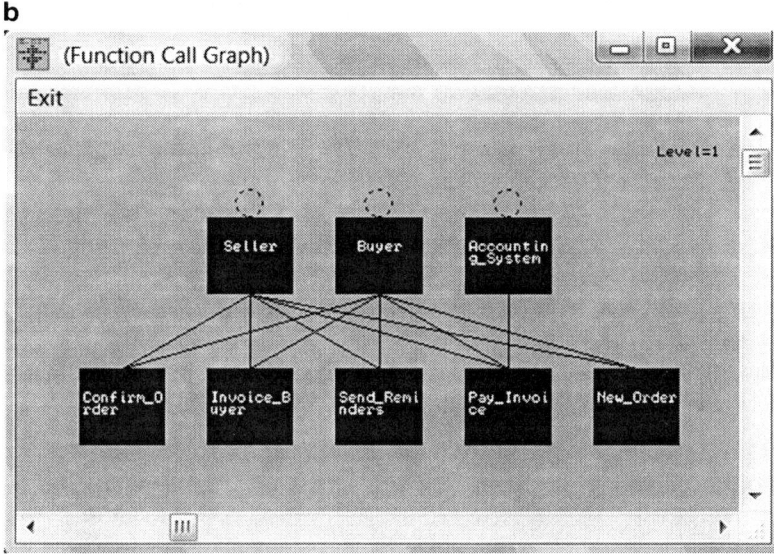

Fig. 15.8 Another example mapping Use Cases (**a**) to HAETVE (**b**)

(c) Within comments, users can use some keywords such as @WORD@, @HTML@, @PDF@, and @BAT@ to indicate the format of a document, followed by the full path name of the document, and a bookmark – for finding inconsistent defects.

(d) Within comments, users can use [path] and [/path] pair to indicate the expected execution path using control flow notation (segment numbers) for a test case – for finding logic defects.

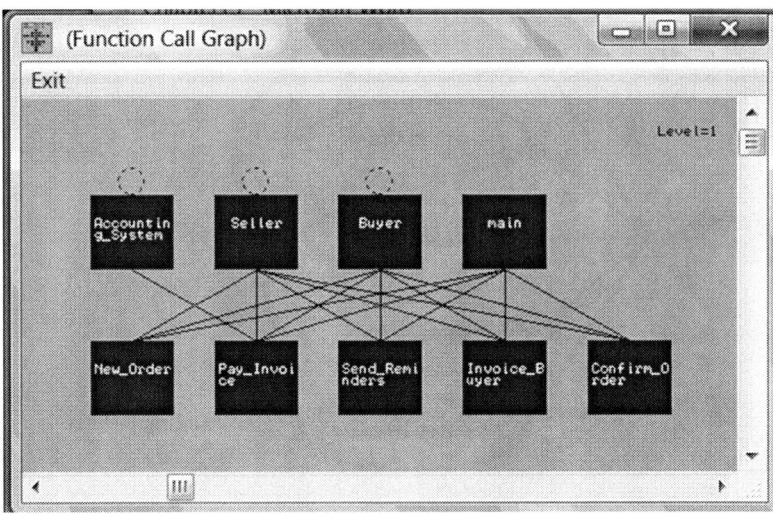

Fig. 15.9 A call graph with a main() module modified from Fig. 15.8b

(e) Within comments, users can use Expected Output to indicate the expected value to be produced – for finding functional defects.
(f) Within comments, users can also use [Not_Hit] and [/Not_Hit] marks to indicate modules or branches (segments) which are prohibited for the related test case execution to enter.
(g) After the comment part, there is a line to indicate the directory for running the corresponding program.
(h) The final line in a test case description is the command line (which may start a program with the GUI) and the options.

Although there is no output from the above simple dummy program, we can design corresponding test cases to not only check whether there are logic errors, but can also check whether there are inconsistent defects through the automatically established traceability among related documents and test cases and the source code.

A simple test script file with five test cases designed for executing the five actions is listed as follows (please pay more attention to what are the related documents used):

```
# @HTML@ C:\Billing_and_Payment\Requirement_specification.htm#New_Order
# @WORD@ C:\Billing_and_Payment2\Prototype_design.doc bmname New_Order
# @WORD@ C:\Billing_and_Payment2\TestRequirements.doc bmname New_Order
# [path] main(int, char**) {s0, s1}New_Order void) [/path]
# Expected output : none
# [Not_Hit] !path [/Not_Hit]
C:\Billing_and_Payment2
Billing_and_Payment.exe New_Order
```

15.6 NSE Test Case Design with HAETVE Technique 401

\# test case 2 for Pay Invoice
\#@HTML@ C:\Billing_and_Payment2\Requirement_specification.htm#Pay_Invoice
\#@WORD@ C:\Billing_and_Payment2\Protorype_design.doc bmname Pay_Invoices
\#[path] main(int, char**) {s0, s4}Pay_Invoice(void) [/path]
\# Expected output : none
\# [Not_Hit] !path [/Not_Hit]
C:\Billing_and_Payment2
Billing_and_Payment.exe Pay_Invoices

\# test case 3 for Confirm_Order
\#@HTML@ C:\Billing_and_Payment2\Requirement_specification.htm#Confirm_Order
\#@WORD@ C:\Billing_and_Payment2\Protorype_design.doc Confirm_Order
\#[path] main(int, char**) {s0 s2} Confirm_Order(void) [/path]
\# Expected output : none
\# [Not_Hit] !path [/Not_Hit]
C:\Billing_and_Payment2
Billing_and_Payment.exe Confirm_Order

\# test case 4 for Invoice_Buyer
\#@HTML@ C:\Billing_and_Payment2\Requirement_specification.htm#Invoice_Buyer
\#@WORD@ C:\Billing_and_Payment2\Protorype_design.doc Invoice_Buyer
\#[path] main(int, char**) {s0, s3} Invoice_Buyer (void) [/path]
\# Expected output : none
\# [Not_Hit] !path [/Not_Hit]
C:\Billing_and_Payment2
Billing_and_Payment.exe Invoice_Buyer

\# test case 5 for Send_Reminders
\#@HTML@ C:\Billing_and_Payment2\Requirement_specification.htm#Send_Reminders
\#@WORD@ C:\Billing_and_Payment2\Protorype_design.doc Send_Reminders
\#[path] main(int, char**) {s0, s5} Send_Reminders (void) [/path]
\# Expected output : none
\# [Not_Hit] !path [/Not_Hit]
C:\Billing_and_Payment2
Billing_and_Payment.exe Send_Reminders

After test execution with the above test script file, we obtained the test result as shown in Fig. 15.10.

Our intention is to execute the five dummy actions of the above dummy program: why has the "Pay_Invoice" action not been executed? Why is the "Send_Reminders" action executed twice?

With NSE, it is the time we should use the automatically established traceability to find the possible logic defects and inconsistent defects among the related documents.

When we clicked test case 1 on the test script window to perform forward tracing through the NSE support platform, Panorama++, we will find nothing wrong as shown in Fig. 15.11.

But when we clicked test case 2, we can easily find two defects/bugs as shown in Fig. 15.12.

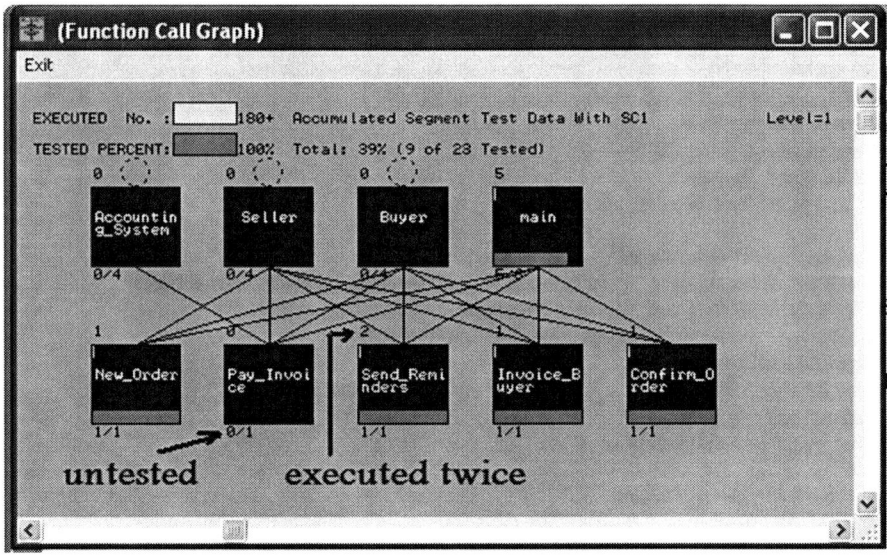

Fig. 15.10 The test coverage measurement result for the sample program and the test cases

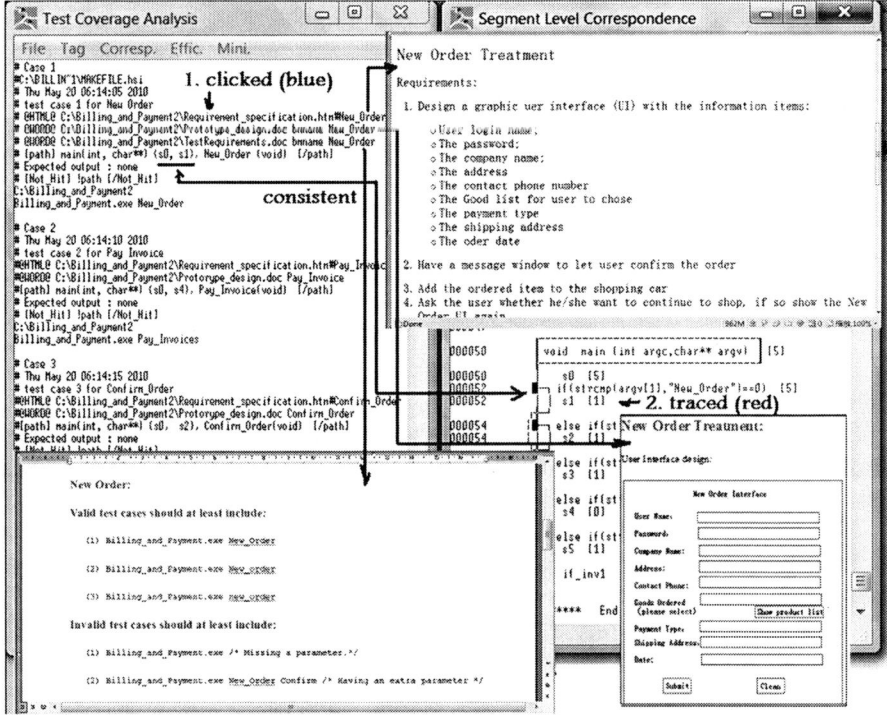

Fig. 15.11 Tracing test case 1 found no defects/bugs

15.6 NSE Test Case Design with HAETVE Technique

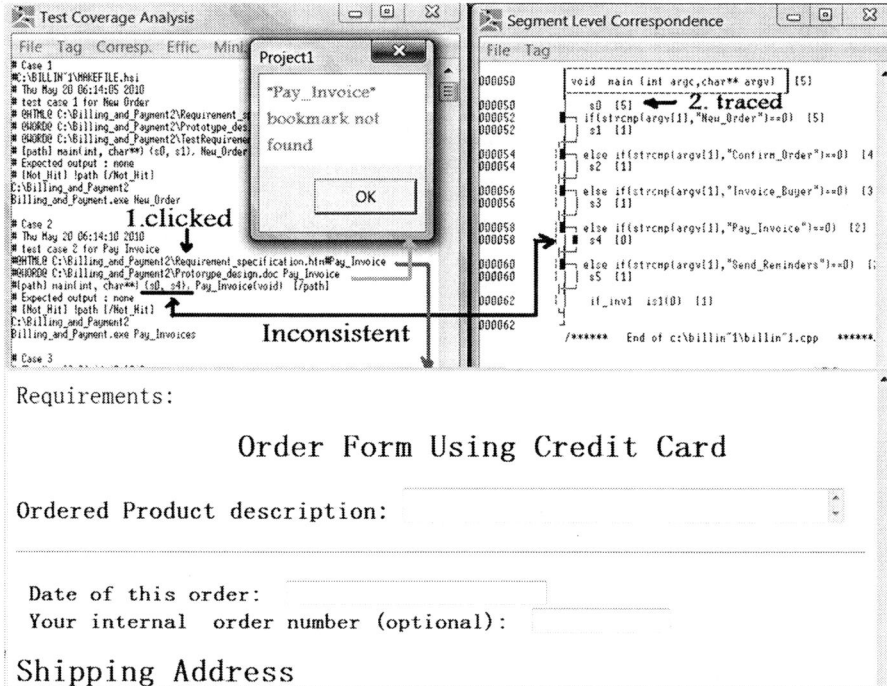

Fig. 15.12 With tracing test case 2, we can easily find two defects/bugs: (1) the specified bookmark is not found and (2) the expected execution path is not covered by the real execution path

As shown in Fig. 15.12, the first defect is due to the inconsistency between the test execution option – "Pay_Invoices" should be "Pay_Invoice" according to the dummy program.

The second defect is from the specified bookmark "Pay_Invoices" which is not found. This defect is due to the name of the bookmark set in the Prototype_design. doc file as "Pay_Invoice," but in the description part of test case 2, the specified bookmark name is "Pay_Invoices" (there is an extra "s" character) – both do not match each other as shown in Fig. 15.13.

Why is the "Send_Reminders" action executed twice? After reading the dummy source code carefully, we can easily find that the defect is from the last statement of the main() function:

"else
 Send_Reminders ();"

which does not check whether the option of the command line is "Send_Reminders."

It means that there are three defects/bugs found, with one logic defect, one inconsistent defect between the test cases and the source code, and one inconsistent defect between the related documents and the test cases.

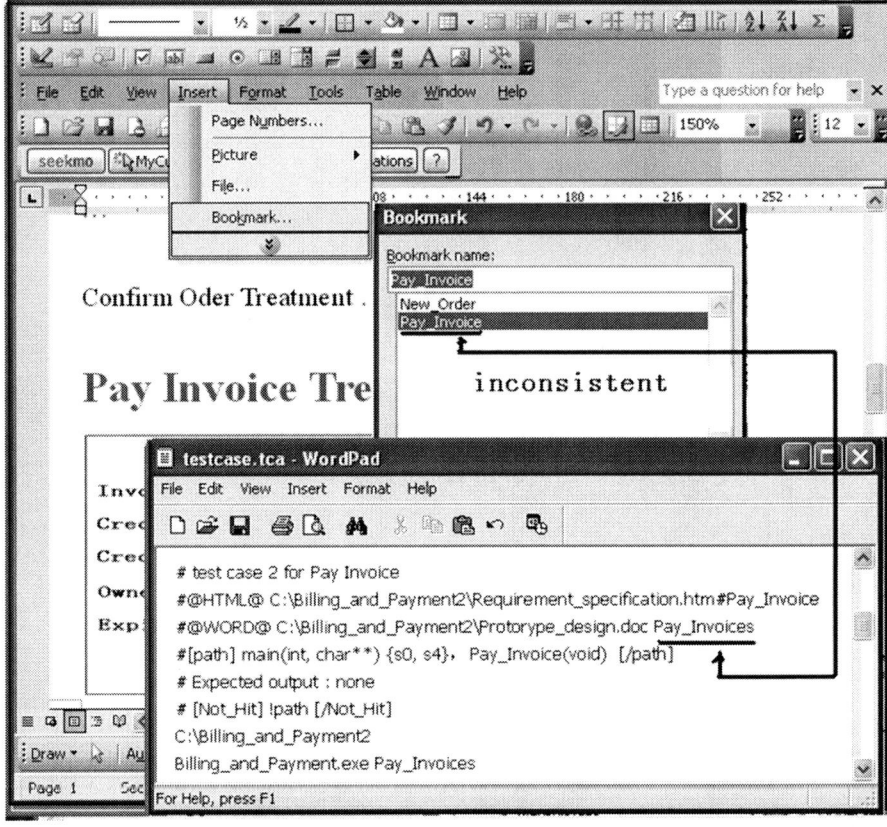

Fig. 15.13 An inconsistent defect found with the use of a bookmark

After fixing the three defects, we obtained the correct result as shown in Fig. 15.14.

We know that the requirement development result is not the real product design; why should we remove the defects introduced in requirement development phase? It is because with NSE the results obtained in the requirement development phase will become the basis for the product design, and the design is precoding as described in Chap. 13.

Of course, the above example just shows the capability of NSE to help software developers remove several logic defects and inconsistent defects introduced in the software requirement development phase. In real applications, most defects introduced in requirement development phases may come from the inconsistency among various different documents.

15.7 Automated Test Case Selection with Automated Test Case Execution

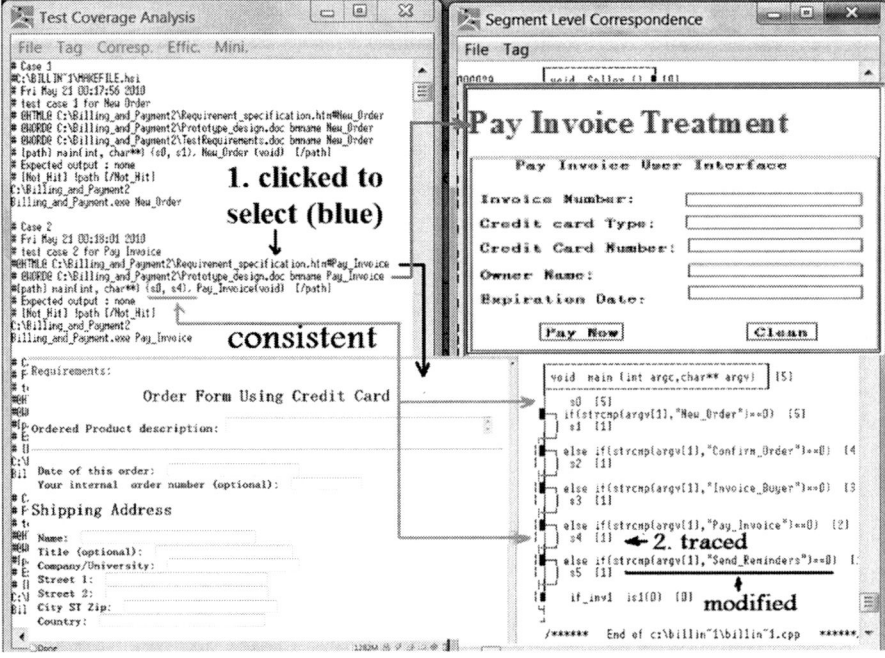

Fig. 15.14 The updated result with the three defects fixed

15.7 Automated Test Case Selection with Automated Test Case Execution

With NSE, when there is only a few modules or branches being modified, users can directly click a modified module or branch on the generated control diagram shown in J-Flow notations to backwardly trace the corresponding test cases or directly execute them through a batch file as shown in Fig. 15.15.

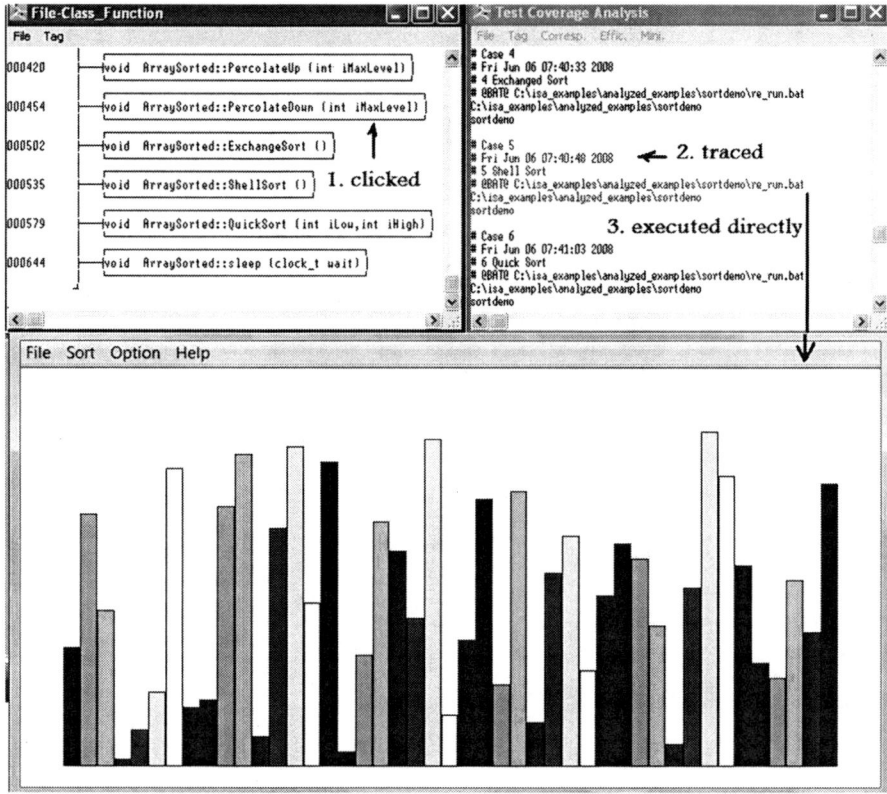

Fig. 15.15 Backwards tracing a modified module to not only automatically select the corresponding test cases through Time Tags, but also automatically execute the test case(s)

15.8 Summary

Software testing is performed using test cases. The test efficiency highly depends on the design of the test cases. There are several test case design approaches. With the old-established software testing methods and technologies, the test cases used for functional testing are different from the test cases used for structural and other testing objectives. Because software testing is dynamically performed after coding, it is too late to be used dynamically to find defects in the requirement development phase and design phase.

With NSE, the software testing paradigm based on the Transparent-box method, functional testing and structural testing are combined together seamlessly, and can be dynamically used in the entire software development lifecycle including the requirement development phase, design phase, coding phase, testing phase, and maintenance phase. The corresponding test case design should indicate the expected value for verifying the functionality, the expected execution path for verifying the program logic, and the related documents for checking the consistency among the related documents, test cases, and the source code.

15.9 Points and Questions to Ponder

(a) What is a test case?
(b) How many basic test case design methods are used today?
(c) How can the NSE software testing paradigm and the NSE software visualization paradigm help users design efficient test cases?
(d) Describe the simple rules for writing test cases for using the Transparent-box software testing method and tools.

15.10 Further Reading and Information Source

(a) Copeland L (2004) A practitioner's guide to software test design. Artech House Publishers, Norwood
(b) Software QA and testing frequently-asked-questions. http://www.softwareqatest.com/qatfaq2.html
(c) Software testing, From Wikipedia, the free encyclopedia. **http://en.wikipedia.org/wiki/Main_Page**

References

[Bei95] Beizer B (1995) Black-Box testing. John Wiley and Sons, New York
[Che06] Chernak Y (2006) Understanding the logic of system testing. Crosstalk, Mar Issue
[Mye79] Myers GJ (1979) The art of software testing. John Wiley and Sons, New York. ISBN 0-471-04328-1

Chapter 16
The NSE Software Testing Paradigm Based on the Transparent-Box Method

> *Software Development: The Need for a New Paradigm ... it recognizes an even stronger need in software development to address quality problems upstream, because that is where almost all software defects are introduced.*
>
> Bijay K. Jayaswal and Peter C. Patton

This chapter introduces another important component of NSE – the NSE software testing paradigm.

The foundation for the establishment of the NSE software testing paradigm is complexity science by complying with the essential principles of complexity science, particularly the Nonlinearity principle and the Holism principle that the whole of a complex system is greater than the sum of its components, and that the characteristics and behaviors of the whole emerge from the interaction of its components, so that with the NSE software testing paradigm almost all software testing engineering tasks/activities are performed holistically and globally to ensure the quality of a software product.

The establishment of the NSE software testing paradigm is done through the use of the FDS framework (the Five-Dimensional Structure Synthesis method – an innovated paradigm-shift framework, see Chap. 4) as shown in Fig. 16.1.

16.1 The Major Existing Software Testing Methods, Techniques, and Tools Are Outdated

Current software quality assurance is mainly based on functional testing using the Black-box testing method being applied after the entire product is produced, structural testing using the White-box testing method after each software unit is coded, and inspection. It violates Deming's Product Quality Assurance Principles, that of "*Cease dependence on inspection to achieve quality. Eliminate the need for inspection on a mass basis by building quality into the product in the first place*" [Dem86].

Both methods are applied separately without any internal logic connection. The White-box testing is mainly performed in unit testing to test an **existing product**

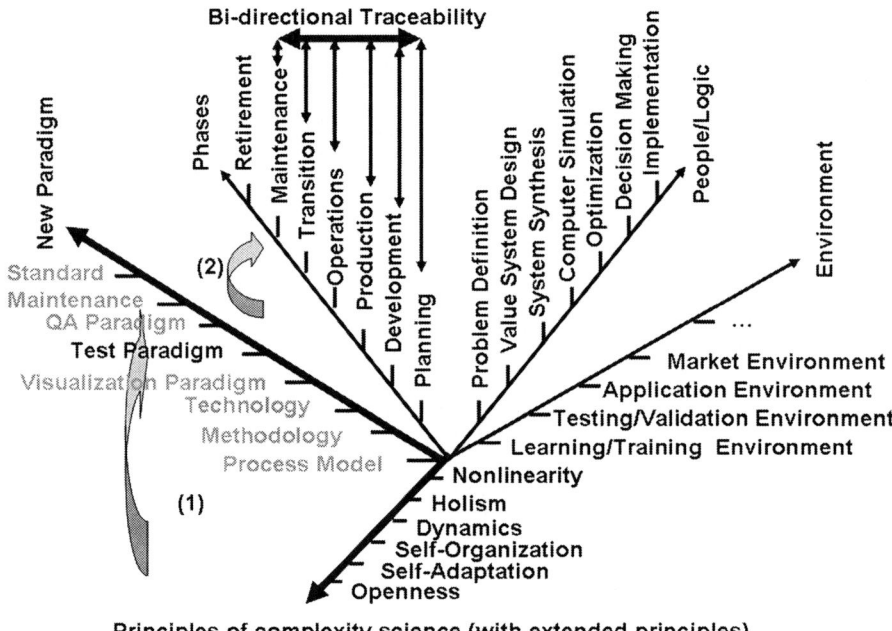

Fig. 16.1 The framework for establishing the NSE software testing paradigm

rather than a **required product**, whereas the Black-box testing is mainly performed in system testing, so that both methods and the corresponding techniques and tools cannot be used dynamically in the requirement development phase and the software design phase.

Even if a requirement development defect or a design defect can be found by both methods after coding, it is too late: the cost for removing the defect will increase tenfold several times.

For those software testing methods, NIST (National Institute of Standards and Technology) concluded that "Briefly, experience in testing software and systems has shown that testing to high degrees of security and reliability is from a practical perspective not possible. Thus, one needs to build security, reliability, and other aspects into the system design itself and perform a security fault analysis on the implementation of the design" ("Requiring Software Independence in VVSG 2007: STS Recommendations for the TGDC," November 2006. http://vote.nist.gov/DraftWhitePaperOnSIinVVSG2007-20061120.pdf).

Those software testing methods and the related techniques and tools are designed to work with the old-established software engineering paradigm based on linear thinking, reductionism, and the superposition principle that **the whole of a system is the sum of its parts**, so that almost all tasks/activities are performed linearly, partially, and locally, making the defects introduced in upper phases easy to propagate to the lower phases to increase the defect removal cost more than 100 times.

As described in Chap. 2, this old-established software engineering paradigm is entirely outdated and should be replaced by a new revolutionary software engineering paradigm based on nonlinear thinking and complexity science.

16.2 The Transparent-Box Testing Method

The innovated Transparent-box testing method is graphically described in Fig. 16.2.

As shown in Fig. 16.2, with the Transparent-box testing method, to each test case, the corresponding tool will not only check whether the output (if any, can be none when it is dynamically used in the requirement development phase and the design phase) is the same as what is expected, but also check whether the execution path covers the expected one specified in the control flow (specified in the description part of each test case after the [path] mark and before the [/path] mark), and whether the execution hits some modules or branches (specified in the description part of each test case after the [Not_Hit] mark and before the [/Not_Hit] mark) which are prohibited for the execution of the corresponding test case, so that it can be used to find functional defects, logic defects, and inconsistency defects. Having an output is no longer a condition to apply this method, so it can be used dynamically in the entire software development lifecycle for defect prevention and defect propagation prevention.

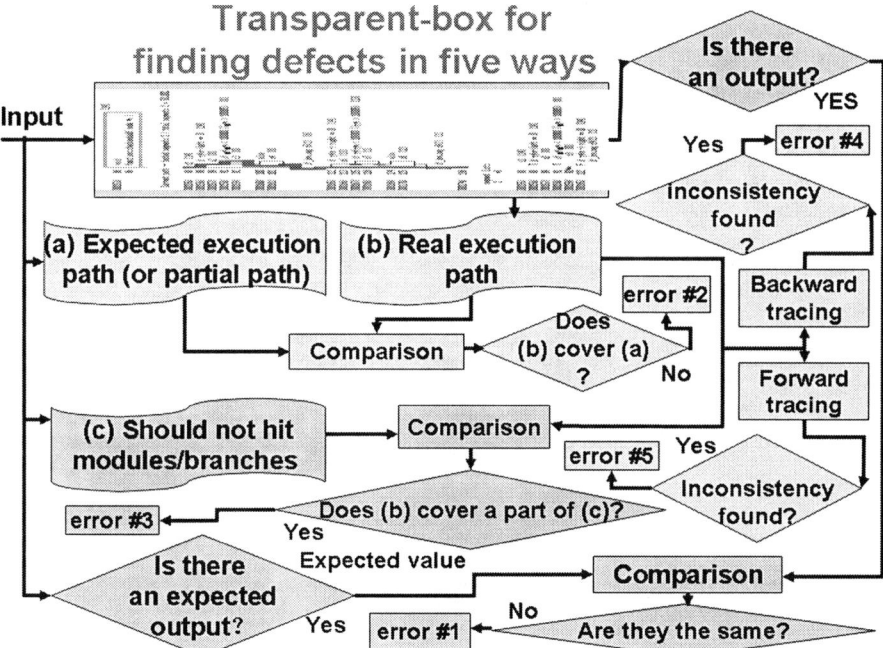

Fig. 16.2 Transparent-box testing method

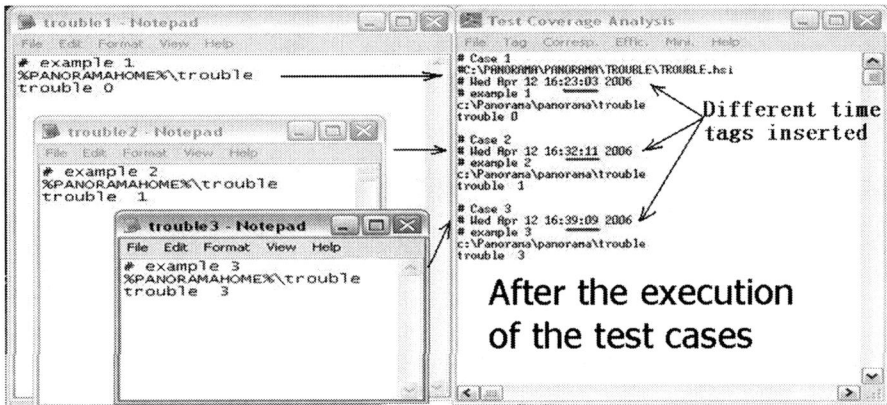

Fig. 16.3 Time Tag examples

The bidirectional traceability between test cases and the source code tested is established through the use of Time Tags that are automatically inserted into the description of the test cases and the database of the source code test coverage analysis for mapping them together accurately. Examples of Time Tags that are automatically inserted into the description path of test cases are shown in Fig. 16.3.

For extending the traceability to include the related documents, some keywords such as @WORD@, @HTML@, @PDF@, and @BAT@ are used for automatically opening the corresponding documents traced at a location specified by a bookmark.

The simple rules for designing a test case have been introduced in Chap. 9 and Sect. 15.6.

An sample test case script file with some test case descriptions is listed as follows (TestScript1):

```
# test case 1 for New Order
# @HTML@ C:\Billing_and_Payment10\Requirement_specification.htm#New_Order
# @WORD@ C:\Billing_and_Payment10\Prototype_design.doc bmname New_Order
# @WORD@ C:\Billing_and_Payment10\TestRequirements.doc bmname New_Order
# [path] main(int, char**) {s0, s1, s9}  [/path]
# [Not_Hit] !path [/Not_Hit]
# Expected output : none
C:\Billing_and_Payment10
Billing_and_Payment.exe new_order Confirm

# test case 2 for Pay Invoice
#@HTML@ C:\Billing_and_Payment10\Requirement_specification.htm#Pay_Invoice
#@WORD@ C:\Billing_and_Payment10\Prototype_design.doc Pay_Invoice
# @BAT@ C:\isa_examples\ganttpro\ganttpr9.bat
#[path] main(int, char**) {s1, s6, s9, }B-Pay_Invoice(void)   [/path]
#[Not_Hit] !path [/Not_Hot]
# Expected output : none
C:\Billing_and_Payment10
Billing_and_Payment.exe Pay_Invoice

......
```

16.3 The New Software Testing Paradigm Based on the Transparent-Box Testing Method

Based on the Transparent-box method, a new revolutionary software testing paradigm is established which offers comprehensive functionalities and capabilities for software testing, including the support not only for Transparent-box testing, but also for MC/DC (Modified Condition/Decision Coverage) test coverage analysis, memory leak and usage violation check, performance analysis, runtime error type analysis and execution path tracing, GUI operation capture and selective playback, test case efficiency analysis and test case minimization for efficient regression testing after code modification, incremental unit testing and integration testing combined together seamlessly, semiautomatic test case design, and so on.

Application examples of this new software testing paradigm in the requirement development phase for finding logic defects and inconsistency defects efficiently with the Holistic, Actor–Action and Event–Response driven, Visual, Traceable, and Executable (HAETVE) software requirement development technique innovated by me to be used to replace the Use Case approach (which is not holistic, not suitable for event–response type applications, not traceable, and not directly executable for defect removal) are shown in Figs. 16.4–16.6.

The dummy programming source code of the main() module is listed as follows:

```
void main(int argc,char** argv)
{
int key;
if(argc==1 /* Missing a parameter * /
    || argc > 2 /* Having an extra parameter */)
    {cout << "Invalid Commands: \n" << argv;
    }
else
{
if(strcmp(argv[1],"New_Order")==0 || strcmp(argv[1],"New_order")==0
    || strcmp(argv[1],"new_order")==0 )
    {
    A_New_Order();
    cout << "*** A_New_Order () called. ***\n";
    }
else if (strcmp(argv[1],"Confirm_Order")==0 ||
  strcmp(argv[1],"Confirm_order")==0
  || strcmp(argv[1],"confirm_order")==0 )
    {
    C_Confirm_Order();
    cout << "*** C_Confirm_Order () called. ***\n";
    }
else if (strcmp(argv[1],"Invoice_Buyer")==0 ||
  strcmp(argv[1],"Invoice_buyer")==0
  || strcmp(argv[1],"Invoice_buyer")==0 )
    {
    D_Invoice_Buyer();
    cout << "*** D_Invoice_Buyer() called. ***\n";
    }
```

```
      else if (strcmp(argv[1],"Pay_Invoice")==0 ||
        strcmp(argv[1],"Pay_invoice")==0
        || strcmp(argv[1],"pay_invoice")==0 )
          {
          B_Pay_Invoice();
          cout << "\n *** B_Pay_Invoice() called. ***\n";
          }
      else if (strcmp(argv[1],"Send_Reminders")==0 ||
        strcmp(argv[1],"Send_reminders")==0
        || strcmp(argv[1],"send_reminders")==0 )
          {
          E_Send_Reminders ();
          cout << "\n *** E_send_Reminders() called. ***\n";
          }
      else
        cout << "Invalid Commands: \n" << (char**) argv <<endl;
        cout << " *** Executed. *** \n" << (char**) argv <<endl;
        }
      }
```

After execution of the test script file, TestScript1, using this new software testing paradigm through the Panorama++ product, one logic defect and another inconsistency defect were found as shown in Fig. 16.5.

After checking the source code, we can easily find that the defect is from an extra space character:

```
                                    | an extra space character
                                    V
    if(argc==1 /* Missing a parameter * /
        || argc > 2 /* Having an extra parameter */)
        {
        cout << "Invalid Commands: \n" << argv;
        }
    else
    {
    if(strcmp(argv[1],"New_Order")==0     ||    strcmp(argv[1],"New_order")==0
        || strcmp(argv[1],"new_order")==0 )
        {
        A_New_Order();
        cout << "*** A_New_Order () called. ***\n";
        }
```

After checking the bookmarks, we found that in the TestRequirements.doc file the bookmark Now_Order is pointing to the Pay_Invoice Treatment position rather than the New_Order Treatment position. After removing the two defects, a correct result is obtained as shown in Fig. 16.6.

When this new software testing paradigm is applied to test a software program without the source code, we can design a virtual main() to indicate the corresponding operations and call the program indirectly through dummy programming too.

16.3 The New Software Testing Paradigm Based on the Transparent-Box Testing Method 415

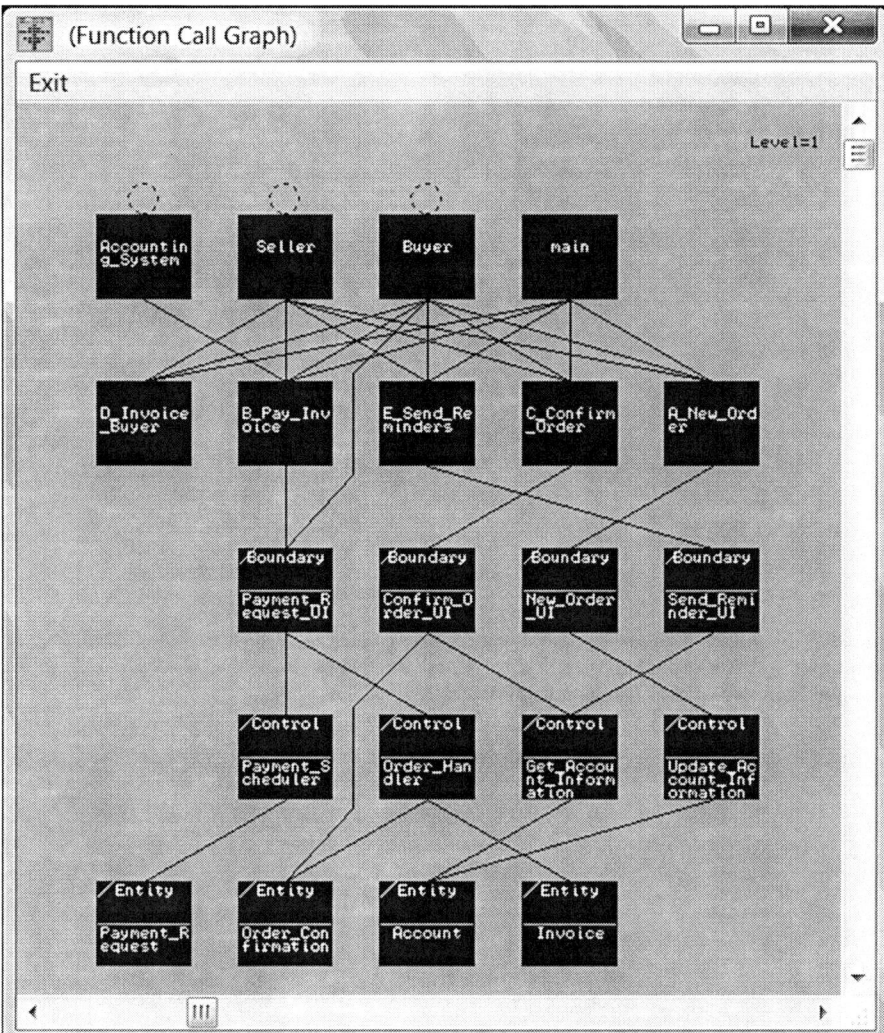

Fig. 16.4 An application result of the HAETVE technique for decomposition of the functional requirements of a Billing_and_Payment product through dummy programming using dummy modules (there are some function call statements in the body of a module (or an empty body) without real program logic)

In this way, the GUI operation can be captured and automatically played back after code modification with the capability to establish bidirectional traceability to find the inconsistency defects among the test cases, the test requirements, the user's manual, and other related documents even if the source code is not available.

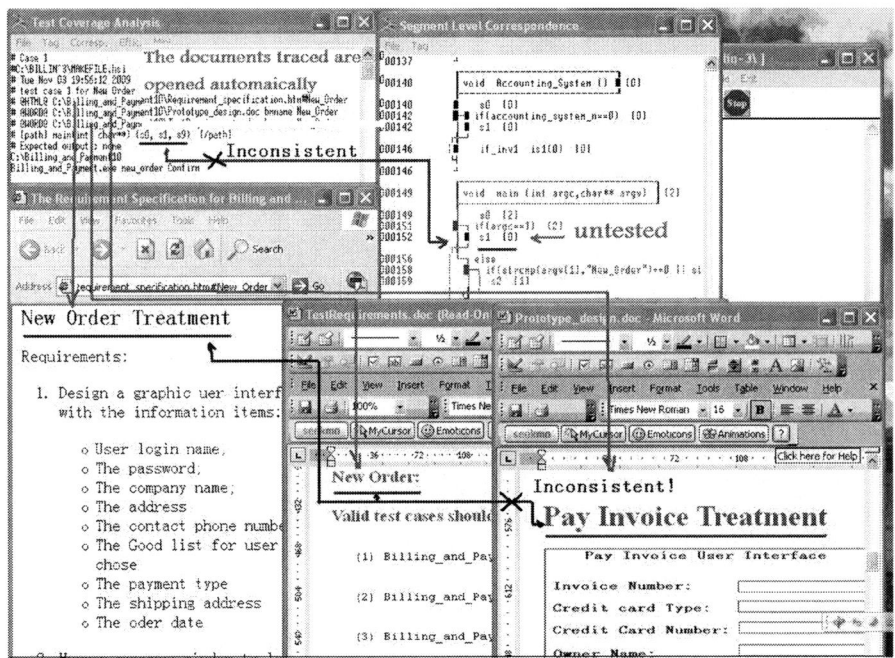

Fig. 16.5 Two defects found through dynamic testing using the Transparent-box method when performing a forward tracing operation (*Note*: all the related documents are opened from the locations indicated by the corresponding bookmarks)

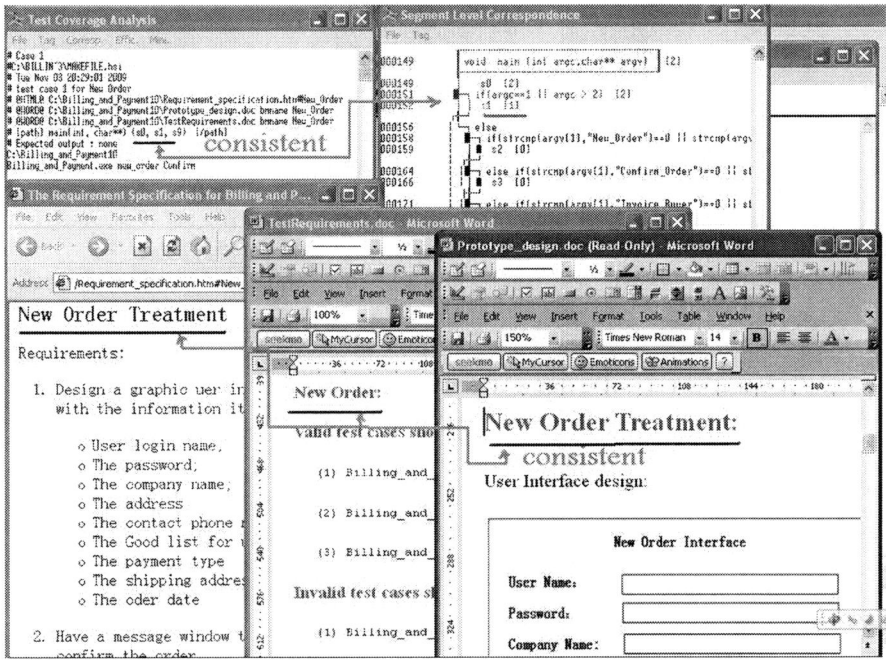

Fig. 16.6 After modification, the two defects shown in Fig. 16.5 are removed

16.4 The Major Features of the New Software Testing Paradigm

The new presented software testing paradigm brings revolutionary changes to software testing. The major features of the new software testing paradigm include:

- It is based on the Transparent-box testing method which combines functional testing and structural testing together seamlessly with close logic connections and a capability to automatically establish bidirectional traceability among the related documents and test cases and the corresponding tested source code, as shown from Figs. 16.4–16.6.
- It can be used dynamically in the entire software development lifecycle, from the requirement development phase down to the maintenance phase.
- It can be used to find functional defects, structural defects, inconsistency defects, memory leaks and memory usage violation defects, and performance bottlenecks.
- It supports MC/DC test coverage analysis required for the RTCA/DO-178B level A standard, being able to show the test coverage analysis results graphically with untested branches and conditions highlighted as shown in Fig. 16.7.
- It supports memory leak analysis and memory usage violation check. An application example is shown in Fig. 16.8.

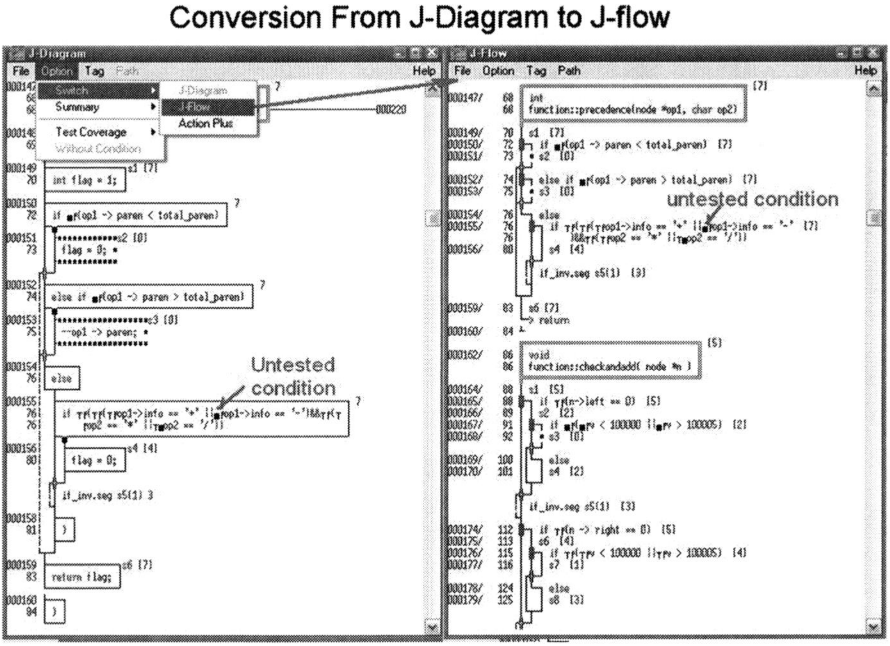

Fig. 16.7 MC/DC test coverage analysis and the analysis results shown graphically

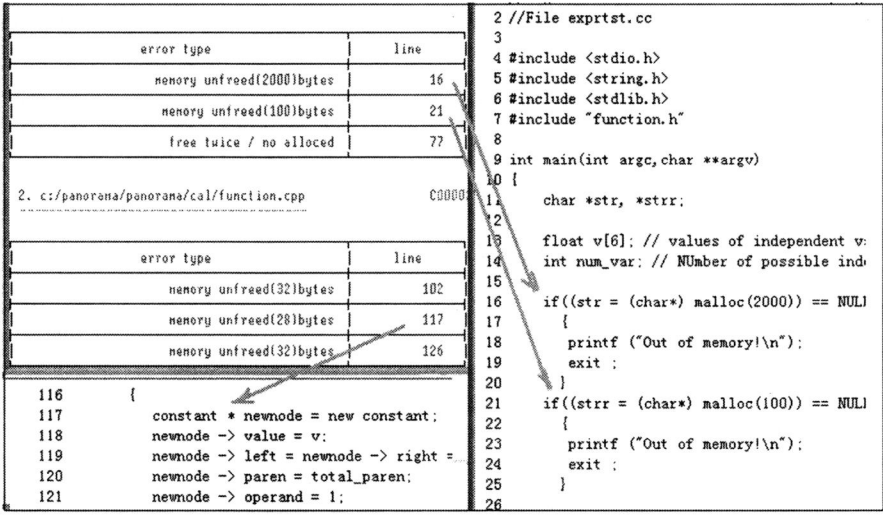

Fig. 16.8 A report on memory leaks and usage violations check

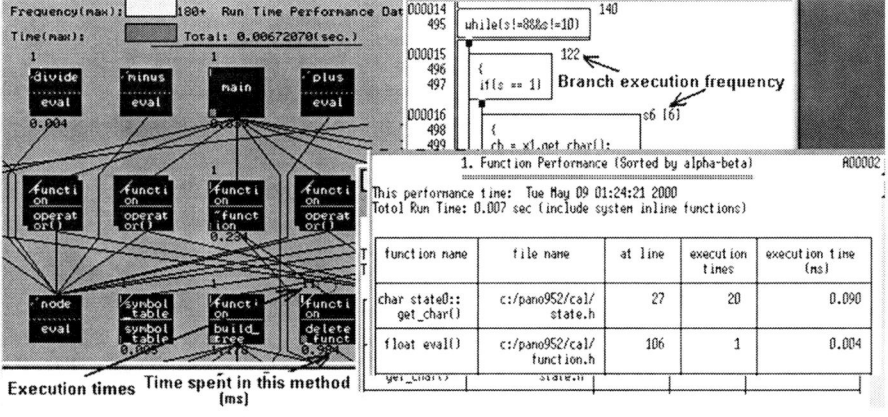

Fig. 16.9 An application example of performance analysis performed by Panorama++

- It supports performance analysis with the capability to report the branch execution frequency to locate performance bottlenecks better as shown in Fig. 16.9.
- It supports efficient test case design by automatically choosing a typical path with the most untested branches and automatically extracting the execution conditions of the chosen path as shown in Fig. 16.10.
- It supports embedded software testing too, as shown in Fig. 16.11.
- It combines software testing and debugging together visually.

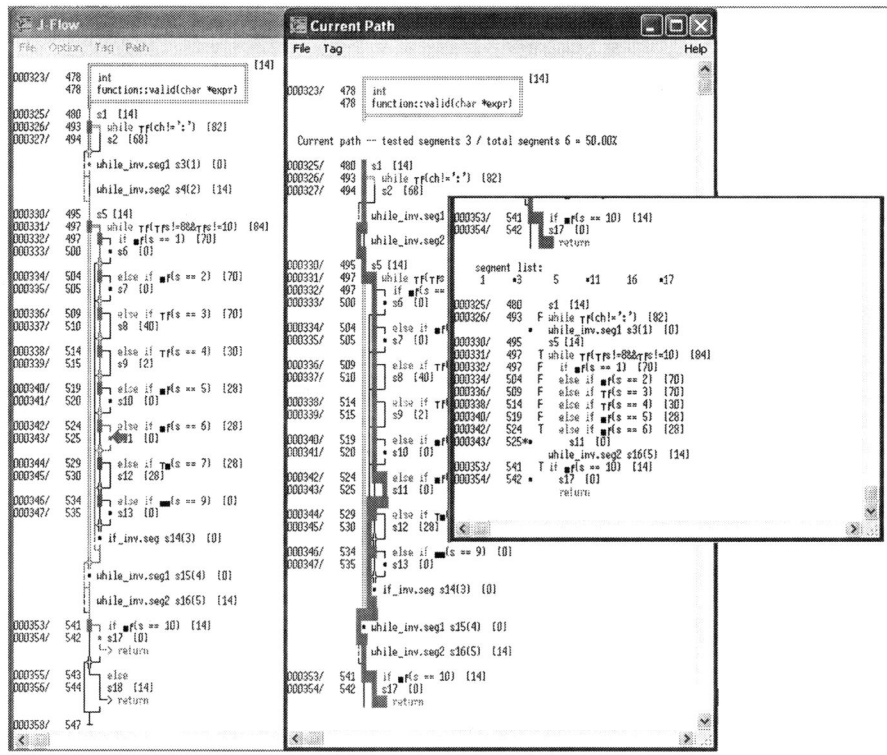

Fig. 16.10 Assisted test case design performed by Panorama++

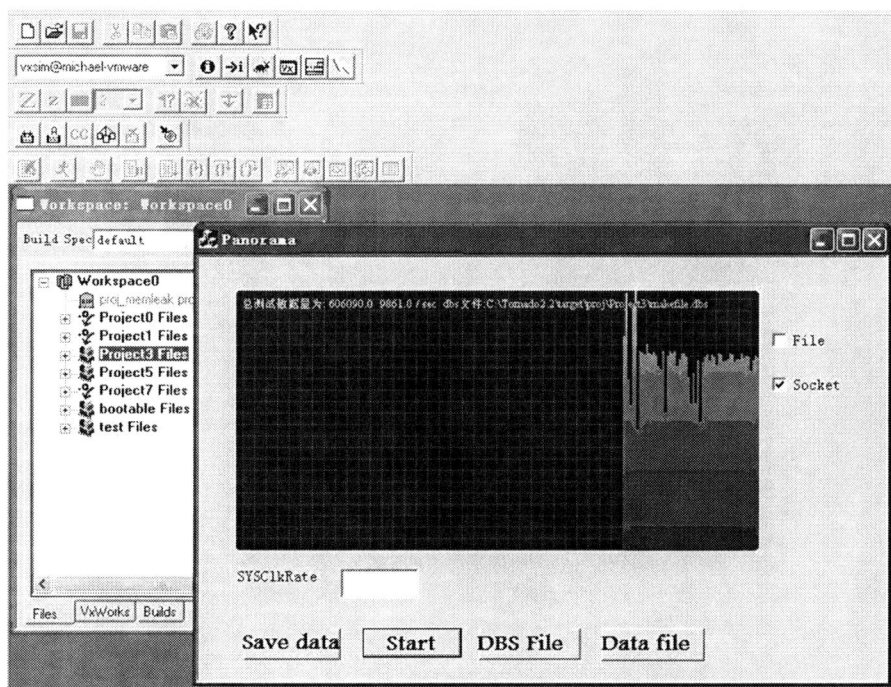

Fig. 16.11 An application example shows that the MC/DC test coverage data are sent from the target to the test server

The NSE software testing paradigm combines software testing and debugging together closely as shown in the following examples:

1. The source code of a sample program module "trouble" with seven defects and the corresponding "main" module is listed as follows:

```
/* File: main.c */
1       #include <stdio.h>
2       static char *tp=NULL;
3       int r=1, x=0, y=1000000, z=0;
4       FILE *fd=NULL;
5       void trouble();
6
7       main(argc, argv)
8       int argc;
9       char **argv;
10      {
11      int  k=0;
12      if(argc>1) trouble(atoi(argv[1]));
13      if(fd) fclose(fd);
14      }
```

```
/* File: trouble.c */

1       /* trouble.c */
2
3       #include <stdio.h>
4       #include <malloc.h>
5
6       #ifdef ERROR_SIMULATION
7       #include "ISA_simu.h"
8       #endif
9       extern int x,y,z;
10      extern FILE *fd;
11      FILE *fi, *fo;
12
13      trouble (x)
14      int x;
15      {
16      int i, t=1;
17      char c,*pc=NULL,ch[10],*p=NULL,*e=NULL;
18      if((e=malloc(4))==NULL)printf("Out of memory,x=%s",x), exit(-1);
19      for(i = x; i <= 8 && t; p=&ch[i++])
20      if(i % 2 ==1) {
21              p=&c; t=0; }
22      ch[0] = *p;      /* seg. fault when x > 8 */
23      i = x ;
24      while (i > -2 && i<=7 ){/*dead loop if x=7 or x=3*/
25      switch ( x + z ) {
26      case  0: case 1: x = z = 1; break;
27      case  2: y = 1; break; }
28      if ( i < 7 )
29           i += 4; }
30      if ( x < 5 )
```

16.4 The Major Features of the New Software Testing Paradigm

```
31        pc = ch;
32      if( x < 6 )
33        fd=fopen("trouble.c", "r");
34        c = getc (fd);           /* seg. fault when x = 6 */
35        strcpy (pc, "ab"); /* seg. fault if x = 5 */
36        c = ch[y]; /* seg. fault when x = 4 */
37        z = x / z; /* Arith. excep. when x = 2 */
38        if((p=malloc(3))!=NULL) strcpy(p,"OK");
39      }
40
```

2. The following shows what is provided by a typical test tool using the statement/ block test coverage metric after the execution of the main() function called the trouble(x) function with x = 0:

```
#include <stdio.h>
    static char *tp=NULL;
    int r=1, x=0, y=1000000, z=0;
    FILE *fd=NULL;
    void trouble();
    main(argc, argv)
    int argc;
    char **argv;
1 -> {
    int  k=0;
    if(argc>1) trouble(atoi(argv[1]));
1 -> if(fd) fclose(fd);
1 -> }

  100.00       Percent of the file executed

    /* trouble.c */
    #include <stdio.h>
    #include <malloc.h>

    #ifdef ERROR_SIMULATION
    #include "ISA_simu.h"
    #endif
    extern int x,y,z;
    extern FILE *fd;
    FILE *fi, *fo;
    trouble (x)
    int x;
1 -> {
    int i, t=1;
          char c,*pc=NULL,ch[10],*p=NULL,*e=NULL;
       if((e=malloc(4))==NULL)printf("Out of memory,x=%s",x), exit(-1);
1,   2 -> for(i = x; i <= 8 && t; p=&ch[i++])
2 ->     if(i % 2 ==1) {
1 ->       p=&c; t=0; }
1 -> ch[0] = *p;        /* seg. fault when x > 8 */
     i = x ;
     while (i > -2 && i<=7 ){/*dead loop if x=7 or x=3*/
```

```
2 ->    switch ( x + z ) {
1 ->      case  0: case 1: x = z = 1; break;
1 ->      case  2: y = 1; break; }
2 ->    if ( i < 7 )
2 ->       i += 4; }
1 -> if ( x < 5 )
1 ->    pc = ch;
1 -> if( x < 6 )
1 ->    fd=fopen("trouble.c", "r");
1 -> c = getc (fd);     /* seg. fault when x = 6 */
1 -> strcpy (pc, "ab"); /* seg. fault if x = 5 */
     c = ch[y]; /* seg. fault when x = 4 */
     z = x / z; /* Arith. excep. when x = 2 */
     if((p=malloc(3))!=NULL) strcpy(p,"OK");
1 -> }

  100.00       Percent of the file executed
```

It means that the tool offering statement test coverage analysis capability reported 100% of the program have been tested without finding any defects.

3. Comments on a typical statement/block test coverage analysis tool:

(a) The analysis result is coding style dependent

```
Suppose there are two statements as follows:

if( 0 ) printf (" Can't be executed. \n");
```

and

```
if( 0 )
        printf ( " Can't be executed. \n");
```

and only the condition parts of them are tested but has never been satisfied, the first statement will report that the entire statement has been tested, but the second one will not.

(b) It cannot identify whether an invisible segment (such as when there is an "if" statement without the "else" part) has been executed or not.

(c) If several "case" statements share an execution body such as

```
case 0: case 1:
    printf(" Less than 2.\n");
        break;
```

but only one of the conditions of the cases is satisfied (such as case 0 is satisfied), it cannot indicate that the other cases are not executed.

(d) It cannot identify whether the high end of a loop boundary is executed or not.

(e) It cannot identify whether a condition outcome or the combination of some condition outcomes is executed or not.

16.4 The Major Features of the New Software Testing Paradigm

4. After compilation, execute the program directly (with X = 6).
   Without using NSE tools, the system shows an error message with no detailed information (see Fig. 16.12).
   In this case, the system debugger can be used to report the related information in object code format as shown in Fig. 16.13.
5. But with NSE, the detailed error information will be reported with the error type and the source code location as shown in Fig. 16.14.
6. Debugging can also be performed visually with the NSE software engineering paradigm as shown in Figs. 16.15–16.19.

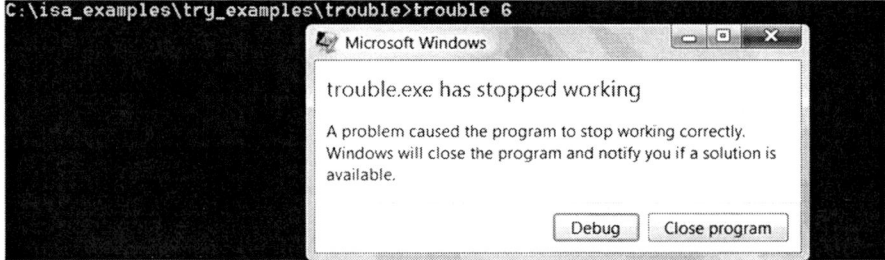

Fig. 16.12 An error message given by the system without showing the error location

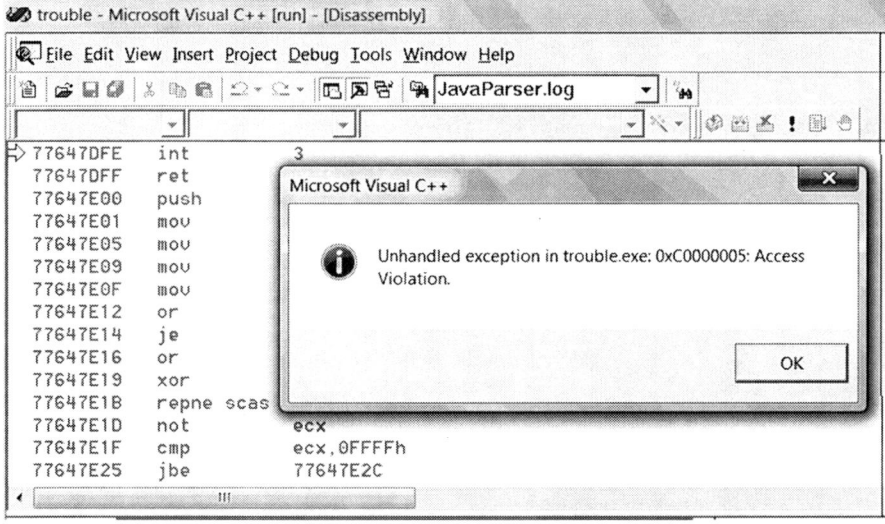

Fig. 16.13 The system debugger can only show the location of the object code which is not very useful

424   16 The NSE Software Testing Paradigm Based on the Transparent-Box Method

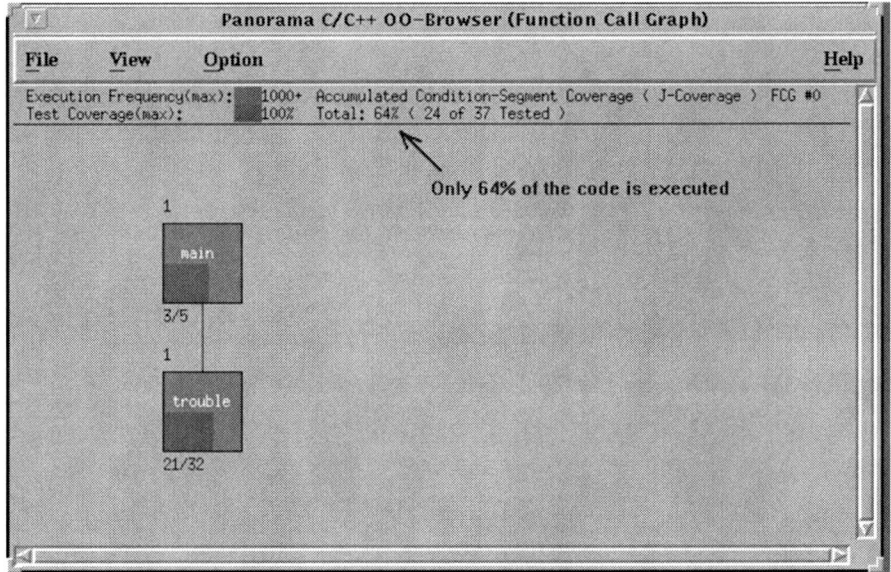

Fig. 16.14 When it is executed under NSE, an error message is given with the error type and the detailed source code location (line 133 in the file trouble.c)

Fig. 16.15 The corresponding program test coverage shown in J-Chart

Figure 16.15 shows that after execution of the main() function called the trouble(x) function with x = 0, NSE's support platform, Panorama++, will report that only 64% of the program have been tested using the MC/DC test coverage metric.

The untested branches/segments and conditions can be highlighted in the J-Diagram as shown in Fig. 16.16.

The untested branches and condition can also be highlighted in a J-Flow diagram as shown in Fig. 16.17.

16.4 The Major Features of the New Software Testing Paradigm

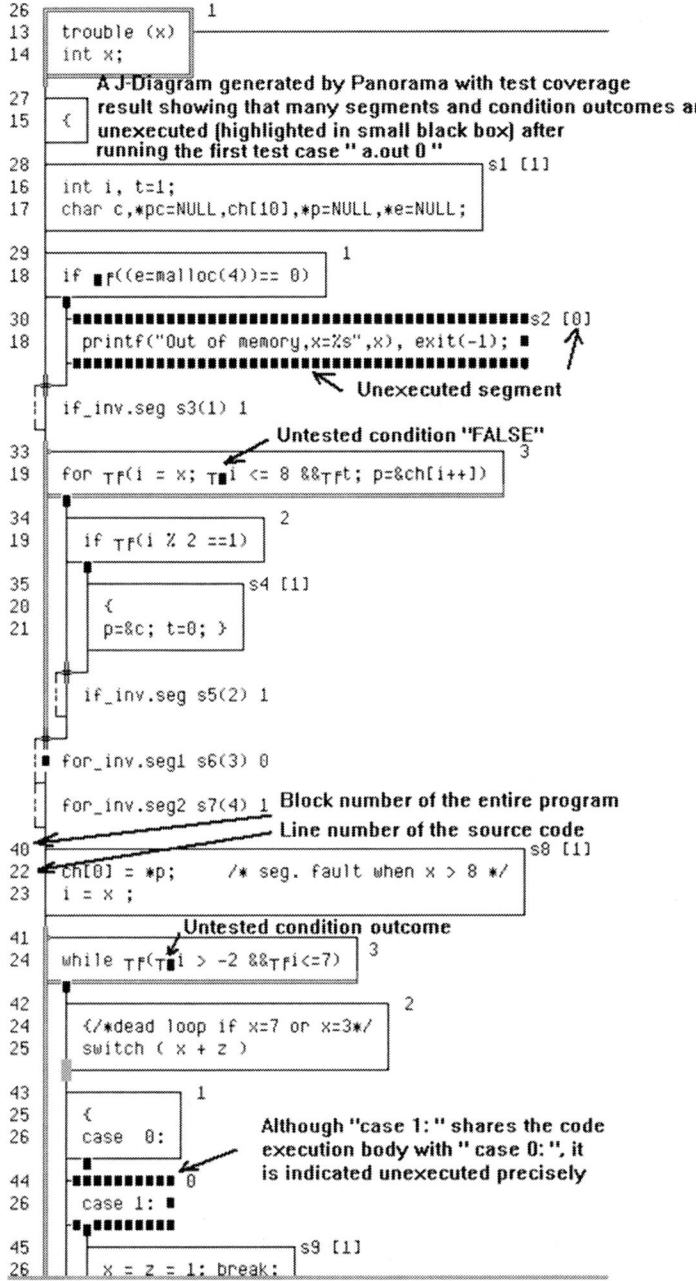

Fig. 16.16 The corresponding logic diagram shown in J-Diagram notation with untested branches and conditions highlighted in *small black boxes*

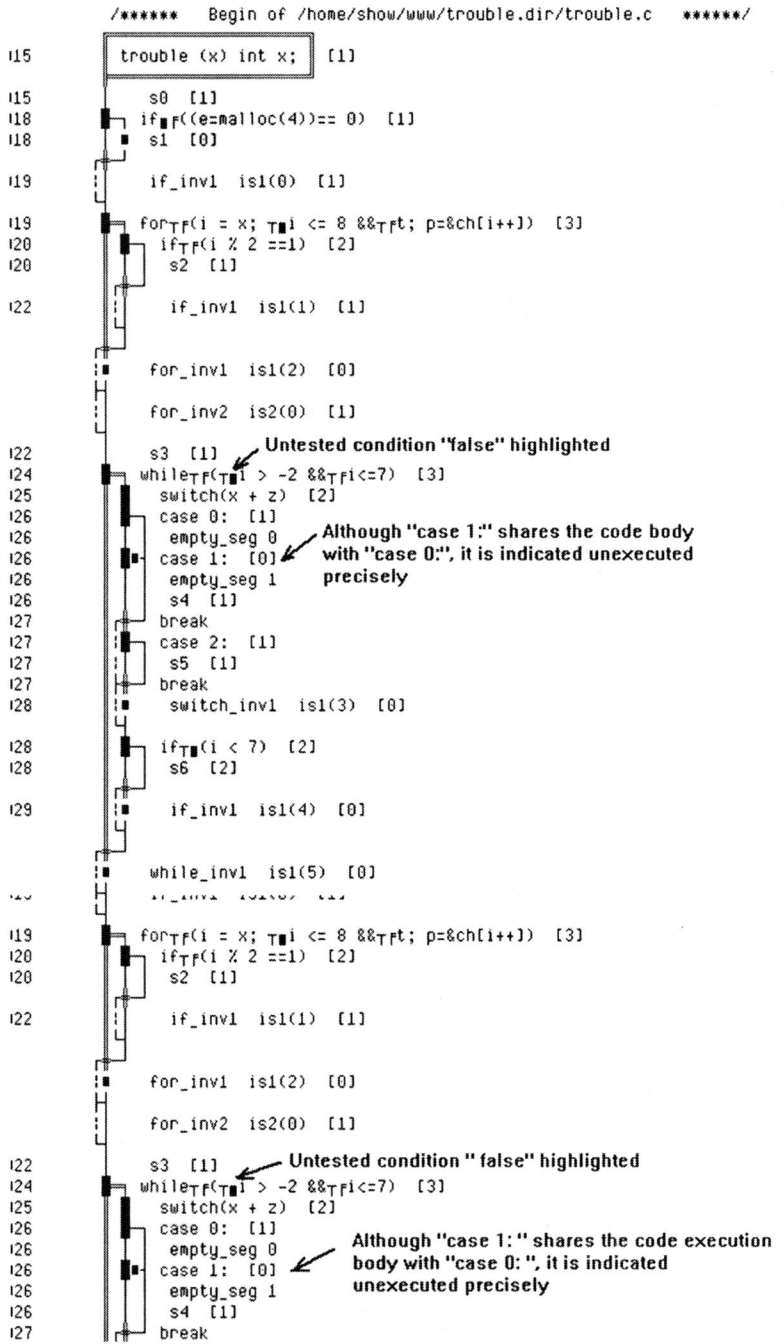

Fig. 16.17 The corresponding J-Flow diagram shown with the untested branches and conditions highlighted

16.4 The Major Features of the New Software Testing Paradigm

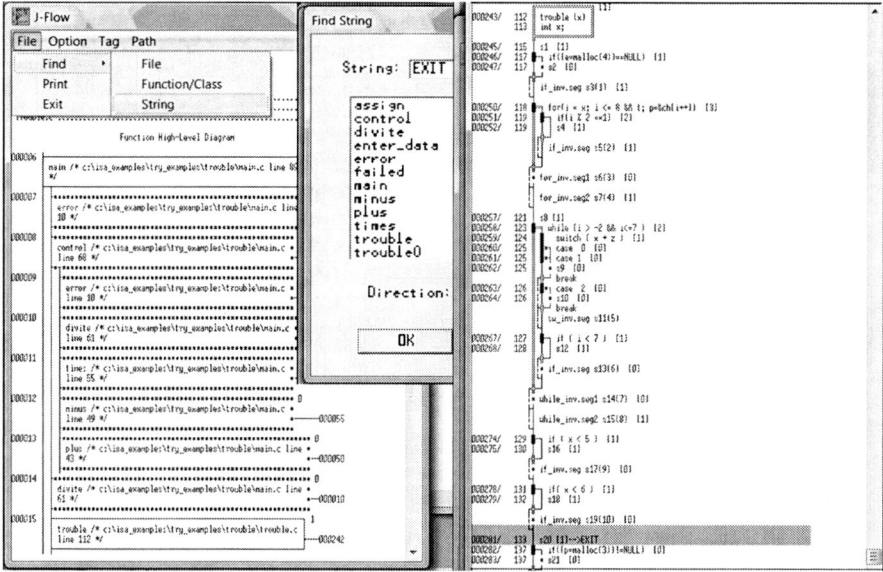

Fig. 16.18 Finding the location where a program terminated unexpectedly using J-Flow diagram through searching the added word "EXIT"

Figure 16.18 shows that when a runtime error happens during the testing process, users can directly find the corresponding source code location using the J-Flow diagram through searching a word "EXIT" which is automatically added into the J-Flow diagram to indicate the error location (sometimes the defect may be introduced earlier but the program is terminated later).

7. With all the untested branches and conditions being tested, seven defects can be found and fixed by modifying the source code. After that, the logic diagram will show that 100% of the branches and the conditions are all tested as shown in Fig. 16.19.

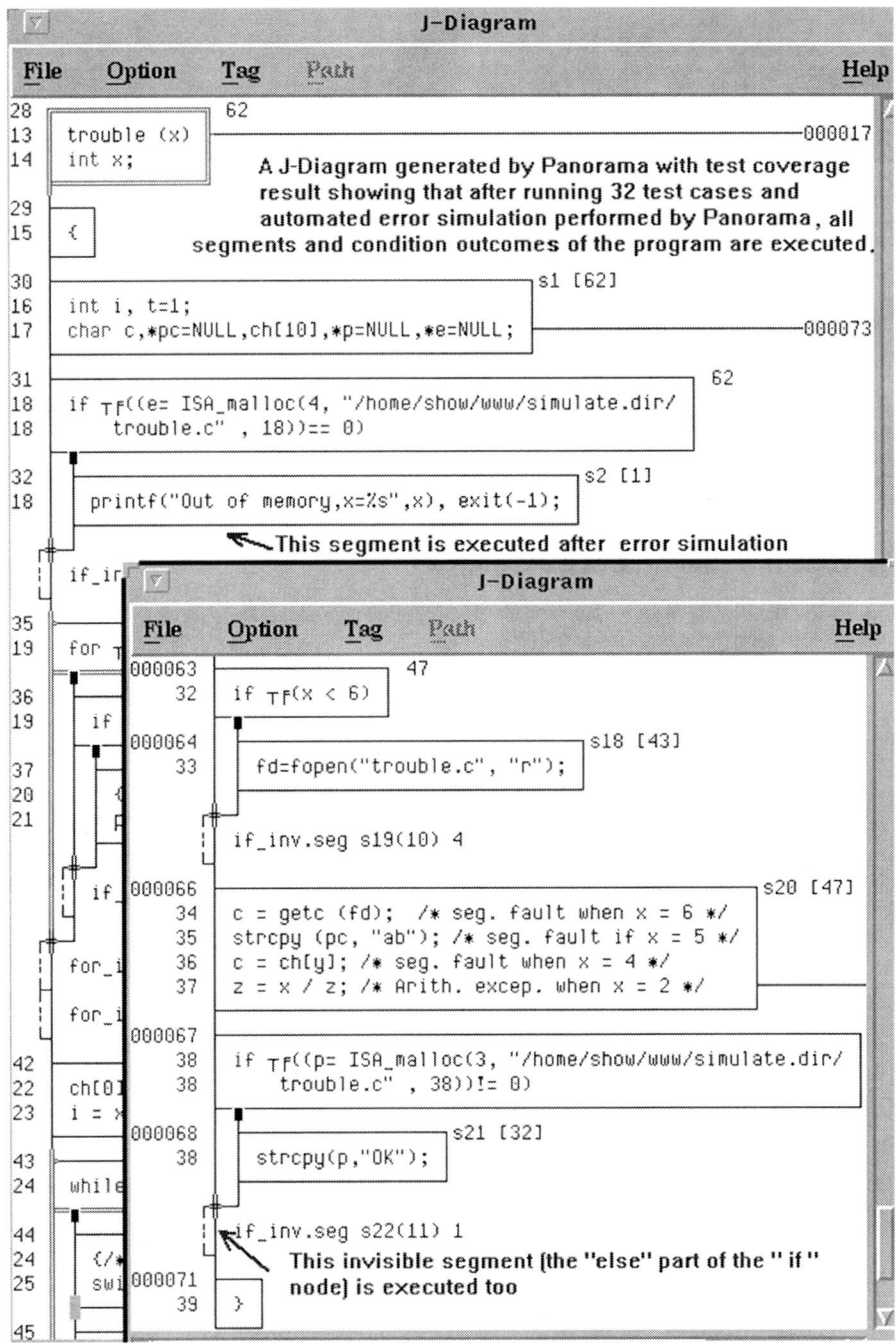

Fig. 16.19 The final result after removing all defects with the trouble module

16.5 A General Comparison Between the New Software Testing Paradigm and the Old One

(a) **The defect finding efficiency**

The old testing paradigm used for incremental software development is shown in Fig. 16.20 [Coc08].

The old testing paradigm used for the iterative software development is shown in Fig. 16.21 [Coc08].

The new presented software testing paradigm used for incremental or iterative software development is shown in Fig. 16.22.

Comparing Figs. 16.20–16.22, it is clear that the new software testing paradigm is much more efficient in finding defects in the software product development process.

(b) **The timing in finding the defects**

The traditional software testing methods can be performed after coding, but it is too late; in comparison, the new presented software testing paradigm can be used in all phases of a software development lifecycle, including the requirement development phase and the design phase.

(c) **The defect types that can be found**

The traditional Black-box method can be used to find functional defects; the traditional structural White-box method can be used to find some structural

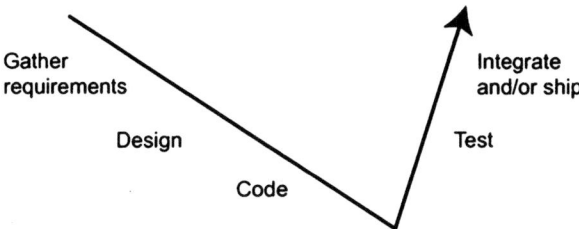

Fig. 16.20 Traditional software testing performed with incremental software development

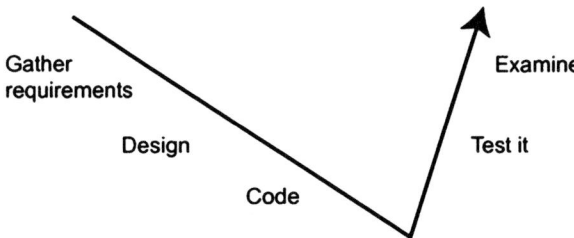

Fig. 16.21 The old testing paradigm used for the iterative software development

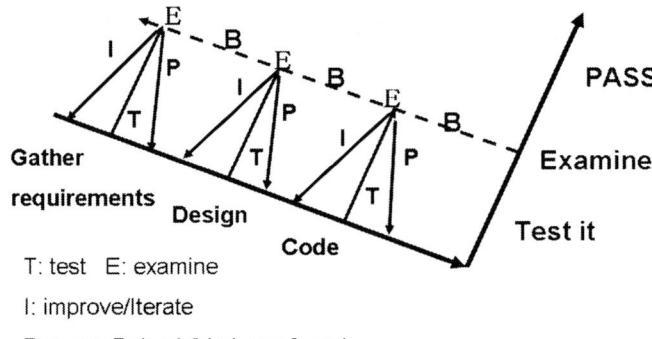

Fig. 16.22 The new presented software testing paradigm used for incremental or iterative software development

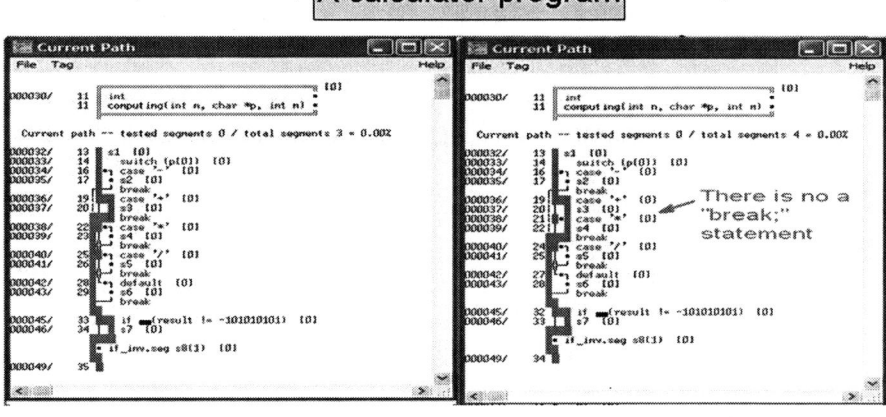

Fig. 16.23 An application example of Transparent-box testing: a bug is found even if the output is the same as what is expected (this defect is from a "break" statement which is missing, so that the result 4 is produced through 2×2 rather than $2+2$)

defects for the existing product no matter if it is the customer-required product or not.

The presented new software testing paradigm can be used to find functional defects, structural defects, logic defects, and inconsistency defects.

Some functional defects cannot be found by the Black-box method, but can be found by the new software testing paradigm as shown in Fig. 16.23.

16.5 A General Comparison Between the New Software Testing Paradigm and the Old One

(d) **The graphical representation techniques for displaying the test results**

The test results obtained from the applications of most traditional software testing methods and tools are shown in textual formats or value tables. But the test results obtained from the applications of the presented new software testing paradigm are graphically shown in the system level and in the detailed source code level in Fig. 16.24.

(e) **The capability to establish automated traceability**

It is only supported by the new presented software testing paradigm.

(f) **The capability to combine software testing and debugging together visually**

It seems that it is only supported by the new presented software testing paradigm.

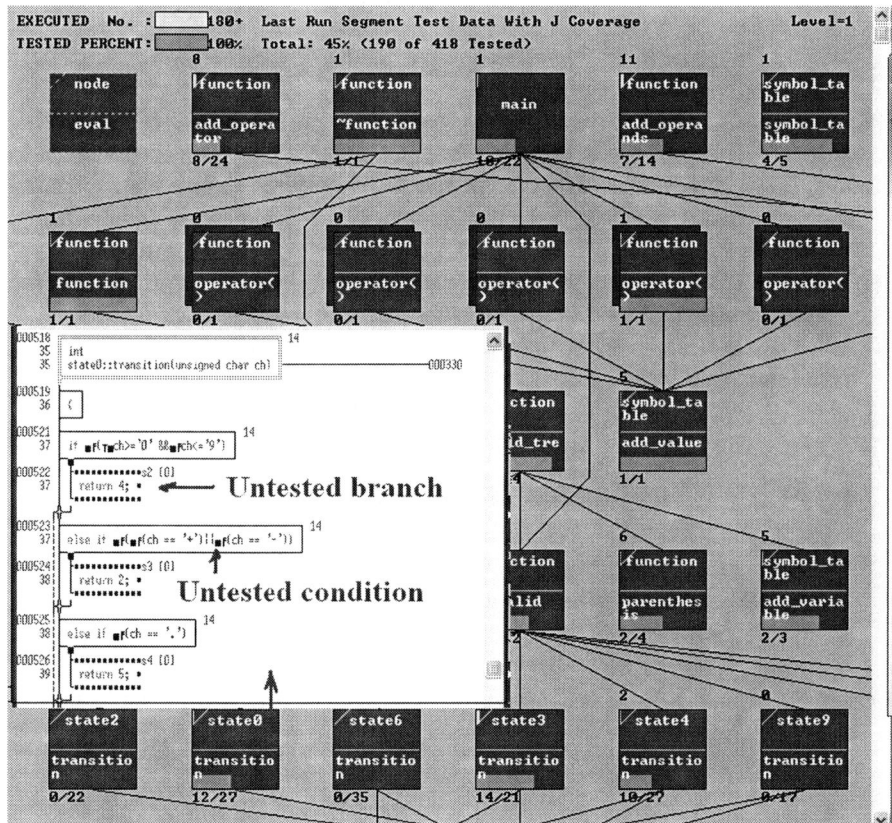

Fig. 16.24 An example of test coverage analysis result obtained using the presented new software testing paradigm (the untested branches and conditions are highlighted with *small black boxes*)

16.6 Summary

This chapter presented a new software testing paradigm based on the Transparent-box testing method which brings revolutionary changes to software testing in the twenty-first century by combining structural testing and functional testing together seamlessly with internal logic connections and the capability to establish automated and self-maintainable traceability among the related documents and test cases and the source code, which can be used dynamically in the entire software development lifecycle from requirement development down to maintenance.

16.7 Points and Questions to Ponder

(a) Why is it that the existing software testing methods, techniques, and tools cannot be dynamically used in the software requirement development phase and the software design phase?
(b) Why are software functional testing and structural testing performed separately with today's software testing paradigm?
(c) Software disasters happen often – is it related to the drawbacks of the existing software testing methods, technologies, and tools? Why?
(d) Why can and should the Transparent-box testing method and the corresponding tools be dynamically used in the requirement development phase and the design phase?
(e) What are the key points in designing test cases for software testing using the Transparent-box method?
(f) What are the major differences between the old-established software testing paradigm and the NSE software testing paradigm?

16.8 Further Reading and Information Source

(a) Mead A. Deming's principles of total quality management (TQM). **http://www.well.com/user/vamead/demingdist.html**, **http://www.ammdoc.com**
(b) Hower R. Software QA and testing frequently-asked-questions. http://www.softwareqatest.com/qatfaq1.html

References

[Dem86]   Deming WE (1982) Out of the crisis. MIT Press, Cambridge
[Coc08]   Cockburn A (2008) Using both incremental and iterative development. CrossTalk, May Issue

Chapter 17
NSE Software Quality Assurance Paradigm Driven by Defect Prevention

> *Cease dependence on inspection to achieve quality. Eliminate the need for inspection on a mass basis by building quality into the product in the first place.*
>
> W.E. Deming

Regarding software quality, Watts S. Humphrey said, "Over the last 50 years there has been very little improvement" [Fry07]. Capers Jones said, "Major software projects have been troubling business activities for more than 50 years. Of any known business activity, software projects have the highest probability of being canceled or delayed. Once delivered, these projects display excessive error quantities and low levels of reliability." [Jon06].

Why? In fact, the quality of a software product cannot be efficiently ensured by quality management and quality assurance visibility only, cannot even be efficiently ensured by general quality assurance methodology and technology and the tools only, because the issue of software quality is strongly related to almost the entire software engineering paradigm, including the foundation of the software engineering, the process model, the software development methodology, the software testing paradigm, the software visualization paradigm, the software documentation paradigm, the software maintenance paradigm, the software project management paradigm, the software support techniques and tools, and the software quality assurance paradigm.

This chapter introduces the NSE software quality assurance paradigm supported by the entire NSE software engineering paradigm with all of its components.

17.1 The Old-Established Software Quality Assurance Paradigm Is Outdated

The old-established software quality assurance paradigm is outdated:

1. **It works with the old-established software engineering paradigm based on reductionism and the superposition principle** that the whole of a complex system is the sum of its components, so that almost all of the tasks

and activities in software quality assurance are performed linearly, partially, and locally, such as the implementation of requirement changes or code modifications.
2. **The corresponding software development process models are linear ones with no upstream movement at all**, making the defects introduced in the requirement development phase and software design phase easily propagate down to the maintenance phase, and the defect removal cost increase tenfold several times.
3. **The corresponding software development methodologies are based on Constructive Holism principle** that the components of a complex system are developed first, then the whole of the system is built from its components – it makes the quality of a software product much more difficult to ensure – for instance, when a runtime error happens in the product integration, it is hard to know where the error comes from.
4. **It is driven by inefficient inspection and testing after coding/production** – current software inspection uses documents and source code without bidirectional traceability, which is highly inefficient; the testing paradigm is mainly based on functional testing using Black-box method being applied after the entire product is produced and on structural testing using the White-box testing method after each software unit is coded.
5. **There is a lack of systematic strategy for quality assurance in the entire software product lifecycle** from the first step down to the retirement of the product.
6. **There is a no systematic, quantifiable, and disciplined method/approach to ensure the quality of a modified product** after requirement changes and/or code modifications.
7. **The quality assurance process and the quality assurance results are almost invisible** – for instance, it is invisible what code branches and condition combinations have not been executed.
8. **The quality management process and the software development process are separated** – for instance, the quality management documents are not traceable with the implementation of the requirements and the source code, and it is hard to update them to maintain consistency with the source code.
9. **The application results show:**
   (a) **About software quality, "Over the last 50 years there has been very little improvement"** (said Watts S. Humphrey, who founded the Software Process Program of the Software Engineering Institute (SEI)) [Fry07].
   (b) **The software project success rate is very low (about 30%).**
   (c) **Software disasters happen often.**

Conclusion: The old-established software quality assurance paradigm is outdated which does not meet the needs for software development in the twenty-first century.

17.2 Outline of NSE Software Quality Assurance Paradigm (NSE-SQA)

The solution offered by NSE for software quality assurance is described in detail in this chapter later. Here is the outline of the solution:

1. **It is based on complexity science by complying with the essential principles of complexity science, particularly the Nonlinearity principle and the Holism principle** that the whole of a complex system is greater than the sum of its components, and that the characteristics and behaviors of the whole emerge from the interaction of its components, so that with NSE almost all of the tasks and activities in software quality assurance are performed nonlinearly, holistically, and globally.
2. **The corresponding software development process model is a nonlinear one with two-way iteration (upstream movement and downstream movement)** for defect prevention and defect propagation prevention through dynamic testing, inspection using traceable documents and the source code, and software visualization.
3. **The corresponding software development methodology is based on the Generative Holism principle** that the whole of a complex system comes first as an embryo, and then grows up with its components – it makes the quality of a software product much easier to ensure [Bro95-p201]. For instance, each time the executable whole system grows up with one module, so that if a runtime error occurs, in most cases the error comes from the newly added module.
4. **It is driven by defect prevention and defect propagation prevention** through dynamic testing using the Transparent-box method combining functional testing and structural testing together with the capability to establish bidirectional traceability among related documents and test cases and source code for efficient inspection and review, and used in the entire software development and maintenance lifecycle.
5. **There is a systematic strategy for the quality assurance in the entire software product lifecycle** from the first step down to the retirement of the product through
    (a) Defect prevention
    (b) Defect propagation prevention (removing defects from the source)
    (c) Refactoring for modules with higher Cyclomatic complexity or performance bottlenecks
    (d) Deeper and broader testing and quality measurement plus quality assurance with side-effect prevention in the implementation of requirement changes and code modification through various traceabilities
6. **There is a systematic, quantifiable, and disciplined method/approach to ensure the quality of a modified product** through side-effect prevention in the

implementation of requirement changes and code modifications supported by various traceabilities.
7. **The quality assurance process and the quality assurance results are visible with the support of the NSE visualization paradigm** – for instance, it is visible what code branches and condition combinations have not been executed.
8. **The quality management process and the software development process are combined together closely** – for instance, the quality management documents are traceable with the implementation of the requirements and the source code, making it easy to update to maintain consistency with the source code.
9. **Preliminary application results show that compared with the old-established software quality assurance paradigm it is possible for NSE to help software development organizations to**
   (a) **Remove more than 99.99% of the defects in their software products**
   (b) **Double their software project success rate (about 60%)**
   (c) **Greatly reduce software disasters**

17.3 Description of NSE Software Quality Assurance Paradigm

17.3.1 The Foundation of NSE-SQA

The foundation for establishing NSE-SQA is complexity science which can efficiently handle the issues of a complex system with many components connected together with dynamic interactions.

17.3.2 The Framework for Establishing NSE-SQA

The establishment of NSE-SQA is done through the use of the FDS (the Five-Dimensional Structure Synthesis method) framework (a paradigm-shift framework, see Chap. 4) as shown in Fig. 17.1.

As shown in Fig. 17.1, the essential principles of complexity science are complied with in the establishment of NSE-SQA, particularly the Nonlinearity principle and the Holism principle that the whole of a complex system is greater than the sum of its components, and that the characteristics and behaviors of the whole emerge from the interaction of its components, so that with NSE-SQA almost all software quality engineering tasks/activities are performed holistically and globally to ensure the quality of a software product. For instance, with NSE-SQA, software maintenance will not be performed linearly, partially, and locally anymore, but nonlinearly, holistically, and globally to prevent the side effects for the implementation of requirement changes and code modifications to ensure the quality of the modified product.

17.3 Description of NSE Software Quality Assurance Paradigm

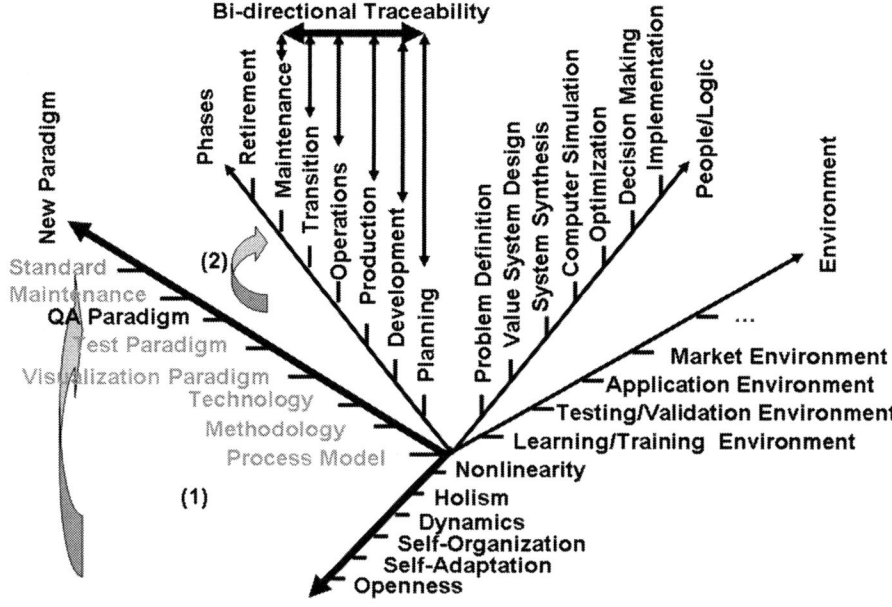

Fig. 17.1 The framework for establishing the NSE software quality assurance paradigm

17.3.3 The Purpose of NSE-SQA

The purpose of NSE-SQA is to revolutionarily solve the quality issues in software product development by applying many software defect prevention techniques, particularly the NSE software testing paradigm based on the Transparent-box method to dynamically test a software product from the first step down to the end step for

(a) Removing more than 99.99% of defects in a software product developed with NSE
(b) Making a software product truly maintainable through side-effect prevention
(c) Working with other efficient quality assurance techniques such as software debugging, Pair Programming, and Joint Application Design (JAD) to realize Six-Sigma quality standards

17.3.4 Definitions

17.3.4.1 Defect

The term defect refers to an error, fault, or failure [Cla01]. The IEEE/Standard defines the following terms as error: a human action that leads to an incorrect result. Fault: an incorrect decision taken while understanding the given information to solve problems or in the implementation of a process. Failure: inability of a function to meet the expected requirements ([Zel03], [Tia01]).

17.3.4.2 Defect Prevention

The Popular Definitions

1. Defect prevention (DP) is a process of identifying defects and their root causes, and taking corrective and preventive measures to prevent them from recurring in the future, thus leading to the production of a quality software product ([Sum08], [Nar08], [Vas05], [Hum89], [Ade05], [Kar07]).
2. The activities involved in identifying defects or potential defects and preventing them from being introduced into a product (SEI).
3. Technologies that minimize the risk of making errors in software deliverables [Jon02].

The New Definition with NSE

Defect prevention is the application process of a set of important software quality assurance techniques and tools for efficiently ensuring the quality of a software product in the entire software development and maintenance lifecycle, from the first step to the retirement of the product, to prevent software defects (majorly in the upstream phases for all kinds of defects including new ones never being found before, minorly in downstream phases for new and repeatable defects) from being introduced into the software product.

With NSE, defect prevention is performed mainly through

1. Dynamic testing using the Transparent-box method combining functional testing and structural testing together seamlessly, can be dynamically used in cases where there is no output (such as the requirement development phase and the software design phase) with the capability to establish automated and self-maintainable traceability to help users remove inconsistent defects among the related documents and test cases and source code.
2. Software visualization.
3. Inspection/review using traceable documents and source code.
4. Side-effect prevention in the implementation of requirement changes or code modifications supported by various traceabilities.
5. Repeatable Defect Prevention through
   (a) Causal analysis
   (b) Preventive actions
   (c) Increase awareness of quality issues
   (d) Data collection
   (e) Improvement of the Defect Prevention Plan
   The key points of the new definition:
(a) Defect prevention should be performed in **the entire software development and maintenance lifecycle.**
(b) It should be performed **from the first step** of the software development mainly through dynamic testing, visualization, and inspection using traceable documents and source code.

17.3 Description of NSE Software Quality Assurance Paradigm 439

(c) It should be performed **until the retirement** of a software product, not only in the product development site, but also in the product maintenance site.
(d) It should be performed **for all kinds of defects** (not only to prevent recurring repeatable defects).

17.3.4.3 Defect Propagation Prevention

The application process of a set of important techniques and tools for removing the defects introduced into a software product from the source.

17.3.5 The Quality Assurance Strategy of NSE-SQA

With NSE, the software quality assurance strategy consists of four major parts with different priorities from higher to lower as follows:

(a) Defect prevention – the top priority
(b) Defect propagation prevention
(c) Refactoring for the modules with higher Cyclomatic complexity or which are the performance bottlenecks (usually 20% of the most-complex modules will have about 80% of the defects)
(d) Deeper and broader software testing, quality measurement, and version comparison

17.3.6 The Implementation of the Quality Assurance Strategy of NSE-SQA

The NSE-SQA strategy has been implemented and commercially supported by the NSE support platform, Panorama++.

17.3.6.1 Defect Prevention

As introduced in Chaps. 8 and 10, with the NSE process model and the NSE software development methodology, defect prevention should be performed in the entire software development and maintenance lifecycle.

1. **In requirement development phase:**
   (a) Helps customer assign priority to requirements according to the importance of the requirements, works with NSE process model to implement the critical requirements (about 20% of the total requirements) first to form an essential version of the product and then incrementally grow the product, delivers all working versions to the customer to review to prevent wrong product development or overuse of the budget.

(b) Works with the HAETVE (Holistic, Actor–Action and Event–Response driven, Traceable, Visual, and Executable) technique and "dummy programming" for requirement development through program execution to prevent possible defects (see Chap. 11) – for instance, if the dummy program cannot be directly executed, there must be something wrong.
(c) Requests prototype design and review for important requirements to prevent the defects of unrealizable requirements.
(d) Provides several standard-based templates to be used to avoid omissions or errors in requirement development, such as the requirement specification template (see Appendix A).
(e) Requests concurrent development of requirement specifications and test scripts with test cases, to avoid untestable functional requirements as shown in Fig. 17.2.
(f) Provides forms for top-down structural documents and test script design using the requirements specification file as the root to assign directories and names and bookmarks for other documents before they have been made or after they have been made, to avoid overlooking any important documents, and then makes the related documents traceable to the test cases and the source code (see Chap. 9).
(g) If the customer requests a requirement change or a new requirement after some versions of a product have been delivered, and the requirement is critical, it is recommended to perform a prototype design and review again to avoid unrealizable requirements.
(h) For the implementation of a requirement change, provides forward traceability to find what documents and code modules need to be modified and backward traceability from each module to be modified to find whether the module is also used for the implementation of other requirement(s), to avoid conflict among different requirements.
(i) After the implementation of a requirement change, finds any inconsistent documents and correct them through bidirectional traceability.

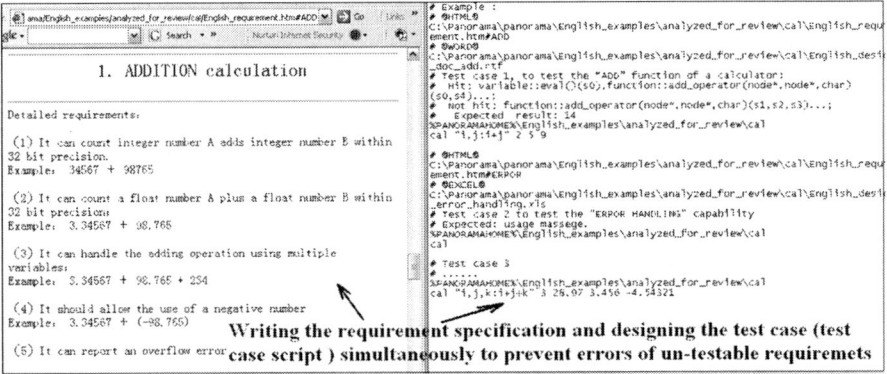

Fig. 17.2 An example of defect prevention in requirement development phase

17.3 Description of NSE Software Quality Assurance Paradigm

(j) For consistent modifications, provides backward traceability to find the related requirement(s), to ensure that the module functionality fulfills the requirements; also provides path traceability to find all related modules calling or called by the module in order to avoid inconsistency, etc.

2. **In software design phase:**

   (a) Combines the product development process and the product maintenance process together, greatly reduces the defects introduced in the product upstream and the propagation of defects down to the maintenance phase, and ensures the quality of a modified product through side-effect prevention supported by various traceabilities, so that it is possible to reduce up to two-third of the total effort and total cost spent in software maintenance to greatly prevent the problems of schedule delay and budget overuse.

   (b) Works with the NSE process model to combine the product development process and project management process together seamlessly to make the project management documents (particularly the product development plan and progress report as well as the cost reports) traceable with the implementation of requirements and the source code, to further prevent the problems of schedule delay and budget overuse – see Fig. 17.3.

   (c) Works with the **Synthesis Design and Incremental growing up** technique (see Chap. 12) and "dummy programming" for software design through program execution to prevent possible defects (see Chap. 12) – for instance, if the program cannot be directly executed, there must be something wrong.

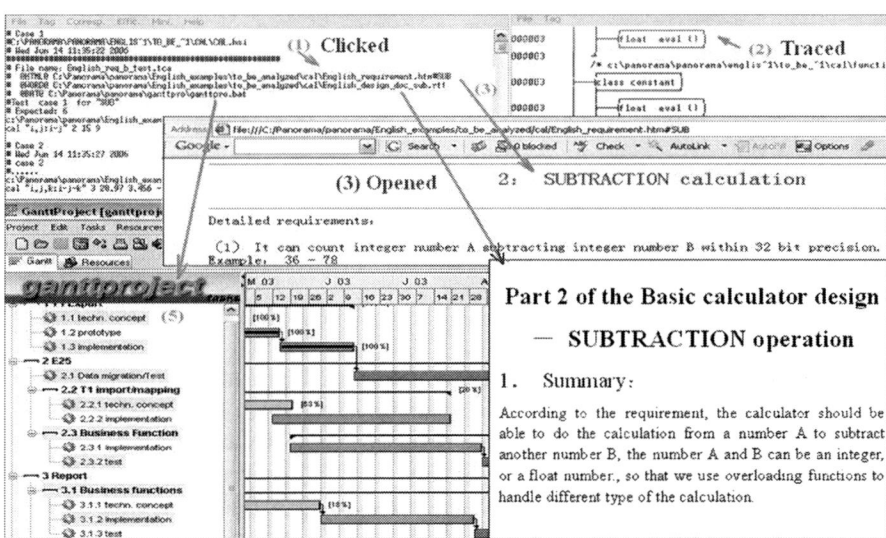

Fig. 17.3 An application example for making project development schedule chart traceable to the implementation of requirements and the source code

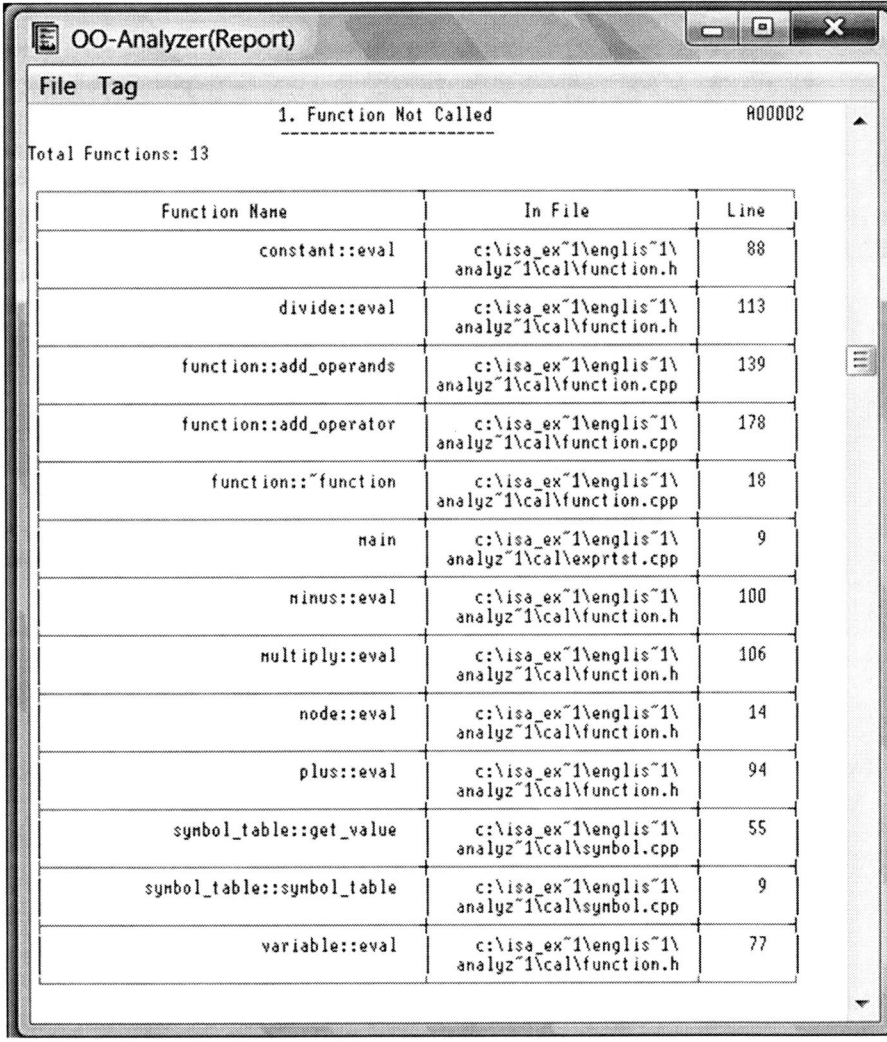

Fig. 17.4 An application example of unused module analysis

(d) Reports unused (uncalled) modules – why are there unused modules in the system? There must be something wrong as shown in Fig. 17.4.
(e) Use the documents including the function decomposition chart of the functional requirements, the description of the nonfunctional requirements, and the Event–Response table (see Table 11.2) as the basis to complete the software design to prevent something missing.
(f) Makes all related design documents and test cases and source code traceable (see Chap. 9) to prevent inconsistency defects.

17.3 Description of NSE Software Quality Assurance Paradigm

(g) After the implementation of requirement changes or code modifications, updates the database automatically to maintain the consistency between the documents and the source code.

3. **In coding phase:**

   (a) Prevents inconsistent defects in the interface coding between the related modules according to the incremental coding order assigned on the call graph generated from the design phase – when writing a function call statement, we can open a new window to view the control flow diagram of the called module (according to the bottom-up coding order, it must have been coded and tested already) to know how many parameters are needed, their types, and their order to prevent the inconsistent defects between the functional call statements and the called modules – see Fig. 17.5.

   (b) When there is a need to modify some data such as a global variable or static variable, performs data analysis to know, for instance, where a global variable is defined, changed, and used to prevent inconsistent defects in data usage; see Fig. 17.6 for an application example.

4. **In software testing phase:**

   (a) Graphically presents the untested modules as shown in Fig. 17.7 – why are they untested? There must be something wrong – either those modules are not needed or there are not enough designed test cases.

5. **In software maintenance phase:**

   (a) See Chap. 18.

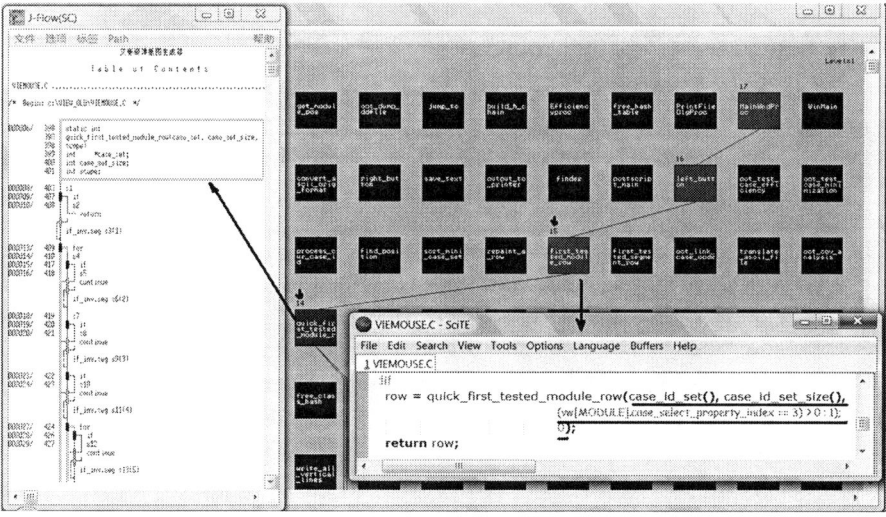

Fig. 17.5 An application example of consistent coding

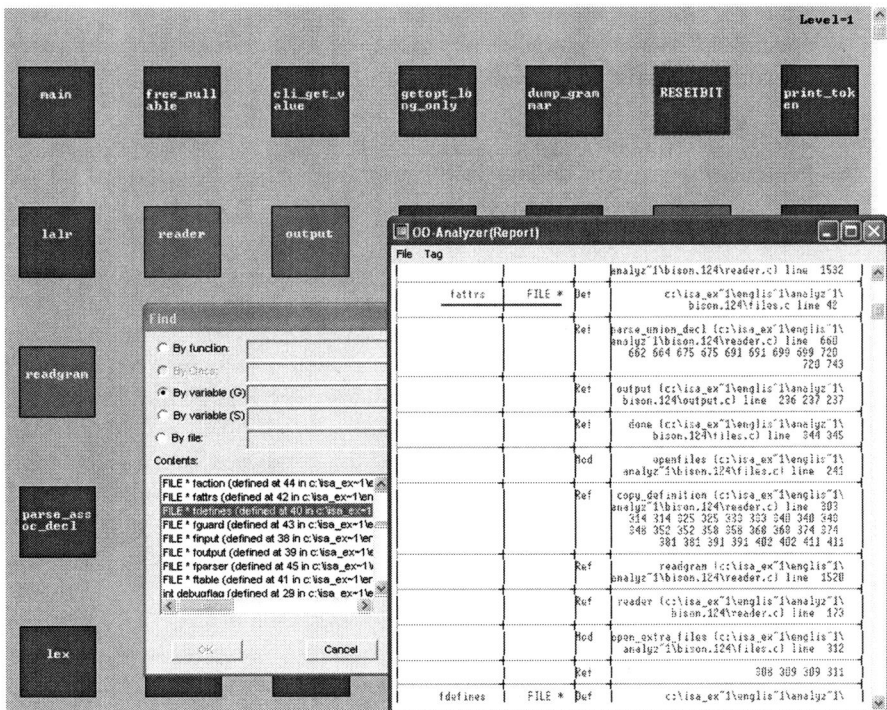

Fig. 17.6 An application example of consistent data (variable) modification support

17.3.6.2 Defect Propagation Prevention

(a) **In requirement development phase:**

Works with the HAETVE (Holistic, Actor–Action and Event–Response driven, Traceable, Visual, and Executable) technique and "dummy programming" for functional requirement decomposition to prevent defects through dynamic testing using the Transparent-box method combining functional and structural testing together seamlessly (which can be used dynamically in requirement development phase because having an output is no longer a condition to use this new software testing method, see Chap. 16) with the capability to establish bidirectional traceability among the related documents and test cases and the dummy source code to help users find and remove logic (structural) defects and inconsistent defects (see Figs. 15.7–15.9). An application example of requirement development and defect propagation prevention is shown in Figs. 17.8 and 17.9.

As shown in Fig. 17.8, there are two defects found.

(b) **In software design phase:**

Similarly, in the software design phase, many defects introduced into the design phase can also be efficiently removed through dynamic software testing supported by the NSE software testing paradigm, inspection using traceable

17.3 Description of NSE Software Quality Assurance Paradigm

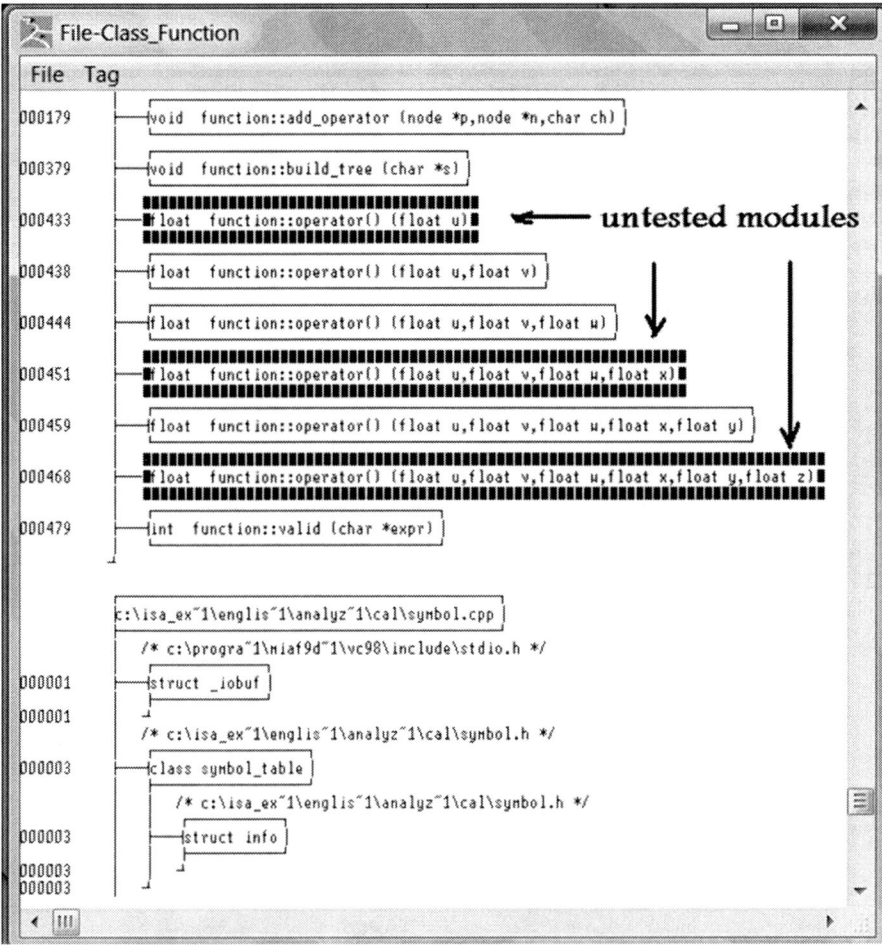

Fig. 17.7 An application example of untested modules report

documents and the source code, dummy programming, and diagram/chart generation supported by the NSE software visualization. An application example is shown in Figs. 17.10–17.12.

(c) **In coding phase:**

With NSE, defect propagation prevention should also be performed in software coding mainly though dynamic testing using the Transparent-box method (see Chap. 16), visualization, and inspection using traceable documents and the source code. Application examples of defect propagation prevention based on code traceability are shown in Figs. 17.13 and 17.14.

Usually, logic defects are hard to find because a program with some logic defect may work normally without providing any error message, but the output could be incorrect. With NSE through software visualization, many logic defects can be found. An application example is shown in Figs. 17.15–17.17.

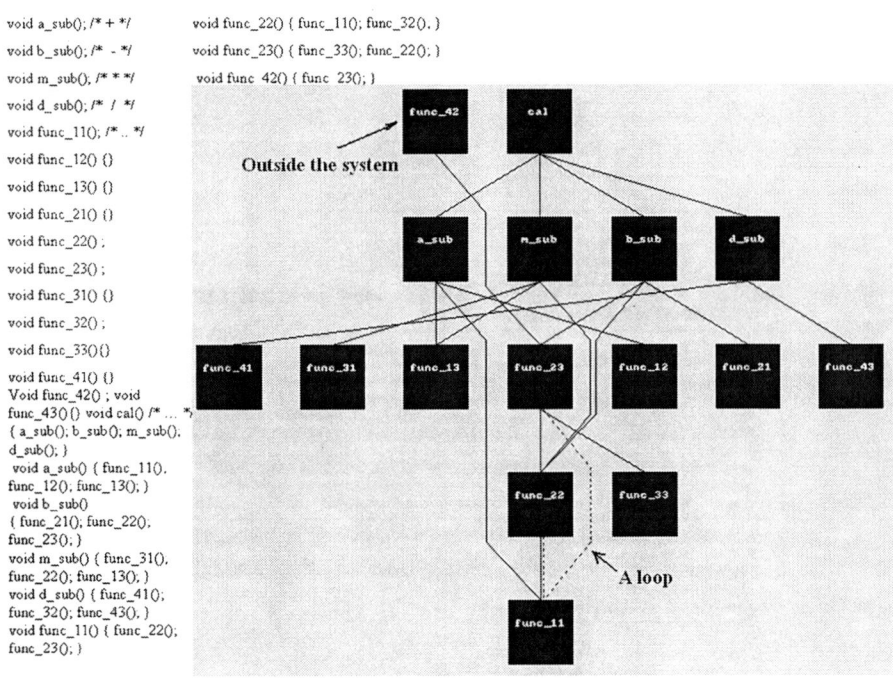

Fig. 17.8 An application example in functional decomposition of functional requirements

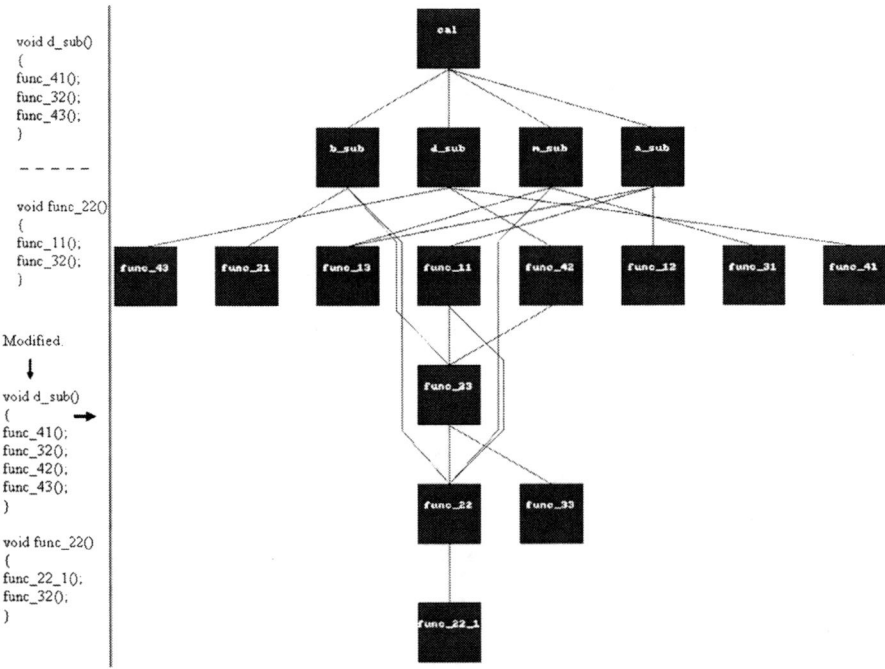

Fig. 17.9 An example of defect propagation prevention in the requirement development phase

17.3 Description of NSE Software Quality Assurance Paradigm

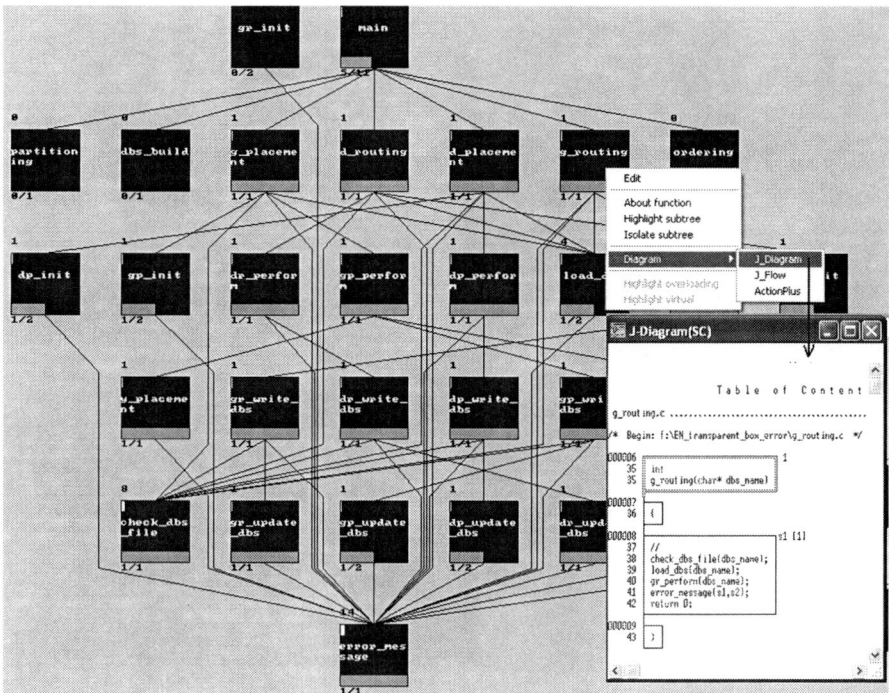

Fig. 17.10 A defect found in the top-down product design

```
int g_routing(char* dbs_name)              int g_routing(char* dbs_name)
{                                          {
                                              gr_init();
   check_dbs_file(dbs_name);     →            check_dbs_file(dbs_name);
   load_dbs(dbs_name);                        load_dbs(dbs_name);
   gr_perform(dbs_name);                      gr_perform(dbs_name);
   error_message(s1,s2);                      error_message(s1,s2);
   return 0;                                  return 0;
}                                          }
```

Fig. 17.11 A simple defect removal process

(d) **In testing phase:**

Many programming defects can be removed through dynamic testing using the Transparent-box testing method (see Chap. 18). With the NSE software testing paradigm, software testing and debugging can be combined together – when a runtime error happens, an extra string "EXIT" or "### Last termination

448   17 NSE Software Quality Assurance Paradigm Driven by Defect Prevention

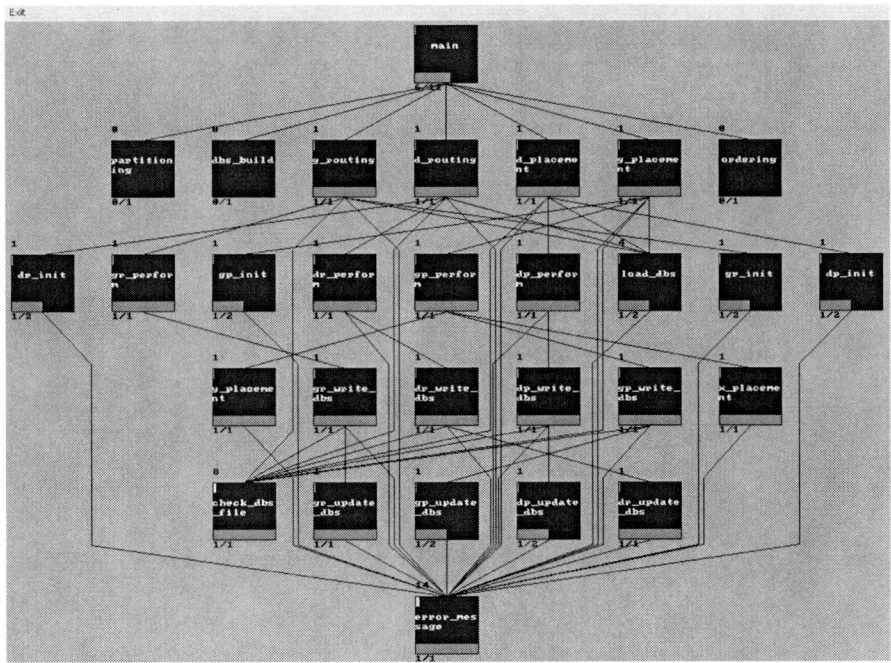

Fig. 17.12 A new system hierarchy modified from that shown in Fig. 17.10

location" will be added into the control flow diagram shown in J-Flow notations to indicate where (the source code location rather than the object code location) the program terminated. An application example is shown in Fig. 17.18.

(e) **In the maintenance phase:**

See Chap. 18.

17.3.6.3 Refactoring

Usually, 20% of modules with high Cyclomatic complexity (the number of decision statements) will have about 80% of the defects in a software product.

Some individual modules, particularly those modules with memory leaks will take more run time than others – being the performance bottlenecks.

With NSE, refactoring is performed for most complex program modules as shown in Fig. 17.19 and the modules that are performance bottlenecks as shown in Fig. 17.20.

With NSE, refactoring is performed with side-effect prevention to ensure the quality of the modules after refactoring (see Chap. 18).

17.3 Description of NSE Software Quality Assurance Paradigm

Fig. 17.13 An application example of static program review through traceability

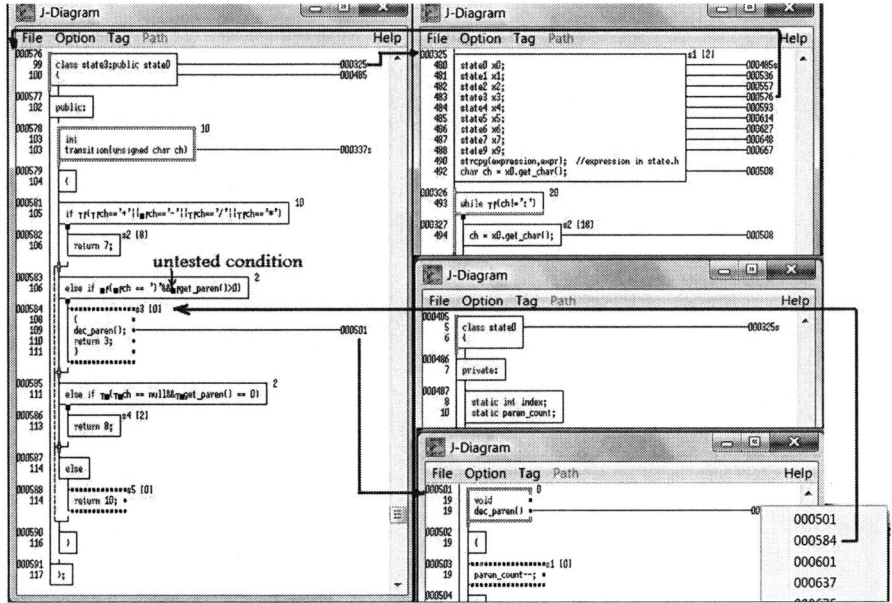

Fig. 17.14 An application example of semiautomated code inspection through bidirectional traceability

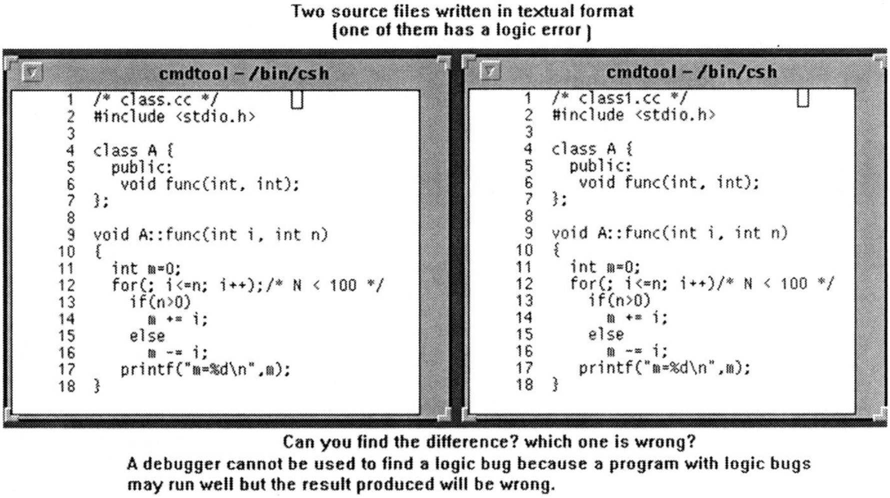

Fig. 17.15 Two similar program modules

17.3 Description of NSE Software Quality Assurance Paradigm

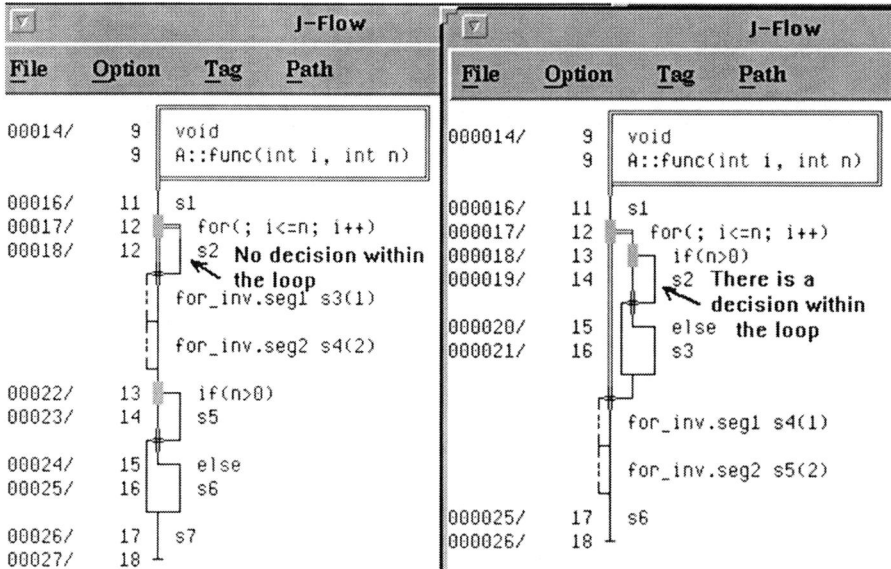

Fig. 17.16 The control flows of the two similar modules shown in Fig. 17.15

17.3.6.4 Deeper and Broader Software Testing, Quality Measurement, and Version Comparison

With NSE-SQA, for ensuring the quality of a software product, various kinds of software testing are performed, including

(a) Unit testing – it is recommended to meet 100% MC/DC test coverage (see Appendix B for an application example).
(b) Functional testing to validate whether the product meets the function requirements – see Fig. 17.21, a C++ program with GUI operation capture and playback.
(c) Structural testing with the capability to highlight untested branches and conditions graphically for testing improvement as shown in Fig. 17.22.
(d) Memory leak and usage violation analysis as shown in Fig. 17.23.
(e) Runtime error type analysis and the execution path tracing – see Fig. 17.24.
(f) Performance testing to check whether the product meets the performance requirement and how much time spent in each module as shown in Fig. 17.25.
(g) Holistic and detailed software quality measurement for an entire software product and each individual module as shown in Figs. 17.26 and 17.27.
(h) Holistic and detailed version comparison for handling "Bad Fixes" (secondary defects) – after fixing some defects, it is still possible to introduce new defects into the product, so that holistic and detailed version comparison is needed to help users to locate the new defects as shown in Fig. 17.28.

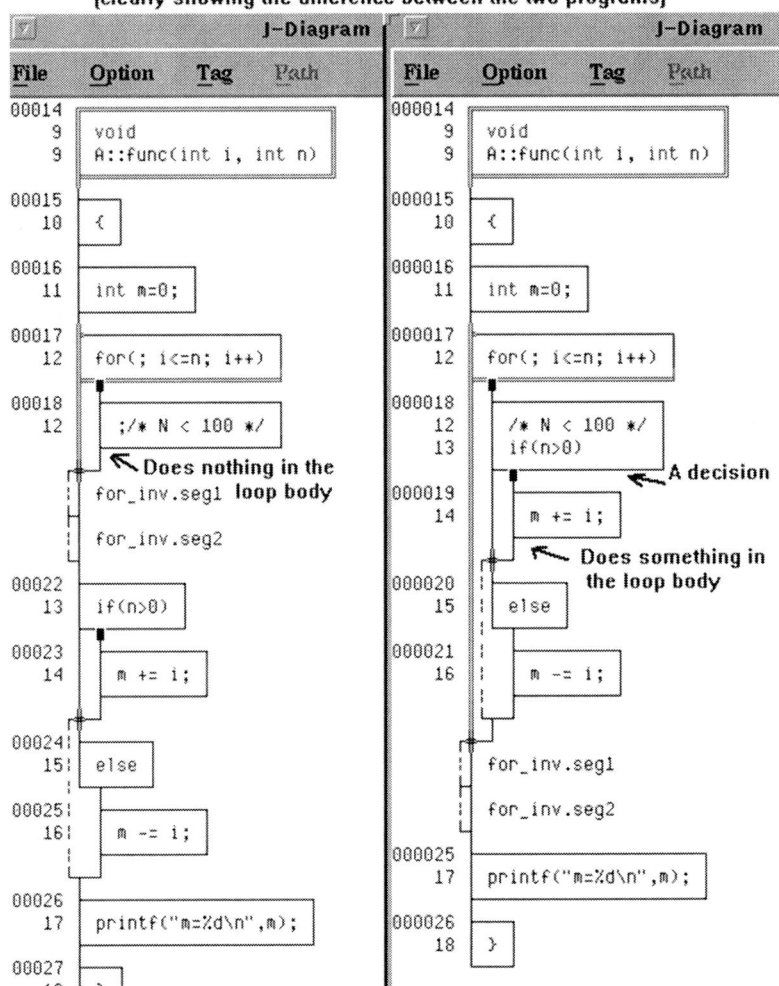

Fig. 17.17 The logic diagrams of the two similar modules shown in Fig. 17.15 (it is easy to find that there is a logic defect with the first module)

17.3 Description of NSE Software Quality Assurance Paradigm

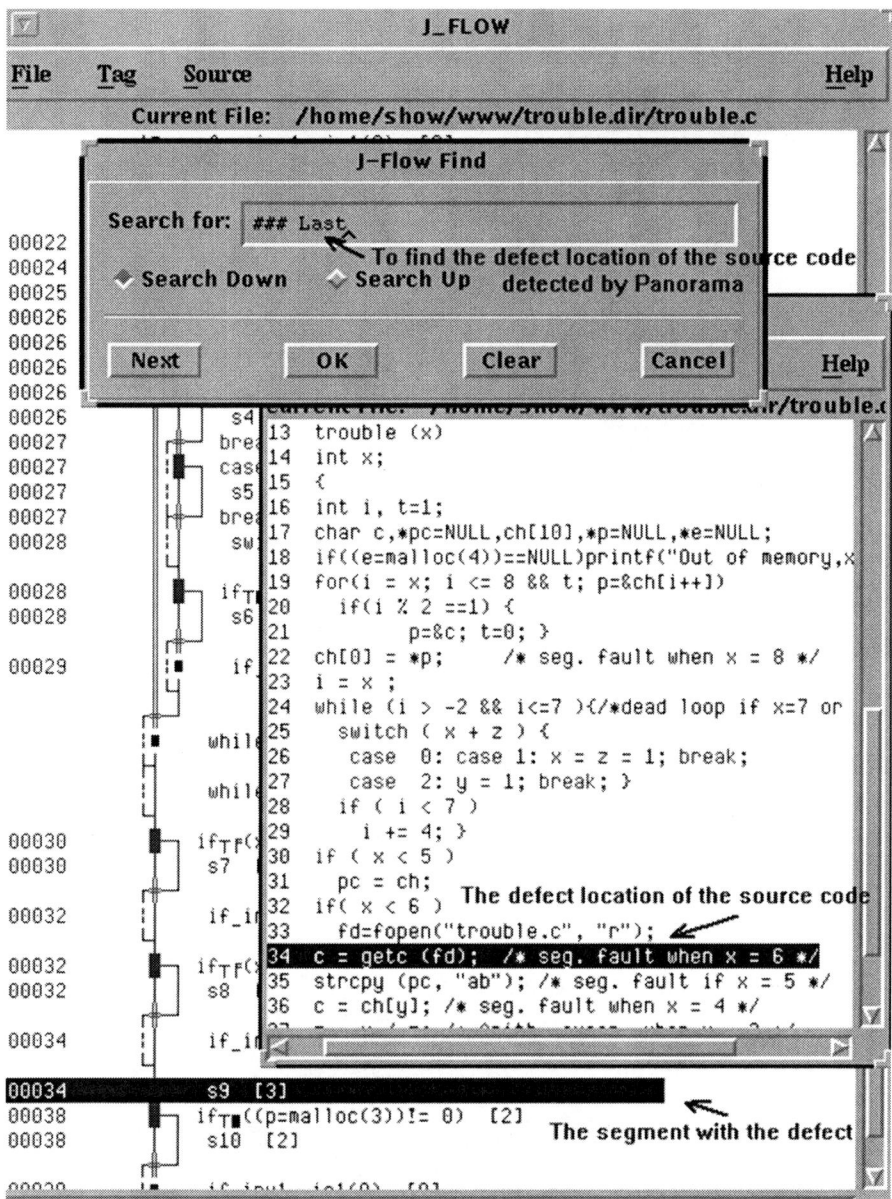

Fig. 17.18 An application example of software testing combined with debugging

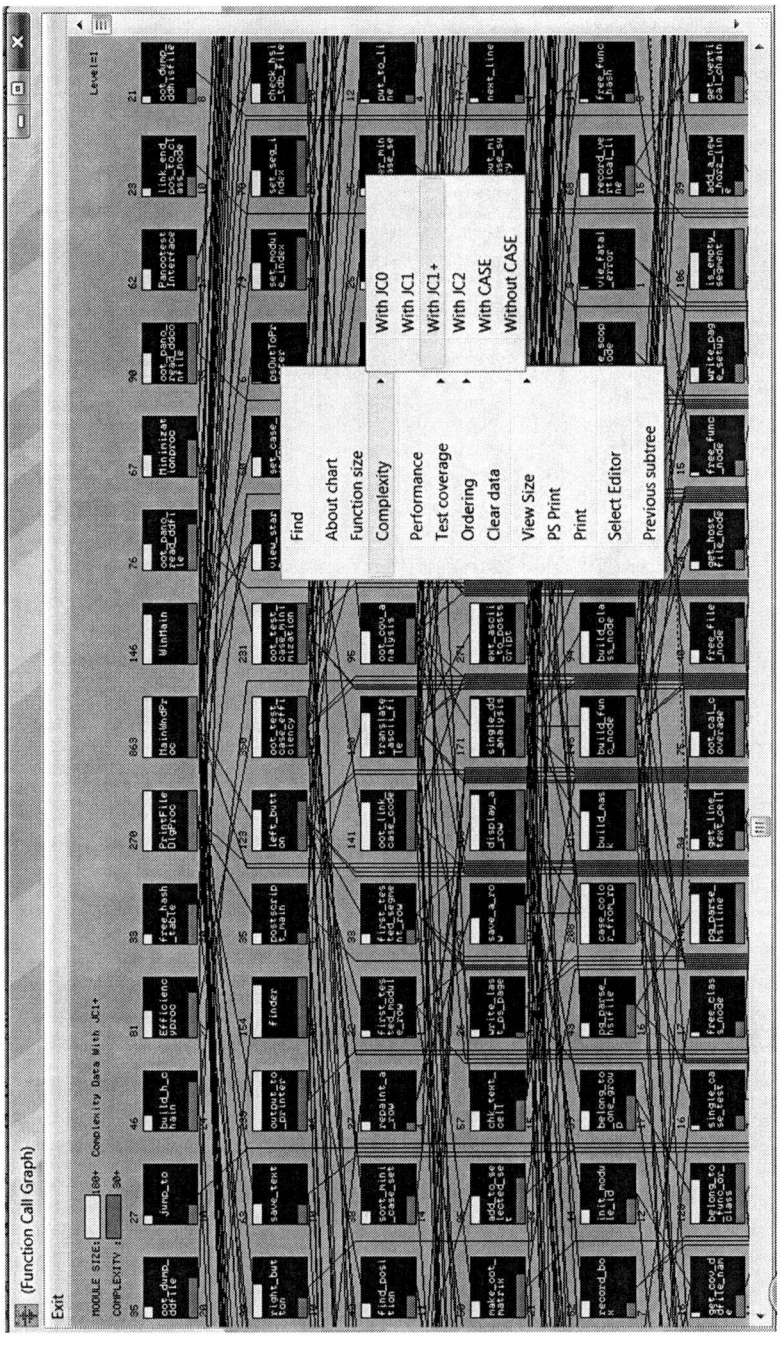

Fig. 17.19 Cyclomatic complexity (the number of decision statement) measurement example (usually module Cyclomatic complexity should be less than 30)

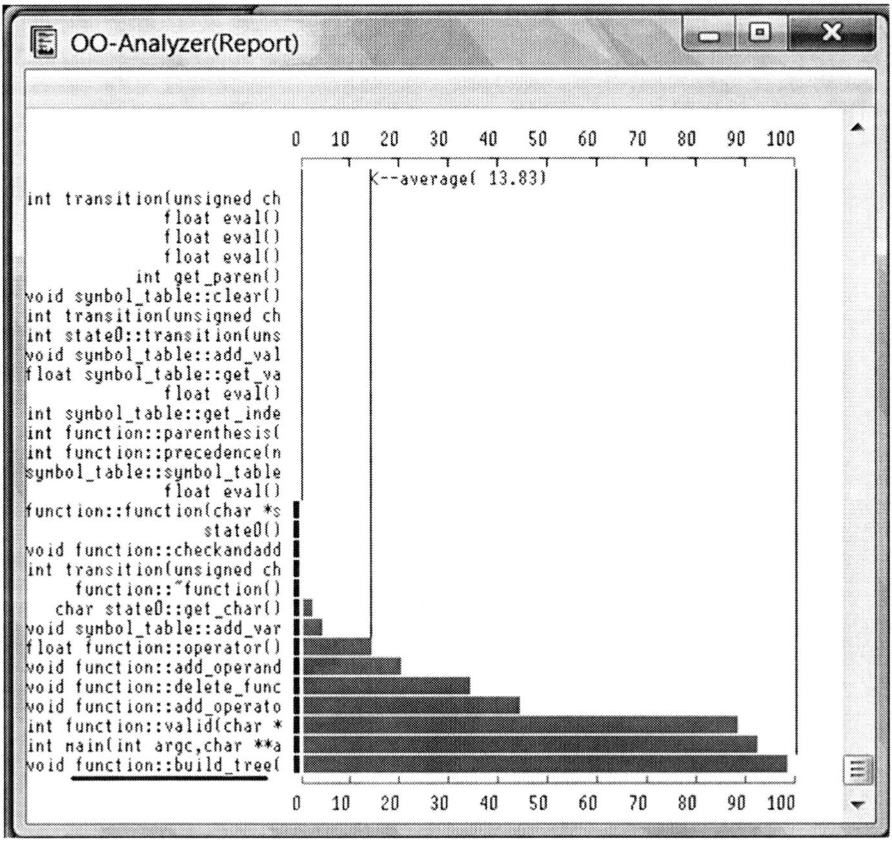

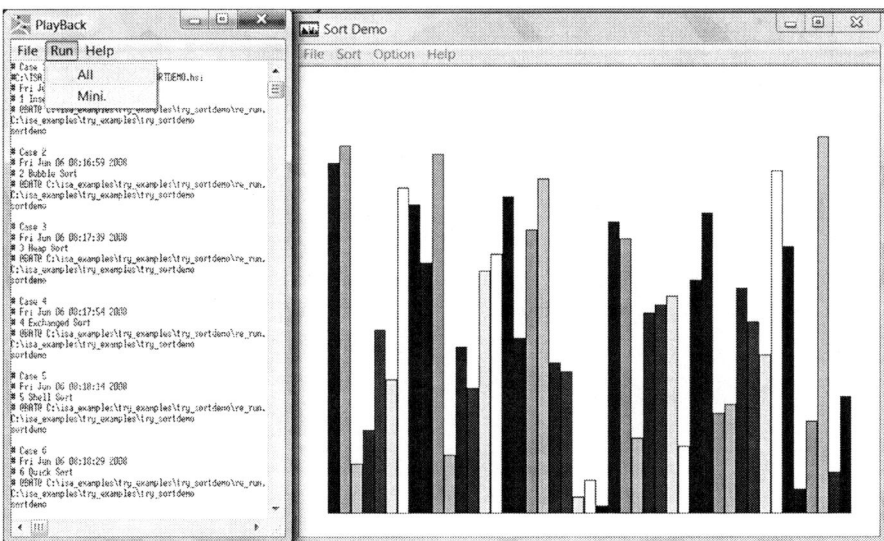

Fig. 17.20 An application example of performance analysis for locating possible performance bottleneck

Fig. 17.21 An application example of functional testing with GUI operation capture and playback

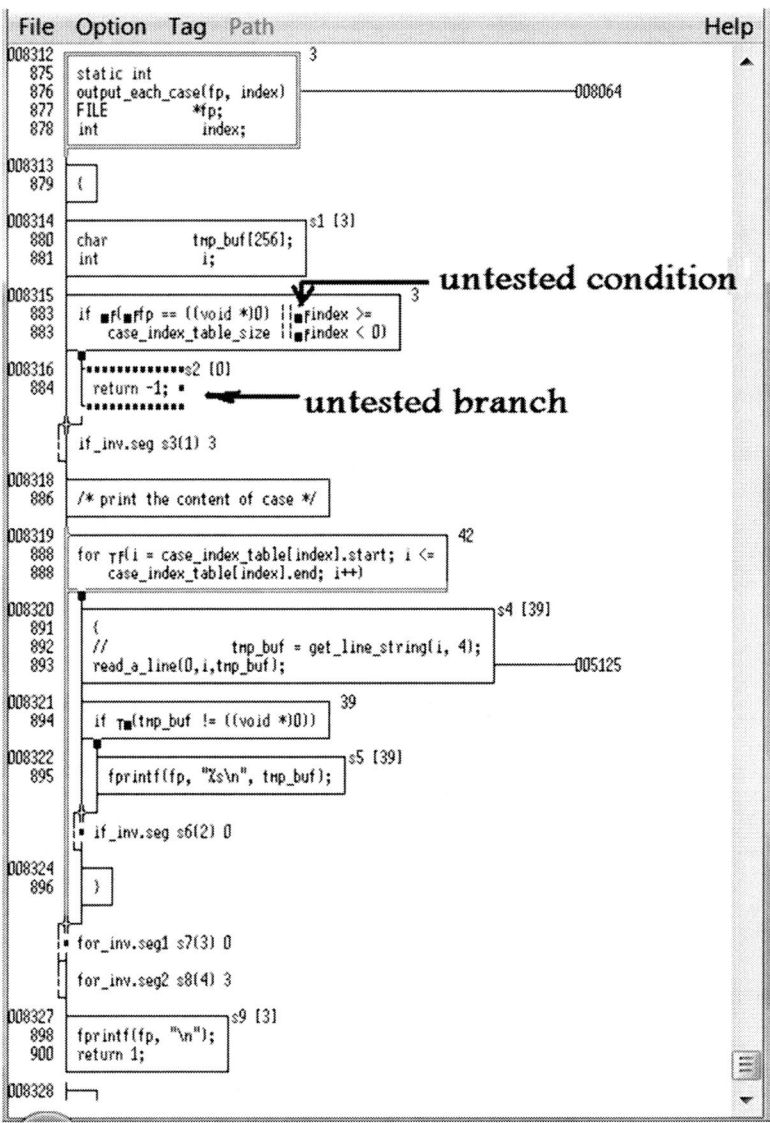

Fig. 17.22 An application example of MC/DC test coverage measurement with the capability to highlight untested branches and conditions

17.3 Description of NSE Software Quality Assurance Paradigm

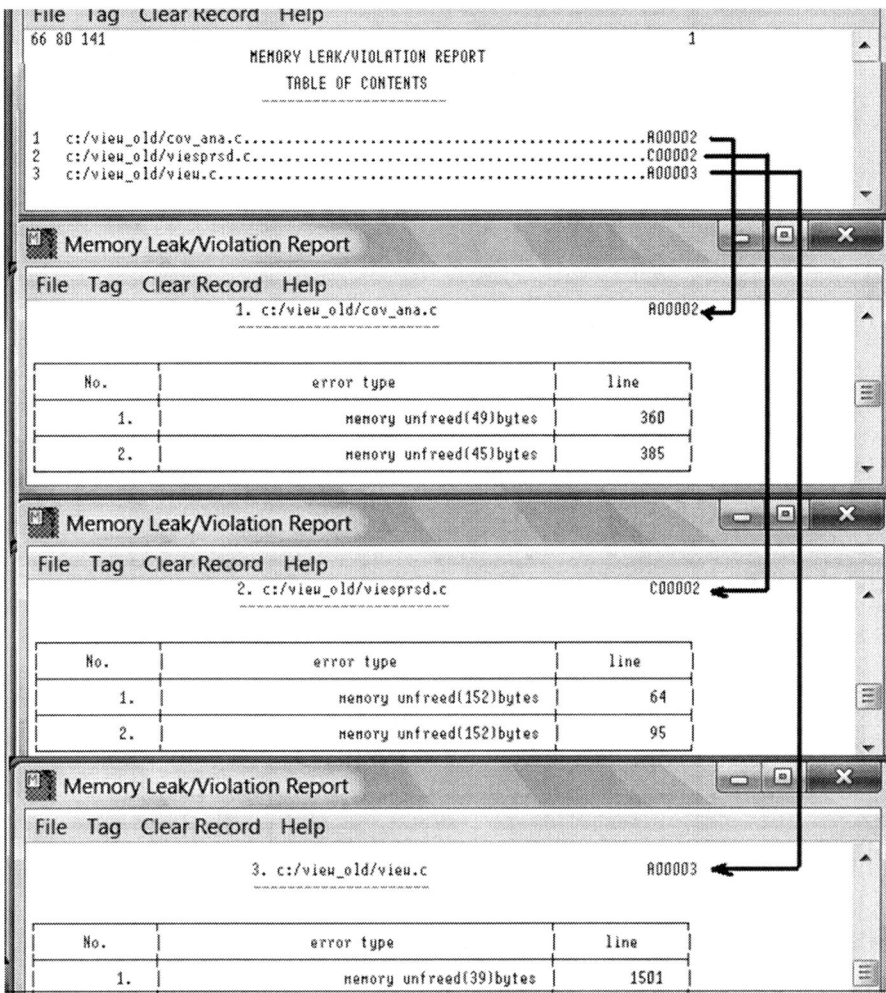

Fig. 17.23 An application example of memory leak and usage violation check

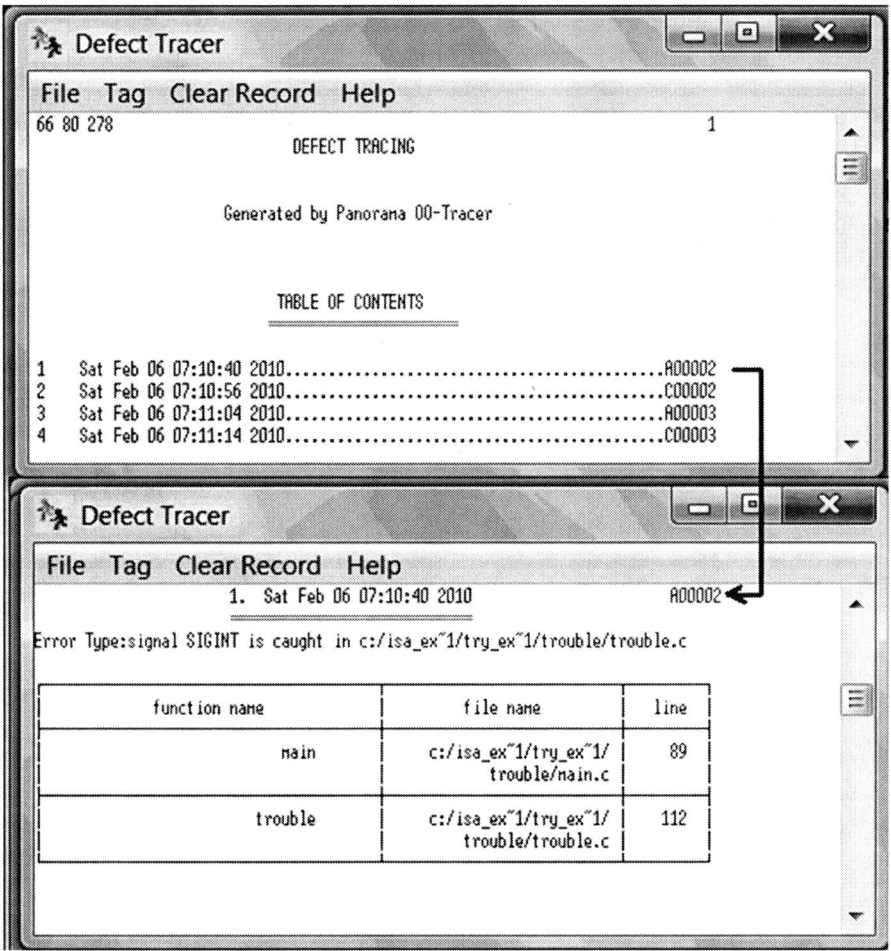

Fig. 17.24 An application example of runtime error type analysis and execution path tracing

17.3 Description of NSE Software Quality Assurance Paradigm

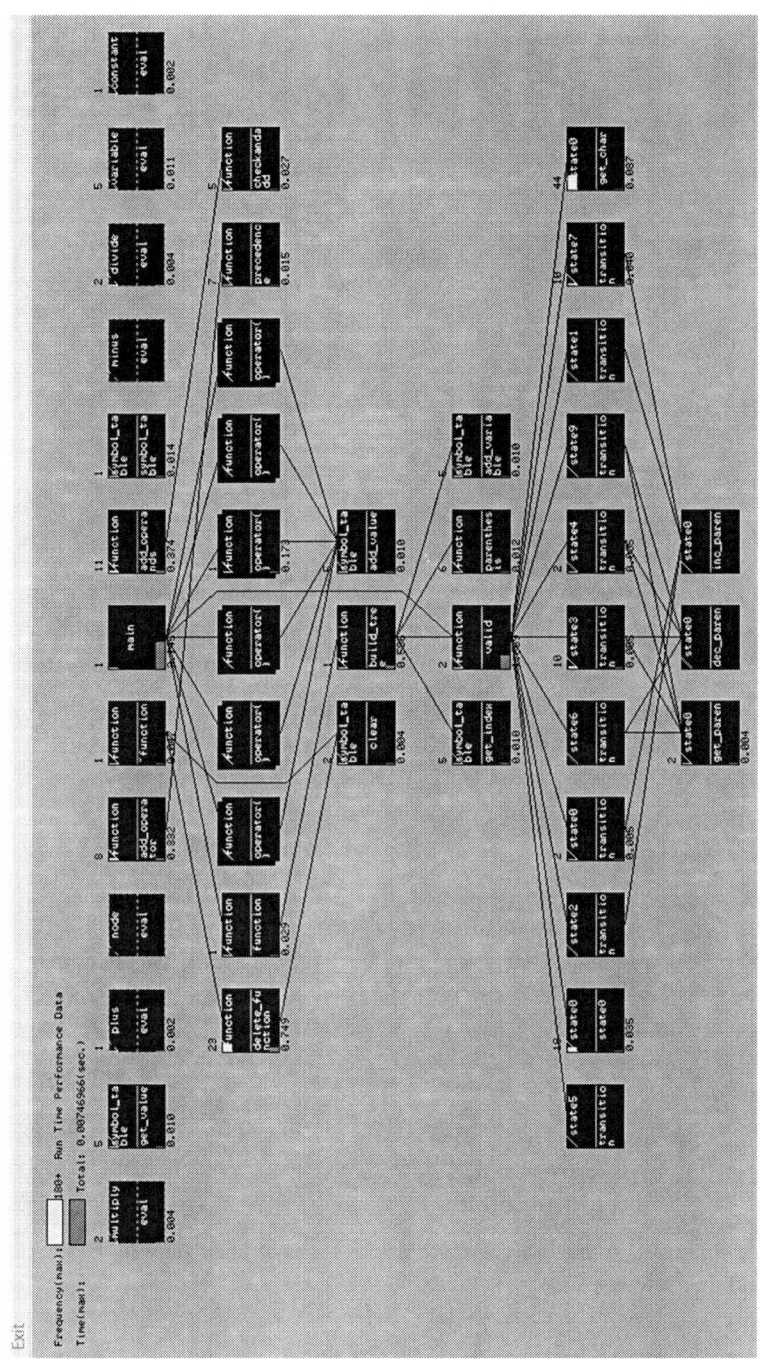

Fig. 17.25 An application example of performance analysis

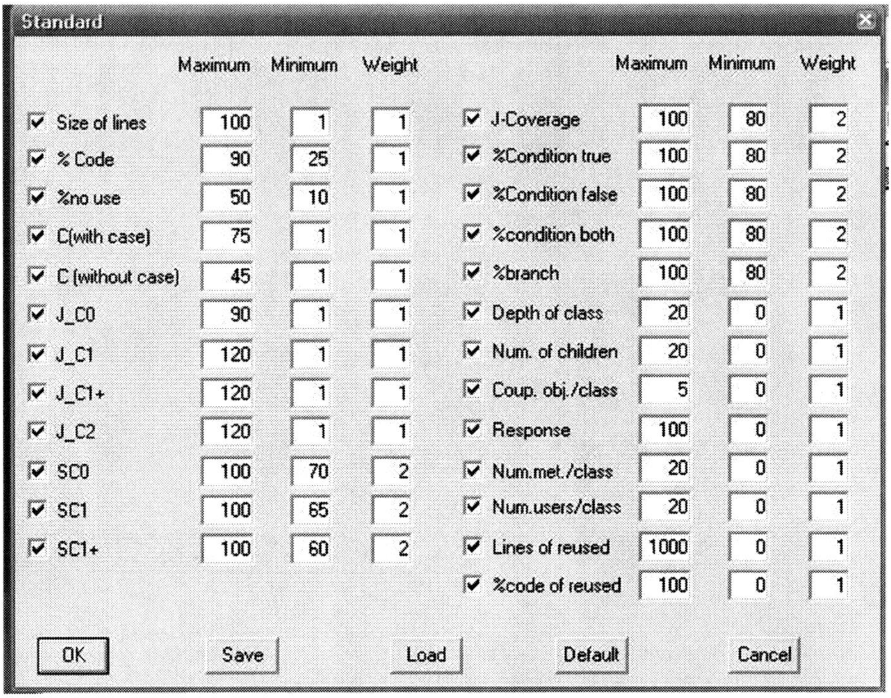

Fig. 17.26 Selecting quality standards and setting required values

17.4 Application of NSE-SQA

NSE-SQA has been preliminarily applied in practice. All screenshots shown in this chapter are taken from real application examples.

With the new revolutionary paradigm for software quality assurance, it is possible to remove 99–99.99% of the defects in a software product. Table 17.1 shows a comparison result in efficiency with various software quality assurance technologies.

17.5 The Major Features of NSE-SQA

The major features of the NSE software quality assurance paradigm are briefly summarized as follows:

(a) Based on complexity science
(b) Performed holistically and globally
(c) Defect prevention driven

17.5 The Major Features of NSE-SQA

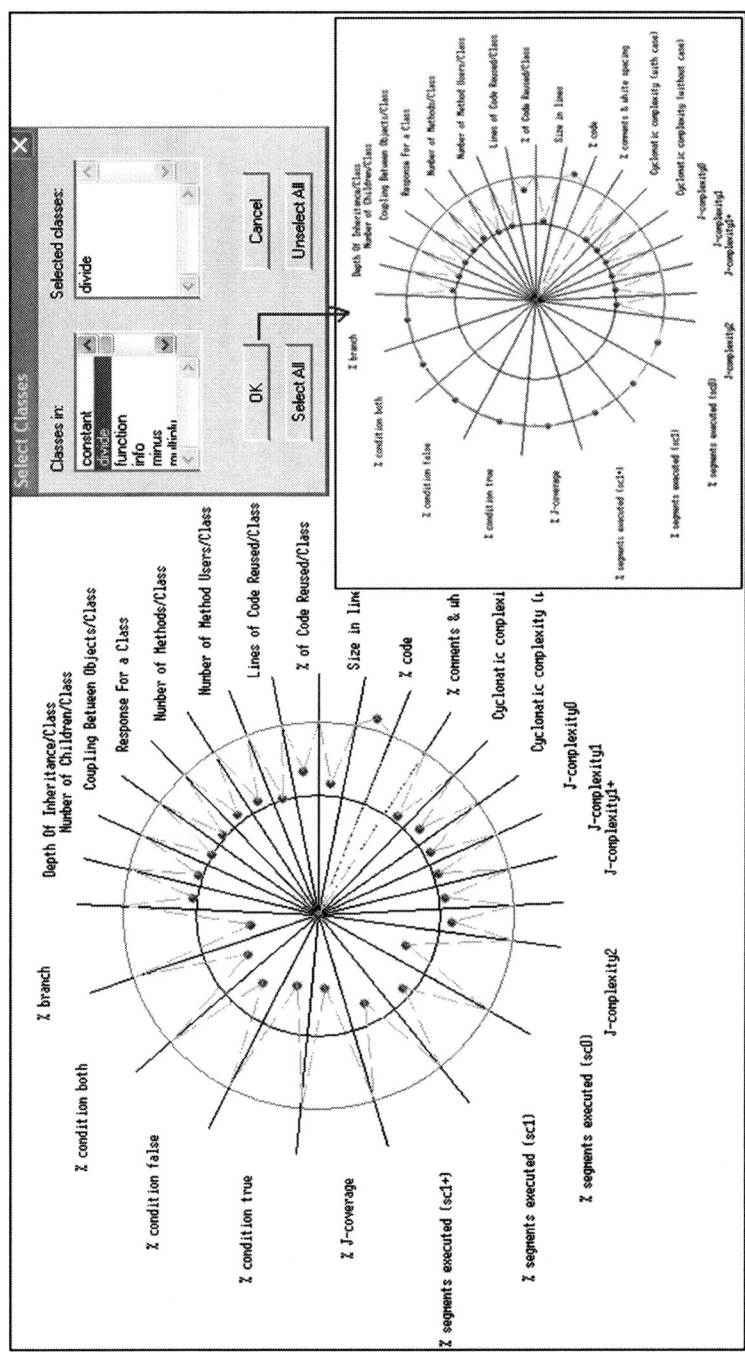

Fig. 17.27 An application example of holistic quality measurement for an entire software product and its individual modules

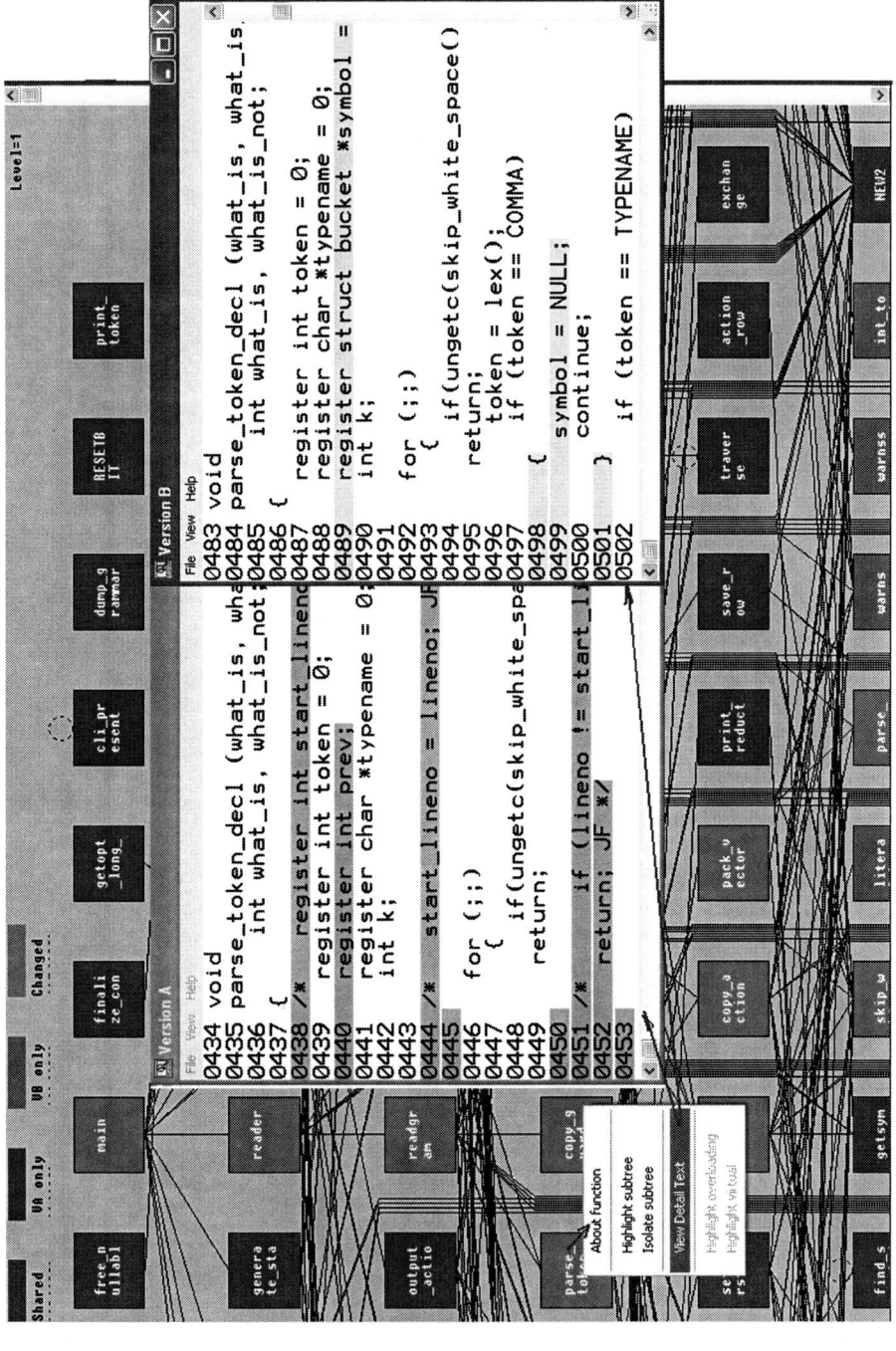

Fig. 17.28 Holistic and detailed version comparison

Table 17.1 SQA technologies and their efficiency

	Defect removal technology	Highest defect removal efficiency (%)
1	*Requirement review*	*50*
	Requirement review with traceable documents	>50
2	*Top level design review*	*60*
	Top level design review using traceable documents and charts	>60
3	*Detailed functional design review*	*65*
	Detailed functional design review using traceable documents	>65
4	*Detailed logic design review*	*75*
	Detailed logic design review using traceable diagrams	>75
5	*Code inspection*	*85*
	Code inspection with bidirectional traceability	>85
6	*Unit testing*	*50*
	Unit testing incrementally according to the assigned bottom-up testing order plus MC/DC test coverage analysis capability and graphical representation of the test result	>50
7	*New function testing*	*65*
	New function testing	>65
8	*Integration testing*	*60*
	Integration testing incrementally	>60
9	*System testing*	*65*
	System testing combining structural and function testing seamlessly	>65
10	*External beta testing*	*75*
	External beta testing with traceable documents and the source code	>75
11	*Cumulative efficiency*	*99.99*
	Cumulative efficiency with Defect prevention in the entire software development life cycle	>99.99

Note: **The item and the data written in italics come from the published reports provided by Software Productivity Research based on the analysis of 12,000 software projects** [Jon02]

(d) Supported by various traceabilities
(e) Visual in the entire software quality assurance process
(f) Systematic, quantifiable, and disciplined
(g) Low cost and high efficiency

17.6 Summary

The old-established software quality assurance paradigm is driven by inefficient inspection without the support of various traceabilities, and testing performed after production. It not only violates Deming's product quality assurance principle that "Cease dependence on inspection to achieve quality. Eliminate the need for inspection on a mass basis by building quality into the product in the first place," but also makes high degrees of software product security and reliability impossible to achieve as pointed by NIST.

With NSE, software quality is ensured through defect prevention, defect propagation prevention, refactoring, deeper and broader testing, plus quality measurement in the entire software development and maintenance process from the first step down to the retirement of a software product, supported by the NSE software testing paradigm based on the Transparent-box method which combines functional testing and structural testing together seamlessly, and can be dynamically used in requirement development (having an output is no longer a condition to dynamically use it), design, coding, testing, and maintenance, and also supported by the NSE software visualization paradigm.

It is possible for NSE-SQA to help software development organizations to remove 99.99% of the defects in their software product development with NSE.

17.7 Points and Questions to Ponder

(a) What is the root cause that regarding software product quality, "Over the last 50 years there has been very little improvement."?
(b) What is defect prevention? Why should it be performed in the entire software development lifecycle from the first step down to the retirement of a software product?
(c) What are the major differences between the old-established software quality assurance paradigm and NSE-SQA?

17.8 Further Reading and Information Source

(a) Rice D (2008) GEEKONOMICS: the real cost of insecure software, 1st edn. Pearson Education, Inc., Publishing as Addison-Wesley, New Jersey
(b) Humphrey WS (2008) The Software Engineering Institute, the software quality challenge. CrossTalk, June Issue
(c) Software QA and testing resource center, FAQ1 – Software QA and testing frequently-asked-questions part 1. http://www.softwareqatest.com/

References

[Ade05] Adeel K, Ahmad S, Akhtar S (2005) Defect prevention techniques and its usage in requirements gathering-industry practices. Paper appears in Engineering Sciences and Technology, SCONEST, August 2005, pp 1–5. ISBN 978-0-7803-9442-1

[Bro95-p201] Brooks FP Jr (1995) The mythical man-month. Addison-Wesley, Reading, p 201

[Cla01] Clark B, Zubrow D (2001) How good is the software: a review of defect prediction techniques. Sponsored by the US Department of Defense, © 2001 by Carnegie Mellon University, version 1.0, p 5

References

[Fry07]	Frye C, News Writer (2007) The state of software quality, part 1: problems remain, but all is not doomed. 16 Feb 2007, SearchSoftwareQuality.com. http://searchsoftwarequality.techtarget.com/news/article/0,289142,sid92_gci1243311,00.html
[Hum89]	Humphrey WS (1989) Managing the software process. In: Defect prevention, Chapter 17. Addison-Wesley, Reading. ISBN-0-201-18095-2
[Jon02]	Jones C. (2002) Software quality in 2002: a survey of the state of the art. Six Lincoln Knoll Lane, Burlington, MA 01803. http://www.SPR.com. Accessed 23 July 2002
[Jon06]	Jones C, Social and Technical Reasons for Software Project Failures, CrossTalk, Jun 2006 Issue
[Kar07]	Karg LM, Beckhaus A (2007) Modelling software quality costs by adapting established methodologies of mature industries. In: Proceedings of 2007 IEEE international conference in industrial engineering and engineering management in Singapore, 2–4 December 2007, pp 267–271. ISBN 078-1-4244-1529-8
[Nar08]	Narayan P (2008) Software defect prevention in a nut shell. Copyright © 2000–2008 iSixSigma LLC. See also http://software.isixsigma.com/library/content/c030611a.asp
[Sum08]	Suma V, Gopalakrishnan Nair TR (2008) Effective defect prevention approach in software process for achieving better quality levels. In: International conference on software engineering (ICSE), WASET, Singapore, vol. 42, pp 258–262
[Tia01]	Tian J (2001) Quality assurance alternatives and techniques: a defect-based survey and analysis. ASQ by Department of Computer Science and Engineering, Southern Methodist University, SQP, vol. 3, no. 3. World Academy of Science, Engineering and Technology 42 2008 260
[Vas05]	Vasudevan S (2005) Defect prevention techniques and practices. In: Proceedings from 5th annual international software testing conference in India
[Zel03]	Zelkowitz MV (2003) The software defect prevention/isolation/detection model. www.cs.umd.edu/~mvz/mswe609/book/chapter2.pdf

Chapter 18
NSE Software Maintenance Paradigm: Systematic, Disciplined, and Quantifiable

> *Over three decades ago, software maintenance was characterized as an 'iceberg'. We hope that what is immediately visible is all there is to it, but we know that an enormous mass of potential problems and cost lies under the surface. In the early 1970s, the maintenance iceberg was big enough to sink an aircraft carrier. Today, it could easily sink the entire navy!*
>
> Roger S. Pressman
>
> *Clearly, methods of developing programs so as to eliminate or at least illuminate side effects can have an immense payoff in maintenance costs.*
>
> Frederick P. Brooks, Jr.

This chapter introduces the NSE software maintenance paradigm established by complying with the essential principles of complexity science through the use of the innovated FDS (Five-Dimension Synthesis Method) framework (**see Chap. 4**) as shown in Fig. 18.1.

18.1 The Existing Software Maintenance Engineering Paradigm Is Outdated

Software products need to be modified for meeting requirement changes, fixing bugs, improving performance, and keeping it usable in a changed or changing environment.

But unfortunately, the old-established software maintenance engineering paradigm is outdated because

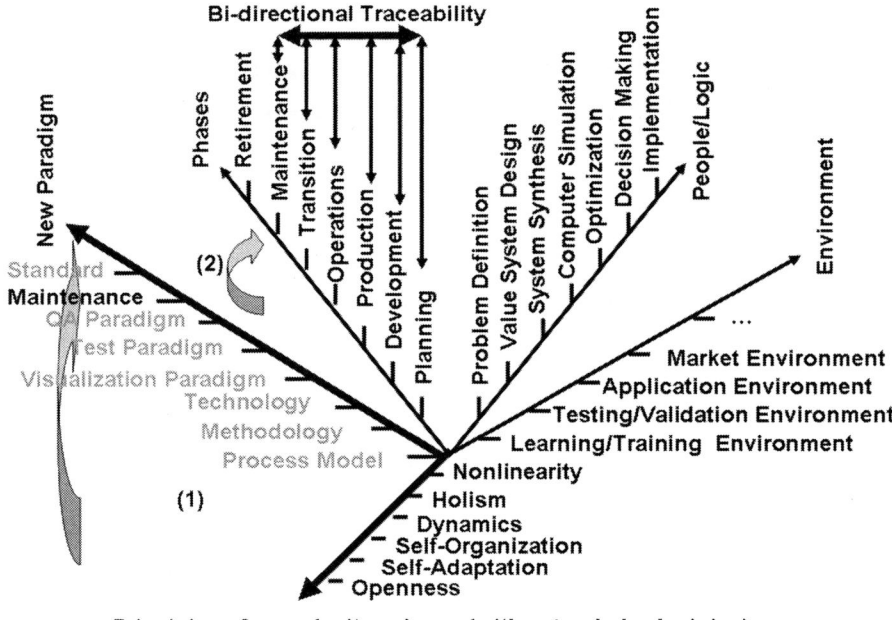

Fig. 18.1 NSE software maintenance paradigm is established by complying with the essential principles of complexity science through the innovated paradigm-shift framework, FDS (Five-Dimension Synthesis Method)

1. **It is based on reductionism and the superposition principle** that the whole of a complex system is the sum of its components, so that almost all of the tasks and activities in software maintenance engineering are performed partially and locally.
2. **The corresponding software development process models are linear ones with no upstream movement at all** – which require software engineers to do all things right at all times without making any mistakes or wrong decisions, but that is impossible.
3. **With the linear process models, the defects brought into a software product in the upper phases easily propagate down to the maintenance phase** to make the maintenance tasks much harder to perform as shown in Fig. 18.2.
4. **The corresponding software development methodologies do not offer "maintainable design"** without the support of various kinds of bidirectional traceabilities.
5. **It is not systematic** – the old-established software maintenance engineering paradigm does not offer systematic approaches for software maintenance: there is no systematic software maintenance process model defined.
6. **It is not quantifiable** – for instance, when a module is modified, there is no facilities provided to get quantifiable data about how many requirements and how many modules may be affected.

18.1 The Existing Software Maintenance Engineering Paradigm Is Outdated

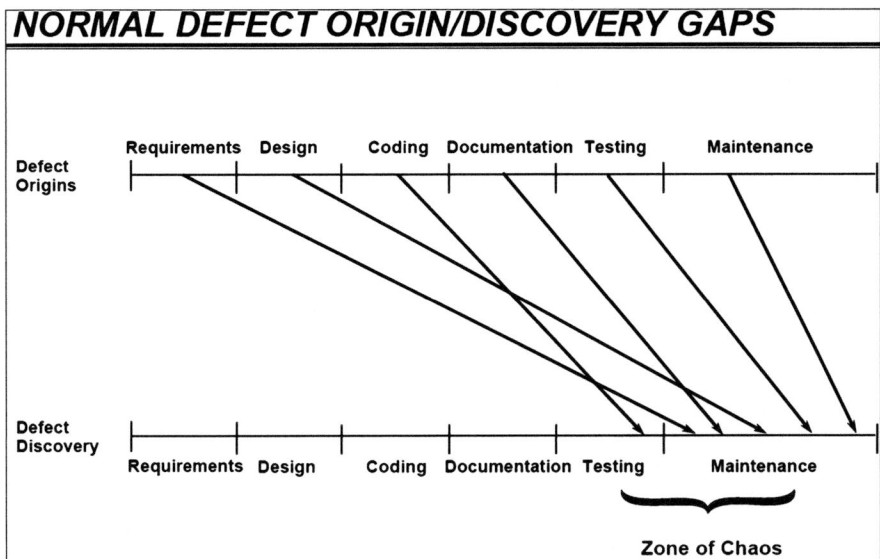

Fig. 18.2 Normal defect origin/discovery gaps [Jon02]

7. **It is not disciplined** – there is no engineering approach and model defined to guide maintainers to perform software maintenance step-by-step to prevent side effects and ensure the quality of the modified products.
8. **It is invisible** – the maintenance engineering process and the results obtained are invisible, making them hard to review and evaluate.
9. **It is blind** – for instance, after the implementation of a requirement change or code modification, it requires the maintainers to use all test cases to perform regression testing blindly, no matter whether a test case is useful or useless to re-test the modified software product.
10. **It is costly** – as pointed out by Scott W. Ambler, "The Unified Process suffers from several weaknesses. First, it is only a development process… it misses the concept of maintenance and support…. It's important to note that development is a small portion of the overall software life cycle. The relative software investment that most organizations make is allocating roughly 20% of the software budget for new development, and 80% to maintenance and support efforts" [Amb05].
11. **It makes a software product being maintained unstable day by day** – as pointed out by Frederick P. Brooks Jr., "*The fundamental problem with program maintenance is that fixing a defect has a substantial (20–50 percent) chance of introducing another. … All repairs tend to destroy the structure, to increase the entropy and disorder of the system*" [Bro95-P120].
12. **It makes a software product developed by others much harder to maintain at the customer site** – today a software product is delivered with the program, the data used, and the documents separated from the source code without bidirectional traceability and intelligent agents (intelligent tools) to support testability, visibility, changeability, conformity, reliability, and maintainability.

13. **It is easy to become a project killer or even a business killer** – as pointed out by Roger S. Pressman, "Over three decades ago, software maintenance was characterized as an 'iceberg'. We hope that what is immediately visible is all there is to it, but we know that an enormous mass of potential problems and cost lies under the surface. In the early 1970s, the maintenance iceberg was big enough to sink an aircraft carrier. Today, it could easily sink the entire navy!" [Pre05-P409].

18.2 Outline of the NSE Software Maintenance Paradigm

The revolutionary solution offered by NSE for software maintenance is described in detail in this chapter later. Here is the outline of the solution:

1. **It is based on complexity science** that the whole of a complex system is greater than the sum of its components – the characteristics and behaviors of the whole emerge from the interaction of its components, so that with NSE almost all of the tasks and activities in software maintenance engineering are performed holistically and globally.
2. **The corresponding software development process model is a nonlinear one with two-way iteration** (see Chap. 8) supported by automated and self-maintainable traceabilities to prevent defects brought into software products by the product developers and the customers. The NSE Process Model includes the preprocess part and the main process part supported by an automated and self-maintainable facility for bidirectional traceability using Time Tags automatically inserted into both the test case description part and the corresponding test coverage database for mapping test cases and the tested source code, and some keywords to indicate the related document types such as @WORD@, @HTML@, @PDF@, @BAT@, and @EXCEL@ written in the test case description part followed by the file paths and the bookmarks to be used to open the traced documents from the specified positions.
3. **With the nonlinear process models, most of the defects brought into a software product can be efficiently removed** through defect propagation prevention mainly by dynamic testing in the entire software development life cycle using the Transparent-box testing method (see Chap. 16) innovated by me to combine functional testing and structural testing together seamlessly: to each test case it not only checks whether the output (if any, can be none when the method is applied in the requirement development phase and the design phase – having an output is no longer a condition to use this software testing method dynamically) is the same as what is expected, but also checks whether the real execution path covers the expected one specified in J-Flow (see Chap. 7), and automatically establishes bidirectional traceability among the related documents, the test cases, and the source code to help the developers remove inconsistency defects. NSE complies with W. Edwards Deming's product quality principle, *"Cease dependence on inspection to achieve quality. Eliminate the need for*

18.2 Outline of the NSE Software Maintenance Paradigm

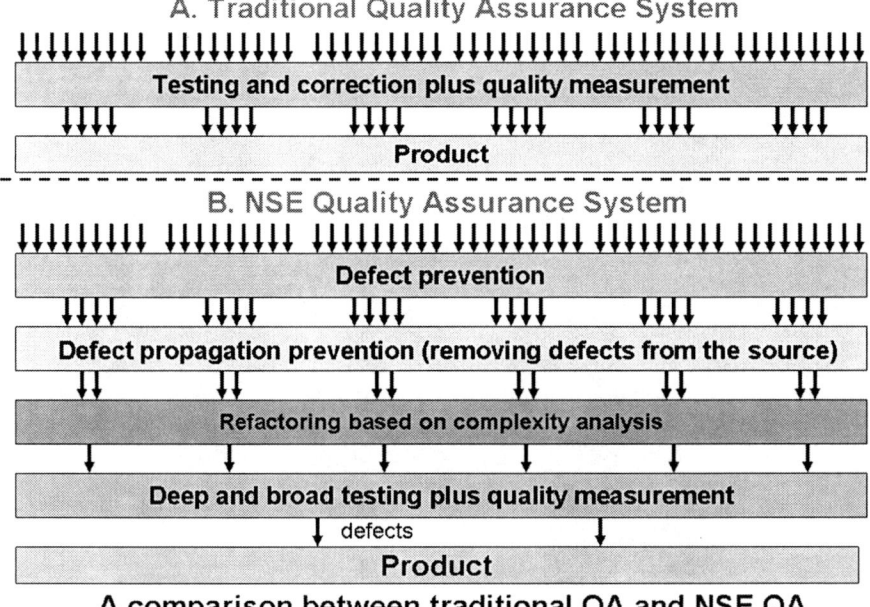

Fig. 18.3 A comparison in quality assurance between the old-established QA paradigm and the NSE QA paradigm

inspection on a mass basis by building quality into the product in the first place" [Dem82]. Figure 18.3 shows the difference in software quality assurance between the old-established software development methodologies and the NSE software development methodology.

4. **The corresponding software development methodology offers "maintainable design"** supported by various kinds of bidirectional traceabilities for defect prevention, defect propagation prevention, and side-effect prevention in the implementation of requirement changes and code modification (see Chap. 10) – as pointed out by Frederick P. Brooks, Jr., "Clearly, methods of designing programs so as to eliminate or at least illuminate side effects can have an immense payoff in maintenance costs" [Bro95-P120].
5. **It is systematic** – the NSE software maintenance engineering paradigm offers systematic approaches for software maintenance: there is a systematic software maintenance process model defined to guide users to perform software maintenance holistically and globally (see Sect. 18.3).
6. **It is quantifiable** – for instance, when a module or even only one statement of the source code is modified, the NSE software maintenance engineering paradigm can help users get quantifiable data on exactly about how many requirements and other modules may be affected. Figure 18.4 shows that if a class member function ArraySorted::Swaps is modified, **seven** related modules may also need to be modified.

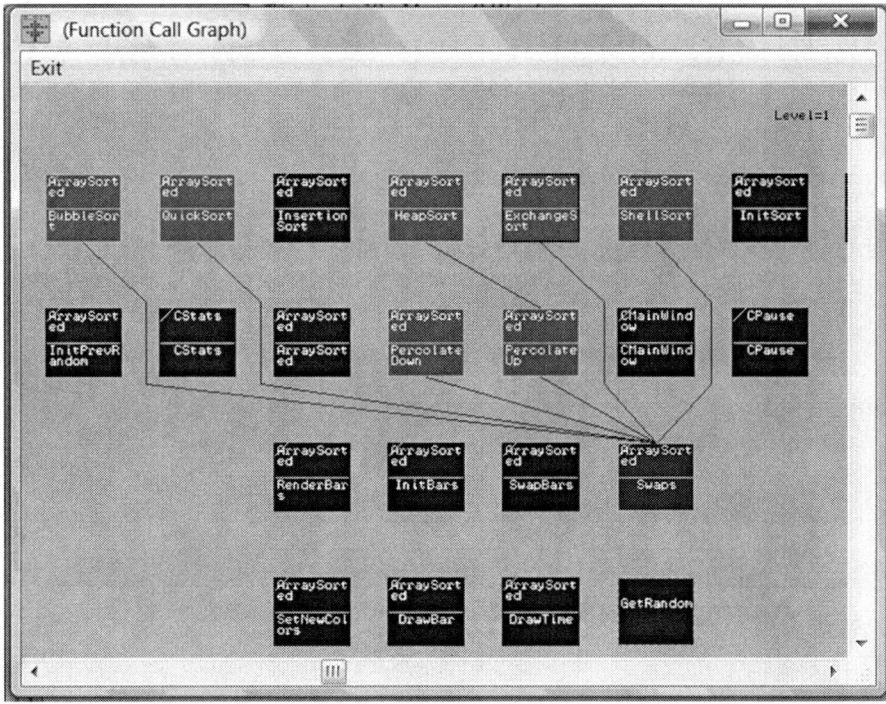

Fig. 18.4 Quantifiable software maintenance example 1: when a class member function ArraySorted::Swaps needs to be modified (e.g., the number of the parameters should be changed), **seven** related modules may also need to be modified

Figure 18.5 shows that correspondingly there are **nine** statements calling the class member function ArraySorted::Swaps – they should be checked to prevent the possible inconsistency defects.

Figure 18.6 shows that if a global variable "vw" is modified, 37 modules need to be checked for consistency.

7. **It is disciplined** – there is a defined engineering approach and model to guide maintainers to perform software maintenance step-by-step to prevent side effects, ensure the quality of the modified products, and perform regression testing efficiently.
8. **It is visible** – with NSE, the maintenance engineering process and the results obtained are visible and easy to review and evaluate, because it is supported with a set of Assisted Online Agents including software visualization tools to automatically generate huge amounts of graphical documents which are interactive and traceable – see Figs. 18.4 and 18.5 again.
9. **It is not blind** – for instance, after the implementation of a requirement change or code modification, it helps the maintainers efficiently select the useful test cases through backward traceability and test case minimization for performing regression testing efficiently.

18.2 Outline of the NSE Software Maintenance Paradigm

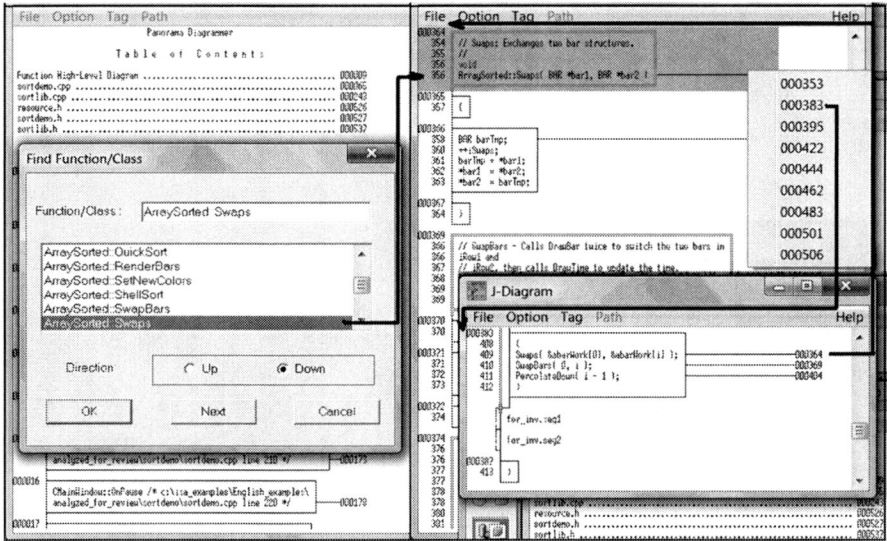

Fig. 18.5 Quantifiable software maintenance example 2: when a class member function ArraySorted::Swaps is modified, **seven** related statements calling it should be checked for consistency

10. **It is not costly** – it is possible for the NSE software maintenance engineering paradigm to help software organizations to greatly reduce the cost and effort spent in software maintenance because:

    (a) With NSE, quality assurance is performed in the entire lifecycle through defect prevention and defect propagation prevention using the Transparent-box testing method dynamically, plus inspection using traceable documents and traceable source code, so that the defects propagated into the maintenance phase are greatly reduced.
    (b) The implementation of requirement changes and code modifications is performed holistically and globally, rather than partially and locally.
    (c) The side effects in the implementation of requirement changes and code modifications are prevented through various kinds of automated and self-maintainable traceabilities.
    (d) Regression testing after software modification is performed efficiently through backward traceability to select the corresponding test cases and perform test case minimization to select the useful test cases to greatly reduce the required time, resources, and cost.

11. **It makes a maintained software product stable** – with NSE, there is no big difference between the product development process and the product maintenance process: in both processes, requirement changes are welcome to support the customers' market strategy, and implemented holistically and globally with side-effect prevention through various kinds of traceabilities.

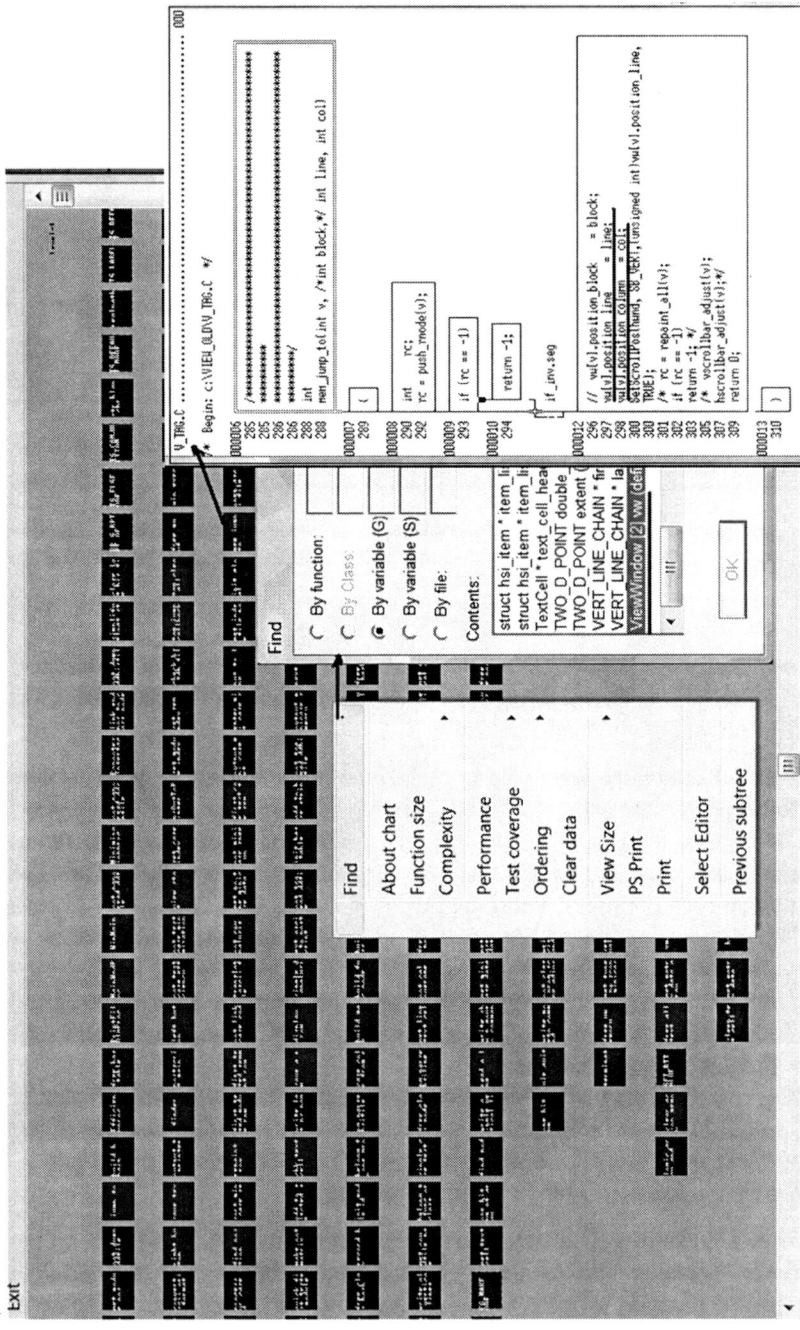

Fig. 18.6 Quantifiable software maintenance example 3: if the variable "vw" is modified, then 37 modules (highlighted in *red* color originally) need to be checked for consistency

18.2 Outline of the NSE Software Maintenance Paradigm

12. **It makes a software product developed by others easy to maintain at the customer site** – even if a software product is maintained at the customer site rather than the product development site, software maintenance engineering can be performed with almost the same conditions as those at the product development site, because with NSE the delivery of a software product includes not only the computer program, the data used, and the documents traceable to and from the source code, but also the database built through static and dynamic measurement of the program, and a set of Assisted Online Agents to make the software adaptive and truly maintainable (see Sect. 18.5 to know how those Assisted Online Agents work together to support testability, reliability, changeability, visibility, conformity, traceability, adaptability, and maintainability).
13. **The NSE software maintenance engineering paradigm becomes a key to make it possible for NSE to help software organization double their productivity and halve their cost in their software product development** – with NSE, not only most defects are removed in the development process through defect prevention and defect propagation prevention, but also new defects are prevented in the maintenance process through various kinds of traceabilities and dynamic testing using the Transparent-box testing method – all software maintenance tasks are performed holistically and globally with side-effect prevention, so that the effort and cost spent in software maintenance will be almost the same as that spent in the software development process – each one takes about 25% of the original cost: about half of the total effort and total cost can be saved as shown in Fig. 18.7.

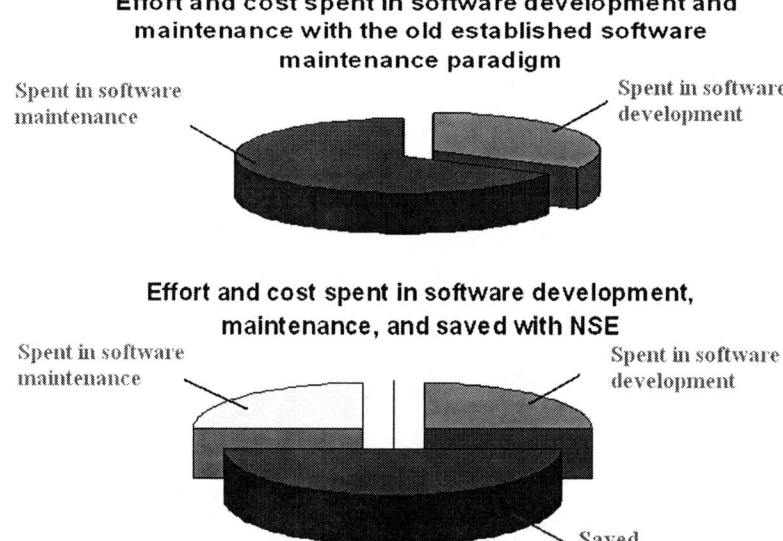

Fig. 18.7 Estimated effort and cost spent in software development and software maintenance

14. **It can be efficiently applied in the worst case scenario where no documents exist at all** – in this case, the NSE software maintenance engineering paradigm will use the Assisted Online Agents to automatically generate huge amounts of various documents through reverse engineering, then help users set bookmarks in the generated documents. After users re-design the test cases with some simple rules and re-test the product, the NSE software maintenance engineering paradigm will automatically establish various automated and self-maintainable traceabilities to make the product adaptive and maintainable.

18.3 Description of NSE Software Maintenance Engineering Paradigm

With NSE, the software maintenance process model is defined as shown in Fig. 18.8.

As shown in Fig. 18.8, the major steps for performing software maintenance engineering are as follows:

Step 1: Begin.
Step 2: Check the maintenance task type. If it is for the implementation of a new requirement, go to step 3; otherwise go to step 4.
Step 3: Perform the implementation of the requirement through the preprocess and the main process regularly as what was performed in the software development process.

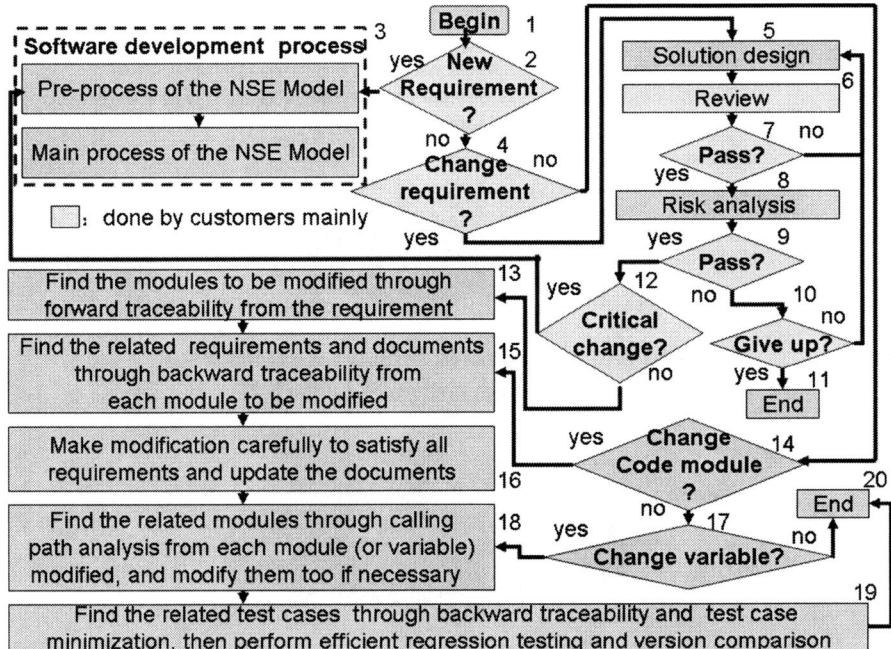

Fig. 18.8 NSE software maintenance process model

Step 4: Is a critical change of the requirement? If not, go to step 14.
Step 5: Perform solution design.
Step 6: Go through the solution review process.
Step 7: If the review result is not good enough, go to step 5.
Step 8: Perform risk analysis.
Step 9: If the risk analysis result is good enough, go to step 12.
Step 10: Give up? If not, go to step 5.
Step 11: End the process without changes.
Step 12: Is a critical change? If so, go to step 3.
Step 13: Find the modules to be modified through forward traceability (from requirement -> the corresponding test cases -> the corresponding source code, see Sect. 18.5 for an application example). Go to step 15.
Step 14: Is it not for changing the source modules? If so, go to step 17.
Step 15: Find the related requirements and documents through backward traceability from each module to be modified.
Step 16: Make modifications carefully to satisfy all of the related requirements (often a module is used for the implementation of more than one requirement) and update the related documents. If necessary, add some new modules and perform unit testing (including memory leak measurement and performance measurement) for the new modules. Go to step 18.
Step 17: Is it to change a global or static variable? If not, go to step 20 (end the process).
Step 18: Find the related modules through calling path analysis from each module/variable that is modified, and modify them too if necessary.
Step 19: Find the related test cases through backward traceability and perform test case minimization, then perform regression testing efficiently (including MC/DC test coverage analysis, memory leak measurement, performance measurement, quality measurement, and runtime error location through execution path tracing, see Sect. 18.4 for an example), and version comparison holistically.
Step 20: End the process.

18.4 Application

As described, with NSE a software product will be delivered with the computer program, the data used, and the documents traceable to and from the source code, plus the database built though static and dynamic measurement of the program, and a set of Assisted Online Agents to support testability, visibility, changeability, conformity, traceability, and maintainability.

The following graphics show an example for maintaining a sample program, a calculator. These graphics are provided by the Assisted Online Agents either in the product development site or in the customer site.

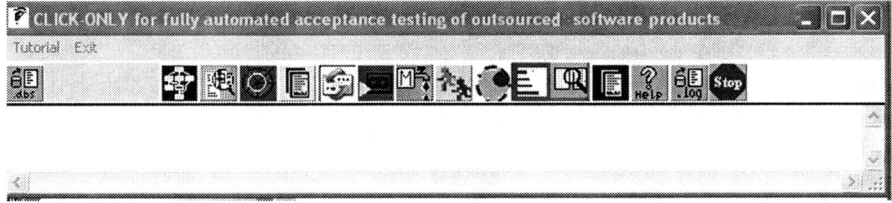

Fig. 18.9 The NSE-CLICK interface (the original icons are shown in different colors)

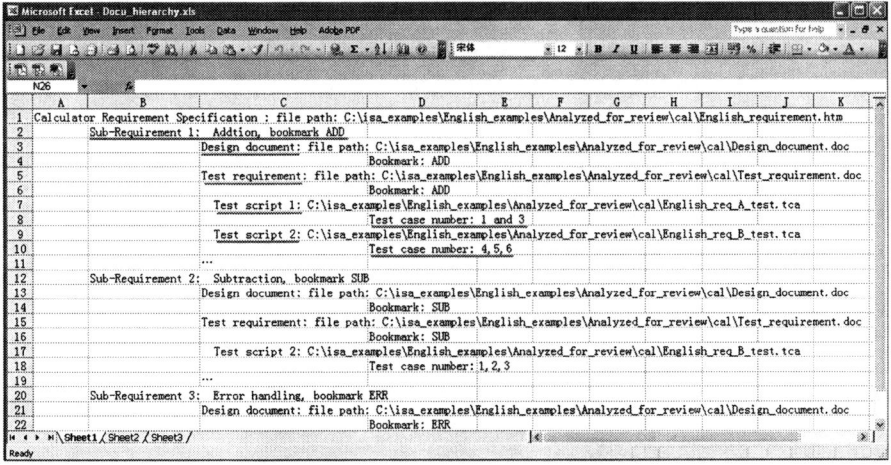

Fig. 18.10 From the requirement(s) to be changed to find the related test cases through the document hierarchy description table

Figure 18.9 shows one of the Assisted Online Agents, NSE-CLICK.

All other Assisted Online Agents are integrated together and can be executed from the interface.

Figure 18.10 shows the document hierarchy description table using bookmarks to indicate the relationship of the documents and the test cases with which we can find what test cases are used for testing the requirement(s) to be changed – for instance, we want to change the ADDITION requirement.

Figure 18.11 shows how we can perform forward tracing from the corresponding test case(s) to find what documents and the source code modules are used for the implementation of the requirement that needs to be modified – click on the test case (automatically shown in blue) in the test case window, then the corresponding source code modules will be highlighted (automatically shown in red) in the control flow window showing the entire product.

Figure 18.12 shows how we can perform backward tracing from the module(s) to be modified to find how many documents and requirements are related (in this case, two requirements are related, so that the modification must satisfy both).

Figure 18.13 shows how we can trace a module to be modified from the call graph shown in J-Chart notation to highlight all of the related modules which may also need to be modified correspondingly.

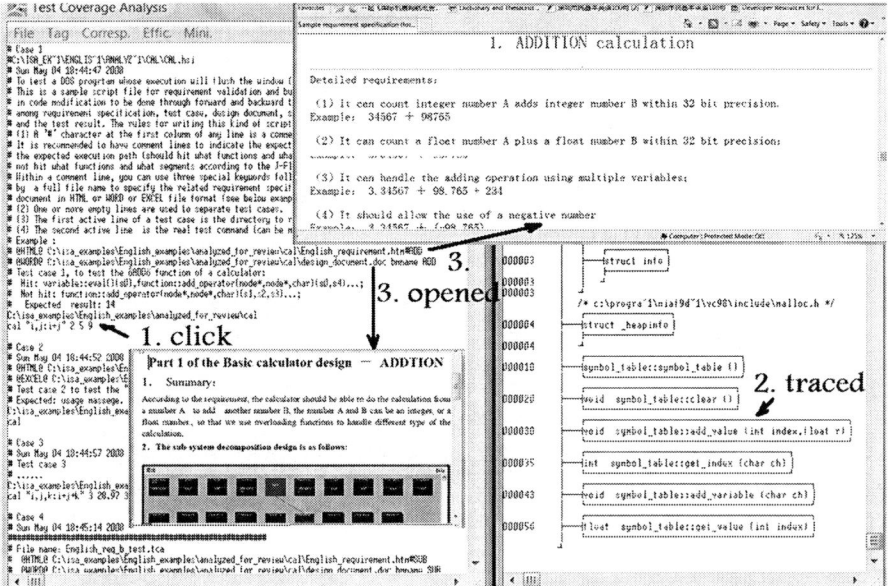

Fig. 18.11 Perform forward tracing from the test case(s) to find the related documents and the program modules (automatically shown in *red* on the screen originally), and then check what module(s) needs to be modified (updated)

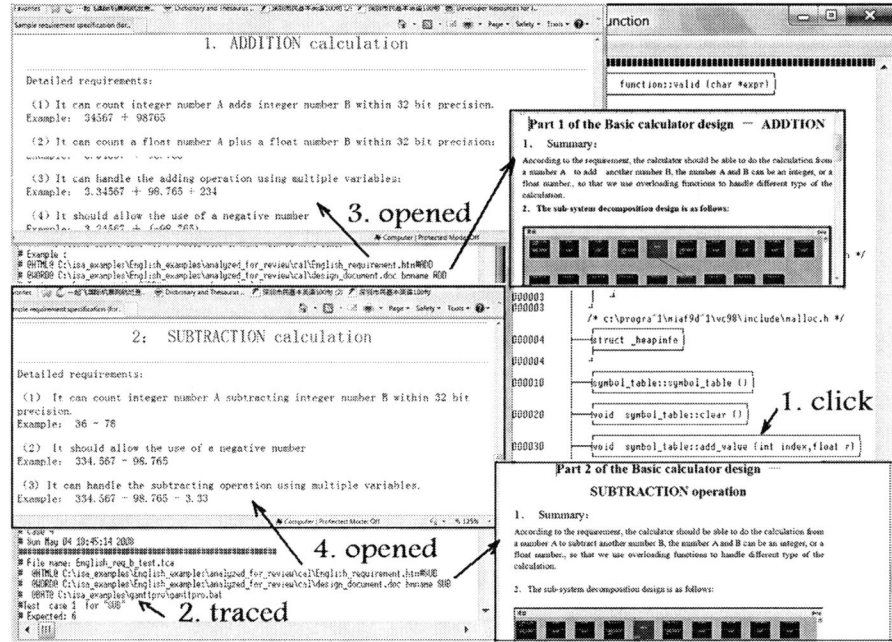

Fig. 18.12 Perform backward tracing from the module(s) to be modified to find how many documents and requirements are related (in this case, two requirements are related, so that the modification must satisfy both)

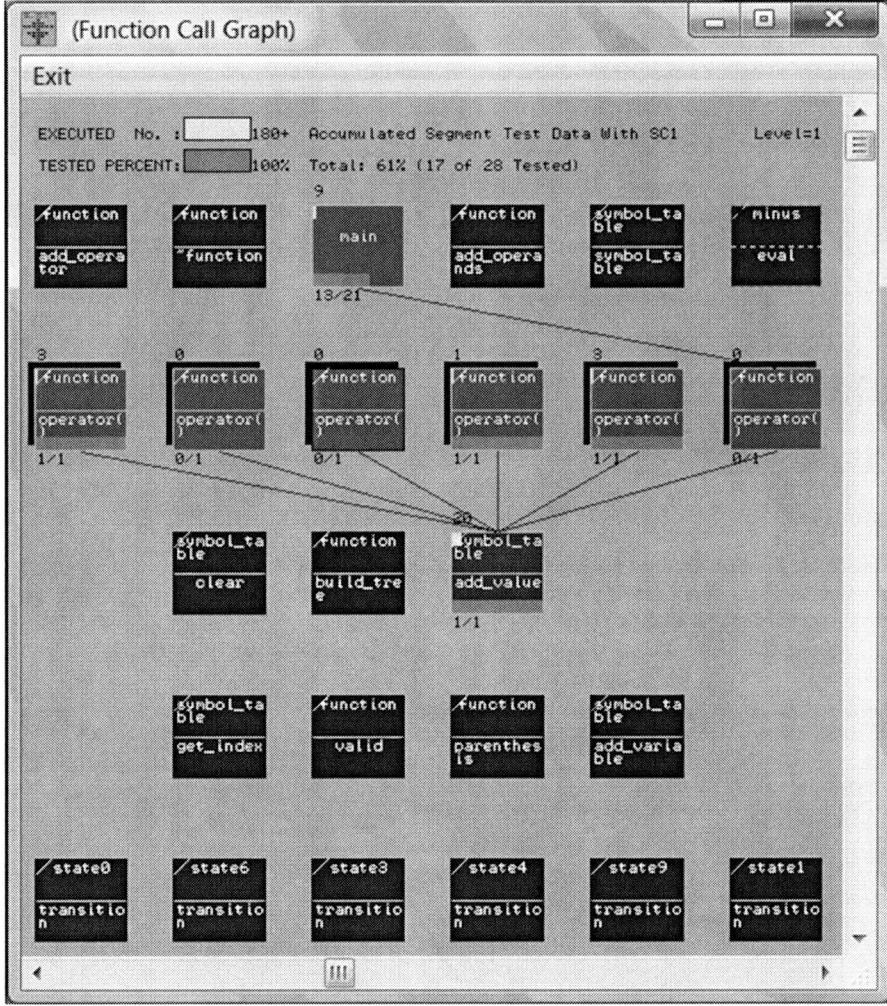

Fig. 18.13 Tracing a module to be modified from the call graph to highlight all the related modules which may need to be modified correspondingly

Figure 18.14 shows how we can find all the related statements calling the module to be modified for ensuring consistency among them from the diagrammed source code shown in J-Diagram

Figure 18.15 shows how to select the related test cases from a modified source code segment (a set of statements with the same execution condition without a decision statement) for efficient regression testing after code modification (in this example, only one test case is selected – it means the other test cases cannot be used to re-test that segment).

Figure 18.16 shows how to update the related documents after software modification through backward traceability.

18.4 Application

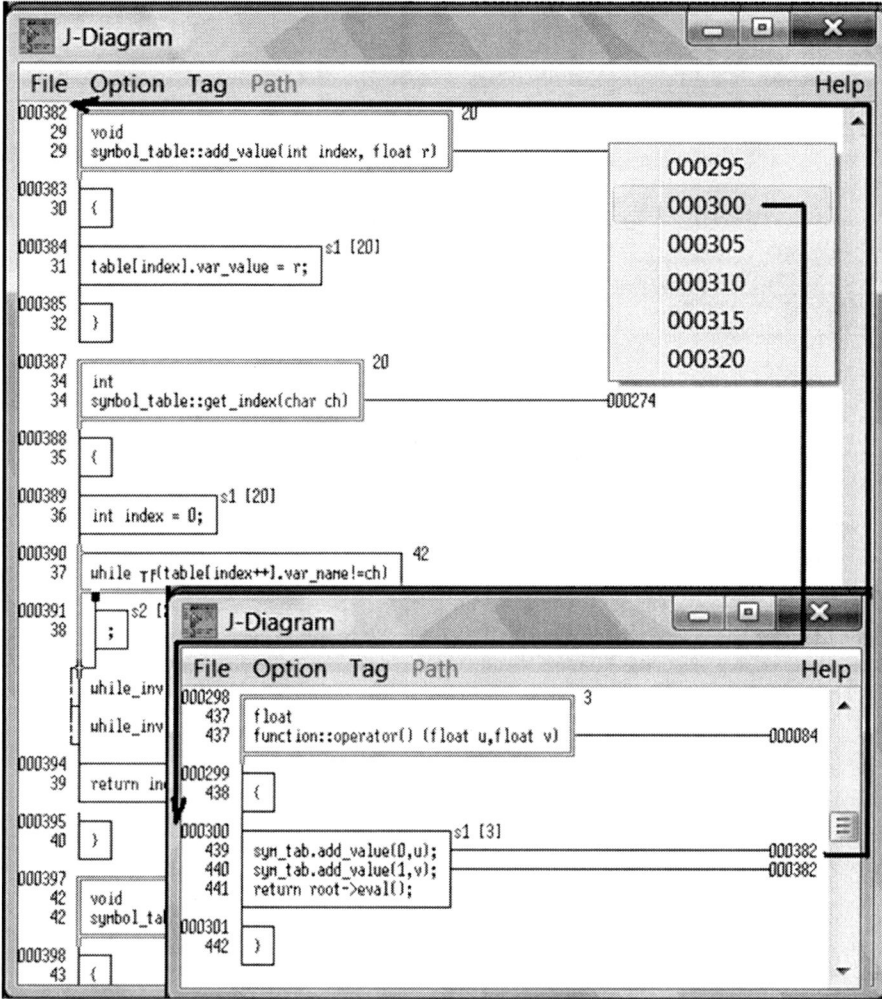

Fig. 18.14 Diagrammed source code shown with a facility used to trace a module to be modified with all the related statements calling that module, for ensuring consistency

If a global or static variable is modified, all modules using that variable must be checked to prevent inconsistency as shown in Fig. 18.17.

Usually less than 20% of the test cases are really useful for retesting a modified software product. Figure 18.18 shows the result of the test case efficiency analysis which will be the basis for test case minimization.

Based on the result of the test case efficiency analysis, test case minimization can be performed to get a minimized set of test cases (see Fig. 18.19) which can be used to obtain the same test coverage results obtained by all test cases (the algorithm is given in Chap. 21) – usually a test which has found a defect will be included into the minimized set of test cases because its execution path will be different from those test cases which have not found a defect.

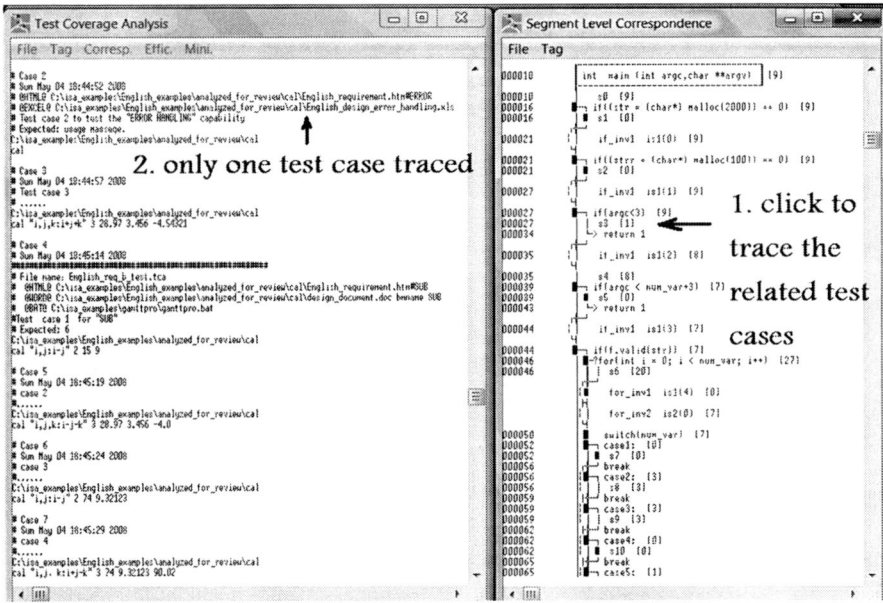

Fig. 18.15 Tracing a source code segment (automatically shown in *blue* on the screen originally) to select the related test case(s) (automatically shown in *red*) for efficient regression testing

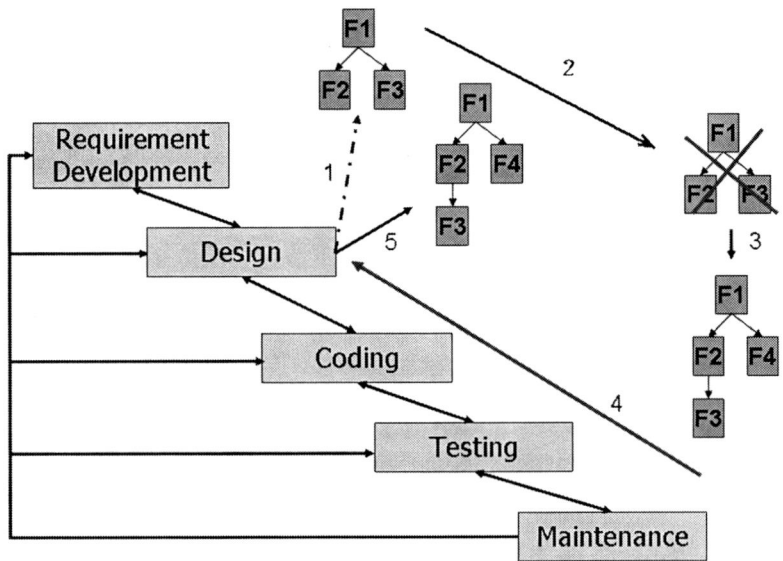

Fig. 18.16 Updating the related documents after software modification through backward traceability

To a software product with a graphical user interface (GUI), we can selectively play the captured GUI operations back automatically through traceability as shown in Fig. 18.20. Although there is only one file used to store the captured GUI opera-

18.4 Application

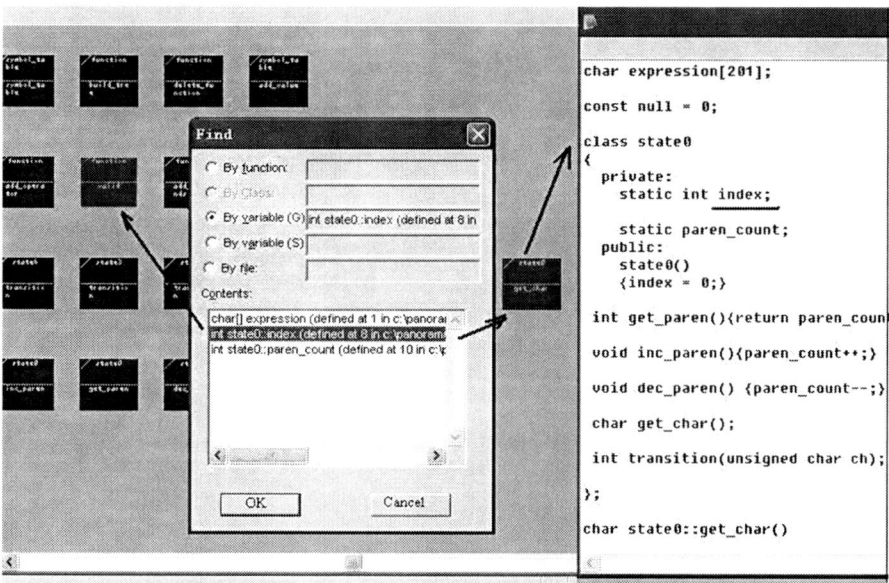

Fig. 18.17 Preventing inconsistent defects in variable modification

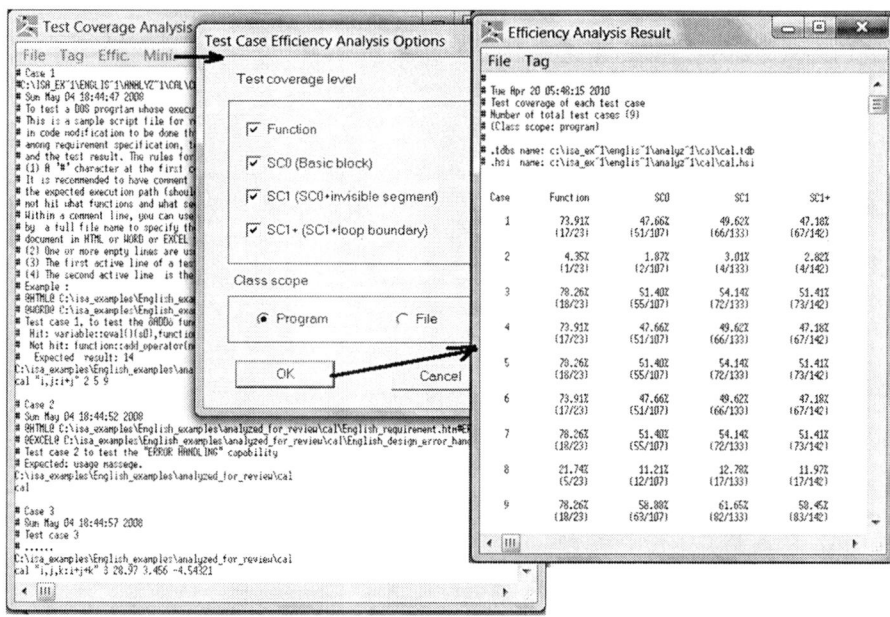

Fig. 18.18 Sample result of test case efficiency analysis

tions for many test cases, and only one batch file used to play the captured GUI operations back, using the Time Tags we can selectively play the corresponding GUI operations back for each test case separately.

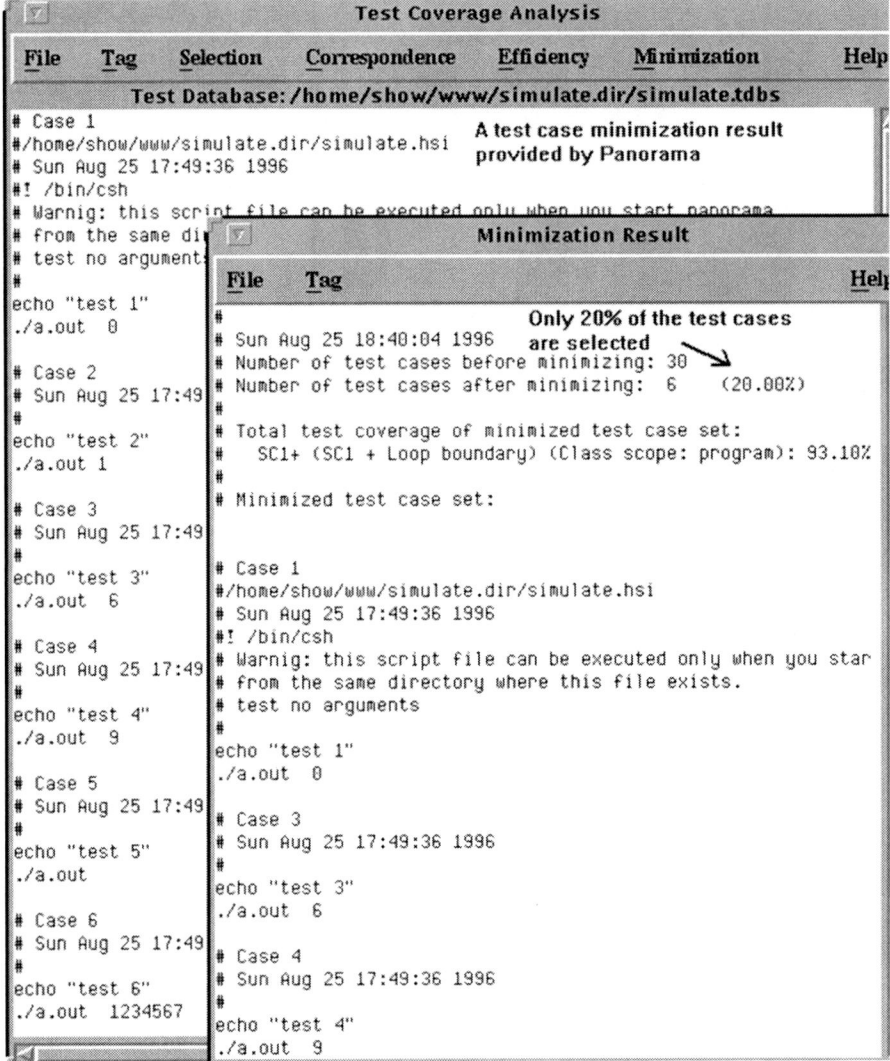

Fig. 18.19 A result of test case minimization, to be used for efficient regression testing

After modification, the following testing should be performed to ensure the quality:

(a) MC/DC test coverage analysis to the modified modules
(b) Memory leak and memory usage violation check
(c) Performance analysis to see whether the modified modules may become performance bottlenecks – if so, perform refactoring with the modified modules
(d) Quality measurement to make sure that the modified modules satisfy the required quality standard

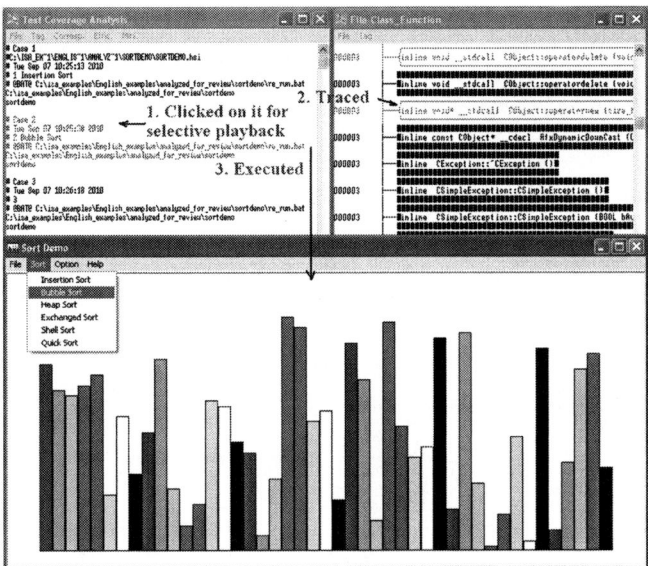

Fig. 18.20 Selectively playing the captured GUI operations back automatically

If new defects are found after the modification, perform runtime error type analysis, trace the runtime error execution path, and locate the detailed error locations (sometime the error may be introduced earlier but are found later).

Figure 18.21 shows the execution path traced from a runtime error.

When some new defects are found after software modification for a complex product, perform holistic version comparison to locate the problems better as shown in Fig. 18.22.

18.5 The Major Features

As described in the NSE software maintenance process model and shown in the application examples, the major features of NSE software maintenance engineering paradigm are briefly summarized as follows:

(a) Based on complexity science
(b) Performed holistically and globally
(c) Side-effect prevention driven
(d) Supported by various traceabilities
(e) Visual in the entire software maintenance process
(f) Intelligent in the test case selection for regression testing through backward traceability
(g) Systematic, quantifiable, and disciplined

486   18 NSE Software Maintenance Paradigm

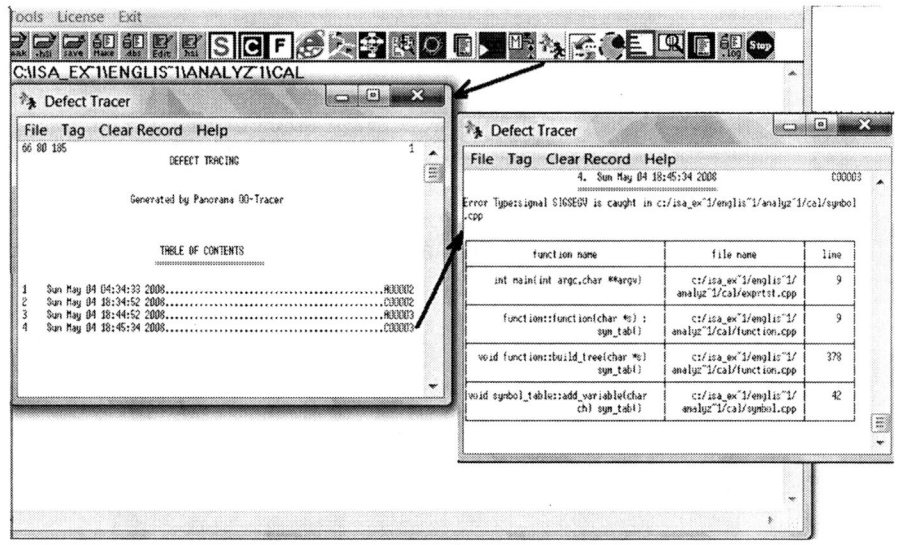

Fig. 18.21 A program execution path traced from a runtime error

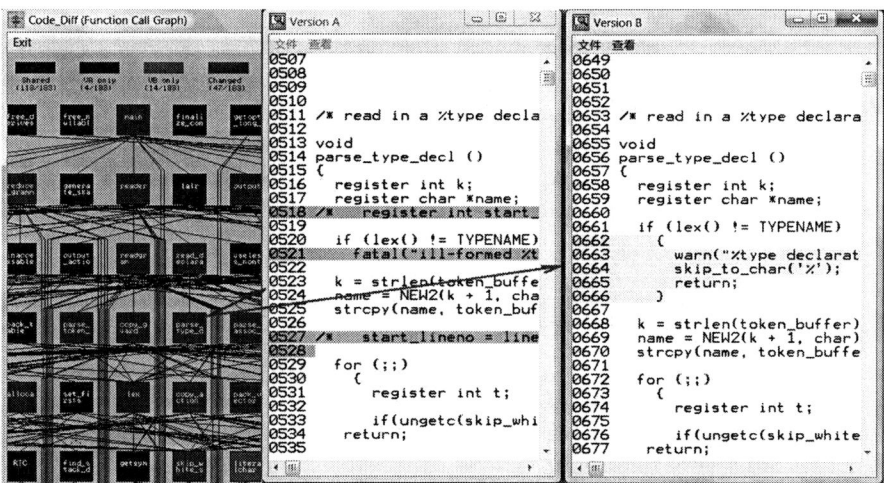

Fig. 18.22 Holistic software version comparison for a complex product (here, with colors on the screen originally, changed modules are shown in *red*, unchanged modules in *blue*, deleted modules in *brown*, and undeleted modules in *green*)

18.6 Summary

Today software maintenance takes 75% or more of the total effort and total cost in software product development, because the existing software maintenance engineering paradigm is based on reductionism and the superposition principle, so that almost all of the tasks and activities in software maintenance engineering are performed partially and locally.

This chapter presented the NSE software maintenance engineering paradigm based on complexity science. With the NSE software maintenance engineering paradigm, almost all software maintenance tasks/activities are performed holistically and globally with side-effect prevention in the implementation of requirement changes and code modifications through various traceabilities. Preliminary applications show that compared with the old-established software maintenance engineering paradigm, it is possible for the NSE software maintenance engineering paradigm to reduce about two-third of the total effort and total cost in software maintenance to help software organizations double their productivity and halve their cost in their software product development.

18.7 Points and Questions to Ponder

(a) Why does software maintenance take 75% or more of the total effort and total cost in software product development today?
(b) What are the major differences between the old-established software maintenance paradigm and the NSE software maintenance paradigm?
(c) How can the side effects in the implementation of requirement changes or code modifications be prevented?
(d) When a software product is made through outsourcing development, what should be provided with the product? Why?

18.8 Further Reading and Information Source

(a) Brooks FP Jr (1995) The mythical man-month. Addison-Wesley, Reading
(b) Ramesh G, Bhattiprolu R (2006) Software maintenance: effective practices for geographically distributed environments. Tata McGraw-Hill, New Delhi. ISBN 9780070483453
(c) Grubb P, Takang A (2003) Software maintenance. World Scientific Publishing, New Jersey. ISBN 9789812384256

References

[Amb05] Ambler SW (2005) A manager's introduction to the rational unified process (RUP). http://www.ambysoft.com

[Bro95-P120] Brooks FP Jr (1995) The mythical man-month. Addison-Wesley, Reading, p 120

[Dem82] Deming WE (1982) Out of the crisis. MIT Press, Cambridge

[Jon02] Jones C (2002) Software quality in 2002: a survey of the state of the art, Six Lincoln Knoll Lane, Burlington, MA 01803. http://www.SPR.com. Accessed 23 July 2002

[Pre05-P409] Pressman RS (2005) Software engineering: a practitioner's approach. McGraw-Hill, New York, p 409

Chapter 19
NSE Documentation Paradigm: Virtual, Traceable, and Consistent with the Source Code

> *The human being, language, reasoning through relationships, and archival representations are universal priors to science (i.e., there can be no science without each of them).*
>
> John N. Warfield

This chapter introduces another component of NSE – the NSE documentation paradigm which helps software organizations document their software products automatically, dynamically, holistically, accurately, precisely, and virtually. Documents for a software product should at least include the software project objectives, product specification, schedule, budget, space allocation, organization chart [Bro95-P110], technical manual, algorithm description, data structure design, testing methods and process and results, user manual, etc. [Woe09]. What is focused on in this chapter is technical documentation engineering.

19.1 The Old-Established Software Documentation Paradigm Is Outdated

The old-established software documentation paradigm is outdated:

(a) The foundation of the old-established software documentation paradigm is reductionism and the superposition principle that the whole of a complex system is the sum of its components, so that with the old-established software documentation paradigm almost all tasks and activities in software documentation are performed linearly, partially, and locally.

(b) It works with the linear process models in which the workflow goes linearly in one way with one track only without upstream movement at all – it requires that the software developers always document the software right without making any mistake and any wrong decision in software documentation, violating nature's laws about people because people are nonlinear and it is easy to make mistakes, although these mistakes can be corrected.

(c) It works with the outdated software development methodologies based on the reductionism principle and Constructive Holism principle that the components of a complex system are completed first, then the whole of the system is assembled from its components.

(d) It is not holistic – with it many small pieces of documents will be created/generated without the capability to document an entire software system – missing the "Big Picture" of a software product. Even if some tools can be used to document an entire software product, without automated and self-maintainable traceability the obtained system-level graphical documents will still be useless, because there will be too many connection lines, making the documents hard to view and hard to understand as shown in Fig. 19.1.

(e) The graphic documents and the source code are separated – hard to keep them consistent.

(f) The obtained documents are not traceable – hard to use.

(g) Often the obtained documents are inconsistent with the source code after product modification.

(h) The documents obtained are stored statically as hard copies in Postscript, XML, or other formats requiring huge amounts of space and long loading times.

(i) Most graphic documents are created manually or using graphic editors, not automatically generated.

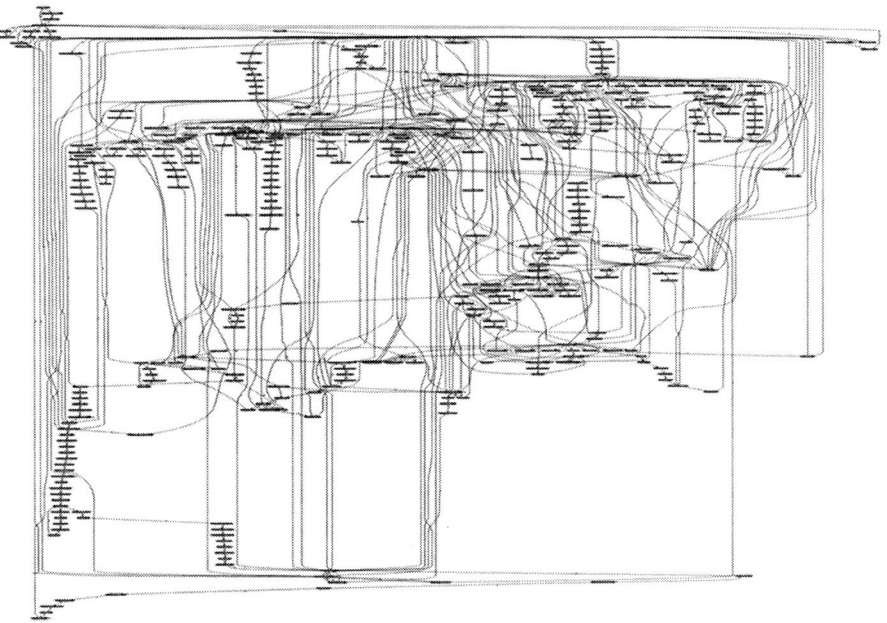

Fig. 19.1 A traditional call graph without traceability (**http://keithcu.com/bookimages/wordpress_html_m1e9af381.jpg**)

(j) The graphic documents are time-consuming to draw, hard to change, and hard to maintain.
(k) Often the obtained documents are not accurate.
(l) Often the obtained documents are not precise – for instance, it cannot directly and graphically show where a code branch or condition combination has been tested or not.

19.2 Outline of NSE Documentation Paradigm

NSE documentation paradigm is introduced in detail later. Here is the outline of the solution:

(a) The foundation of the NSE documentation paradigm is complying with the essential principles of complexity science, particularly the Nonlinearity principle and the Holism principle, so that with NSE almost all software documentation tasks and activities are performed holistically and globally.
(b) It works with the NSE process model in which the workflow goes nonlinearly through two-way iteration with multiple tracks – supporting upstream movement and downstream movement.
(c) It works with NSE software development methodology based on complexity science and the Generative Holism principle that the whole of a complex system comes first as an embryo, then grows up with its components.
(d) It is holistic – many holistic documents for an entire software product will be automatically generated to make the documents easy to view and easy to understand as shown in Figs. 19.2 and 19.3.
(e) With NSE, source code (either a dummy program or a regular program) is the source for most graphical document generation, whereas the graphic documents are the visual representation of the corresponding source code – so that they are always consistent to each other.
(f) The obtained documents are traceable to and from the source code – easy to use.
(g) The obtained documents are consistent with the source code after product modification – only need to update the database.
(h) The documents are generated directly from the source code and the corresponding database virtually exists without storing hard copies in memory or hard disk (unless the users want them) to greatly save the space and make the display speed about 1,000 times faster.
(i) With NSE, almost all graphic documents are automatically generated.
(j) The graphic documents are easy to generate, change, and maintain.
(k) The generated graphical documents are accurate and consistent with the code.
(l) The generated graphical documents are precise – for instance, it can directly and graphically show where a code branch or condition combination has been tested or not.

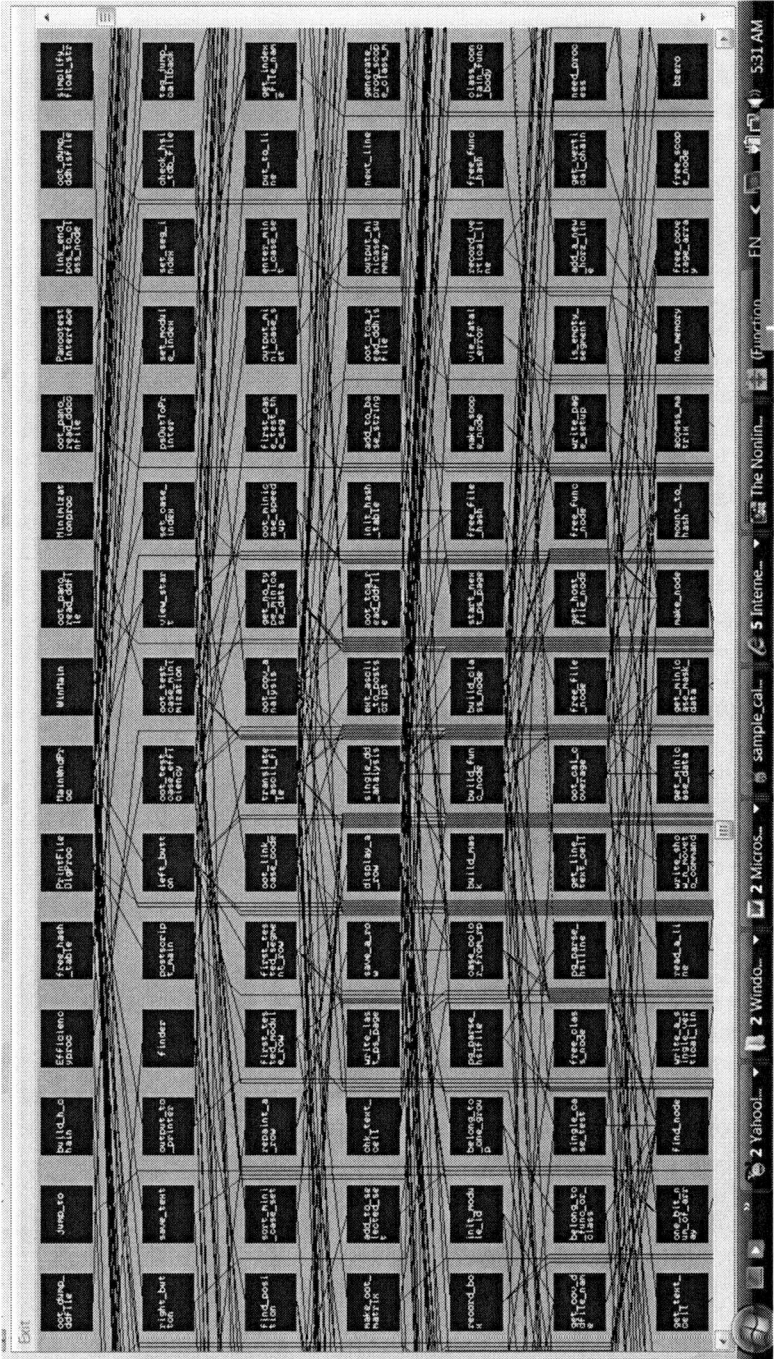

Fig. 19.2 An application example for NSE to document a complex software holistically

19.2 Outline of NSE Documentation Paradigm

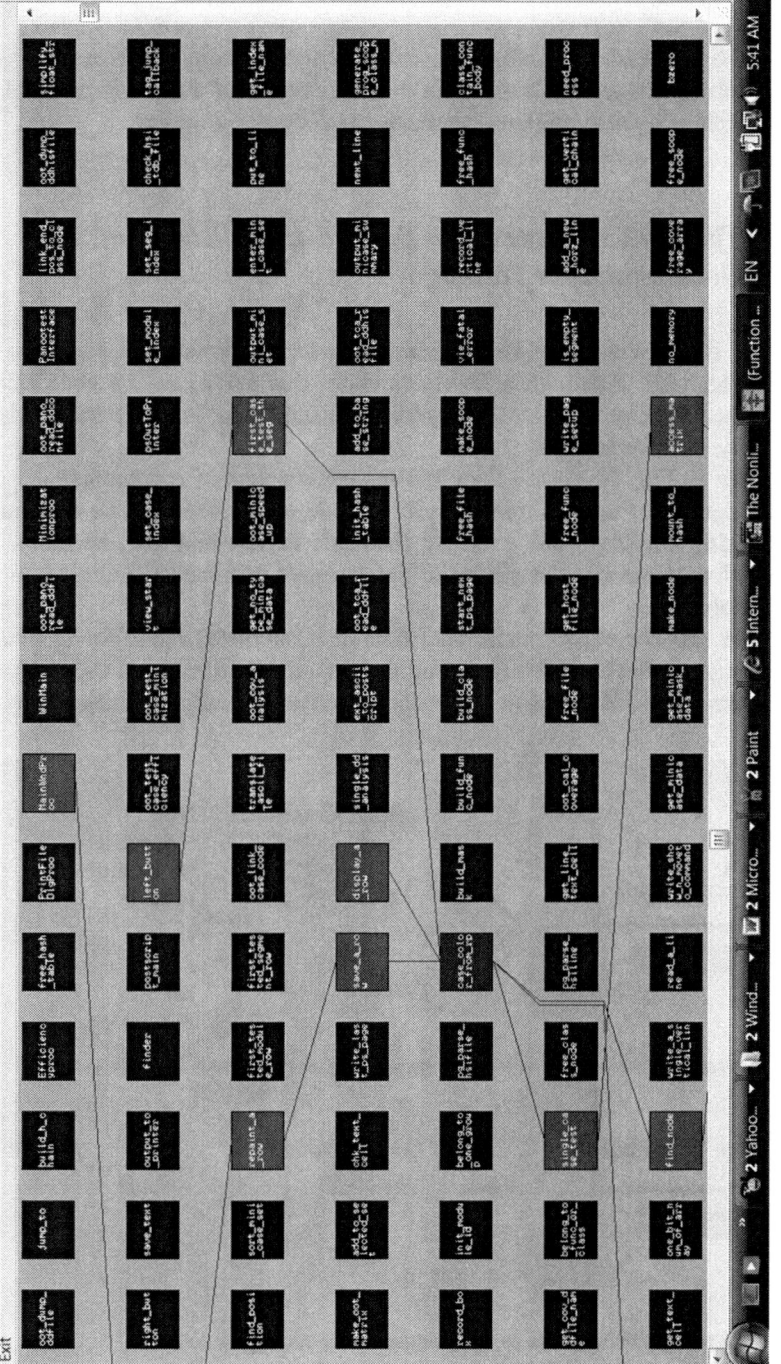

Fig. 19.3 An application example for NSE to trace a module with all the related modules

19.3 Description of the NSE Documentation Paradigm

The NSE documentation paradigm generates holistic, graphical, interactive, and traceable documents automatically from the source code of a dummy program for requirement development and product design or a regular program.

19.3.1 The Critical Issues with the Old-Established Software Documentation Paradigm

The critical issues with the old-established software documentation paradigm are shown in Fig. 19.4. Although a few tools claim that partial source code can be directly generated from the UML diagrams [Arl05], there are still many related issues that need to be solved.

As shown in Fig. 19.4, with the old-established software documentation paradigm, the design documents and the source code are separated. The documents designed using UML are not traceable for static review and not executable for dynamic defect removal – the quality of the designed documents is almost impossible to ensure.

The issues also come from the fact that there is no automated and self-maintainable facility to support the traceability among the design documents and the test cases and the source code – it is almost impossible to make the design documents and the

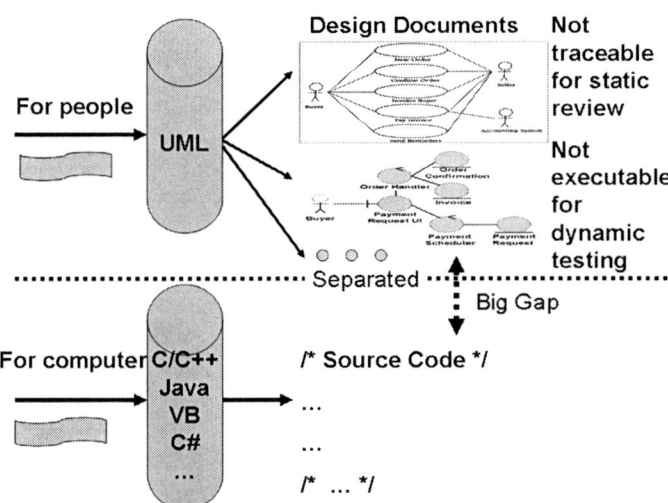

Fig. 19.4 The big gap between the design documents and the source code

19.3 Description of the NSE Documentation Paradigm

source code consistent, so that after the source code is changed again and again with the implementation of requirement changes and code modifications, most design documents will become useless garbage.

19.3.2 The Solution Offered with NSE

The solution offered with NSE is shown in Fig. 19.5.

As shown in Fig. 19.5, with NSE the source code (either a dummy program or a regular program) is also the source for automatically generating the graphic design documents, while the graphical design documents become the visual face of the source code – design becomes precoding, and coding becomes further design. The designed graphical documents are traceable for static review as shown in Fig. 19.3 and executable for dynamic defect removal (see Figs. 11.22 and 11.23).

With NSE, there is an automated and self-maintainable facility established through dynamic testing using the Transparent-box method to support traceability among the design documents and the test cases and the source code as shown in

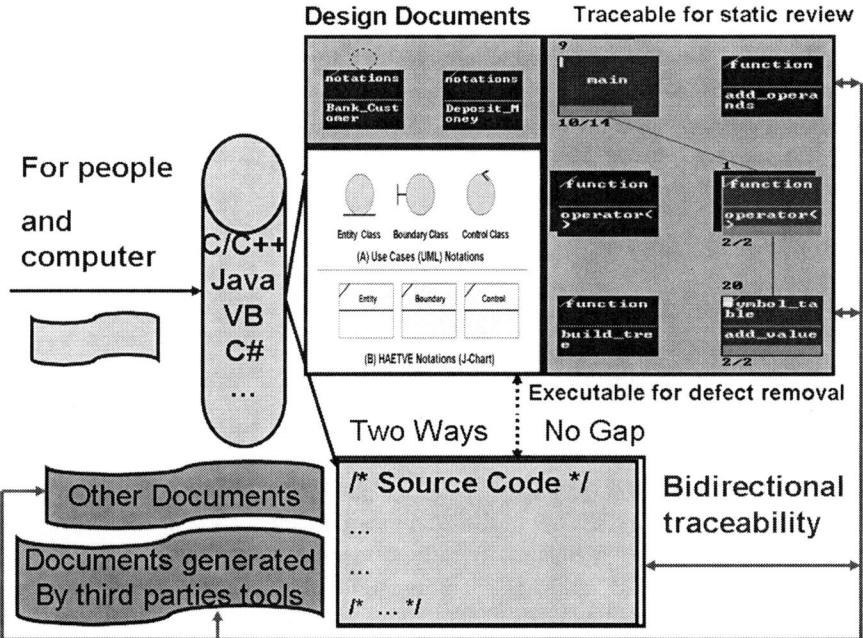

Fig. 19.5 The solution offered with NSE

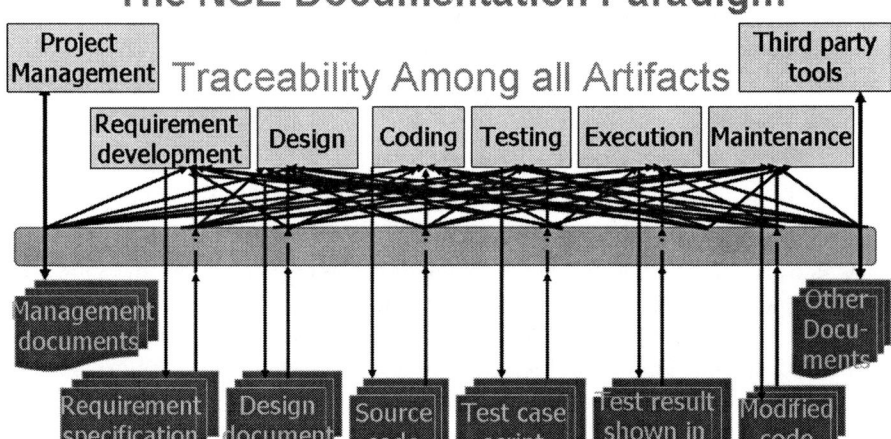

Fig. 19.6 Traceability among all related documents, test cases, and source code

Fig. 19.6 – it makes the design documents and the source code consistent. After the source code is changed with the implementation of requirement changes and code modifications, most design documents can be automatically and incrementally (only the modified source files need to be re-analyzed) updated to maintain consistency with the source code.

19.3.3 The Objectives of the NSE Documentation Paradigm

The objectives of the NSE documentation paradigm are given as follows:

(a) Combining software programming and graphical software documentation together seamlessly.
(b) Making one source for both people understanding and computer "understanding" – through static people review of the graphical documents and dynamic program execution to ensure the upstream quality of a software product.
(c) Making all kinds of documents (including those manually drawn and those generated by third party tools) traceable to the source code to keep them consistent with each other through automated and self-maintainable traceability established by dynamic testing using the Transparent-box method combining functional testing and structural testing together seamlessly with a capability to establish bidirectional traceability.

19.3 Description of the NSE Documentation Paradigm

(d) Generating most software documents automatically as much as possible.
(e) Making software documents visible as much as possible.
(f) With the graphical documents consistent with the source code, making a software product truly maintainable and adaptive to the changed or changing environment.

19.3.4 Working with Dummy Programming

The NSE software documentation paradigm works with dummy programming using dummy modules consisting of an empty body or only simple function call statements – any software professional can write the dummy programs easily without extra learning.

19.3.5 Working with NSE Software Visualization Paradigm

The NSE documentation paradigm works closely with the NSE Software Visualization Paradigm which mainly generates interactive and traceable 3J graphics (J-Chart, J-Diagram, and J-Flow diagram). As described above, making software documents visible as much as possible is one of the objectives of NSE software documentation paradigm.

19.3.6 Working with HAETVE Requirement Development Technique

The NSE software documentation paradigm works with the HAETVE (Holistic, Actor–Action and Event–Response driven, Traceable, Visual, and Executable) requirement development technique and the corresponding tools. As described in Chap. 11, the notations of HAETVE can map with most of the UML notations as follows:

(a) Figure 19.7 shows the sample notations for representing an actor and an action.

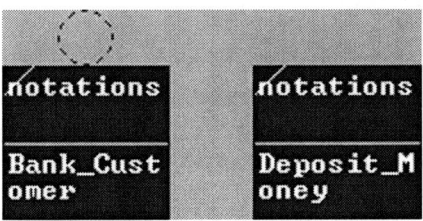

Fig. 19.7 HAETVE notations for representing an actor and an action

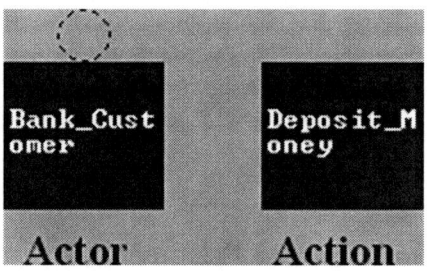

Fig. 19.8 The notations for representing an actor and the action for C/C++

The dummy Java program corresponding to the notations shown in Fig. 19.7 is as follows:

```
public class notations {
   public static void Bank_Customer ()
     {
        Bank_Customer () ;
     }
   public static void Deposit_Money ()
     {
     }
}
```

The sample notations for representing an actor and the action for C/C++ are shown in Fig. 19.8.

The corresponding dummy source code written in C/C++ is listed separately as follows:

```
Bank_Customer ()
{
Bank_Customer ();
}
   Void Deposit_Money ()
     {
     }
```

(b) HAETVE notations mapping to Use Cases analysis notations are shown in Fig. 19.9.
(c) Graphical representation of class.
   With HAETVE, classes are represented in several graphical notations as shown in Fig. 19.10.

19.3 Description of the NSE Documentation Paradigm

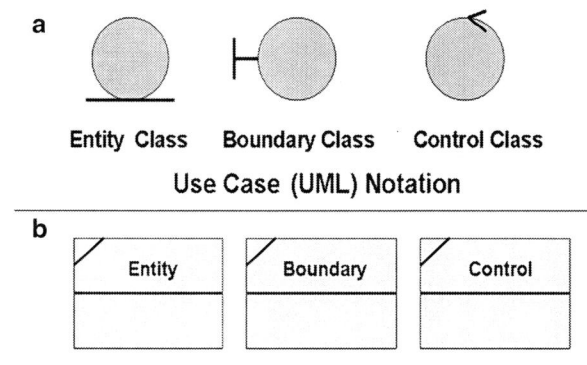

Fig. 19.9 Analysis notation mapping between Use Cases (UML) and HAETVE

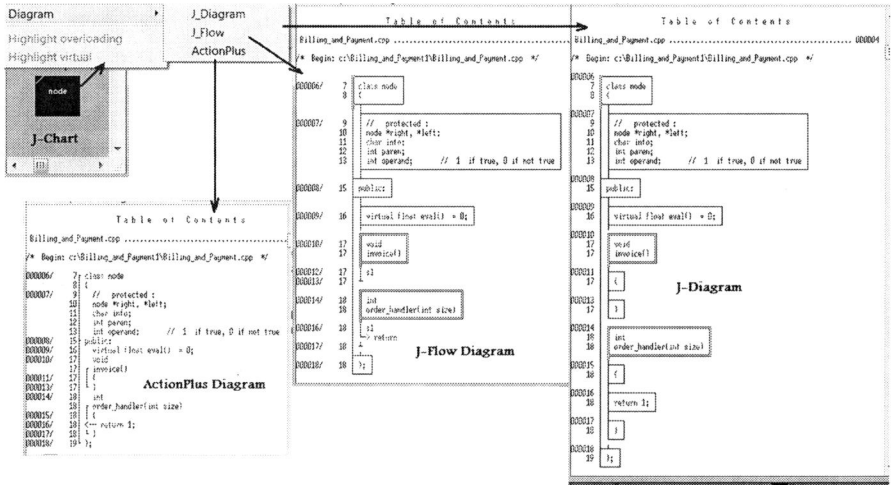

Fig. 19.10 NSE graphical representation for a class

(d) Time-Event table.

With NSE, the time-event table is written in the comment part of a dummy program or a regular program. An example is listed as follows:

```
/* Time-Event table:
|  Timing  |    t1    |    t2    |    t3    |    t4    |
|  Events  |  Event1  |          |          |          |
|          |          |  event2  |          |          |
|          |          |          |  event3  |          |
|          |          |          |          |  event4  |

|  Timing  |    t5    |    t6    |    t7    |    t8    |
|  Events  |  Event5  |          |          |          |
|          |          |  event6  |          |          |
|          |          |          |  event7  |          |
|          |          |          |          |  event8  |

...
*/
```

(e) Mapping to Activity diagram.
With NSE, a new type logic diagram, J-Diagram, is used to map to Activity diagrams. An application example is shown in Figs. 12.9 and 12.10.
(f) Method for graphically representing message sending and receiving.
With NSE, message sending and receiving are represented with the automatically established "click-to-jump" facility as shown in Fig. 19.11.

19.3.7 How It Works

Figure 19.12 shows the workflow for the NSE software documentation paradigm.

19.3.8 Making a Software Product Visible in Multiple-Views

1. **Static + dynamic**

    (a) Static analysis of a program's Cyclomatic complexity – see Fig. 19.13.
    (b) **Dynamic analysis of program performance** – see Fig. 19.14.

2. **Macro + micro**

    (a) **Holistic MC/DC test coverage analysis for an entire software product** – see Fig. 19.15.
    (b) **Detailed MC/DC test coverage analysis for a individual class/function** – see Fig. 19.16.

19.3 Description of the NSE Documentation Paradigm

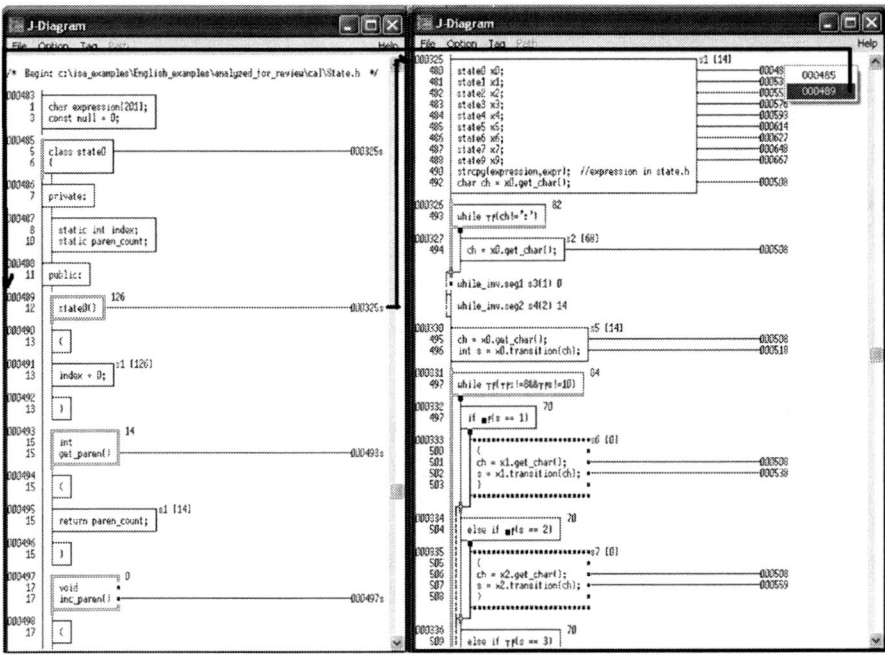

Fig. 19.11 Click-to-jump facility automatically established for showing message sending and receiving

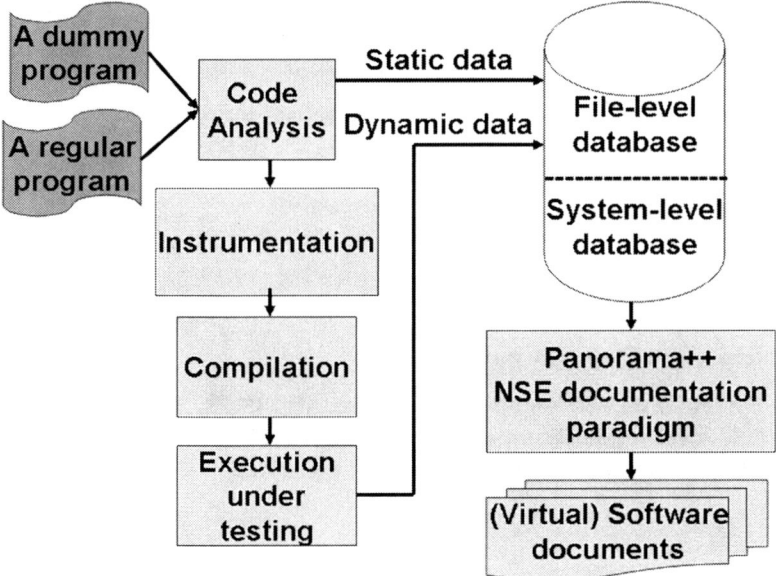

Fig. 19.12 The workflow of NSE documentation paradigm

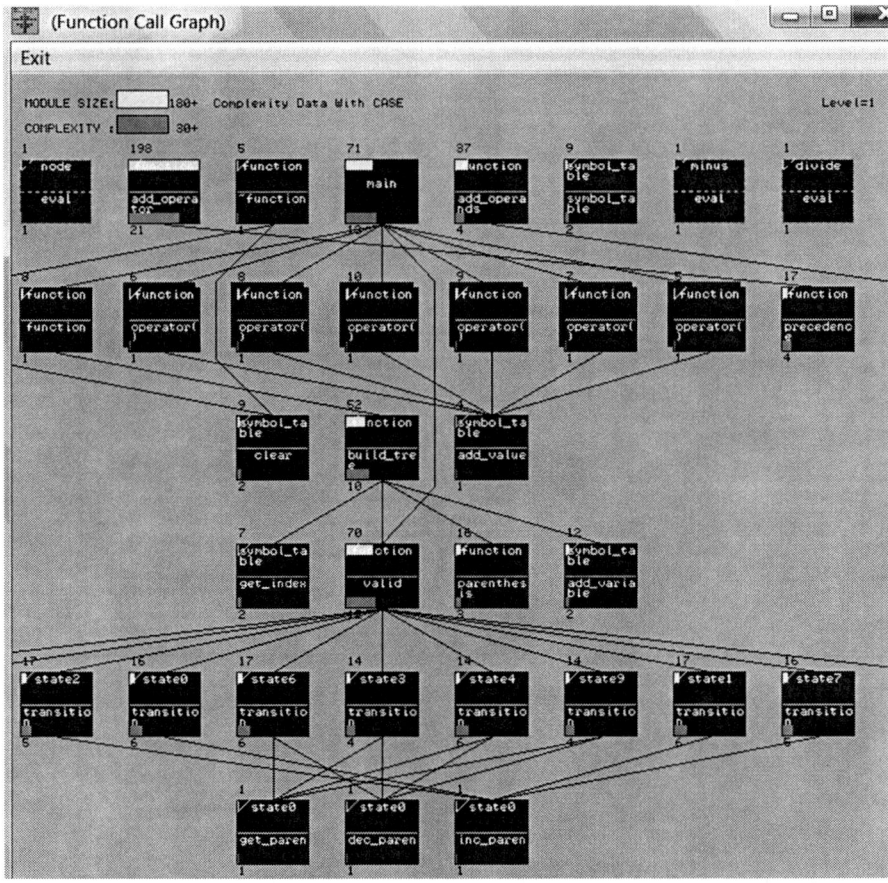

Fig. 19.13 An application example of Cyclomatic complexity analysis

3. **Procedure + data**

   (a) **Function cross-reference analysis** – see Fig. 19.17.
   (b) **Data analysis** – see Fig. 19.18.

4. **System level + file level + statement level**

   (a) **System-level version comparison** – see Fig. 19.19.
   (b) **File-level version comparison** – see Fig. 19.20.
   (c) **Statement version comparison** – see Fig. 19.21.

19.3 Description of the NSE Documentation Paradigm 503

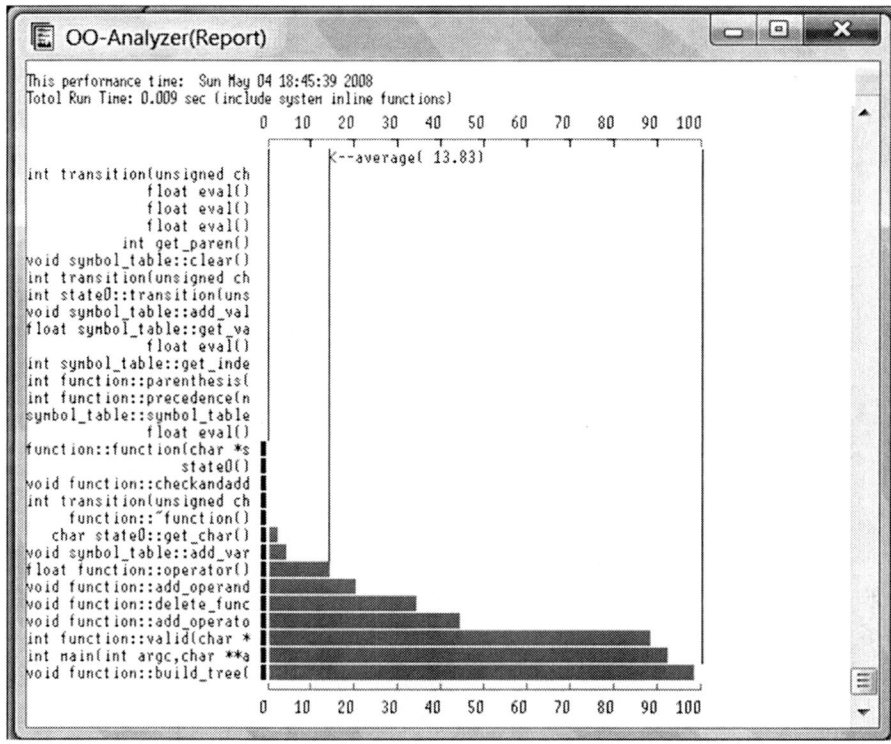

Fig. 19.14 An application example of performance analysis

5. **Static visibility + dynamic visibility**

    (a) **Forward tracing from a test case to find what modules can be tested** – see Fig. 19.22.
    (b) **Dynamic visibility – tracing a test case to not only find what modules can be tested, but also directly play the captured test operations back through the batch file (.bat) specified in the @BAT@ keyword within the test case description part** – see Fig. 19.23.

6. **Integrative + traceable**

    (a) **With NSE, the generated documents are interactive – for instance, users can click on a module box to use that module as the root to generate a sub call graph** – see Fig. 19.24.
    (b) With NSE, most of the generated documents are traceable – see Fig. 19.25.

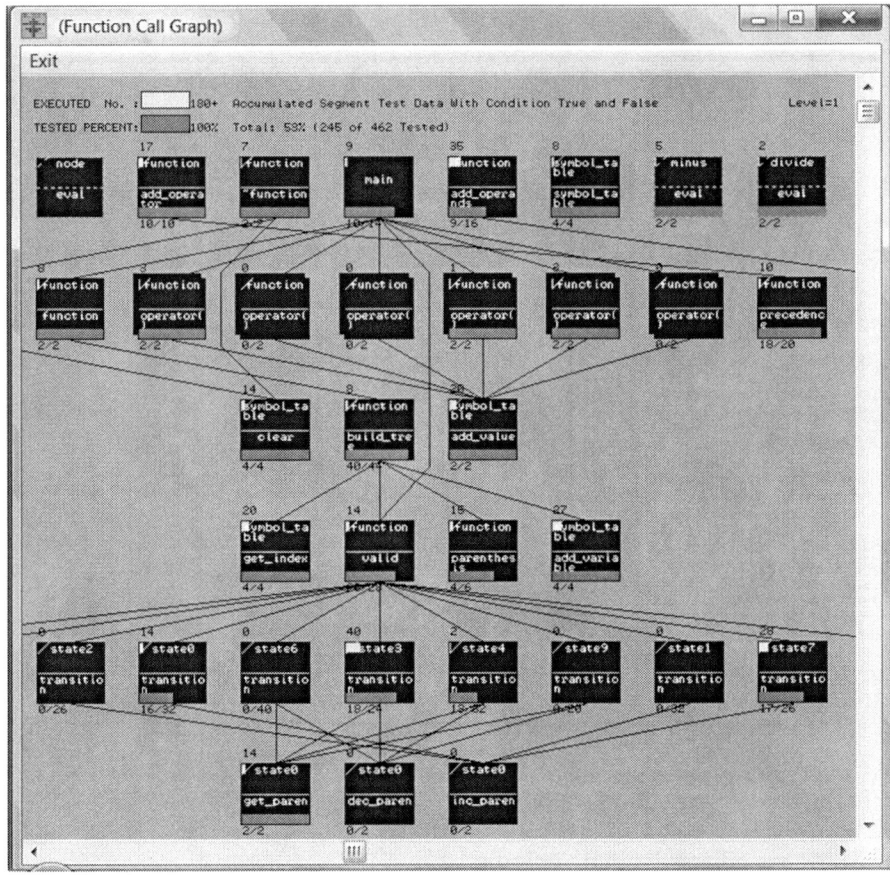

Fig. 19.15 Holistic MC/DC (Modified Condition/Decision Coverage) test coverage analysis

7. **Linkable + convertible**

   (a) **With NSE, different graphical documents can be linked together** – see Fig. 19.26.

   (b) **Converting between the generated logic diagram and the control flow diagram** – see Fig. 19.27.

8. **Local + internet**

   (a) **With NSE, many static and dynamic analysis reports can be automatically generated** – see Fig. 19.28.

   (b) **With NSE, the generated reports for static and dynamic program analysis can be saved in the HTML format to be used as Web pages** – see Fig. 19.29.

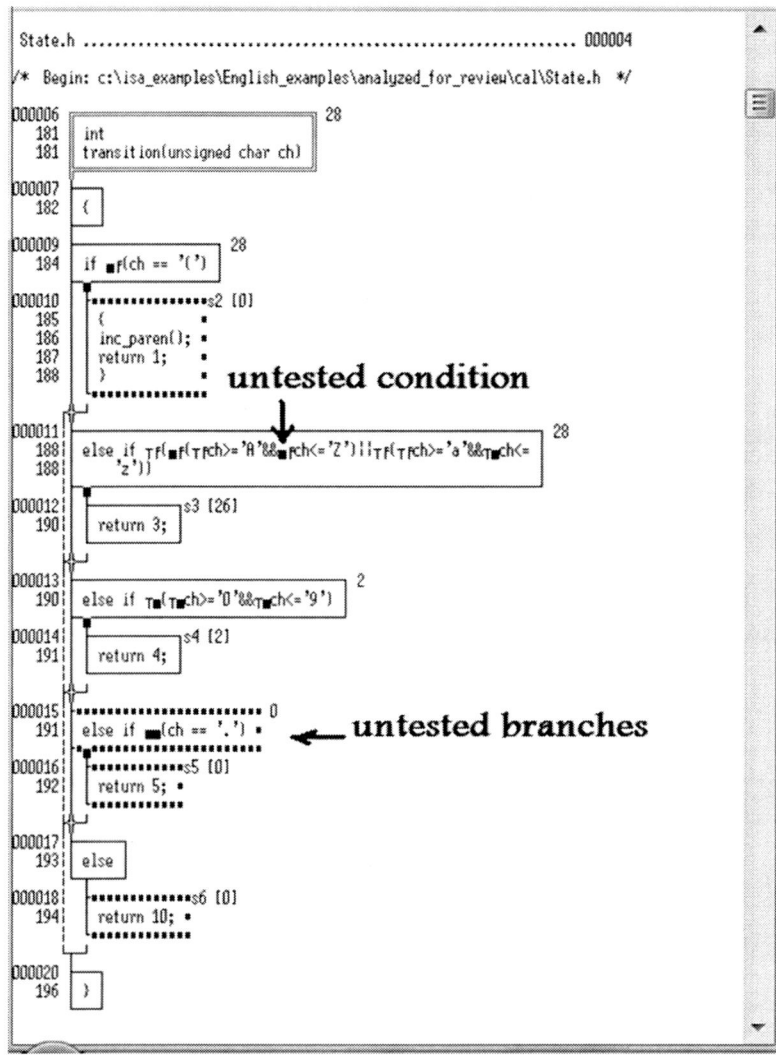

Fig. 19.16 An application example of detailed test coverage analysis of a module

19.4 The Major Features of NSE Documentation Paradigm

The graphical documents generated by the NSE documentation paradigm are given as follows:

- **Holistic** – NSE documentation paradigm generates holistic charts and diagrams to document an entire software product.
- **Interactive** – the generated graphical documents are interactive, the generated charts/diagrams themselves are also the interfaces to accept the user's commands/operations.

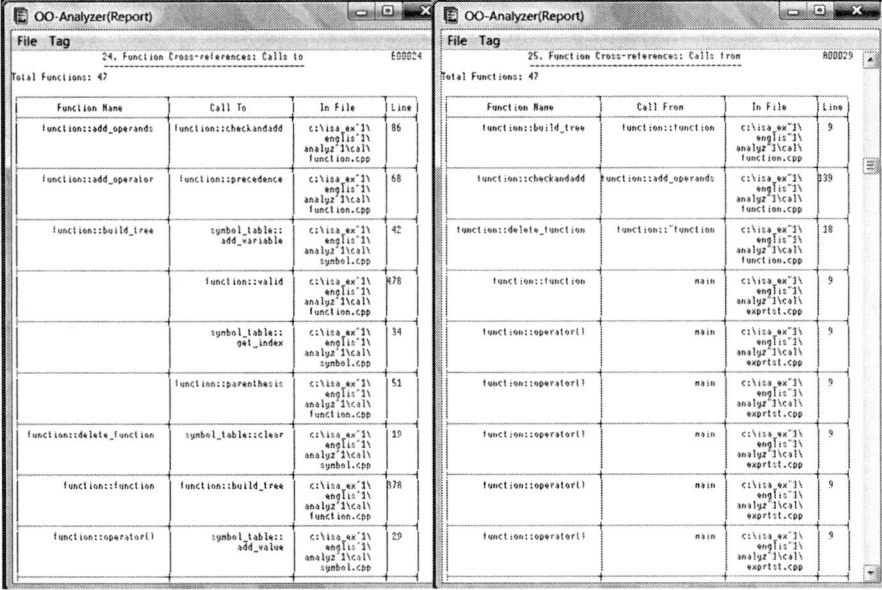

Fig. 19.17 An application example of functional cross-reference analysis

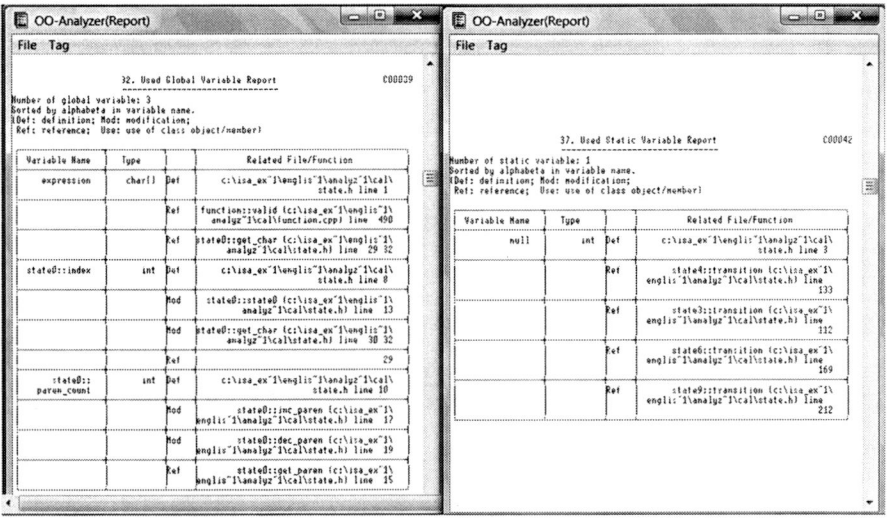

Fig. 19.18 An application example of variable analysis

19.4 The Major Features of NSE Documentation Paradigm

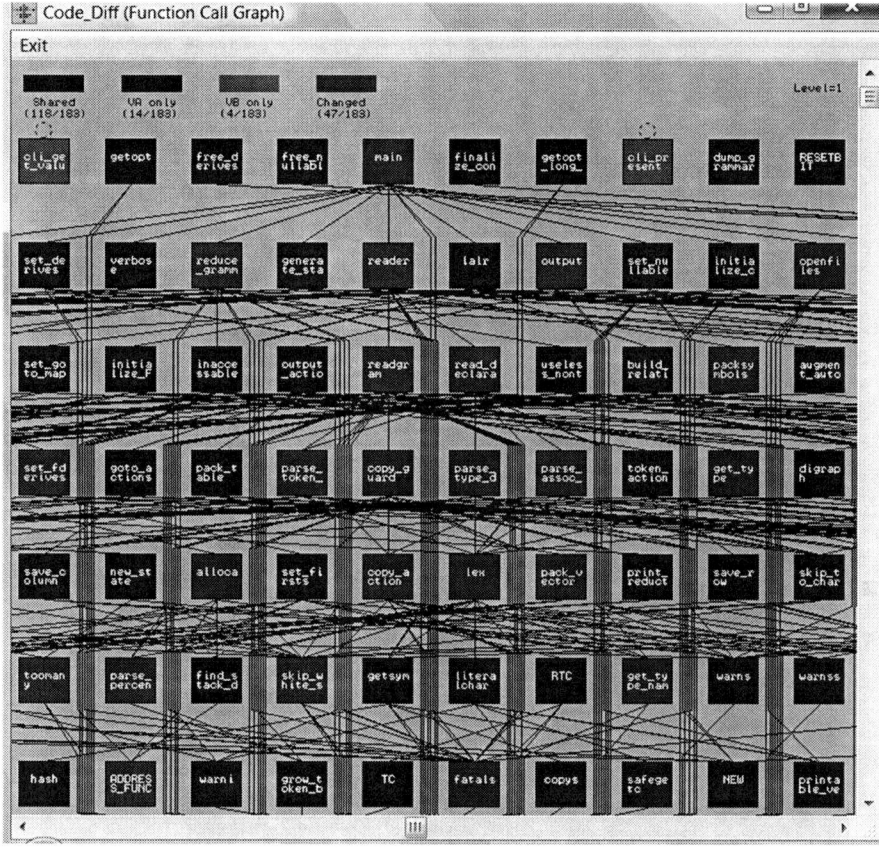

Fig. 19.19 An application example of holistic version comparison in system level

- **Traceable** – with NSE, most of the generated documents are traceable, useful for validation, verification, and semiautomated inspection and walkthrough.
- **Accurate** – with NSE, the source code of a dummy program or a regular program is also the source to automatically generate most graphical documents, so that the generated documents are accurate and consistent with the source code.
- **Precise** – the generated graphical documents are precise; for instance, the corresponding documents can show how many times a branch is executed, and what code branches and conditions have not been executed.
- **Virtual** – with NSE, most graphical documents are dynamically generated from the source code, there is no need to save their hard copies in memory or disk, so that a huge amount of space can be saved, and the display speed is about 1,000 times faster. The generated holistic charts and diagrams are shown within a window, no more or less. When a chart or diagram needs to move around, a new one will be regenerated dynamically without the real movement of the chart or diagram, so that the display

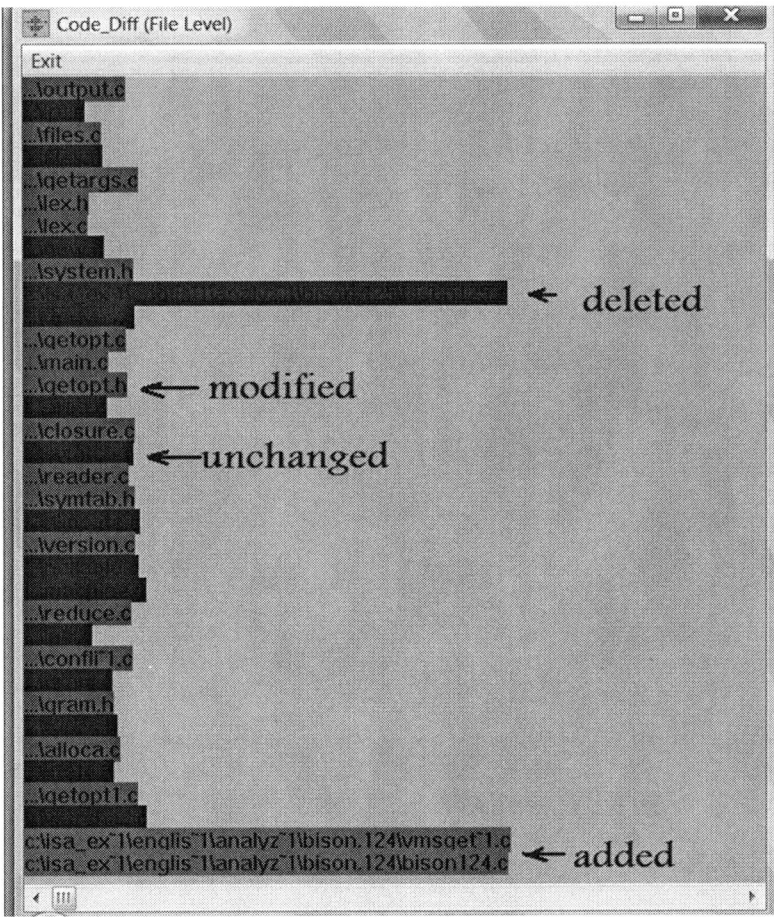

Fig. 19.20 An application example of file-level version comparison

speed is very fast, but from a users' point of view, there is no difference between the virtual charts/diagrams and the regular charts/diagram needing huge amounts of space to store in hard disk and computer memory.
- **Massive** – the graphical documents with the size being more than 100 times the size of the source code (if the documents are stored in disk regularly) can be automatically generated in system level, file level, and module level. For instance, for each class/function, the NSE documentation paradigm can automatically generate the logic diagram shown in J-Diagram notations with the untested branches and untested conditions highlighted, control flow diagram shown in J-Flow diagram notations, the quality measurement result shown in Kiviat diagram, etc. – massive documentation.

19.4 The Major Features of NSE Documentation Paradigm

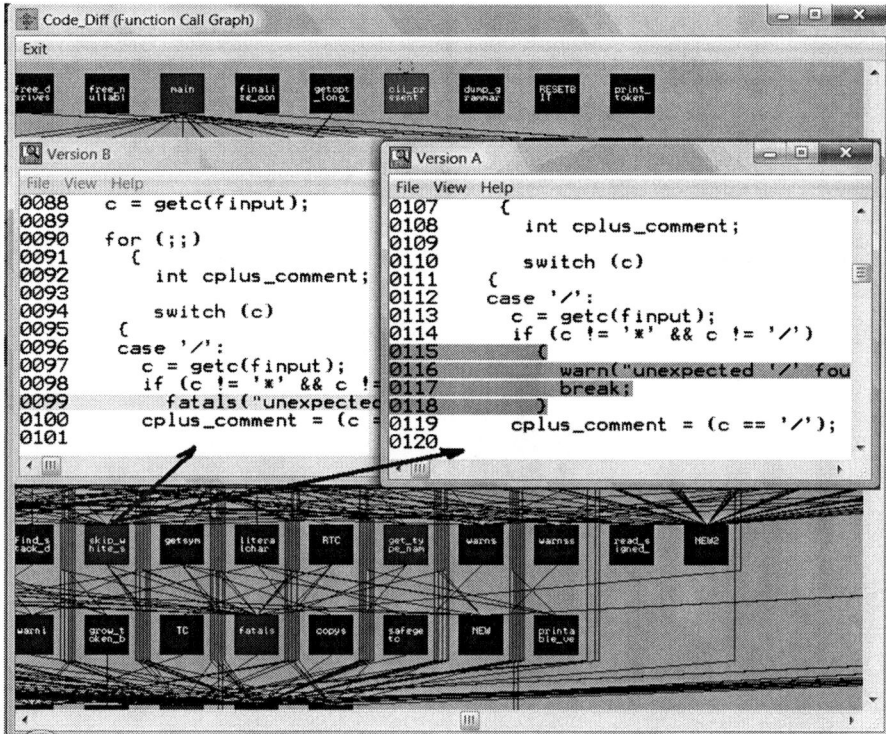

Fig. 19.21 An application example of statement version comparison

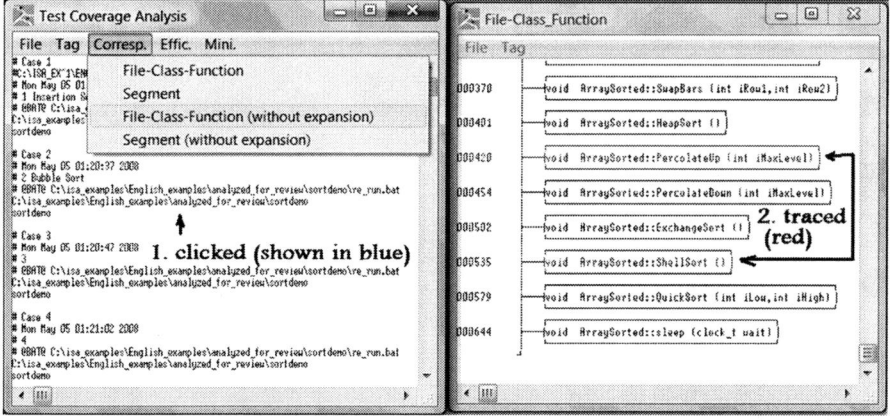

Fig. 19.22 Example of static visibility – tracing a test case to view what modules can be tested

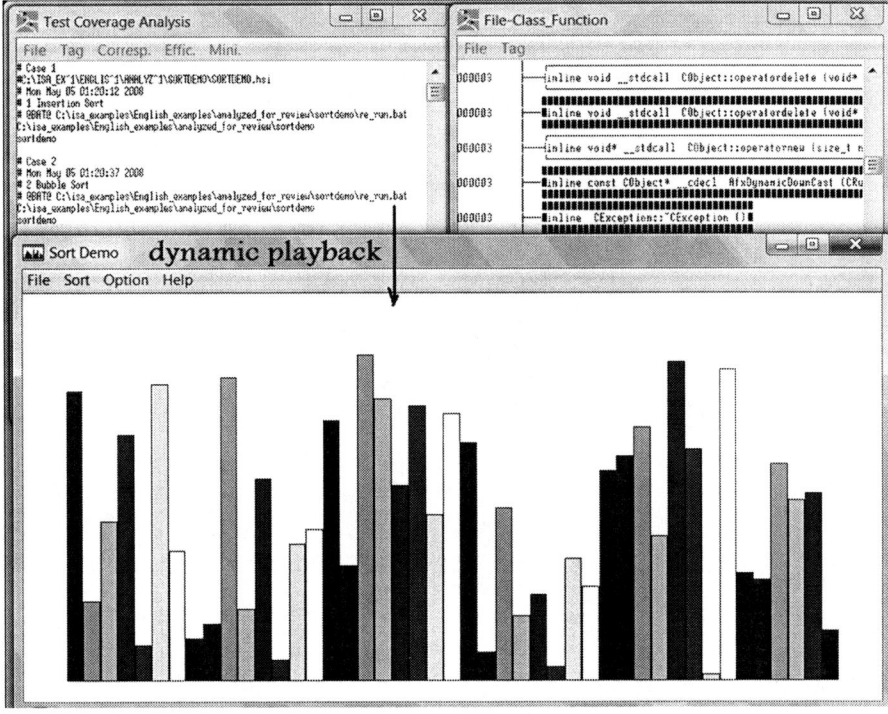

Fig. 19.23 Dynamic visibility – tracing a test case to play the captured operations back

19.5 Application

NSE documentation paradigm has been commercially implemented and supported by Panorama++. All the screenshots shown in this chapter are taken from real application examples.

19.6 Summary

The old-established software documentation paradigm is outdated because it is based on reductionism and the superposition principle that the whole is the sum of its components, so that with it almost all software documentation tasks and activities are performed linearly, partially, and locally. The sources used to generate/create

19.6 Summary

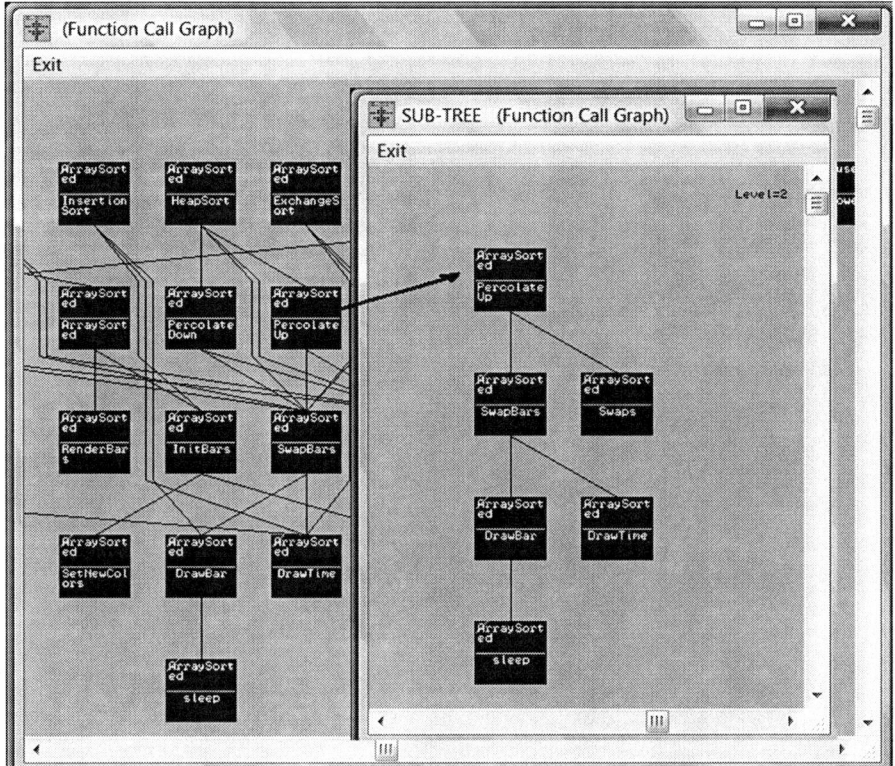

Fig. 19.24 Interaction example: click on a module box to generate an isolated sub call graph

software documents are different from the source code of the software product. The generated/created graphical documents are not traceable for static review and not executable for dynamic testing, so that the quality of the documents is hard to ensure and the documents are hard to maintain consistency with the source code after code modifications.

The NSE software documentation paradigm is based on complexity science by complying with the essential principles of complexity science, particularly the Nonlinearity principle and the Holism principle that the whole of a complex system is greater than its components, and that the characteristics and behaviors of the whole emerge from the interaction of its components, so that with NSE software documentation paradigm almost all software documentation tasks and activities are performed holistically and globally. The sources used to generate most graphical software documents are also the source code of the dummy programs or regular programs. The generated graphical documents are traceable for static review, and

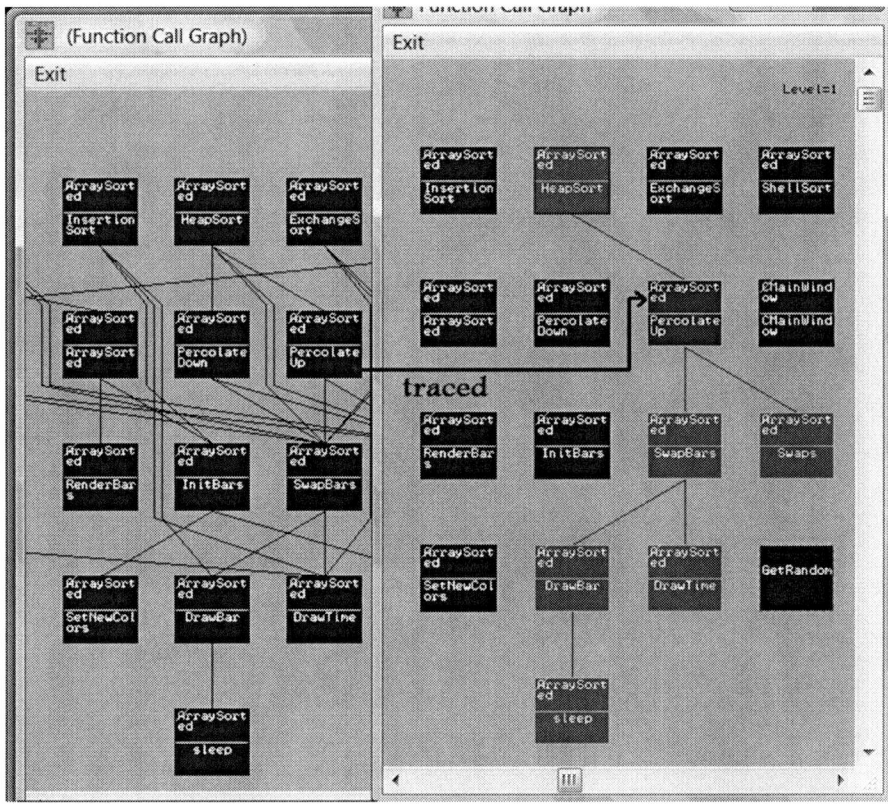

Fig. 19.25 An application example – tracing a module to see all the related modules

the corresponding source code is executable for dynamic testing, so that the quality of the documents is easy to ensure and the documents are easy to maintain consistency with the source code after code modifications – with NSE design is precoding, while coding is further design.

Source code is not the best documentation of a software product, but **source code is the best source to directly and automatically generate holistic, interactive, traceable, consistent, accurate, precise, massive, and virtual documents of the software product**.

19.7 Points and Questions to Ponder

(a) What are the major issues existing with the old-established software documentation paradigm?
(b) Is source code the best documentation for a program? Why?
(c) What are the major differences between the old-established software documentation paradigm and the NSE software documentation paradigm?

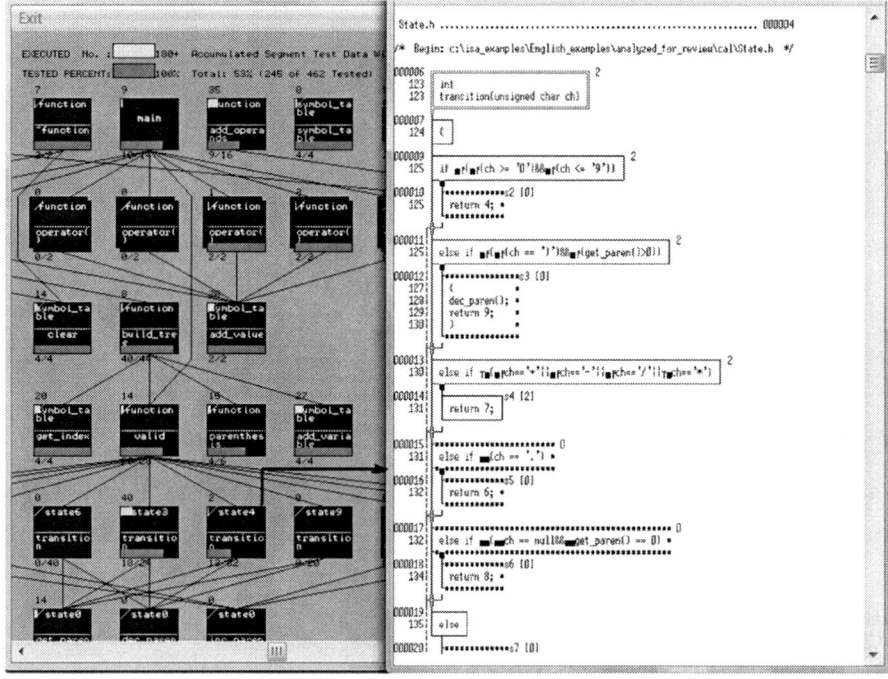

Fig. 19.26 An application example – linking a call graph to the logic diagram

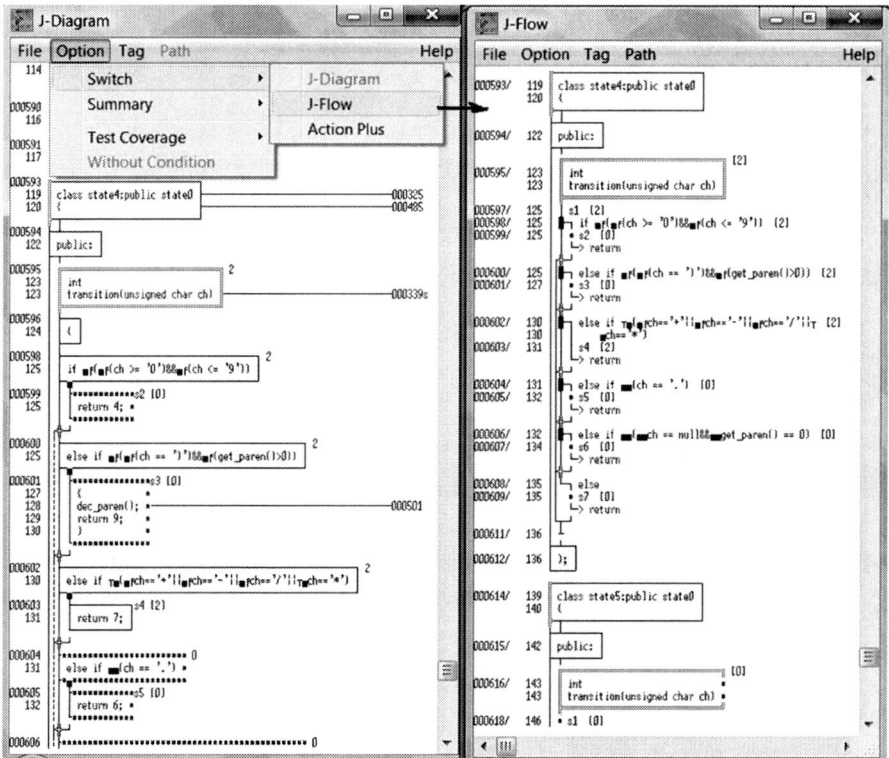

Fig. 19.27 An application example of diagram conversion from a logic diagram to the control flow diagram

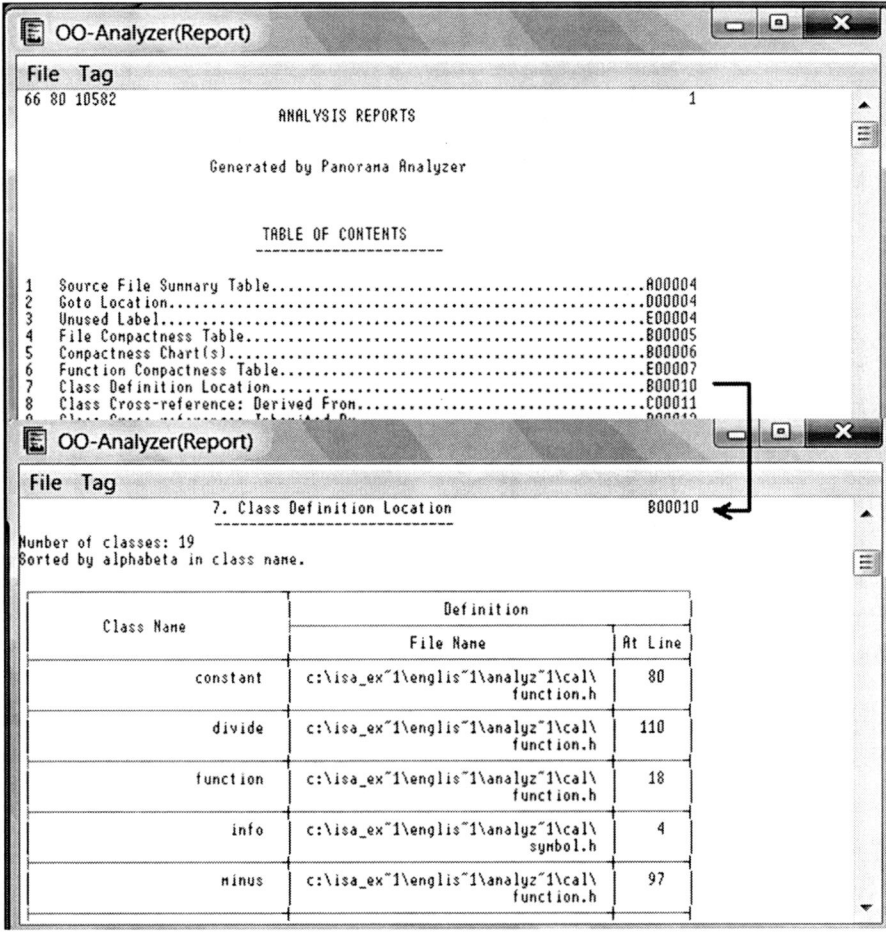

Fig. 19.28 An application example of static and dynamic program analyses and reporting

19.8 Further Reading and Information Source

(a) Problems of bad software documentation. http://www.software-documentation.co.uk/software-documentation-problems.html, © TechScribe, UK. Page updated 29 December 2008

(b) *Dennis Crane*, Writing cost-effective documentation for software systems. White Papers (Download as Acrobat PDF file). http://www.drexplain.com/press/cost-effective-documentation-for-software-systems/

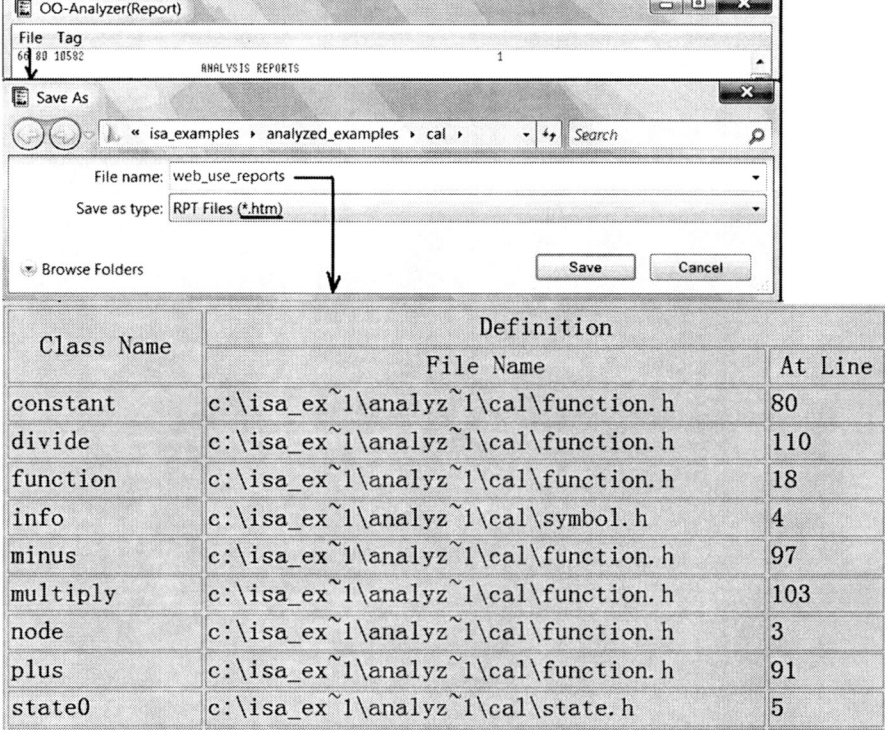

Fig. 19.29 Code analysis reports saved in HTML format to be used as Web pages

(c) Software documentation. From Wikipedia, the free encyclopedia. **http://en.wikipedia.org/wiki/Software_documentation**
(d) Software documentation resource portal. http://www.softwaredocumentation.info/Default.aspx

References

[Arl05]	Arlow J, Neustadt I (2005) UML 2 and the unified process: practical object-oriented analysis and design. Pearson Education, Inc., Boston
[Bro95-P110]	Brooks FP Jr (1995) The mythical man-month. Addison-Wesley, Reading, p 182
[Woe09]	Woelz C (2009) The KDE documentation primer. http://i18n.kde.org/docs/doc-primer/index.html. Retrieved 15 June 2009.

Chapter 20
NSE Project Management Paradigm: Seamlessly Combined with the Project Development Process

> *There is nothing in the world constant but inconstant.*
>
> Jonathan Swift

This chapter introduces the NSE project management paradigm with which the software development process and project management process are combined together seamlessly, so that the documents for project management are traceable with the implementation of the requirements and the source code to help the project management team and the project development teams find possible problems quickly and solve the problems in time.

20.1 The Old-Established Software Project Management Paradigm Is Outdated

Software projects need to be managed according to the project development plan, budget, and functions. Usually software project management tasks include:

1. Project planning/scheduling
2. Project monitoring
3. Risk management
4. Project cost estimation
5. Process management
6. Project documentation
7. Unexpected event handling
8. People/team management

Since the term *software engineering* first appeared in the 1968 NATO Software Engineering Conference, it has been more than 40 years past. Many books on software project management are published. But unfortunately, the project success rate is still very low – only about 30%. Why?

The root reason is that not only the old-established software engineering paradigm based on reductionism and the superposition principle is outdated, the old-established software project management paradigm is also outdated because

1. It focuses on process rather than people as the first-order effect on software development, violating John N. Warfield's "**Twenty laws of complexity**" that "The human being, language, reasoning through relationships, and archival representations are universal priors to science (i.e., there can be no science without each of them)" [War98]. As pointed by Alistair Cockburn that "The fundamental characteristics of 'people' have a first-order effect on software development" [Coc99]. Even if there are some models claiming people is the first-order effect on software development, they handle people having positive effects only to offer better working conditions and tool support for the software development team – they ignore the negative effects from people. But the fact is, almost all defects are introduced into a software product by people – the customers and the developers.
2. It does not seize the principal contradiction – in most software organizations, 75% or more of effort and cost are spent in software maintenance, but often the management team does not give more importance to software maintenance – why? They know it is a critical issue, but they feel powerless.
3. It cannot efficiently handle the issue of changeability.
4. There is a lack of support methods and tools.
5. With it, software development process and project management process are separated.
6. The project management is always half a beat behind – hard to find problems in time and hard to solve problems in time.
7. With it, the project success rate is still very low – only about 30%, not acceptable in any other industry.
8. The root cause for the issues in software project management is that the old-established software development paradigm and the old-established software project management paradigm are based on reductionism and the superposition principle that the whole of a complex system is the sum of its components, so that with the old-established software project management paradigm almost all software management tasks are performed partially and locally.

20.2 Outline of the NSE Project Management Paradigm

The revolutionary solution offered by NSE for software project management is described in detail in this chapter later. Here is the outline of the solution:

1. It focuses on people rather than process. The NSE software project management paradigm treats people with two side impacts: the positive side and the negative side that almost all defects are introduced into a software product by people – the customer and the developers, so that it forces the management team not only to offer better working conditions and support to the software development team,

but also to provide many efficient methods and tools to prevent people from introducing defects into a software product.
2. It seizes the principal contradiction – software maintenance. It makes both the software development process and the software maintenance process be managed together and forces side-effect prevention in the implementation of requirement changes and code modifications to greatly reduce the effort and cost spent in software maintenance. It is performed with the support of various automated and self-maintainable traceabilities.
3. It can handle the issue of changeability better by preventing the side effects in the implementation of requirement changes to ensure the quality of a modified product.
4. With NSE, there is a lot of methods and tools provided to support software project management, such as the method and tool to make the project management documents traceable to the implementation of requirements and source code.
5. With it, the software development process and project management process are combined together.
6. The project management documents such as the schedule chart, project progress reports, and cost reports are traceable with the implementation of requirements and source code, so that the project management team can find the problems early and solve the problems in time.
7. With the NSE software engineering paradigm including the NSE project management paradigm, it is possible for NSE to help software organizations double their productivity and double their project success rate (see Chap. 24 for more information about it), compared with the old paradigms.
8. The foundation for establishing the NSE software project management paradigm is complexity science. The NSE software project management paradigm complies with the essential principles of complexity science, particularly the Nonlinearity principle, the Holism principle, the Self-Adaptability principle, and the Self-Organizing principle, so that with the NSE software project management paradigm almost all software project management tasks are performed holistically and globally, such as the cost estimation is done with the decomposition result of an entire software product preliminarily designed.
9. Real-time communication support: with the NSE software project management paradigm, a project Web site and the corresponding BBS are traceable to and from the implementation of requirements and source code is required for real-time communication support.

20.3 The Foundation of NSE Project Management Paradigm

The foundation for establishing NSE software project management paradigm is complexity science. It is established through the application of FDS (Five Dimension Synthesis Method) framework as shown in Fig. 20.1.

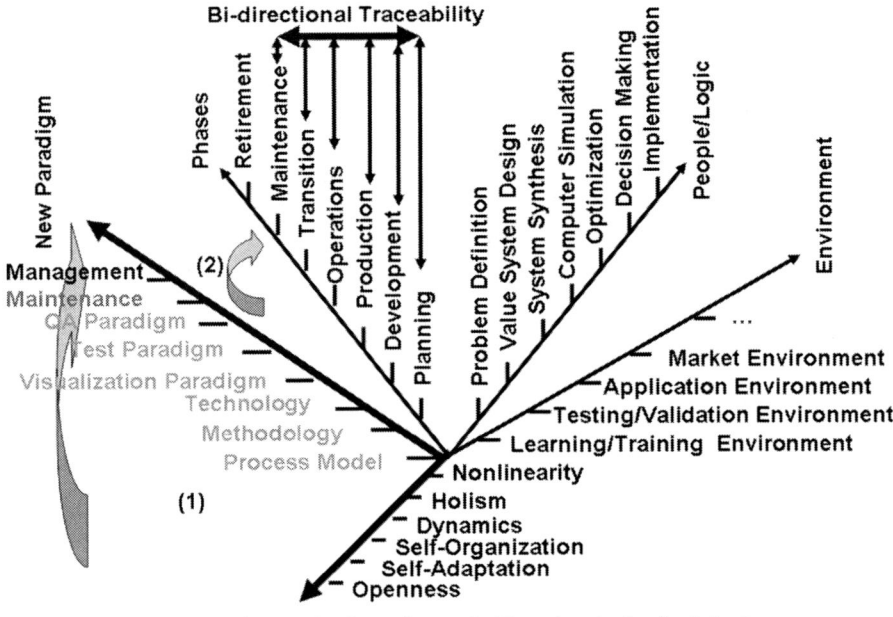

Fig. 20.1 The paradigm-shift framework, FDS

As shown in Fig. 20.1, the NSE project management paradigm complies with the essential principles of complexity science.

The NSE project management paradigm consists of project management strategy, methods, tools, and templates for project planning/scheduling, project monitoring, risk management, project cost estimation, process management, project documentation, unexpected event handling, and people/team management including training.

20.4 The Strategy of NSE Project Management Paradigm

The NSE project management paradigm emphasizes on self-organization, self-adaptation, and self-maintenance.

Self-organization relies on feedback (positive and negative), interaction, and balance of exploitation and exploration [Bon99], so that, for instance, with the NSE project management paradigm, all working product versions, even if it is a dummy whole system (as an embryo), will be provided to customers for review to obtain customer's feedback.

Self-adaptation emphasizes on the support for software changeability and maintainability. With NSE, requirement changes are welcome and implemented with side-effect prevention through various traceabilities. With NSE, a software product is maintainable not only in the produce development site, but also in the customer site – with NSE "Software" is redefined as and delivered to the customer with

1. Instructions (computer programs) that when executed provide desired features, function, and performance
2. Data structures that enable the programs to adequately manipulate information
3. Documents that describe the operation and use of the programs (including the test case script files too) **plus**
4. **The database built through static and dynamic measurement of the programs**
5. **A set of Assisted Online Agents (AOA, artificial intelligence tools working with the database) for supporting testability, reliability, visibility, changeability, conformity, and traceability to make the software program maintainable and adaptive**

20.5 People Oriented

With NSE, people-oriented management emphasizes on

1. Innovation and continuous improvement in the existing products and services to match the fast changing demands of the market.
2. Establishing a set-up with an environment to enhance operational efficiency of the organization.
3. Developing human resources, by taking care of the needs and aspirations relating to career progression and job satisfaction through involvement, participation, training, and commitment [Kha02].
4. Understanding of that people are nonlinear and it is easy to make mistakes in reading, writing, thinking, making decisions, communication, etc., – almost all defects are introduced into a software product by people (customers and developers), so that with NSE a set of methods and tools are developed to prevent the defects introduced into a software product by people (see Chap. 17). For instance, we know that the obtained function decomposition result using Use Cases is not traceable and not directly executable, then how do we know where the defect exists? With NSE, the HAETVE (Holistic, Actor–Action and Event–Response, driven Traceable, Visual, and Executable) requirement development technique and the tool are applied to prevent defects introduced into a software by people through traceability for static review and program execution for dynamic testing – see Figs. 20.2 and 20.3.

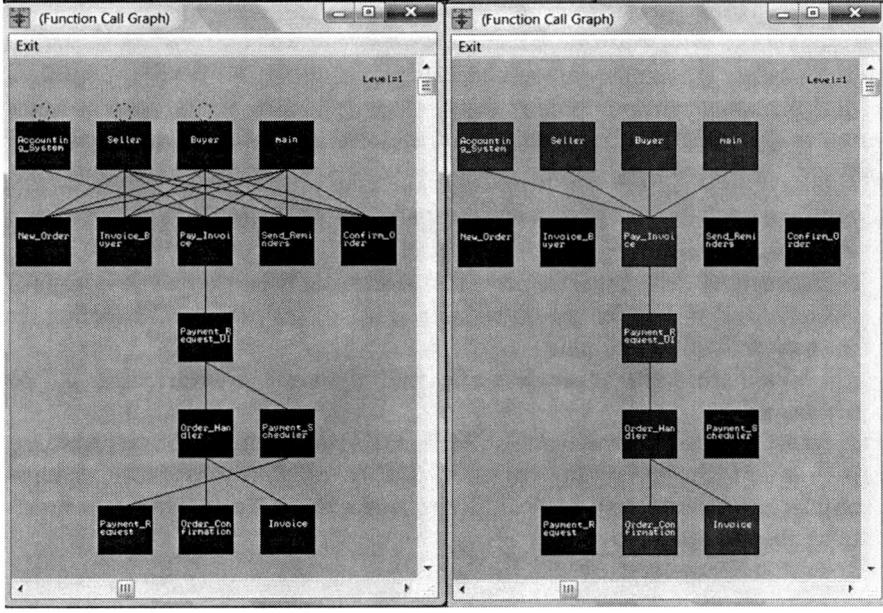

Fig. 20.2 The requirement development result using the HAETVE technique is traceable for defect finding and removing

20.6 Focusing on Maintenance

With the old-established software engineering paradigm based on reductionism and the superposition principle, linear process models are used, making defects easy to propagate from upstream to downstream, and software maintenance is performed partially and locally, so that 75% or more of the total effort and total cost are spent in software maintenance. It is clear that to be able to double software productivity and halve software development cost, we must solve the issues with software maintenance.

With NSE, the solution is simple:

1. Combining the software development process and maintenance process together closely, supporting requirement changes at any stage through side-effect prevention.
2. Greatly reducing the defects introduced into a software product and the defects propagated into software maintenance through defect prevention and defect propagation prevention (see Chap. 17).
3. Greatly reducing the new defects introduced into a software in the maintenance phase by performing the implementation of requirement changes and code modifications holistically and globally with side-effect prevention supported by various traceabilities (see Chap. 18).

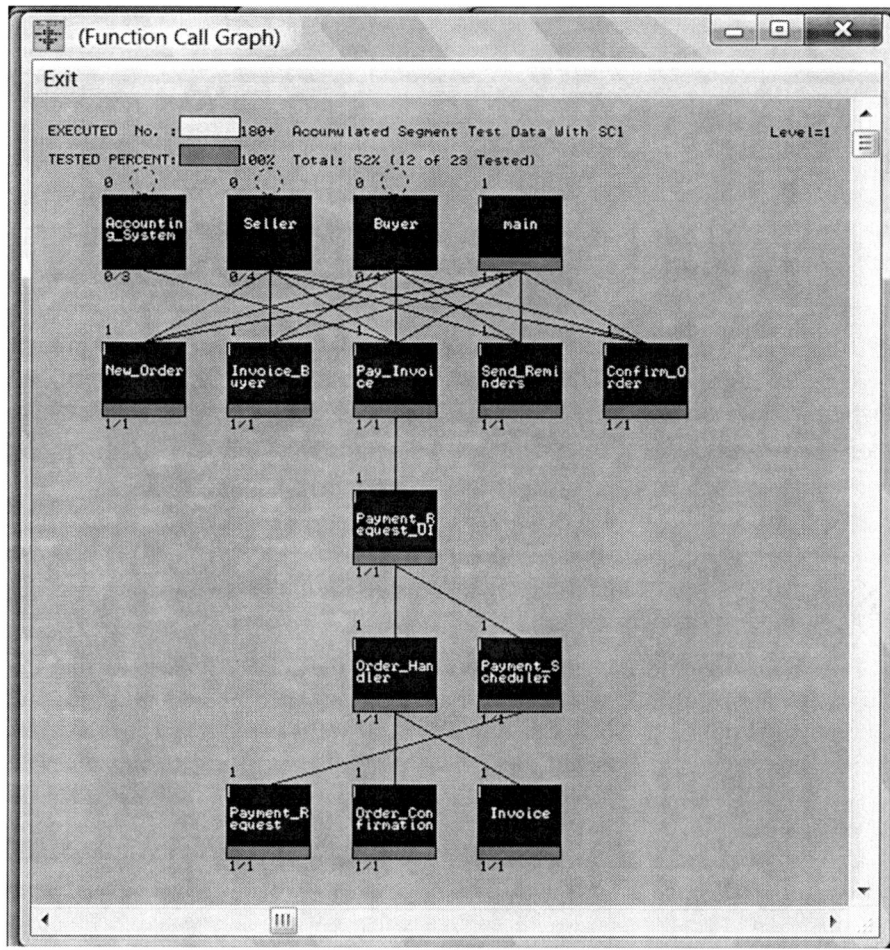

Fig. 20.3 The requirement development result using the HAETVE technique is executable for defect removing (as shown in this figure, all "Actions" are executed except the "Actors")

20.7 More Method and Tool Support

Almost all required methods and tools for supporting software project management are developed and provided (see Chap. 22), particularly

1. The method and tool for making project management documents such as the schedule charts and cost reports traceable with the implementation of the requirements and the source code
2. The methods and tools of the NSE software visualization paradigm to make the entire software development process and management process visible, and the work products visible

3. The methods and tools for cost estimation using call graphs shown in J-Chart notations, see Fig. 20.4
4. Precise productivity measurement methods and tools – see an application example shown in Fig. 20.5

20.8 Combination of Product Development and Project Management

One of the root causes of software failures is that the project management process is separated from the product development process. In the article "Social and Technical Reasons for Software Project Failures," Capers Jones pointed that there are five root causes for software failures:

1. Root causes of inaccurate estimating and schedule planning
2. Root causes of incorrect and optimistic status reporting
3. Root causes of unrealistic schedule pressures
4. Root causes of new and changing requirements during development
5. Root causes of inadequate quality control [Jon06]

I think the fundamental root cause for software project failures is that the old-established software engineering paradigm is based on reductionism and superposition principle, so that with it almost all software engineering tasks and project management tasks are performed partially and locally. But the root causes pointed out by Capers Jones are also existing with today's software development.

How can we solve these issues? First, these problems should be handled holistically and globally; the project management process and the product development process should be combined together to make the project management documents such as the schedule and cost reports traceable with the requirement implementation and the source code.

"So it is today. Schedule disaster, functional misfits, and system bugs all arise because the left hand doesn't know what the right hand is doing" [Bro95-p74] – by combining the project management process and the product development process together, and make the work products of project management and the work products of the product development traceable, and set up a project/product Web site and BBS for real-time communication, will **make the left hand know what the right hand is doing, and the right hand know what the left hand is doing** to solve those issues efficiently. A schedule chart traced and opened when performing forward tracing from a requirement/test case is shown in Fig. 20.6.

A Web page traced and opened when performing forward tracing from a requirement/test case is shown in Fig. 20.7.

20.8 Combination of Product Development and Project Management

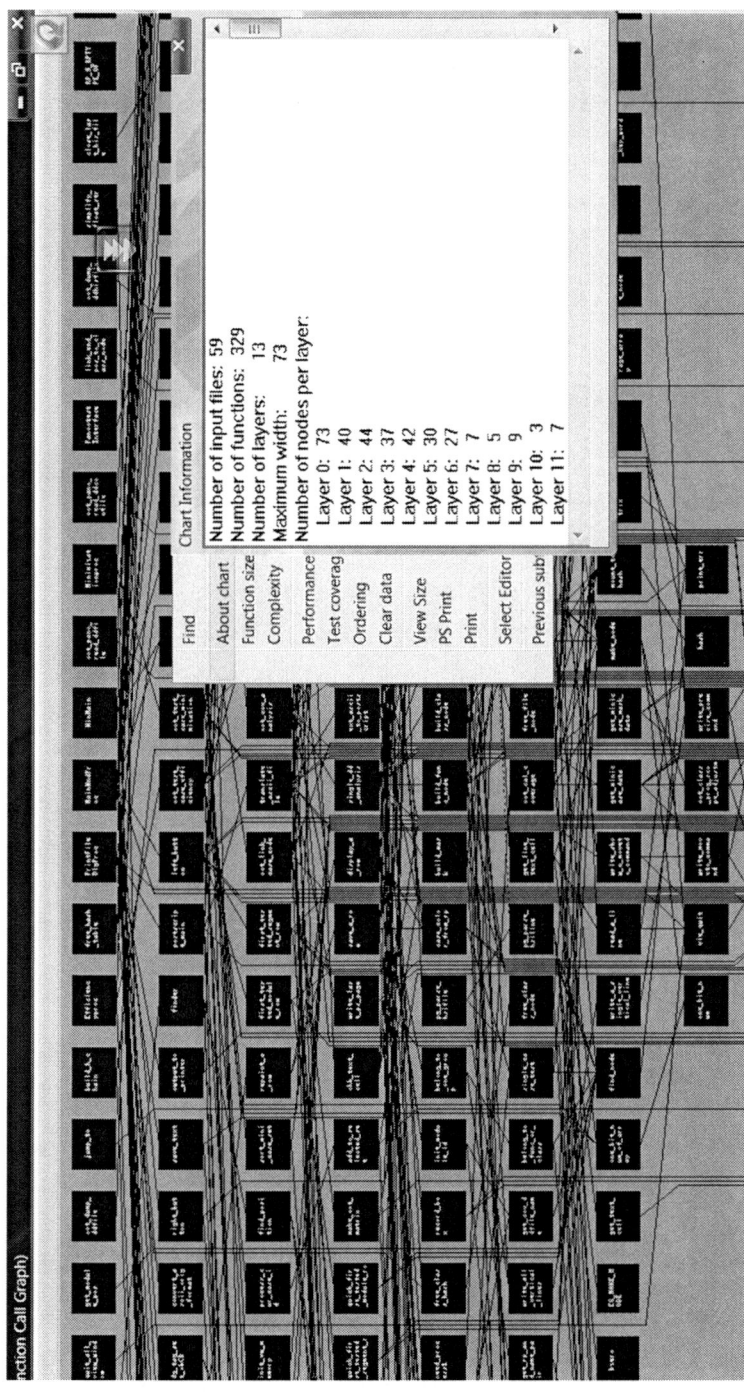

Fig. 20.4 A system call graph shown in J-Chart for cost estimation

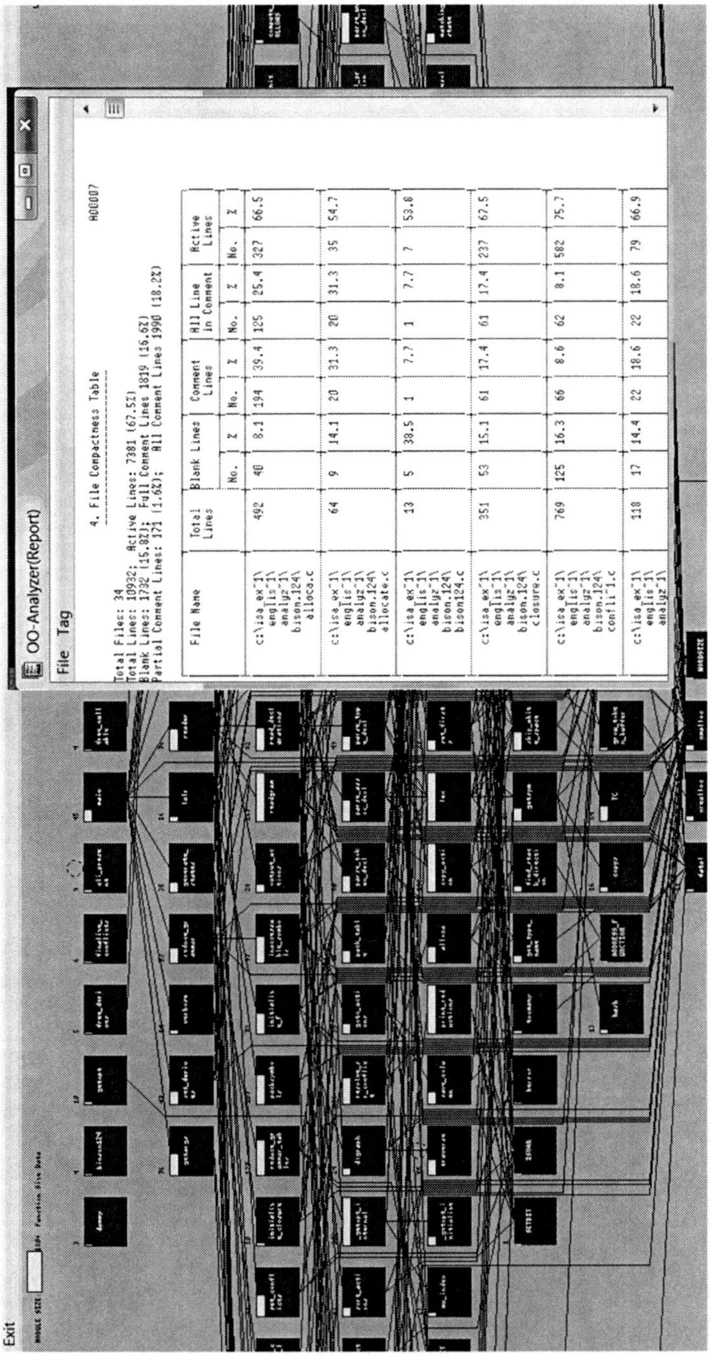

Fig. 20.5 Precise productivity measurement support

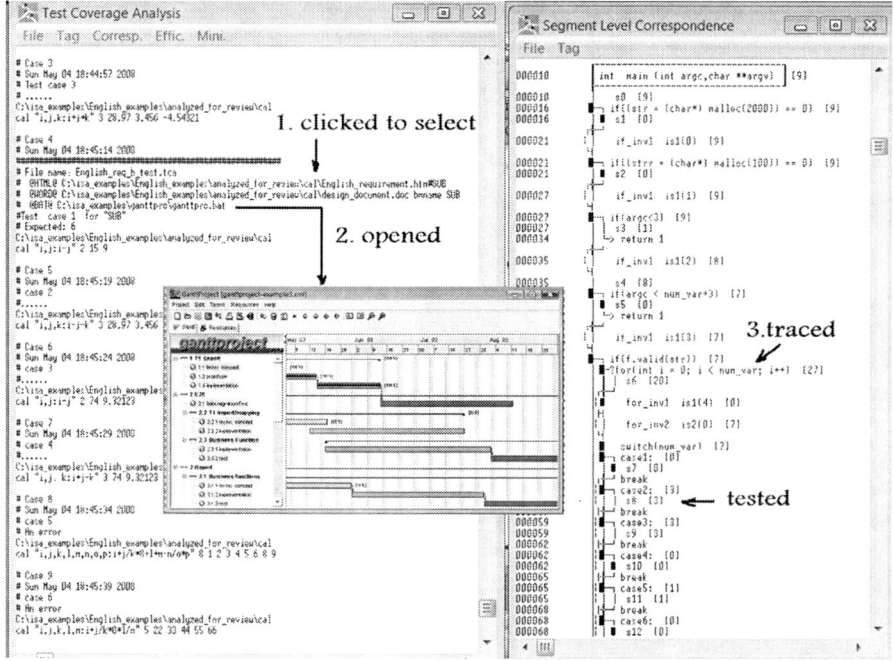

Fig. 20.6 A schedule chart traced and opened when performing forward tracing for a requirement/test case

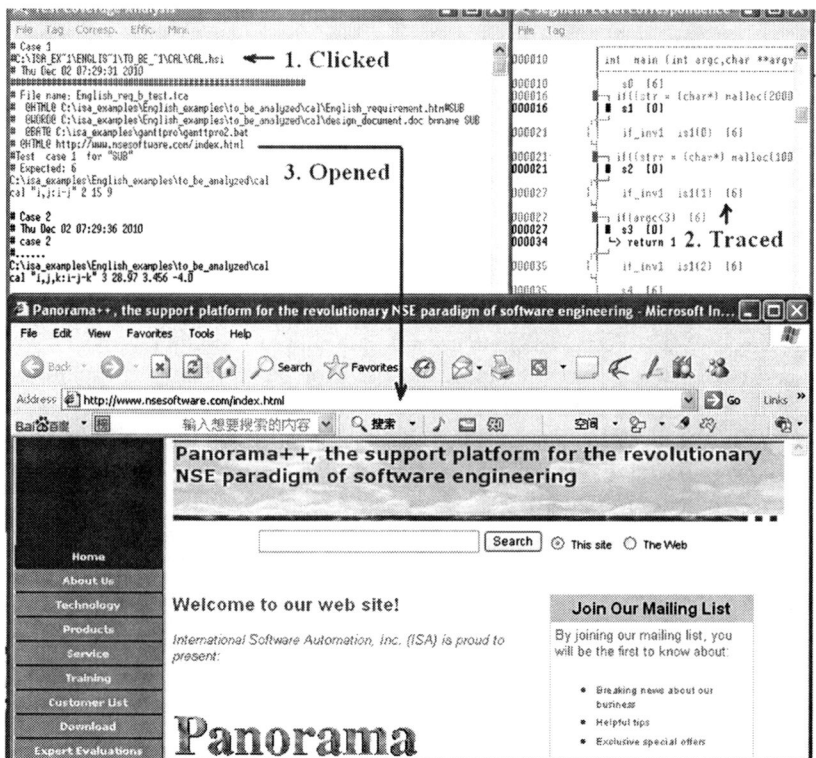

Fig. 20.7 An application example of tracing a requirement/test case to open a related Web page

20.9 Finding Problems Early and Solving the Problems in Time

As pointed by Frederick P. Brooks, Jr., "When one hears of disastrous schedule slippage in a project, he imagines that a series of major calamities must have befallen it. Usually, however, the disaster is due to termites, not tornadoes; and the schedule has slipped imperceptibly but inexorably. Indeed, major calamities are easier to handle; one responds with major force, radical reorganization, the invention of new approaches. The whole team rises to the occasion." "But the day-by-day slippage is harder to recognize, harder to present, harder to make up"[Bro95-P154].

The benefits of combining software development process and software project management process together, and making the work products of software development and the work products of project management traceable, is that one will be able to find problems early and solve the problems in time.

20.10 Quality Management

As pointed by Roger S. Pressman, "Software engineering will change – of that we can be certain. But regardless of how radical the changes are, we can be assured that quality will never lose its importance and that effective analysis and design and competent testing will always have a place in the development of computer based systems" [Pre05-p867].

With NSE, software quality assurance and quality management is performed from the first step down to the final step – the retirement of a software product – see Chap. 17.

20.11 Multiple-Project Management

The NSE software project management paradigm supports multiple-project development and management – two or more related projects' documents, including the management documents and the progress reports, can be traced to each other as shown in Fig. 20.8. With the traceability, events, progress, and issues in one project can be viewed by the management team in another project to take corresponding actions and to help each other.

20.12 Summary

Software project management is an important factor for project success. The old-established software project management paradigm is outdated which is based on reductionism and the superposition principle so that with it almost all project

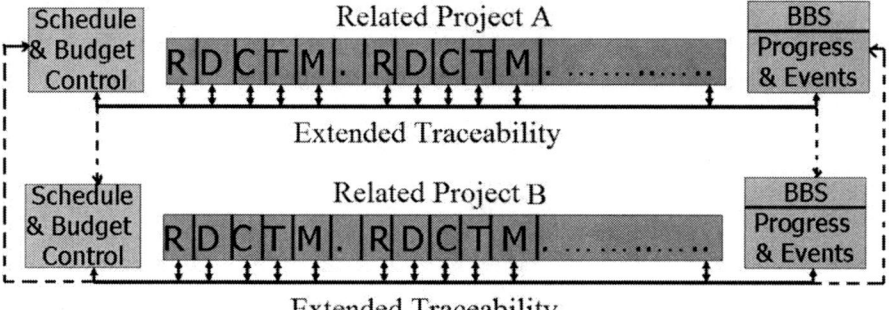

Fig. 20.8 Multiple-project management

management tasks are performed partially and locally, such as the management for software changes.

The NSE software project management paradigm is based on complexity science by complying with the essential principles of complexity science, particularly the Nonlinearity principle, the Holism principle, the Self-Organization principle, and the Self-Adaptation principle, so that with the NSE software project management paradigm almost all software project management tasks are performed holistically and globally.

The most important feature of the NSE project management paradigm is that the software project management process is combined with software development process – the management materials such as the schedule charts, cost reports, progress reports, and unexpected event reports are traceable with the implementation of requirements and the source code, so that the management team can find possible problems early and solve the problems in time.

People oriented and maintenance focused are also the important features of the NSE project management paradigm for efficiently increasing the software project success rate.

20.13 Points and Questions to Ponder

(a) What are the benefits of combining the project management process and product development process together to make their work products traceable?
(b) Why should a project Web site and BBS be established and the related Web pages or BBS title pages be made traceable with the related requirements and test cases and source code?

20.14 Further Reading and Information Source

(a) Software project management, Wikipedia, the free encyclopedia. **http://en.wikipedia.org/wiki/Software_project_management**
(b) Farthing DW Software project management. University of Glamorgan. http://www.comp.glam.ac.uk/staff/dwfarthi/projman.htm

References

[Bon99]   Bonabeau E, Dorigo M, Theraulaz G (1999) Swarm intelligence: from natural to artificial systems. Oxford University Press, New York, pp 9–11
[Bro95-p74]   Brooks FP Jr (1995) The mythical man-month. Addison-Wesley, Reading, p 74
[Bro95-P154]   Brooks FP Jr (1995) The mythical man-month. Addison-Wesley, Reading, p 154
[Coc99]   Cockburn AAR (1999) Characterizing people as non-linear, first-order components in software development. Humans and Technology, HaT Technical Report 1999.03, 21 October 1999
[Jon06]   Jones C (2006) Social and Technical reasons for software project failures. CrossTalk, June Issue
[Kha02]   Khatoon A (2002) People-oriented management. DAWN. http://www.dawn.com/2002/11/11/ebr19.htm. Accessed November 2002
[Pre05-p867]   Pressman RS (2005) Software engineering: a practitioner's approach. McGraw-Hill, New York, p 867
[War98]   Warfield JN (1998) Twenty laws of complexity: science applicable in organizations, Wiley InterScience (http://www.interscience.wiley.com). http://www3.interscience.wiley.com/journal/71007260/abstract?CRETRY=1&SRETRY=0

Chapter 21
Algorithms Innovated for Establishing NSE

The algorithm is the soul of software.

Jay Xiong

The algorithm is the soul of software.

An algorithm is an effective method for solving a problem expressed as a finite sequence of steps/instructions.

Because the size limitation of this book, I will not introduce all of the detailed steps of an algorithm innovated by me and used in the establishment of NSE, but I will introduce the idea and key steps in the implementation of the algorithm.

Here is a list of the algorithms to be introduced in this chapter:

1. The algorithm for realizing MC/DC (Modified Condition/Decision Coverage) test coverage analysis
2. The algorithm for test case efficiency analysis and test case minimization
3. The algorithm for performance analysis
4. The algorithm for Cyclomatic complexity analysis
5. The algorithm for tracing the execution path of a runtime error
6. The algorithm for the layout of the call graph of a program
7. The algorithm for holistic version comparison of a software product
8. The algorithm for memory leak and usage violation analysis
9. The algorithm for realizing the traceability of the diagrammed source code
10. The algorithm for dynamic traceability

21.1 The Algorithm for Realizing Modified Condition/Decision Coverage Test Coverage Measurement

21.1.1 The Requirements

The requirements for implementing MC/DC test coverage measurement [RTCA], [DO-178B] include:

1. Realize the function/capability of MC/DC test coverage measurement for large software products
2. Do not affect the performance too much (less than one-fifth of the time spent in the instrumentation method using object code)
3. Do not require too much space
4. Support MC/DC test coverage measurement for classes which cannot be directly executed
5. Can show the accumulated result or the last-run result
6. Can show the results graphically, holistically, and precisely with untested branches and conditions highlighted
7. Support incremental update of the test coverage measurement results

21.1.2 The Basic Idea

For meeting the above listed requirements, the proposed solution methods include:

1. Perform code implementation with the source code rather than object code
2. To C/C++ program, for instance, use "?:" statement structure to replace the method using function call
3. Use only one bit to record whether a branch/condition is tested or not
4. Design a special preprocessor to replace the system preprocessor for mapping the test coverage results from class instances to the corresponding class
5. Record both the accumulated and last-run results
6. Use J-Chart to show the overall result and J-Diagram or J-Flow diagram (see Chap. 7) to show the detailed measurement results with the capability to highlight the untested elements
7. Store the test coverage measurement results in system level and file level for supporting incremental database update – for instance, if only one source file is modified, then only this file should be re-measured without affecting the results of other files

21.1.3 The Major Steps

The major steps for realizing MC/DC test coverage measurement are as follows:

1. Perform source code analysis to make sure the program itself works
2. Perform code instrumentation precisely, such as that
   to a statement as:
   if (a && b) printf ("OK\n");
   change it to:

   if (((a) ? (aisai_rp -> con[0] |= excc, 1): (aisai_rp -> con[0] |= 0x33, 0)) && ((b)?
   (aisai_rp -> con[1] |= excc, 1): (aisai_rp -> con[1] |= 0x33, 0))) ? (aisai_rp ->
   con[2] |= excc, 1): (aisai_rp -> con[2] |= 0x33) printf("OK\n");
   or (for embedded systems):

   if ((a)? (va = con |= excc, 1): (va = con |= 0x33, 0)) && ...
   (Use a variable rather than a data array to record the test coverage data, which will be read in real time by a corresponding tool)
3. Re-compile the program
4. Run the compiled program with the test cases under the control of Panorama++

21.1.4 Application

A sample MC/DC test coverage measurement for classes is shown in Fig. 21.1.

21.2 The Algorithm for Test Case Efficiency Analysis and Test Case Minimization

21.2.1 The Requirements

The requirements for test case efficiency analysis include:

1. Realize the function/capability of test case efficiency analysis and test case minimization
2. Combine this function with the MC/DC test coverage measurement
3. Can show the analysis results according to different test coverage metrics
4. Support incremental update of the test case efficiency measurement results
5. Make the minimized test cases useful in regression testing after software modification

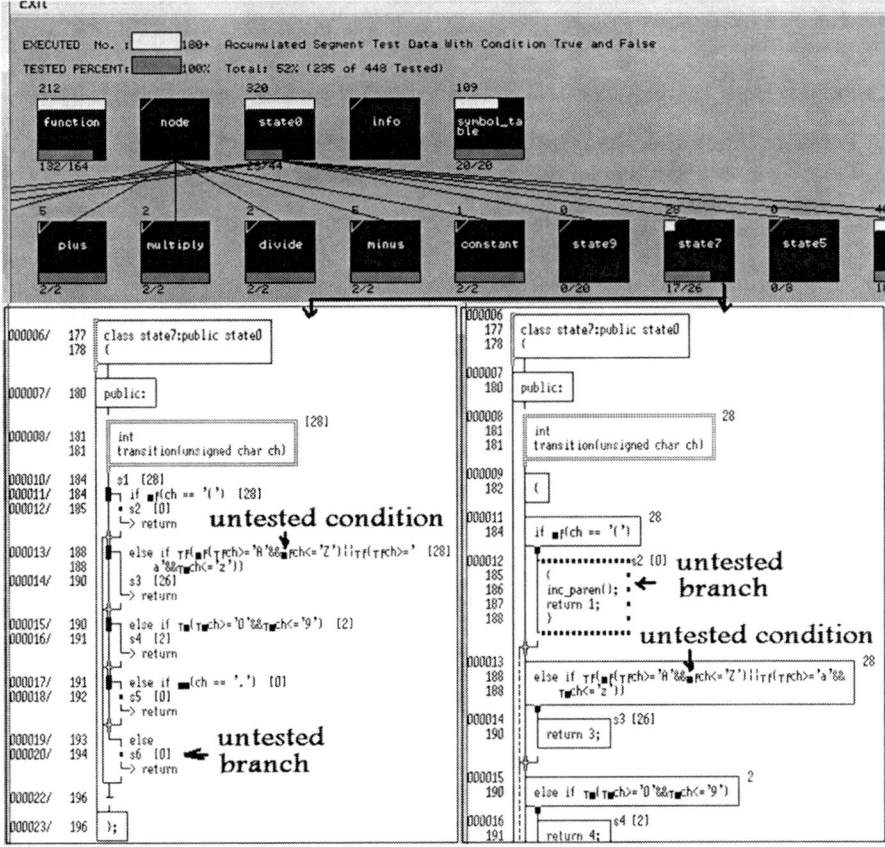

Fig. 21.1 MC/DC test coverage measurement for classes

21.2.2 The Basic Idea

The basic ideas for realizing this function include:

1. Use Time Tags (when a test case is executed) automatically inserted into both the test case description part and the test coverage database for data mapping
2. Perform logic "and" or "or" operations to separate the test results according to different metrics (such as statement test coverage or branch test coverage)
3. Set the test case selection rules according to the net contribution rather than absolute test coverage contribution (see Sect. 21.2.3)
4. Store the test coverage measurement results in system level and file level for supporting incremental database update
5. Make it possible to store the minimized test cases which can be used in regression testing after code modification

21.2.3 The Major Steps

The major steps for test case efficiency analysis and test case minimization are as follows:

1. After the test case execution, support users to select different metrics through an interface as shown in Fig. 21.2.
2. According to the Time Tags and users' selection, get the test coverage results for each test case used.
3. Support users to select different metrics through an interface for test case minimization as shown in Fig. 21.3.
4. According to the users' selection, perform test case minimization according to the net contribution rather than absolute test coverage contribution based on the test case efficiency analysis result:

   (a) Make two sets, A and B
   (b) Put all test cases into set A, let set B be empty
   (c) From A, select the test case with

   - Biggest test coverage result, or
   - A test case which has been used to find a defect
   - then move it from A to B

   (d) From A, select a test case which has the biggest net test coverage contribution with B (covers most elements which are not covered by all test cases in set B), then move it from A to B
   (e) Repeat step (d) until no test case in A can make any net test coverage contribution with B

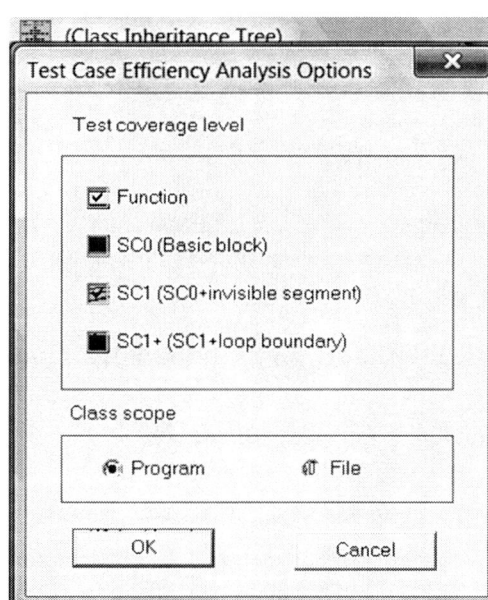

Fig. 21.2 The interface for selecting the metrics for test case efficiency analysis (about the meaning of SC0, SC1, etc., see "Glossary" of this book)

Fig. 21.3 The interface for users to select the metric for test case minimization

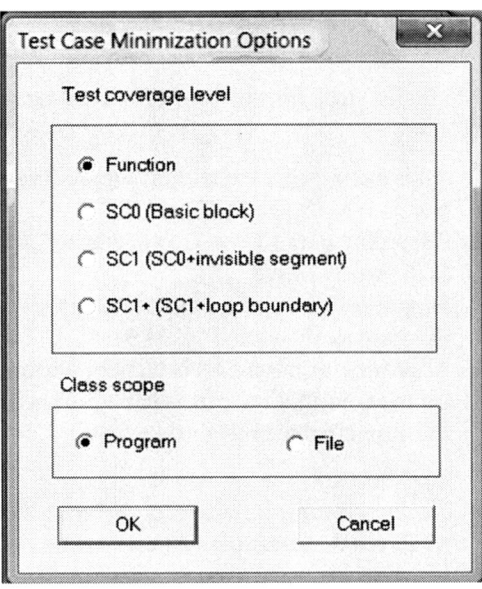

(f) Set B is the minimized set of test cases
Usually, with NSE, a test case which has found a defect will be selected into the minimized set of test cases, because the execution path for that test case is different from other test cases (with NSE support platform Panorama++, there are some extra statements to be executed to record the defect and analyze the type of the defect, etc.)
(g) Show the minimized test cases, then support users to decide whether they want to save the result for regression testing

21.2.4 Application

A sample result of test case efficiency analysis is shown in Fig. 21.4.
A sample result of test case minimization is shown in Fig. 21.5.

21.3 The Algorithm for Performance Analysis

21.3.1 The Requirements

The requirements for performance analysis are as follows:

1. Implement the function/capability for performance analysis
2. Combine performance analysis and MC/DC test coverage together as an option

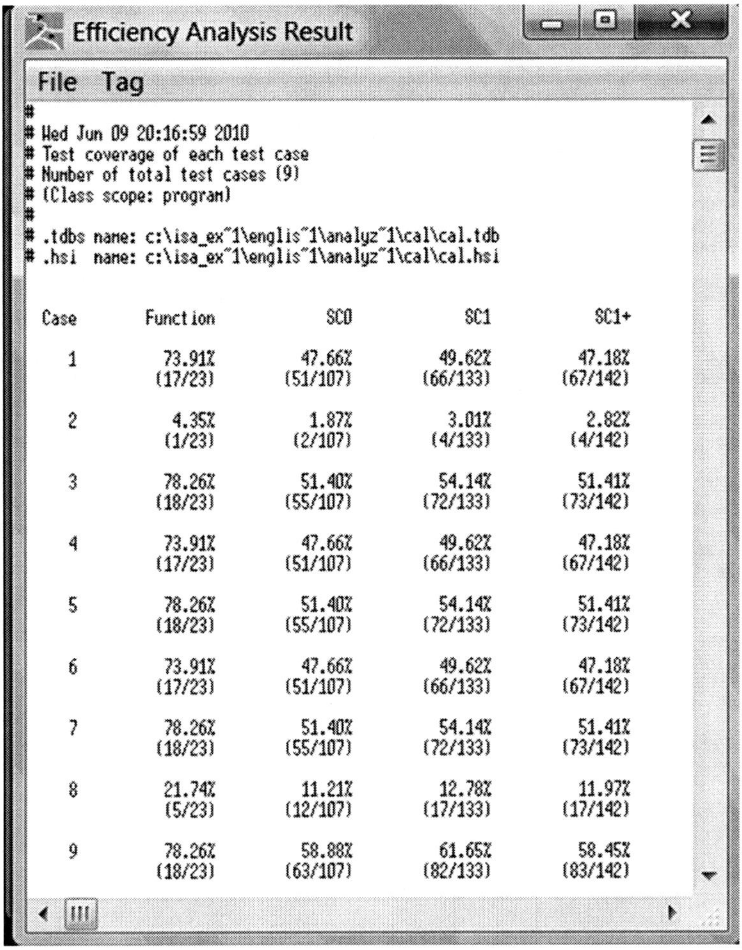


```
#
# Wed Jun 09 20:16:59 2010
# Test coverage of each test case
# Number of total test cases (9)
# (Class scope: program)
#
# .tdbs name: c:\isa_ex~1\englis~1\analyz~1\cal\cal.tdb
# .hsi  name: c:\isa_ex~1\englis~1\analyz~1\cal\cal.hsi
```

Case	Function	SC0	SC1	SC1+
1	73.91% (17/23)	47.66% (51/107)	49.62% (66/133)	47.18% (67/142)
2	4.35% (1/23)	1.87% (2/107)	3.01% (4/133)	2.82% (4/142)
3	78.26% (18/23)	51.40% (55/107)	54.14% (72/133)	51.41% (73/142)
4	73.91% (17/23)	47.66% (51/107)	49.62% (66/133)	47.18% (67/142)
5	78.26% (18/23)	51.40% (55/107)	54.14% (72/133)	51.41% (73/142)
6	73.91% (17/23)	47.66% (51/107)	49.62% (66/133)	47.18% (67/142)
7	78.26% (18/23)	51.40% (55/107)	54.14% (72/133)	51.41% (73/142)
8	21.74% (5/23)	11.21% (12/107)	12.78% (17/133)	11.97% (17/142)
9	78.26% (18/23)	58.88% (63/107)	61.65% (82/133)	58.45% (83/142)

Fig. 21.4 A sample result of test case efficiency analysis

3. Record the execution frequency for each branch for locating the performance bottlenecks better
4. Sort the functions according to the time spent
5. Show the overall result of performance analysis with the capability to also show the branch execution frequency

21.3.2 The Basic Idea

The basic idea for performance analysis is about how to record the time spent in each function and all functions. The method used with NSE is to use a stack to record when a function is called and when the function returns the control to the

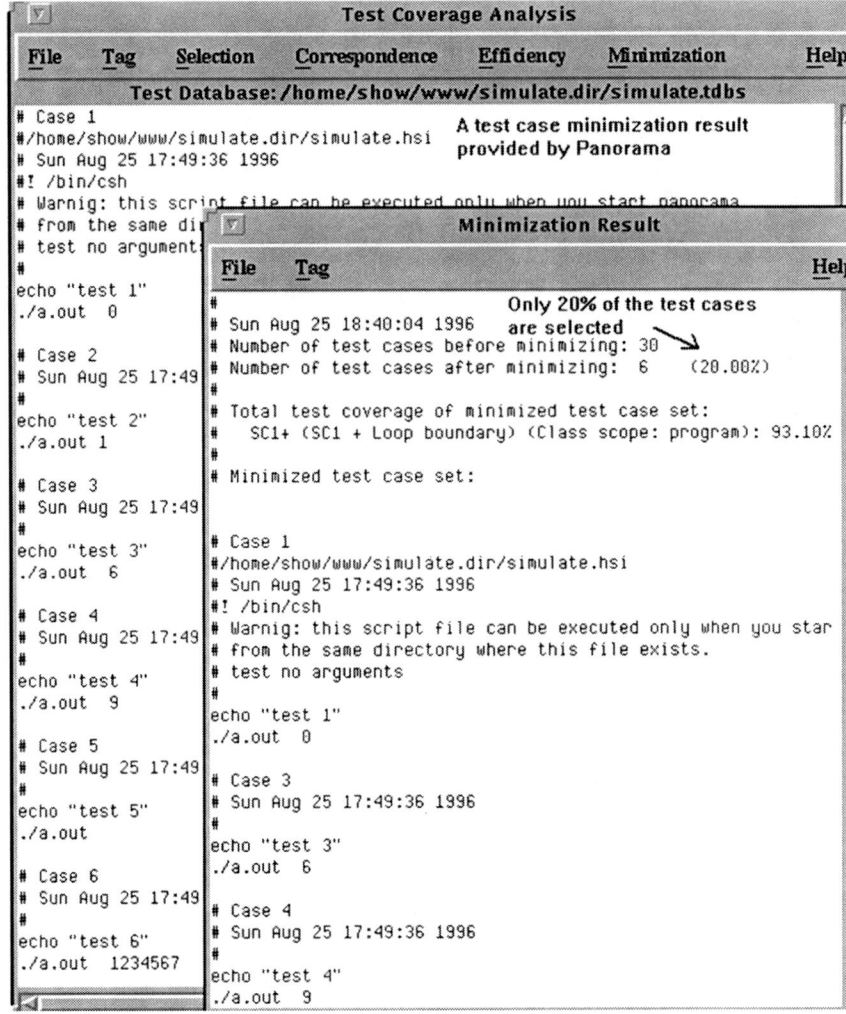

Fig. 21.5 Sample result of test case minimization

caller. If there are nested calls, count the time spent in the deepest function first, and so on, then count the net time spent in each function.

21.3.3 The Major Steps

1. Analyze the source code of a software product.
2. Perform instrumentation with capability to record the execution frequency of each branch.

21.3 The Algorithm for Performance Analysis

3. Set a stack to record what function and the time Tin – when it is called.
4. When the program control is returned from a function called by others, record the time Tout, remove it from the stack, and count the time used by that function: if it does not call other functions, then

$$T = Tout - Tin$$

is the time spent in that function; if the function calls other functions, for instance, it calls function1 and function2, then count T1 and T2 using the same method, then the net time

$T - T1 - T2$ is the net time spent in that function, and record it. Later on if that function is called again by other functions, accumulate the total time spent in that function.
5. After the program execution, show the overall time spent in all functions and the percentage of the time spent in each function using J-Chart.
6. Show the performance analysis results through different sorting orders.

21.3.4 Application

Figure 21.6 shows the overall performance analysis results in J-Chart notation with the capability to show the branch execution frequency of a function.

Figure 21.7 shows a sample sorted performance analysis result.

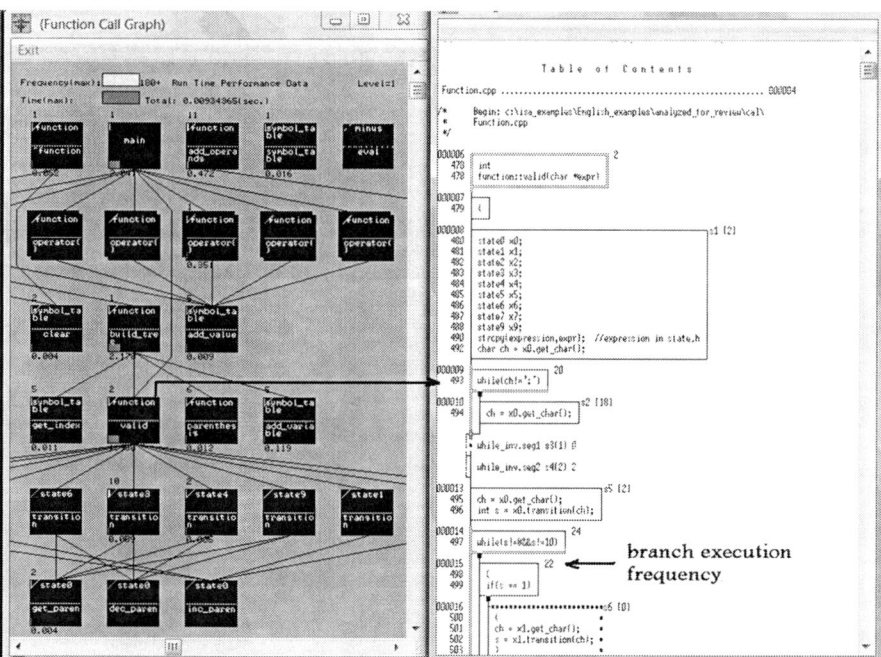

Fig. 21.6 An application example of performance analysis

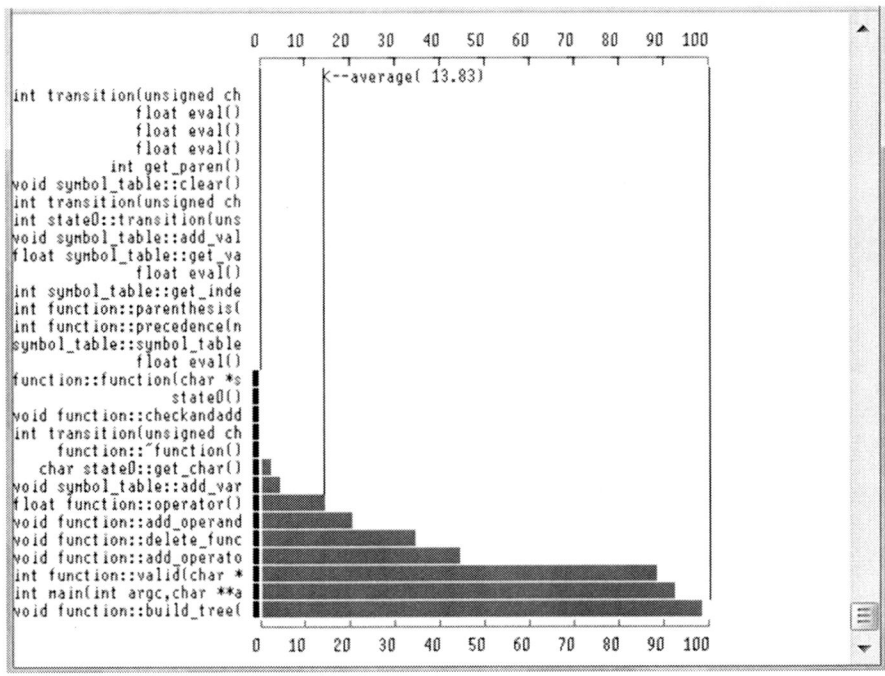

Fig. 21.7 Sample sorted performance analysis result

21.4 The Algorithm for Cyclomatic Complexity Analysis

21.4.1 The Requirement

The requirements for Cyclomatic complexity [Mcc76] analysis are as follows:

1. Offer the function/capability for Cyclomatic complexity (the number of decision statements) analysis
2. Count the Cyclomatic complexity with and without including the "case" statements
3. To classes, count the complexity of the classes with and without the parent classes
4. Support incremental update
5. Show the overall complexity analysis results with the capability to show the control flow of a class or a function for users to understand the complexity of the class/function better

21.4.2 The Basic Idea

With NSE, it is performed through source code static analysis to count the numbers of keywords of the decision statements.

21.4.3 The Major Steps

The major steps for Cyclomatic complexity analysis are as follows:

1. Perform code static analysis.
2. Count the keywords of decision statements with and without including "case" statements.
3. Analyze the control flow for each class/function.
4. Store the complexity analysis results in system level and file level to support incremental update of the results.
5. Show the overall Cyclomatic complexity analysis results in J-Chart with a bar graph at the bottom of each module box to indicate the Cyclomatic complexity level (a full bar graph means the complexity is equal or bigger than 30), as well as the capability to display the control flow diagram using J-Flow notations.

21.4.4 Application

An application example is shown in Fig. 21.8.

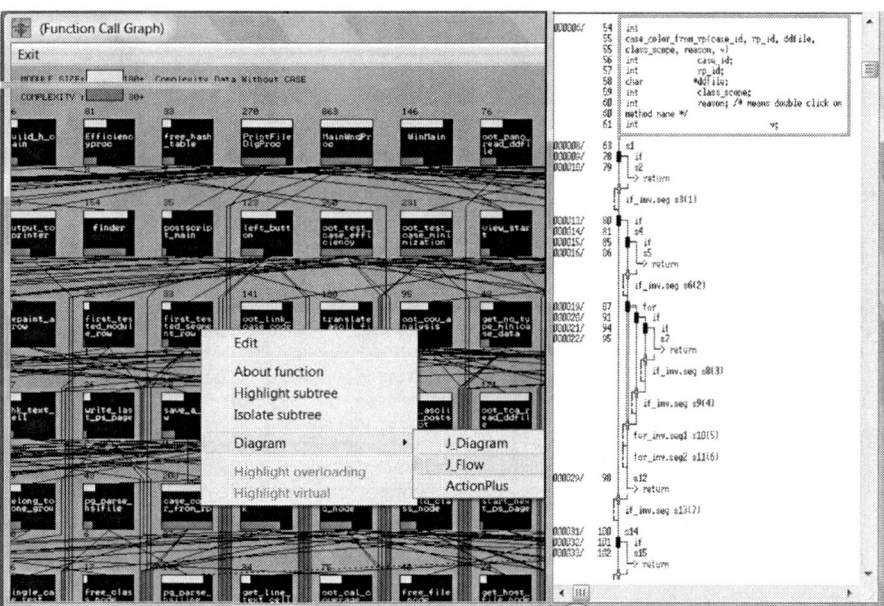

Fig. 21.8 An application example of Cyclomatic complexity analysis

21.5 The Algorithm for Tracing the Execution Path of a Runtime Error

The algorithm and the major steps are similar to that used for performance analysis but there is no need to count the time spent. The major difference between performance analysis and tracing the execution path of a runtime error is that when the program is unexpectedly terminated, we need to get the information and the control earlier than the system. It is realized through the replacement of the on_exit() function with isa_exit() function in a header file.

An application example is shown in Fig. 21.9.

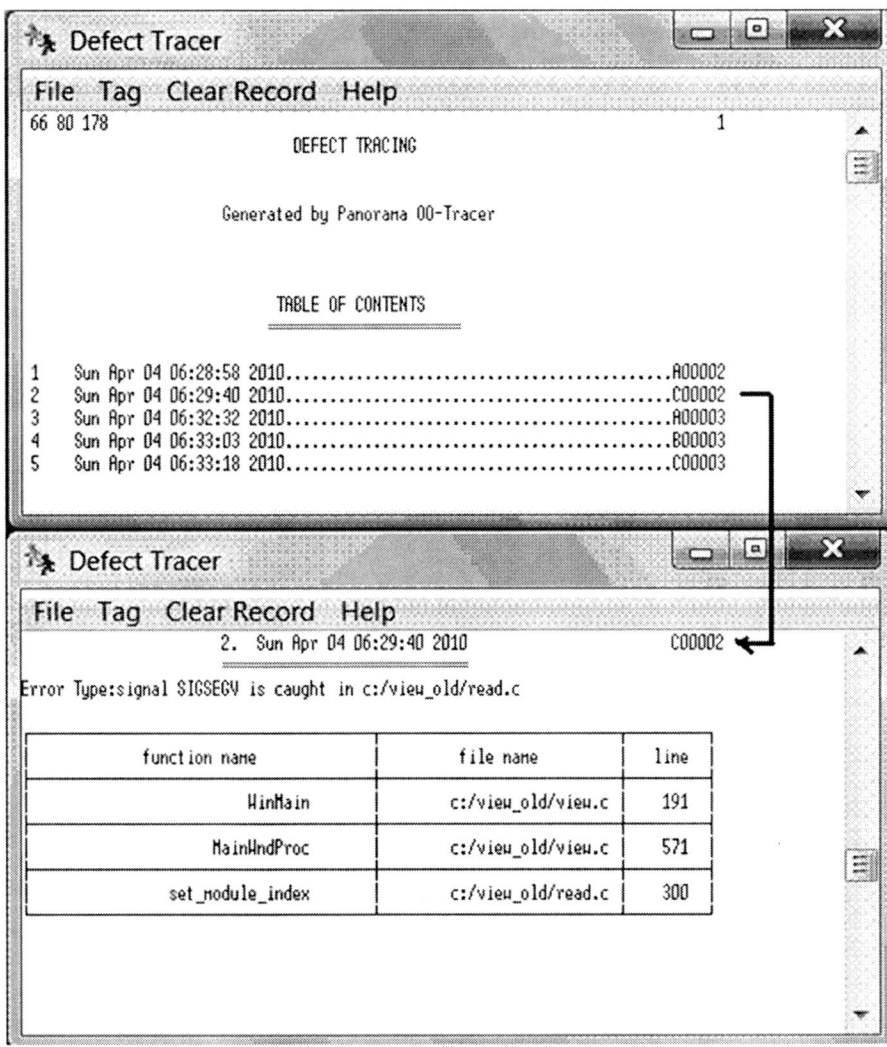

Fig. 21.9 An application example of tracing the execution path for a runtime error

21.6 The Algorithm for the Layout of the Call Graph of a Program Using J-Chart Notations

The major steps of the algorithm for the layout of a call graph of a program are as follows:

1. Put the functions not called by any function at the first level.
2. Put the functions called by only one function at the next level of the caller.
3. If a function is called by more than one other functions, put this function at the next level lower than the function which is at the lowest level among all the callers.
4. If a function calls itself (recursive), put a small circuit on the top of the function box.
5. If there is a loop such as that function A calls function B, function B calls function C, but function C calls function A, then use a dotted line to link them from the lowest level to the highest level.

   An application example is shown in Fig. 21.10.

21.7 The Algorithm for Holistic Version Comparison of a Software Product

The major steps of the algorithm are as follows:

1. To the version A and version B of a software product, remove all the extra space characters.
2. Perform static analysis to generate the two databases for the version A and version B.
3. Merge the two databases together by marking unchanged modules in blue, changed modules in red, deleted modules in brown, and added modules in green.
4. To changed modules, identify the statement differences.

   An application example of holistic version comparison is shown in Fig. 21.11.

21.8 The Algorithm for Memory Leak and Usage Violation Analysis

The major steps of the algorithm are as follows:

1. Analyze the source code of a program.
2. Replace all the memory handling system functions with a new one, such as replace "new" to "isa_new," and "malloc" to "isa_malloc."
3. To a memory assignment statement, such as malloc(n), use isa_malloc to call malloc(n+m+m); here "m" is an integer, can be 1 or 2 or more.

544 21 Algorithms Innovated for Establishing NSE

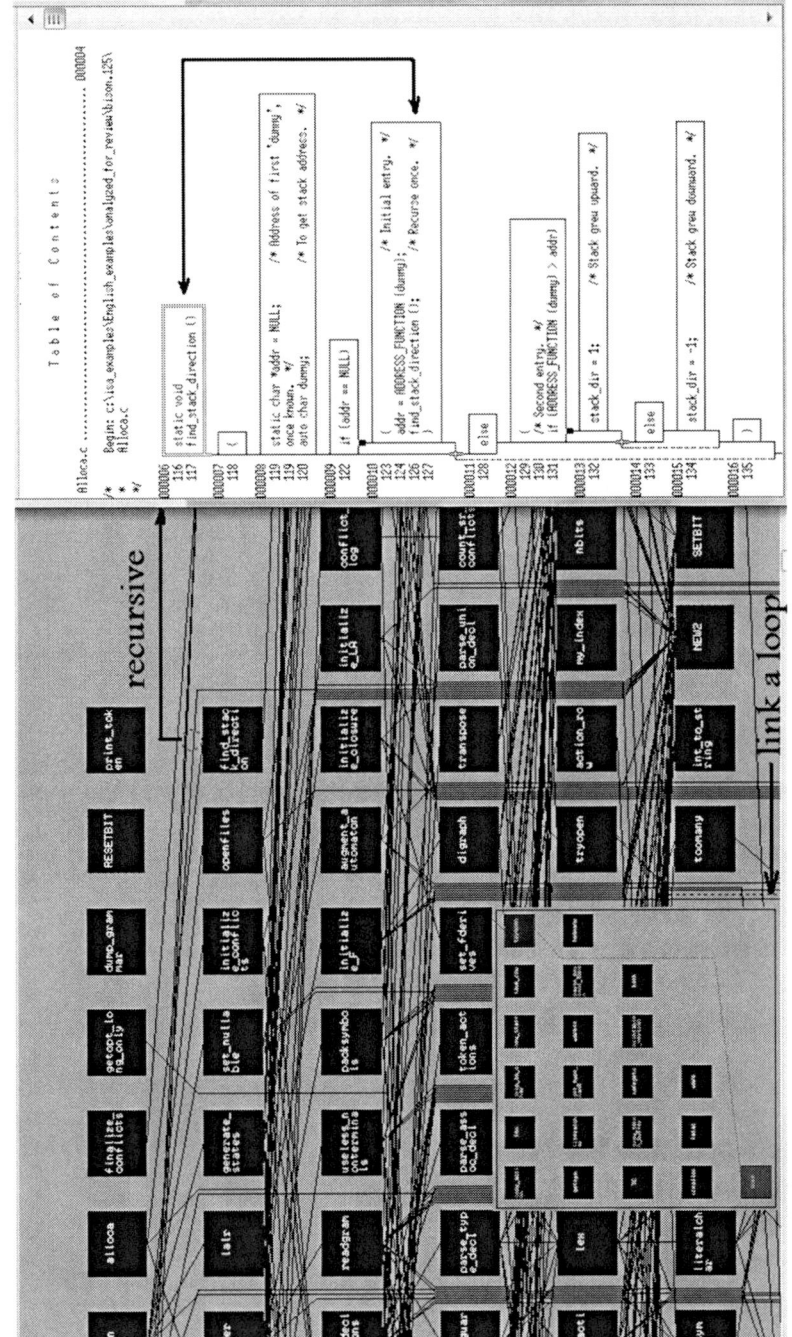

Fig. 21.10 An application example of the layout of a call graph

21.8 The Algorithm for Memory Leak and Usage Violation Analysis

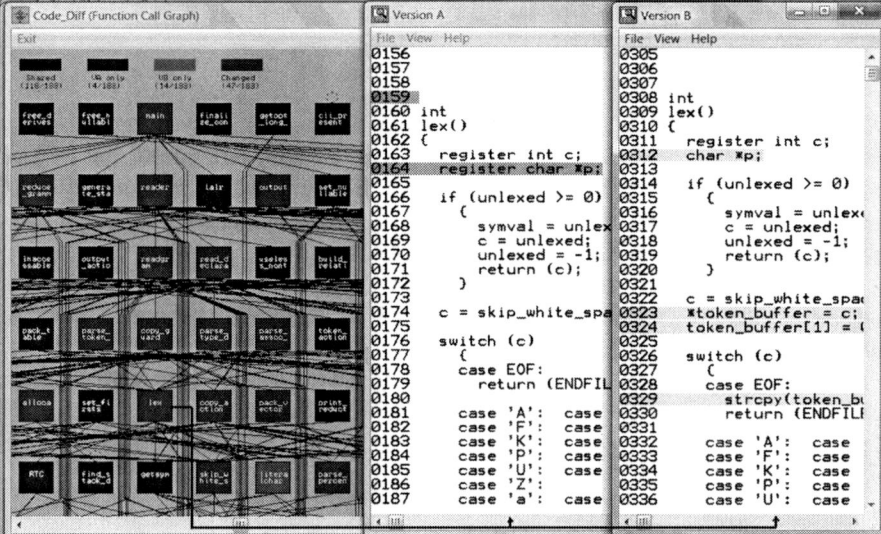

Fig. 21.11 An application example of holistic version comparison

Fig. 21.12 An application example of memory leak and usage violation analysis

4. To the above example, put special values to the first m memory units and the last m memory units.
5. Return the required memory to the memory assignment statement from the m + 1 location, so that if at the end of the program execution the assigned memory has not been freed, or in the program running time the memory unit within the first m units or the last m units is used, a memory leak issue or a memory usage violation issue will be detected.

An application example is shown in Fig. 21.12.

21.9 The Algorithm for Realizing the Traceability of the Diagrammed Source Code

With NSE, various bidirectional traceabilities for source code are established by diagramming the source code and showing it with J-Chart, J-Diagram, and J-Flow, including the traceability between a class and its parent classes, a source file and the included file, a label and the goto statements, a function and the callers, a class instance and the class definition and the constructor.

The algorithms used for various traceabilities are almost the same, so that we only need to know a typical one used to realize the traceability between a function and the callers/callees.

The related major data structures for realizing the traceability between a function and the callers/callees are as follows:

```
struct func_def {
  unsigned            id;              /* numbered from 1 */
  unsigned char       code;            /* identifier diff funcs */
  PANO_STR_INDEX_T    name;
  PANO_STR_INDEX_T    orig_name;
  PANO_STR_INDEX_T    ret_type;
  PANO_STR_INDEX_T    host_file;
  long                base_ddindex;    /* the first ddindex of the func */
                                       /* useless when incl_list not null */
  struct incl_node    *incl_list;      /* incl info of some funcs */
  struct class_def    *host_class;     /* for member func */
  struct call_node    *callers;
  struct call_node    *callees;
  struct func_ref     *overloadings;   /* will be linked after loading */
  struct func_ref     *virtuals;
  struct var_ref      *vars;
  unsigned long       first_lineno;
  unsigned long       last_lineno;
  struct class_ref    *local_classes;
  struct class_ref    *friend_to_classes;
  unsigned long       fst_blk;         /* 1st blk's seq no the file */
  unsigned long       blk_num;         /* blk num of the func */
  struct label_node   *labels;
  unsigned short      comment_lines;
  unsigned short      partial_lines;
  unsigned short      blank_lines;
```

21.9 The Algorithm for Realizing the Traceability of the Diagrammed Source Code

```
  unsigned short        complexity_w;
  unsigned short        complexity_o;
  unsigned short        vis_num;        /* visible segment number */
  unsigned short        invs0_num;      /* IF, SWITCH's invisible segs */
  unsigned short        invs1_num;      /* low end of loop cond */
  unsigned short        invs2_num;      /* high end of loop cond */
  unsigned short        cond_num;       /* cond num of all expressions */
  unsigned short        branch_num;     /* branch number */
#ifdef   FOR_TEMPLATE
  int             is_template;          /* template flag */
#endif
  struct func_def       *next_hsitem;   /* next hash item in the same bucket */
  struct func_def       *next_infile;   /* next func in the same file */
};

struct call_node {
  struct func_def       *func;
  struct line_node      *lineno_list;
  struct call_node      *next;
};

struct func_ref {
  struct func_def       *func;
  struct func_ref       *next;
};

typedef struct jump_info
{
  unsigned long         jump2_blockseq_dia;
  unsigned long         jumpfrom_sourceline;      /* source code line */
  struct jump_info      *next;
}JumpInfo;


typedef struct fc_head_block
{
  unsigned long         type;

  unsigned short                fc;
}FcHeadBlock;

typedef struct detail_block
{
  unsigned long         type;

  long                  ddindex;
  off_t                 text_offset;
  int                   text_len;
  int                   key_offset;
  int                   cond_len;
  unsigned long         first_line;
  int                   blockseq_dbs;
  InactiveNode          *inactive_list;
  JumpInfo              *jump_list;

  char                  *cond_str;     /* after cpp */
  int                   cond_str_len;    /* after cpp */
  int                   test_data;
  int                   segment_no;
  int                   host_file;
  int                   host_fc;       /* func/class */
  int                   is_template; /* add by clq */

}DetailBlock;
```

```
typedef struct fc_info
{
  void              *fc_def;
  unsigned short    type; /* FUNC_INFO/CLASS_INFO/
                            DUMB_FUNC_INFO/DUMB_CLASS_INFO */
  char              *name; /* no free */
  unsigned long     first_blockseq_dia;
  unsigned long     begin_line;
  FcNode            *fathers;
  FcNode            *sons;
  unsigned short    host_file;

  void              *aux_p;
  int               test_data;   /* aux_int */
  Coverage          cov[2][NCOVS]; /* T_SC0/T_SC1/... */
  int               complexity[2][NCOMS]; /* C_JC0/C_JC1/... */
                                 /* FUNC/WITH_BASE/WITHOUT_BASE*/
  int               is_template; /* add by clq */
}FcInfo;

/* ... */
```

The major steps of the algorithm for realizing the traceability of the diagrammed source code are as follows:

1. Analyze users' program, fill the Hash table for functions according to the above data structures, including the information about the callers and the callees.
2. Put all the source files of the program together to count the relative global block number and the local line number in the corresponding source file according to the order of code static analysis, and fill the JumpInfo for the function traceability.
3. Count the corresponding JumpInfo as follows:
   For instance, a function FC1 is defined at line 20 of source file SF1, local block 5, the global block number is also 5; a caller of the function is located in source file SF5, local block 6, local line 33; if the source file SF1 has 20 blocks, SF2 has 30 blocks, SF3 has 22 blocks, SF4 has 26 blocks, then the global block number of the caller is

   $$20 + 30 + 22 + 26 + 6 = 104.$$

4. Show the generated logic diagram with the JumpInfo – for instance, in global block 5, the location where the function FC1 is defined, add a jump number 104 on the right side of the diagram.
5. Make the generated diagram interactive to accept users' command for realizing the traceability – for instance, while the user is viewing the diagram of FC1 located in global block 5 in the generated diagram and makes a click on the number 104 added on the right side of the function FC1, the diagram generator should jump to global block 104 – there is a caller calling to function FC1.
6. Correspondingly, after the diagram jumped to global block 104 and the diagram generator shows the diagram from global block 104, it also adds a jump number 5 on the right side to indicate where the called function is located.

21.10 The Algorithm for Dynamic Traceability

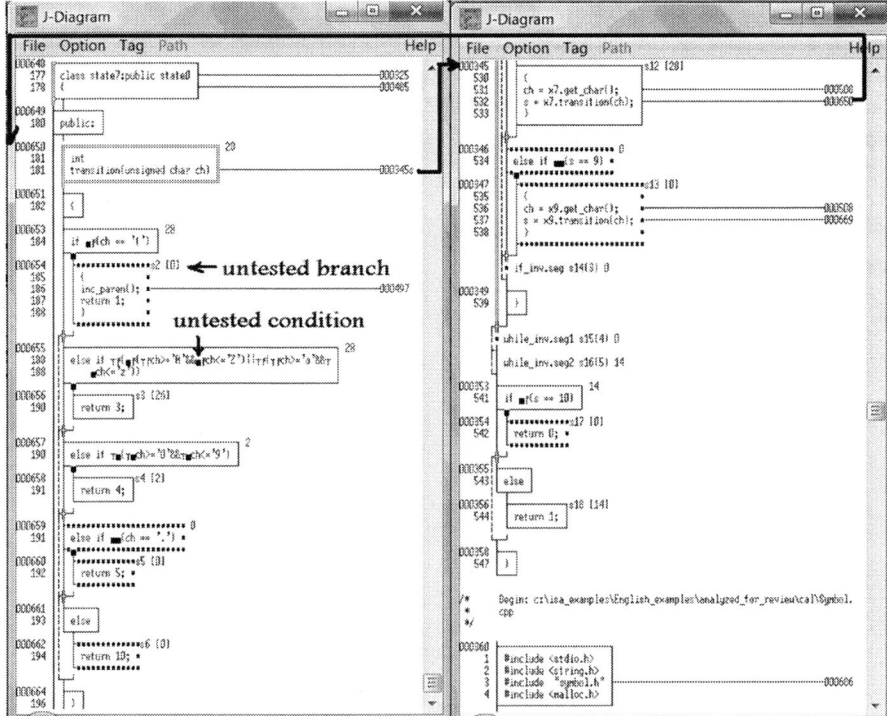

Fig. 21.13 Automated traceability established with diagrammed source code

Note: In the real tool development, the holistic diagram for an entire software product is virtually existing – there is no big diagram stored in the computer memory or hard disk at all! All diagrams are generated dynamically from the database with several Hash tables only – each time only a small diagram with the size same as that of Windows used to show the diagram, so that when a diagram jumps to a new location, there is no real diagram movement at all, instead, a new one is dynamically generated. In this way, the display speed is about 1,000 times faster than the traditional approaches. But as an option, users can also require the tool to save (to a Postscript file) or print out the entire diagram which can be as large as 100,000 pages or more, depending on the original size of the source program.

An application example is shown in Fig. 21.13.

21.10 The Algorithm for Dynamic Traceability

With NSE, "dynamic traceability" means that when performing forward tracing from requirements/test cases to find the corresponding source code, or performing backward tracing from a program module or a segment (branch) to find the corresponding test cases/requirements, if a batch file specified in the description part of a test case using the keyword @BAT@ is selected or traced, the batch file will be executed dynamically. Dynamic traceability is useful for making a document created by a third party also traceable with the implementation of requirements and test cases and the source code, or automatically playing back the captured GUI testing operations back for automated acceptance testing, etc.

Now, for instance, let us consider how to automatically play the captured GUI operation back through dynamic traceability. The major steps of the algorithm/process are as follows:

1. Design the corresponding test case script file as follows:
   sortdemo.tca:

   ```
   # 1 Insertion Sort
   # @BAT@ C:\isa_examples\English_examples\analyzed_for_review\sortdemo\re_run.bat
   C:\isa_examples\English_examples\analyzed_for_review\sortdemo
   sortdemo

   # 2 Bubble Sort
   # @BAT@ C:\isa_examples\English_examples\analyzed_for_review\sortdemo\re_run.bat
   C:\isa_examples\English_examples\analyzed_for_review\sortdemo
   sortdemo

   # 3
   # @BAT@ C:\isa_examples\English_examples\analyzed_for_review\sortdemo\re_run.bat
   C:\isa_examples\English_examples\analyzed_for_review\sortdemo
   sortdemo

   # 4
   # @BAT@ C:\isa_examples\English_examples\analyzed_for_review\sortdemo\re_run.bat
   C:\isa_examples\English_examples\analyzed_for_review\sortdemo
   Sortdemo

   ...
   ```

2. Run the test cases with Panorama++ to capture the GUI test operations as shown in Figs. 21.14 and 21.15.
3. Design a batch file for playing the captured operations back as follows:
   re_run.bat:

   ```
   C:\isa_examples\play -hsi=C:\isa_examples\English_examples\analyzed_for_review\sortdemo\sortdemo.hsi
   -dbspath=C:\isa_examples\English_examples\analyzed_for_review\sortdemo\dbs\
   -tdb=C:\isa_examples\English_examples\analyzed_for_review\sortdemo\sortdemo.tdb
   -tdb=C:\isa_examples\English_examples\analyzed_for_review\sortdemo\playout.tdb
   ```

   Here "play" (play.exe) is the tool name, others are the parameters required by the tool, sortdemo.tdb is the file storing the captured GUI operations.
4. Although there is only one file (sortdemo.tab) storing the captured GUI test operations for all test cases in a test script file, clicking on a test case will only play the corresponding GUI operation for that test case back selectively without playing other operations captured for other test cases back – it is done through the Time Tag.
5. How it works – through an environment variable TIME_TAG to map the corresponding GUI operations captured and play them back selectively.

21.10 The Algorithm for Dynamic Traceability

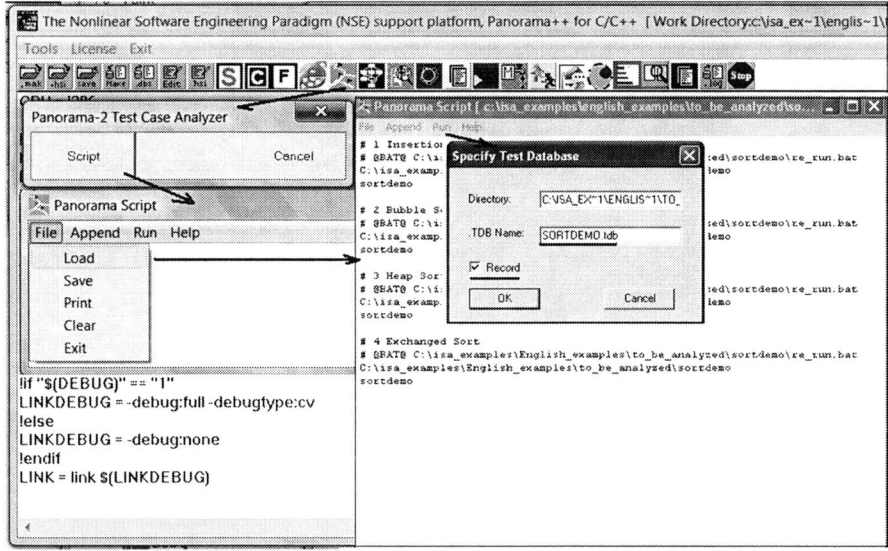

Fig. 21.14 Running the test cases to capture the GUI test operations with Panorama++

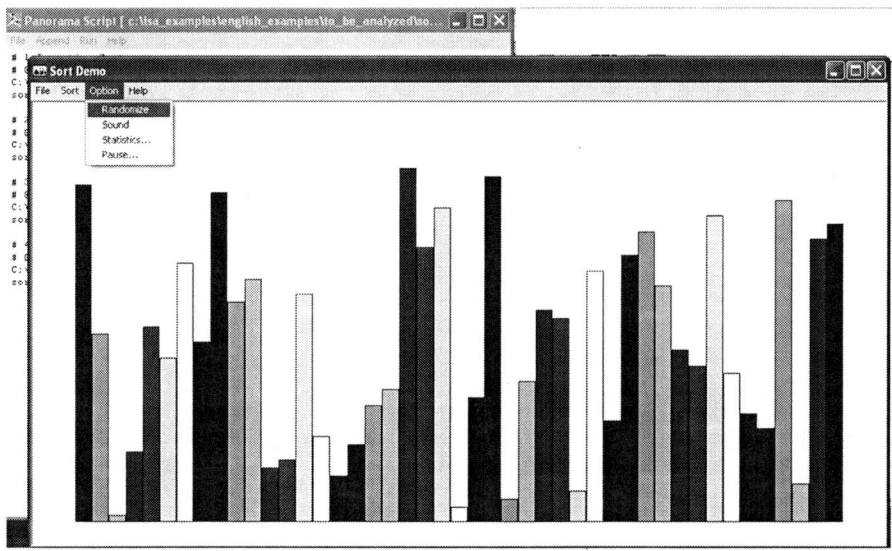

Fig. 21.15 The test process with GUI operations

(a) When a user click on a test case, set the environment variable TIME_TAG before running the batch file:

```
...
if(case_and_code_only == 0 && check_color != 0 && flag==0 && ( strstr(text, "# Mon " ) != NULL || strstr(text, "# Tue ") != NULL || strstr(text, "# Wen ") != NULL ||
strstr(text, "# Tur ") != NULL || strstr(text, "# Fri ") != NULL || strstr(text, "# Sat ") != NULL || strstr(text, "# Sun ") != NULL ))
{
int  stat ;
char *env;
int len ;
int k = 0;
char *pp, cmd [1024];
strcpy(time_tag, text);
len = strlen (time_tag);
while (k < len-2)
{
time_tag[k] = time_tag [k+2];
++k;
}
time_tag[k] = '\0';

sprintf (cmd, "TIME_TAG=%s", time_tag);
stat =  putenv(cmd);
...
```

(b) Run the batch file

```
void run_bat_file (char * file_name)
{
char cmd_line[1024];
STARTUPINFO si;
PROCESS_INFORMATION pi;
FILE *fp, *ffpp;
char *panohome;
int ln;

  ZeroMemory( &si, sizeof(si) );
  si.cb = sizeof(si);
  ZeroMemory( &pi, sizeof(pi) );
  if ((panohome=getenv("PANORAMAHOME")) == NULL)
    MessageBox (NULL, "Can get the environment variable of PANORAMAHOME", "ERROR", MB_OK );
  else
  {
// Jay 2007, 1, 29, check to see whether there is a need to play back a test case
// If it is a Code_to_case action, there is no a need.
// But we should check the batch file first to see whether there is a playback command.
```

```
        sprintf(cmd_line,
            "%s\\run_bat\.exe %s", panohome, file_name);

        // MessageBox (NULL, cmd_line, "The cdm line = ",  MB_OK );
        CreateProcess( NULL,
        cmd_line,
          NULL,
          NULL,
          FALSE,
          0,
          NULL,
          NULL,
          &si,
          &pi);
            }
      }
```

(c) In the play.exe file used for playing the captured GUI operations back, get the environment variable TIME_TAG used to selectively play the corresponding GUI operations back:

if(runbackground)

{

time_tag=getenv("TIME_TAG");

...

}

6. Show dynamic traceability through forward tracing (or backward tracing) operation using the OO-Validate tool of Panorama++ – see Figs. 21.16 and 21.17.

21.11 Summary

In this chapter, the major steps of ten algorithms are introduced. In the NSE support platform, Panorama++, much more algorithms are innovated and applied.

Compared with a poor algorithm, the efficiency of an excellent algorithm can increase more than 1,000 times such as the algorithm example of the virtual diagram generation and display.

Algorithms are the soul of software.

21.12 Points and Questions to Ponder

(a) Why are software algorithms so important?
(b) What is a hash table? Where do we need hash tables?

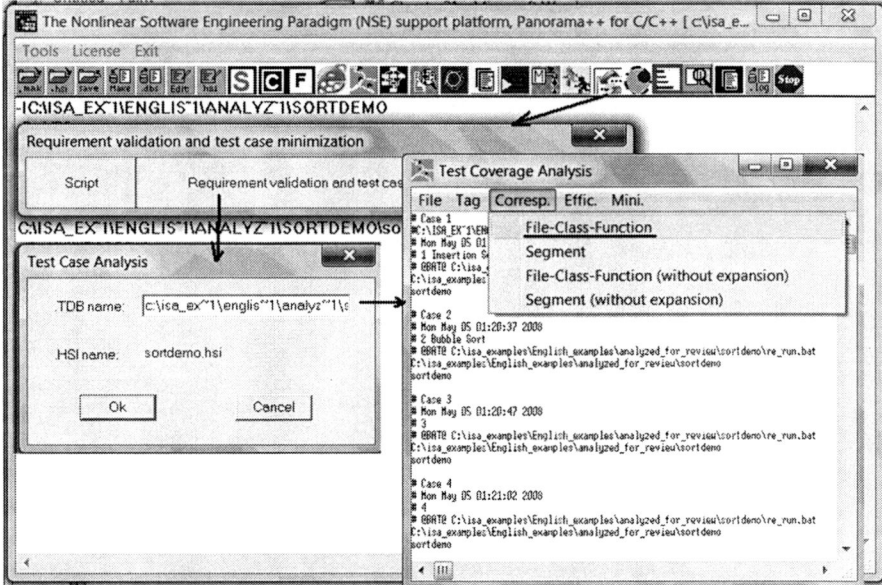

Fig. 21.16 Process I for dynamic traceability

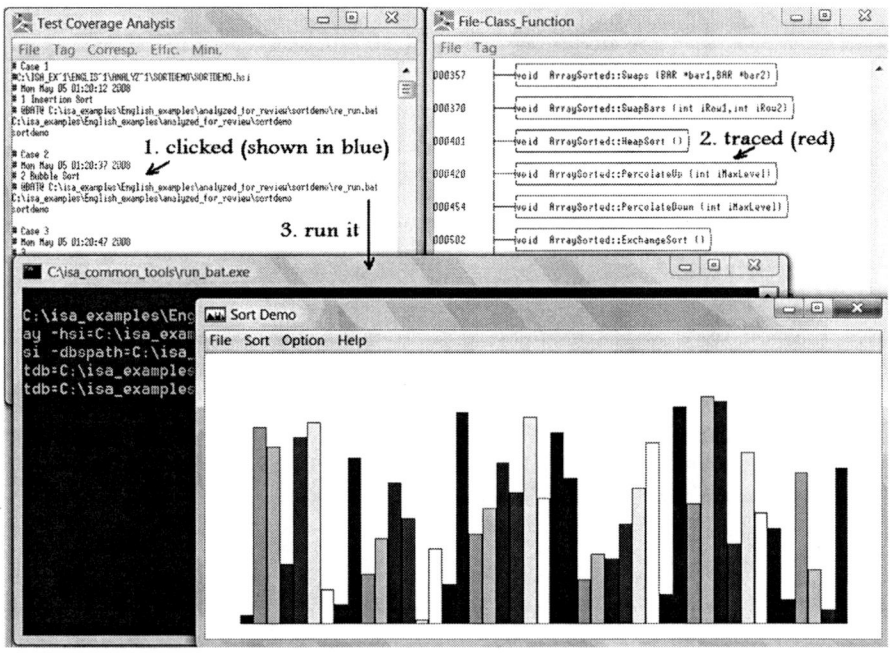

Fig. 21.17 Process II for dynamic traceability – selectively playing the captured GUI operations back through forward tracing: click on a test case (automatically shown in *blue* on the screen originally) to see the source code modules tested (automatically shown in *red* on the screen originally), and run the batch file traced

21.13 Further Reading and Information Source

(a) Algorithm, **Wikipedia, the free encyclopedia.** http://www.en.wikipedia.org/wiki/Algorithm
(b) What is Algorithm, whatis.com. http://www.whatis.techtarget.com/definition/0,,sid9_gci211545,00.html

References

[Mcc76]   McCabe T (1976) A complexity measure. IEEE Trans Software Eng, December 1976
[DO-178B] DO-178B, DO-254 Questions & answers. http://www.highrely.com/do178b_questions.php
[RTCA]   http://www.rtca.org/

Chapter 22
NSE Support Tools and NSE Support Platforms

> *Right tool for the right job.*
>
> At-Risk Survivors

This chapter introduces NSE support tools and NSE support platform. These tools are developed mainly from the innovated techniques described in Chap. 7.

22.1 Full Software Development Lifecycle Support

NSE support tools and NSE support platforms, Panorama++ and Silver Bullet++, fully support the entire software development and maintenance process as shown in Fig. 22.1.

22.2 The Product Development History

The history of Panorama++ development

- First generation: Hindsight
- Second generation: Panorama
- Third generation: Panorama++

22.2.1 The First Generation: Hindsight

In 1989, I founded Advanced Software Automation, Inc. in Silicon Valley. Our first product Hindsight designed by me and implemented by me and my colleagues was chosen by Sun Microsystems as the internal test suite for almost all software products except the operating systems – see Fig. 22.2 for a related article.

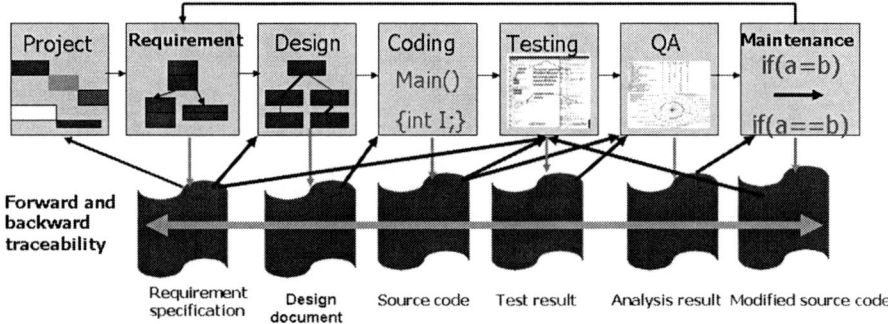

Fig. 22.1 NSE integrated tools and platforms fully support the software development lifecycle

Hindsight was also selected as one of the typical reverse engineering tools by CASE OutLook – see Fig. 22.3.

It is pointed in the article shown in Fig. 22.3 that "ASA's Hindsight The aptly named *Hindsight* from Advanced Software Automation (ASA) is a program understanding and maintenance workbench for C and Fortran. The company was founded by Mr. Jay Xiong who formerly worked in the field of CAD for integrated circuit (microchip) development at Hitachi and U.C. Berkeley. Xiong's vision in founding ASA was to address the problems associated with testing and maintaining very large programs (one million or more lines of code). Hindsight uses algorithms derived from Xiong's work with very large IC layout tools to represent large program structure."

22.2.2 Second Generation: Panorama

In 1992, I founded International Software Automation, Inc. (ISA) in Silicon Valley. Our Panorama product designed by me and implemented by me and my colleagues was introduced in 1994. In 1995, Panorama was selected as one of the best software analysis and testing products by Edward Kit in his book "Software Testing in the Real World" [Kit05] – see Fig. 22.4.

The following contents are a comment from Professor ROGER S. PRESSMAN:

> Panorama: developed by International Software Automation, Inc. (http://www.software-automation.com. Note: now http://www.nsesoftware.com) encompasses a complete set of tools for object-oriented software development including tools that assist test case design and test planning [Pre05].

22.2 The Product Development History

057 OCTOBER 29, 1990   The Newspaper Of Open Systems Computing   A CMP Publication®

Fixing C Code

BY PAUL KRILL

Santa Clara—Advanced Software Automation (ASA) has released its Hindsight software maintenance tool, designed to cut the time programmers spend fixing existing C code.

Through the use of interactive structure charts and active logic diagrams, ASA's Hindsight software maintenance environment speeds understanding of code modification effects on the rest of a program, according to ASA. The product also locates test coverage deficiencies and updates documentation directly from code.

Sun Microsystems has adopted Hindsight as a software development suite testing tool. Hindsight will be used to evaluate test suites on non-OS software, such as compilers, programming environments and network software, said Charles Miller, a Sun software technology engineer.

The product works on major Unix platforms, including Hewlett-Packard/Apollo, Sun, Digital Equipment and IBM. It is priced from $12,000, for the structure chart feature only, to $23,000 for the structure chart plus all modules, including structural complexity and test coverage. Shipments begin by mid-November.

Although Hindsight currently supports only C, ASA is working on COBOL, FORTRAN and C++ versions, with no release dates set.

ASA senior vice president Tom McHugh said the tool helps programmers understand code that might be obscure to anyone but the original programmer. "Once you understand the code, it helps you evaluate performance, [reliability] and quality," he said.

Operating under X-Window-based interfaces such as Motif and Open Look, Hindsight covers the spectrum of maintenance tasks without requiring new methodologies or source code changes, the company said. Hindsight's automatically generated structure chart provides a graphic overview of a program structure.

Hindsight can overlay bar graphs on each function in the structure chart to depict test coverage, run-time performance, module size and complexity data. Using a mouse, programmers, including engineers and managers, navigate through structure charts and diagrams, pop-up logic flow or complexity diagrams and select feature buttons.

"You can do in maybe two or three minutes with [Hindsight] what would take hours or days in terms of understanding how the code works," McHugh said.

Miller said Hindsight builds upon standard interfaces. The product uses interfaces similar to *tcove*, which is a standard Unix utility, and *prof*, a run-time profiler, to generate results, he said.

"We've evaluated multiple test coverage tools," such as Verilog Logiscope and TCAT, from Software Research, Miller said. "Hindsight seems to provide the best graphical interface [and] the most usable graphical interface, as well as an integrated environment."

The product also features assisted code tracing, allowing users to follow function calls and references by pointing to a reference, then jump to the referenced line. This ability is included in Hindsight's block-flow logic diagram, the J-Diagram, a graphic representation of both high-level and detailed procedural logic, according to ASA.

Documentation updates also are generated after a change is made to the source code. More than 20 tables and charts document function cross-references, size, compactness, complexity, test coverage, run-time performance and other traits.

ASA can be reached at 408-492-1668.

Copyright© 1990 by CMP Publications, Inc., 600 Community Drive, Manhasset, NY 11030. Reprinted from UNIX Today! with permission.

Fig. 22.2 An article on Unix Today

Expert David Spuler pointed that:

> The front end is a GUI that provides many useful reports and code views. ... There is explicit support for navigating through the source code for assisted code inspections and walkthroughs, which are an important bug reduction coding technique. Using the coverage data from OO-Test, it can show analysis of path conditions covered and unexecuted segments [Spu96].

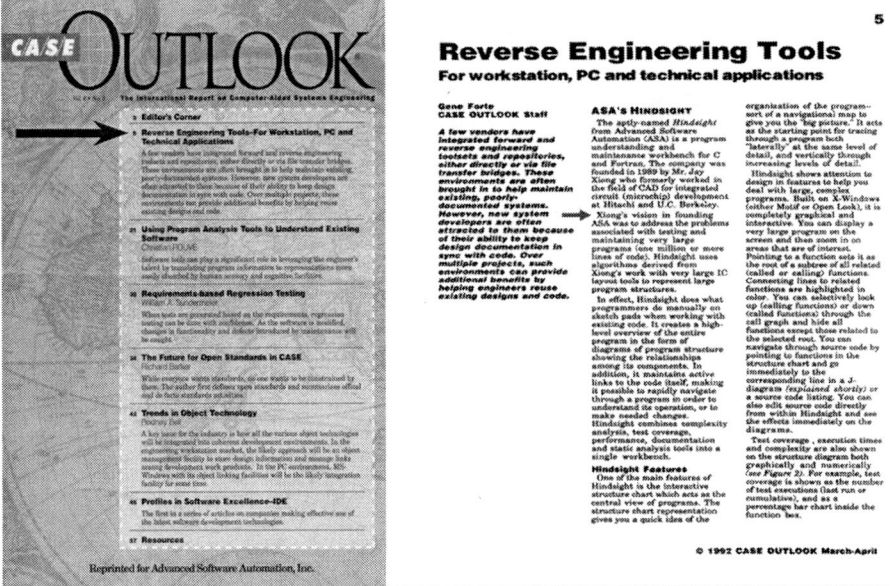

Fig. 22.3 An article introducing Hindsight as one of the typical reverse engineering products

Our Web site introducing Panorama got an 5 star award form ItmWEB as shown in Fig. 22.5.

22.2.3 Panorama++

Panorama++, the third generation of the major product family, is designed by me and implemented by me and my colleagues for fully supporting NSE nonlinear software engineering paradigm. The interface of Panorama++ is shown in Fig. 22.6.

The most tools/functions of Panorama++ for C/C++ have been ported to Panojava for handling Java programs. The Panojava interface is shown in Fig. 22.7.

22.3 Automated Tools Integrated with Panorama++

There are 15 automated tools plus 4 third party tools (open source tools) integrated into the Panorama++ product family (including Panojava for the java platform) as shown in Table 22.1.

22.4 Panorama++ Product Installation

Panorama++ is one of the "green software products" without complicated installation operations.

22.4 Panorama++ Product Installation

Type	MS Windows Tool		Vendor
Reviews & Inspections	Complexity Analysis	Panorama	International Software
		Analysis of Complexity Tool	McCabe & Associates
		Design Complexity Tool	
	Code comprehension	Panorama	International Software Automation
		Battlemap Analysis Tool	McCabe & Associates
Test Design & Development	Test Data Generator	TDGEN	Software Research, Inc
	Test Design	SoftTest	Bender and Associates
Test Execution & Evaluation	Capture/Playback	WinRunner	Mercury Interactive Corporation
		QA Partner	
		Ferret	Segue Software, Inc. Azor Inc.
	Client/Server	LoadRunner/PC	Mercury Interactive Corporation
	Coverage Analysis	Instrumentation Tool	McCabe & Associates
		Panorama	International Software Automation
Test Support	Problem Management	Defect Control System	Software Edge

Fig. 22.4 The information about Panorama from the book "Software Testing in the Real World"

Requirements

Machine Requirements: **A PC 486/586 or up running Windows NT/ Windows XP**
Disk space required to load Panorama++ C/C++: **500 megabytes**
Main Memory required: **640+ megabytes**

The itmWEB□ Five Star Selection Award

An exclusive award which recognizes useful, high quality Information Technology related web sites.

United States Five Star Award Winners

- ✓ International Software Automation

Fig. 22.5 Award issued by ItmWEB

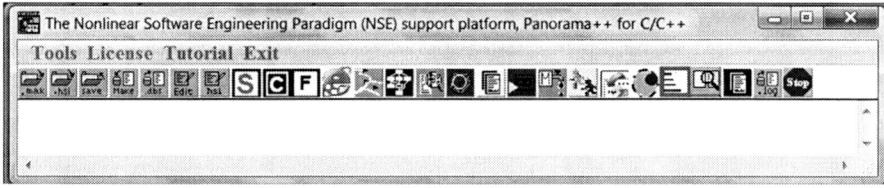

Fig. 22.6 The interface of Panorama++ for C/C++

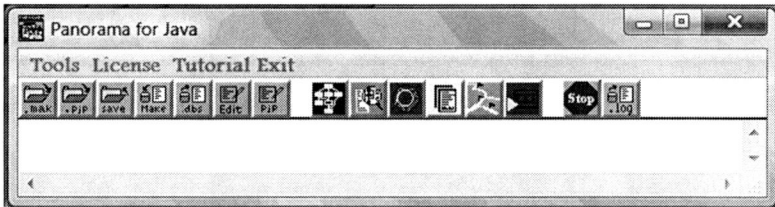

Fig. 22.7 The interface of Panojava

Operations

1. Copy the isa_examples directory to C: disk to form the C:\isa_examples directory for the example programs provided with the product for trial or directly view the results in the isa_examples\analyzed_for_review sub-directory.
2. Copy the isa_common_tools directory to any disk where you have enough space, such as the F: disk.
3. Copy the isa_NSE directory to any disk where you have enough space, such as the G: disk.
4. Set an environment variable PANORAMAHOME to point to the isa_common_ tools directory. For instance, if it is located in F: disk, then set

    PANORAMAHOME=F:\isa_common_tools

Table 22.1 Tools integrated in Panorama++ for C/C++

Support tools	Project management	Preprocess/ prototyping	Requirement development	Product design	Coding	Testing	Maintenance	Configuration (version comparison)
OO-Browser	●	●		●	●	●	●	●
OO-Diagrammer	●	●		●	●	●	●	●
OO-Test		●		●	●	●	●	
OO-SQA		●		●	●	●	●	●
OO-V&V	●	●	●	●	●	●	●	
OO-Analyzer	●	●	●	●	●	●	●	
OO-Performance						●	●	
OO-Playback						●	●	
OO-MemoryCheck							●	
OO-MiniCase							●	
OO-DefectTracer						●	●	
OO-CodeDiff								●
OO-DiffReporter								●
OO-BlackBox						●	●	
OO-ToolIntegration								
Open source products from third party								
Fujaba			●	●	●			
Gantpro	●							
OpenLoad						●		
CVS							●	●

(continued)

Table 22.1 (continued)

The tool function/capability:

OO-Browser for generating interactive and traceable call graphs or Class Inheritance Charts shown in J-Chart notations (see Chap. 7) innovated by me

OO-Diagrammer for generating interactive and traceable logic diagrams shown in J-Diagram notations or control flow diagram in J-Flow notations (see Chap. 7) innovated by me

OO-Test for performing software testing using Transparent-box method combining functional testing and structural testing [for Modified Condition/Decision (MC/DC) test coverage analysis] together seamlessly with the capability to establish bidirectional traceability among the related documents and test cases and source code (see Chap. 16) innovated by me

OO-V&V for Requirement Validation and Verification through bidirectional traceability

OO-SQA for software quality measurement

OO-Analyzer for dynamic and static program measurement

OO-MemoryCheck for checking memory leaks and usage violations

OO-Performance for performance measurement

OO-DefectTracer for tracing each runtime error to the execution path

OO-MiniCase for test case efficiency analysis and test case minimization in order to perform regression testing efficiently after code modification

OO-Playback for GUI operation capture and playback after code modification

OO-CodeDiff for holistic and intelligent software version comparison, etc.

OO-DiffReporter for providing reports on software version comparison

OO-BlackBox for functional testing with the capability for GUI operation capture and playback without the source code

OO-ToolIntegration for running a third party tool without presetting. There are two icons used to run the server-side program and the client-side program (some tools need both, some tools only need one of them). Panorama++ supports deeper tool integration: not only can it run third party tools without presetting, but it can also make the work products (or documents) generated be traceable with the implementation of requirements and test cases and source code through the use of a batch file and a special keyword @BAT@ to indicate the path of the batch file in the test case description. See application examples shown in Figs. 9.6 and 9.11

Note:

1. Different from other software testing tools, OO-Test using the Transparent-box method can be dynamically used in the entire software development lifecycle including the requirement development phase and the design phase for defect prevention and defect propagation prevention
2. Although the integrated third party products are originally open source products, users still need to check whether the policies of those products have been changed or not

5. Set an environment variable PANORAMATMPDIR to a working area (a directory exists). It is recommended to set it to the existing isa_common_tools\panotemp subdirectory. For instance, if the isa_common_tools directory is copied to the F: disk, then set

   PANORAMATMPDIR=F:\isa_common_tools\panotemp
6. Set an environment variable ISA_NSE_HOME to point to the isa_NSE directory. For instance, if it is located in G: disk, then set
   ISA_NSE_HOME=G:\isa_NSE

Verification

After that please click e_panorama from the isa_NSE directory to start the toolkit to see whether the settings are correct through examples (see the User's Manual).

22.5 A Guided Tour of Panorama++ for C/C++

The Panorama++ C/C++ software testing, validation, maintenance, and re-engineering environment is designed to simplify and speed up the tasks of understanding, evaluating, testing, validating, and maintaining a software product. It is easy to use.

Panorama++ fully supports the NSE nonlinear software engineering paradigm, including requirement development by working with the **HAETVE** technique (see Chap. 11), software design and coding by working with the **Synthesis Design and Incremental growing up (Implementation and Integration)** technique (see Chaps. 12 and 13), software testing using the Transparent-box method (see Chap. 16), quality assurance driven by defect prevention and defect propagation prevention, software maintenance with side-effect prevention in the implementation of requirement changes and code modifications, etc.

This demo operation guide uses one of the demo programs provided by Panorama++ C/C++, a calculator program, and for Panorama++ C the provided demo program is SORTDEMO.

Make sure that your compiler (such as cl.exe) and linker utility (such as link.exe) can be found in your path. If you use Microsoft Visual C++ 6.0 on Windows NT or Windows XP, you should select an option to register the environment variable to run the compiler in command line when you install the Visual C++ – see Fig. 22.8:

If you have not registered it, please ask your system manager to install VC++ again to register the environment variable.

The major operations:

A. Specifying the source program

Panorama++ C/C++ generates information about a software program directly from its source code. To specify the source files, the user only needs to identify the

Fig. 22.8 Register environment variable needed

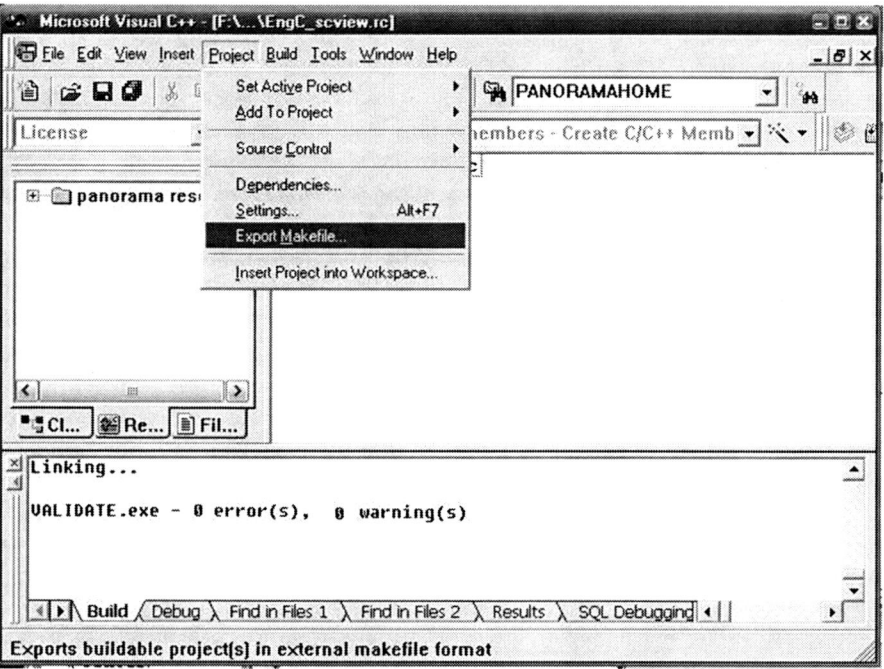

Fig. 22.9 Exporting a Makefile

makefile of the program. If you do not have a makefile for your VC project, you can get it from the VC interface – see Fig. 22.9.

You may use a batch file (such as xxx.bat) to replace a Makefile.

22.5 A Guided Tour of Panorama++ for C/C++

The major steps:

1. Open the Panorama++ C/C++ toolkit from the HOME directory pointed by the environment variable, PANORAMA_HOME, and click on the Panorama++ icon. The Panorama++ MAIN menu will pop up. It contains a tool bar with some buttons for dealing with input/output (makefile, hsi and database), and the individual tools in Panorama++: OO-Test, OO-Browser, OO-Diagrammer, OO-SQA, OO-Analyzer, and OO-Validate.
2. Click on .MAK from the tool bar. The LOAD.MAK FILE menu will come up.
   For a sample of Panorama++ C, click on the directory
   C:\isa_examples\English_examples\to_be_analyzed\SORTDEMO
   For a sample of Panorama++ C++, click on the directory
   C:\isa_examples\English_examples\to_be_analyzed\CAL
3. Select the makefile of program.
   For Panorama++ C, click on the file sortdemo.mak, and press OK.
   For Panorama++ C++, click on the file cal.mak, and press OK.
   The file will be loaded into the white area below the tool bar.

B. Creating an analysis database

Panorama++ will analyze the makefile and build a static and dynamic database. The analysis results can then be viewed on charts, diagrams, and reports. It also creates an .hsi file that lists all the source files and the correct CPP options for your reference. If your program does not contain a makefile such as in the case that your program has not been completed, you need to manually create a .hsi file specifying all the files of the program to allow Panorama++ to analyze your program.

1. Click on MAKE. The CREATE.DBS FILE dialog box will come up.
2. Click OK to accept the default database file name, for Panorama++ C is **sortdemo.dbs** and for Panorama++ C++ is **cal.dbs**.
   The ANALYSIS.MAK FILE menu will come up.
3. Click DYNAMIC. This will capture dynamic test coverage data.
4. Click OK.

Panorama++ will analyze the source program to build a database of information. A new window will be opened to display information. When **sortdemo.hsi** and **sortdemo.dbs** (for Panorama++ C) or **cal.hsi** and **cal.dbs** (for Panorama++ C++) have been generated, you will be notified. You can then close the window.

Follow Step C now to collect dynamic test coverage data or skip to Step D to directly view only the static analysis information in charts, diagrams, reports, and metrics diagrams.

C. Getting dynamic test coverage data

You need to run the demo program to obtain its test coverage data. This is done using a test script or a batch file containing the test cases. A test script sortdemo.tca (for Panorama++ C) or English_req_a_test.tca and English_req_b_test.tca (for Panorama++ C++) have been provided in the related directory. Just run it from Panorama++ OO-Test to obtain test data.

1. On the tool bar of the MAIN menu, click the green OO-TEST button. The Panorama++ OO-Test menu bar will pop up.
2. Press the SCRIPT button. The SCRIPT window will come up.
3. Click FILE, then select LOAD from the submenu. A LOAD CASE FILE dialog box will pop up.
4. Load the corresponding test case script file mentioned in the previous step, and press OK to load the script file. The file will show up in the SCRIPT window.
5. Click RUN from the menu bar of the SCRIPT window. The SPECIFY TEST DATABASE dialog box will come up showing **sortdemo.tdb** (for Panorama++ C) or **cal.tdb** (for Panorama++ C++) in the .TDB NAME field.
6. Press OK to run the script and accumulate data in **sortdemo.tdb** (for Panorama++ C) or **cal.tdb** (for Panorama++ C++). A command window will pop up showing the results of the execution. Then a notification window will come up informing you that the execution has been completed.
7. Click OK to close the notification window.
8. Click FILE then EXIT to quit the SCRIPT window.

D. Obtaining a system-level overview using J-Charts

Panorama++ OO-Browser generates three kinds of J-Charts from your source code: Function Calling Graph, Class Inheritance Chart, and Class-Function Coupling Chart. They give a quick, graphic overview of program hierarchies.

1. On the tool bar of the MAIN menu, click the blue OO-BROWSER button. The Panorama++ OO-Browser dialog box will pop up.
2. Select FUNCTION CALL GRAPH.
3. Select CHART ONLY and press OK. The Function Calling Graph will come up. Each box represents a function, connecting lines represent function calling relationships.
4. Click the left mouse button on a function box in the J-Chart to highlight the related functions.
5. Click the right mouse button on a function box will activate the FUNCTION pop-up menu. For example, select DIAGRAM then J-DIAGRAM to view the J-Diagram for the function.
6. Click the right mouse button anywhere outside the function boxes to open the CHART pop-up menu. For example, select COMPLEXITY then WITH CASE to overlay complexity information on the chart.
7. If you have collected test coverage data (step 3), you can select TEST COVERAGE then ACCU RUN then SC0 from the CHART pop-up menu. Then the test coverage data will be shown on the charts.
8. You can repeat D-1–D-7 for Class Inheritance Chart and Class-Function Calling Graph for Panorama++ C++.

22.5 A Guided Tour of Panorama++ for C/C++

9. Click on HELP from the chart menu bar. A Panorama++ C/C++ Help window will come up, showing an Index. Text in green can be expanded to show more information.
10. Click on the green "Description of Structure Chart" a few lines down from the top. The Help window will now display the description of the Structure Chart.
11. Click the BACK from the menu bar to jump back to the Index display.
12. Click EXIT from the Chart window to exit OO-Browser.

E. Understanding detailed code logic using logic diagrams

Panorama++ OO-Diagrammer generates three kinds of logic diagrams from your source code: J-Diagram, J-Flow, and ActionPlus. They give insights into the detailed logic of a program.

1. On the tool bar of the MAIN menu, click the orange OO-DIAGRAMMER button. The Panorama++ OO-Diagrammer menu will pop up.
2. Select J-DIAGRAM and press OK. The J-Diagram will come up. After the table of contents is the Function High Level Diagram. It shows the calling relationship between functions. At the right of each function name is a corresponding active number.
3. Click the left mouse button on an active number at the right side of the diagram to jump to detailed diagram for that function.
4. Code in diagrammed format is much easier to read through. You can scroll through it using the keyboard? or what key?
5. If there is a function call or other reference in the code, just click the left mouse button on the active number at the right side to jump to the corresponding part of the diagram. Press again to jump back.
6. If you have collected test coverage data (Step C), you can display test coverage data on the diagram. From the menu bar, select OPTION, then TEST COVERAGE, then ACCUMULATED. The test frequencies will be displayed on the top right of each segment, and the untested segment will be highlighted with black boxes.
7. Select OPTION, then SWITCH, then J-FLOW to switch to the J-Flow Diagram. In this control flow diagram, you can highlight an untested path and view information about it to help you test an untested segment.
8. Scroll down to a function that contains untested segments and is fairly complex. Each vertical line from the left represents a new control level. So one vertical line that has many levels is more complex in logic.
9. Double-click the left mouse button on an untested segment (a black box). A path containing the untested segment will be highlighted in magenta.
10. From the menu bar, select PATH then CURRENT PATH INFO to view the logic conditions which when satisfied will test this path.
11. Unselect the path by double-clicking the left mouse button away from the logic diagram (such as on the line/block number in the left).
12. Select OPTION, then SWITCH, then ACTIONPLUS to switch to the ActionPlus Diagram, an enhanced version of the Action Diagram.
13. Click FILE then EXIT to exit OO-Diagrammer.

F. Measuring program quality using metrics diagrams

Panorama++ OO-SQA allows you to set practical quality standards for your object-oriented program, then collect quality data from your source code and see how it compares to the standards you have set. The quality data is shown in four easy-to-see formats: Bar Graph, Kiviat Diagram, MultiMetrics Diagram, and Reports.

1. On the tool bar of the MAIN menu, click the magenta OO-SQA button. The Panorama++ OO-SQA menu bar will pop up.
2. Press CLASSES. The METRICS window will come up.
3. From the menu bar, click TYPE then BAR GRAPH. The Bar Graph will be displayed. For each of the quality assurance metrics listed on the left, a bar shows how well the classes in the program satisfy the metric: the blue part represents the passed classes and the red part the failed classes.
4. Click STANDARD. The STANDARD menu will come up. You can change the minimum/maximum acceptable value or the weight for each metric, then press OK to observe the change in the metrics diagram.
5. Press TYPE, then KIVIAT DIAGRAM to view a different metrics diagram. In this diagram, each radius represents a metric. The inner circle represents the minimum acceptable value and the outer circle the maximum acceptable value. Each function's metric values are connected to form a polygon, if the polygon falls entirely within the two circles, the function satisfy all the metrics.
6. You can also view quality assurance REPORTS and MULTIPLE METRICS DIAGRAM by selecting them from the TYPE menu.
7. Click FILE then EXIT to exit OO-SQA.

G. Viewing online program documentation

Panorama++ OO-Analyzer automatically generates reports from the source code to fully document a software program. More than 100 reports provide information on program compactness, function/class structure, special functions, global/static variable, complexity, and test coverage.

1. On the tool bar of the MAIN menu, click the yellow OO-ANALYZER button. The Panorama++ OO-Analyzer menu will pop up.
2. Press OK on the dialog box. The 44 default reports will be generated. The REPORT window will come up, showing the Table of Contents.
3. To the right of each report title is an active number in red. Click the left mouse button on an active number to view the specific report.
4. Read the report.
5. Jump back to the table of contents by pressing the "enter" or "return" key.
6. Select FILE then EXIT from the report window to exit OO-Analyzer.

H. Regression testing: test case playback

Playback function is very useful for rerunning test cases, especially for user's interface testing (GUI playback). You must have done Step C (Script action), on Step

22.5 A Guided Tour of Panorama++ for C/C++

C-5 you must select Record button as TRUE in the SPECIFY TEST DATABASE dialog box.

1. On the tool bar of the MAIN menu, click the red OO-Playback button. A Playback window pops up.
2. Select FILE then LOAD menu to load the sortdemo.tdb file, and the sortdemo. tdb file will be shown on the playback window.
3. Select RUN on the menu bar, then you can select the ALL or MINI submenu. If you select ALL, A SPECIFY TEST DATABASE dialog box pops up, press OK and then all test cases in sortdemo.tdb will be played back and all of the test database will be added to the playout.tdb file.
4. Select RUN then MINI menu. TestCase Minimization Options dialog box pops up, select options you want then press OK button, the test cases of minimization will be shown in Minimization Result window. Select Run then ALL menu and SPECIFY TEST DATABASE dialog box pops up, press OK. The test cases of minimization result will be played back and all of the test database will be added to the playout.tdb file.
5. You can press the FILE menu then SAVE submenu, a SAVE dialog box will pop up, type file name sortmini.tdb then press OK, the minimized test cases will be saved in sortmini.tdb.
6. Modifying the source code of sortdemo (copy sortdemo.c to sortdemo.old, then copy sortdemo.new to sortdemo.c or directly modify sortdemo.c) then building a new database. Running step 1 and step 2 with loading sortmini.tdb to playback (select RUN and ALL this time rather than MINI). The test cases of minimization result will be played back and all of the test database will be added to the playout.tdb file.
7. View the new result after playing back using minimized test cases by repeating steps from Sects. 22.4 to 22.7. You will find that the test coverage result obtained using the minimized test cases through an automated playing back of the operations is the same or almost the same as that obtained using all the test cases. It means that with the capture/playback tool seamlessly integrated with test coverage analysis, test case minimization, and analysis of program structure, logic, control flow, complexity, compactness, and data, Panorama++ can bring you great savings for your software testing and re-testing after code modification.

Note

1. As you can see from the execution of the sortdemo program, its window position is different each time as controlled by the Windows Manager. It is recommended not to move the window (the title) of the sortdemo program; otherwise the window may be moved completely or partially out of the screen in the automatic playback process (if it happens, press Alt-Esc to recover it).
2. Please make the beginning environment for playback be the same as that for capturing the operations. If the environment is different such as that a file name to be used to save a file does not exist in the capturing process, but after that it exists in the playback process, then the message window shown by the Windows

Manager will be different so that the playback process may not be successful (for solving the problem, you may create an empty file before capturing or rename the file capturing).
3. The source file names and the dbs names for capture and for playback after code modification should be the same for correctly running the playback and obtaining the correct test coverage data.

I. Validation of requirement implementation and minimization of the test cases

You need to have done Step C to capture test coverage data before running this step. Panorama++ provides an ideal environment for validating the implementation of the requirements through bidirectional traceability built automatically. In the following description, we use the CAL example written in C++ programming language.

1. On the tool bar of the MAIN menu, click the OO-Validate button (next to the OO-Analyzer icon, with a white background). The **Requirement validation and test case minimization** menu bar will pop up. Click on "Requirement validation and test case minimization," a "Test Case Analyzer" window will pop up. Click "OK" to accept the default setting.
2. The **Test Coverage Analysis** window will pop up. Use the mouse to change the width of the window until you can see all the contents.
3. Click on "Corresp.," and then click on "File.Class.Function" from the menu bar. A "File.Class.Function" window will pop up. Use mouse to move this window to a suitable location without overlaying the "Test Coverage Analysis" window.
4. Validation through forward traceability: move the mouse to point to a test case from the "Test Coverage Analysis" window, and make a mouse click, then the description part of the test case selected will be shown in blue color automatically. The corresponding modules that can be tested by that test case will be shown in red color automatically in the "File.Class.Function" window. At the same time, all of the related requirement specification file, the design document file, and other related documents specified using a keyword to identify the file type followed by a full path and the corresponding bookmark will be opened automatically from the location pointed by the bookmark, so that you can validate the requirement with the test case and all related documents and the corresponding source modules to check the consistency to see whether the requirement has been fully implemented. If there are several test cases used for validating the requirement, all of them need to be checked.
5. Validation through backward traceability: move the mouse to point to a tested module (not highlighted in black boxes) from the "File.Class.Function" window, and make a mouse click, then the module will be shown in blue color automatically. The corresponding test cases that can be used to test that module will be shown in red color automatically in the "Test Case Analysis" window. At the same time, all of the related requirement specification file, the design document file, and other related documents specified using a keyword to identify the file type followed by a full path and the corresponding bookmark will be opened automatically from the

22.5 A Guided Tour of Panorama++ for C/C++

location pointed by the bookmark, so that you can validate how many requirements are related to the module and how many test cases can be used to test that module. You can also check the consistency among all of the related documents and the module.

6. After you try some operations, please close the "File.Class.Function" window.
7. The keywords available include @WORD@, @HTML@, @EXCEL@, @PDF@, and @BAT@. They should be used within the comment line (with a "#" character at the beginning of the line) in the test case script file.
8. You can view the bidirectional traceability at the code statement level for requirement validation too: from the "Test Coverage Analysis" window, click on "Corresp.," and then click on "Segment," a "Segment Level Correspondence" window will show up.
9. Repeat step 4 and step 5 using "Segment" to replace "Module" in the step description.
10. After you try some operations, please close the "Segment Level Correspondence" window.

J. Analyzing test cases efficiency for more efficient testing

Panorama++ provides an ideal environment for analyzing the test cases efficiency and performing test case minimization to reduce re-testing effort after code modification.

1. Test cases have different test efficiencies. Adopting the highly efficient test cases results in great savings of test effort. Click EFFICIENCY on the TEST COVERAGE ANALYSIS window.
The TEST CASE EFFICIENCY ANALYSIS OPTIONS dialog box pops up.
2. Press OK. The EFFICIENCY ANALYSIS RESULT window will pop up. It shows all the test cases and their test coverage results for different levels (here "Function" is used for module coverage; "SCO" for segment (statement) test coverage; "SC1" for branch test coverage; "SC1+" for branch test coverage plus loop boundary coverage).
3. In a large set of test cases, many test cases merely duplicate the test coverage results obtained by the previous test cases run. A lot of test efforts thus are wasted. To minimize test cases, press MINIMIZATION on the menu bar of TEST COVERAGE ANALYSIS window.
The TEST CASE MINIMIZATION OPTIONS dialog box pops up.
4. Press OK. The MINIMIZATION RESULT window will pop up. It shows the minimized set of test cases obtained from all the test cases in the **sortdemo.tca** (for Panorama++ C) or **testcase.cal** (for Panorama++ C++) script.
5. Press FILE then EXIT on the TEST COVERAGE ANALYSIS to exit OO-Test.

K. Memory leak/violation check

If you want to get memory leak and violation information, use the sample program in the "leak" subdirectory. You must select "Check Memory Leak/Violation" as

TRUE in "Panorama++-2 Analysis.mak file" dialog box in Step B. In the Memory Checker Report, it will show all of the memory leaks and violations to you after you have run the program.

1. On the tool bar of the MAIN menu, click the OO-MemoryChecker button. A Memory Leak/Violation Report window will open. You can see all of the memory leaks and violation in your program which you have run earlier.
2. Scroll through the report using the "page down" key.
3. If you want to clear the earlier record, you can simply click "Clear Record" on the menu bar.
4. Press File then Exit to exit Memory Checker Report.

L. Defect Tracer (Optional)

Defect Tracer can trace the execution path from the beginning to the end of an execution of a program when a problem exists with the execution. It can also identify the type of the problem in most cases. Use the sample program for program tracing in the "trouble" subdirectory. You must select "Record Problem Tracing" as TRUE in "Panorama++-2 Analysis.mak file" dialog box in Step B.

1. On the tool bar of the MAIN menu, click the OO-DefectTracer button. A Defect Tracer Report window will open. You can see all of the execution paths of a program which has some problems you have run earlier.
2. To the right of each case's running time is an active number in red. Click the left mouse button on an active number to view the specific defect tracing report.
3. Scroll through the report using the "page down" key.
4. Jump back to the table of contents by pressing the "return" key.
5. If you want to clear the earlier record, you can simply click "Clear Record" on the menu bar.
6. Select FILE then EXIT from the window to exit OO-DefectTracer.

M. Panorama++ Log File

If you want to know the information of making the database, you can click the "Open Log" button. The Open File dialog box will open. You can select a Panorama++.log file for reading.

N. Exiting Panorama++

Click FILE then QUIT from the Panorama++ C/C++ MAIN menu.

22.6 Network Floating License Support

With NSE support platform, Panorama++, not only computer-specific licenses are available, but network floating licenses are also available. The licenses are counted for each individual tool. Users can select different numbers of licenses for different tools.

22.7 The Major Features of Panorama++

The major features of Panorama++ are as follows:
- **Highly automated** – all integrated tools are automated ones
- **Highly integrated** – all tools are integrated closely to share the same small database
- **Intelligent to some degree** – for instance, the version comparison tool CodeDiff
- **No size limitation** – there is no limitation to the tools, depends on the systems
- **Small database** – there are only six hash tables
- **High speed display** – the generated holistic results are virtually existing, the display speeds are very fast
- **Incremental update** – after the database is built the first time, it can be easily updated incrementally if a few source files are modified – only the modified files need to re-analyzed
- **Easy to use – this highly automated software development and maintenance platform is very easy to use with** application examples and tutorials provided.

22.8 Applications

All screenshots provided in this chapter and all chapters in this book are application examples of Panorama++ for C/C++ or Panojava for Java programs. For more information about the application, please see Chap. 23.

22.9 Summary

In this chapter, many software productivity and quality tools are introduced, which are integrated into Panorama++, the NSE support platform. Those tools are highly automated and easy to use.

22.10 Points and Questions to Ponder

(a) Why do we need to use software tools?
(b) Why should software tools be automated?

22.11 Further Reading and Information Source

(a) CVS – Concurrent Versions System. http://www.nongnu.org/cvs/
(b) The Fujaba Project. http://www.fujaba.de/

(c) OpenLoad. http://www.opendemand.com/
(d) Ganttpro/Ganttproject. http://sourceforge.net/projects/ganttproject/

References

[Kit05]   Kit E (2005) Software testing in the real world. Addison-Wesley, New York
[Pre05]   Pressman RS (2005) Software engineering: a practitioner's approach. McGraw-Hill, New York
[Spu96]   Spuler D (1996) C++ & C tools, utilities, libraries, and resources. Prentice Hall PTR, Upper saddle River, pp 233–234. ISBN 0-13-226697-0

Chapter 23
NSE Applications

> *Practical wisdom is only to be learned in the school of experience.*
>
> Samuel Smiles (1812–1904)

This chapter introduces the applications of NSE and its support platform, Panorama++, for both new software product development and a product being developed using other approaches – in this case, users only need to rewrite their test cases according to some simple rules and set bookmarks to the related documents; the other work can be performed automatically by Panorama++. It means that NSE can be added on to any approach at any stage.

23.1 The Whole and Its Components: A General Comparison Between NSE and Other Approaches

Both software and the software engineering paradigm are complex systems with many components connected closely with strong interactions. According to the Holism principle of complexity science, the whole of a complex system is greater than the sum of its components, the characteristics and behaviors of the whole emerge from the interaction of its components, and cannot be determined or explained by its components alone, and cannot be inferred simply from the behavior of its individual components.

The old-established software engineering paradigm is based on reductionism and the superposition principle that the whole of a complex system is the sum of its components, so that with the old-established software engineering paradigm almost all tasks and activities in software engineering are performed linearly, partially, and locally, such as the implementation of requirement changes and code modification.

Low quality and productivity, high cost and risk – those critical issues exist for more than 40 years with the old-established software engineering paradigm. Almost all of its components make bad "contributions" to those issues, including the linear process models which always go forward in one direction with only one track without upstream movement at all – it forces software developers to always do all things right without making any mistakes, violating the nature of human beings; the software development methodologies based on the Constructive Holism principle that the components of a software are developed first, then the whole of the product is assembled from the components, making the quality hard to ensue; the test paradigm mainly based on Black-box testing after production, and cannot be used dynamically to remove the critical defects introduced into a software product in requirement development phase and design phase; the quality assurance paradigm based on inspection and testing after coding; the maintenance paradigm where the implementation of requirement changes and code modifications are performed partially and locally without means to prevent the side effects; the documentation paradigm separated from the source code; the project management paradigm where the project management process is separated from the product development process, etc. – it means that **any partial solution or improvement in any individual part(s) of software engineering will not be able to solve the critical issues with software development today: low quality and productivity, high cost and risk.**

Unfortunately, all existent software development approaches and the new ones being developed such as MDD (Model-Driven Development)/MDA (Model-Driven Architecture) are partial solution approaches.

NSE is different. NSE offers a holistic and global solution for software engineering by

1. Using complexity science as the sharp weapon to complete the revolutionary change for each component of the software engineering paradigm, including the process model (see Chap. 8), software development methodology (see Chap. 10), software testing paradigm (see Chap. 16), software quality assurance paradigm (see Chap. 17), software visualization paradigm (see Chap. 7), software documentation paradigm (see Chap. 19), software maintenance paradigm (see Chap. 18), software management paradigm (see Chap. 20), etc.
2. Making the desired characteristics and behavior of the whole of the new software engineering paradigm emerge from the interaction of its new components – let them work together closely and support each other. For instance, the NSE software testing paradigm based on the Transparent-box testing method can be used dynamically in the requirement development phase with the HAETVE technique and dummy programming technique. But the dummy programs do not provide real outputs. In this case, the NSE testing tool will check whether the execution path covers the expected path shown in J-Flow notations which needs the support from NSE visualization paradigm; NSE testing paradigm supports MC/DC test coverage analysis, it also needs support from the NSE visualization paradigm to show the test results with the capability to highlight the untested branches and untested conditions graphically.
3. Making almost all software engineering tasks and activities be performed holistically and globally.

23.2 What Makes NSE Special?

About software development strategy, NSE offers a different solution:

1. **One source for both** – for human understanding (graphic diagrams) and computer understanding (source code).
2. **Design becomes precoding, coding becomes further design** – top-down plus bottom-up. The graphic design diagrams are not only generated through reverse engineering, but also generated from forward engineering using dummy programs written in the same language as that used for target coding or different programming language (such as Java for better portability) – in this case, a language transformer/translator is needed.
3. **All tasks and activities are performed holistically and globally** – to avoid partial and local solutions and side effects in the implementation of requirement changes and code modifications.
4. **Dynamic testing using the Transparent-box method to be performed from the first step down to the delivery of a software product** – to ensure the quality of the product and extend the product life.
5. **Any change, no matter it is in the product development process or maintenance process, is implemented with side-effect prevention through various traceabilities** – to ensure the quality of the product being developed or maintained.

23.3 Applications in New Software Development

23.3.1 Benefits

Applying NSE for a new software product development can bring the maximum benefits to the product, including:

1. **Low risk** – through preprocess (see Chap. 8) to perform prototype design, testing, and review for critical and not familiar requirements.
2. **High quality** – following the NSE process model and NSE methodology, the product quality can be ensured through defect prevention and defect propagation prevention in the entire software development and maintenance lifecycle, particularly in the requirement development phase and the design phase by applying the Transparent-box testing method dynamically – having an output is no longer a condition to use this testing method dynamically.
3. **High productivity** – high quality also means high productivity (requiring less time and effort to find and fix defects), plus less defects propagate to the maintenance phase, and the side effects for the implementation of requirement changes and code modifications can also be prevented through various traceabilities, so about two-third of the total effort spent in software maintenance can be saved – it equals to double the productivity.

4. **Low cost** – following the NSE process model and NSE methodology, less time and resources need to be used for finding and fixing defects, plus less defects propagate to the maintenance phase, and the side effects for the implementation of requirement changes and code modifications can also be prevented through various traceabilities, so about two-third of the total cost spent in software maintenance can be saved – it equals to half the cost.
5. **Easy to meet the schedule** – with NSE, the project management process and product development process are combined together closely – product plans, schedule charts, and progress reports are traceable with the implementation of requirements and the source code, plus the project Web site and the BBS are also traceable with the implementation of requirements and the source code, so that any schedule issues can be found early and solved in time.
6. **Easy to meet the budget** – complying with the Generative Holism principle, the whole of a software product will be formed first, so that the required budget can be estimated better; in the requirement development phase, requirements are ordered according to the importance: the more important ones will be implemented earlier to meet the market needs within the budget – if necessary some optional requirements or not important requirements can be temporarily ignored or implemented in the next round; high quality and productivity and low cost also means easy to meet the budget.
7. **Easy to maintain** – with NSE, not only the defects introduced into a software product and the defect propagated into the maintenance phase can be greatly reduced, but the new defects introduced in the implementation of requirement changes and code modifications can also be greatly reduced through side-effect prevention supported by various traceabilities.

23.3.2 Recommended Process

To fully benefit from applying NSE in new software development, it is strongly recommended to:

1. **In the preprocess phase:**
   - Assign priority to the requirements according to the importance of the requirements.
   - Perform prototype design, testing, and review for important and unfamiliar requirements to reduce the risks.
   - Build a project Web site and the corresponding BBS.
2. **In the requirement development phase (see Chap. 11):**
   - Use dummy programming for requirement development and modeling to generate design documents as much as you can, because graphic documents that are manually generated or designed using graphic editors are time-consuming to draw, hard to review, hard to change, and hard to maintain consistency with the source code.

23.3 Applications in New Software Development

- Use the HAETVE technique for requirement development (including the function decomposition of the functional requirements); with it, the automatically generated graphical documents are traceable for static review for defect removal and dynamic execution for dynamic defect prevention and defect propagation prevention.
- Preliminarily design your requirement specification file (which should be improved in the implementation of the requirements) using the template shown in Appendix A.
- According to the simple rules for writing test cases (see Chap. 9), design the corresponding test cases to dynamically execute your dummy program for defect removal. Do not forget to use the special keywords to specify the related documents.
- Run your dummy programs using the Transparent-box testing method – to each test case it not only checks whether an output (if any, can be none) is the same as what expected, but also checks whether the execution path is the same as what expected with the capability to establish bidirectional traceability to help you find and remove the inconsistent defects.

3. **In the preliminary system design phase:**

- Refer to the requirement development results, perform the system preliminary design of the product to form the whole of the system as an embryo.
- Perform defect prevention and defect propagation prevention using the Transparent-box testing method.
- Estimate your project cost according to your system design results.
- Make your product development plan and schedule.
- Preliminarily design your document hierarchy using bookmarks (see Chaps. 8 and 9).

4. **In the implementation process:**

- Select a set of (or one) requirements according to the assigned priority to implement it – it is recommended to implement a essential version of the product (about 20% of the total requirement).
- Further improve the requirement specification and design the test requirements and the test scripts.

5. **In the design phase (see Chap. 12):**

- Use the **Synthesis Design and Incremental growing up (Implementation and Integration)** technique to improve the preliminary product design, form the system call graph, and then complete the detailed product design for the selected requirements.
- Dynamically test the designed product version for defect prevention and defect propagation prevention – with NSE, before coding, all designed versions should be executable. If something unexpected is found, go back to improve the design; if it is found that the solution method did not meet the requirement(s), go back to the preprocess to select a new solution method.
- Use the designed system call graph to assign a bottom-up incremental coding order.

6. **In the coding phase (see Chap. 13):**
    - Follow the coding order to perform incremental coding.
    - Insist on performing MC/DC test coverage analysis for each program unit (see Appendix B).
    - Prevent possible defects between the interfaces.
    - If there is a need for a called function to return special values, use the technique introduced in Appendix C.
    - Diagram the source code and use the automated traceability to perform code inspection.
    - Coding can be parallelly performed, but the integration should be done by adding modules one by one.
    - If something critical is found, go upstream; if the solution method does not meet the requirement(s), go back to the preprocess to try a better solution method.

7. **In the testing phase (see Chaps. 14–16):**
    - Perform system testing using the Transparent-box method.
    - Perform performance measurement, memory leak, and usage violation check.
    - Perform GUI testing operation capture and playback after code modification.
    - Perform code static and dynamic measurement to provide various reports for documenting the product.
    - Measure the quality of the executable product and each module.
    - If something critical is found, go upstream; if the solution method does not meet the requirement(s), go back to the preprocess to try a better solution method.

8. **In the maintenance phase (see Chap. 18):**
    - Perform implementation of requirement changes and code modifications with side-effect prevention through various traceabilities.
    - Perform regression testing using minimized test cases (sometimes there is a need to design some new test cases), and using backward traceability to find the corresponding test cases to save time and resources.
    - If after the implementation of requirement changes and code modifications some new defects exist, perform holistic and detailed version comparison to locate the defects.

9. **Others:**
    - Frequently deliver all working products to customers for review, and get feedback to improve the product.
    - Make the project Web site and the BBS as a real-time communication channel to share information and perform problem solving.
    - Always combine the software development process and the project management process together, and make the project plan, schedule, cost report, progress report traceable with the implementation of requirements and the source code, for better project management and budget control.
    - Perform tasks holistically and globally as much as possible.
    - Always use the traceability among all related documents and test cases and the source code to ensure product quality.

23.4 Applications in a Software Product Being Developed Using Other Approaches

NSE can be applied to a product being developed or tested or maintained at any stage to help it (currently Panorama++ supports C, C++, Visual Basic, and Java on Windows. Linux versions of Panorama are under testing). The major work that users need to do are:

1. Set bookmarks to your documents.
2. Use the bookmark to form your documents hierarchy, particularly to indicate what test script and test cases are used for what requirements.
3. Redesign your test cases using the basic format specified in Chap. 9.

Almost all other work can be performed automatically by the NSE support platform, Panorama++. As described in Chap. 19, with NSE the source code is also the source to automatically generate the graphical documents.

23.5 Possible Combination with UML

23.5.1 About the Future of UML

In Chap. 11, I mentioned that regarding the future of UML, Jim Arlow and Ila Neustadt pointed that:

> "MDA – the future of UML
> The future of UML may be a recent OMG initiative called MDA... MDA defines a vision for how software can be developed based on models... In MDA software is produced through a series of model transformations aided by an MDA modeling tool. An abstract computer-independent model (CIM) is used as basis for a platform-independent model (PIM). The PIM is transformed into a platform-specific model (PSM) that is transformed into code" [Arl06].

23.5.2 Question to the Future of UML

About MDA, Harry Sneed pointed that:
"Model driven considered harmful

- Model-driven tools magnify the mistakes made in the problem definition.
- Model-driven tools create an additional semantic level to be maintained.
- Model-driven tools distort the image of what the program is really like.
- The model cannot be directly executed. It must first be transformed into code which may behave other than expected.
- Model-driven tools complicate the maintenance process by creating redundant descriptions which have to be maintained in parallel.

- Model-driven tools are designed for top-down development.
- Top-down functional decomposition creates maintenance problems."

"Summary:

- If a UML design can really replace the programming code as envisioned by Jacobson in his paper, 'UML all the way down,' then it becomes just another programming language.
- The question then comes up as to what is easier to change
    - The design documents or
    - The programming language
- This depends on the nature of the problem and the people trying to solve it. If they are more comfortable with diagrams, they can use diagrams. If they are more comfortable with text, they should write text.
- Diagrams are not always the best means of modeling a solution. A solution can also be described in words. The important thing is that one model is enough – either the code or the diagrams. They should be reproducible from one another" [Sne07].

23.5.3 *Possible Combination with UML (NSE-UML?)*

It is possible to combine NSE and UML together using platform-independent programming language such as Java programming language as the original programming language for not only 3J graphics (J-Chart, J-Diagram, and J-Flow) generation, but also for UML diagram generation. It may need a new programming language which can be easily used for both graphic document generation and source code design.

As described in Chap. 11, with the Source Code Driven Dynamic Software Modeling and Engineering using the HAETVE technique, not only the models/diagrams can be automatically generated from the source code through forward engineering using dummy programs or reverse engineering using regular source code, but the generated models/diagrams can also work together with the source code dynamically to help users understand a software product better, test the product better, and maintain the product better. Different from traditional software models which only represent some static properties of a software product, with NSE the models generated from the source code also represent the dynamic properties of a software product, such as the overall test coverage, the performance measurement result, the execution path traced for a runtime error, etc. Furthermore, the generated models/diagrams can also dynamically respond to users' requests (see Figs. 11.31 and 11.32).

Instead of making UML diagrams executable (executable UML), I think it will be better to have one kind of source for both graphic document generation and source code.

A dummy program written in Java for representing an actor and an action with the corresponding graphical document automatically generated by NSE visualization paradigm is shown in Fig. 23.1.

23.5 Possible Combination with UML

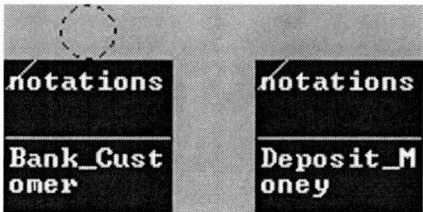

Fig. 23.1 The notations for representing an actor and the action for Java

```
public class notations {

  public static void Bank_Customer ()
  {
        Bank_Customer ();
  }
  public static void Deposit_Money ()
  {
  }
}
```

A Java class inheritance chart with the branch test coverage measurement result and the logic diagram as well as the control flow diagram of a class is shown in Fig. 23.2.

In my opinion, the future of UML should offer full automation for dynamic software modeling. The concept using graphic editors to draw graphical software documents is outdated. I absolutely agree with Harry Sneed's idea that "The important thing is that one model is enough – either the code or the diagrams. They should be reproducible from one another."

A virtual comparison of UML and NSE-UML proposed is shown in Table 23.1.

23.5.4 Possible Combination with CMMI (NSE-CMMI?)

CMMI is a great invention innovated by Watts S. Humphrey and his colleagues. In my opinion, the great contribution of CMMI is not only in helping DoD to choose qualified vendors and qualified products, and helping software organizations to improve their product quality and their management capability, but also in helping the entire software industry in the world to understand the importance of software quality improvement and software process improvement much better.

As described in Sect. 23.1 for solving the critical issues (low quality and productivity, high cost and risk) existing with software development today, partial solutions without bringing revolutionary changes to all areas of software engineering will not work well, so I believe in the future that CMMI will no longer focus on

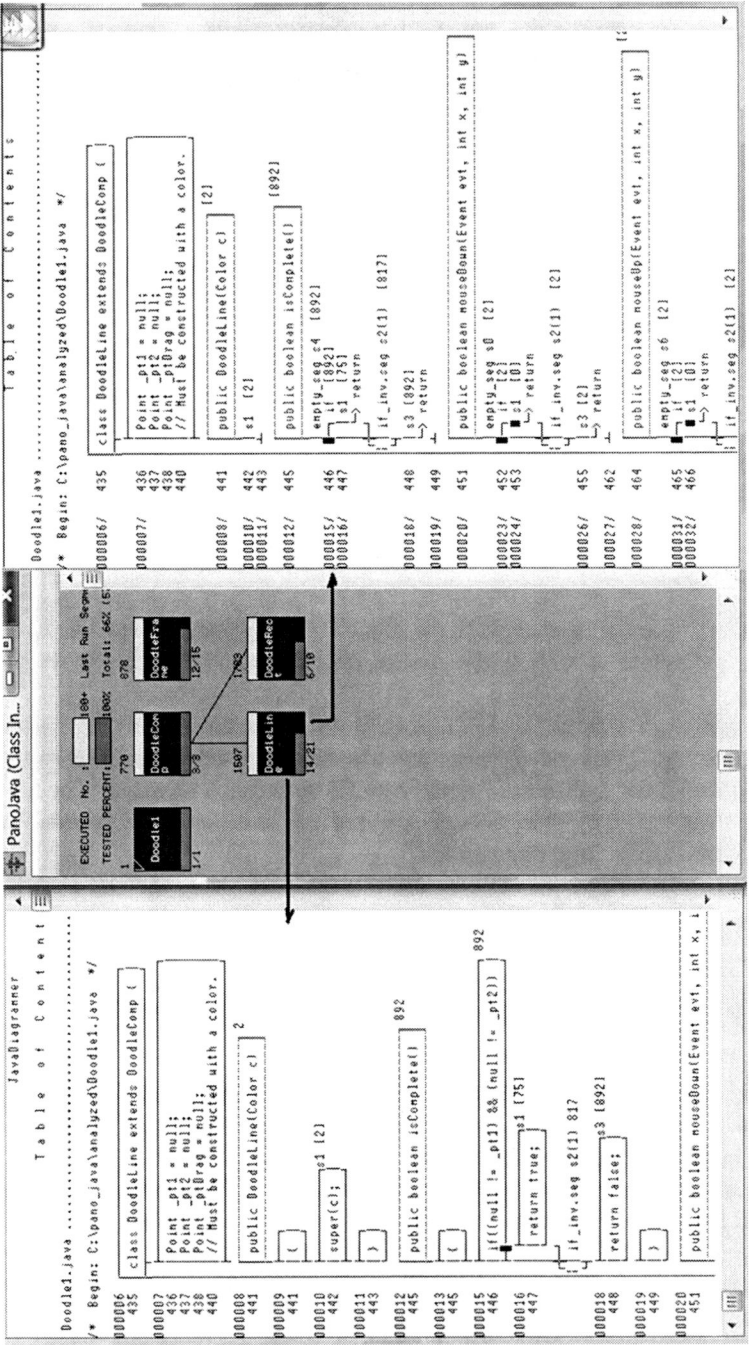

Fig. 23.2 A class inheritance chart with test coverage measurement result and the logic diagram as well as the control flow diagram of a class directly generated from a Java program

23.6 Possible Combination with Agile Software Development Approaches

Table 23.1 Virtual comparison between UML and NSE-UML proposed

Description	UML	NSE-UML
How many kinds of sources are used	Two: one in diagrams for human understanding of a software system; another one in textual format for computer understanding of a software system	One kind of source for both human understanding (in diagrams automatically generated from the source) and computer understanding of a software product (in textual format)
Modeling type	Static modeling: • Not traceable • Not executable • Not testable • The obtained models are statically existing	Dynamic modeling: • Traceable • Executable • Testable • The obtained models are dynamically existing – when a model/diagram is shown, the corresponding generator is always working and waiting for users' commands through the interface (using the model/diagram itself) for dynamically responding to users' requests
Supported software development methods	Top-down	Top-down and bottom-up: the models generated from dummy programs through forward engineering can be automatically updated through reverse engineering from the regular programs

software process improvement and management improvement only, but cover all the areas of software development and management.

The purpose of CMMI is to guide software organizations on "What to do" or "What should be done" rather than "How to do," while the purpose of NSE is for both – "What to do" and "How to do," so that NSE can be chosen by software development organizations as the powerful means for the implementation of the updated CMMI framework.

The combination of NSE and CMM – NSE-CMMI will bring revolutionary changes to CMMI as shown in Table 23.2.

23.6 Possible Combination with Agile Software Development Approaches

Advanced concepts and excellent ideas can be found from the Manifesto for Agile Software Development:

Individuals and interactions over processes and tools
Working software over comprehensive documentation

Table 23.2 A comparison between CMMI and NSE-CMMI

Comparison item	CMMI	NSE-CMMI
The software development foundation	Reductionism and the superposition principle that the whole of a complex system is the sum of its components, so that with it almost all software engineering tasks are performed partially and locally	Complexity science by complying with the essential principles of complexity science, particularly the Nonlinearity principle and the Holism principle, so that with NSE-CMMI almost all software development tasks will be performed holistically and globally
The first-order effect on software development	Process	People ["The fundamental characteristics of 'people' have a first-order effect on software development" (Alistair Cockburn)]
Does it intend to attack the essential issues in software development (complexity, conformity, changeability, and invisibility)?	No: 1. It does not really attack the essential issues related to software complexity, changeability, invisibility, and conformity. The process improvement suggested by CMMI misses the most important part: the improvement of the process models themselves 2. It does not really hit the software maintenance issue which takes about 75% of the software development effort	Yes: 1. It attacks the essential issues related to software complexity, changeability, invisibility, and conformity 2. It makes revolutionary changes to the software engineering paradigm, including the NSE process model 3. It hits both the software development issues and the maintenance issue, with side-effect prevention in the implementation of requirement changes or code modifications through a set of techniques and tools for many types of traceability to greatly reduce the effort and cost spent in software maintenance
Can the design documents keep consistency with the source code after software modification?	No: It is very difficult to perform without automated and self-maintainable facility for various traceabilities	Yes: It is supported by automated and self-maintainable facility for various traceabilities
Does it offer not only "what to do" but also "how to do?"	No: It mainly offers "what to do" only	Yes: It offers both
Does it provide a partial and local solution or a holistic and global solution for solving the major critical issues (low quality and productivity, high cost and risk) existing with today's software engineering?	Partial and local solution mainly in software development process improvement and project management improvement	Holistic and global solution by bringing revolutionary changes to almost all areas in software engineering based on complexity science, including the process model, the software development methodology, the software testing paradigm, software quality assurance paradigm, software visualization paradigm, software documentation paradigm, software project management paradigm, and software maintenance paradigm

(continued)

23.6 Possible Combination with Agile Software Development Approaches

Table 23.2 (continued)

Comparison item	CMMI	NSE-CMMI
Is the implementation of this model expensive?	Yes: There is a lack of suitable models and tools to support the implementation	No: No longer. With NSE-CMM, many suitable models and automated tools are available to support the implementation – for instance, almost all graphic design documents will be automatically generated and maintained. Rough estimation shows that it may take only 25% of the cost spent in the implementation of CMMI

Customer collaboration over contract negotiation
Responding to change over following a plan

XP (Extreme programming) is one of the most popular Agile software development approaches. About XP, Kent Beck pointed that:
"Here is a quick summary of each of the major practices in XP.

Planning game. Customers decide the scope and timing of releases based on estimates provided by programmers. Programmers implement only the functionality demanded by the stories in this iteration.

Small releases. The system is put into production in a few months, before solving the whole problem. New releases are made often anywhere from daily to monthly.

Metaphor. The shape of the system is defined by a metaphor or set of metaphors shared between the customer and programmers.

Simple design. At every moment, the design runs all the tests, communicates everything the programmers want to communicate, contains no duplicate code, and has the fewest possible classes and methods. This rule can be summarized as, 'Say everything once and only once.'

Tests. Programmers write unit tests minute by minute. These tests are collected and they must all run correctly. Customers write functional tests for the stories in the iteration. These tests should also run, although practically speaking, sometimes a business decision must be made comparing the cost of shipping a known defect with the cost of delay.

Refactoring. The design of the system is evolved through transformations of the existing design that keep all the tests running.

Pair programming. All production code is written by two people at one screen/keyboard/mouse.

Continuous integration. New code is integrated with the current system after no more than a few hours. When integrating, the system is built from scratch and all tests must pass or the changes are discarded.

Collective ownership. Every programmer improves any code anywhere in the system at any time if they see the opportunity.

Onsite customer. A customer sits with the team full time.

40-h Weeks. No one can work a second consecutive week of overtime. Even isolated overtime used too frequently is a sign of deeper problems that must be addressed.

Open workspace. The team works in a large room with small cubicles around the periphery. Pair programmers work on computers set up in the center.

Just rules. By being part of an Extreme team, you sign up to follow the rules. But they're just the rules. The team can change the rules at any time as long as they agree on how they will assess the effects of the change" [Bec01].

About the advantages and disadvantages, the creator of CMMI, Watts Humphrey, concluded that:

"Advantages for using XP:

1. Emphasis on customer involvement: a major help to projects where it can be applied.
2. Emphasis on teamwork and communication: as with the TSP, this is very important in improving the performance of just about every software team.
3. Programmer estimates before committing to a schedule: this helps to establish rational plans and schedules and to get the programmers personally committed to their schedules – a major advantage of XP and TSP.
4. Emphasis on responsibility for quality: unless programmers strive to produce quality products, they probably would not.
5. Continuous measurement: since software development is a people-intensive process, the principal measures concern people. It is therefore important to involve the programmers in measuring their own work.
6. Incremental development: consistent with most modern development methods.
7. Simple design: though obvious, worth stressing at every opportunity.
8. Frequent redesign or refactoring: a good idea but could be troublesome with any but the smallest projects.
9. Having engineers manage functional content: should help control function creep.
10. Frequent, extensive testing: cannot be overemphasized.
11. Continuous reviews: a very important practice that can greatly improve any programming team's performance (few programmers do reviews at all, let alone continuous reviews)."

Disadvantages for using XP:

1. "Code-centered rather than design-centered development: although the lack of XP design practices might not be serious for small programs, it can be disas-

23.6 Possible Combination with Agile Software Development Approaches

trous when programs are larger than a few thousand lines of code or when the work involves more than a few people.

2. Lack of design documentation: limits XP to small programs and makes it difficult to take advantage of reuse opportunities.
3. Producing readable code (XP's way to document a design) has been a largely unmet objective for the last 40-plus years. Furthermore, using source code to document large systems is impractical because the listings often contain thousands of pages.
4. Lack of a structured review process: when engineers review their programs on the screen, they find about 10–25% of the defects. Even with pair programming, unstructured online reviews would still yield only 20–40%. With PSP's and TSP's structured review process, most engineers achieve personal review yields of 60–80%, resulting in high-quality programs and sharply reducing test time.
5. Quality through testing: a development process that relies heavily on testing is unlikely to produce quality products. The lack of an orderly design process and the use of unstructured reviews mean that extensive and time-consuming testing would still be needed, at least for any but the smallest programs.
6. Lack of a quality plan: we have found with the TSP that quality planning helps properly trained teams produce high-quality products, and it reduces test time by as much as 90%. XP does not explicitly plan, measure, or manage program quality.
7. Data gathering and use: we have found with the TSP that, unless the data are precisely defined, consistently gathered, and regularly checked, they will not be accurate or useful. The XP method provides essentially no data-gathering guidance.
8. Limited to a narrow segment of software work: since many projects start as small efforts and then grow far beyond their original scope, XP's applicability to small teams and only certain kinds of management and customer environments could be a serious problem.
9. Methods are only briefly described: while some programmers are willing to work out process details for themselves, most engineers will not. Thus, when engineering methods are only generally described, practitioners will usually adopt the parts they like and ignore the rest. Kent Beck notes that, when the XP method fails in practice, this is usually the cause.
10. Obtaining management support: the biggest single problem in introducing any new software method is obtaining management support. The XP calls for a family of new management methods but does not provide the management training and guidance needed for these methods to be accepted and effectively practiced.
11. Lack of transition support: transitioning any new process or method into general use is a large and challenging task. Successful transition of any technology requires considerable resources, a long-term support program, and a measurement and analysis effort to gather and report results. I am not aware of such support for the XP" [Hum01].

23.6.1 Possible Combination with XP (NSE-XP?)

It is possible for NSE to combine with XP together to form NSE-XP to remove the disadvantages of XP pointed by Watts Humphrey.

1. Design-centered development: using the HAETVE technique for requirement development and the Synthesis Design and Incremental growing up (Implementation and Integration) technique for design for program development of all sizes.
2. Huge amounts of design documentation: automatically generated from dummy programs and regular programs.
3. Automatically generate the call graph, the logic diagram, and the control flow diagram to document large systems with bidirectional traceability to make the programs much easier to understand, test, and maintain. The generated graphical documents are virtually existing, often containing thousands of pages with almost no extra space needed.
4. Fully supported structured review process: when engineers review their programs on the screen, they will find that the generated holistic and traceable graphics are much more useful and efficient than PSP's and TSP's structured review.
5. Quality through defect prevention and defect propagation prevention: a development process that relies heavily on traditional testing is unlikely to produce quality products, so that with NSE-XP, the quality of a software product is ensured through defect prevention and defect propagation prevention mainly using the Transparent-box testing method which can be dynamically used in the entire software development lifecycle to combine functional testing and structural testing together with the capability to establish bidirectional traceability among the related documents and test cases and the source code.
6. Plan, measure, and manage program quality better: with NSE-XP, quality planning, quality measurement, and quality management are performed holistically for the entire product and in detail for each individual program module, from the first step down to the retirement of a software product, through defect prevention and side-effect prevention in the implementation of requirement changes and code modifications.
7. Data gathering and use: with NSE-XP, detailed information about data (such as the global variables and static variables) are collected and measured, including where they are defined, used, referred, changed, never been used, how many of them are used in each file or each class/module, etc. for helping users to check and keep their consistency.
8. Remove the limitation on program size: in fact with the virtual documentation and virtual diagramming techniques, NSE-XP can easily handle very big programs without size limitation.
9. Methods are clearly described: with NSE-XP, the NSE software development methodology based on the Generative Holism principle will be used to cooperate with the NSE nonlinear process model, the HAETVE technique for requirement development, and the Synthesis Design and Incremental growing up (Implementation and Integration) technique for product design.

10. Obtaining management support: remove the biggest single problem – NSE-XP can obtain management support by combining the product development process and the project management process together and making the management documents traceable with the implementation of requirements and the source code, so that managers can directly get first-hand information easily.
11. Transition support: transitioning NSE-XP into general use is easy because NSE-XP can be applied for new software development or added on to a product being developed using any other approach – only need to rewrite the test cases and set bookmarks to the related documents, all the other work can be performed automatically. With the support platform, Panorama++ that is integrated with many automated tools, almost no extra resources are needed. Since NSE has brought revolutionary changes to almost all areas of software engineering, it is possible for NSE-XP to help software organizations double their productivity and project success rate, halve their cost, and remove about 99.99% of defects. It will be easy for NSE-XP to obtain the necessary transition support.

23.7 Possible Combination with RUP (NSE-RUP?)

Many advanced concepts and ideas have been implemented into RUP (Rational Unified Process), particularly the concepts of Use Case driven, Architecture-centric, iterative, and incremental development.

But as Scott Ambler pointed out, "The Unified Process suffers from several weaknesses. First, it is only a development process… it misses the concept of maintenance and support… It's important to note that development is a small portion of the overall software life cycle. The relative software investment that most organizations make is allocating roughly 20% of the software budget for new development, and 80% to maintenance and support efforts" [Amb05].

For overcoming this major drawback, a possible combination (NSE-RUP?) is proposed and shown in Fig. 23.3.

As shown in Fig. 23.3: (1) the proposed combination model (NSE-RUP) not only has the inception phase, elaboration phase, construction phase, and transition phase, but also has the maintenance phase and (2) the proposed combination model supports two-way iteration with possible upstream movement through various traceabilities, such as refactoring for a highly complex module to further divide it to several small modules performed with side-effect prevention through various traceabilities (of course, refactoring can also be done in a forward engineering approach to modify the design first).

23.8 Support for CBSE

According to the Generative Holism principle, the whole of a complex system should come earlier (as an embryo) than its components, then it grows up with its components – the integration should be performed by adding one module at a time (see Chap. 12).

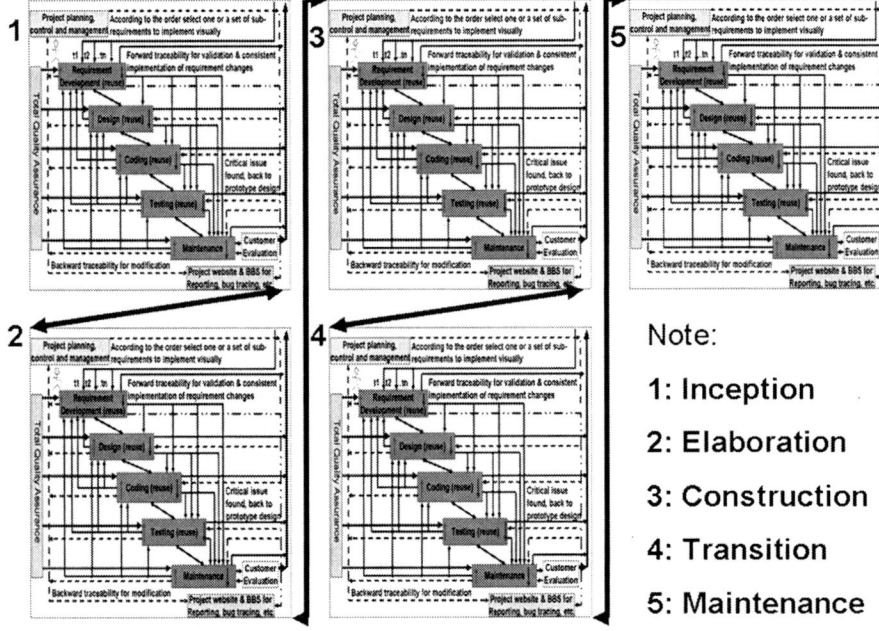

Fig. 23.3 A proposed combination of NSE and RUP (NSE-RUP?)

Computer software products are nonlinear systems, so that any local and small change may bring unexpected effects to the entire system – "Butterfly-effects", so that a reusable software component should be designed like "Broken Limbs" rather than "Artificial Limbs" – a reusable component must be qualified as a Broken Limb with self-adaptive capability – at least no negative effects on the system quality, no overuse of the system memory, no memory leaks, no negative effects on the performance, fully tested with test cases for verification, and fully fulfills the functionality required.

A reusable component should be tested using the MC/DC metric with memory leak check, performance measurement, and quality measurement.

23.9 Summary

It is recommended to apply NSE for new software product design to get the maximum benefit. But NSE can also be applied to a software product being developed using other approaches by setting bookmarks in the related documents and rewriting the test case files – the other work can be performed by the NSE support platform Panorama++.

NSE can be applied by itself to make it possible for NSE to help users double their productivity, halve their cost, and remove 99.99% defects in their products. NSE can also be combined with UML, CMMI, XP, and RUP to benefit their users greatly.

23.10 Points and Questions to Ponder

(a) Why is it that "The important thing is that one model is enough – either the code or the diagrams. They should be reproducible from one another"?
(b) How to realize that "One model is enough – either the code or the diagrams. They should be reproducible from one another"?
(c) Complete a small software project with NSE and the NSE support platform Panorama++.

23.11 Further Reading and Information Source

(a) Wikiversity. Unsolved problems in software engineering. http://en.wikiversity.org/wiki/Unsolved_problems_in_software_engineering
(b) Scientific Research Publishing (SRP: http://www.scirp.org). J Software Eng Appl. http://www.scirp.org/journal/jsea/

References

[Amb05] Ambler S W (2005) A Manager's Introduction to The Rational Unified Process (RUP), Ambysoft.
[Arl06] Arlow J, Neustadt I (2006) UML 2 and the unified process: practical object-oriented analysis and design, 2nd edn. Pearson Education, Upper Saddle River
[Hum01] Humphrey W (2001) "Comments on eXtreme Programming", eXtreme Programming Pros and Cons: what questions remain? IEEE Computer Society Dynabook. **http://www.computer.org/SEweb/Dynabook/HumphreyCom.htm**
[Sne07] Sneed H (2007) The drawbacks of model driven software evolution. In: IEEE CSMR 07- Workshop on Model-Driven Software Evolution (MoDSE2007), Amsterdam, 20 Mar 2007. http://www.sciences.univ-nantes.fr/MoDSE2007/; http://www.cs.vu.nl/csmr2007/workshops/I-%20Summary%20Description.pdf
[Bec01] Beck K (2001) "XP practices", eXtreme Programming Pros and Cons: what questions remain? IEEE Computer Society Dynabook. http://www.computer.org/SEweb/Dynabook/XPPracSdb.htm

Chapter 24
Candidates of "Silver Bullet"

> *Let us consider the inherent properties of this irreducible essence of modern software systems: complexity, conformity, changeability, and invisibility... Therefore it appears that the time has come to address the essential parts of the software task, those concerned with fashioning abstract conceptual structures of great complexity.*
>
> Frederick P. Brooks, Jr.

Today software has become the driving force for the development of all kinds of businesses, engineering, sciences, and the global economy – software reliability affects not only our lives and the global economy today but also affects the future of mankind. As pointed by David Rice, like cement, software is everywhere in modern civilization. One cannot live in modern civilization without touching, being touched by, or depending on software in one way or another [Ric08].

But unfortunately, software itself is not well engineered. For instance, the total economic cost of insecure software is very high: $180 billion a year in the USA [Ros08].

For addressing the essential issues in software engineering, NSE is established and introduced in this book from Chaps. 3 to 23.

This chapter summarizes the NSE nonlinear software engineering paradigm: the whole and its components, the main features, the major differences between it and the old-established software engineering paradigm, and the qualification as candidates of "Silver Bullets" to slay software "werewolves" – a monster of missed schedules, blown budgets, and flawed products.

24.1 Is "The Mythical Man-Month" an Outcome of Linear Thinking, Reductionism, and Superposition Principle?

"The Mythical Man-Month" is a popular book written by Frederick P. Brooks, Jr.

24.1.1 A Great book

"The Mythical Man-Month" is a great book from which I learned a lot on software engineering. I will continuously learn more from it.

What I have particularly learned from it include the following:

- "Plan the system for change"
- "Plan the organization for change"
- "Testing the specification"
- "Control changes"
- "Add one component at a time"
- "Quantize updates"
- "Self-documenting programming techniques find their greatest use and power in high-level languages used with online systems, which are the tools one should be using"
- "Incremental development – grow, not build, software"
- "Much of software architecture, implementation, and realization can proceed in parallel"
- "Timely updating is of critical importance"
- "There has to be upstream movement"
- "To keep documentation maintained, it is crucial that it be incorporated in the source program, rather than kept as a separate document"
- "People are everything (well, almost everything)"
- "A program should be shipped with a few test cases, some for valid input data, some for borderline data, and some for invalid data"
- "The fundamental problem with program maintenance is that fixing a defect has a substantial (20–50%) chance of introducing another. So the whole process is two steps forward and one step back"
- "Why aren't defects fixed more cleanly? First, even a suitable defect shows itself as a local failure of some kind. In fact it often has system-wide ramifications, usually nonobvious. Any attempt to fix it with minimum effort will repair the local and obvious, but unless the structure is pure or the documentation very fine, the far reaching effects of the repair will be overlooked. Second, the repairer is usually not the man who wrote the code"
- "All repairs tend to destroy the structure, to increase the entropy and disorder of the system"
- "Less and less effort is spent on fixing original design flaws; more and more is spent on fixing flaws introduced by early fixes. As time passes, the system becomes less and less well-ordered. Sooner or later the fixing ceases to gain any ground. Each forward step is matched by a backward one"
- "Clearly, methods of designing programs so as to eliminate or at least illuminate side effects can have an immense payoff in maintenance cost" [Bro95].

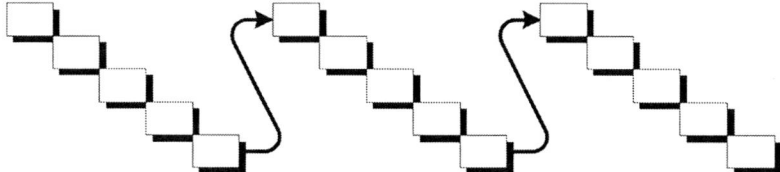

Fig. 24.1 Incremental model [GSAM03]

24.1.2 Limitation

Unfortunately, the old-established software engineering paradigm is based on linear thinking, reductionism, and the superposition principle that the whole of a system is the sum of its components, so that with it almost all software engineering tasks and activities are performed linearly, partially, and locally. It impacts and limits the book "The Mythical Man-Month" too – the book itself is also an outcome of linear thinking, reductionism, and the superposition principle.

In the 1995 edition of "The Mythical Man-Month" book, Frederick P. Brooks, Jr. criticized his 1975 edition of the book that **"Don't Build One to Throw Away – The Waterfall Model Is Wrong! …The biggest mistake in the 'Build one to throw away' concept is that it implicitly assumes the classical sequential or waterfall model of software construction. …Chapter 11 is not the only one tainted by the sequential waterfall model; it runs through the book, beginning with the scheduling rule in Chapter 2."**

Unfortunately, in the 1995 edition of the book, it also assumes sequential model – "An Incremental – Build Model" **which is "a series of Waterfalls"** [GSAM03] as shown in Fig. 24.1.

Comparing the "Incremental – Build Model" with the traditional waterfall model, besides that the Incremental – Build Model can help in reducing risk and waiting time in some degree, it keeps all of the major drawbacks of the one-time waterfall model – complying with the linear sequence, the defects introduced into a software product in the upper phases can easily propagate to the lower phases, making the final defect removal cost increase more than 100 times; the requirement changes and code modifications are implemented locally and blindly without support of bidirectional traceability, making software maintenance take 75% or more of the total effort and total cost in a software product development, etc.

24.2 Is the "No Silver Bullet" Conclusion Outdated?

As Brooks' law states as "No Silver Bullet": "There is no single development, in either technology or management technique, which by itself promises even one order-of-magnitude improvement within a decade in productivity, in reliability, in simplicity." [Bro95-P179] "Adding manpower to a late software project makes it later." [Bro95-P274]

"Of all the monsters who fill nightmares of our folklore, none terrify more than werewolves, because they transform unexpectedly from the familiar into horrors. For these, we seek bullets of silver can magically lay them to rest. The familiar software project has something of this character (at least as seen by the nontechnical manager), usually innocent and straightforward, but capable of becoming a monster of missed schedules, blown budgets, and flawed products." [Bro95-P180]

"Not only are there no silver bullet now, the very nature of software makes it unlikely that there will be any – no inventions that will do for software productivity, reliability, and simplicity what electronics, transistors, and large-scale integration did for computer hardware. **We cannot expect ever to see twofold gains every 2 years.**" [Bro95-P181].

Brooks' law and his related conclusions are only suitable to the old-established software engineering paradigm (based on linear process, reductionism, and superposition principles) where

- "Testing the specification" – is not dynamically supported/implemented.
- "Control changes" – is performed partially and locally.
- "Add one component at a time" – is not systematically supported.
- "Quantize updates" – is not fully supported.
- "Self-documenting programming techniques find their greatest use and power in high-level languages used with online systems, which are the tools one should be using." – it is not well supported.
- "Incremental development – grow, not build, software" – is not really supported.
- "Timely updating is of critical importance," is not systematically supported.
- "There has to be upstream movement," but people still insist "no upstream movement at all."
- "People are everything (well, almost everything)," but some models still focus on process improvement.
- "A program should be shipped with a few test cases, some for valid input data, some for borderline data, and some for invalid data" – is not really supported.
- "The fundamental problem with program maintenance is that fixing a defect has a substantial (20–50%) chance of introducing another. So the whole process is two steps forward and one step back." – is true because the implementation of requirement changes and code modifications are performed partially and locally.
- "Why aren't defects fixed more cleanly? First, even a suitable defect shows itself as a local failure of some kind. In fact it often has system-wide ramifications, usually non-obvious. Any attempt to fix it with minimum effort will repair the local and obvious, but unless the structure is pure or the documentation very fine, the far reaching effects of the repair will be overlooked. Second, the repairer is usually not the man who wrote the code" – is still true.
- "All repairs tend to destroy the structure, to increase the entropy and disorder of the system." – is true because the repairing process is performed blindly and locally.

24.2 Is the "No Silver Bullet" Conclusion Outdated? 601

- "Less and less effort is spent on fixing original design flaws; more and more is spent on fixing flaws introduced by early fixes. As time passes, the system becomes less and less well-ordered. Sooner or later the fixing ceases to gain any ground. Each forward step is matched by a backward one." – is true too.
- "Clearly, methods of designing programs so as to eliminate or at least illuminate side effects can have an immense payoff in maintenance cost." – but it has not been realized with the old-established software engineering paradigm.

But Brooks' law ("No Silver Bullet") is no longer suitable for NSE, the nonlinear software engineering paradigm based on complexity science, where

- "Testing the specification" is dynamically supported by NSE software testing paradigm.
- "Control changes" is performed holistically and globally with side-effect prevention.
- "Add one component at a time" is systematically supported by complying with the Generative Holism principle.
- "Quantize updates" is realized through incremental development and incremental integration.
- "Self-documenting programming techniques find their greatest use and power in high-level languages used with online systems, which are the tools one should be using." – with NSE, a set of Assisted Online Agents (AOA) will be delivered with the corresponding software product, including the tools for automatically generating many graphical documents directly from the source code or the dummy program used in high-level product design.
- "Incremental development – grow, not build, software" is fully supported by complying with the Generative Holism principle.
- "Timely updating is of critical importance" – now it is one of the key features of NSE.
- "There has to be upstream movement" – it is the key feature of NSE process model supported by various traceabilities.
- "People are everything (well, almost everything)" – NSE treats people as the first positive order factor in software engineering to fully support them with many advanced techniques and tools, but also treat people as the first negative order factor that almost all defects are introduced into a software product by people, so that NSE provides many techniques and tools to prevent people from introducing defects into a software product.
- "A program should be shipped with a few test cases, some for valid input data, some for borderline data, and some for invalid data" – is fully supported with NSE.
- "The fundamental problem with program maintenance is that fixing a defect has a substantial (20–50%) chance of introducing another. So the whole process is two steps forward and one step back." – it is not suitable for NSE now, because the implementation of requirement changes is performed holistically and globally with side-effect prevention supported by various traceabilities.
- "Why aren't defects fixed more cleanly? First, even a suitable defect shows itself as a local failure of some kind. In fact it often has system-wide ramifications, usually non-obvious. Any attempt to fix it with minimum effort will repair the

local and obvious, but unless the structure is pure or the documentation very fine, the far reaching effects of the repair will be overlooked. Second, the repairer is usually not the man who wrote the code" – it is not suitable for NSE now, because the implementation of requirement changes is performed holistically and globally, no matter who maintains the product.
- "All repairs tend to destroy the structure, to increase the entropy and disorder of the system." – it is no longer true because with NSE all repairs are performed with side-effect prevention through various traceabilities.
- "Less and less effort is spent on fixing original design flaws; more and more is spent on fixing flaws introduced by early fixes. As time passes, the system becomes less and less well-ordered. Sooner or later the fixing ceases to gain any ground. Each forward step is matched by a backward one." – it is no longer true because with NSE software changes are performed holistically and globally with side-effect prevention.
- "Clearly, methods of designing programs so as to eliminate or at least illuminate side effects can have an immense payoff in maintenance cost." – with NSE, it is not good enough, because eliminating or illuminating side effects still requires time, resources and costs! The better solution offered by NSE is to directly prevent the side effects through various traceabilities.

24.3 The First Candidate of "Silver Bullet"

Here the first proposed candidate of "Silver Bullet" is the various NSE automated and self-maintainable traceability techniques, including the most important one shown in Fig. 24.2 – the traceability among related documents and test cases and source code.

As shown in Fig. 24.2, the traceability between test cases and the source code is established through the use of Time Tags automatically inserted into both the test case description part and the program test coverage database after test case execution for mapping test cases and the corresponding source code; the traceability extended to include the related documents is established using some keyword such as @WORD@, @HTML@, @EXCEL@, @PDF@, @BAT@ written in the comment part of a test case to indicate the format of a document, followed by the file path of the document and the corresponding bookmark for automatically showing the document traced from the specified location. The keyword @BAT@ is used for dynamic traceability to directly run a batch file for special applications such as playing the captured GUI testing operations back or running a third-party tool for handling the corresponding document generated by that third-party tool.

> "Software traceability can help bring software development into the twenty-first century. It reduces costs, gives better visibility and adequate test coverage, and helps software engineers meet customer needs. Changes can be implemented much faster and new projects can be estimated more accurately."
>
> Rick Coffey, Document Control Supervisor, Tyco Healthcare/Mallinckrodt

24.3 The First Candidate of "Silver Bullet"

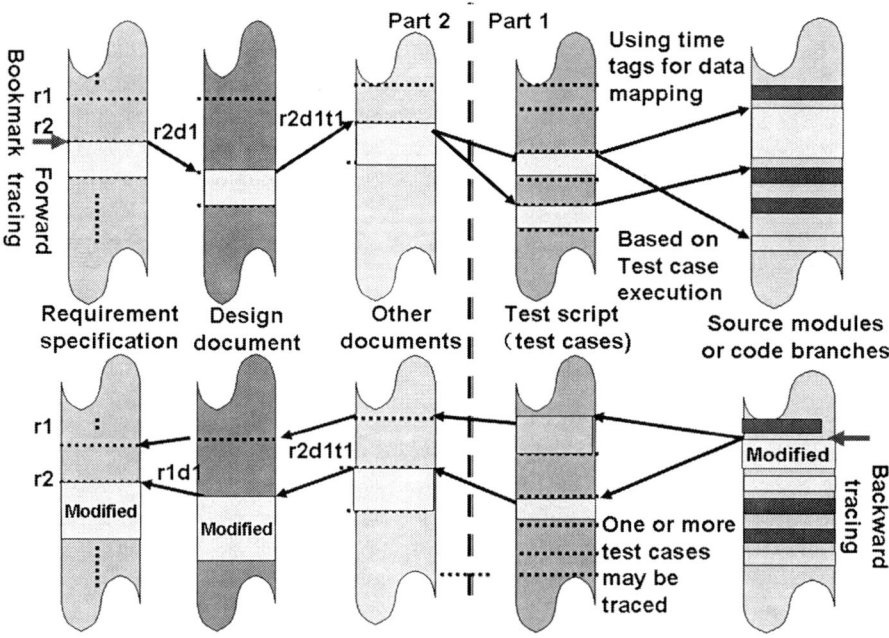

Fig. 24.2 Automated, bidirectional, and self-maintainable traceability among the documents and the test cases and the source code of a software product

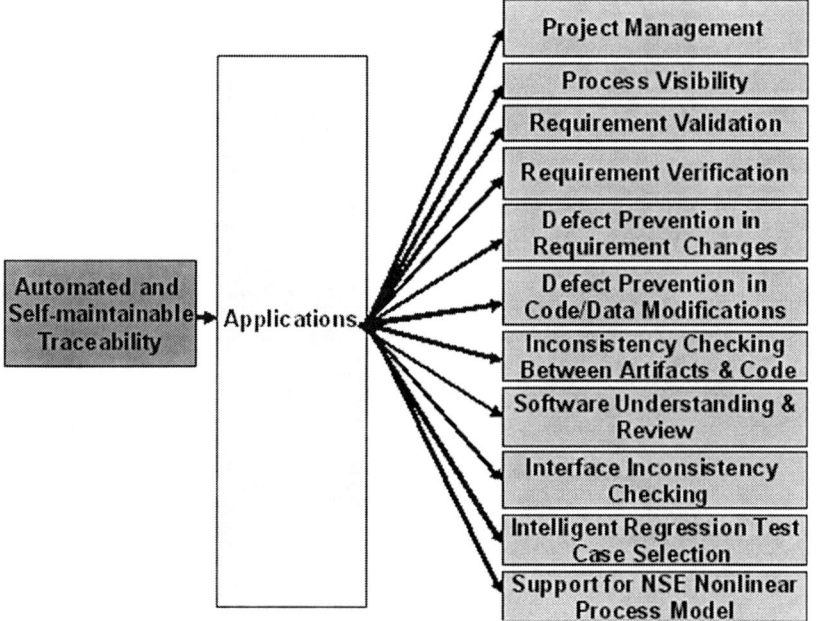

Fig. 24.3 Sample applications of the automated and self-maintainable traceability

The automated and bidirectional traceability technique, "which by itself promises even one order-of-magnitude improvement within a decade in productivity, in reliability, in simplicity" – Software traceability can be applied widely in the entire software development process as shown in Fig. 24.3, particularly in requirement validation and verification, defect propagation prevention, side-effect prevention in the implementation of requirement changes and code modifications, semiautomated inspection. Regarding reliability, NSE traceability techniques can realize one order-of-magnitude improvement immediately without waiting for a decade.

24.4 The Second Candidate of "Silver Bullet"

Here the second proposed candidate of "Silver Bullet" is the NSE defect prevention and defect propagation prevention technique mainly based on inspection using traceable documents and the source code, and the Transparent-box testing method combining functional testing and structural testing together as shown in Fig. 24.4, which can be used in the entire software development lifecycle. After the execution of test cases using the Transparent-box method, a facility for bidirectional traceability will be automatically established for helping users remove the inconsistent defects.

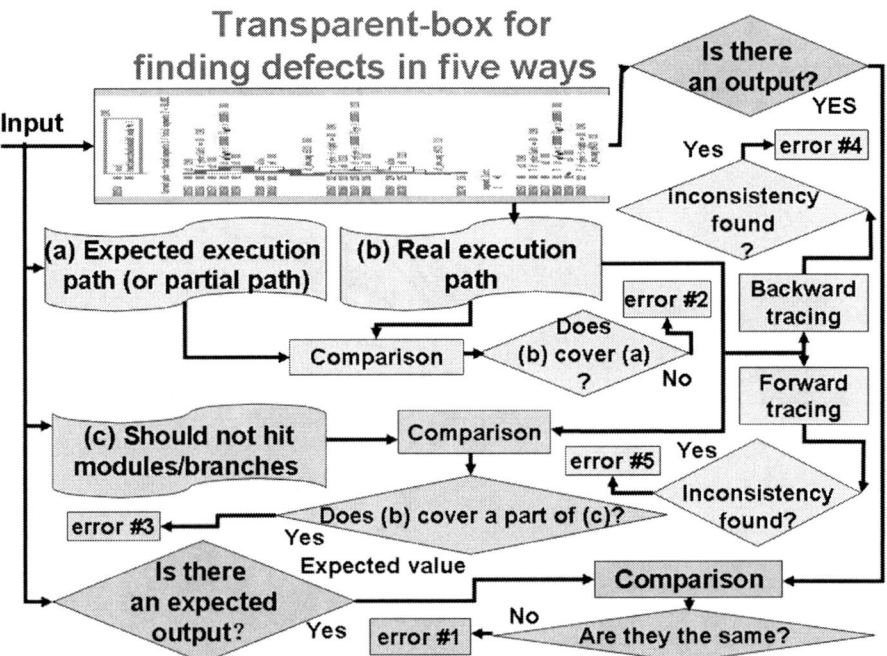

Fig. 24.4 The Transparent-box software testing method

The reasons are as follows:

1. The Transparent-box testing method combines functional and structural testing together seamlessly, for each input it not only checks whether the output (if any, can be none) is the same as what is expected but also checks whether the execution path is the same as the expected one indicated in J-Flow – so it can find more defects efficiently, including logic defects.
2. Having a real output is no longer a condition to dynamically use this technique, so that it can be dynamically used in the requirement development phase and the design phase to find out defects introduced in those phases before coding to ensure the quality of a software product from the first step.
3. It can also establish automated and bidirectional traceability among the related documents, the test cases, and the source code to help software developers to check and remove inconsistent defects.
4. Based on the traceability established by this technique, software maintenance can be performed holistically and globally with side-effect prevention – a key to ensure the quality of a software product being maintained.

Regarding reliability, NSE defect prevention and defect propagation prevention techniques can realize one order-of-magnitude improvement immediately without waiting for a decade. Figure 24.5 shows the differences of the defect finding efficiency between NSE testing paradigm using the Transparent-box testing method and the old-established software testing paradigm.

24.5 Can the "Silver Bullet" Defined by Brooks Slay the "Werewolves" Defined by Him?

No! It is impossible. The reasons are:

1. The software "Werewolves" is defined by Brooks as that: "Of all the monsters who fill nightmares of our folklore, none terrify more than werewolves, because they transform unexpectedly from the familiar into horrors" and "The familiar software project has something of this character (at least as seen by the non-technical manager), usually innocent and straightforward, but capable of becoming a monster of **missed schedules, blown budgets, and flawed products.**"
2. The "Silver Bullet" is defined by Brooks that "a **single development, in either technology or management technique, which by itself promises even one order-of-magnitude improvement within a decade in productivity, in reliability, in simplicity.**"

Here it is clear that, **the "werewolves" is a monster of missed schedules, blown budgets, and flawed products** – these issues relate to the entire software engineering paradigm, including the process models, the software development methodology, the quality assurance paradigm, the software testing paradigm, the

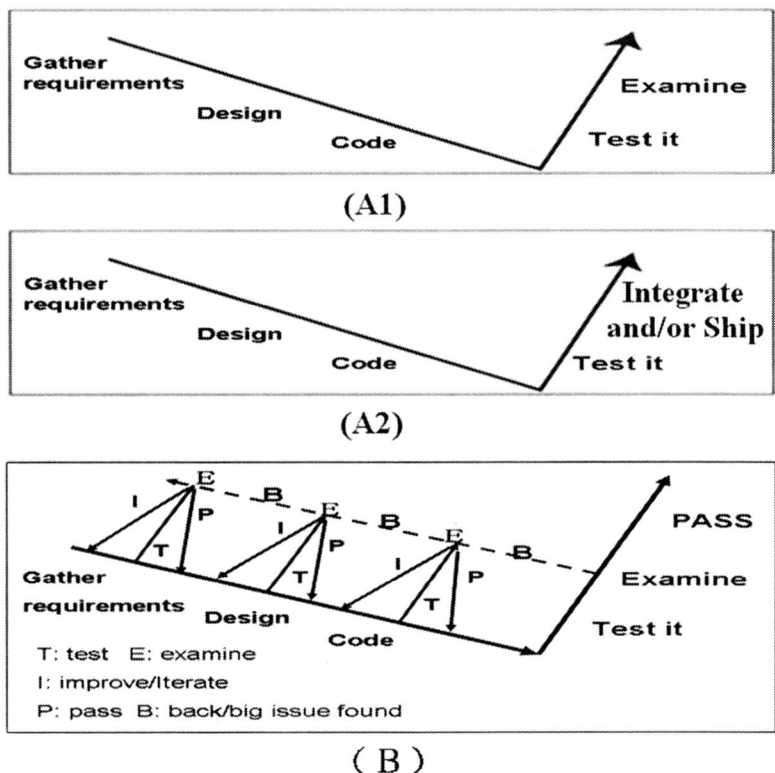

Fig. 24.5 A comparison of defect finding efficiency between the old-established software testing paradigm (A1 for iterative development, A2 for incremental development [Coc08]) and NSE software testing paradigm (B)

project management paradigm, the software documentation paradigm, the self-organization capability, the Capability Maturity of the organization and the team, and more.

But the "Silver Bullet" defined by Brooks **is a "single development, in either technology or management technique,** which by itself promises even one order-of-magnitude improvement within a decade in productivity, in reliability, in simplicity." – **how can a single technology or management technique solve the issues of missed schedules, blown budgets, and flawed products** which are not only technology or technique issues but strongly related to people and project management?

3. **In theory, it is impossible**

   According to complexity science, the whole of a complex system is greater than the sum of its parts, the characteristics and behaviors of the whole of a complex system emerge from the interaction of its components and cannot be inferred simply from the behavior of its individual components. It means that no matter how excellent is the single development, in either technology or management technique, the characteristics and behaviors of any individual part cannot be inferred simply by the whole of the software engineering paradigm, so that it is impossible for the single development, in either technology or management technique to slay the software monster of missed schedules, blown budgets, and flawed products – those problems come from the whole of the old-established software engineering paradigm.

4. **From practices, it is impossible**

   After analyzing more than 12,000 software projects, Capers Jones pointed out in his article titled "Social and Technical Reasons for Software Project Failures" that "Major software projects have been troubling business activities for more than 50 years. Of any known business activity, software projects have the highest probability of being canceled or delayed. Once delivered, these projects display excessive error quantities and low levels of reliability. Both technical and social issues are associated with software project failures. Among the social issues that contribute to project failures are the rejections of accurate estimates and the forcing of projects to adhere to schedules that are essentially impossible. Among the technical issues that contribute to project failures are the lack of modern estimating approaches and the failure to plan for requirements growth during development. However, it is not a law of nature that software projects will run late, be canceled, or be unreliable after deployment. A careful program of risk analysis and risk abatement can lower the probability of a major software disaster." [Jon06] – it means that the issues of missed schedules, blown budgets, and flawed products are not only technology issues but also social issues, can never be solved by a single development, in either technology or management technique.

   With the same reasons, focusing on Software Process Improvement and project management improvement only without changing the entire software engineering paradigm will not be able to slay the software "werewolves" too.

24.6 What Kind of "Silver Bullet" Can be Used to Slay the "Werewolves" Defined by Brooks?

Slaying the software "werewolves" (a monster of missed schedules, blown budgets, and flawed products) is equal to solving the most critical problems existing with today's software development – low quality and productivity, and high cost and risk, so before answering the question, let us consider what in the old-established software engineering paradigm allow the software "werewolves" to exist:

(a) The existing process models (no matter if they are waterfall models, incremental development models **which is "a series of Waterfalls"** [GSAM03], or iterative development models on which each time of an iteration is a waterfall) which are based on linear thinking, reductionism, and the superposition principle that the whole of a complex system is the sum of its components, so that with them almost all software process tasks and activities are performed linearly, partially, and locally, making the defects introduced into a software product upstream easy to propagate down to the maintenance phase and the final defect removal cost increase tenfold many times.

(b) The software development methodologies which are based on linear process, reductionism, superposition, and constructive holism principles, so that with them almost all software development tasks and activities are performed linearly, partially, and locally for the components of a software product first, then the components are "assembled" to form the whole of the software product – this makes the quality of the software product very hard to ensure, and software maintenance much harder to perform.

(c) The software testing paradigm which ignores the fact that most critical software defects are introduced to a software product in the requirement development phase and the product design phase, and can only be dynamically used after production, so that NIST (National Institute of Standards and Technology) concluded that "Briefly, experience in testing software and systems has shown that testing to high degrees of security and reliability is from a practical perspective not possible. Thus, one needs to build security, reliability, and other aspects into the system design itself and perform a security fault analysis on the implementation of the design." (**"Requiring Software Independence in VVSG 2007: STS Recommendations for the TGDC," November 2006**, http://www.vote.nist.gov/DraftWhitePaperOnSIinVVSG2007-20061120.pdf.). Even if a defect has been found through dynamic software testing, the defect removal cost will increase tenfold several times.

(d) The quality assurance paradigm based on inspection and software testing after production, which violates W. Edwards Deming's product quality principle, *"Cease dependence on inspection to achieve quality. Eliminate the need for inspection on a mass basis by building quality into the product in the first place."* [Dem86].

(e) The software visualization paradigm mainly supports visual modeling only and does not make the entire software development and maintenance process visible, so that software engineers and maintainers need to spend much more time to understand and maintain a software product.

(f) The software documentation paradigm which are not traceable with the source code, and often are not consistent with the source code after code modification, making the software hard to understand and hard to maintain.

(g) The software maintenance paradigm with which the implementation of requirement changes and code modifications are performed partially and locally, so that fixing a defect has a substantial (20–50%) chance of introducing another [Bro95], making a software product unstable day by day.
(h) The project management paradigm with which software project management process and the software development process are separated, the software management documents are not traceable to the implementation of requirements and the source code, making the schedule hard to meet, and the budget hard to control.
(i) The corresponding software development techniques and tools are designed to work with the linear process models; it is hard to use them to handle a complex software which is nonlinear.
(j) The entire software engineering paradigm is based on linear thinking, reductionism, and the superposition principle, and it is hard to efficiently handle a nonlinear software system.

It means that almost all parts of the old-established software engineering paradigm are allowing the possibility for the software werewolves to exist.

Now it is the time we can answer the question: only such a Silver Bullet can be used to slay software werewolves:

1. It is based on complexity science, complying with the essential principles of complexity science, particularly the Nonlinearity principle and the Holism principle, so that with it almost all software development tasks and activities are performed holistically and globally.
2. It not only can bring revolutionary changes to all parts of the software engineering paradigm, but also can make the required characteristics and behaviors emerge from the interaction of all of its parts.

In fact, the qualified "Silver Bullet" being able to slay software "werewolves" (a monster of missed schedules, blown budgets, and flawed products) means a complete revolution in software engineering through a paradigm-shift from the old one based on linear thinking, reductionism, and the superposition principle to a new one based on nonlinear thinking and complexity science.

24.7 The Third Candidate of "Silver Bullet": The Entire NSE Paradigm

As discussed in Sect. 24.5 that the "Silver Bullet" defined by Brooks cannot slay the "Werewolves" defined by him, so that we need another Silver Bullet which is not "a single development, in either technology or management technique," but a qualified Silver Bullet being able to slay the software werewolves – a monster of missed schedules, blown budgets, and flawed products.

24.7.1 What Is NSE: The Whole and Its Components

As described in Chap. 1, NSE (Nonlinear Software Engineering paradigm) is a new revolutionary software engineering paradigm based on complexity science by complying with the essential principles of complexity science, particularly the Nonlinearity principle and the Holism principle, so that with NSE almost all software engineering tasks and activities are performed holistically and globally.

NSE is established with the objectives to revolutionarily solve the critical problems existing with the old-established software engineering paradigm. Those critical problems can be summarized as follows:

(a) **Incomplete** – For instance, there is no defined process model and support for software maintenance which takes 75% or more of the total effort and cost for a software product.
(b) **Unreliable** – The quality of a software product mainly depends on inspection and testing after production which has been proven impossible to ensure high quality.
(c) **Invisible** – The existing visualization methods, techniques, and tools do not offer the capability to make the entire software development lifecycle visible, the generated charts and diagrams are not holistic and not traceable.
(d) **Inconsistent** – The documents and the source code are not traceable to each other and not consistent after code modification again and again.
(e) **Unchangeable** – The implementation of requirement change or code modification is performed locally and blindly with high risks.
(f) **Not maintainable** – Software maintenance is performed partially and locally without support for bidirectional traceability to prevent side effects, so that each code modification will have a 20–50% of chance to introduce new defects into the software product being maintained.
(g) **Low productivity and quality** – Most resources are spent in inefficient software maintenance, the quality cannot be ensured with the blind and local implementation of software changes.
(h) **High cost and risk** – Most costs are spent in blind and local maintenance of the software products, which makes software product unstable day by day in responding to needed changes.
(i) **Low project success rate** – It is still less than 30% for projects with budgets over $1 million.
(j) **Often the software projects developed with the old-established software engineering paradigm are capable of becoming a monster of missed schedules, blown budgets, and flawed products** – because the old-established software engineering paradigm is based on linear thinking, reductionism, and the superposition principle.

It is clear that those problems are related to the entire software engineering paradigm with all of its components, including the process models, the software development methodologies, the visualization paradigm, the software testing

24.7 The Third Candidate of "Silver Bullet": The Entire NSE Paradigm

paradigm, the quality assurance paradigm, the documentation paradigm, the maintenance paradigm, the project management paradigm, and the related techniques and tools. It means that a local and partial solution will not work – we need a holistic and global solution in almost all aspects of software engineering: a complete revolution.

For solving those critical problems existing with today's software development efficiently, NSE is established. The essential difference between the old-established software engineering paradigm and NSE is how to handle the relationship between the whole and its parts of a software system. **The former adheres to the reductionism principle and superposition principle that the whole is the sum of its parts,** so that nearly all software development tasks/activities are performed locally, such as the implementation of requirement changes. **The latter complies with the Holism principle of complexity science, that a software product is a Complex Adaptive System (CAS [Hol95]) having multiple interacting agents (components), of which the overall behavior and characteristics cannot be inferred simply from the behavior of its individual agents but emerge from the interaction of its parts**, so that with NSE nearly all software development tasks/activities are performed globally and holistically to prevent defects in the entire software lifecycle[Xio09-1], [Xio09-2].

NSE brings revolutionary changes to almost all aspects in software engineering, including the following:

- **The foundation** (see Chaps. 3 and 4)
  From: that based on linear thinking and the reductionism principle and superposition principle that the whole is the sum of its parts, so that nearly all software development tasks/activities are performed linearly, partially, and locally, such as the implementation of requirement changes.
  To: that based on nonlinear thinking and complexity science – to comply with the essential principles of complexity science, particularly the Nonlinearity Principle and the Holism Principle that the whole of a complex system is greater than the sum of its parts – the characteristics and the behavior of a complex system is an emergent property of the interactions of its components (agents), so that with NSE nearly all software development tasks/activities are performed nonlinearly, holistically, and globally to prevent defects in the entire software lifecycle – for instance, if there is a need to change a requirement, with NSE and the support platform Panorama++ the implementation of the change will be performed nonlinearly, holistically, and globally through various bidirectional traceabilities: (1) Performs forward tracing for the requirement change (through the corresponding test cases) to determine what modules should be modified. (2) Performs backward tracing to check the related requirements of the modules to be modified for preventing requirement conflicts – sometimes a module is used for the implementation of multiple requirements. (3) Checks what other modules may also need to be changed with the modification by tracing the modules to find all related modules on the corresponding call graph shown

in J-Chart. (4) Checks where the global variables and static variables may be affected by the modification. (5) After modification, checks all related statements calling the modified module for preventing inconsistency defects between them. (6) Performs efficient regression testing through backward tracing from the modified module to find the related test cases. (7) Performs backward tracing to find and modify inconsistent documents after code modification.

- **The process model(s)** (see Chap. 8)
  From: linear ones based on linear thinking and the reductionism principle and superposition principle, including the waterfall model, the incremental development models, the iterative development models, or the incremental and iterative development models with which there is only one track in one direction – no upstream movement at all, always going forward from the upper phases to the lower phases, so that defects introduced in the upper phases will easily propagate to the lower phases to make the defect removal cost greatly increase.
  To: a nonlinear one (called the NSE process model) based on nonlinear thinking and complexity science with which there are multiple tracks in two directions through various traceabilities to prevent defects and defect propagation, so that experience and ideas from each downstream part of the construction process may leap upstream, sometimes more than one stage, and affect the upstream activity. With NSE, the software development process and software maintenance process are combined together closely, the software development process and the project management process are also combined together closely so that the project management documents are traceable with the implementations of software requirements and the source code. With the NSE process model, requirement validation and verification can be done easily through forward traceability in parallel, and code modification can be done with side-effect prevention through backward traceability in parallel too.
- **The software development methodologies** (see Chap. 10)
  From: the software development methods based on Constructive holism – "**building**" a software system with its components – the components are developed first, then the system of a software product is built through the integration of the components developed. From the point of view of quality assurance, those methodologies are test-driven but the functional testing is performed after coding; it is too late. These methodologies consider a software product as a machine rather than a logical product created by human beings. They all comply with the reductionism principle and superposition principle.
  To: the software development method (NSE software development method, innovated by the me) based on Generative Holism of complexity science – having the whole dummy system first, then "**growing up**" with its components.

24.7 The Third Candidate of "Silver Bullet": The Entire NSE Paradigm

- **The software testing paradigm** (see Chap. 16)

  From: that mainly based on functional testing using the Black-box testing method being applied after the entire product is produced, structural testing using White-box testing method being applied after each software unit is coded for the incremental software development and iterative software development [Coc08]. Both methods are applied separately without internal logic connections.

  To: that mainly based on the Transparent-box method to combine functional testing and structural testing together seamlessly: to each set of inputs, it not only verifies whether the output (if any, can be none) is the same as the expected value but also helps users to check whether the execution path covers the expected path with the capability to automatically establish bidirectional traceability among all of the related documents and the source code for inconsistent defect checking.

- **The quality assurance paradigm** (see Chap. 17)

  From: a test-driven approach, mainly using Black-box testing method plus structural testing method and code inspection after coding.

  To: NSE-SQA – defect prevention-driven approach mainly using the Transparent-box testing method in all phases of a software development lifecycle from the first step to the end, because having an output is no longer a condition to use the Transparent-box testing method dynamically. The priority of NSE-SQA for assuring the quality of a software being developed is ordered as (1) defect prevention; (2) defect propagation prevention; (3) Refactoring applied to highly complex modules and module(s) that are performance bottlenecks; (4) Deep and broad testing.

- **The software visualization paradigm** (see Chap. 7)

  From: that drawing the diagrams manually or using graphic editors or using a tool to generate partial charts/diagrams which are neither interactive nor traceable in most cases. Even if some charts/diagrams for an entire software system can be generated, they are still not useful because there are too many connection lines to make the charts/diagrams hard to view and hard to understand without a capability to trace an element with all the related elements.

  To: a holistic, interactive, traceable, and virtual software visualization paradigm to make an entire software development lifecycle visible. The charts/diagrams are dynamically generated from several hash tables from the database and the source code through dummy programming or reverse engineering virtually without storing the hard copies in hard disk or memory to greatly reduce the space. The generated charts/diagrams are interactive and traceable between related elements – for instance, users can highlight an element with all of the related elements easily.

- **The documentation paradigm** (see Chap. 19)

  From: (a) separated from the source code without bidirectional traceability; (b) inconsistent with the source code after code modifications; (c) requiring

huge disk space and memory space to store the graphical documents; (d) the display and operation speed is very slow; (e) hard to update; (f) not very useful for software product understanding, testing, and maintenance.

To: (a) managed together with the source code based on bidirectional traceability; (b) consistent with the source code after code modification; (c) most documents are dynamically generated from several hash tables and exist virtually without huge storage space; (d) the display and operation speed is very fast; (e) most documents can be updated automatically; (f) very useful for software product understanding, testing, and maintenance.

- **The software maintenance paradigm** (see Chap. 18)

  From: that performed blindly, partially, and locally without the capability to prevent the side effects for the implementation of requirement changes or code modifications, and takes about 75% of the total effort and cost in the software system development in most software organizations.

  To: that performed visually, holistically, and globally using a systematic, disciplined, quantifiable approach to prevent the side effects for the implementation of requirement changes or code modifications through various automated traceabilities; and takes only about 25% of the total effort and total cost in software system development, because with NSE there is no big difference between the software development process and the software maintenance process – both support requirement changes or code modification with side-effect prevention.

- **The software project management paradigm** (see Chap. 20)

  From: that performed separately from the software product development process, and often makes the necessary actions being done too late.

  To: that performed closely with the software development process and makes the project management documents such as the product development schedule, the cost reports, and the progress reports traceable with the requirement implementation or the corresponding test cases or the source code, making the necessary actions being done in time.

24.7.2 The Components of NSE

As described in Chap. 5, NSE consists of the following components:

1. **The NSE process model** – It is the core part of NSE, a roadmap of the NSE paradigm. The NSE process model is nonlinear, through two way iteration with multiple tracks (see Chap. 8) supported by automated and self-maintainable traceabilities (see Chap. 9).
2. **The NSE software development methodology** – It is based on **Generative Holism** and driven by defect prevention and traceability, different from the existing software development methodology based on **Constructive Holism** and driven by testing (see Chap. 10).

24.7 The Third Candidate of "Silver Bullet": The Entire NSE Paradigm

3. **The NSE visualization paradigm** – It makes the entire software engineering process visible from the first step down to the maintenance phase using interactive and traceable 3J graphics by generating the overall charts/diagrams for an entire software system and detailed logic diagrams and control flow diagrams for each file/class/function, with the capability to highlight untested conditions and branches when working with the MC/DC test coverage measurement tools integrated into the NSE support platforms (see Chaps. 7 and 22).
4. **The NSE testing paradigm** – It is based on the Transparent-box testing method which combines functional testing and structural testing together seamlessly; to each test case it not only checks whether the output (if any, can be none) is the same as what is expected, but it also helps users to check whether the real execution path covers the expected one specified in control flow diagram, and then it automatically establishes bidirectional traceability among the related documents and test cases and the source code through the use of bookmarks and Time Tags automatically inserted into both the test case description and the test coverage database for mapping the test cases and the tested source code together, so that it can be used dynamically in the entire software development and maintenance process, including the requirement development phase and the design phase, to greatly reduce the amount of defects introduced into a software product developed with NSE (see Chap. 16).
5. **The NSE quality assurance paradigm** – It is based on defect prevention and defect propagation prevention from the first step down to the maintenance phase (see Chap. 17).
6. **The NSE documentation paradigm** – It makes the documents traceable to and from the source code to keep consistency with the source code at all times. The generated documents exist virtually to greatly reduce the required space and to speed up the display much faster (see Chap. 19).
7. **The NSE maintenance paradigm** – It helps users perform software maintenance holistically and globally with side-effect prevention for the implementation of requirement changes or code modifications supported by various traceabilities to ensure the product quality and greatly reduce the cost in regression testing after code modification through the use of test case minimization and intelligent test case selection (see Chap. 18).
8. **The NSE project management paradigm** – It combines the software development process and project management process together, making the project management documents (such as the schedule chart, the project development plan, and the cost estimation tables) traceable with the implementation of requirements and the source code for finding and fixing management problems in time (see Chap. 20).
9. **The NSE support techniques** – They are the driving force for the establishment of NSE: 14 advanced techniques are innovated and applied to NSE and the support platforms (see Chap. 6).
10. **The NSE support tools and support platforms** – They help software organizations to apply NSE in their software product development easily, no matter if it is used for new software development or to test or maintain an existing software product (see Chap. 22).

24.7.3 The Major Features and Characteristics of NSE

The major features and characteristics of NSE are listed as follows:

- **It is based on a solid foundation – complexity science:** The entire NSE paradigm is established by complying with the essential principles of complexity science, particularly the Nonlinearity principle and the Holism principle.
- **It is complete** – NSE itself is complete, including its own process model, software development methodology, visualization paradigm, testing paradigm, QA paradigm, documentation paradigm, maintenance paradigm, management paradigm, etc.
- **It brings revolutionary changes to almost all aspects in software engineering** – It makes them change from the old one based on linear processes and the superposition principle to the new one based on complexity science.
- **It offers both "what to do" and "how to do"** – different from some popular models which only offer "what to do" but ignore "how to do," NSE offers both.
- **With it almost all software engineering tasks/activities are performed holistically and globally** – With NSE, from requirement development down to maintenance, all tasks/activities are performed holistically and globally with defect prevention including side-effect prevention for the implementation of requirement changes and code modifications.
- **It combines the software development process and software maintenance process together closely** – With NSE, requirement changes are welcome at any stage and implemented with side-effect prevention though various bidirectional traceabilities (see Chaps. 8 and 18).
- **It combines the software development process and software management process together closely** – It makes all documents including the management documents such as the schedule chart and the cost reports traceable to the implementation of requirements and the source code to control a software project better and to find and fix the related issues in time (see Chaps. 8 and 20).
- **It ensures software product quality from the first step to the final step through defect prevention and dynamic testing using the Transparent-box testing method** – NSE offers many means to prevent defects that are introduced into a software product by people (the customers and the developers) with dynamic testing using the Transparent-box testing method which combines functional testing and structural testing seamlessly and can be dynamically used in cases where there is no real output from the software system such as a dummy system with dummy modules only without detailed program logic (see Chaps. 11, 17, and 18).
- **With NSE, the design becomes precoding (top-down), and the coding becomes further design (bottom-up)** – With NSE, in most cases the design through dummy programming using dummy modules becomes precoding, and the coding becomes further design through reverse engineering (see Chaps. 12 and 13).
- **It makes software documents traceable to and from source code** – With NSE, all related documents and test cases and the source code are traceable forwards or backwards through automated and self-maintainable traceabilities.

24.7 The Third Candidate of "Silver Bullet": The Entire NSE Paradigm

- **It supports real-time communication through traceable Web pages and traceable technical forums** – With NSE, the bidirectional traceability is extended to include Web pages and BBS for real-time communication.
- **It makes the entire software development process visible from first step down to the final step** – The NSE visualization paradigm is capable of making the entire software development process visible through dummy programming and reverse engineering.
- **It makes a software product much easier to read, understand, test, and maintain** – With NSE, a software is represented graphically and shown in both the overall structure of the entire product and the detailed logic diagram and control flow diagram with various traceabilities and where the untested conditions and branches are highlighted.
- **It can be applied at any time in any stage for a software product development using any other method originally** – NSE can be added onto a software product being developed using any other approach by adding bookmarks in the related documents and modifying the test cases to use some key words to indicate the format of a document and the file path plus the bookmark, then the other work can be performed by the NSE support platform automatically.
- **It requires much less time, resources, and manpower to apply, compared with other existing approaches** – One just needs to reorganize the document hierarchy using bookmarks and modify the test case descriptions using some simple rules; all of the other work can be performed automatically by the NSE support platform with many automated and intelligent tools integrated together, including the creation of huge amounts of traceable and virtual documents based on static and dynamic measurement of the software, the diagramming of the entire software product to generate holistic and detailed system call graphs and class inheritance charts, the holistic and detailed test coverage measurement results shown in J-Chart and J-Diagram or J-Flow diagram with untested conditions and branches highlighted, the holistic and detailed quality measurement results shown in Kiviat diagram for the entire software product and each class or function, the holistic and detailed performance measurement results shown in J-Chart and bar chart with branch execution frequency measurement result shown in J-Diagram or J-Flow Diagram to locate the performance bottleneck better, the software logic analysis results shown in J-Diagram with various kinds of traceability for semiautomated code inspection and walk through, the software control flow analysis results shown in J-Flow with untested conditions and branches highlighted, the GUI test operation capture and selective playback for regression testing after code modification, the test case efficiency analysis and test case minimization to form a minimized set of test cases to replace all the test cases to speed up the regression testing process and greatly save the required time and resources, the establishment of bidirectional traceability among all related documents and the test cases and the source code, the generation of more than 100 reports based on the static and dynamic measurement of the software and the reports can be stored in HTML format for being used on the internet, the Cyclomatic complexity measurement results shown in J-Chart and J-Flow diagram for performing refactoring for the over complicated modules to reduce possible defects, and more.

- **It is possible for NSE to help software organizations double their productivity, halve their cost, and reduce 99–99.99% of the defects in their software products** – With NSE, the quality of a software product is ensured from the first step through defect prevention and defect propagation prevention rather than testing after coding, so that the amount of defects introduced into a software product is greatly reduced, and that the defects propagating to the maintenance phase are also greatly reduced; software maintenance is performed holistically and globally with side-effect prevention; the regression testing after software modification is performed using a minimized test case set and some test cases selected through backward traceability from the modified modules and branches; software testing is performed in the entire software development process dynamically using the Transparent-box method which combines functional testing and structural testing together seamlessly and can be dynamically used in the case that there is no real output in running some test cases, when it is used in the requirement development phase and the software design phase.

Is there any weaknesses with NSE? Yes. We know that nothing in the world is completely perfect. For instance, the waterfall model can be applied in practice with or without tool support, the RUP (Rational Unified Process) can be applied with mainly static tool support, but NSE can only be applied with mainly dynamic tool support, such as those tools offered by Panorama++ to perform defect prevention and defect propagation prevention, and to establish bidirectional traceabilities. Of course, this disadvantage of NSE can also be considered as an advantage, because in the twenty-first century dynamic tools should be used in every software development company; otherwise, the company may lose its competition power.

24.7.4 The Major Differences Between NSE and the Old-Established Software Engineering Paradigm

The essential difference between the old-established software engineering paradigm and NSE is how to handle the relationship between the whole and its parts of a software system. **The former adheres to the reductionism principle and superposition principle that the whole is the sum of its parts,** so that nearly all software development tasks/activities are performed locally, such as the implementation of requirement changes. **The latter complies with the Holism principle of complexity science, that a software product is a Complex Adaptive System (CAS [Hol95]) having multiple interacting agents (components), of which the overall behavior and characteristics cannot be inferred simply from the behavior of its individual agents but emerge from the interaction of its parts**, so that with NSE nearly all software development tasks/activities are performed globally and holistically to prevent defects in the entire software lifecycle[Xio09-1], [Xio09-2]. A comparison between traditional software engineering paradigm and NSE is shown in Table 24.1.

24.7 The Third Candidate of "Silver Bullet": The Entire NSE Paradigm

Table 24.1 A general comparison between the traditional software engineering paradigm and NSE (nonlinear software engineering paradigm)

Comparison item	Traditional software engineering paradigm	NSE (nonlinear software engineering paradigm)
The definition of software (software products)	Software is (1) instructions (computer programs) that when executed provide desired features, function, and performance; (2) data structures that enable the programs to adequately manipulate information; and (3) documents that describe the operation and use of the programs [Pre05-P4]	Software is (1) instructions (computer programs) that when executed provide desired features, function, and performance; (2) data structures that enable the programs to adequately manipulate information; and (3) documents that describe the operation and use of the programs (including the test case script files too); plus (4) **the database built though static and dynamic measurement of the programs**; and (5) **a set of Assisted Online Agents (AOA, artificial intelligence tools working with the database) for supporting testability, reliability, visibility, changeability, conformity, and traceability to make the software program maintainable, adaptive, and that the static and dynamic measurement results can be viewed easily, and the requirement validation and the acceptance testing can be dynamically done in a fully automated way through mouse clicks only** (when a software is delivered to an end user rather than the customer, it may (for open source products) or may not include the part (4) and the part (5)
The foundation in software development	Linear thinking and simplistic science complying with the superposition principle that **the whole of a system is equal to the sum of its parts**, so that almost all tasks/activities are performed partially and locally through a linear process	Nonlinear thinking and complexity science with a set of essential principles including the **Nonlinearity** principle, the **Holism** principle that **a whole is greater than the sum of its parts – the characteristics and the behavior of a complex system is an emergent property of the interactions of its components (agents)**, the **Dynamics** principle, the **Openness** principle, the **Self-adaptation** principle, the **Self-organization** principle, the **Initial Condition Sensitivity** principle, the **Sensitivity to Change** principle, the **Complexity Arises From Simple Rules** principle, etc... so that with NSE, almost all tasks/activities are performed globally and holistically through a nonlinear process

(continued)

Table 24.1 (continued)

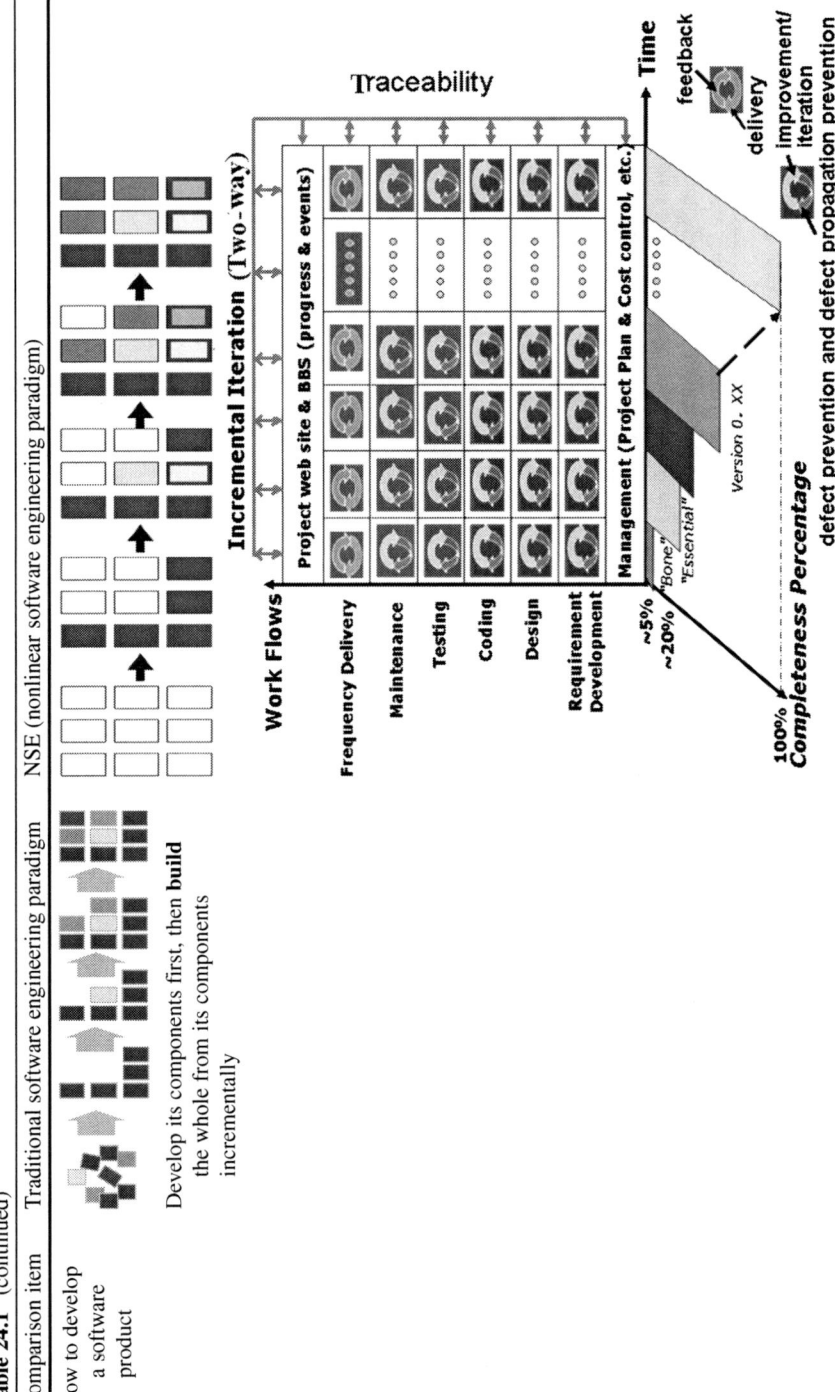

24.7 The Third Candidate of "Silver Bullet": The Entire NSE Paradigm   621

How to capture customers' requirements	Captures customers' requirements mainly using the Use Case approach: • Used with linear process models • Complying with the superposition principle that the whole of a system is the sum of its parts, so that many small pieces are obtained • Hard to get the big picture of a software product being developed • Even if a big picture of the entire software product can be obtained, it is still useless because of the lack of traceability and the lack of the capability to highlight a unit and the related units (so there will be too many connection lines to make the entire system diagram hard to read and understand) • The result obtained is not executable directly, so that it is hard to check whether the result obtained is correct or not	Captures customers' requirements mainly using Holistic, Actor–Action and Event–response driven, Traceable, Visual, and Executable technique (HAETVE): • Used with a nonlinear process model • Based on complexity science, complying with the Nonlinearity and the Holism principles • Easy to get the big picture of the entire software product • Make the charts of an entire system useful with traceabilities and the capability to highlight a unit and all of the related units • Not only useful for actor–action type applications but also useful for event–response type applications, or for both • Not only useful for the decomposition of functional requirements but also useful for nonfunctional requirements through the definition and use of a SuperActor that can request what functions or tools are needed for the interface design, the product performance, the quality level, and more • Easy to map the result obtained to the real product design, because it is done mainly through dummy programming • Easy to check whether the result obtained is correct and consistent or not through dynamic execution using the Transparent-box method
How to ensure the quality of a software product	**Test-driven:** Finds and fixes the defects after production (coding) through testing, inspection, and debugging	**Defect prevention driven:** according to Dr. W. Edwards Deming's principles for product quality control – *"Cease dependence on inspection to achieve quality. Eliminate the need for inspection on a mass basis by building quality into the product in the first place."* – NSE ensures the software product quality in the entire software system development lifecycle, particularly in the requirement development and design phases before coding though dynamic execution of test cases using the Transparent-box method combining functional testing and structural testing together seamlessly. To each test case, it checks whether the output (if any, can be none) is the same as what is expected, checks whether the execution path covers the expected path specified, and then establishes bidirectional traceability to help users remove the inconsistency defects among the requirement specification and all related documents, plus many other ways for defect prevention and inspection using traceable documents and traceable source code

(continued)

Table 24.1 (continued)

Comparison item	Traditional software engineering paradigm	NSE (nonlinear software engineering paradigm)
How to dynamically test a software product	Mainly performs functional testing using the Black-box testing method after coding, and structural testing separately	Performs functional testing and structural testing together seamlessly using the Transparent-box method in the entire software development lifecycle from the requirement development phase down to the maintenance phase with the capability to establish bidirectional traceability to help developers remove inconsistency defects
How to document a software product	Documents a software product with a man-made traceability-matrix which is time consuming to build and very hard to maintain, so that often the designed documents are not consistent with the source code after code modification. Although some tools may be used to establish bidirectional traceability, it still needs manual work to maintain	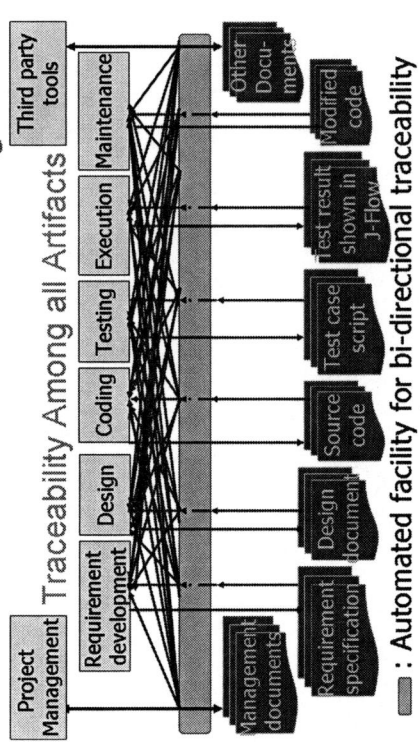

The NSE Documentation Paradigm

Traceability Among all Artifacts

■ : Automated facility for bi-directional traceability
Any related documents can be viewed anywhere in any phase for anybody to perform any task.

Documents a software product with a traceability facility which is automatically built through dynamic testing and is self-maintainable, so that the designed documents and the source code can be managed together – when the source code is changed, the developers can perform backward tracing to find the related documents, check the consistency, and remove the inconsistency defects. Besides this, more traceability facilities are provided to make the documents traceable with related source code, and the source code traceable with related source code, and so on |

24.7 The Third Candidate of "Silver Bullet": The Entire NSE Paradigm

| How to manage a software product development process and control the schedule and the budget | The project management processes are separated from the product development processes – the project plan/schedule information and the cost information are not traceable with the requirement implementation, so that often a software implementation becomes a monster of missed schedules and blown budgets | 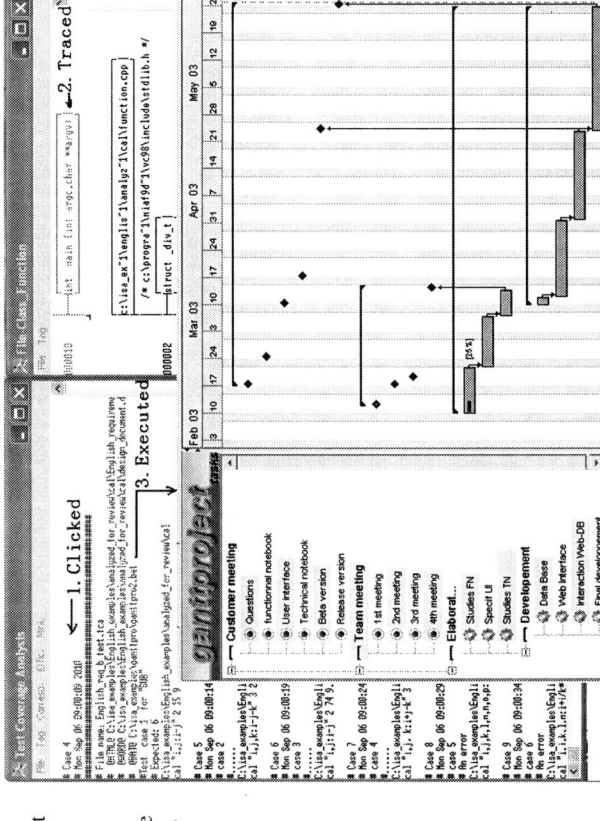 The project management processes and the product development processes are combined together, making the project plan/schedule information and the cost information traceable with the requirement implementation and the source code, assigning implementation priorities to the requirements according to the importance and market needs, so that the schedules and budgets can be controlled better. Particularly, the NSE nonlinear process model is used with defect prevention for the implementation of requirement changes or code modification to greatly reduce the cost spent in the software development process and the software maintenance process, and ensure the quality from the first step to the end of a software development project |

(continued)

Table 24.1 (continued)

Comparison item	Traditional software engineering paradigm	NSE (nonlinear software engineering paradigm)
How to maintain a software product and handle the issue of changeability	Based on linear process models without facilities for various bidirectional traceabilities or very limited traceability made manually, software maintenance is performed locally and partially with no way to prevent the side effects for the implementation of requirement changes or code modifications, so that often when a bug is fixed, there is a 20–50% change to introduce a new one to the software product. Often the regression testing is performed by reusing all test cases – it is time consuming and costly. It is why software maintenance takes more than 75% of the total cost and total effort in software system development	Based on the NSE nonlinear process model with the support of facilities for various bidirectional traceabilities that are automatically established, software maintenance is performed globally and holistically with side-effect prevention. There is no big difference between the software development process and the maintenance process, because with NSE requirement changes are welcome at any time to support the customer's market competition strategy, and responded to in real time where the side effects for the implementation of requirement changes or code modifications can be prevented to assure the quality through various bidirectional traceabilities. The regression testing after code modification can be performed with minimized test cases to greatly save the cost and time. In the case that only a few code branches are modified, only some related test cases will be selected for regression testing through backward tracing from the modified branches to the test case scripts. The regression testing will use the Transparent-box method which combines functional testing and structural testing together seamlessly with the capability to establish the new bidirectional traceabilities, and the capability to perform performance measurement, memory leak and usage violation check, and MC/DC (Modified Condition/Decision Coverage) test coverage measurement. If something wrong is found after the code modification, a global and holistic version comparison will be performed for helping users to fix the problem quickly
How to handle the issues of conformity	Without bidirectional traceability, often the documents and the modified source code are inconsistent. It is very hard to handle the issue of conformity	With NSE process model, the issue of conformity can be solved easily through the use of various traceabilities established automatically, particularly the traceabilities among the related documents, the test cases, and the source code

24.7 The Third Candidate of "Silver Bullet": The Entire NSE Paradigm

| How to handle the issue of invisibility | It can be solved partially in the modeling process using UML and the support tools |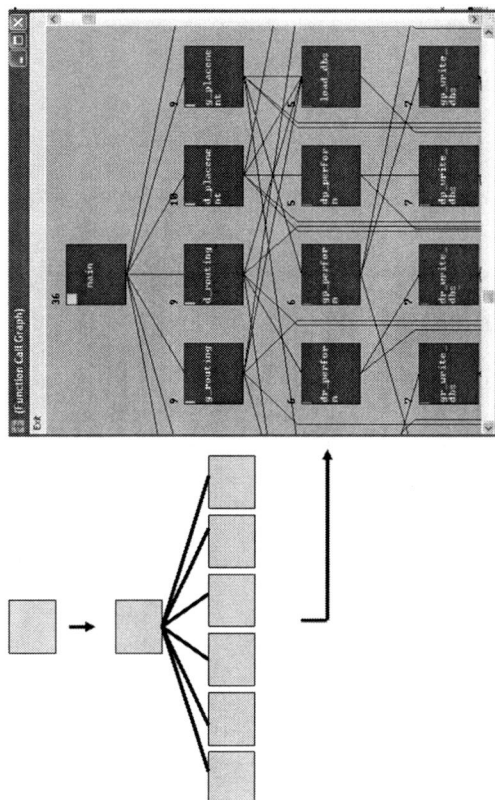

Hierarchical System Design Supported by NSE

With the NSE process model and the support platforms, the entire software development process is visible from the first step to the maintenance phase using integrative and traceable 3J graphics and the corresponding diagramming tools, which generate all charts and diagrams globally and holistically with various kinds of traceabilities to make the software product being developed much easier to understand, test, and maintain |
|---|---|---|

(continued)

Table 24.1 (continued)

Comparison item	Traditional software engineering paradigm	NSE (nonlinear software engineering paradigm)
The graphical presentation of the process models		

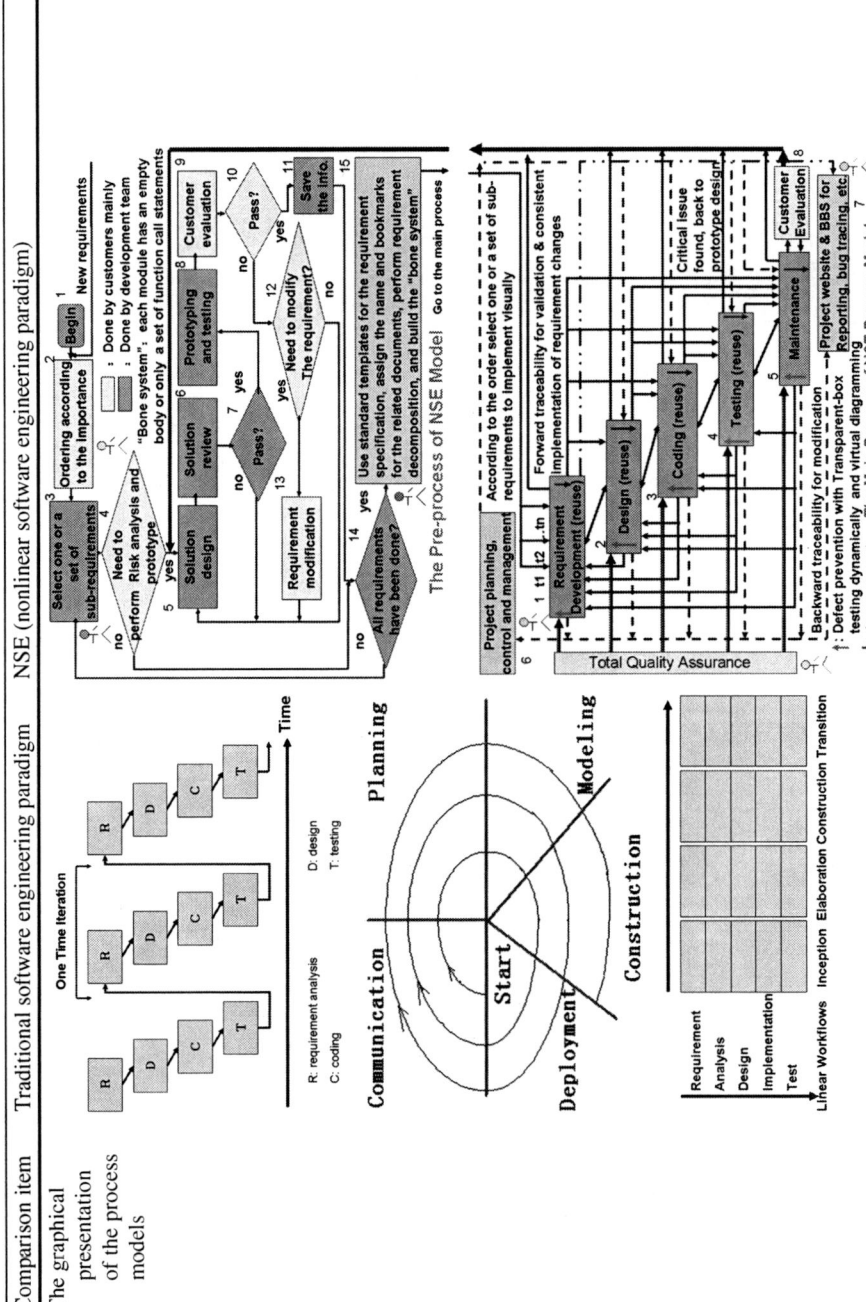

24.7 The Third Candidate of "Silver Bullet": The Entire NSE Paradigm

The major characteristics of the process models	1. Linear 2. Test-driven 3. Iteration in one direction 4. There is no preprocess, but some of models include a prototyping process 5. There is no self-maintainable facility to truly support automated and bidirectional traceability 6. There is no defined process or systematic method for software maintenance 7. The software development process and the project management process are separated, the cost reports and schedule charts and other management material are not directly traceable with the requirement implementation and the source code 8. Dynamic software testing is performed after coding 9. Almost all tasks are performed locally and partially according to the superposition principle that the whole of a system is the sum of its parts	1. Nonlinear 2. Defect prevention and traceability driven 3. Bidirectional iteration 4. Divided into two parts: the preprocess and the main process 5. There are many self-maintainable facilities to support bidirectional traceabilities 6. Combining the software development process and the software maintenance process together, responding to software changes in real time with side effects prevented 7. Combining the software development process and the project management process together closely to make the project management materials (cost reports, schedule charts, etc.) traceable with the requirement implementation and the source code 8. Dynamic software testing is performed in the entire software development process and the maintenance process from the first step to the end using the Transparent-box testing method which combines functional and structural testing together seamlessly, with the capability to establish a self-maintainable facility to help users check and remove inconsistency defects among all related artifacts and the source code 9. Almost all tasks are performed globally and holistically according to the holism principle of complexity science

24.7.5 Qualification as a Candidate of "Silver Bullet" for Slaying Software "Werewolves"

In this section, we will discuss the qualification of NSE as a candidate of the **"Silver Bullet" to slay software "werewolves."**

1. Efficiently Solving the Issue of Missed Schedules

(a) Helping the project development team and the customer work together closely to assign priority to requirements according to the importance, so that the important requirements will be implemented early to meet the market needs. If necessary, some optional requirements can be temporally ignored (see Chap. 8).

(b) Making the project plan, the schedule chart, and other related documents traceable with the implementations of requirements and the source code as shown in Fig. 24.6, so that the management team can find and solve the schedule issues in time (see Chap. 9).

(c) Helping the software development team set up a project Web site and technical forum, and making the Web pages and the topic pages of the technical forum traceable to the implementation of requirements and the source code, so that any schedule delay will be known by the members of the team, and

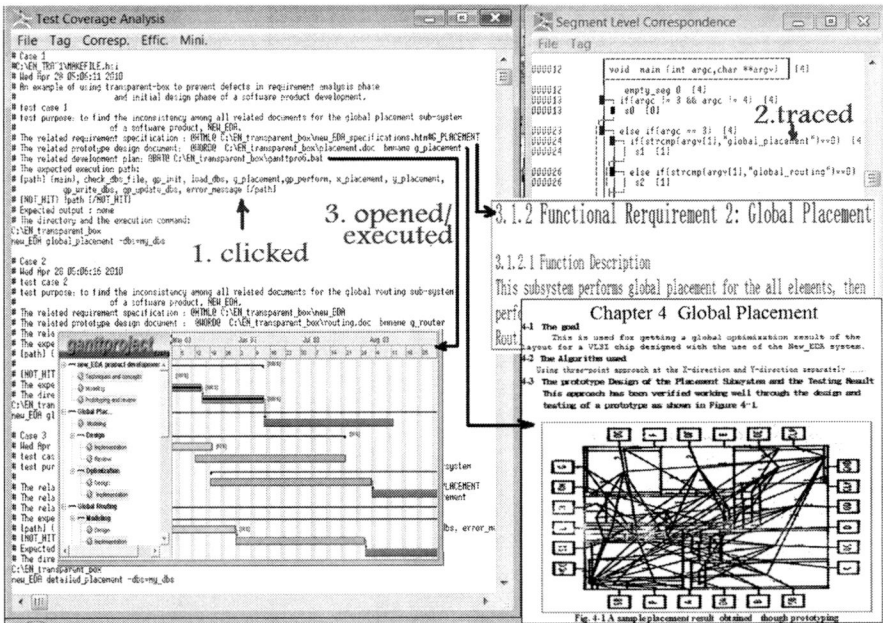

Fig. 24.6 An application example to make the project development schedule chart traceable with the implementation of requirements and the source code

24.7 The Third Candidate of "Silver Bullet": The Entire NSE Paradigm

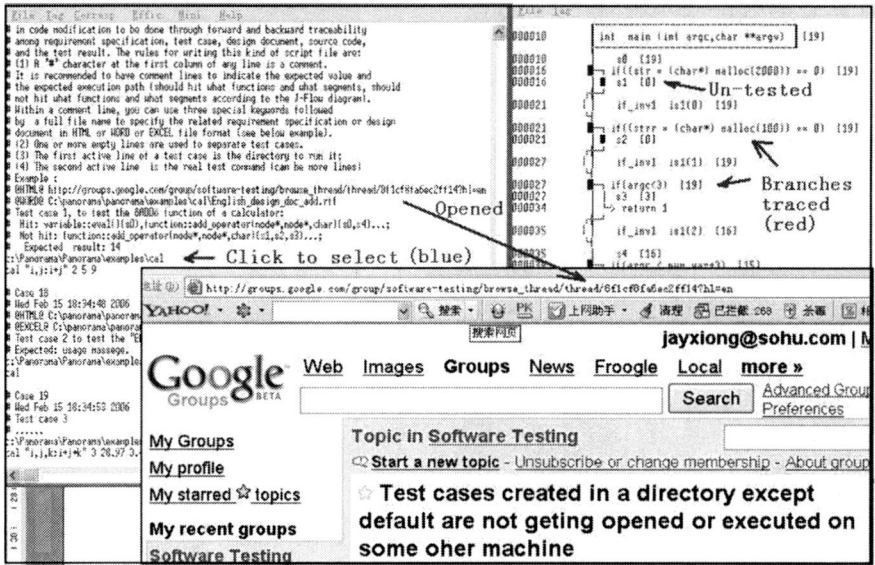

Fig. 24.7 An example of making Web pages traceable to the implementation of requirements and the source code

each member may make his/her contribution to solve the issue quickly – see Fig. 24.7 for an application example.

(d) See (3) "Efficiently Solving the Issue of Flawed Products – Removing More Than 99.99% of the Defects" – through greatly reducing the amount of defects to help the development team meet the project development schedule much more easily.

(e) See (4) "How Is It Possible for NSE to Help Users Double Their Productivity" – through defect prevention and defect propagation prevention upstream to greatly reduce the defects propagated downstream and side-effect prevention in the implementation of requirement changes and code modifications to make it possible to reduce two-thirds of the total effort spent in software changes and maintenance to help the development team meet the project development schedule better.

2. **Efficiently Solving the Issue of Blown Budgets**

(a) Assigning priority to the requirements according to their importance ((1) must have, (2) should have, (3) better to have, (4) may have or optional…) to make the critical and important requirements be implemented early to form an essential working version (about 20% of the requirements) first, then making the working product grow up incrementally according to the assigned priority (see Figs. 24.8 and 24.9), to avoid the issue of blown budgets – if necessary, some optional requirements can be ignored or implemented in the future (see Chap. 8).

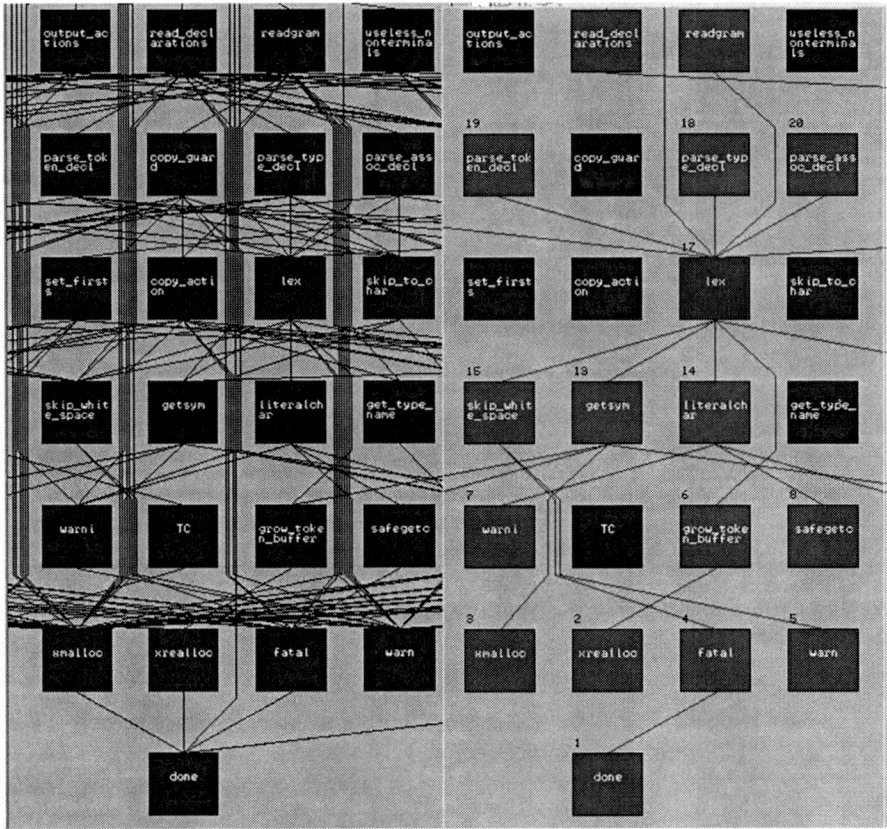

Fig. 24.8 Incremental development support with assignment of bottom-up coding order

(b) Complying with the Generative Holism principle of complexity science, helping users to form the whole of a software product first through dummy programming (using dummy modules with an empty body or only some function call statements) as an embryo through the use of the HAETVE (Holistic, Actor–Action and Event–Response driven, Traceable, Visual, and Executable) technique (see Chap. 11) for requirement development and the **Synthesis Design and Incremental growing up (Implementation and Integration)** technique (see Chap. 12) for product design, to help users estimate the cost/budget better.

(c) Making the cost estimation chart, the budget plan, and other related documents traceable with the requirement implementation and the source code, so that the management team can know the situation in time and control the budget better (see Chap. 9).

24.7 The Third Candidate of "Silver Bullet": The Entire NSE Paradigm

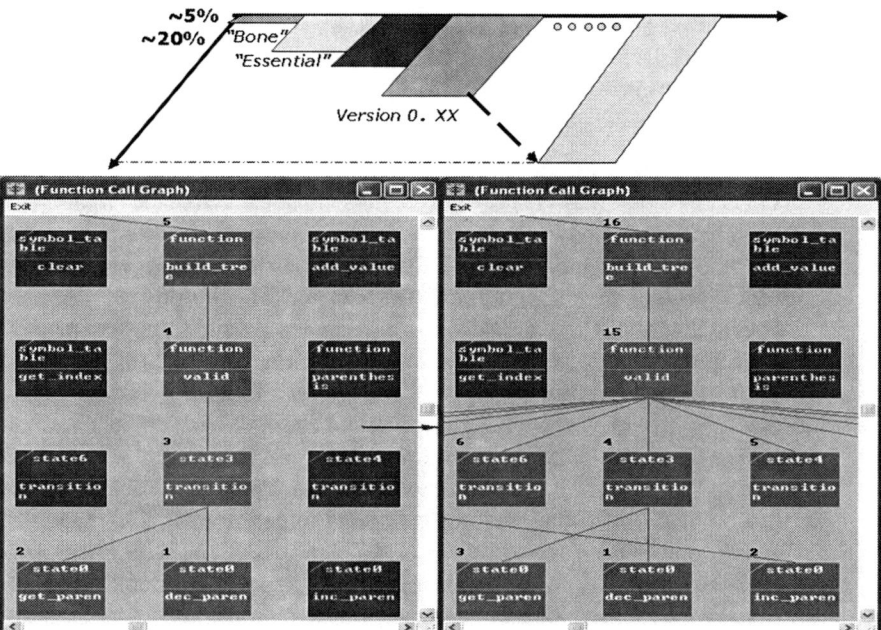

Fig. 24.9 Incremental development support

(d) Making the Web pages or topic pages of the technical forum traceable to the implementations of requirements and the source code, so that any budget issue can be known by the members of the team early, and each member may make his/her contribution to solve the issue quickly.

(e) Helping users to make the product grow up incrementally, according to the requirement priority (see Chap. 10).

(f) See (3) "Efficiently Solving the Issue of Flawed Products – Removing More Than 99.99% of the Defects" – through greatly reducing the amount of defects to help the development team to develop the product within the budget much more easily.

(g) See (4) "How Is It Possible for NSE to Help Users Double Their Productivity" – through defect prevention and defect propagation prevention upstream to greatly reduce the defects propagated downstream, and side-effect prevention in the implementation of requirement changes and code modifications to make it possible to reduce two-thirds of the total effort spent in software changes and maintenance to help the development team to develop the product within the budget better.

(h) See (5) "How Is It Possible for NSE to Help Users Halve Their Cost" – through greatly reducing the cost to further ensure the product be developed under the budget.

3. **Efficiently Solving the Issue of Flawed Products – Removing More Than 99.99% of the Defects mainly through Defect Prevention and Defect Propagation Prevention**
   (a) Helping users efficiently remove defects, particularly upstream defects, through

   - Defect prevention by (1) providing some templates such as the requirement specification template (see Appendix A) to prevent something missing; (2) helping users apply the HAETVE technique for requirement development through dummy programming and making the dummy program executable under dynamical testing using the Transparent-box method combining functional and structural testing together seamlessly, which can be used dynamically in the entire software development lifecycle (see Chap. 16); (3) supporting incremental coding to prevent inconsistencies between the interfaces (see Chap. 13).
   - Defect propagation prevention mainly through dynamic testing using the Transparent-box testing with the capability to perform MC/DC (Modified Condition/Decision Coverage) test coverage measurement, memory leak and usage violation check, performance analysis, and the capability to automatically establish bidirectional traceability to help users check and remove the inconsistency defects among the related documents and the source code, plus inspection using traceable documents and source code (see Chap. 7).
   - Refactoring for those modules with higher Cyclomatic complexity (the number of decision statements) and performance bottleneck modules with side-effect prevention – often 20% of the highest complex modules have about 80% of the defects.

   (b) Supporting quality assurance from the first step to the end through dynamic testing using the Transparent-box method.
   (c) Providing techniques and tools for quality measurement to the entire software product and each component for finding and solving the quality problems in time.
   (d) Helping users perform software maintenance holistically and globally with side-effects prevention though various bidirectional traceabilities.
   (e) See (6) "How Is It Possible for NSE to Help Users Reduce the Risk" and (7) "Efficiently Handling the Issue of Changeability" for more information about quality assurance with NSE.
   (f) See Table 17.1 about how it is possible for NSE to help users remove 99.99% or more of the defects for a software product (see Chap. 17).

4. **How Is It Possible for NSE to Help Users Double Their Productivity**

   (a) With the old-established software engineering paradigm, linear process models are used and dynamic testing is performed after coding, so that defects are easily introduced into a software product upstream, and the defects easily propagate to the maintenance phase in which the implementation of requirement changes and code modifications are performed partially and locally, so that software maintenance takes 75% or more of the total

24.7 The Third Candidate of "Silver Bullet": The Entire NSE Paradigm

effort in software development; but with NSE, the nonlinear NSE process model is used which combines the software development process and maintenance process together, ensuring software quality from the first step down to the final step through defect prevention, defect propagation prevention, refactoring, and software testing dynamically using the Transparent-box method in the entire software system development lifecycle, so that the defects propagated into maintenance phase are greatly reduced, plus that the implementation of requirement changes and code modifications are performed holistically and globally with side-effect prevention – the result is that the effort spent in software maintenance will be almost the same as that spent in the software development process. It means about two-thirds of the effort originally spent in software maintenance can be saved – about half of the total effort can be saved (equal to double the productivity).

(b) As described in (3), with NSE about 99.99% of the defects can be removed. So that as Capers Jones pointed, "Focus on quality, and productivity will follow" [Jon94].

(c) NSE also supports the reuse of qualified components (see Chap. 8, Sect. 8.10 (5)) to increase software productivity.

(d) With NSE, the NSE software documentation paradigm (see Chap. 19) and NSE software visualization paradigm support traceability between the software documents and source code, making a software product much easier to read, understand, test, and maintain to increase the productivity.

(e) With NSE, there are more means to help users increase their productivity:

- Provides techniques and automated tools to help users manage and control their software projects better.
- Provides automated tools and templates to help users execute their project development plan easily.
- Provides techniques and visual tools to help users perform requirement development, product design, and bug fixing quickly.
- Supports reverse engineering to generate a lot of design documents automatically.
- Supports incremental and visual coding.
- Provides techniques and automated complexity analysis tools to help users design their test plan quickly.
- Provides techniques and tools to help users perform test case design efficiently through unexecuted path analysis.
- Provides techniques and tools for capturing GUI operations and playing them back automatically.
- Provides techniques and automated tools for test case efficiency analysis and test case minimization, to help users perform regression testing quickly (at least five times faster, in general).
- Provides techniques and automated tools for incremental database management, so that unchanged source files do not need to be analyzed twice to speed up the regression process (ten times faster than other tools without incremental database management capability).

- Provides techniques and automated tools to analyze the system structure, data usage, and logic flow of a users' software product to help them manage the product better.
- Provides intelligent version comparison tools to help users maintain their product versions easier.

5. **How Is It Possible for NSE to Help Users Halve Their Cost**

    (a) All of the techniques and tools used for helping users double their productivity are also useful for reducing the software development cost.
    (b) All techniques and tools provided for reducing 99.99% of the bugs are also useful for reducing the software development cost.
    (c) With the old-established software engineering paradigm, software maintenance takes 75% or more of the total cost in software development; but with NSE, the nonlinear NSE process model is used which combines the software development process and maintenance process together, ensuring software quality from the first step down to the final step through defect prevention, defect propagation prevention, refactoring, and software testing dynamically using the Transparent-box method in the entire software system development lifecycle, so that the defects propagated into the maintenance phase are greatly reduced, plus the implementation of requirement changes and code modifications are performed holistically and globally with side-effect prevention – the result is that the effort spent in software maintenance will be almost the same as that spent in the software development process, meaning that about two-thirds of the cost originally spent in software maintenance can be saved – about half of the total cost can be saved as shown in Fig. 24.10.
    (d) Provides techniques and tools to diagram the entire system of a user's product and links the related parts to each other, making code inspection and walk-through much easier to perform.
    (e) Supports efficient regression testing using minimized test cases.
    (f) Provides techniques and tools to capture users' GUI operations and play them back to reduce regression test costs, plus

    - Provides techniques and visual tools to help users quickly perform requirement development, functional decomposition, and bug fixing
    - Supports reverse engineering to automatically generate design documents
    - Supports incremental and visual coding
    - Provides automated tools for complexity analysis to help users design their test plan rapidly
    - Provides tools to help users perform efficient test case design
    - Provides techniques and tools for capturing GUI operations and playing them back
    - Provides techniques and automated tools for test case efficiency analysis and test case minimization
    - Provides techniques and tools to diagram the entire system of a user's software product for immediate product comprehension and understanding

24.7 The Third Candidate of "Silver Bullet": The Entire NSE Paradigm

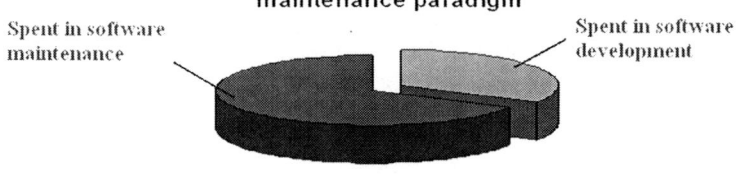

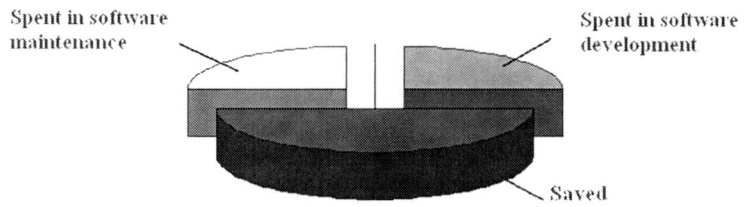

Fig. 24.10 Estimated effort and cost spent in software development and software maintenance

- Provides techniques and automated tools to analyze the system structure, data usage, and logic flow of users' software products for better product management
- Provides intelligent version comparison tools to help users maintain their products effortlessly
- Provides automatic forward and backward traceability among requirement specifications, design documents, test cases, source code, and tests, making the software product easier to understand, test, and maintain

6. **How Is It Possible for NSE to Help Users Reduce the Risk**

    (a) Helping users work with the customer to assign a priority order to requirements according to the importance for implementing the important requirements earlier.
    (b) Helping users perform prototype design and testing for important and unfamiliar requirements to prevent unrealizable requirements.
    (c) Helping users estimate the cost better using the designed dummy system through dummy programming.
    (d) Making it possible to help users remove 99.99% of the defects in the designed product, double their productivity, and halve their cost – further reducing the risk.

7. **Efficiently Handling the Issue of Changeability**

    (a) Responds to requirement changes in "real-time" without waiting for a milestone.

(b) Helps users communicate about the changes through a corresponding title in a Project BBS with detailed information that is traceable with the requirements and source code (also with face-to-face meetings).
(c) Supports side-effect prevention for the implementation of requirement changes through various traceabilities:

- Helps users perform forward tracing from the test cases(s) related to the requirement to be changed to determine what modules should be modified for a requirement change – see Fig. 24.11.
- Helps users perform backward tracing to check related requirements of the modified modules for preventing requirement conflicts (in this example, two requirements are related) – see Fig. 24.12.
- Helps users check what other modules may also need to change with the modification – see Fig. 24.13.
- After modification, helps users check all related call statements for defect prevention – see Fig. 24.14.
- Helps users perform efficient regression testing through related test case collection based on backward traceability – see Fig. 24.15.
- Helps users check the consistency of global variables or static variables – see Fig. 24.16.

8. **Efficiently Handling the Issues of Complexity**

(a) "Complexity is by levels
- Hierarchically, by layered modules or objects
- Incrementally, so that the system always works" [Bro95-P211]

– With NSE, HAETVE (Holistic, Actor–Action and Event–Response driven, Traceable, Visual, and Executable) technique for top-down function decomposition of functional requirements

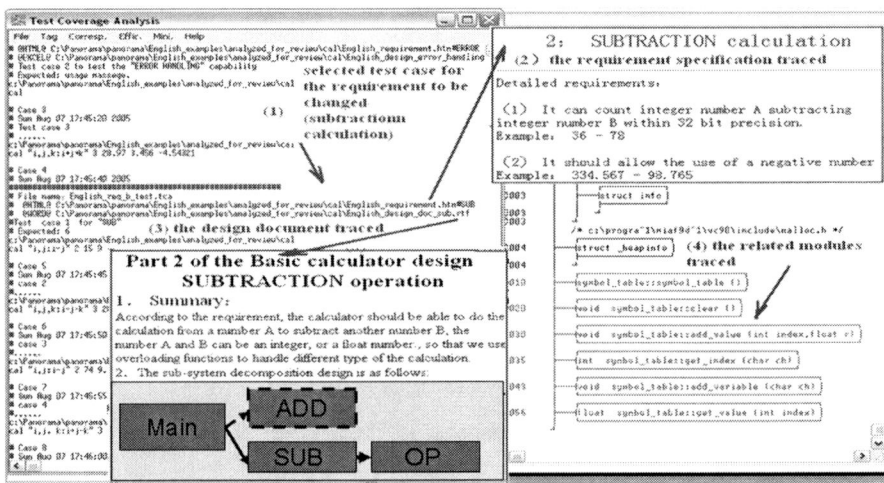

Fig. 24.11 Forward tracing for finding the modules to be modified

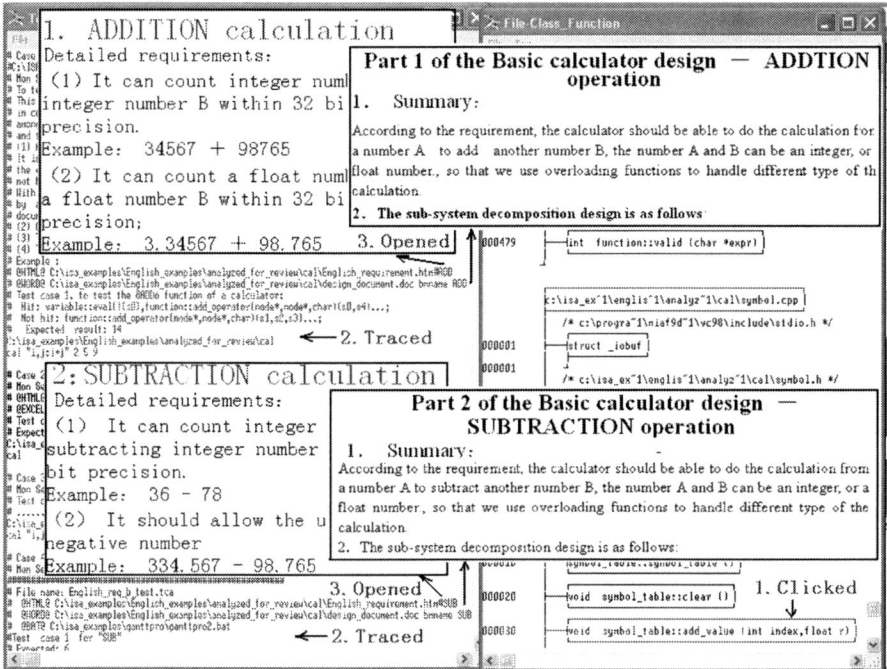

Fig. 24.12 Backward tracing from the module(s) to be modified to see how many requirements are related (if more than one requirement is related, the modification must satisfy all of them)

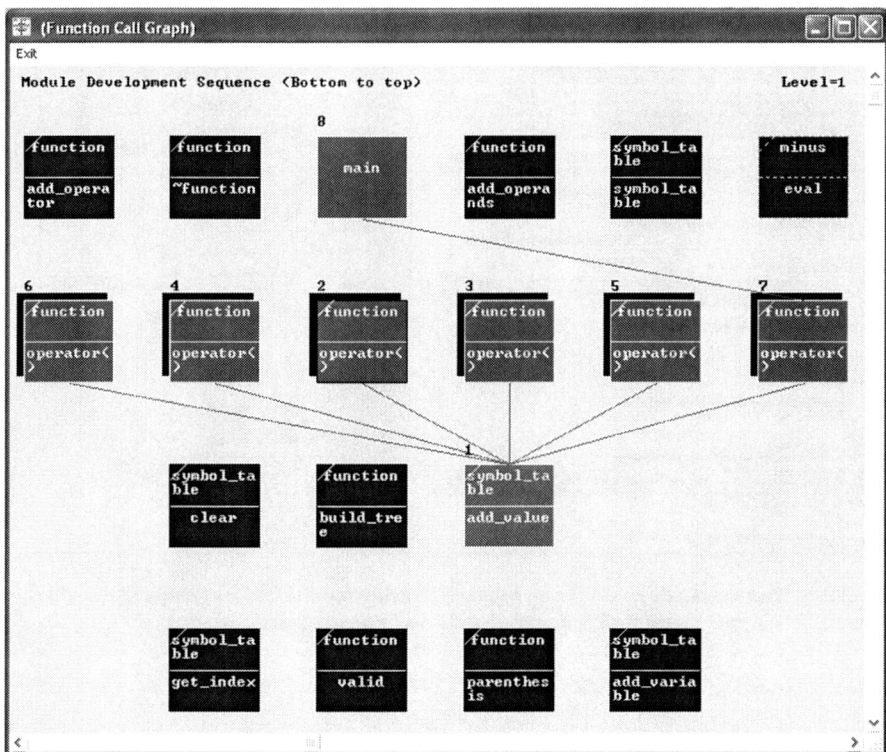

Fig. 24.13 Helps users checks what other modules may also need to be changed with the modification

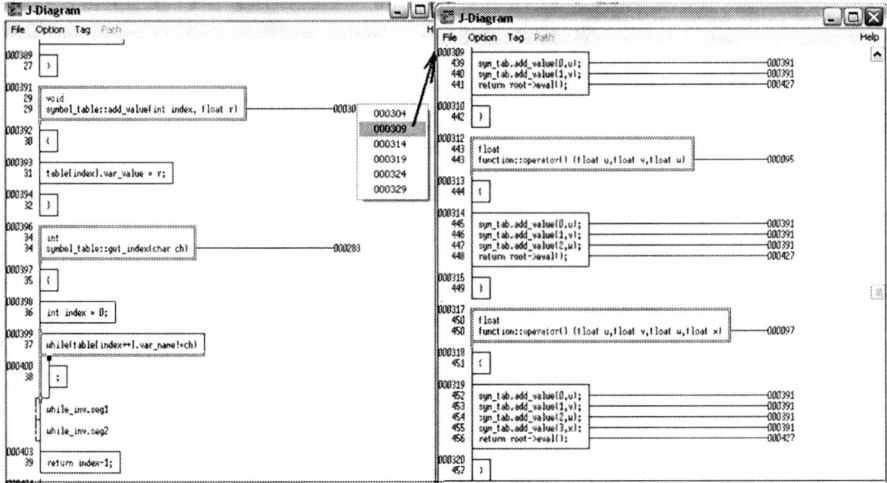

Fig. 24.14 Checking all related call statements for defect prevention

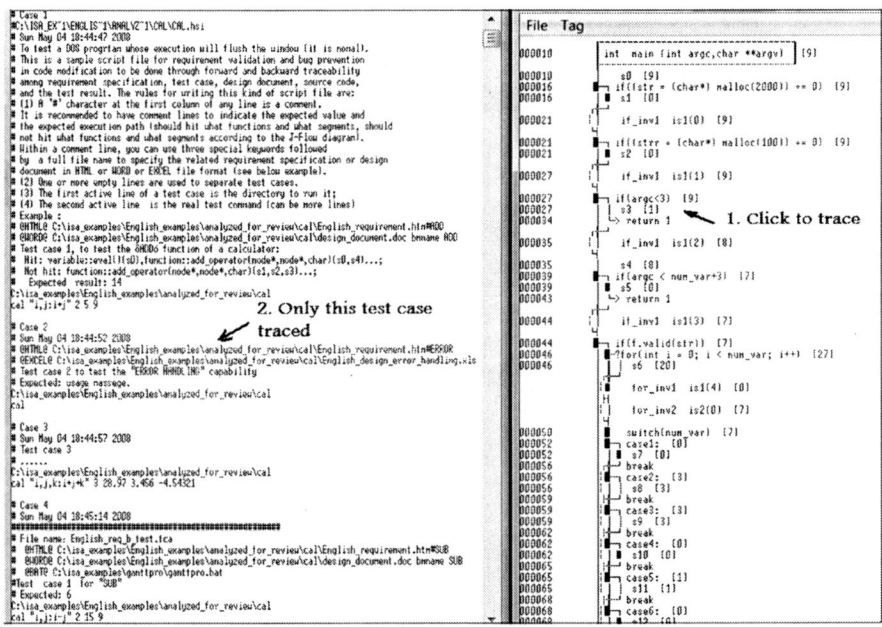

Fig. 24.15 Test cases selection through backward tracing for efficient regression testing (in this example, when code segment s3 is modified, only one test case is needed)

24.7 The Third Candidate of "Silver Bullet": The Entire NSE Paradigm

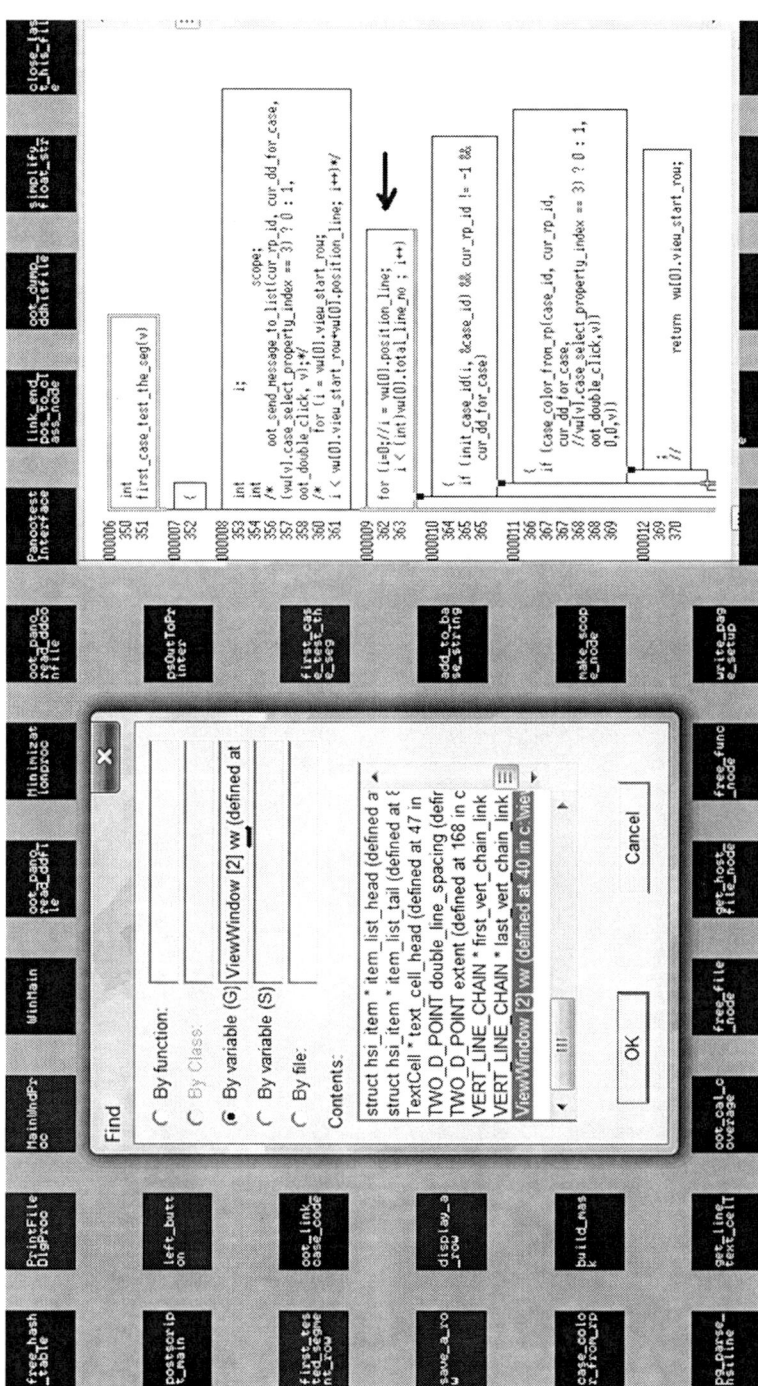

Fig. 24.16 Checking variable consistency

- With NSE, the software development methodology is based on the Generative Holism principle that the whole of a software system is formed first as an embryo, then it grows up with its components incrementally. A system leveling process is shown in Fig. 24.17.
(b) Complying with the Nonlinearity principle and the Holism principle, with NSE almost all software development tasks and activities are performed holistically and globally. For instance, the implementation of requirement changes or code modifications are performed holistically and globally with side-effect prevention supported by various traceabilities.
(c) Making the entire software development process and the work products visible through the applications of the NSE software visualization paradigm.
(d) Helping users perform refactoring for the program modules with higher Cyclomatic complexity.
(e) For dynamic traceability-based understanding of the complexity and structure of software and its ecosystem, see Fig. 24–18 to Fig. 24–20.
(f) For comprehensive (static+dynamic) program element analysis, see Fig. 24.21 and Fig. 24.22.

9. Efficiently Handling the Issues of Invisibility

With the NSE software visualization paradigm, NSE makes the entire software development process and the work products visible.

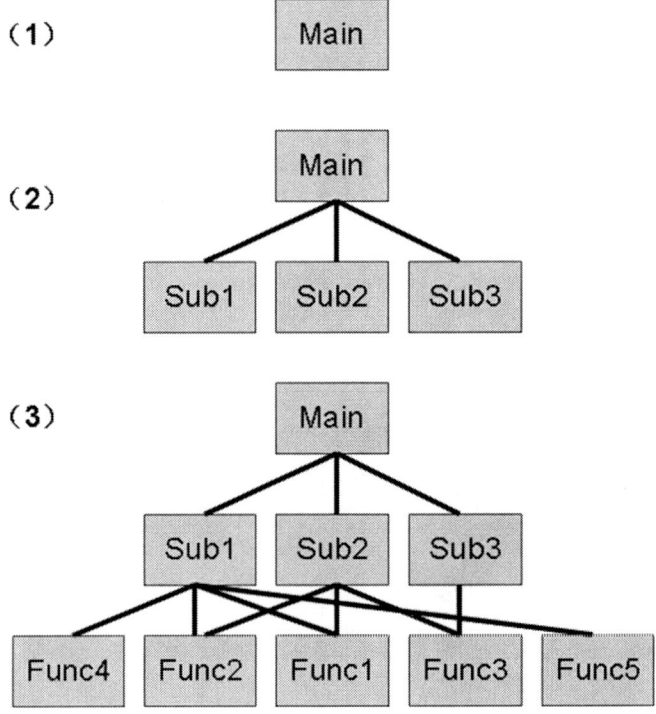

Fig. 24.17 An application example of Top-Down system leveling

24.7 The Third Candidate of "Silver Bullet": The Entire NSE Paradigm

Fig. 24.23 and Fig. 24.24 show the types of graphics provided/supported.

With NSE, software charts and diagrams can be automatically generated from both dummy programs and regular programs as shown in Fig. 24.25.

About the detailed application examples for making the entire software development process and work products visible, see Chap. 7, Sect. 7.9.

10. Efficiently Handling the Issue of Conformity

(a) Making all documents and test cases and source code traceable forwards and backwards as shown in Fig. 24.26 through the execution of test cases and the Time Tags automatically inserted into both the description part of a test case and the corresponding test coverage database, and some special keywords such as @WORD@, @HTML@, @PDF@, @EXCEL@, and @BAT@ to indicate the format of a document followed by the file path and bookmark to open the document form the bookmark location.

(b) After the implementation of requirement changes or code modifications, solving the inconsistency problems between design documents and the modified source code through bidirectional traceability – see Fig. 24.27.

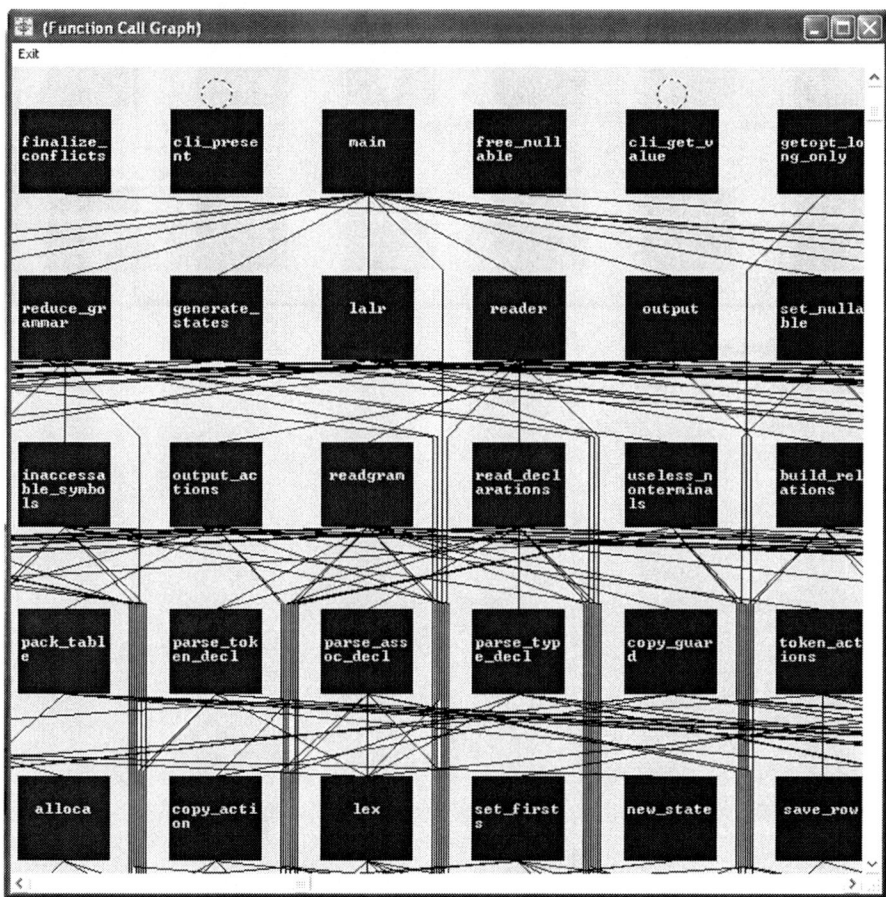

Fig. 24.18 A call graph without and with dynamic traceability

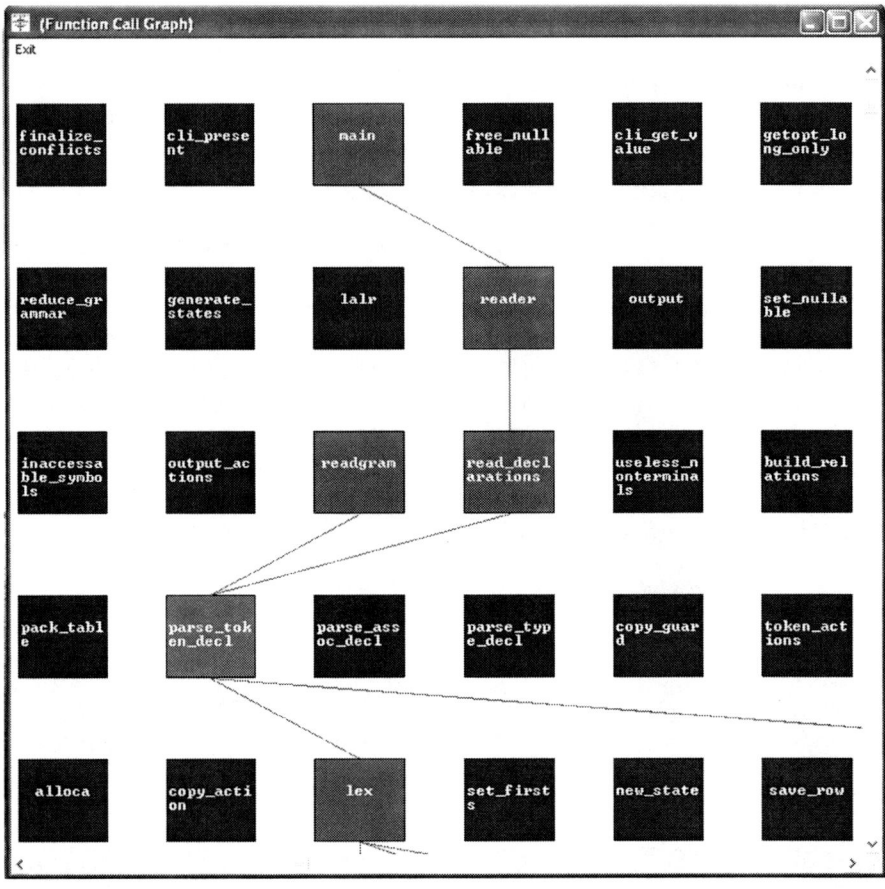

Fig. 24.18 (continued)

24.7 The Third Candidate of "Silver Bullet": The Entire NSE Paradigm

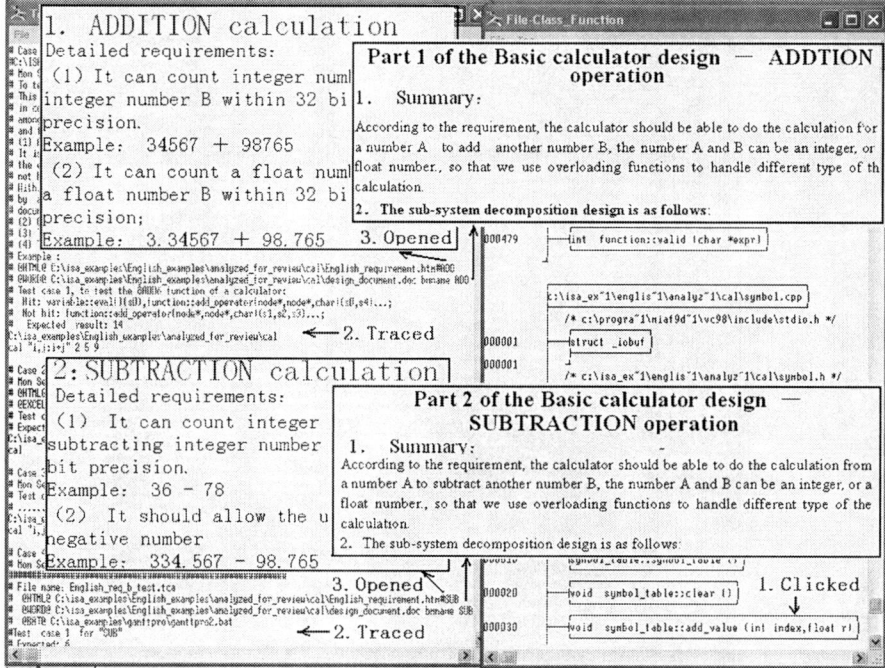

Fig. 24.19 Tracing a module backwardly to highlight the related test cases and the related requirements (in this example, two sub-requirements are traced, so that the modification of this module must satisfy both requirements to prevent inconsistent defects)

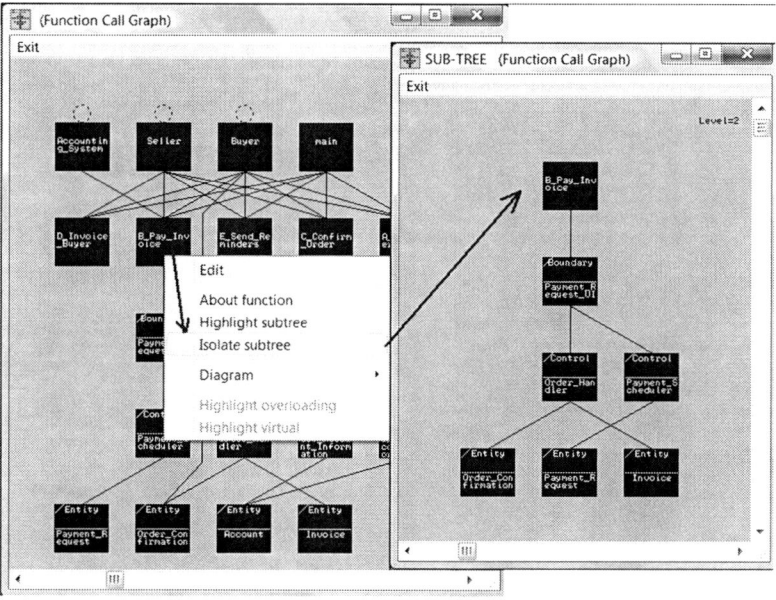

Fig. 24.20 Program structure analysis based sub-system isolation

644　24 Candidates of "Silver Bullet"

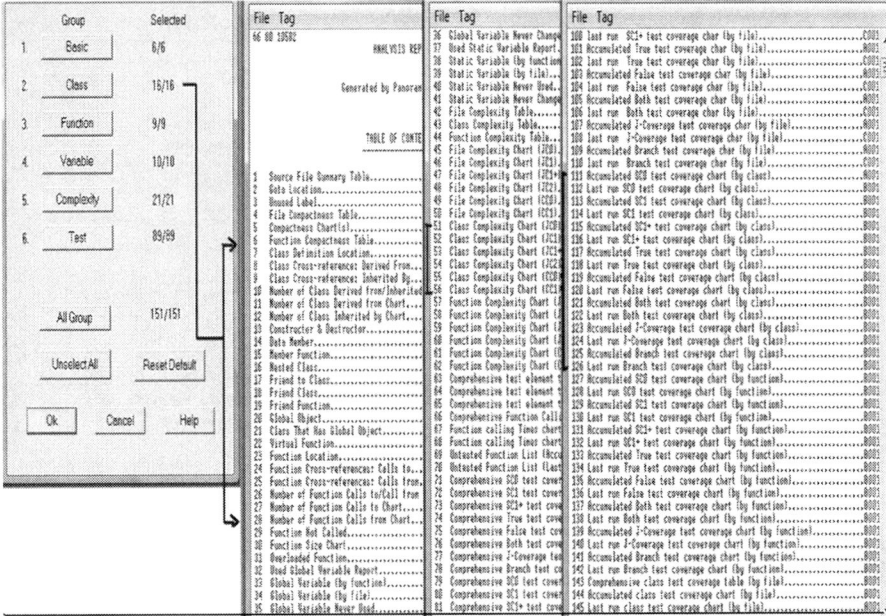

Fig. 24.21 A report list of comprehensive (static + dynamic) class and other program element analysis

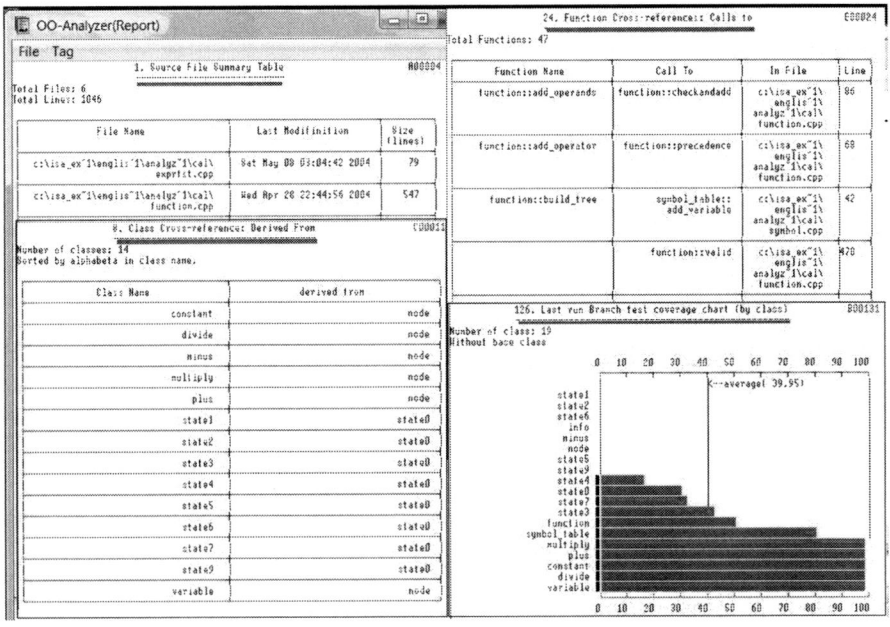

Fig. 24.22 Sample reports of comprehensive (static + dynamic) program element analysis

24.7 The Third Candidate of "Silver Bullet": The Entire NSE Paradigm

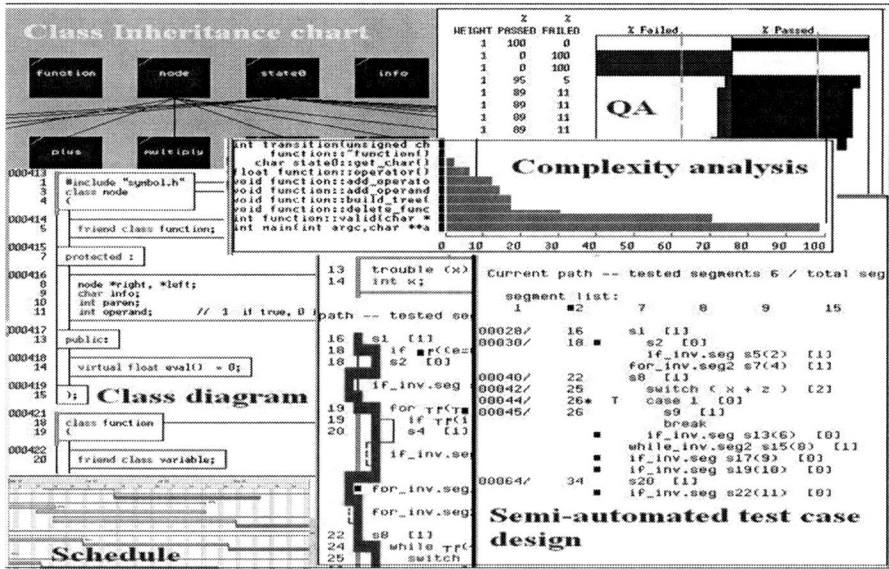

Fig. 24.23 Graphic types provided with NSE

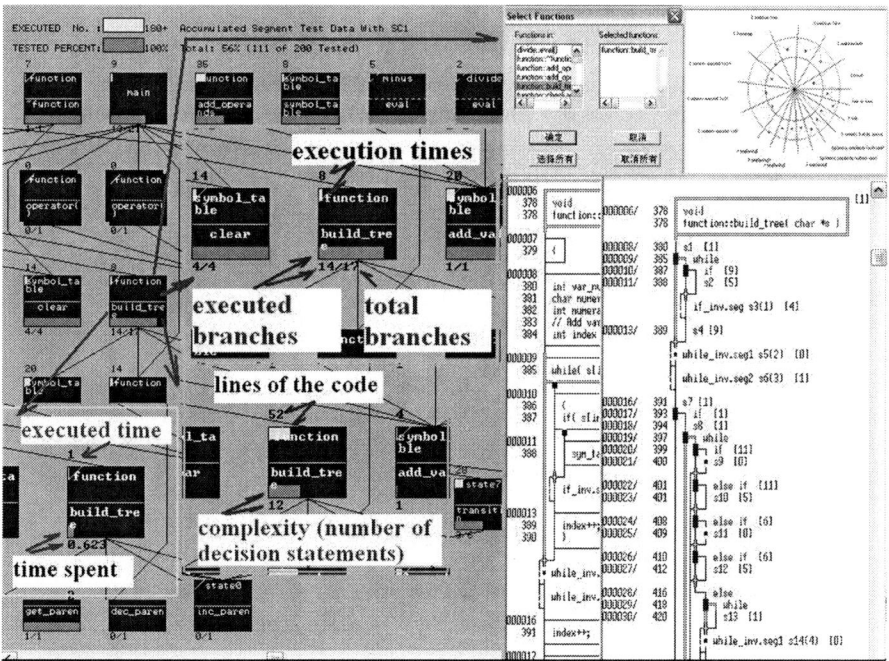

Fig. 24.24 More graphic types provided with NSE

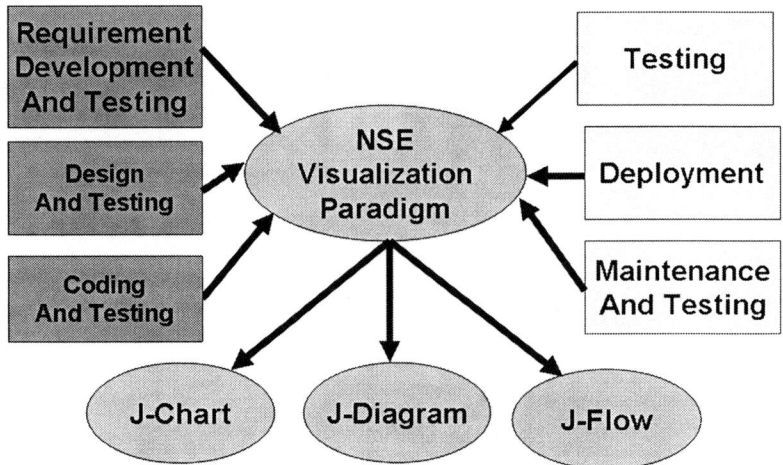

Note: With NSE software testing is performed in the entire software product development life cycle dynamically using Transparent-Box method combining functional testing and structural testing together

Fig. 24.25 Entire software development lifecycle visualization support

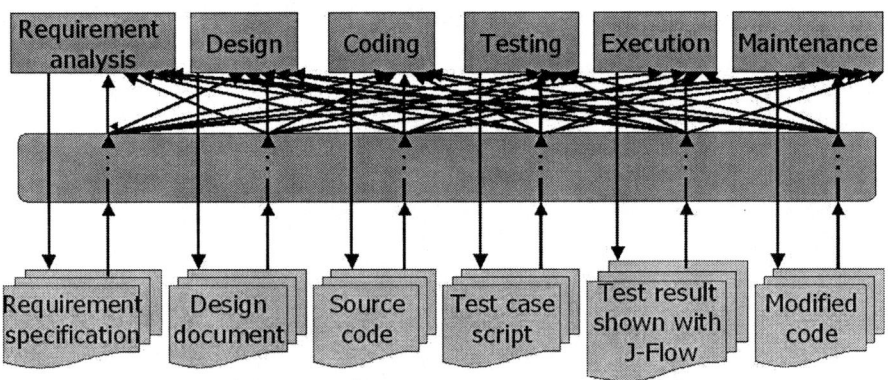

Fig. 24.26 With NSE all related documents and test cases and source code are traceable

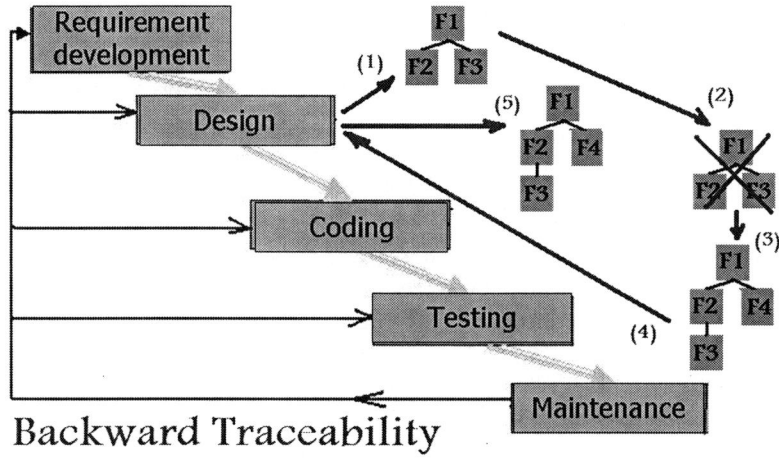

Fig. 24.27 Keeping consistency between documents and the source code through traceability

24.8 Summary

Software has become the driving force for the development of all kinds of businesses, engineering, sciences, and the global economy. But unfortunately, software itself is not well engineered – unreliable and almost unmaintainable. The root cause is that the old-established software engineering paradigm based on linear thinking, reductionism, and the superposition principle is entirely outdated – with it, almost all software development tasks and activities are performed linearly, partially, and locally, making software "Werewolves" (a monster of missed schedules, blown budgets, and flawed products) exist for more than 50 years and hard to find "Silver Bullets" to slay them.

For efficiently slaying the software werewolves, a new software engineering paradigm, the NSE paradigm is established which is based on complexity science by complying with the essential principles of complexity science, particularly the Nonlinearity principle and the Holism principle that the whole of a complex system is greater than the sum of its components, and the characteristics and behaviors emerge from the interaction of its components, so that with NSE, almost all software development tasks and activities are performed holistically and globally.

Preliminary applications show that compared with the old-established software engineering paradigm, it is possible for NSE to help software development organizations double their productivity and their project success rate, halve their cost, remove 99.99% of defect in their software products, and greatly reduce the software development risks.

Bringing revolutionary changes to almost all aspects in software engineering, NSE could be a qualified candidate of the "Silver Bullet" to slay the software "Werewolves" – a monster of missed schedules, blown budgets, and flawed products.

24.9 Points and Questions to Ponder

(a) Can any single development, in either technology or management technique efficiently solve the critical problems existing with today's software development: low quality and productivity, high cost and risk? Why?
(b) Does a qualified "Silver Bullet" which is able to slay software "Werewolves" (a monster of missed schedules, blown budgets, and flawed products) mean a complete revolution in software engineering through a paradigm shift from the old one based on reductionism and the superposition principle to a revolutionary one based on complexity science? Why?
(c) What are the major differences between the old-established software engineering paradigm and NSE?
(d) Refer to Chap. 1, use the NSE-CLICK tool to view the examples as described in Chap. 1; then use S_Panorama (for C/C++) or S_Panojava (for Java) product included in the CD attached with this book for your small project development through all major processes, including the following:

1. The preprocess to assign priority to the requirements.
2. Perform prototype design, testing, and review.
3. Through dummy programming, use the HAETVE technique (see Chap. 11) to perform function decomposition of functional requirements, write the requirement specification using the corresponding template (see Appendix A), use bookmarks to specify the relationship of the requirement specification, the prototype design documents, the testing requirement specification, the project plan, etc.
4. Design the corresponding test script and test cases according to the simple rules (see Chap. 9) – must indicate the expected execution path, and the related documents using the corresponding keyword for each test case.
5. Use the OO-Browser tool (see the corresponding user manual) to generate the call graph, then use the traceability to review the work products for static defect removal.
6. Use the OO-Test tool to execute your test cases using the Transparent-box method and meet the MC/DC test coverage requirement (see Appendix B).
7. Use OO-Browser and OO-Diagrammer to graphically show your testing results, then check whether there are logic defects or inconsistent defects – if there are, remove them; If necessary, go back to the upstream phase.
8. Use the function decomposition result and the "Synthesis Design and Incremental growing up" Technique to perform preliminary system design, and then perform detailed system design. After that repeat steps steps 4–7 and 9 or if something unexpected is found, go back to the upstream phases.

9. On the generated call graph, assign the bottom-up coding order.
10. According to the coding order, perform unit programming the testing incrementally; if something unexpected is found, go back to the upstream phases.
11. Use OO-Test to perform system testing; repeat steps (d)–(g), then go to step (l).
12. Use OO-Diagrammer to generate the logic diagram of your entire product, perform static review using the traceable diagrammed code.
13. Use the OO-Validate tool to remove inconsistent defects through bidirectional traceability.
14. Use other tools of S_Panorama or S_Panojava to perform quality measurement, static and dynamic program measurement, memory leak checking, performance measurement, tracing the execution path for each runtime error, and more.
15. Try side-effect prevention for requirement changes or code modification.
16. Make a summary of your project development, then answer the following questions:

    - Is the entire development process of your project visible?
    - Is all of your work products for your project visible?
    - Have you made your documents traceable to and from the source code?
    - Have your project satisfied 100% of the MC/DC test coverage? (If it is not 100% satisfied, then answer the percentage and state the reasons).
    - Have you prevented the side-effects in the implementation of requirement changes and code modifications? How did you do it?
    - What have you learned from this small project?

24.10 Further Reading and Information Source

(a) David Rice (2008) GEEKONOMICS the real cost of insecure software. Pearson Education Publishing as Addison-Wesley, Reading
(b) Pressman RS (2005) The Road Ahead (Chap. 32). In: Software engineering: a practitioner's approach. McGraw-Hill, New York

References

[Bro95] Brooks FP Jr (1995) The mythical man-month. Addison–Wesley, Reading
[Bro95-P179] Brooks FP Jr (1995) The mythical man-month. Addison-Wesley, Reading, p 170
[Bro95-P180] Brooks FP Jr (1995) The mythical man-month. Addison-Wesley, Reading, p 180
[Bro95-P181] Brooks FP Jr (1995) The mythical man-month. Addison-Wesley, Reading, p 181

[Bro95-P211]	Brooks FP Jr (1995) The mythical man-month. Addison-Wesley, Reading, p 211
[Bro95-P274]	Brooks FP Jr (1995) The mythical man-month. Addison-Wesley, Reading, p 274
[Coc08]	Alistair C (2008) Using both incremental and iterative development, CrossTalk, May Issue
[Dem86]	Deming WE (1986) Out of the crisis. MIT Press, Cambridge, MA
[GSAM03]	Department of the Air Force Software Technology Support Center, Condensed GSAM handbook, chapter 2 (2003). CrossTalk
[Hol95]	Holland JH (1995) Hidden order: how adaptation builds complexity. Addison-Wesley, Reading
[Jon06]	Jones C (2006) Social and technical reasons for software project failures. CrossTalk, June Issue
[Jon94]	Jones C (1994) Assessment and control of software risks. Prentice Hall, Englewood Cliffs, p 619
[Pre05-P4]	Pressman RS (2005) Software engineering: a practitioner's approach. McGraw-Hill, New York, p 4
[Ric08]	Rice D (2008) Geekonomics the real cost of insecure software. Pearson, Reading
[Ros08]	Dave R (2008) Total economic cost of insecure software: $180 billion a year in the U.S. http://www.news.cnet.com/8301-13846_3-9978812-62.html
[Xio09-1]	Jay X (2009) Tutorial, a complete revolution in software engineering based on complexity science, WORLDCOMP'09, Las Vegas, July 13–17, 2009
[Xio09-2]	Jay X, Jonathan X (2009) A complete revolution in software engineering based on complexity science, WORLDCOMP'09 – SERP (Software Engineering Research and Practice 2009), pp 109–115

Appendix A
Software Requirements Specification Template To Be Used with NSE

> Items that are intended to stay in as part of your document are in **bold**; explanatory comments are in *italic* text. Plain text is used where you might insert wording about your project.

This document is an outline for specifying software requirements with NSE, adapted from the IEEE Guide to Software Requirements Specifications (Std 830-1998 and Std 830-1993). The corresponding file name of this document in the provided disk (or virtual "disk" held on a Web site) of this book is NSE_SRS.doc with bookmarks inserted, so that the information can be easily used in the description part of the related test cases for establishing automated and self-maintainable traceability among all the related documents and test cases and the source code (see Chap. 9 for guidance and examples). It is recommended for the customer and the software development organization to work closely together to write and maintain the software requirement specification document.

With NSE, it is recommended to use the HAETVE (Holistic, Actor–Action and Event–Response driven, Traceable, Visual, and Executable) technique (see Chap. 11) through dummy programming using dummy modules (only providing the dummy source code is good enough, providing both the source code and corresponding J-Chart is better) to replace the Use Case approach in the requirement specification.

NSE Software Requirements Specification Template
For
<Project>
Version 1.0 approved
Prepared by <author>
<organization>
<date created>
Revision History

Name	Date	Reason For change	Approval	Version

Table of Contents
1 Introduction
    1.1   Purpose
    1.2   Scope
    1.3   Definitions, Acronyms, and Abbreviations
    1.4   References
    1.5   Overview
2 The Overall Description
    2.1   Product Perspective
        2.1.1   System Interfaces
        2.1.2   User Interfaces
        2.1.3   Hardware Interfaces
        2.1.4   Software Interfaces
        2.1.5   Communications Interfaces
        2.1.6   Memory Constraints
        2.1.7   Operations
        2.1.8   Site Adaptation Requirements
    2.2   Product Functions
    2.3   User Characteristics
    2.4   Constraints
    2.5   Assumptions and Dependencies
    2.6   Apportioning of Requirements
3 Specific Requirements
    3.1   External Interfaces
    3.2   Functions
    3.3   Performance Requirements
    3.4   Logical Database Requirements
    3.5   Design Constraints
        3.5.1   Standards Compliance
    3.6   Software System Attributes
        3.6.1   Reliability
        3.6.2   Availability
        3.6.3   Security
        3.6.4   Maintainability
        3.6.5   Portability
    3.7   Organizing the Specific Requirements
        3.7.1   System Mode
        3.7.2   User Class
        3.7.3   Objects

        3.7.4   Feature
        3.7.5   Stimulus
        3.7.6   Response
        3.7.7   Functional Hierarchy
  3.8 Additional Comments
4 Supporting Information
5 Change Management Process
6 Document Approvals
7 Delivery
8 Appendix

1 Introduction

The introduction of the SRS should provide an overview of the entire SRS. It should contain the following subsections:

(a) *Purpose*
(b) *Scope*
(c) *Definitions, acronyms, and abbreviations*
(d) *References*
(e) *Overview*

1.1 Purpose

This subsection should

(a) *Delineate the purpose of the SRS*
(b) *Specify the intended audience for the SRS*

1.2 Scope

In this subsection:

(a) *Identify the software product(s) to be produced by name*
(b) *Explain what the software product(s) will, and, if necessary, will not do*
(c) *Describe the application of the software being specified, including relevant benefits, objectives, and goals*
(d) *Be consistent with similar statements in higher-level specifications if they exist*

1.3 Definitions, Acronyms, and Abbreviations

This subsection should provide the definitions of all terms, acronyms, and abbreviations required to properly interpret the SRS. This information may be provided by reference to one or more appendixes in the SRS or by reference to other documents.

1.4 References

In this subsection:

1. *Provide a complete list of all documents referenced elsewhere in the SRS*
2. *Identify each document by title, report number (if applicable), date, and publishing organization*
3. *Specify the sources from which the references can be obtained*

This information may be provided by reference to an appendix or to another document.

1.5 Overview

In this subsection:

1. *Describe what the rest of the SRS contains*
2. *Explain how the SRS is organized*

2 The Overall Description

This section of the SRS should describe the general factors that affect the product and its requirements. This section does not state specific requirements. Instead, it provides a background for those requirements, which are defined in detail in Sect. 3 of the SRS, and makes them easier to understand. This section usually consists of six subsections as follows:

(a) *Product perspective*
(b) *Product functions*
(c) *User characteristics*
(d) *Constraints*
(e) *Assumptions and dependencies*
(f) *Apportioning of requirements*

2.1 Product Perspective

This subsection of the SRS should put the product into perspective with other related products. If the product is independent and totally self-contained, it should be so stated here. If the SRS defines a product that is a component of a larger system, as frequently occurs, then this subsection should relate the requirements of that larger system to functionality of the software and should identify interfaces between that system and the software. A block diagram showing the major components of the larger system, interconnections, and external interfaces can be helpful.

This subsection should also describe how the software operates inside various constraints. For example, these constraints could include

(a) *System interfaces*
(b) *User interfaces*
(c) *Hardware interfaces*
(d) *Software interfaces*
(e) *Communications interfaces*
(f) *Memory*
(g) *Operations*
(h) *Site adaptation requirements*

2.1.1 System Interfaces

This should list each system interface and identify the functionality of the software to accomplish the system requirement and the interface description to match the system.

2.1.2 User Interfaces

This should specify the following:

(a) *The logical characteristics of each interface between the software product and its users.* This includes those configuration characteristics (e.g., required screen formats, page or window layouts, content of any reports or menus, or availability of programmable function keys) necessary to accomplish the software requirements.
(b) *All the aspects of optimizing the interface with the person who must use the system.*

This is a description of how the system will interact with its users. Is there a GUI, a command line or some other type of interface? Are there special interface requirements? If you are designing for the general student population, for instance, what is the impact of ADA (American with Disabilities Act) on your interface?

2.1.3 Hardware Interfaces

This should specify the logical characteristics of each interface between the software product and the hardware components of the system. This includes configuration characteristics (number of ports, instruction sets, etc.). It also covers such matters as what devices are to be supported, how they are to be supported, and protocols. For example, terminal support may specify full-screen support as opposed to line-by-line support.

2.1.4 Software Interfaces

Specify the use of other required software products and interfaces with other application systems. For each required software product, include:

1. *Name*
2. *Mnemonic*
3. *Specification number*
4. *Version number*
5. *Source*

   For each interface, provide:

1. *Discussion of the purpose of the interfacing software as related to this software product*
2. *Definition of the interface in terms of message content and format*

2.1.5 Communications Interfaces

Specify the various interfaces to communications such as local network protocols, etc. These are protocols you will need to directly interact with.
   With NSE, for improving the communication capability and efficiency, besides the developer's internal Web site, it is recommended *to share an extra Project Web Site and the BBS held in the customer site or the developer site, then make the Web pages and the index pages of the BBS traceable to the requirements and the source code (see an application example shown in Fig. 8.17).*
   List:

1. *The URL for the index page of the shared Web site*
2. *The URL for the index page of the BBS*
3. *The URL for the feedback page*
4. *Other important URLs*

2.1.6 Memory Constraints

Specify any applicable characteristics and limits on primary and secondary memory. Do not just make up something here. If all the customer's machines have only

Appendix A

128 KB of RAM, then your target design has got to come under 128 KB so there is an actual requirement. You could also cite market research here for shrink–wrap type applications "Focus groups have determined that our target market has between 256 and 512 MB of RAM, therefore the design footprint should not exceed 256 MB." If there are no memory constraints, so state.

With NSE, memory leak and usage violation should be checked and reported (see Fig. 16.8).

2.1.7 Operations

Specify the normal and special operations required by the user such as:

1. The various modes of operations in the user organization
2. Periods of interactive operations and periods of unattended operations
3. Data processing support functions
4. Backup and recovery operations

(Note: This is sometimes specified as part of User Interfaces) If you separate this from the UI stuff earlier, then cover business process type stuff that would impact the design. For instance, if the company brings all their systems down at midnight for data backup that might impact the design. These are all the work tasks that impact the design of an application, but which might not be located in software.

2.1.8 Site Adaptation Requirements

In this section:

1. Define the requirements for any data or initialization sequences that are specific to a given site, mission, or operational mode
2. Specify the site or mission-related features that should be modified to adapt the software to a particular installation

If any modifications to the customer's work area would be required by your system, then document that here. For instance, "A 100 KW backup generator and 10,000 BTU air conditioning system must be installed at the user site prior to software installation."

This could also be software-specific like, "New data tables created for this system must be installed on the company's existing DB server and populated prior to system activation." Any equipment the customer would need to buy or any software setup that needs to be done so that your system will install and operate correctly should be documented here.

2.2 Product Functions

This subsection of the SRS should provide a summary of the major functions that the software will perform. For example, an SRS for an accounting program may use this part to address customer account maintenance, customer statement, and invoice preparation without mentioning the vast amount of detail that each of those functions requires. Sometimes the function summary that is necessary for this part can be taken directly from the section of the higher-level specification (if one exists) that allocates particular functions to the software product. Note that for the sake of clarity:

(a) The functions should be organized in a way that makes the list of functions understandable to the customer or to anyone else reading the document for the first time.
(b) Textual or graphical methods can be used to show the different functions and their relationships. Such a diagram is not intended to show a design of a product, but simply shows the logical relationships among variables.

2.3 User Characteristics

This subsection of the SRS should describe those general characteristics of the intended users of the product including educational level, experience, and technical expertise. It should not be used to state specific requirements, but rather should provide the reasons why certain specific requirements are later specified in Sect. 3 of the SRS.

2.4 Constraints

This subsection of the SRS should provide a general description of any other items that will limit the developer's options. These include

(a) Regulatory policies
(b) Hardware limitations (e.g., signal timing requirements)
(c) Interfaces to other applications
(d) Parallel operation
(e) Audit functions
(f) Control functions
(g) Higher-order language requirements
(h) Signal handshake protocols (e.g., XON-XOFF and ACK-NACK)
(i) Reliability requirements
(j) Criticality of the application
(k) Safety and security considerations

2.5 Assumptions and Dependencies

This subsection of the SRS should list each of the factors that affect the requirements stated in the SRS. These factors are not design constraints on the software

Appendix A 659

but are, rather, any changes to them that can affect the requirements in the SRS. For example, an assumption may be that a specific operating system will be available on the hardware designated for the software product. If, in fact, the operating system is not available, the SRS would then have to change accordingly.

2.6 Apportioning of Requirements

This subsection of the SRS should identify requirements that may be delayed until future versions of the system.

3 Specific Requirements

This section of the SRS should contain all of the software requirements to a level of detail sufficient to enable designers to design a system to satisfy those requirements, and testers to test that the system satisfies those requirements. Throughout this section, every stated requirement should be externally perceivable by users, operators, or other external systems. These requirements should include at a minimum a description of every input (stimulus) into the system, every output (response) from the system, and all functions performed by the system in response to an input or in support of an output. As this is often the largest and most important part of the SRS, the following principles apply:

(a) Specific requirements should be stated in conformance with all the characteristics described in Sect. 4.3.
(b) Specific requirements should be cross-referenced to earlier documents that relate.
(c) All requirements should be uniquely identifiable.
(d) Careful attention should be given to organizing the requirements to maximize readability.

Before examining specific ways of organizing the requirements, it is helpful to understand the various items that comprise requirements as described in Sects. 3.1–3.7.

3.1 External Interfaces

This contains a detailed description of all inputs into and outputs from the software system. It complements the interface descriptions in Sect. 2 but does not repeat information there. Remember Sect. 2 presents information oriented to the customer/ user while Sect. 3 is oriented to the developer.

It contains both content and format as follows:

- *Name of item*
- *Description of purpose*
- *Source of input or destination of output*
- *Valid range, accuracy, and/or tolerance*
- *Units of measure*
- *Timing*
- *Relationships to other inputs/outputs*
- *Screen formats/organization*
- *Window formats/organization*
- *Data formats*
- *Command formats*
- *End messages*

3.2   Functions

Functional requirements define the fundamental actions that must take place in the software in accepting and processing the inputs and in processing and generating the outputs. These are generally listed as "shall" statements starting with "The system shall..."
   These include:

- *Validity checks on the inputs*
- *Exact sequence of operations*
- *Responses to abnormal situations, including*
    - *Overflow*
    - *Communication facilities*
    - *Error handling and recovery*
- *Effect of parameters*
- *Relationship of outputs to inputs, including*
    - *Input/Output sequences*
    - *Formulas for input to output conversion*

It may be appropriate to partition the functional requirements into subfunctions or subprocesses. This does not imply that the software design will also be partitioned that way.

3.3   Performance Requirements

This subsection specifies both the static and the dynamic numerical requirements placed on the software or on human interaction with the software, as a whole. Static numerical requirements may include:

(a) *The number of terminals to be supported*
(b) *The number of simultaneous users to be supported*
(c) *Amount and type of information to be handled*

Appendix A

Static numerical requirements are sometimes identified under a separate section entitled capacity.

Dynamic numerical requirements may include, for example, the numbers of transactions and tasks and the amount of data to be processed within certain time periods for both normal and peak workload conditions.

All of these requirements should be stated in measurable terms.

For example,

 95% of the transactions shall be processed in less than 1 s

rather than,

 An operator shall not have to wait for the transaction to complete.

(Note: Numerical limits applied to one specific function are normally specified as part of the processing subparagraph description of that function.)

3.4 Logical Database Requirements

This section specifies the logical requirements for any information that is to be placed into a database. This may include:

- *Types of information used by various functions*
- *Frequency of use*
- *Accessing capabilities*
- *Data entities and their relationships*
- *Integrity constraints*
- *Data retention requirements*

If the customer provided you with data models, those can be presented here. ER diagrams (or static class diagrams) can be useful here to show complex data relationships. Remember a diagram is worth a thousand words of confusing text.

3.5 Design Constraints

Specify design constraints that can be imposed by other standards, hardware limitations, etc.

3.5.1 Standards Compliance

Specify the requirements derived from existing standards or regulations. They might include:

1. *Report format*
2. *Data naming*

3. Accounting procedures
4. Audit tracing

For example, this could specify the requirement for software to trace processing activity. Such traces are needed for some applications to meet minimum regulatory or financial standards. An audit trace requirement may, for example, state that all changes to a payroll database must be recorded in a trace file with before and after values.

3.6 Software System Attributes

There are a number of attributes of software that can serve as requirements. It is important that required attributes be specified so that their achievement can be objectively verified. The following items provide a partial list of examples. These are also known as nonfunctional requirements or quality attributes.

These are characteristics the system must possess, but that pervade (or crosscut) the design. These requirements have to be testable just like the functional requirements. It is easy to start philosophizing here, but keep it specific.

3.6.1 Reliability

Specify the factors required to establish the required reliability of the software system at the time of delivery. If you have MTBF requirements, express them here. This does not refer to just having a program that does not crash. This has a specific engineering meaning.

3.6.2 Availability

Specify the factors required to guarantee a defined availability level for the entire system such as checkpoint, recovery, and restart. This is somewhat related to reliability. Some systems run only infrequently on-demand (like MS Word). Some systems have to run 24/7 (like an e-commerce Web site). The required availability will greatly impact the design. What are the requirements for system recovery from a failure? "The system shall allow users to restart the application after failure with the loss of at most 12 characters of input".

3.6.3 Security

Specify the factors that would protect the software from accidental or malicious access, use, modification, destruction, or disclosure. Specific requirements in this area could include the need to:

Appendix A 663

- *Utilize certain cryptographic techniques*
- *Keep specific log or history data sets*
- *Assign certain functions to different modules*
- *Restrict communications between some areas of the program*
- *Check data integrity for critical variables*

3.6.4 Maintainability

Specify attributes of software that relate to the ease of maintenance of the software itself. There may be some requirement for certain modularity, interfaces, complexity, etc. Requirements should not be placed here just because they are thought to be good design practices.

3.6.5 Portability

Specify attributes of software that relate to the ease of porting the software to other host machines and/or operating systems. This may include:

- *Percentage of components with host-dependent code*
- *Percentage of code that is host dependent*
- *Use of a proven portable language*
- *Use of a particular compiler or language subset*
- *Use of a particular operating system*

3.7 *Organizing the Specific Requirements*

For anything but trivial systems, the detailed requirements tend to be extensive. For this reason, it is recommended that careful consideration be given to organizing these in a manner optimal for understanding. There is no one optimal organization for all systems. Different classes of systems lend themselves to different organizations of requirements in Sect. 3. Some of these organizations are described in the following subclasses.

3.7.1 System Mode

Some systems behave quite differently depending on the mode of operation. When organizing by mode, there are two possible outlines. The choice depends on whether interfaces and performance are dependent on mode.

3.7.2 User Class

Some systems provide different sets of functions to different classes of users.

3.7.3 Objects

Objects are real-world entities that have a counterpart within the system. Associated with each object is a set of attributes and functions. These functions are also called services, methods, or processes. Note that sets of objects may share attributes and services. These are grouped together as classes.

3.7.4 Feature

A feature is an externally desired service by the system that may require a sequence of inputs to effect the desired result. Each feature is generally described in a sequence of stimulus–response pairs.

3.7.5 Stimulus

Some systems can be best organized by describing their functions in terms of stimuli.

3.7.6 Response

Some systems can be best organized by describing their functions in support of the generation of a response.

3.7.7 Functional Hierarchy

When none of the above organizational schemes prove helpful, the overall functionality can be organized into a hierarchy of functions organized by either common inputs, common outputs, or common internal data access. Data flow diagrams and data dictionaries can be used to show the relationships between and among the functions and data.

With NSE, it is recommended to form a document hierarchy description table using bookmarks to indicate what requirements are related to what design documents and other documents and the test scripts as well as the test case numbers, so that when users want to modify a requirement or perform requirement validation, it is easy to locate the related test cases to perform forward tracing to identify the corresponding source code.

3.8 Additional Comments

Whenever a new SRS is contemplated, more than one of the organizational techniques given in Sect. 3.7 may be appropriate. In such cases, organize the specific requirements for multiple hierarchies tailored to the specific needs of the system under specification.

There are many notations, methods, and automated support tools available to aid in the documentation of requirements. For the most part, their usefulness is a function of organization. For example, when organizing by mode, finite state machines or state charts may prove helpful; when organizing by object, object-oriented analysis may prove helpful; when organizing by feature, stimulus–response sequences may prove helpful; when organizing by functional hierarchy, data flow diagrams and data dictionaries may prove helpful.

In any of the outlines below, those sections called "Functional Requirement i" may be described in native language, in pseudo code, in a system definition language, or in four subsections titled: Introduction, Inputs, Processing, and Outputs.

4 Supporting Information

The supporting information makes the SRS easier to use. It includes the following:

(a) *Table of contents*
(b) *Index*
(c) *Appendixes*

4.1 Table of Contents and Index

The table of contents and index are quite important and should follow general compositional practices.

4.2 Appendixes

The appendixes are not always considered part of the actual SRS and are not always necessary. They may include

(a) *Sample input/output formats, descriptions of cost analysis studies, or results of user surveys*
(b) *Supporting or background information that can help the readers of the SRS*
(c) *A description of the problems to be solved by the software*
(d) *Special packaging instructions for the code and the media to meet security, export, initial loading, or other requirements*

When appendixes are included, the SRS should explicitly state whether or not the appendixes are to be considered part of the requirements.

5 Change Management Process

Identify the change management process to be used to identify, log, evaluate, and update the SRS to reflect changes in project scope and requirements. How are you going to control changes to the requirements? Can the customer just call up and ask for something new? Does your team have to reach consensus? How do changes to requirements get submitted to the team? Formally in writing, email, or phone call?

6 Document Approvals

Identify the approvers of the SRS document. Approver name, signature, and date should be used.

7 Delivery

Indicate what should be delivered – with NSE, it is recommended to deliver:
1. ***The computer program (source code and executable product)***
2. ***The data used***
3. ***The documents traceable to and from the source code***
4. ***The database built through static and dynamic measurement of the program***
5. ***A set of Assisted Online Agents (AOA) to support testability, visibility, reliability, traceability, changeability, conformity, and maintainability, including a set of AOA of NSE if the product is developed with NSE***

8 Appendix

This section is optional.
   Appendices may be included if any, either directly or by reference, to provide supporting details that could aid in the understanding of the Software Requirements Specifications.

Outline for SRS Section 3
Organized by mode: Version 1

3 Specific Requirements
   3.1 External Interface Requirements
       3.1.1 User Interfaces
       3.1.2 Hardware Interfaces
       3.1.3 Software Interfaces
       3.1.4 Communications Interfaces
   3.2 Functional Requirements
       3.2.1 Mode 1
           3.2.1.1 Functional Requirement 1.1
           ...
           3.2.1.n Functional Requirement 1.n
       3.2.2 Mode 2
       ...
       3.2.m Mode m
           3.2.m.1 Functional Requirement m.1
           ...
           3.2.m.n Functional Requirement $m.n$
   3.3 Performance Requirements
   3.4 Design Constraints
   3.5 Software System Attributes
   3.6 Other Requirements

Outline for SRS Section 3
Organized by mode: Version 2

3 Specific Requirements
   3.1 Functional Requirements
       3.1.1 Mode 1
           3.1.1.1 External Interfaces
               3.1.1.1.1 User Interfaces
               3.1.1.1.2 Hardware Interfaces
               3.1.1.1.3 Software Interfaces
               3.1.1.1.4 Communications Interfaces
           3.1.1.2 Functional Requirement
               3.1.1.2.1 Functional Requirement 1
           ...
               3.1.1.2.n Functional Requirement n
           3.1.1.3 Performance
       3.1.2 Mode 2
       3.1.m Mode m
   3.2 Design Constraints
   3.3 Software System Attributes
   3.4 Other Requirements

Outline for SRS Section 3
Organized by user class (i.e. different types of users → System Administrators, Managers, Clerks, etc.)

3 Specific Requirements
   3.1 External Interface Requirements
      3.1.1 User Interfaces
      3.1.2 Hardware Interfaces
      3.1.3 Software Interfaces
      3.1.4 Communications Interfaces
   3.2 Functional Requirements
      3.2.1 User Class 1
          3.2.1.1 Functional Requirement 1.1
          ...
          3.2.1.n Functional Requirement 1.n
      3.2.2 User Class 2
      ...
      3.2.m User Class m
          3.2.m.1 Functional Requirement m.1
          ...
          3.2.m.n Functional Requirement $m.n$
   3.3 Performance Requirements
   3.4 Design Constraints
   3.5 Software System Attributes
   3.6 Other Requirements

Outline for SRS Section 3
Organized by object (Good if you did an object-oriented analysis as part of your requirements)

3 Specific Requirements
   3.1 External Interface Requirements
      3.1.1 User Interfaces
      3.1.2 Hardware Interfaces
      3.1.3 Software Interfaces
      3.1.4 Communications Interfaces
   3.2 Classes/Objects
      3.2.1 Class/Object 1
          3.2.1.1 Attributes (Direct or Inherited)
              3.2.1.1.1 Attribute 1
              ...
              3.2.1.1.n Attribute n
          3.2.1.2 Functions (Services, Methods, Direct, or Inherited)
              3.2.1.2.1 Functional Requirement 1.1
          ...
              3.2.1.2.m Functional Requirement 1.m

Appendix A 669

           3.2.1.3   Messages (Communications Received or Sent)
     3.2.2   Class/Object 2
           ...
     3.2.p   Class/Object p
  3.3   Performance Requirements
  3.4   Design Constraints
  3.5   Software System Attributes
  3.6   Other Requirements

Outline for SRS Section 3
Organized by feature (Good when there are clearly delimited feature sets)

3   Specific Requirements
  3.1   External Interface Requirements
     3.1.1   User Interfaces
     3.1.2   Hardware Interfaces
     3.1.3   Software Interfaces
     3.1.4   Communications Interfaces
  3.2   System Features
     3.2.1   System Feature 1
           3.2.1.1   Introduction/Purpose of Feature
           3.2.1.2   Stimulus/Response Sequence
           3.2.1.3   Associated Functional Requirements
                 3.2.1.3.1   Functional Requirement 1
                 ...
                 3.2.1.3.n   Functional Requirement n
     3.2.2   System Feature 2
           ...
     3.2.m   System Feature m
  3.3   Performance Requirements
  3.4   Design Constraints
  3.5   Software System Attributes
  3.6   Other Requirements

Outline for SRS Section 3
Organized by stimulus (Good for event-driven systems where the events form logical groupings)

3   Specific Requirements
  3.1   External Interface Requirements
     3.1.1   User Interfaces
     3.1.2   Hardware Interfaces
     3.1.3   Software Interfaces
     3.1.4   Communications Interfaces
  3.2   Functional Requirements
     3.2.1   Stimulus 1

                3.2.1.1   Functional Requirement 1.1
                ...
                3.2.1.n   Functional Requirement 1.*n*
        3.2.2   Stimulus 2
                ...
    3.2.m   Stimulus *m*
                3.2.m.1   Functional Requirement *m*.1
                ...
                3.2.m.n   Functional Requirement *m.n*
    3.3   Performance Requirements
    3.4   Design Constraints
    3.5   Software System Attributes
    3.6   Other Requirements

Outline for SRS Section 3
Organized by response (Good for event-driven systems where the responses form logical groupings)

3   Specific Requirements
    3.1   External Interface Requirements
        3.1.1   User Interfaces
        3.1.2   Hardware Interfaces
        3.1.3   Software Interfaces
        3.1.4   Communications Interfaces
    3.2   Functional Requirements
        3.2.1   Response 1
                3.2.1.1   Functional Requirement 1.1
                ...
                3.2.1.n   Functional Requirement 1.*n*
        3.2.2   Response 2
                ...
    3.2.m   Response *m*
                3.2.m.1   Functional Requirement *m*.1
                ...
                3.2.m.n   Functional Requirement *m.n*
    3.3   Performance Requirements
    3.4   Design Constraints
    3.5   Software System Attributes
    3.6   Other Requirements

Outline for SRS Section 3
Organized by functional hierarchy (Good if you have done structured analysis as part of your design)

3 Specific Requirements
   3.1 External Interface Requirements
      3.1.1 User Interfaces
      3.1.2 Hardware Interfaces
      3.1.3 Software Interfaces
      3.1.4 Communications Interfaces
   3.2 Functional Requirements
      3.2.1 Information Flows
         3.2.1.1 Data Flow Diagram 1
            3.2.1.1.1 Data Entities
            3.2.1.1.2 Pertinent Processes
            3.2.1.1.3 Topology
         3.2.1.2 Data Flow Diagram 2
            3.2.1.2.1 Data Entities
            3.2.1.2.2 Pertinent Processes
            3.2.1.2.3 Topology
         ...
         3.2.1.n Data Flow Diagram *n*
            3.2.1.n.1 Data Entities
            3.2.1.n.2 Pertinent Processes
            3.2.1.n.3 Topology
      3.2.2 Process Descriptions
         3.2.2.1 Process 1
            3.2.2.1.1 Input Data Entities
            3.2.2.1.2 Algorithm or Formula of Process
            3.2.2.1.3 Affected Data Entities
         3.2.2.2 Process 2
            3.2.2.2.1 Input Data Entities
            3.2.2.2.2 Algorithm or Formula of Process
            3.2.2.2.3 Affected Data Entities
         ...
         3.2.2.m Process *m*
            3.2.1.m.1 Input Data Entities
            3.2.1.m.2 Algorithm or Formula of Process
            3.2.1.m.3 Affected Data Entities
      3.2.3 Data Construct Specifications
         3.2.3.1 Construct 1
            3.2.3.1.1 Record Type
            3.2.3.1.2 Constituent Fields
         3.2.3.2 Construct 2
            3.2.3.2.1 Record Type

                3.2.3.2.2   Constituent Fields
            ...
        3.2.3.p   Construct *p*
            3.2.3.p.1   Record Type
            3.2.3.p.2   Constituent Fields
    3.2.4   Data Dictionary
        3.2.4.1   Data Element 1
            3.2.4.1.1   Name
            3.2.4.1.2   Representation
            3.2.4.1.3   Units/Format
            3.2.4.1.4   Precision/Accuracy
            3.2.4.1.5   Range
        3.2.4.2   Data Element 2
            3.2.4.2.1   Name
            3.2.4.2.2   Representation
            3.2.4.2.3   Units/Format
            3.2.4.2.4   Precision/Accuracy
            3.2.4.2.5   Range
        ...
        3.2.4.q   Data Element *q*
            3.2.4.q.1   Name
            3.2.4.q.2   Representation
            3.2.4.q.3   Units/Format
            3.2.4.q.4   Precision/Accuracy
            3.2.4.q.5   Range
3.3   Performance Requirements
3.4   Design Constraints
3.5   Software System Attributes
3.6   Other Requirements

Outline for SRS Section 3
Showing multiple organizations (Can't decide? Then glob it all together)

3   Specific Requirements
    3.1   External Interface Requirements
        3.1.1   User Interfaces
        3.1.2   Hardware Interfaces
        3.1.3   Software Interfaces
        3.1.4   Communications Interfaces
    3.2   Functional Requirements
        3.2.1   User Class 1
            3.2.1.1   Feature 1.1
                3.2.1.1.1   Introduction/Purpose of Feature
                3.2.1.1.2   Stimulus/Response Sequence
                3.2.1.1.3   Associated Functional Requirements

3.2.1.2   Feature 1.2
    3.2.1.2.1   Introduction/Purpose of Feature
    3.2.1.2.2   Stimulus/Response Sequence
    3.2.1.2.3   Associated Functional Requirements
...
3.2.1.m   Feature 1.*m*
    3.2.1.m.1   Introduction/Purpose of Feature
    3.2.1.m.2   Stimulus/Response Sequence
    3.2.1.m.3   Associated Functional Requirements

- 3.2.2 User Class 2
  ...
- 3.2.n User Class *n*
- 3.3 Performance Requirements
- 3.4 Design Constraints
- 3.5 Software System Attributes
- 3.6 Other Requirements

Outline for SRS Section 3
Organized by HAETVE Applications

3 Specific Requirements
  3.1 External Actor Descriptions
    3.1.1 Human Actors
    3.1.2 Hardware Actors
    3.1.3 Software System Actors
  3.2 HAETVE Application Descriptions
    3.2.1 (Dummy Source Code for HAETVE) 1
    3.2.2 (Dummy Source Code for HAETVE) 2
    ...
    3.2.n (Dummy Source Code for HAETVE) *n*
  3.3 Performance Requirements
  3.4 Design Constraints
  3.5 Software System Attributes
  3.6 Other Requirements

Appendix B
An Example About How to Realize 100% MC/DC (Modified Condition/Decision Coverage) for a Program Unit

In this appendix, an example is used for illustrating the test coverage analysis metrics of Panorama C/C++ for Windows XP.

SUM_PRODUCT is a sample program which requests the input of three integers: LOW, HIGH, and MAX. The integers should not be negative, otherwise an error message will be given.

The source code of SUM_PRO.cpp is listed below:

```
#include <stdio.h>
main(void)
// This program prints for each k in the range LOW to HIGH
// k + k and k * k. No more than MAX number of k's are used.
{
  int low, high, max, k, n=0;
  printf("Enter positive integers LOW, HIGH, and MAX:");
  scanf("%d %d %d", &low, &high, &max);
  printf(" LOW = %d  HIGH = %d  MAX = %d \n",low,high,max);
  if ( low >= 0 && high >=0 && max >= 0)
          for (k=low; k<=high; k++)
          {
                  ++n;
                  if (n > max)
                      break;
                  printf(" %d + %d = %d   %d * %d = %d\n",
                  k, k, k+k, k, k, k*k);
          }
  else
          printf("Error! The input data are incorrect! \n");
}
```

The Makefile of SUM_PRO.exe is listed below:

```
#"Makefile"
LINK = link32
CC = cl
SUM_PRO.exe: SUM_PRO.cpp
        $(CC) -c SUM_PRO.cpp
        $(LINK) -out: sum_pro.exe -subsystem:console sum_pro.obj libc.lib kernel32.lib
```

Note: if it is for Panorama C, the file name SUM_PRO.cpp must be renamed to SUM_PRO.c.

A SUM_PRO.hsi file is generated from the Makefile of SUM_PRO.exe and loaded into the main menu of Panorama. Then, a .dbs file is created for SUM_PRO.exe. To capture the dynamic test coverage data, SUM_PRO.exe is executed with several groups of integers as listed below:

LOW	HIGH	MAX
2	8	0
10	20	12
10	1	11
2	8	−2
2	−2	8
−2	2	8

A series of J-Flow and J-Diagrams in OO-Diagrammer are listed to show the changes of accumulated test coverage each time when SUM_PRO.exe is executed.

Note: in this Appendix, the test coverage refers to the accumulated test coverage in order to show the result of all the executions.

Before the execution of SUM_PRO.exe, the test coverage of the code is zero. This is reflected in the Bar graph and diagrams given below (Figs. B.1–B.3):

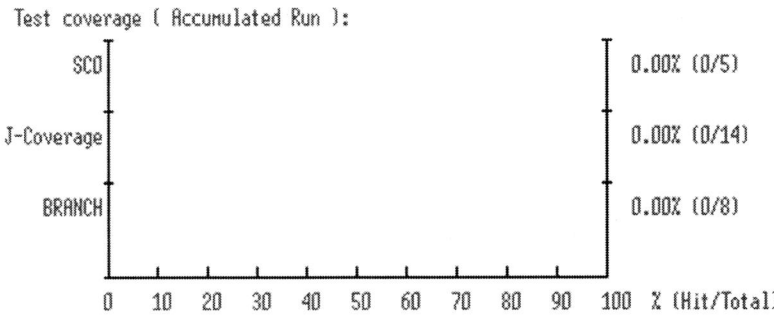

Fig. B.1 Bar graph in OO-Diagrammer:
the test coverage data are all zeros (here, J-Coverage means MC/DC)

Appendix B

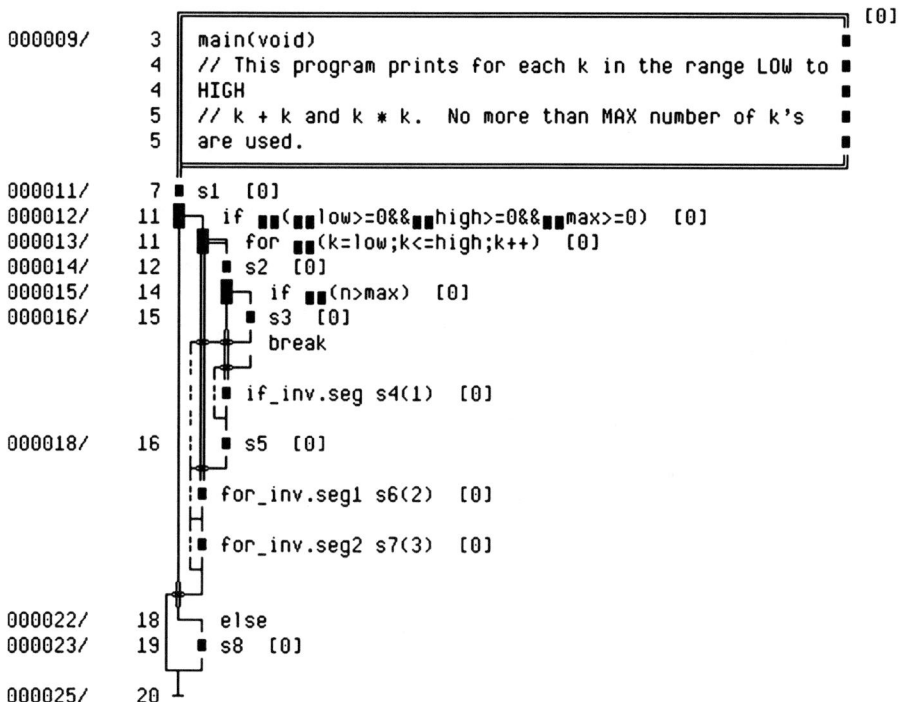

Fig. B.2 J-Flow in OO-Diagrammer:
all the elements are untested and highlighted

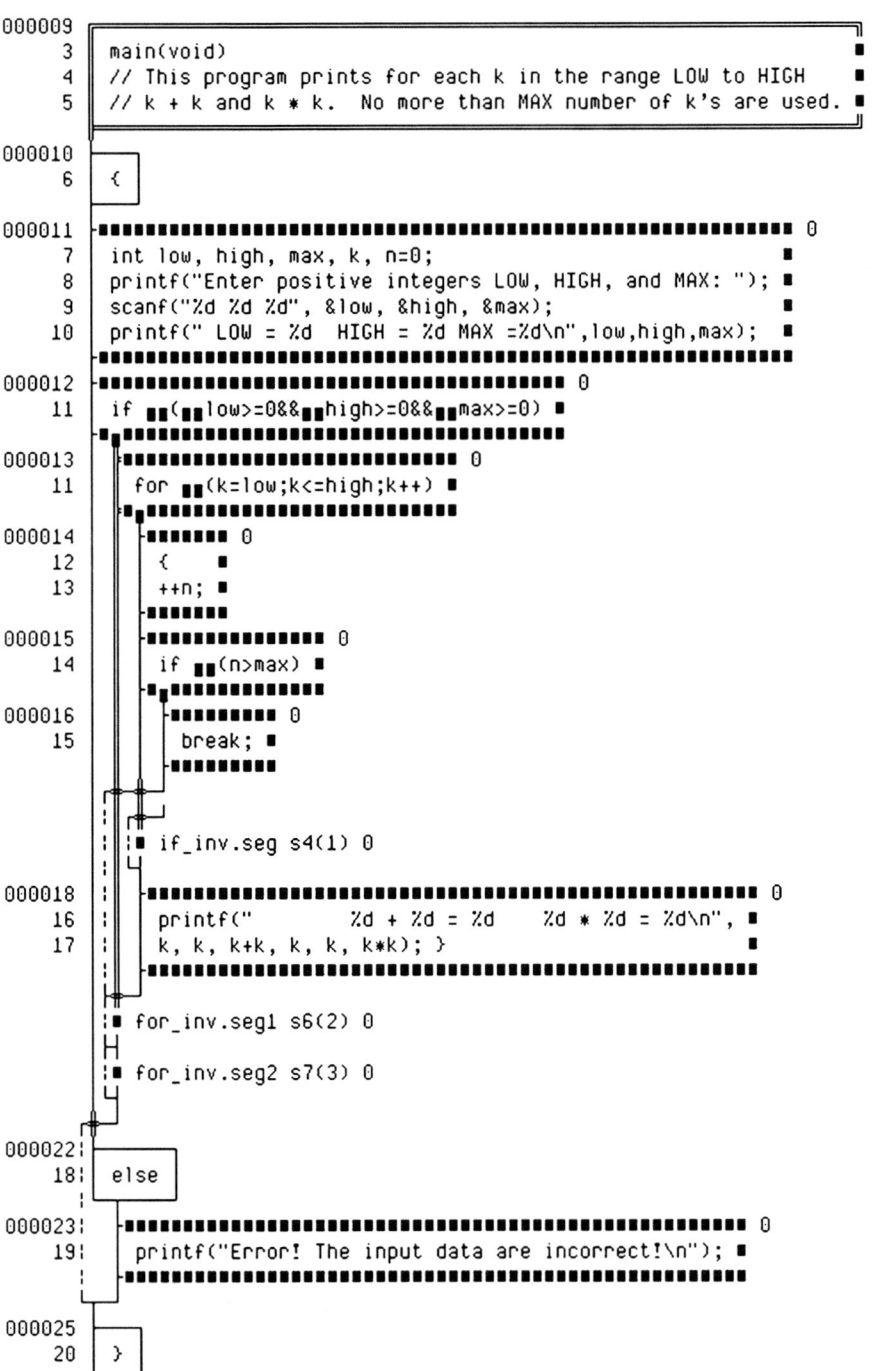

Fig. B.3 J-Diagram in OO-Diagrammer:
accumulated test coverage: all the elements are untested and highlighted

Appendix B

To execute the sample program, type SUM_PRO.exe in the appropriate directory at the prompt:

C: >\Func\SUM_PRO\sum_pro.exe

Enter positive integers LOW, HIGH, and MAX: **2 8 0**

LOW = 2 HIGH = 8 MAX = 0

The bold characters above are typed in at the prompts, while the italic characters are displayed by the sample program SUM_PRO.

Then check the Bar graph, J-Flow, and J-Diagram in OO-Diagrammer. Select the accumulated test coverage on the corresponding Options dialog box, then click OK. The test coverage data are automatically updated (Figs. B.4–B.6):

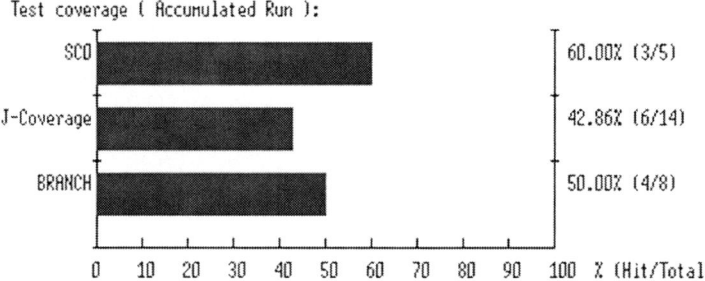

Fig. B.4 Bar graph in OO-Diagrammer:
after the first execution of sum_pro.exe, the test coverage results are to be improved

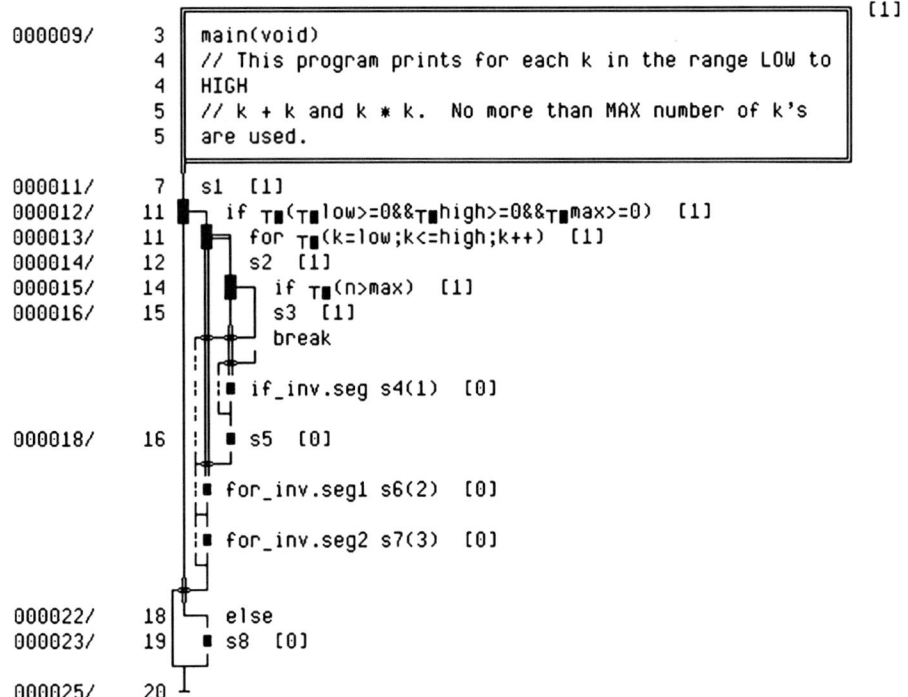

Fig. B.5 J-Flow in OO-Diagrammer: **after the first execution of sum_pro.exe**

Appendix B 681

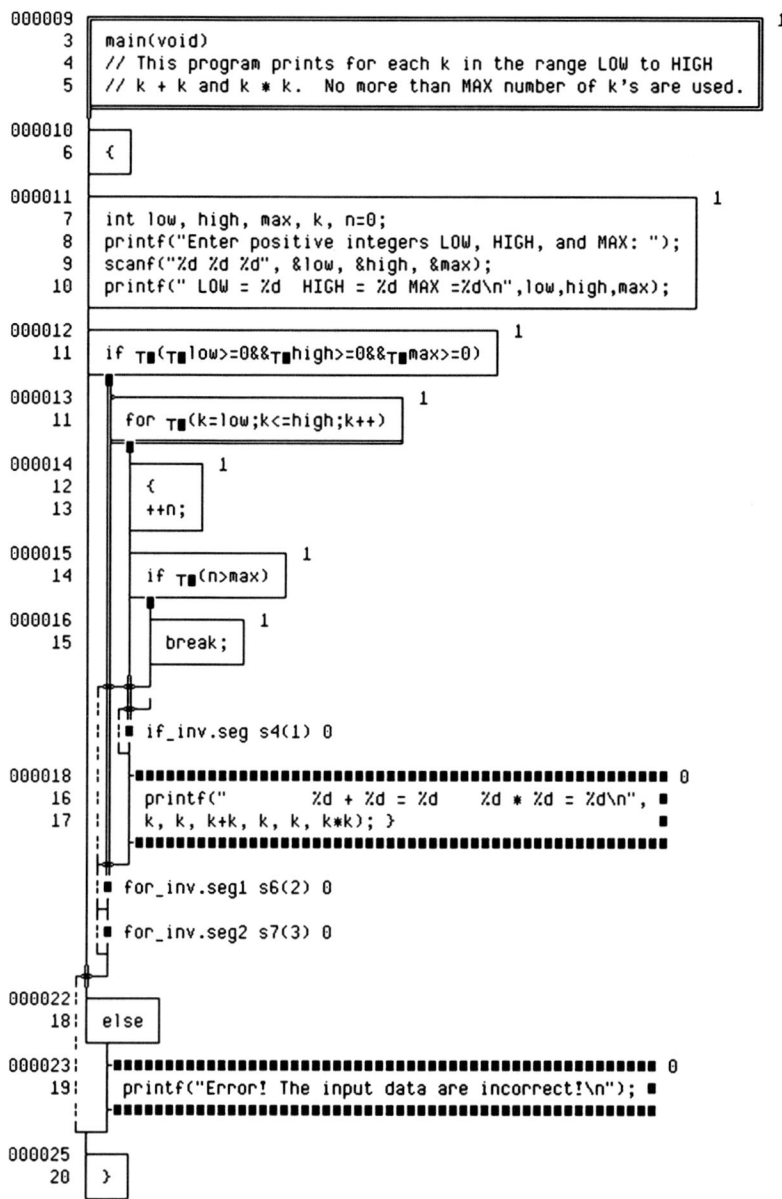

Fig. B.6 J-Diagram in OO-Diagrammer:
after the first execution of sum_pro.exe

Now, execute SUM_PRO.exe again. This time three integers 10, 20, and 12 are inputted. SUM_PRO.exe outputs, from 10 to 20, 11 groups of equations:

C: >\Func\SUM_PRO\sum_pro.exe

Enter positive integers LOW, HIGH, and MAX: **10 20 12**

LOW = 10 HIGH = 20 MAX = 12
*10 + 10 = 20 10 * 10 = 100*
*11 + 11 = 22 11 * 11 = 121*
*12 + 12 = 24 12 * 12 = 144*
*13 + 13 = 26 13 * 13 = 169*
*14 + 14 = 28 14 * 14 = 196*
*15 + 15 = 30 15 * 15 = 225*
*16 + 16 = 32 16 * 16 = 256*
*17 + 17 = 34 17 * 17 = 289*
*18 + 18 = 36 18 * 18 = 324*
*19 + 19 = 38 19 * 19 = 361*
*20 + 20 = 40 20 * 20 = 400*

The bold characters above are typed in at the prompts, while the italic characters are displayed by the sample program SUM_PRO.exe.

Then check the Bar graph, J-Flow, and J-Diagram in OO-Diagrammer. Select the accumulated test coverage data on the corresponding Options dialog box, then click OK. The test coverage data on the diagrams are automatically updated (Figs. B.7–B.9):

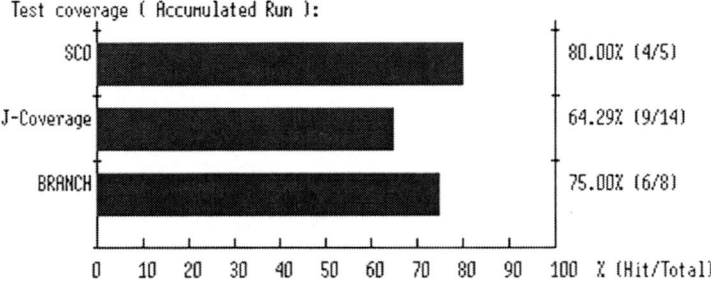

Fig. B.7 Bar graph in OO-Diagrammer:
the test coverage data have increased significantly

Appendix B   683

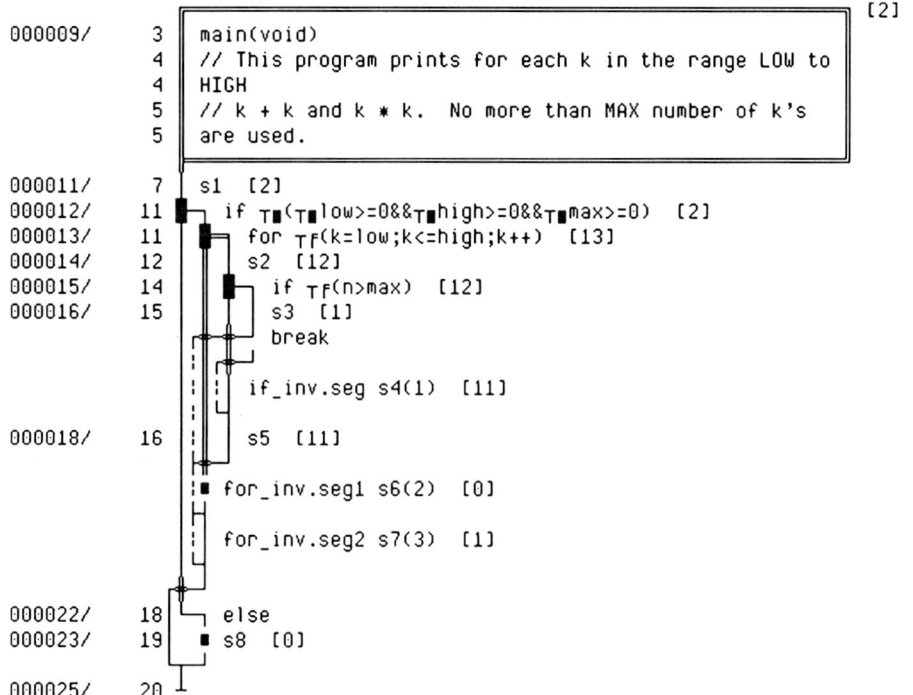

Fig. B.8 J-Flow in OO-Diagrammer:
accumulated test coverage after the second execution

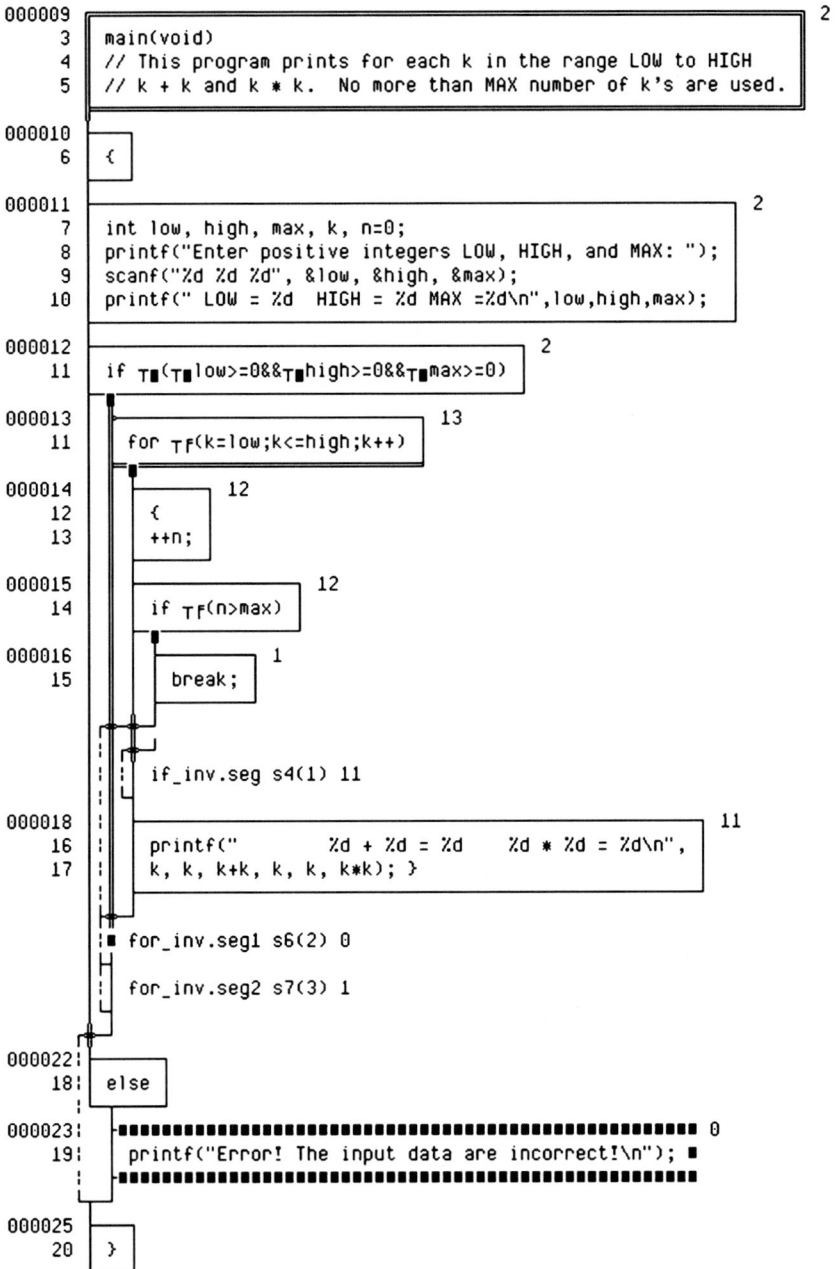

Fig. B.9 J-Diagram in OO-Diagrammer: accumulated test coverage:
the number of unexecuted elements highlighted has been greatly decreased compared with the diagrams before

Appendix B

Now, execute SUM_PRO.exe again to increase its test coverage further. This time integers 10, 1, 11 are inputted.

C: >\Func\SUM_PRO\sum_pro.exe

Enter positive integers LOW, HIGH, and MAX: **10 1 11**

LOW = 10 HIGH = 1 MAX = 11

The bold characters above are typed in at the prompts, while the italic characters are displayed by the sample program SUM_PRO.exe.

Since Low=10>High=1, no equation is outputted this time.

Then check the Bar graph, J-Flow, and J-Diagram in OO-Diagrammer. Select the accumulated test coverage on the corresponding Options dialog box, then click OK. The test coverage data are automatically updated (Figs. B.10–B.12):

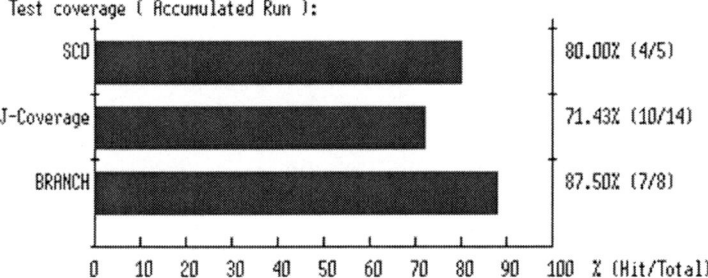

Fig. B.10 Bar graph in OO-Diagrammer: accumulated test coverage: compared with Fig. B.7, one more branch and one more segment are tested. Consequently, J-Coverage is increased by one too

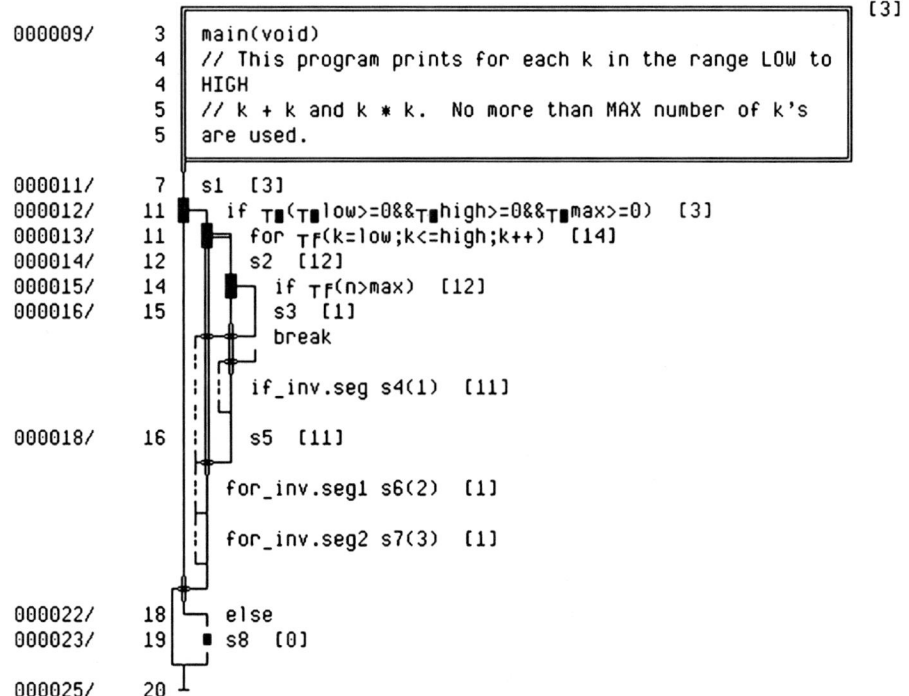

Fig. B.11 J-Flow in OO-Diagrammer: accumulated test coverage: compared with Fig. B.8, one more segment (branch) is tested

Appendix B

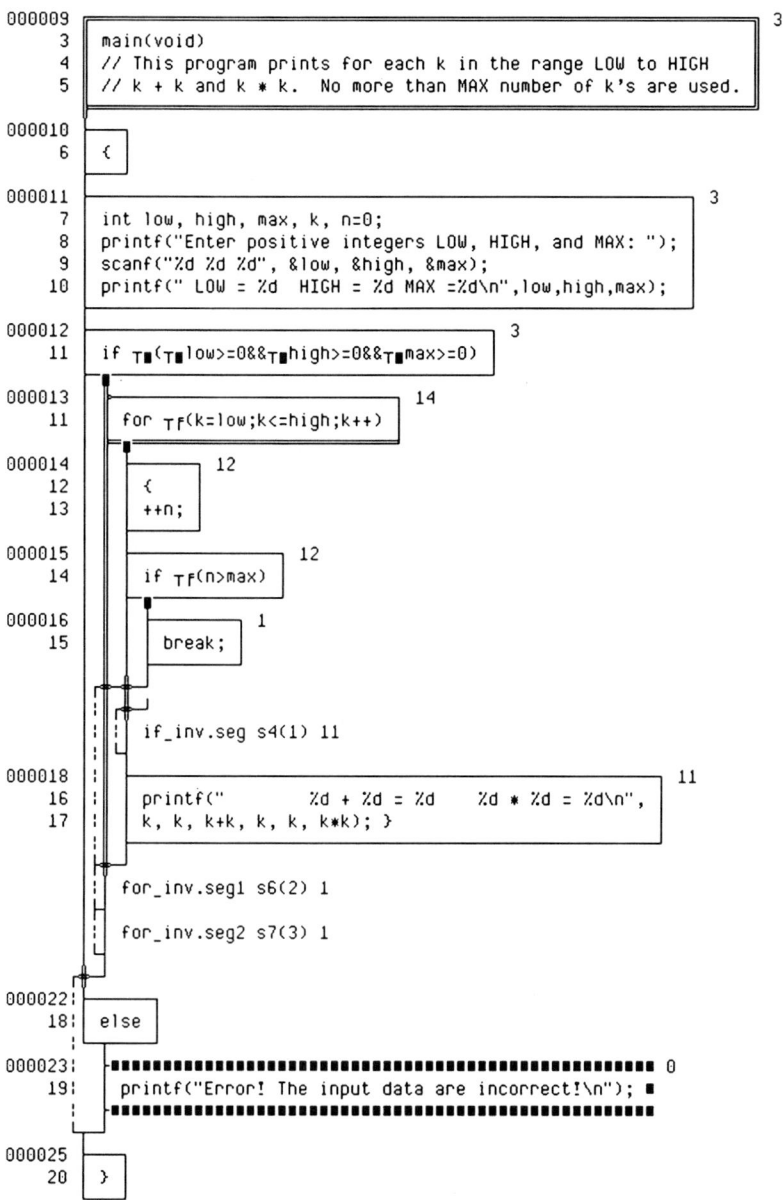

Fig. B.12 J-Diagram in OO-Diagrammer: accumulated test coverage: compared with Fig. B.9, one more segment (branch) is tested

Now, carefully observe the J-Flow or J-Diagram, you may find out that the condition test coverage should be increased. Since Condition True has reached 100% coverage, the Condition False needs to be increased.

C: >\Func\SUM_PRO\sum_pro.exe

Enter positive integers LOW, HIGH, and MAX: **2 8 −2**

LOW = 2 HIGH = 8 MAX = −2

Error! The input data are incorrect!

The bold characters above are typed in at the prompts, while the italic characters are displayed by the sample program SUM_PRO.exe.

Since a negative integer is inputted, an error message is given this time.

Then check the Bar graph, J-Flow, and J-Diagram in OO-Diagrammer. Select the accumulated test coverage on the corresponding Options dialog box, then click OK. The test coverage data are automatically updated (Figs. B.13–B.15):

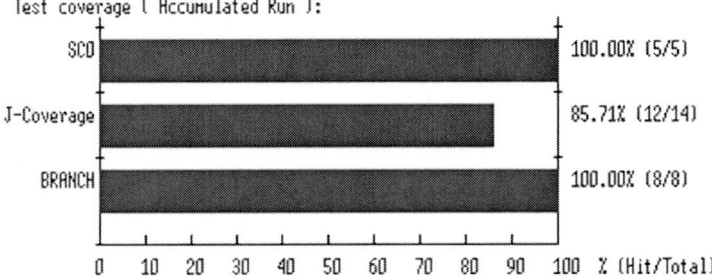

Fig. B.13 Bar graph in OO-Diagrammer:
the accumulated test coverage of SC0 branch has reached 100%. J-Coverage is increased too

Appendix B

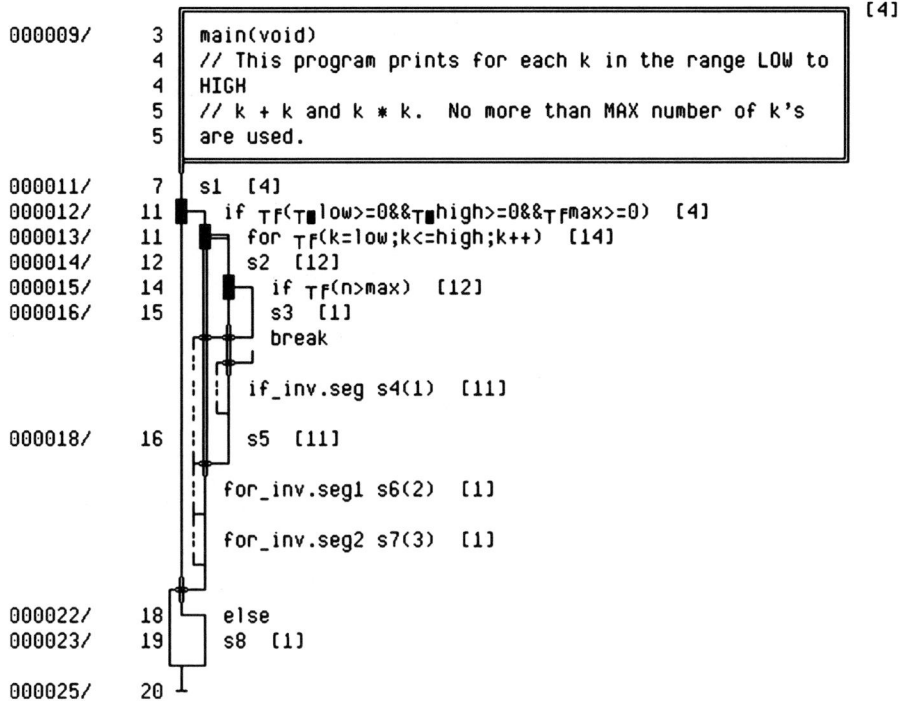

Fig. B.14 J-Flow in OO-Diagrammer:
accumulated test coverage: only two conditions are untested

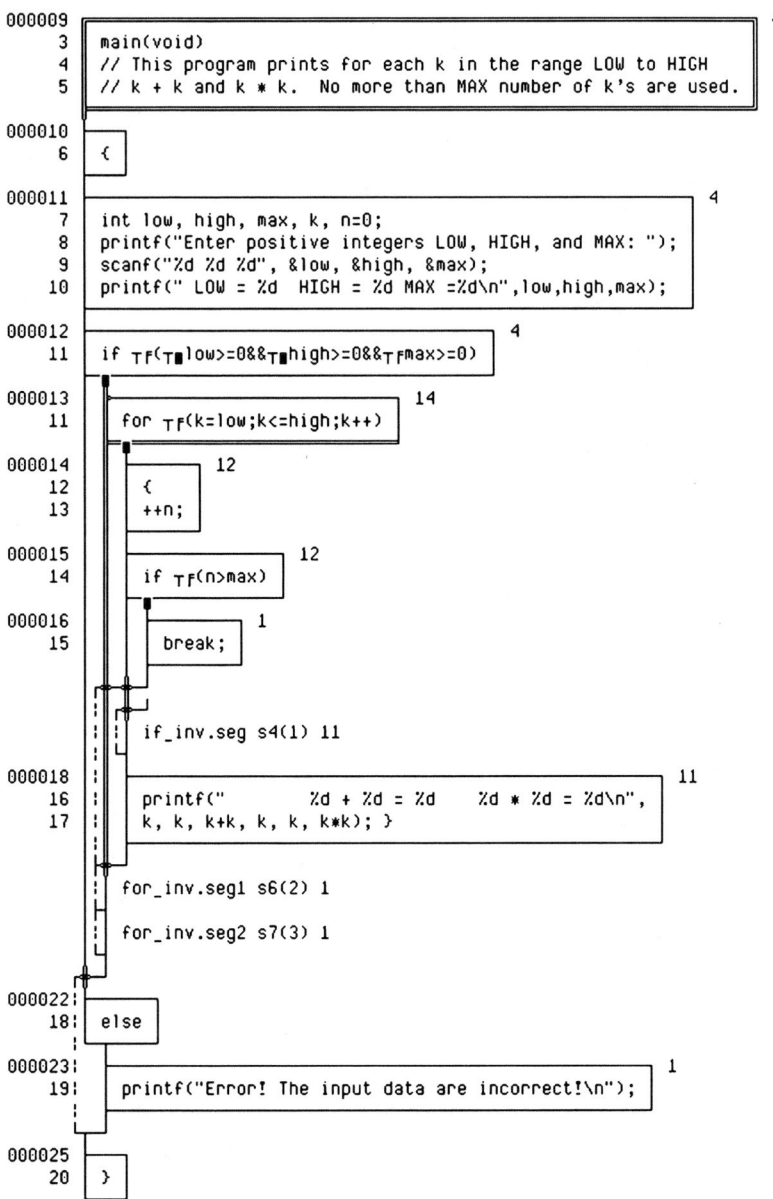

Fig. B.15 J-Diagram in OO-Diagrammer:
accumulated test coverage: only two conditions are untested

Appendix B

To increase the coverage of Condition False, run SUM_PRO.exe again and input another group of integers. This time, integer High is negative.

C: >\Func\SUM_PRO\sum_pro.exe

*Enter positive integers LOW, HIGH, and MAX:***2 −2 8**

LOW = 2 HIGH = −2 MAX = 8

Error! The input data are incorrect!

The bold characters above are typed in at the prompts, while the italic characters are displayed by the sample program SUM_PRO.exe.

Since a negative integer High is inputted, an error message is given too.

Then check the Bar graph, J-Flow, and J-Diagram in OO-Diagrammer. Select the accumulated test coverage in the corresponding Options dialog box, then click OK. The test coverage data are automatically updated (Figs. B.16–B.18):

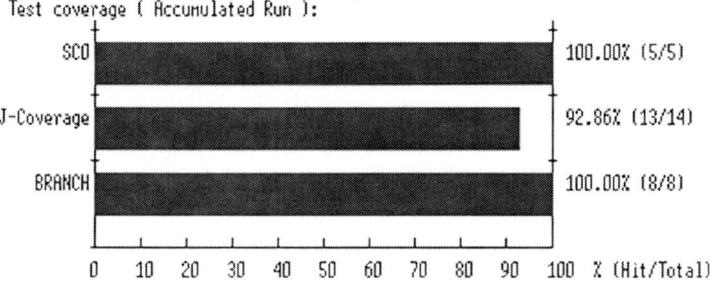

Fig. B.16 Bar graph in OO-Diagrammer:
J-Coverage has been increased

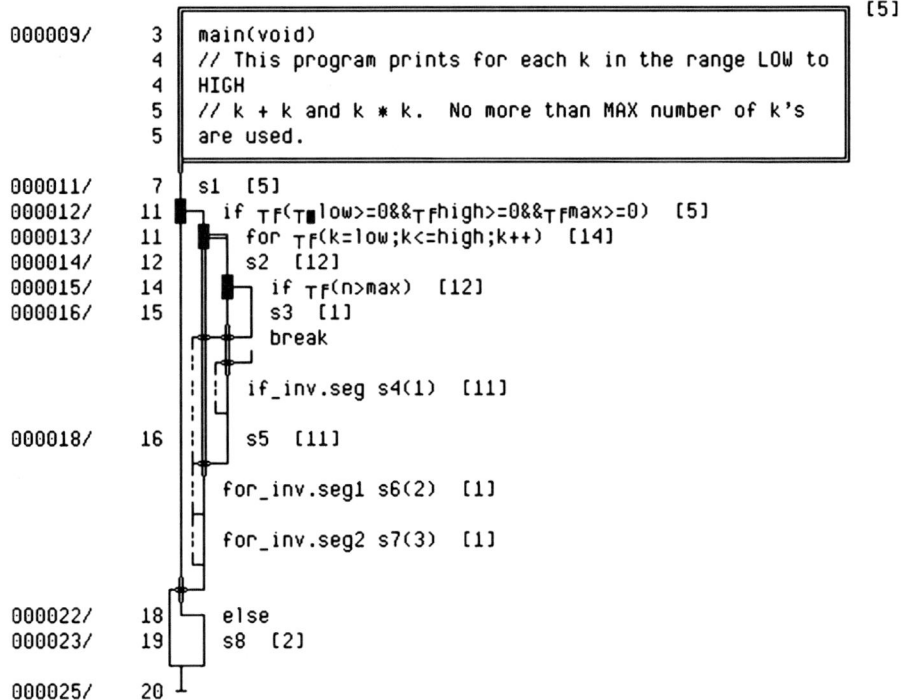

Fig. B.17 J-Flow in OO-Diagrammer:
accumulated test coverage: only one False condition is untested

Appendix B

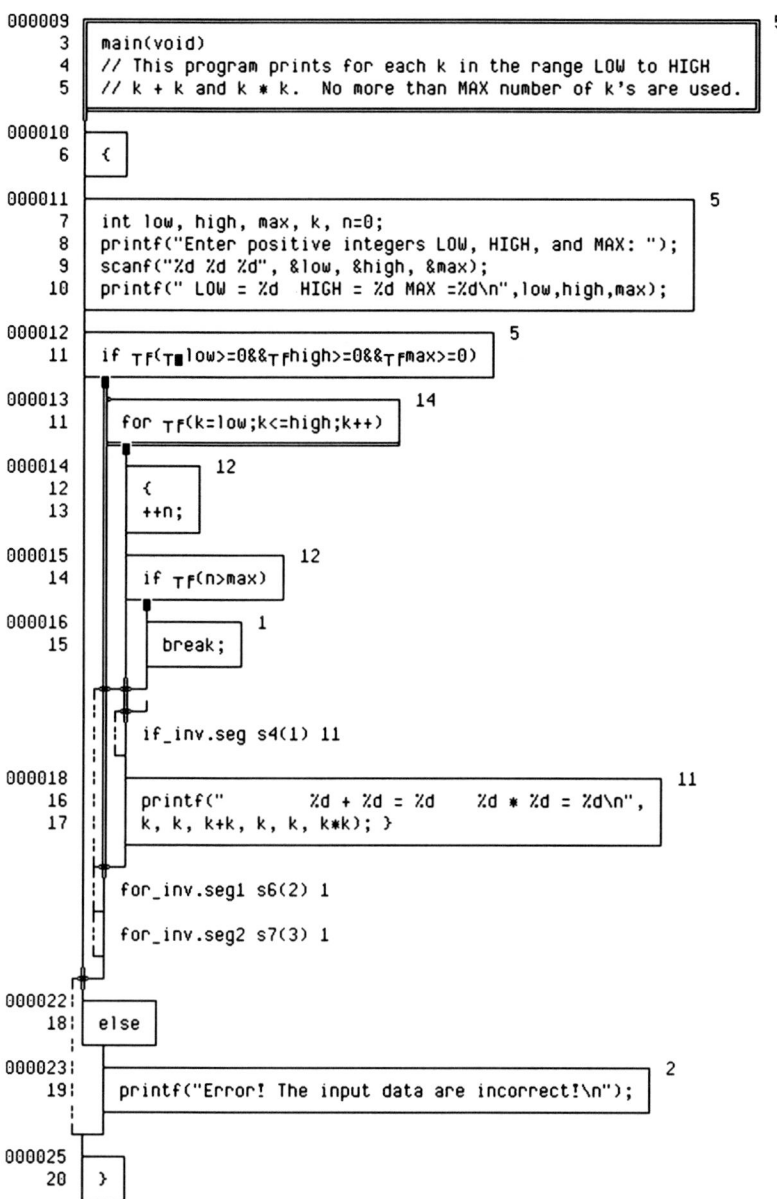

Fig. B.18 J-Diagram in OO-Diagrammer:
accumulated test coverage: only one False condition is untested

To cover all the conditions, run SUM_PRO.exe again and input another group of data with negative Low integer.

C: >\Func\SUM_PRO\sum_pro.exe

Enter positive integers LOW, HIGH, and MAX: **−2 2 8**

LOW = −2 HIGH = 2 MAX = 8

Error! The input data are incorrect!

The bold characters above are typed in at the prompts, while the italic characters are displayed by the sample program SUM_PRO.exe.

Since a negative integer Low is inputted, an error message is given too.

Then check the Bar graph, J-Flow, and J-Diagram in OO-Diagrammer. Select the accumulated test coverage on the corresponding Options dialog box, then click OK. All the conditions should have been covered (Figs. B.19–B.21):

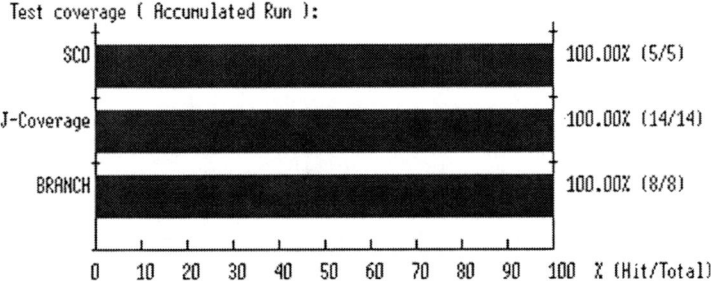

Fig. B.19 Bar graph in OO-Diagrammer:
accumulated test coverage: all the test coverage metrics have been reached 100%

Appendix B

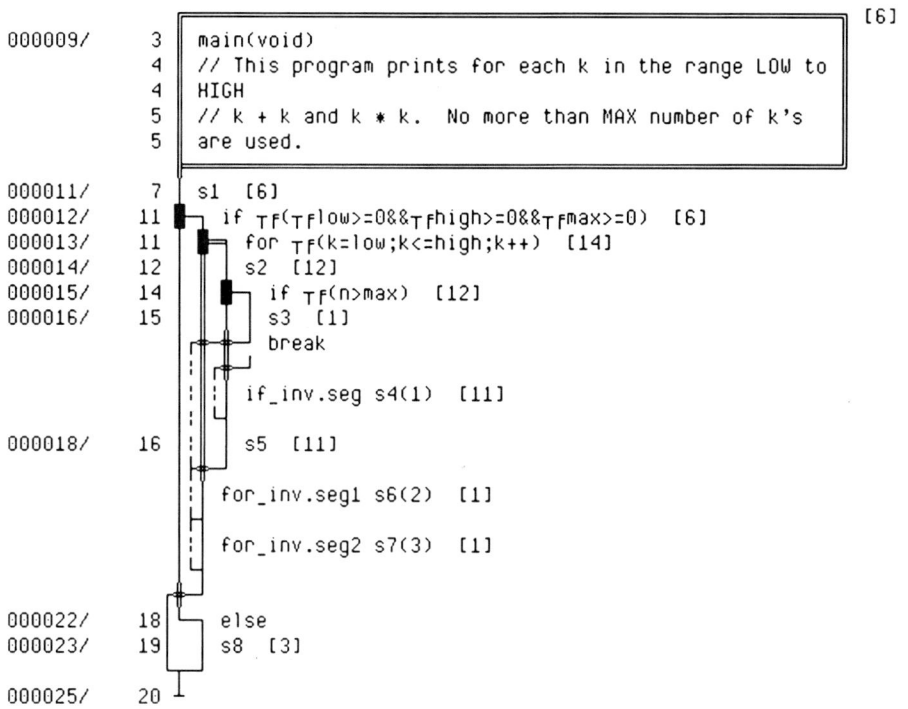

Fig. B.20 J-Flow in OO-Diagrammer:
accumulated test coverage: the program sum_pro.exe is completely tested

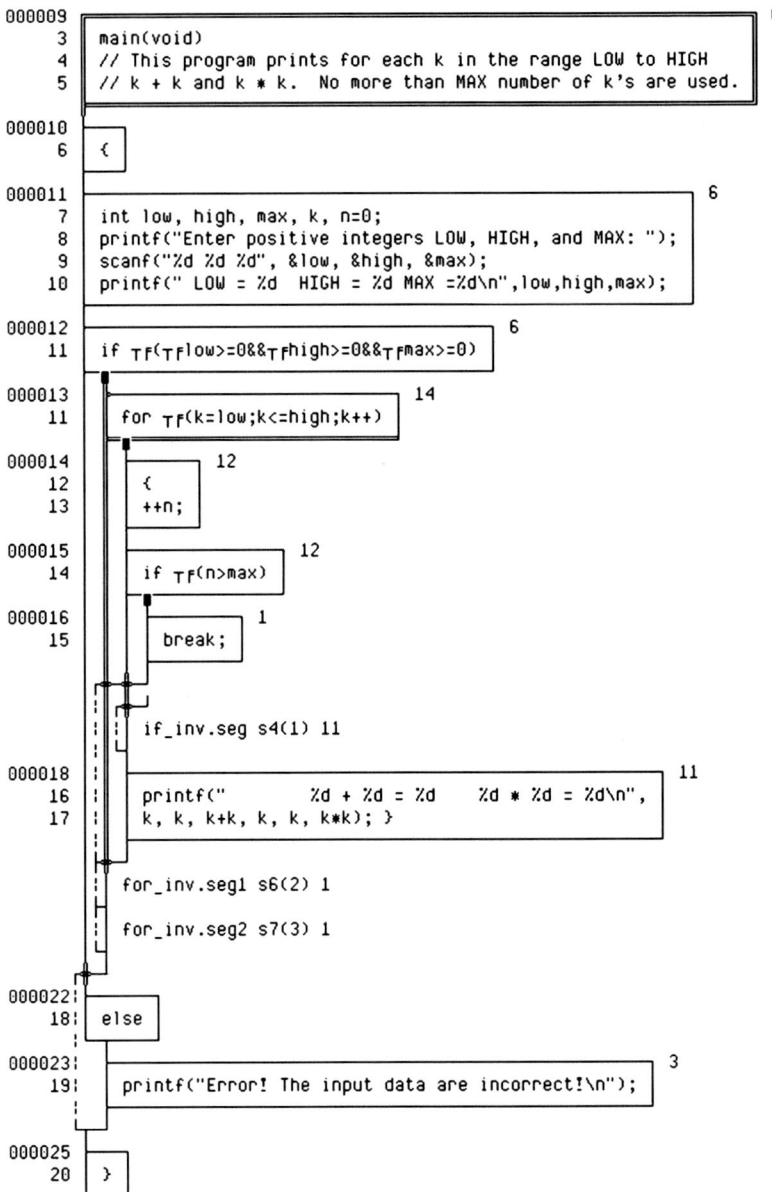

Fig. B.21 J-Diagram in OO-Diagrammer:
accumulated test coverage: the program sum_pro.exe is completely tested

From the example above, it is clearly shown how test coverage data are displayed on J-Flows and J-Diagrams and how the result shown may help you to increase the coverage of your program.

Similarly, other tools of Panorama C/C++, such as the structure charts, software metrics diagrams, reports, ActionPlus diagrams, etc., can also show the dynamic test data vividly and help you successfully plan any further testing.

Appendix C
How to Control/Simulate the Return Values of a Program Unit Being Tested

There are two ways provided: (1) using a global/static variable to control the return values from a called program unit and (2) using files to control the return value of a program unit according to the execution times.

1. The source code (here the statements shown in bold are inserted for getting/controlling the return value in a function call statement)

```
(1) main.c
#include <stdio.h>
extern int sub1();
extern int sub2();


void main(argc, argv)
int argc;
char **argv;
{
int i, returned_value1, returned_value2;
if(argc==1)
   {
    printf("Order  returned_value \n");
    for (i=1; i<7; i++)
       {
        returned_value1=sub1();
        printf("%d \t %d\n", i, returned_value1);
       }
   }
 else
   {
       returned_value2 = sub2();
//    printf ("#################################################\n");
    printf("The retuned value from a unit called by the unit being tested: %d\n", returned_value2);
   }
 printf("\n");
}
(2) trouble.c (a sample application program)
```

```c
/* trouble.c */
#include <stdio.h>
static int flag=0;
int send_value1()
{
++ flag;
  switch (flag)
  {
    case 1: return -1;
    case 2: return 0;
    case 3: return 1;
    case 4: return 32797;
    case 5: return 32798;
    case 6: return 32799;
  }
/* The original statements. */
}
int send_value2()
{
char order_file_name[]="order_file.txt";
char data_file_name[]="data_file.txt";
char line_buf[1024];
FILE *f_order, *f_data;
int order, new_order, line_number, value;
if ((f_order=fopen(order_file_name, "r"))==NULL)
  {
    printf("Error in opening the file of %s \n", order_file_name);
    return -1;
  }
if((fgets(line_buf, sizeof (line_buf), f_order))!=NULL)
  {
    sscanf(line_buf, "%d", &order);
    fclose (f_order);
    new_order=order +1;
    if((f_order=fopen(order_file_name, "w"))==NULL)
      {
        printf("Error in reading the file of %s \n", order_file_name);
        return -1;
      }
    fprintf(f_order, "%d", new_order);
    fclose (f_order);
  }
else
  {
    printf("Error in reading data from the file of %s \n", order_file_name);
    return -1;
  }
if((f_data = fopen(data_file_name,"r"))==NULL)
  {
    printf("Error in reading the data file of %s \n", data_file_name);
```

```
        return -1;
    }
    line_number=1;
    while ((fgets(line_buf, sizeof (line_buf), f_data))!=NULL)
    {
        if(line_number == order) /* It is the right value to be used. */
        {
            sscanf (line_buf, "%d", &value);
            fclose(f_data);
            return value;
        }
        ++line_number;
    }
    fclose(f_data);
/* The original code statements. */
}
int sub1()
{
/* Original code statements */
return send_value1();
}

int sub2()
{
/* Original code statements */
return send_value2();
}
```

2. Other files
(1) order_file.txt
```
1
```
(2) save_order_file.txt
```
1
```
(3) data_file.txt
```
-1
0
1
32797
32798
32799
```

3. The Makefile
```
# Some n make macros for building Win32 applications
CPU=i386
cc=cl
link=link

cflags=-c

all: trouble.exe

OBJS=main.obj trouble.obj
```

```
trouble.obj: trouble.c
   $(cc) $(cflags) $*.c
main.obj: main.c
   $(cc) $(cflags) $*.c
panounit.exe: $(OBJS)
   $(link) -out:panounit.exe -subsystem:console main.obj trouble.obj
libc.lib kernel32.lib
```

—

4. The batch file: run_panounit.bat
copy save_order_file.txt order_file.txt
panounit
panounit 1
panounit 2
panounit 3
panounit 4
panounit 5
panounit 6

Appendix D
Hints for Answering the "Points and Questions to Ponder" in Each Chapter

Note: hints are not real answers, but something to help you make your answers.

Chapter 1

Points and Questions to Ponder

(a) What are the major differences between the traditional **software** definition and the new one defined with NSE? Do you think it is necessary to provide a software product to the customer (not the end-user) with the database built through static and dynamic measurement of the product, and a set of Assistant Online Agents? Why?

Hints: Traditional: **Software**=program+data+document, but the program and the documents are separated without traceability that is established automatically.

With NSE: **Software**=program+data+documents traceable to and from the source code, plus the **database** built through static and dynamic measurement of the program, and a set of **Assisted Online Agents** (automated and intelligent tools working with the program and the database) for handling the issue of complexity and supporting the testability, visibility, changeability, conformity, reliability, and traceability – making the software product adaptive and truly maintainable in the new working environment at the customer site, and that the requirement validation and the acceptance testing can be done dynamically in a fully automated way with mouse clicks only.

(b) Are today's software products sufficiently engineered? Why?

Hints: Today's software products are not sufficiently engineered, because the old-established software engineering paradigm is based on reductionism and the super-position principle that the whole of a complex system is the sum of its components, so that almost all software engineering tasks and activities are performed partially and locally; many critical problems exist such as … (Please complete it).

(c) What are the common limitations existing with current software process models?

Hints: About the common limitations with the current software process models, refer to Section 1.4.5.2.

(d) For efficiently supporting software maintenance, what conditions do you think a process model or software development approach should satisfy?

Hints: The following conditions:

1. Being able to help users perform software maintenance holistically and globally.
2. Being able to greatly reduce the amount of defects introduced into the software product and propagated to the software maintenance phase through defect prevention and defect propagation prevention performed from the first step to the entire software development process.
3. Being able to help users prevent the side effects for the implementation of requirement changes or code modifications.
4. Being able to provide the necessary means to help users greatly reduce the time, resources requested, and cost in regression testing after the implementation of requirement changes or code modifications, such as the capability for test case efficiency analysis and test case minimization, or automated, efficient, and intelligent test case selection.
5. Being able to help the customer side to maintain a software product with almost the same conditions as if the software product is maintained by the product development side.

(e) Although the software engineering paradigm itself is a complex system consisting of many related parts which are connected closely and interactively, some people still believe that only improving one or two parts of the software engineering paradigm without improving its other parts can still dramatically improve the overall characteristics, performance, behavior, and the problem-solving capability of the software engineering paradigm – do you agree with their conclusion? Why?

Hints: No. According to the Holism principle of complexity science, the characteristics and behaviors of the whole of a complex system emerge from the interaction of its parts, and cannot be inferred simply from the behavior of its individual components …(please complete it).

Chapter 2

(a) How is a successful project defined?

Hints:

    The definition of a successful project is one that completed within 10% or so of its committed cost and schedule and delivered all of its intended functions.

(b) What is the root cause that about 70% of software projects are failures?

Appendix D 705

Hints:

1. The foundation of the old-established software engineering paradigm is based on linear thinking, reductionism, and the superposition principle.
2. The old-established software engineering paradigm is outdated including the process models, the software development methodologies, the software testing paradigm, the quality assurance paradigm, the maintenance paradigm, etc. (please read Sects. 2.5–2.12, write down your notes, then close this book and make your answer)

Chapter 3

(a) What is complexity science?

Hints:

Read Sect. 3.1, write down your notes, then close this book and make your answer.

(b) What are the major differences between Reductionism and Holism?

Hints:

Compare the difference: the superposition principle and the Holism principle (about how to handle the relationship between the whole of a complex system and its parts).

(c) What are the essential principles of complexity science? How are they related to the establishment of NSE?

Hints:

Read Sect. 3.2, write down your notes, then close this book and make your answer.

The essential principles of complexity science are the foundation for establishing NSE.

Chapter 4

(a) What are the major differences between Hall's framework and FDS?

Hints:

Consider the differences in the four aspects:
1. The objectives
2. The phases being performed; follow or do not follow a linear order
3. The contents of the axes
4. The use of computer simulation

(b) Why is it recommended to apply complexity science to solve the problems of a complex system in an industry through two major steps (the first one is to complete the paradigm shift by the organization performing the tasks or a tool vendor, then the second one is to handle the detailed tasks by applying the corresponding new paradigm established in the first step)?

Hints:

> The "Sunlight" of complexity science cannot directly "Reach" the target without removing the big "Umbrella" in the middle – the old-established paradigm… (please complete it)

Chapter 5

(a) What are the major problems existing with today's software development? Why are those problems so hard to solve?

Hints:

> 1. Consider the issues of quality, productivity, cost, risk, missed schedules, blown budgets, and flawed products …
> 2. Those problems cannot be solved by a single development, in either technology or management technique – they are caused by the entire existing software engineering paradigm based on reductionism and superposition principle including the linear process models …

(b) Why does today's software maintenance take 75% or more of the total effort and total cost in software product development?

Hints:

> 1. With the linear process models, huge amounts of defects will be introduced into a software product
> 2. The defects easily propagate down to the maintenance phase
> 3. The implementation of requirement changes and code modifications is performed partially and locally without the means to prevent the side-effects …
> 4. The process and the result in software maintenance are invisible.

(c) What is NSE?

Hints:

> Consider:
> 1. The objectives
> 2. The foundation
> 3. The major features
>
>> (Read Sect. 1.7, write down your notes, then close this book and make your answer.)

Appendix D

Chapter 6

(a) What are the driving forces for the establishment of NSE (Nonlinear Software Engineering paradigm)? Describe them in as much detail as possible.

Hints:

Describe each technique listed in Fig. D.1:

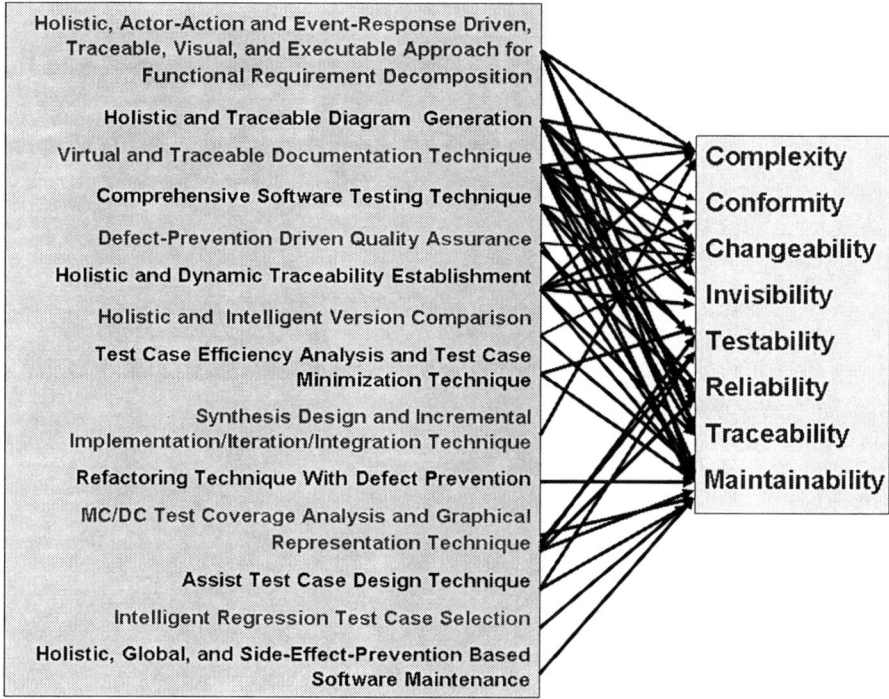

Fig. D.1 NSE techniques

(b) Which principles do the techniques introduced in this chapter comply with? Why?

Hints:

The essential principles of complexity science, including the Nonlinearity principle, the Holism principle, the ... (read Chap. 3 for the detailed description of the principles).

Chapter 7

(a) What are the major differences between the NSE software visualization paradigm and the traditional software visualization paradigm?

Hints:

> Briefly compare Sects. 7.1 and 7.2.

(b) What are the major benefits of virtually existing charts and diagrams without storing hard copies in the hard disk and the memory of a computer?

Hints:

> Consider: (1) the space saved; (2) the time spent in loading the graphics and in displaying the graphics.

(c) Point out the reasons why a system-level call graph or diagram should be made interactive and traceable.

Hints:

Consider:
1. The difference between static graphics and dynamic graphics – which one is more useful?
2. About traceability, compare the following Figs. D.2 and D.3, then make your answer

Appendix D 709

Fig. D.2 A call graph shown in J-Chart

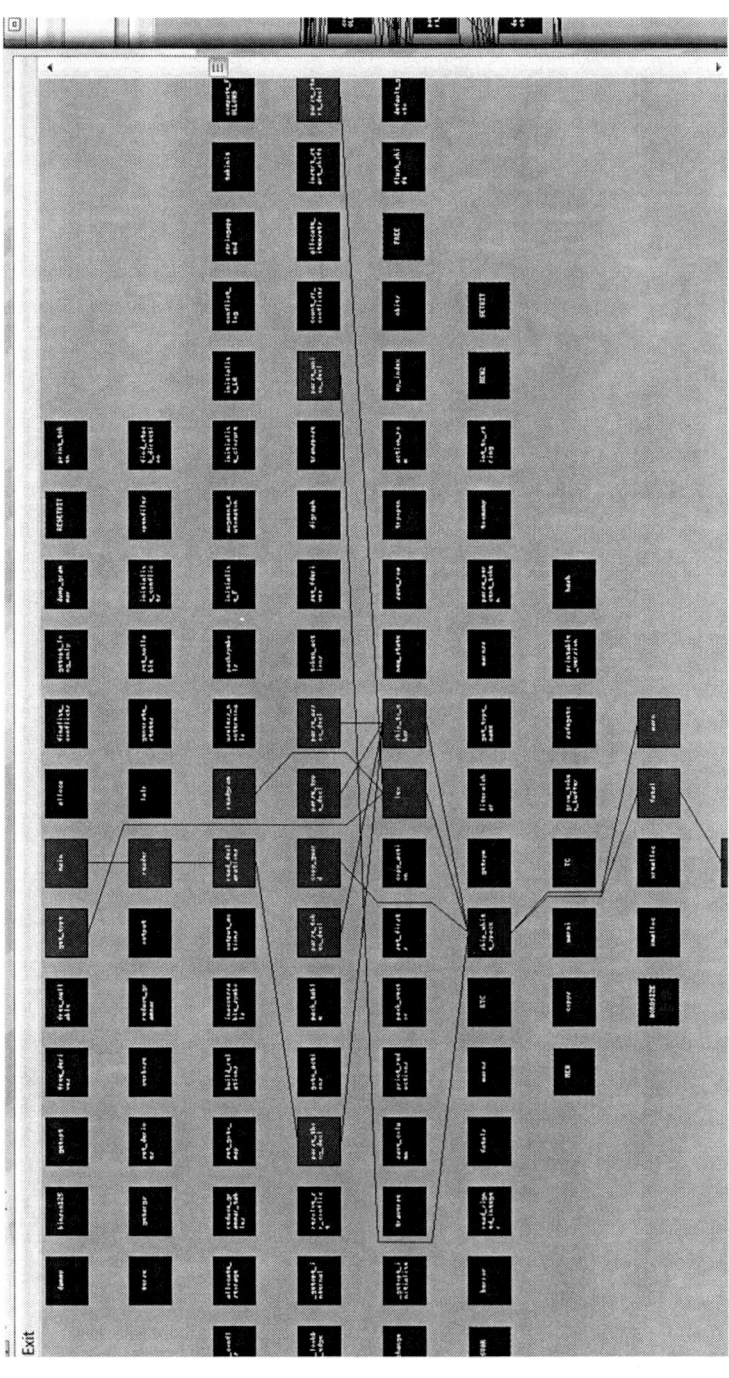

Fig. D.3 A module and the related modules highlighted

Appendix D

(d) Write three small programs for generating the three charts shown in Fig. D.4 separately through dummy programming, then compile them and run the executable programs to correct possible defects.

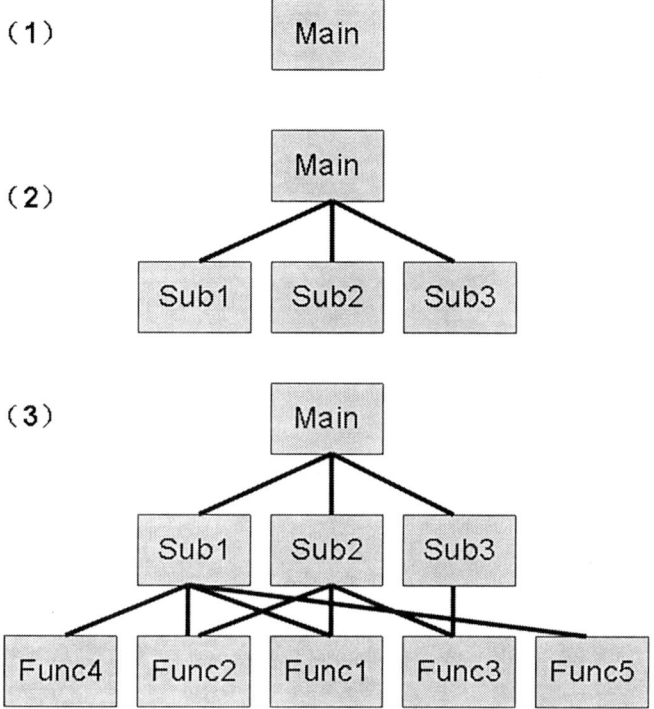

Fig. D.4 Three charts

Hints:

Refer to Fig. D.5, design your three dummy programs, then use S_Panorama for C/C++ or S_Panojava for Java (see Preface) to generate the call graphs and correct possible bugs:

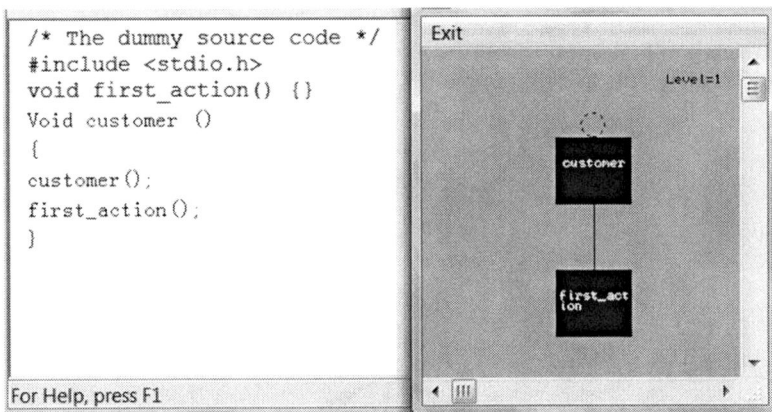

Fig. D.5 A dummy program and the graph generated

Chapter 8

(a) About the software process model, "There has to be upstream movement" – why?

Hints:

Without upstream movement, defects introduced in the upper phases will easily propagate to the lower phases, and the defect removal cost will increase tenfold several times – now software maintenance takes 75% or more of the total effort and total cost of software development.

(b) Why is there no upstream movement at all in all the existing software process models (excluding the NSE process model)?

Hints:

Please consider what is the foundation of those models.

(c) Why should software maintenance be performed globally and holistically? How can software maintenance be performed globally and holistically?

Hints:

1. Find out what are called "Butterfly-effects"
2. Consider what kinds of traceability are needed for performing software maintenance holistically and globally.

(d) Is a modified waterfall model with feedback as shown in the following figure (Fig. D.6) a linear model or not? Why?

Appendix D

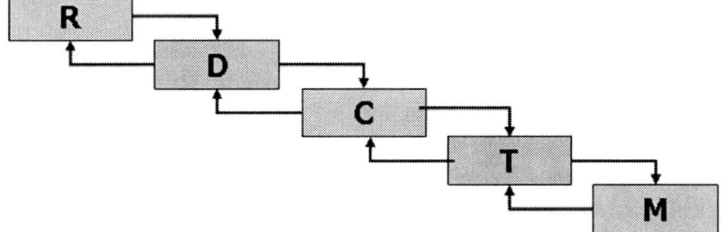

Fig. D.6 A waterfall model with feedback

Hints:

Read Sect. 1.4 for answering this question.

(e) List the drawbacks of a linear life cycle model without upstream movement.

Hints:

Read Sects. 2.5 and 8.2 carefully, then make your answer.

(f) What are the major differences between the NSE process model and the existing process models?

Hints:

Read Sect. 8.10, then make your answer.

Chapter 9

(a) Why is software traceability, particularly requirement traceability, so important?

Hints:

Please read

1. "Requirement Traceability", http://www.en.wikipedia.org/wiki/Requirements_traceability
2. Andrew K, et al (2009) Why software requirements traceability remains a challenge. CrossTalk, Jul/Aug Issue
3. Ramesh B, Matthias J (2001) Toward reference models for requirements traceability. IEEE Trans Software Eng 1:58–93
4. Juergen R, et al (2007) CASCON 2007 Workshop Report, Traceability in software engineering – past, present and future. IBM Technical Report: TR-74-211, 25 Oct 2007

After that, make your answer.

(b) Why should a bookmark be used to open a related document that is traced automatically?

Hints:

1. Usually a requirement specification file will include the descriptions for all requirements, so that if we do not use bookmarks, all requirements traced will be shown from the beginning location of the requirement specification file – it will cause confusion about exactly which requirement is traced.
2. Using bookmarks, we can open the related document and show it from the location indicated by a bookmark.
3. Bookmarks will not affect the contents.
4. When the contents of a document are modified, in most cases the bookmarks will automatically point to the new locations without manual modification.
5. Try to set some bookmarks in a document, then modify the contents, and view the documents again using the bookmarks to see what happens.

(c) What are the benefits to use Time Tags for implementing the bidirectional traceability between the test cases and the source code?

Hints:

Consider the following points to make your answer:
1. Automation
2. Accuracy
3. Self-maintainability

(d) What are the major features of this automated and self-maintainable traceability?

Hints:

Read Sect. 9.5 and the article "*Software Requirements Traceability Remains a Challenge*" [Kan09], then make your answer.

(e) Where do you think this automated and self-maintainable traceability can be efficiently used in software engineering?

Hints:

Consider the following points to make your answer:
1. Identify and fix the inconsistency defects among documents, and between documents and source code
2. Prevent side effects in the implementation of requirement changes and code modification
3. Perform regression testing efficiently
4. Validate and verify the product efficiently
5. Make the documents generated by third party tools also traceable
6. Automate the acceptance testing

(f) How can this automated and self-maintainable traceability be used to make a document produced by a third party tool traceable with the requirements of a project being developed using this technique and tools?

Appendix D 715

Hints:

Using batch files – try your own examples before making your answer.

Chapter 10

(a) What are the differences in software development methodology between that based on Constructive Holism and that based on Generative Holism?

Hints:

Consider the following points to make your answer:

1. The whole and its parts of a complex system, which one comes first?
2. With the software development method based on Generative Holism, we can begin user testing very early, and we can adopt a build-to-budget strategy that protects absolutely against schedule or budget overrun (at the cost of possible functional shortfall).

(b) What are the major differences between RUP (Rational Unified Process) and the NSE software development methodology?

Hints:

Compare Figs. 1.57 and 1.58, and Sect. 23.7, then make your answer.

(c) How can the NSE software testing paradigm be dynamically used in upstream quality assurance for defect prevention and defect propagation prevention?

Hints:

Read Chap. 16, then consider the following points to make your answer:

1. Having an output is no longer a condition to dynamically use the Transparent-box method for software testing.
2. In the case where there is no output in the execution of a test case, we can specify the expected execution path in the test case description part, and then check whether the real execution path covers the expected path to find logic defects.
3. With the HAETVE technique (see Chap. 11), the requirement development work products and graphic design documents are generated from dummy programs which are executable.
4. After the execution of the test cases, the bidirectional traceability facility will be established for checking the consistency among the documents and test cases and source code.

(d) How can the NSE software visualization paradigm be used in software defect prevention, defect propagation prevention, software understanding, testing, and maintenance?

Hints:

Read Chap. 7 and see Figs. 10.10 and 10.11 to make your answer.

Chapter 11

(a) What are the major differences between the Use Case approach and the HAETVE technique?

Hints:

Consider the following points to make your answer:

1. Holism
2. Visibility
3. Maintainability
4. Traceability for static review and defect removal
5. Execution for dynamic defect prevention and defect propagation prevention
6. Whether it is suitable for Event–Response type applications

(b) Why should the graphical result of the function decomposition of the functional requirements of a product be made traceable?

Hints:

Consider the following points to make your answer:

1. Visibility
2. Big Picture for program understanding
3. Static review for defect removal
4. Support for incremental development (to assign orders for incremental unit coding and testing)

(c) Why do we need not only static review, but also dynamic testing in the software requirement development phase?

Hints:

1. Read Capers Jones' article "Software quality in 2002: a survey of the state of the art" (Six Lincoln Knoll Lane, Burlington, Massachusetts 01803 http://www.SPR.com 23 July 2002) to know
   - Usually how many percent of the defects are introduced in the requirement development phase
   - The impact of those defects introduced in the requirement development phase
   - The difficulty to remove the defects introduced in the requirement development phase

Appendix D   717

2. Many defects are hard to find without dynamic execution of the program
3. Without dynamic testing, it is impossible to establish automated and self-maintainable traceability for detecting the inconsistent defects among the related documents and test cases and source code.

Chapter 12

(a) What are the major problems with today's software design?

Hints: Read Sect. 12.1 carefully to make your answer.

(b) What are the benefits to use the **Software Synthesis Design (and Incremental growing up) technique** for software design?

Hints: Read Sects. 12.2 and 12.5 carefully to make your answer.

(c) Complete the dummy program for generating the top-down design result shown in Fig. D.7:

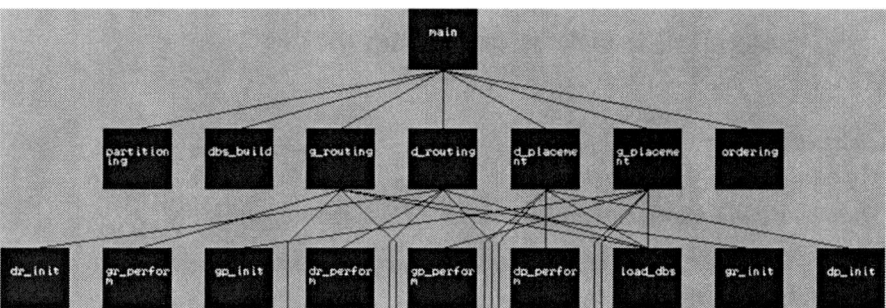

Fig. D.7 A call graph through top-down design

Hints: Read Sect. 12.3 carefully, particularly the corresponding dummy programs, to complete your dummy program design, then use the S_Panorama or S_Panojava (see Preface) tools to verify your dummy program design.

Chapter 13

(a) What are the major problems existing with today's software programming?

Hints: Read Sect. 13.1 carefully (write down your notes, then close this book) to make your answer.

(b) Why, with NSE, does software design become precoding and coding become further design?

Hints: With NSE, the dummy programs used for design to generate graphical design documents can be directly extended to become the source code in the coding process, and the extended source code can be directly used to generate the graphical design documents to update the design.

(c) Why should all commercial software products satisfy 100% MC/DC test coverage?

Hints: Read Sects. 13.4 and 16.4 carefully, and consider the following points to make your answer:

1. If the MC/DC test coverage is low, or if we only obtain high branch-level test coverage or high statement-level test coverage, many logic paths may not be tested.
2. Once the execution conditions are satisfied for those untested paths in the customer site, software disasters may happen.
3. As introduced in Appendix B, with NSE it is not very difficult to get 100% MC/DC test coverage result for program units, because NSE visualization tools can highlight any untested branches and conditions graphically to help users design the necessary test cases.

Chapter 14

(a) Why should a software product be tested before its application?

Hints: Read Sect. 14.1 carefully (write down your notes and then close this book) to make your answer.

(b) How many kinds of tests are needed?

Hints: Read Sects. 14.3–14.13 carefully (write down your notes and then close this book) to make your answer.

(c) Can a software product be tested manually only, without using tools?

Hints: Consider the following points to make your answer:

1. Can structural testing be performed manually?
2. Can performance testing be performed manually?
3. Can memory leak be completely checked manually?
4. Can the execution path of a runtime error be traced manually?
5. Can GUI test operations be captured and played back manually?
6. Can load testing for Web applications be performed manually?

(d) Who should test a software product – the product developers, other teams or groups but not the developers, or both? Why?

Appendix D

Hints: Consider the following points to make your answer:
1. The types of defects
2. The required knowledge to test a software product
3. The timing and the difference of the cost

Chapter 15

(a) What is a test case?

Hints: Read Sect. 15.1 then make your answer.

(b) How many basic test case design methods are used today?

Hints: Read Sect. 15.2, write down your notes, then close this book and make your answer.

(c) How can the NSE software testing paradigm and the NSE software visualization paradigm help users design efficient test cases?

Hints: Read Sect. 15.3, write down your notes, then close this book and make your answer.

(d) Describe the simple rules for writing test cases for using the Transparent-box software testing method and tools.

Hints: Read Sect. 15.6, write down your notes, then close this book and make your answer.

Chapter 16

(a) Why is it that the existing software testing methods, techniques, and tools cannot be dynamically used in the software requirement development phase and the software design phase?

Hints:

1. With the old-established software engineering paradigm, there is nothing executable in the requirement development phase and the design phase – people will think there is no need to perform software testing dynamically.
2. The existing software testing methods, techniques, and tools are mainly based on Black-box testing which is used to compare the output with the expected value, so that it can only be used dynamically after production.

(b) Why are software functional testing and structural testing performed separately with today's software testing paradigm?

Hints: Consider

1. Today, in functional testing people mainly use the Black-box testing method.
2. Most people think the purposes of the two kinds of testing are different.
3. It is difficult to combine them together without automatic and visual tools.

(c) Software disasters happen often – is it related to the drawbacks of the existing software testing methods, technologies, and tools? Why?

Hints: Consider

1. Today, in what phases is software testing dynamically performed?
2. Why did NIST (National Institute of Standards and Technology) conclude that "Briefly, experience in testing software and systems has shown that testing to high degrees of security and reliability is from a practical perspective not possible. Thus, one needs to build security, reliability, and other aspects into the system design itself and perform a security fault analysis on the implementation of the design." ("Requiring Software Independence in VVSG 2007: STS Recommendations for the TGDC," November 2006 http://www.vote.nist.gov/DraftWhitePaperOnSIinVVSG2007-20061120.pdf)?

(d) Why can and should the Transparent-box testing method and the corresponding tools be dynamically used in the requirement development phase and the design phase?

Hints:

1. Different from the old-established software engineering paradigm with which there is nothing executable in the requirement development phase and the design phase (except for prototype design), with NSE the executable dummy programs are used for requirement development and product design, so that dynamic testing is needed for ensuring the quality of the product.
2. To the Transparent-box testing method, having an output is no longer a condition to dynamically use the method and the corresponding tool – in the cases where there is no output, it will check whether the execution path covers the expected path specified using J-Flow notations for helping users remove logic defects, and it can also establish the bidirectional traceability facility to help users remove inconsistent defects.
3. The forward and backward tracing processes are supported by a set of visual tools.

(e) What are the key points in designing test cases for software testing using the Transparent-box method?

Hints: Read Chap. 9 and Sect. 15.6 carefully (then close this book) to make your answer.

(f) What are the major differences between the old-established software testing paradigm and the NSE software testing paradigm?

Appendix D

Hints: Read Sect. 16.5 carefully (and then close this book) to make your answer.

Chapter 17

(a) What is the root cause that regarding software product quality, "Over the last 50 years there has been very little improvement?"

Hints: Read Sects. 2.3 and 17.1 carefully, write down your notes, then close this book and make your answer.

(b) What is defect prevention? Why should it be performed in the entire software development lifecycle from the first step down to the retirement of a software product?

Hints: Read Sect. 17.3.4, then consider the following points to make your answer:

1. The timing
2. The cost savings
3. The support for changeability
4. The possibility to extend the life time of a software product
5. "An ounce of prevention is worth a pound of cure!"

(c) What are the major differences between the old-established software quality assurance paradigm and NSE-SQA?

Hints: Compare Sects. 17.1 and 17.2, write down your notes, then close this book and make your answer.

Chapter 18

(a) Why does software maintenance take 75% or more of the total effort and total cost in software product development today?

Hints: Read Sect. 18.1, write down your notes, then close this book and make your answer.

(b) What are the major differences between the old-established software maintenance paradigm and the NSE software maintenance paradigm?

Hints: Compare Sects. 18.1 and 18.2, write down your notes, then close this book and make your answer.

(c) How can the side effects in the implementation of requirement changes or code modifications be prevented?

Hints:
1. Perform software maintenance holistically and globally.
2. Use various traceabilities.
3. Read Sect. 18.4, write down your notes, then close this book and make your answer.

(d) When a software product is made through outsourcing development, what should be provided with the product? Why?

Hints: Read Sect. 1.1, particularly about the new definition of software, write down your notes, then close this book and make your answer.

Chapter 19

(a) What are the major issues existing with the old-established software documentation paradigm?

Hints: Read Sect. 19.1, write down your notes, then close this book and make your answer.

(b) Is source code the best documentation for a program? Why?

Hints: Consider the following points to make your answer:
1. Text and graphics – which one do people like more for understanding a complex system?
2. Traceable and not traceable – which documents are more useful in software understanding?

(c) What are the major differences between the old-established software documentation paradigm and the NSE software documentation paradigm?

Hints: Compare Sects. 19.1 and 19.2, write down your notes, then close this book and make your answer.

Chapter 20

(a) What are the benefits of combining the project management process and product development process together to make their work products traceable?

Hints: Consider the following points to make your answer:
1. Timing in finding the possible problems
2. Timing in solving the problems found
3. The importance of getting first hand information

Appendix D

(b) Why should a project Web site and BBS be established and the related Web pages or BBS title pages be made traceable with the related requirements and test cases and source code?

Hints: Consider the following points to make your answer:

1. Besides regular short meetings, how can the members of the product management team and product development team communicate efficiently?
2. In case unexpected events occur, how can everybody contribute to solving the problems quickly?
3. In case a project is developed by a distributed network of teams, how can they share information better?

Chapter 21

(a) Why are software algorithms so important?

Hints: According to the popular algorithms textbook Introduction to Algorithms (Second Edition by Thomas H. Cormen, Charles E. Leiserson, Ronald L. Rivest, Clifford Stein), "an algorithm is any well-defined computational procedure that takes some value, or set of values, as input and produces some value, or set of values as output." In other words, algorithms are like road maps for accomplishing a given, well-defined task (lbackstrom, *Algorithm Tutorial,* http://www.topcoder.com/tc?module=Static&d1=tutorials&d2=importance_of_algorithms).

So the efficiency of problem-solving greatly depends on the algorithm used.

(b) What is a hash table? Where do we need to use hash tables?

Hints: A **hash table** or **hash map** is a data structure that uses a hash function to map identifying values, known as keys (e.g., a person's name) to their associated values (e.g., their telephone number) (Wikipedia, Hash Table, **http://www.en.wikipedia.org/wiki/Hash_table**).

Hash tables are used for sorting items such as the names of students.

Chapter 22

(a) Why do we need to use software tools?

Hints: Please consider the following points to make your answer:

1. Can every software engineering task be performed manually?
2. If a task can be done by both people and tools, which one can be used to save time and resources?

(b) Why should software tools be automated?

Hints: Please consider the following points to make your answer:

1. The efficiency
2. The maintainability of the work products

Chapter 23

(a) Why is it that "The important thing is that one model is enough – either the code or the diagrams. They should be reproducible from one another"?

Hints: Read Harry Sneed's article, *"The Drawbacks of Model driven Software Evolution"*, IEEE CSMR 07-workshop on model-driven software evolution (MoDSE2007) Amsterdam, 20 March 2007

http://www.sciences.univ-nantes.fr/MoDSE2007/

http://www.cs.vu.nl/csmr2007/workshops/I-%20Summary%20Description.pdf, then make your answer.

(b) How to realize that "One model is enough – either the code or the diagrams. They should be reproducible from one another"?

Hints: Consider the following points to make your answer:

1. Diagrams and source code – which one is easier to change?
2. From code to diagram, and from diagram to code – which one is easier to do precisely?
3. What is offered by NSE?
4. Do you have a better solution? If you do, describe your solution in detail.

(c) Complete a small software project with NSE and the NSE support platform Panorama++.

Hints: It is important for you to test yourself how well you have learned from this book and how well you can apply NSE in practice. Please use the learning versions of the Panorama++ and Panojava product available to handle your small projects – try the provided application example (see Chap. 1) first.

Chapter 24

(a) Can any single development, in either technology or management technique efficiently solve the critical problems existing with today's software development: low quality and productivity, high cost and risk? Why?

Appendix D

Hints: Read Sects. 24.5 and 24.6, write down your notes, then close this book and make your answer.

(b) Does a qualified "Silver Bullet" which is able to slay software "Werewolf" (a monster of missed schedules, blown budges, and flawed products) mean a complete revolution in software engineering through a paradigm shift from the old one based on reductionism and the superposition principle to a revolutionary one based on complexity science? Why?

Hints:

1. I recommend reading one or two books introducing complexity science, or visit the Web site of complexity science map, http://www.art-science-factory.com/complexity-map_feb09.html.
2. Read Sect. 24.6, write down your notes, then close this book and make your answer.

(c) What are the major differences between the old-established software engineering paradigm and NSE?

Hints: Read Sect. 24.7.4, write down your notes, then close this book and make your answer.

MACRO Representation in Diagrams

For code inspection and walk through, users often prefer to having the code of a class/function diagrammed with the original source code locations (line numbers) shown before preprocessing. Panorama C/C++/OO-Browser and Panorama C/C++/OO-Diagrammer provide the logic and control flow diagrams of a class/function, with or without MACRO definition, before preprocess, thus to satisfy these users' requirements.

For code test coverage analysis, users often want to have the code being diagrammed after preprocessing and have the unexecuted logic elements highlighted. Panorama C/C++/OO-Test provides the control flow diagram of the class/function after preprocessing with the unexecuted elements highlighted, thus to satisfy those users' demands.

If in the case that the existence of some macro functions in the code makes it hard to identify the corresponding program logic before preprocessing, a group of lines of the code with macro functions used will be merged into one block in the diagrams.

Glossary

Block A group of contiguous computer program statements that are treated as a unit.

Both See "Condition Both."

Branch (1) A computer program construct in which one of two or more alternative sets of program statements is selected for execution; or (2) a point in a computer program at which one of two or more alternative sets of program statements is selected for execution; or (3) any of the alternative sets of program statements in (1).

Branch testing Testing designed to execute each outcome of each decision point in a computer program.

Class test coverage Class test coverage is defined as the ratio of the tested classes to the total number of classes. When one function within a class is tested, this class is considered to be tested.

Condition (predicate) coverage Condition (predicate) coverage is defined as the percentage of both simple and compound conditions that have been tested. It is further defined as follows:

- Condition True: **the percentage of true conditions and function entry points that have been tested. It can also be presented as a ratio.**
- Condition False: **the percentage of false conditions and function entry points that have been tested. It can also be presented as a ratio.**
- Condition Both: **the percentage of true and false conditions and function entry points (2 for each function) that have been tested. It is equal to the ratio of the sum of the numerators for Condition True and Condition False Coverage to the sum of their denominators.**

Note: Function entry points are used in calculating condition coverage, because for an executed function with no conditions, its condition coverage should be 1/1 or 100%.

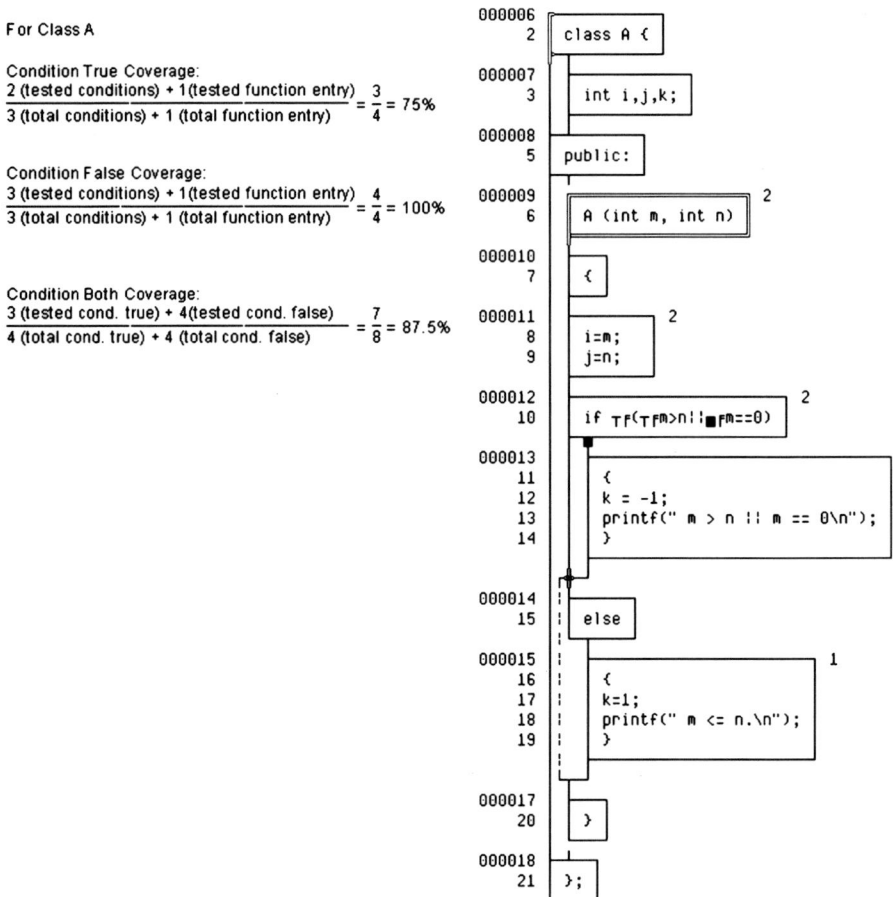

For Class A

Condition True Coverage:
$$\frac{2 \text{ (tested conditions)} + 1 \text{ (tested function entry)}}{3 \text{ (total conditions)} + 1 \text{ (total function entry)}} = \frac{3}{4} = 75\%$$

Condition False Coverage:
$$\frac{3 \text{ (tested conditions)} + 1 \text{ (tested function entry)}}{3 \text{ (total conditions)} + 1 \text{ (total function entry)}} = \frac{4}{4} = 100\%$$

Condition Both Coverage:
$$\frac{3 \text{ (tested cond. true)} + 4 \text{ (tested cond. false)}}{4 \text{ (total cond. true)} + 4 \text{ (total cond. false)}} = \frac{7}{8} = 87.5\%$$

Control level	Program control is leveled. Level 1 begins from an entry of a program and ends at the end of the program. Each time that the control flow reaches a decision statement and its condition is satisfied, a new control level is begun. This new control level ends upon reaching the end of the executable part of the decision statement.
Cyclomatic complexity	Panorama uses a complexity metric adapted from Cyclomatic complexity metrics. The algorithm is as follows:

- **Each function has a base complexity of 1 and each decision or loop statement adds a complexity of 1 to the base complexity.**

Glossary

- **For switch statements, the user has the option of including or not including case statements as part of the calculation. When case statements are included each N-way switch will add a complexity of (N – 1). When case statements are not included, the added complexity is 2.**

Empty segment	(1) If there is no statement between two conjunctive nodes and the second node is a label, we define that there is an empty segment between the two nodes. (2) If the statement preceding the case statement is not an unconditional escape statement, we define that there is an empty segment between the case statement label and the conditional part of the case statement.
False	See "Condition False."
High-level logic	The high-level logic is the table presentation of all functions or all classes within a program. The high-level logic can be displayed with the detailed logic and serves as a table in the diagrams.
Invisible segment	For each decision statement, if there is no executable statement associated with the decision statement when its condition is unsatisfied, we define there should be an invisible segment next to the decision statement (e.g., any if statement lacking an else part has an invisible segment by definition). For each repetition statement, there are two invisible segments. One of them will be executed when the program control reaches the statement but its condition is never satisfied. The other will be executed if the condition is satisfied at least once, and the program control exits the repetition body normally when the condition is no longer satisfied (vs. exiting directly from the body).
J-Coverage (condition-segment coverage)	J-Coverage metric is defined as the ratio of the number of executed visible and invisible segments plus executed outcomes of conditions to the number of all visible and invisible segments plus all outcomes of conditions in a program or program module. It is equal to MC/DC test coverage.
J-Complexity (test complexity)	JC0 (Block Test Complexity): Block Test Complexity is defined as the minimal instrumentation points required for recording all block (visible segment) test coverage data.

JC1 (Segment Test Complexity): Segment Test Complexity is defined as the minimal instrumentation points required for recording all segment test coverage data; the value of the Segment Test Complexity is equal to the sum of visible segments (including empty segments) plus invisible segments.

JC2 (Condition-Segment Test Complexity): Condition-Segment Test Complexity is defined as the minimum instrumentation points required for recording all Condition-Segment Test Coverage (J-Coverage) data. The value of JC2 is equal to the sum of all outcomes of conditions in all decision statements plus all visible segments and invisible segments. J-Complexity used alone usually denotes the Condition-Segment Test Complexity (JC2).

Node   The condition part of a decision statement, an **else** clause, a junction such as a label, or an entry or exit point of a program unit is called a node.

Path   A sequence of instructions that may be performed in the execution of a computer program.

Path condition A set of conditions that must be met in order for a particular program path to be executed.

Path testing Testing designed to execute all or selected paths through a computer program.

SC0   The Basic Segment Test Coverage (also known as the Block Test Coverage). A set of test cases of a program satisfies SC0 if all nodes and visible segments of the program have been executed at least once.

SC1   The Standard Segment Test Coverage. A set of test cases of a program satisfies SC1 if it satisfies SC0 coverage and all invisible segments of the program have been executed at least once.

SC1+   The Standard Segment Test Coverage. A set of test cases of a program satisfies SC1+ if it satisfies SC1 coverage and all the low-end invisible segments of the loops in the program have been executed at least once.

Index

A
Abstraction, 64, 238
Acceptance testing, 2, 3, 17, 57, 112, 224, 238, 252, 255, 383, 550, 699, 710
Action
 actor-, 31, 130, 206, 213, 231, 273, 277, 279, 280, 282, 293, 295, 306, 311, 397, 413, 440, 445, 447, 521, 630, 640, 649
 diagram, 569
 type, 282
Activity
 business, 92, 433, 607
 diagram, 323, 324, 500
 linear, 199
 processing, 660
 prototyping, 274
 upstream, 51, 97, 200, 612
Actor
 action, 31, 130, 206, 213, 231, 273, 277, 279, 280, 282, 293, 295, 306, 311, 397, 413, 440, 445, 447, 521, 630, 640, 649
 descriptions, 671
 external, 671
 special, 280
Adaptation, 86, 229–230
Adaptive software development, 88
Agile
 manifesto, 40
 methods, 42
 models, 57
 processes, 41
 software development, 40–41, 88, 230, 587–593
Agility, 41
Agent-oriented software engineering, 88, 236

Algorithm, 17, 31, 62, 66, 85–86, 103, 105, 126, 131, 132, 182, 183, 313, 317, 373, 391, 395, 481, 489, 531–555, 558, 669, 719, 723
Analysis
 class, 192, 389
 complexity, 126, 194, 378, 502, 540–541, 633, 635,
 control flow, 111, 120, 188, 194, 617
 correspondence, 148
 data, 443, 502
 dynamic, 177, 500, 504
 function cross reference, 120, 502
 logic, 111, 120, 182, 188, 617
 memory leak, 127, 216, 413, 417, 477
 object-oriented, 64, 274, 666–667
 path, 123, 148, 183, 232, 309, 391, 477, 633
 performance, 120, 123, 127, 132, 185, 216, 219, 341, 413, 418, 454, 459, 484, 503, 536, 537, 539, 540, 542, 632
 requirement, 31, 33, 34, 273
 risk, 68, 131, 208, 269, 477, 607
 source code, 533
 static, 500, 540, 541, 543, 548, 567
 structured, 62, 63, 669–670
 systems, 188
 test coverage, 3, 120, 127–128, 133, 149, 155, 165, 168, 170, 178, 181, 182, 194, 216, 217, 219, 232, 241, 242, 271, 294, 298, 374, 375, 379, 380, 382, 412, 413, 417, 422, 431, 477, 484, 500, 504, 505, 571–573, 578, 582, 672, 721
 test case efficiency, 3, 12, 16, 30, 76, 111, 125–126, 134, 219, 232, 233, 373, 378, 413, 481, 483, 533–537, 573, 617, 633, 635, 700
 variable, 120, 127, 188, 506

Application
    examples, 2, 51, 94, 99, 101, 106, 122, 124, 126, 129, 130, 132, 145, 153, 166, 168, 170, 177, 178, 182, 206, 207, 210, 216, 222–224, 227, 246–249, 253, 270, 271, 282, 284, 286, 303–305, 326, 341, 342, 361, 366, 367, 374, 376, 382, 391, 395, 396, 413, 418, 419, 430, 441–445, 447–450, 452, 454–461, 485, 492, 493, 500, 502, 503, 505–510, 512–514, 524, 527, 539, 541–545, 549, 575, 628, 629, 640, 642, 643, 654, 720
    process, 306, 438, 439
    results, 92, 135, 258, 314, 315, 317, 415, 434, 436
    systems, 263, 654
Architecture, 31, 38, 41, 42, 64, 69, 132, 137, 138, 178, 200, 227, 274, 313, 317, 340, 348, 578, 593, 598
Assertion, 64, 217, 353
Assisted Online Agents (AOA), 2, 30, 76, 270, 472, 475–478, 601, 664, 699
Assisted test case design, 128, 129, 419
Attributes, 39, 68, 194, 650, 660–662, 665–668, 670, 671
Automated and self-maintainable traceability, 56, 97, 140, 237–256, 260, 261, 263, 265, 267, 304, 323, 362, 432, 438, 490, 496, 519, 602, 603, 649, 710
Automation
    full, 122, 585
    insufficient, 69
    partial, 62
Availability, 650, 653, 660

B
Bandwidth, 377
Batch file, 28, 189, 237, 251, 252, 318, 405, 482, 503, 550, 552, 554, 566, 567, 602, 698
Black box testing, 52, 117, 231, 234, 373, 375, 409, 410, 578, 613, 715, 716
Bookmarks, 21, 108, 110, 111, 209, 210, 225, 237, 240–242, 246, 249, 251, 254, 255, 261, 264, 270, 303, 304, 318, 331–333, 387, 399, 403, 404, 412, 414, 416, 440, 470, 476, 478, 572, 573, 581, 583, 593, 594, 602, 615, 617, 642, 647, 649, 662, 709, 710
Boundary value analysis, 388–389

Builds, 38, 41, 73, 75, 245, 257, 259, 277, 389, 410, 567, 571, 580, 599, 608, 664, 716
Business processes, 68, 655
Butterfly-effect, 69, 81, 87, 129, 226, 238, 594

C
C++ programming language, 64, 243, 572
Candidates, 43, 48, 75, 95, 106, 597–650
Catastrophe theory, 83
CBSD (Component-Based Software Development), 72, 229, 258, 259, 262
Cellular automata, 84–85
Challenges, 45, 87, 380, 389
Change control, 45, 46, 598, 600, 601, 664
Change management, 664
Changeability, 2, 3, 44, 50, 92, 94, 104, 105, 134, 202, 203, 224, 229, 234, 270, 337, 469, 477, 518, 519, 521, 632, 635–640, 664, 699, 717
Chaos theory, 80–81, 99
Class
    definition, 117, 546
    diagrams, 659
    entity, 662
    hierarchy, 140
    inheritance, 138
    inheritance chart, 3, 8, 111, 162, 172, 188, 323, 340, 568, 585, 586, 617
    instance, 264, 532, 546
    member function, 140, 378, 471–473
    object, 353
    overloading member function, 140, 378, 471–473
    relationship, 138, 323
    structure, 132, 140, 178, 317, 570
    test coverage, 8, 181, 182, 316, 376, 380, 722
Client, 564
CMMI (Capability Maturity Model Integration)
    framework, 587
Coding
    engineering, 137, 138, 177, 339–369
    order, 124, 132, 177, 215, 232, 341, 346, 347, 443, 581, 582, 630, 649
    phase, 105, 260, 270, 341, 406, 443, 447, 582
    process, 124, 216, 340, 342, 344, 368

Index   733

standards, 369
style, 232, 422
Complex adaptive system (CAS), 49, 84, 92,
    611, 618
Complex systems, 44, 50, 55, 57, 58, 69, 76,
    80, 81, 84, 85, 87–89, 92, 95, 96,
    99, 101, 103, 104, 106, 112,
    200–202, 207, 214, 226, 238, 258,
    259, 262, 275, 277, 313, 314, 316,
    336, 337, 434–436, 468, 470,
    489–491, 511, 518, 577, 593,
    607, 608, 611, 648, 699, 700,
    702, 711
Complexity arises from simple rules principle,
    87, 95, 203
Complexity science
    essential principles of, 50, 69, 75,
        87–89, 92, 98, 101, 106, 109,
        112, 133, 135, 162, 199, 200,
        202, 203, 258, 259, 277, 279,
        314, 337, 409, 435, 436, 468,
        491, 511, 519, 520, 529, 609–611,
        616, 648, 701
    generative Holism principle of, 104, 201,
        259, 277, 314, 316, 336, 342, 435,
        491, 630, 640
    holism principle of, 49, 87, 104, 201, 259,
        271, 577, 611, 618
    manifestation of the Essential Principles
        of, 225–226
Communication, 63, 69, 76, 110, 126, 207,
    223, 271, 380, 519, 521, 524, 582,
    590, 617, 654, 658, 661
Component
    important, 199, 237, 257, 409
    reusable, 229, 259, 594
    reuse, 229, 633
Computer aided software engineering
    (CASE), 31, 62, 63, 558,
    568, 573
Computer software, 92, 374, 594
Concept, 29, 74, 75, 83, 86, 258,
        314, 383, 469, 585, 587,
        593, 597
Condition outcomes, 140, 148, 422
Conditions combination method,
    389–390
Configuration management, 32, 38
Conformity, 2, 3, 44, 76, 92, 104, 105, 131,
    134, 202, 203, 208, 224, 229, 234,
    255, 270, 337, 469, 475, 521,
    642–643, 664, 699
Consistency, 2, 109, 117, 134, 136, 137, 140,
    167, 214, 240, 253, 377, 407, 434,

436, 443, 472–474, 480, 481, 511,
572, 573, 592, 615, 636, 639, 647,
711
Constraints, 68, 83, 86, 652–656, 659–660
Construction, 31, 42, 51, 75, 97, 200, 593,
    599, 612
Control flow diagram, 3, 4, 110, 117, 126,
    138, 145, 148, 153, 154, 183, 186,
    194, 215, 225, 237, 240, 249, 375,
    391, 443, 450, 504, 508, 513, 585,
    586, 617, 721
Conversion, 389, 513, 658
Cost estimation, 62, 109, 207, 213, 223,
    229, 251, 517, 519, 520, 524,
    525, 615, 630
Coupling, 132, 138, 317, 367, 568
CRC, 99
Critical issues, 92, 105, 216, 229, 239, 267,
    317, 342, 494–495, 518, 577, 578,
    585
Critical path, 8, 341, 342, 347, 348
Customers, 2, 30–32, 34, 35, 40, 41, 45,
    58, 75, 110, 130, 132, 206,
    208, 209, 213, 214, 223, 224,
    229–231, 267, 269, 277, 317,
    362, 371–374, 378, 380, 383,
    430, 439, 440, 469, 470, 473,
    475, 477, 518, 520, 521, 582,
    589–591, 602, 616, 628, 635,
    654–657, 659, 664, 699,
    700, 714
Cyclomatic complexity, 6, 126, 142, 155, 160,
    317, 378, 379, 439, 450, 453, 500,
    502, 531, 540–541, 632, 641,
    723–724
Cyclomatic complexity measurement, 111,
    140, 162, 170, 175, 176, 181, 182,
    232, 343, 617

D

Data dictionary, 662, 663
Data element, 670
Data flow diagram, 662, 663
Data types, 372
Debugging, 72, 137, 138, 148, 179, 180, 263,
    264, 276, 277, 279, 311, 418, 420,
    423, 431, 437, 450, 452
Design
    approach, 316, 406
    automation, 93
    changes, 263
    constraints, 656, 659–660
    defect, 105, 250, 410

description, 237
documents, 99, 127, 207, 223, 239, 248, 253, 271, 275, 314, 315, 317, 342, 345, 368, 442, 494–496, 572, 580, 584, 633
efficiency, 633
engineering, 31, 137, 138, 177, 215, 313–338
example, 323
flaws, 29, 598, 601, 602
graphics, 318, 495, 579, 711
methodologies, 61–62
optimization, 231, 317
phase, 46, 72, 100, 105, 108, 111, 117, 124, 155, 207, 231, 260, 270, 275, 314, 315, 406, 410, 411, 429, 432, 434, 438, 441–443, 445, 470, 578, 579, 581, 605, 608, 615, 618, 715, 716
process, 179, 215, 318, 342, 344, 591
purpose, 97
result, 211, 317, 318, 333, 334, 338, 342, 581, 713
review, 463
rules, 302
strategy, 260
tasks, 313, 314, 337
Documentation
design, 304–306, 591, 592
engineering, 489
paradigm, 44, 49, 54, 69, 73, 76, 103–105, 107, 109, 112, 234, 260, 262, 489–515, 578, 606, 608, 611, 613–616, 633, 718
tasks, 491, 510, 511
technique, 119–121, 134, 231
Document
approvals, 664
creation, 111, 318–320, 617
generation, 348, 491, 584
hierarchy, 111, 209, 213, 241, 242, 304, 305, 318, 478, 581, 617, 662
information, 317, 387
template, 338
types, 238, 261, 470
Decomposition, 31, 132, 177, 190, 206, 207, 211–213, 215, 231, 266, 269, 273, 274, 278, 286, 289, 292–295, 303, 307, 311, 317, 415, 442, 445, 519, 521, 581, 584, 635, 640, 649, 712
Defect prevention and defect propagation prevention, 30, 32, 43, 58, 100, 109, 111, 133, 207, 254, 260, 261, 270–272, 279, 290, 306, 311, 315, 318, 321–323, 337, 347, 364, 368, 398, 411, 435, 471, 473, 475, 522, 565, 581, 592, 604, 605, 615, 618, 631, 633, 634, 700, 711, 712
Defect tracer, 3, 574
Definition of a successful project, 67, 700
Deterministic chaos, 81
Dissipation structure, 82
Downstream, 51, 200, 270, 383, 384, 435, 438, 491, 522
Drawbacks, 35, 44, 75, 76, 234, 274, 275, 374–375, 383–384, 432, 593, 599
Driving forces, 61, 79, 89, 92, 99, 109, 134, 203–204, 263–265, 597, 615, 648
Dummy module, 110, 115, 165, 207, 266, 269, 271, 277, 306, 317, 397, 415, 497, 616, 630, 649
Dummy programming technique, 31, 207, 397, 578
Dummy system, 52, 110, 115, 132, 207, 211–214, 223, 230, 231, 234, 317, 612, 616, 635
Downstream movement, 435, 491
Dynamic defect prevention, 285, 306, 321–323, 581, 712
Dynamic modeling, 278, 279, 309–310
Dynamic traceability, 122, 133, 253, 306, 318, 363, 531, 549–554, 602
Dynamic testing, 110, 124, 132, 207, 214, 227, 231, 261, 263, 270, 271, 276, 278, 315, 317, 327, 329, 337, 347, 364, 368, 416, 438, 447, 450, 470, 475, 495, 496, 511, 512, 521, 579, 616, 632, 633, 712, 713, 715, 716
Dynamical system, 81–84
Dynamics principle, 88, 95, 203

E
EDA (Electronic Design Automation), 93, 277, 326, 336
Effort, 29, 30, 34, 47, 48, 54, 62, 68, 74, 75, 85, 104, 112, 133, 202, 207, 219, 230, 235, 239, 266, 441, 469, 473, 475, 487, 518, 519, 522, 573, 579, 591, 593, 598–602, 610, 614, 629, 631–634, 702, 708, 717

Index 735

Entity, 659, 662
Equivalence class partition,
    388–389
Errors, 3, 12, 14, 31, 65, 66, 92, 123, 176–179,
    182, 215, 223, 232, 259, 266, 275,
    282, 294, 295, 316, 353, 364, 365,
    372–374, 378, 383, 388, 389, 400,
    413, 423, 424, 427, 434, 435, 437,
    438, 440, 447, 450, 455, 458, 477,
    485, 486, 531, 542, 584, 607, 649,
    658, 672, 685, 688, 691
Estimation, 62, 68, 109, 207, 213, 223, 229,
    251, 317, 517, 519, 520, 524, 525,
    615, 630
Extreme programming (XP), 40, 41,
    589–593, 595

F
Facility, 70, 117, 122, 138, 160, 164–166, 204,
    206, 208, 224–225, 231, 237–256,
    265, 470, 481, 494, 495, 500, 501,
    604, 711, 716
Feasibility, 213
Feedback, 32, 33, 35, 82, 87, 100, 106, 131,
    224, 235, 267, 520, 582, 654, 709
Five-dimensional structure synthesis method
    (FDS), 79, 89, 91–101, 106, 107,
    112, 115, 133, 135, 136, 199, 200,
    257, 258, 409, 436, 467, 468,
    519, 520
Floating license, 574
Flow chart, 142, 145, 146
Forms, 62, 82, 83, 560, 636
Formal methods, 62, 63
Foundation, 44–47, 50, 57, 69, 79–89, 91, 109,
    234, 262, 409, 433, 436, 489, 491,
    519–520, 611, 616, 701, 702, 708
Foundation of modern civilization, 1, 57
Fractal, 82
Fractal dimension, 82
Framework
    adaptive, 97
    FDS, 79, 89, 91–101, 106, 107, 112, 115,
        133, 135, 136, 199, 200, 257, 258,
        409, 436, 467, 468, 519, 520
    general, 101
    Hall's, 92, 93, 96, 97, 101, 701
    Paradigm-shift, 79, 89, 91–101, 106, 107,
        112, 115, 135, 136, 200, 257, 258,
        409, 436, 468, 520, 609
    systems engineering, 92–93
Function call graph, 138, 162, 172, 178,
    190, 568

Function decomposition, 31, 177, 206,
    211–213, 266, 269, 273, 278, 289,
    292–295, 311, 317, 442, 445, 521,
    581, 640, 649, 712
Function points, 105, 277, 336
Fundamental problem, 29, 76, 200, 469, 598,
    600, 601

G
Genetic algorithm, 85–86
Group, 42, 46, 84, 86, 246, 371, 372, 378, 388,
    655, 673, 679, 688, 691, 721

H
Hall's systems engineering framework, 92, 93
Hardware, 373, 377, 380, 381, 600, 653, 654,
    656, 657, 659
Hierarchical design, 210, 212, 213, 321,
    328, 581
Hindsight, 557–558, 560
Holism principle of complexity science,
    49, 87, 104, 201, 259, 271, 577,
    611, 618

I
Idea, 51, 74, 81, 87, 97, 117, 200, 258, 315,
    377, 531, 532, 534, 537–538, 540,
    585, 587, 590, 593, 612
Incremental development, 34–35, 51, 70, 201,
    213, 229, 259, 271, 277, 590, 593,
    598, 600, 601, 606, 612, 620, 630,
    631, 712
Incremental and iterative development, 51, 70,
    77, 201, 612
Incremental unit coding, 8, 132, 341, 712
Index, 569, 654, 663
Industrial revolution, 91–101
Information
    flows, 79, 669
    hiding, 61
    processing, 63, 552
    sharing, 233
    source, 58, 77, 89, 101, 112, 134, 196–197,
        236, 256, 272, 311, 338, 369, 384,
        407, 432, 464, 487, 514, 530, 555,
        575–576, 595, 650
    systems, 236
    technology, 274
Inheritance, 3, 8, 111, 138, 162, 172,
    188, 323, 340, 367, 568, 585,
    586, 617

Initial Condition Sensitivity principle, 87, 88, 95, 203
Innovation, 79, 93, 136, 199, 263–265, 521
Installation, 3, 560–565, 655
Instrumentation, 241, 242, 374–375, 532, 533, 538, 724, 725
Integration, 52, 63, 71, 131–133, 214, 215, 231, 275, 313, 341, 348, 384, 434, 565, 581, 582, 592, 593, 600, 612, 630
Integration testing, 123, 132, 216, 219, 232, 341, 349–353, 373, 378, 413
Integrity, 659, 661
Integer, 86, 371, 372, 388, 543, 672, 673, 676, 679, 682, 685, 686, 691
Intelligent agents, 2, 88, 469
Intelligent test case selection, 22, 30, 76, 109, 219, 615, 700
Intelligent version comparison, 32, 121–122, 134, 222, 233, 634, 635
Interface, 31, 35, 68, 132, 138, 155, 158, 160, 162, 166, 200, 215, 232, 259, 303, 309, 313, 316, 341, 349, 382, 443, 478, 482, 505, 535, 560, 566, 582, 632, 653, 654, 656–658, 661
Interface testing, 535
Interoperability, 372, 380
Invisible segment, 381, 422, 724, 725
ISO 9001:2000, 237

J

Java, 64, 280, 309, 378, 498, 560, 575, 579, 583–586, 648, 707
3J graphics (J-Chart, J-Diagram, and J-Flow), 100, 117, 138, 155–160, 195, 497, 584
Joint Application Design (JAD), 437

K

Keywords, 237, 241, 246, 254, 261, 264, 318, 387, 399, 412, 470, 540, 541, 573, 581, 642
Knowledge, 67, 84, 97, 101, 208, 371, 375, 715

L

Lateral thinking, 86, 87
Li-Yorke Theorem, 83
Linear system, 44, 81, 104, 201, 226

Linear thinking, 44, 45, 49, 50, 57, 66, 69, 72, 74–76, 80, 86–88, 91–94, 97, 101, 103, 112, 135, 199, 201, 232, 233, 258, 313, 410, 597–599, 608–612, 648
Load testing, 72, 377, 380, 381, 714
LOC, 367
Logic defects, 133, 182, 185, 187, 215–217, 232, 254, 364, 366, 399, 401, 403, 404, 411, 413, 414, 430, 447, 451, 605, 649, 711, 716
Logic diagram, 4, 53, 110, 120, 136, 138, 140, 149, 162, 166–168, 170, 176, 179, 182, 183, 194, 215, 232, 340, 373, 375, 425, 427, 500, 504, 508, 513, 548, 569, 585, 586, 592, 617

M

Manpower, 76, 111, 373, 600, 617
Maintainability, 3, 76, 104, 115, 203, 469, 475, 477, 521, 650, 661, 664, 712, 720
Maintenance
  approach, 97, 235
  capability, 279
  cost, 29, 471, 598, 601, 602
  engineering, 467–473, 475–477, 485, 487
  example, 472–474
  Iceberg, 467, 470
  issue, 588
  model, 30, 58, 471, 476, 485, 700
  paradigm, 44, 54, 56, 69, 73–74, 76, 88, 103–105, 107, 109, 112, 235, 304, 433, 467–487, 578, 609, 611, 614–616, 717
  phase, 30, 103, 108, 109, 153, 233, 260, 264, 270, 275, 315, 406, 417, 434, 441, 443, 450, 468, 473, 522, 579, 580, 582, 593, 608, 615, 618, 632, 633, 700, 702
  problems, 274, 584
  process, 30, 51, 54, 100, 108, 110, 190, 207, 219, 229, 233, 235, 269, 274, 279, 311, 318, 464, 468, 471, 473, 475, 476, 485, 519, 522, 557, 579, 583, 608, 612, 614–616, 623, 632, 634
  site, 106
  support, 57, 222
  tasks, 468, 475, 476, 487
  team, 30
  technique, 133, 134, 219, 230
  visualization,

MC/DC test coverage analysis, 3, 127–128, 133, 149, 170, 178, 182, 216, 217, 219, 271, 340, 379, 382, 417, 477, 484, 500, 578, 582
MDA, 274, 578, 583
Measurement
  branch execution frequency, 111, 617
  Cyclomatic complexity, 111, 140, 162, 170, 175, 176, 181, 182, 232, 343, 617
  database, 12, 261
  memory leak, 12, 232, 353, 477, 582, 632, 649
  performance, 3, 4, 6, 11, 111, 162, 176, 232, 341, 377, 477, 582, 584, 594, 617, 649
  productivity, 349, 524, 526
  program, 1, 3, 181, 224, 649
  quality, 3, 12, 14, 111, 120, 121, 132, 138, 162, 176, 194, 262, 341, 366–367, 435, 439, 454–461, 477, 484, 508, 592, 594, 617
  reports, 12, 194, 377
  results, 4–6, 9–12, 14–16, 111, 120, 121, 156, 160, 162, 170, 175, 176, 179, 181, 182, 188, 231, 237, 241, 249, 254, 260, 299, 300, 327, 330, 357–359, 361, 376, 402, 508, 532–534, 584–586, 617
  size, 194
  support, 349, 353–362, 526
  test case efficiency, 391, 395
  test coverage, 4, 5, 9, 10, 108, 111, 136, 142, 145, 156, 160, 162, 176, 179, 181, 182, 237, 241, 249, 250, 254, 261, 299, 300, 327, 330, 353–362, 376, 402, 456, 532–536, 585, 586, 615, 617, 632
Memory leak, 3, 12, 123, 127, 132, 216, 229, 232, 259, 262, 341, 353, 372, 413, 417, 418, 450, 455, 457, 477, 484, 531, 543–545, 573–574, 582, 594, 632, 649, 655, 714
Meta-synthesis, 84
Metrics, 262, 367, 374, 421, 424, 533–536, 567, 570, 672, 691, 694, 723, 724
Milestone, 267, 269, 635
Model
  capability maturity, 39, 63
  computer-independent (CIM), 274, 583
  incremental, 75, 105, 595
  integration, 32
  iteration, 99, 105
  life cycle, 36, 57, 105, 235, 709
  linear, 32, 33, 70, 201, 227, 235, 708
  micro-waterfall, 35, 42, 43, 45, 92, 99
  modified waterfall, 33
  NSE, 37, 99, 227
  original waterfall, 33
  platform-specific, 274, 583
  prototype, 36, 37, 57
  quality, 278, 279
  requirements, 51, 612
  RUP, 39
  sequential, 75, 599
  SPIN, 61
  spiral, 36, 37, 57
  transformations, 274, 583
  XP, 40, 41
Model-driven tools, 274, 583, 584
Modularity, 661
Monster of missed schedules, blown budges, and flawed products, 49, 597, 600, 605, 607, 609, 610, 641, 642, 721
Multi-agent systems, 201

N

NIST (National Institute of Standards and Technology), 73, 104, 257, 383, 410, 463, 608, 716
No Silver Bullet (NSB), 63, 74–76, 96, 599–602
Nonlinear complex system, 69, 81, 88, 93, 94, 99, 100, 226, 238
Nonlinear system, 81, 83, 87, 129, 594
Nonlinear thinking, 30, 50, 57, 75, 76, 80, 86–88, 91–94, 97, 101, 137, 162, 234, 314, 411, 609, 611, 612
Nonlinearity principle, 69, 75, 87, 89, 95, 101, 106, 109, 112, 133, 162, 195, 199, 202, 203, 259, 277, 279, 314, 336, 409, 436, 491, 511, 519, 529, 610, 616, 640, 648, 703
Notations
  class, 287
  control, 399
  event-response relationship, 280
  function, 287
  HAETVE, 279, 497, 498
  J-Chart, 3, 53, 118, 128, 132, 138, 139, 190, 232, 478, 524, 539, 543

J-Diagram, 3, 4, 53, 143, 145, 189, 215–216, 232, 425, 508
J-Flow, 3, 148, 150, 189, 215, 225, 246, 405, 450, 541, 578, 716
mapping, 284, 398, 499
UML, 497
use case, 282, 397
use cases analysis, 498
NSE
  applications, 270, 577–595
  components of, 107–109, 614–615
  documentation paradigm, 107, 109, 112, 260, 489–515, 615, 616, 633, 718
  establishment of, 75, 79, 89, 101, 105–107, 115, 134, 409, 531, 701, 703
  outline of, 103–112, 202, 259–263, 276–279, 314–315, 435–436, 472–476, 491–493, 518–519
  process model, 49, 51, 71, 107, 112, 115, 130, 199–235, 238, 273, 275, 277, 279, 315, 316, 336, 337, 439, 441, 470, 491, 580, 601, 612, 614, 634, 708, 709
  software development methodology, 107, 112, 183, 257–273, 279, 315, 337, 368, 439, 471, 491, 592, 614, 640, 711
  software engineering visualization paradigm, 135–196
  software quality assurance paradigm, 433–464
  software testing paradigm, 270, 272, 278, 285, 306, 316, 384, 406, 407, 409–432, 437, 445, 450, 578, 601, 606, 711, 715, 716
  structure of, 107, 108
  support platform, 2, 30, 50, 108–111, 129, 134, 156, 179, 202, 216, 220, 222, 230–233, 245, 270, 272, 280, 303, 361, 367, 375, 378, 379, 382, 401, 424, 439, 536, 553, 557–575, 577, 583, 593–595, 611, 615, 617, 720
  support techniques, 107, 109, 112, 183, 234, 615
  support tools, 107, 109, 112, 557–575, 615
NSE-CMMI, 585–589
NSE-RUP, 593, 594
NSE-UML, 584–585, 587
NSE-XP, 592–593

O
Object Oriented analysis and design, 64, 274
Object-Oriented modeling, 64
Object-Oriented software engineering 64
Object oriented technologies, 63
Objectives, 2, 31, 48, 56, 57, 85, 93–94, 103, 105–106, 206–208, 214, 227, 239, 387, 406, 489, 496–497, 591, 610, 651, 701, 702
Obstacles, 92, 94
Open source software development, 63, 64
Openness principle, 88, 95, 96, 203
Operating system, 377, 380, 657, 661
Organization, 33, 47, 50, 54, 67, 83, 94, 97, 98, 101, 105, 106, 111, 134, 203, 207, 219, 235, 238, 262, 272, 315, 374, 436, 464, 469, 473, 475, 487, 489, 518, 519, 521, 585, 587, 593, 598, 606, 614, 618, 649, 652, 655, 658, 661, 663, 670, 702
Outsourcing, 2, 230, 487, 718

P
Package, 64, 258
Pair programming, 339, 437, 589, 591
Panorama product, 170, 558
Panorama++ product, 378, 414, 560–565
Paradigm
  existing software engineering, 1, 29, 32–46, 57, 59, 77, 103–105, 702
  nonlinear software engineering, 2, 48, 57, 89, 101, 107, 134, 199, 560, 565, 597, 601, 610, 619–627, 703
  software documentation, 44, 55, 73, 103–105, 234, 262, 433, 489–491, 494–495, 497, 500–512, 606, 608, 633, 718
  software maintenance, 44, 54, 56, 73–74, 103–105, 235, 304, 467–487, 578, 609, 614, 717
  software project management, 44, 54, 74, 103, 104, 235, 433, 517–519, 528, 529, 614
  software quality assurance, 44, 55, 72–73, 104, 105, 234, 433–464, 578, 717
  software testing, 44, 49, 52, 55, 72, 103–105, 234, 270, 272, 278, 285, 306, 316, 383–384, 406, 407, 409–433, 437, 445, 450, 578, 601, 605, 606, 608, 610, 613, 711, 715, 716

Index   739

software visualization, 44, 55, 73,
    103–105, 135, 136, 160–180,
    183, 188–190, 195, 196, 260,
    272, 285, 391, 433, 464, 497,
    523, 578, 608, 613, 633, 640,
    641, 643, 704, 711
traditional software engineering,
    618–627
Paradigm-shift framework, 79, 89, 91–101,
    106, 107, 112, 115, 135, 136, 200,
    257, 258, 409, 468, 520
Parameter, 83, 85, 215, 225, 232,
    237, 250, 413, 414, 443, 472,
    550, 658
Patterns, 131
Payoff, 29, 200, 471, 598, 601, 602
Performance measurement, 3, 4, 6, 11, 111,
    162, 176, 232, 377, 477, 582, 584,
    594, 617, 649
Portability, 579, 650, 661
Prediction, 80, 91–101
Preliminary design, 117, 214, 215, 266, 269,
    317
Pre-process, 37, 204, 206–214, 216, 225, 227,
    229, 265–267, 269, 317, 341, 342,
    470, 476, 579–582, 649, 725
Priority, 41, 52, 132, 206, 208, 213, 229, 230,
    265, 285, 341, 439, 580, 581, 613,
    628, 629, 631, 635, 649
Procedure, 62, 316, 660, 683, 719
Process
    business, 68, 655
    coding, 124, 216, 340, 342, 344, 368, 714
    construction, 51, 97, 200, 612
    design, 179, 215, 318, 342, 344, 591
    improvement, 44, 105, 600
    leveling, 640
    linear, 34–36, 44, 46, 55, 71, 91, 93, 98,
        101, 104, 109, 195, 227, 233, 275,
        276, 314, 337, 339, 383, 468,
        489, 522, 600, 608, 609, 616,
        632, 706
    maintenance, 30, 51, 54, 100, 108, 110,
        190, 207, 219, 229, 233, 235, 269,
        274, 279, 311, 318, 464, 468,
        471, 473, 475, 476, 485, 519,
        522, 557, 579, 583, 608, 612,
        614–616, 634
    management, 32, 51, 104, 109, 110, 223,
        229, 235, 251, 434, 436, 441,
        517–520, 523, 524, 528, 529, 578,
        580, 593, 609, 612, 615, 616, 651,
        664, 718

nonlinear, 30, 32, 33, 101, 109, 315, 342,
    368, 470, 592
playback, 571, 572
quality assurance, 434, 436, 463
review, 477, 591, 592
solution review, 477
testing, 111, 216, 218, 271, 361, 382, 617
tracing, 716
unified, 29, 42, 274, 593, 618
Production, 43, 46, 48, 55, 104, 105, 123, 434,
    438, 463, 578, 589, 608, 610, 715
Productivity
    measurement, 349, 524, 526
    precise, 524, 526
    programming, 340, 348
    research, 463
    software, 463, 522, 575, 600, 633
    software development, 47
Program module, 126, 149, 157, 185, 242,
    279, 353–355, 361, 371, 374, 375,
    420, 449, 450, 479, 549, 592,
    641, 724
Project failure rate, 67–69
Project management, 32, 35, 44, 46, 49, 51,
    54, 57, 62, 69, 74, 76, 97, 103, 104,
    107, 109, 112, 207, 223, 229, 231,
    235, 238, 251, 268, 270, 271, 433,
    441, 517–529, 578, 580, 582, 593,
    606, 609, 611, 612, 614, 615, 718
Project manager, 230–231
Project plan, 207, 223, 235, 582, 628, 649
Project planning, 263, 517, 520
Project scope, 42, 664
Property, 50, 87, 89, 106, 131, 309, 372,
    584, 611
Prototyping
    activity, 274
    design and evaluation, 206
    document, 303, 328, 331
    model, 36
    technologies, 62

Q

Quality assurance
    defect-prevention, 711
    driven, 565
    methodology, 433
    metrics, 570
    paradigm, 44, 49, 52, 55, 69, 72–73, 76,
        88, 94, 103–105, 107, 109, 112,
        234, 433–464, 578, 605, 608, 611,
        613, 615, 717

principles, 55, 279, 311, 328, 409, 463
process, 434, 436, 463
reports, 570
requirements, 353
results, 434, 436
standards, 92, 97
strategy, 260, 261, 270, 279, 311, 314, 315, 337, 439–460
technique, 123–125, 133, 437, 438
visibility, 433
Quality measurement, 3, 12, 14, 111, 120, 121, 132, 138, 162, 176, 194, 262, 341, 366–367, 435, 439, 454–461, 477, 484, 508, 592, 594, 617, 632, 649

R

Rapid prototyping, 62, 231
Rational unified process (RUP), 29, 39, 42, 43, 57, 272, 593–595, 618, 711
Re-engineering, 138, 565
Realization, 200, 227, 598
Record, 28, 243, 253, 532, 533, 536–539, 571, 574
Reductionism, and superposition principle, 44, 49–52, 55, 69, 72, 74, 80, 88, 91, 98, 101, 103, 104, 112, 135, 201, 257, 258, 262, 275, 311, 313, 336, 337, 339, 383, 410, 433, 468, 487, 489, 510, 517, 518, 522, 524, 528, 577, 597–600, 608–612, 618, 648, 699, 702, 721
Refactoring, 52, 111, 126–127, 134, 271, 340, 342, 435, 439, 450–454, 464, 484, 589, 590, 593, 613, 617, 632–634, 641
Relationship, 20, 70, 97, 98, 138, 190, 194, 280, 282, 318, 323, 389, 478, 518, 568, 569, 611, 618, 649, 656, 658, 659, 662, 701
Reliability, 68, 73, 75, 76, 92, 115, 203, 224, 229, 234, 257, 258, 270, 377, 410, 433, 463, 469, 475, 521, 597, 599, 600, 604–608, 650, 656, 660, 699, 716
Requirement
analysis, 31, 33, 34, 273
change, 29, 30, 32, 34, 49, 50, 54, 73, 75, 100, 104, 109, 110, 178, 207, 208, 219, 220, 222, 226, 228–231, 233, 235, 254, 255, 259, 265, 269, 434–436, 438, 440, 443, 467, 469, 471–473, 487, 495, 496, 519, 521, 522, 565, 577–580, 582, 592, 599–602, 604, 609–611, 614–616, 618, 629, 631–636, 640, 642, 649, 650, 700, 702, 710, 717
conflicts, 50, 220, 611, 636
decomposition, 132, 231, 286, 294, 303, 317, 445
development phase, 46, 72, 100, 105, 108, 111, 117, 153, 207, 214, 224, 231, 253, 260, 270, 275, 311, 404, 406, 410, 411, 413, 417, 429, 432, 434, 438–441, 445, 446, 470, 564, 578–581, 605, 608, 615, 618, 712, 715, 716
documents, 303–306
elicitation/gathering, 31, 178, 182
engineering, 62, 273–311
implementation, 35, 54, 55, 133, 207, 223, 229, 235, 303, 316, 572–573, 614, 630
management, 265
modeling, 31
priority, 132, 341, 631
specification, 20, 21, 214, 237, 239, 241, 248, 253, 273, 313, 328, 331, 371, 396, 440, 581, 632, 649–671
traceability, 208, 238–240, 255, 709
validation and verification, 3, 12, 18, 51, 238, 240, 242, 254, 255, 304, 604, 612
Resources, 30, 34, 38, 47, 49, 57, 76, 83, 111, 373, 391, 473, 521, 580, 582, 591, 593, 602, 610, 617, 700, 719
Retirement, 434, 435, 438, 439, 464, 528, 592, 717
Reusable components, 229, 259, 594
Reverse engineering, 8, 110, 137, 138, 155, 179, 207, 271, 277, 340, 348, 476, 558, 560, 579, 584, 613, 616, 617, 633, 635
Reviews, 35, 45, 46, 68, 70, 130, 166, 182, 207–209, 213, 214, 223, 230–232, 267, 270, 271, 273, 275, 278, 279, 306, 311, 314, 315, 318, 321, 329, 337, 340, 435, 438–440, 448, 469, 472, 477, 494–496, 511, 520, 521, 579–582, 590–592, 649, 712
Revolutionary solution, 137–138, 202, 259–263, 276–279, 470, 518
Risk, 34, 35, 38, 42, 44, 45, 48–50, 56, 57, 68, 69, 75, 131, 134, 202, 206, 208, 266, 269, 274, 317, 362, 388, 438, 477, 517, 520, 577–580, 585, 599, 607, 610, 632, 635, 702, 720

Index   741

Risk analysis, 68, 131, 208, 269, 477, 607
Risk management, 274, 517, 520
Road map, 72, 107, 199, 276, 311, 719
Root cause for software disasters, 67–69
Rule, 75, 83–85, 87, 93, 95, 111, 189, 203,
    254, 270, 302, 336, 399, 412, 476,
    534, 577, 581, 589, 590, 599, 617,
    649, 715
RUP. *See* Rational unified process (RUP)

S
Scheduling, 34, 36, 75, 517, 520, 599
Scope, 42, 388, 589, 591, 650,
    651, 664
Security, 73, 257, 258, 410, 463, 608, 650,
    656, 660–661, 663, 716
Self-adaptation principle, 88, 95, 97, 203, 529
Self-documenting, 191–194, 320–321, 598,
    600, 601
Self-organization principle,
    88, 95, 203, 529
Semi-automated test case design, 156, 217,
    390–391
Sensitivity to Change principle, 88, 95, 203
Side-effects, 29, 30, 48, 54, 73, 99, 104, 125,
    126, 133, 182, 200, 208, 214, 219,
    226, 229–231, 238, 239, 436, 469,
    471–473, 519, 578–580, 598, 601,
    602, 610, 614, 632, 650, 700, 702,
    710, 717
Side-effect prevention, 2, 30, 32, 51, 54, 57,
    76, 109–111, 122, 134, 207, 219,
    229–231, 233, 235, 251, 254, 255,
    259, 269, 435, 437, 438, 441, 450,
    471, 473, 475, 485, 487, 519, 521,
    522, 565, 579, 580, 582, 592, 593,
    601, 602, 604, 605, 612, 614–616,
    618, 629, 631–634, 636, 640, 649
Silver bullet, 75, 106, 107, 270, 361, 557,
    597–650, 721
Simultaneous users, 658
Six sigma, 437
Smalltalk, 64
Software architectures, 31, 137, 138, 178, 227,
    313, 598
Software definition
    traditional, 58, 699
    new, 2, 30
Software configuration management, 32
Software deliverables, 438
Software design engineering, 31, 215, 313,
    337, 341
Software development lifecycle, 31, 46, 49,
    54, 117, 122, 133, 135, 207, 214,
    223, 230, 231, 233, 254, 255, 260,
    261, 270, 337, 406, 411, 417, 429,
    432, 464, 557, 558, 592, 604, 610,
    613, 632, 646, 717
Software development methodologies
    test-driven, 52, 71, 100, 104, 257, 612
    defect prevention driven, 52, 123–125,
        133, 230, 234, 364–366, 460, 613
Software disasters, 47, 65–69, 374, 434, 436,
    607, 714, 716
Software documentation paradigm, 44, 55, 73,
    103–105, 234, 262, 433, 489–491,
    494–495, 497, 500, 510–512, 606,
    608, 633, 718
Software engineering, 1, 61, 80, 92, 103–105,
    115, 135, 199, 239, 257, 275, 313,
    339, 383, 410, 433, 517, 560, 577,
    597, 699
Software inspection, semi-automated, 166
Software lifecycle, 50, 611, 618
Software maintenance paradigm, 44, 54, 56,
    73–74, 103–105, 235, 304,
    467–487, 578, 609, 614, 717
Software metrics, 694
Software process improvement, (SPI), 55, 63,
    104, 585, 587, 607
Software process model, 44, 57, 58, 63, 199,
    200, 202, 235, 699, 700, 708
Software project management, 44, 46, 54, 74,
    103, 104, 235, 251, 270, 271, 433,
    517–519, 523, 528, 529, 609, 614
Software quality assurance (SQA), 31–32,
    44–46, 55, 72–73, 104, 105, 123,
    234, 255, 270, 409, 433–464, 471,
    528, 578, 717
Software requirements engineering, 31
Software requirements specification, 649–671s
Software testing. *See also* Testing
    embedded, 381–382, 418
    engineering, 409
    method, 117, 260, 375, 383, 406, 407,
        409–411, 429, 431, 432, 445, 470,
        604, 715, 716
    object-oriented, 378–380
    objectives, 406
    paradigm, 44, 49, 52, 55, 72, 103–105,
        234, 270, 272, 278, 285, 306, 316,
        383–384, 406, 407, 409–433, 437,
        445, 450, 578, 601, 605, 606, 608,
        610–611, 613, 711, 715, 716
    performance, 3, 111, 429, 454, 618, 715
    phase, 270, 716
    support, 219
    system, 100
    tasks, 383

technique, 122–123, 133
tools, 340, 381, 383
Software understanding, 145, 272, 711, 718
Software process model, 44, 57, 58, 63, 199, 200, 202, 235, 699, 700, 708
Soliton, 86
Solution method, 131, 133, 206, 208, 209, 213, 214, 216, 229, 230, 267, 317, 341, 532, 581, 582
Source code
 analysis, 533
 design, 584
 location, 148, 177, 423, 424, 427, 450, 721
 module, 115, 176, 178, 179, 245–246, 249, 251, 263, 478, 554
 segment, 480, 482
 test coverage, 125, 242, 412
 writing style, 166, 168, 179, 182
Spiral model, 36–37, 57
Stage, 51, 68, 97, 110, 200, 270, 374, 522, 577, 583, 612, 616, 617
Stakeholders, 31, 68, 274
Standards, 31, 32, 63, 92, 94, 127, 206, 209, 257, 367, 371, 417, 437, 440, 460, 484, 570, 608, 650, 659–660, 725
State transition, 389
Strategy, 34, 36, 106, 260, 261, 270, 277, 279, 311, 314, 315, 337, 378, 434, 435, 439–460, 473, 520–521, 579, 711
Structured analysis, 62, 63, 669
Structured programming, 61
Sub-call-graph, 7, 166, 169, 179, 182, 503, 511
Support
 center, 59
 efforts, 29, 469, 593
 facility, 208, 224–225, 265
 functions, 655
 MC/DC test coverage analysis, 340, 353–362, 532
 methods, 518
 multi-project development, 251, 252, 272
 people, 230
 platform, 2, 49, 103, 105, 106, 108–112, 129, 134, 156, 179, 220, 230–234, 245, 270, 274, 280, 303, 361, 367, 375, 378, 379, 382, 557–575, 577, 583, 595, 611, 615, 617
 platform Panorama++, 30, 50, 202, 216, 231, 374, 378, 401, 424, 439, 536, 553, 593, 594, 611, 720
 requirement changes, 30, 54, 235, 473, 614
 reverse engineering, 8, 340, 348

software maintenance, 29, 48, 222, 610
software project management, 519
 techniques, 107, 109, 112, 183, 203–204, 234, 433, 615
 testability, 229, 469, 475, 477, 664
 tools, 62, 103, 107, 109, 112, 232, 433, 518, 523–524, 557–575, 615, 618, 663
 traceability, 73, 75, 239, 347, 463, 468, 494, 495, 610
 two-way iteration, 593
 users, 372, 535, 536
 visibility, 76
Synergetics, 83
Synthesis design, 96, 98, 100, 131–133, 207, 214–216, 313–324, 336–339, 341–349, 441, 565, 581, 592, 630, 649, 713
System
 activation, 655
 actors, 671
 administrators, 666
 attributes, 650, 660–661, 665–668, 670, 671
 call graphs, 111, 525, 581, 617
 categories, 81
 chart, 45
 debugger, 177, 423
 definition language, 663
 design, 73, 92, 98, 177, 179, 182, 212, 226, 258, 410, 581, 608, 649, 716
 development, 54, 98, 133, 614, 633, 634
 engineering, 92–94
 feature, 667
 functions, 543
 hierarchy, 132, 188, 211, 212, 321, 447
 interfaces, 650, 653
 level, 16, 23, 32, 46, 121, 160, 165, 172, 182, 188, 196, 233, 264, 342, 343, 378, 431, 490, 502, 507, 508, 532, 534, 541, 568–569, 704
 leveling, 640
 linear, 44, 81, 104, 201, 226
 manager, 565
 mode, 650, 661
 nonlinear, 81, 83, 87, 88, 93, 94, 99, 100, 129, 594
 preprocessor, 532
 problems, 84, 88
 quality, 229, 594
 recovery, 660
 requirement, 653
 structure, 83, 120, 633, 635

Index 743

testing, 100, 123, 216, 218, 219, 340, 342, 353, 378, 384, 410, 582, 649
theory, 79, 83

T
Table, 42, 46, 85, 191, 194, 203, 209, 213, 241, 280–281, 283, 304, 305, 318, 320, 389, 390, 442, 460, 463, 478, 499, 548, 553, 560, 563, 564, 569, 570, 574, 585, 587–589, 618–627, 632, 662, 719, 724
Task, 31–32, 38, 44, 49, 50, 57, 69, 70, 72, 74, 75, 80, 89, 94, 95, 98, 99, 101, 103, 106, 109, 110, 112, 135, 195, 199, 227, 257, 259, 279, 306, 313, 314, 323, 337, 339, 342, 383, 409, 410, 433, 435, 436, 468, 470, 475, 476, 487, 489, 491, 510, 511, 517–519, 524, 529, 565, 577–579, 582, 591, 599, 608–611, 616, 618, 640, 655, 659, 699, 702, 719
Technical forum, 76, 110, 223, 617, 628, 630
Technique
    assisted test case design, 128, 129, 203, 204, 419
    comprehensive software testing, 122–123, 133, 203, 204
    defect-prevention driven quality assurance, 123–125, 133, 364–366, 565
    defect propagation prevention, 30, 32, 43, 52, 57, 100, 109, 111, 133, 207, 254, 260, 261, 270–272, 290, 306, 311, 315, 318, 321, 322, 337, 339, 347, 364, 368, 398, 411, 435, 439, 445–450, 471, 473, 475, 522, 565, 581, 592, 604, 605, 615, 618, 631–634, 700, 711, 712
    HAETVE, 31, 130, 133, 183, 189, 211, 231, 279–286, 306–309, 311, 396–399, 413, 415, 440, 445, 497–500, 521–523, 565, 578, 581, 584, 592, 630, 632, 640, 649
    holistic and dynamic traceability, 122, 133, 203, 204, 253, 306, 318, 363, 531, 549–554, 602
    holistic and intelligent version comparison, 121–122, 134, 222, 233
    holistic, virtual, and traceable diagram generation, 117–119, 133, 203, 204, 214, 231
    intelligent regression test case selection, 109, 128–130, 134, 203, 204
    intelligent version comparison technique, 32, 121–122, 134, 203, 233
    refactoring, 52, 111, 126–127, 134, 203, 204, 271, 340, 342, 435, 439, 450–454, 464, 484, 589, 590, 593, 613, 617, 632–634, 641
    software maintenance, 2, 29, 30, 32, 44, 46, 48, 51, 54, 56–58, 73–76, 100, 103–105, 110, 111, 122, 125, 133, 134, 138, 178, 204, 207, 219, 222, 229, 230, 233, 235, 238, 239, 251, 269–270, 304, 436, 441, 467–487, 518, 519, 522, 565, 578–580, 599, 605, 608–610, 612, 614, 616, 618, 632–634, 700, 702, 708
    synthesis design and incremental growing up, 96, 98, 100, 131–133, 203, 214, 313–324, 336–339, 341–349, 367, 368, 441, 565, 581, 592, 630, 649, 713
    test case efficiency analysis, 3, 12, 16, 76, 111, 125–126, 134, 204, 219, 232, 233, 373, 378, 413, 481, 483, 533–537, 573, 617, 633, 635
    test case minimization, 3, 12, 17, 30, 76, 109, 111, 125–126, 134, 204, 219, 232, 373, 378, 391–396, 413, 472, 473, 477, 481, 484, 531, 533–536, 538, 571–573, 615, 617, 633, 635, 700
    virtual and traceable documentation, 119–121, 134, 203, 204, 231
Technology, 38, 64, 66, 73, 75, 84, 88, 97, 99, 100, 104, 116, 121, 208, 230, 257, 274, 371, 382, 383, 410, 433, 591, 599, 605–609, 648, 716, 720
Test case
    analysis, 148, 219, 572
    analyzer, 572
    collection, 636
    description, 108, 111, 115, 225, 240, 249, 251, 254, 264, 318, 400, 470, 503, 534, 602, 615, 617, 711
    design, 128, 129, 156, 160, 182, 203, 204, 217, 302, 387–407, 413, 418, 419, 633, 635
    efficiency, 3, 12, 16, 76, 111, 125–126, 134, 204, 219, 232, 233, 373, 378, 391, 395, 413, 481, 483, 533–537, 573, 617, 633, 635
    execution, 128, 237, 254, 387, 400, 405–406, 535
    generation, 232, 349, 352
    inputs, 373

manager, 373
minimization, 3, 12, 17, 30, 76, 109, 111, 125–126, 134, 204, 219, 232, 373, 378, 391–396, 413, 472, 473, 477, 481, 484, 531, 533–536, 538, 571–573, 615, 617, 633, 635, 700
numbers, 241, 304, 662
parameter, 225, 250
playback, 570–571
script, 115, 117, 122, 225, 240, 241, 302, 304, 412, 521, 550, 568, 573
selection, 22, 30, 76, 109, 128–130, 134, 203, 204, 219, 405–406, 485, 534, 615, 638, 700
set, 111, 618
window, 246, 478
Test coverage measurement, 4, 5, 9, 10, 108, 111, 136, 156, 160, 162, 176, 179, 181, 182, 237, 241, 249, 250, 254, 261, 299, 300, 327, 330, 353–362, 376, 402, 456, 532–534, 585, 586, 615, 617, 632
Test plan, 633, 635
Testability, 2, 3, 76, 115, 203, 229, 234, 270, 469, 475, 477, 521, 664, 699
Testing. *See also* Software testing
approach, 100, 123–125, 231, 232, 261, 294
black-box, 52, 117, 231, 234, 373, 375, 410, 578, 613, 715, 716
efficiency, 573
effort, 29, 30, 34, 48, 54, 74, 75, 85, 104, 112, 133, 202, 207, 219, 230, 235, 239, 269, 441, 473, 475, 487, 518, 519, 573, 579, 591, 598–600, 610, 614, 629, 631, 633, 634, 702
engineer, 232–233
functional, 3, 52, 71, 72, 104, 105, 108, 110, 111, 117, 125, 127, 132, 133, 207, 234, 260, 280, 341, 373, 375, 378, 380, 384, 396–406, 409, 417, 432, 434, 454, 455, 464, 470, 496, 592, 604, 613, 615, 616, 618, 715
load, 377, 381, 714
methods, 52, 97, 108, 110, 117, 122, 123, 214, 234, 260, 261, 263, 270, 271, 278, 285, 318, 323, 337, 339, 347, 364, 368, 372, 373, 375, 376, 381, 383, 387, 406, 409–417, 429, 431, 432, 434, 445, 450, 470, 473, 475, 489, 578, 579, 581, 592, 604, 605, 613, 615, 616, 715, 716
operations, 550, 582, 602
order, 124, 156, 183, 216, 232, 349
performance, 376–377, 380, 384, 455, 714
phase, 260, 270, 406, 450, 582
plan, 378
process, 111, 216, 218, 271, 361, 382, 617
purpose, 371–373
regression, 3, 12, 16, 22, 30, 50, 76, 109, 111, 126, 128, 134, 208, 219, 221, 232, 233, 242, 254, 255, 263, 373, 378, 391, 413, 469, 472, 473, 477, 480, 482, 484, 485, 533, 534, 536, 570–572, 582, 612, 617, 618, 633, 635, 636, 638, 700, 710
reliability, 377
resource, 76, 83, 311, 391, 464, 580, 602
requirements, 373, 643
result, 381
stress, 72, 377, 384
structural, 3, 31, 52, 72, 104, 108, 110, 111, 117, 122, 123, 125, 127, 132, 133, 207, 216, 234, 260–261, 278, 340, 341, 373–375, 378, 380, 383, 384, 396–406, 417, 432, 434, 435, 438, 445, 455, 464, 470, 496, 592, 604, 605, 613, 615, 618, 632, 714, 715
support, 219, 232, 349–353
technique, 122–123, 133, 203, 204
tool, 578
transparent-box, 52, 108, 110, 117, 122–124, 133, 214, 230–232, 234, 261, 263, 270, 271, 278, 280, 285, 318, 323, 337, 339, 347, 364, 387, 411–417, 430, 450, 470, 475, 578, 579, 581, 592, 604, 605, 613, 615, 616, 632, 716
visualization, 44, 48, 49, 55, 68, 69, 73, 103–105, 108–110, 112, 135–195, 261, 270, 277, 287, 317, 318, 337, 339, 347, 364–366, 368, 391, 433, 435, 436, 438, 445, 447, 464, 472, 497, 523, 578, 608, 610, 613, 615–617, 633, 640, 641, 643
white-box, 52, 373–375, 389, 409, 429, 434, 613
web, 380–381
The mythical man-month, 29, 74–76, 97, 200–201, 597–599
Thinking
linear, 44–45, 49, 50, 57, 66, 69, 72, 74–76, 80, 86–88, 91–94, 97,

101, 103, 112, 135, 199, 201, 233, 234, 258, 313, 410, 597–599, 608–612
nonlinear, 30, 50, 57, 75, 76, 80, 86–88, 91–94, 97, 101, 137, 162, 234, 314, 411, 609, 611, 612
Time tags, 12, 28, 108, 122, 208, 225, 240–242, 245, 246, 249, 250, 252, 254, 255, 261, 264, 318, 412, 470, 482, 534, 535, 602, 615, 642, 710
Tool
  Adobe, 156, 303
  analysis, 422
  bar, 567–572, 574
  chain, 137, 138, 179
  development, 549
  freeware, 381
  function/capability, 564
  integration, 63, 137, 138, 179, 564
  modeling, 276, 583
  OO-Analyzer, 3, 12, 563, 564, 567, 570, 572
  OO-Browser, 3, 5, 155, 563, 564, 567–569, 649, 721
  OO-CodeDiff, 3, 22, 563, 564
  OO-DefectTracer, 3, 563, 564, 574
  OO-Diagrammer, 3, 9, 160, 563, 564, 567, 569, 649, 673–693, 721
  OO-MemoryChecker, 574
  OO-MiniCase, 3, 15, 563, 564
  OO-Test, 3, 559, 563, 564, 567, 568, 573, 649, 721
  OO-Validate, 553, 567, 572, 649
  Panounit, 216–219, 349, 352, 353, 698
  performance, 11
  playback, 3, 123, 189, 232, 382–383, 413, 454, 455, 563, 564, 570–572, 582, 617
  software engineering, 137, 138, 179
  support, 43, 237–239, 518, 523–524, 618
  test coverage, 3–5, 8–10, 16, 75, 108, 111, 115, 120, 122, 125–128, 133, 136, 138, 148, 149, 155, 156, 159–162, 165, 168, 170, 176, 178, 179, 181, 182, 194, 208, 216, 217, 219, 225, 237, 240–242, 245, 249, 250, 254, 262, 264, 271, 294, 298–300, 309, 316, 318, 327, 329, 330, 340, 347, 353–362, 367, 374–376, 378–380, 382, 391, 402, 412, 413, 417, 419, 421, 422, 424, 431, 454, 456, 463, 477, 481, 484, 500, 505, 532–536, 567–573, 578, 582, 584–586, 602, 615, 617, 632, 649, 673, 676, 679–685, 688–694, 714, 721, 722, 724, 725
  third party, 106, 251, 252, 255, 305, 564, 602, 710
  vendor, 94, 97, 98, 101, 702
  version comparison, 575, 634, 635
Traceability
  backward, 51, 111, 189, 193, 219, 221, 227, 228, 248, 250, 271, 378, 440, 441, 472, 473, 477, 480, 482, 485, 572, 582, 612, 618, 635, 636
  bi-directional, 1–3, 12, 49, 52, 54, 55, 73, 75, 76, 97, 99, 100, 104, 108, 110, 111, 115, 117, 122, 133, 204, 206, 208, 214, 216, 219, 224–225, 227, 231, 233, 240, 241, 245–249, 252, 264, 278, 348, 412, 417, 435, 440, 449, 469, 470, 496, 564, 572, 573, 592, 599, 604, 605, 610, 613, 615, 617, 632, 642, 647, 711, 716
  code, 447
  facility, 117, 208, 231, 238, 239, 711, 716
  forward, 31, 227, 247, 304, 440, 477, 572, 612
  requirement, 208, 238–240, 255, 305
  software, 75, 122, 255, 602, 604
Transaction, 659
Transparent-box testing method, 3, 31, 52, 100, 108, 110, 111, 117, 122–125, 132, 133, 207, 214, 216, 227, 230–232, 234, 260, 261, 263, 270, 271, 277, 278, 285, 294, 306, 315, 317, 318, 323, 337, 339, 347, 364, 368, 375, 384, 387, 406, 407, 409–432, 435, 437, 438, 445, 447, 450, 464, 470, 473, 475, 495, 496, 565, 578, 579, 581, 582, 592, 604, 605, 613, 615, 616, 618, 632–634, 649, 711, 715, 716

U

Unit testing, 8, 53, 123, 124, 132, 156, 171, 183, 216, 232, 271, 340–342, 347, 349–353, 361, 378, 390–391, 409, 413, 454, 477
UML
  charts, 138
  design, 275, 584

diagrams, 494, 584
notations, 497
Upstream movement, 33, 51, 70, 97, 103, 200, 201, 235, 268, 275, 276, 340, 434, 435, 468, 489, 578, 593, 598, 600, 601, 612
Usability, 44, 159
Use case
analysis, 286, 287
approach, 31, 130, 133, 274, 282, 413, 649
drawbacks, 274
notations, 282
User interface, 303, 382, 482
User manual, 237, 489, 649

V
Viewpoints, 100, 237
Virtual, 54, 111, 117–122, 133, 134, 137, 168, 170, 179, 195, 214, 231, 277, 341, 507, 508, 512, 553, 587, 592, 613, 617, 649
Virtual function, 414
Visibility, 2, 3, 68, 73, 75, 76, 238, 270, 433, 469, 475, 477, 503, 509, 510, 521, 602, 664, 699

W
Waterfall model, 32–36, 42, 43, 45, 51, 75, 92, 105, 599, 608, 612, 618
Waterfall model with feedback, 32, 33, 708, 709
Web, 68, 76, 95, 110, 156, 207, 223, 224, 231, 237, 238, 251, 252, 268, 271, 303, 377, 380–381, 504, 515, 519, 524, 527, 529, 560, 580, 582, 617, 628–630, 649, 654, 660, 714, 719, 721
Web engineering, 63
Werewolves, 597, 600, 605–609, 628–647
White box testing, 52, 375, 389, 409, 434, 613
Work flow, 206, 267, 268
Work product, 641–643, 649
Workflows, 97, 101

X
XP. *See* Extreme programming

Y
Y2K, 66